**Soil Me**

**Unsatu**

# Soil Mechanics for Unsaturated Soils

*D. G. Fredlund, Ph.D.*

Professor of Civil Engineering
University of Saskatchewan
Saskatoon, Saskatchewan

*H. Rahardjo, Ph.D.*

Senior Lecturer
School of Civil and Structural Engineering
Nanyang Technological University

A Wiley-Interscience Publication
**JOHN WILEY & SONS, INC.**
New York • Chichester • Brisbane • Toronto • Singapore

This text is printed on acid-free paper.

Copyright © 1993 by John Wiley & Sons, Inc.

All rights reserved. Published simultaneously in Canada.

Reproduction or translation of any part of this work beyond that permitted by Section 107 or 108 of the 1976 United States Copyright Act without the permission of the copyright owner is unlawful. Requests for permission or further information should be addressed to the Permissions Department, John Wiley & Sons, Inc., 605 Third Avenue, New York, NY 10158-0012.

This publication is designed to provide accurate and authoritative information in regard to the subject matter covered. It is sold with the understanding that the publisher is not engaged in rendering legal, accounting, or other professional services. If legal advice or other expert assistance is required, the services of a competent professional person should be sought.

*Library of Congress Cataloging in Publication Data:*

Fredlund, D. G. (Delwyn G.), 1940–
    Soil mechanics for unsaturated soils / D. G. Fredlund and H. Rahardjo
      p.  cm.
    Includes bibliographical references and index.
    ISBN 0-471-85008-X
    1. Soil mechanics.  2. Soil moisture.  3. Soil—Testing.
I. Rahardjo, H. (Harianto)  II. Title.
TA710.5.F73  1993
624.1'5136—dc20                      92-30869

Printed in the United States of America

10 9 8 7 6 5 4 3 2

Dedicated

to our parents

George and Esther Fredlund

and

Sugiarto and Pauline Rahardjo

who taught us that the fear of the Lord was the beginning of
wisdom and that the love of the Lord makes life worth living

Delwyn G. Fredlund
Harianto Rahardjo

# FOREWORD

The appearance of a new book on Geotechnical Engineering is always an important occasion; but the appearance of the first book on an important aspect of Soil Mechanics is especially noteworthy. In this volume, Professor Fredlund and Dr. Rahardjo present the first textbook solely concerned with the behavior of unsaturated soils. The timing is particularly propitious.

It is evident that since much of the developed world enjoys a temperate climate, resulting in primarily saturated soil conditions, the literature has been biased toward problems involving saturated soils. Moreover, the theoretical understandings and associated experimental procedures required for an understanding of unsaturated soil behavior are intrinsically more complex than those required for saturated soil behavior. As a result, the ability to synthesize unsaturated soil mechanics has lagged behind its saturated counterpart. This has been to the detriment of both students and practitioners alike.

The climatic conditions that give rise to unsaturated soils can be found on every continent. Indeed, in some countries, unsaturated soil conditions dominate. The engineering problems associated with unsaturated soil mechanics extend over an enormous range. The requirements for design and construction of low-cost lightly loaded housing on expansive soils have been with us for a long time. More financial losses arise annually from damages due to unsaturated expansive soil behavior than from any other ground failure hazard. At the other extreme, unsaturated soils are used as a buffer material in almost every proposal for the underground storage of nuclear waste. Hence, the need to understand the mechanics of unsaturated soil behavior extends from concerns for low cost housing to some of the most complex environmental issues of our time.

I expect that this volume will quickly become the classic reference in its field. It will not be possible to teach, conduct research, or undertake modern design related to unsaturated soils without reference to Fredlund and Rahardjo. The authors have wisely maintained the framework of classical soil mechanics and sought to extend it in order to incorporate soil suction phenomena as an independent variable that is amenable to measurement and calculation. This will greatly facilitate the use of this comprehensive volume and quickly result in a more profound understanding of unsaturated soil behavior.

The road to this volume has been a difficult one. Many early leaders of Soil Mechanics pointed in the right direction, but it has taken more than thirty years of sustained effort to reach the end of the journey marked by this publication. All those who participated in the voyage should share pleasure in the outcome.

N. R. MORGENSTERN
University Professor of Civil Engineering

*University of Alberta*
*April 1993*

# PREFACE

Numerous textbooks have been written on the subject of soil mechanics. The subject matter covered and the order of presentation vary somewhat from text to text, but the main emphasis is always on the application of the principles of soil mechanics to problems involving *saturated* soils.

A significant portion of the earth's surface is subjected to arid and semi-arid climatic conditions, and as a result, many of the soils encountered in engineering practice are unsaturated. This textbook addresses the subject of soil mechanics as it relates to the behavior of *unsaturated* soils. More specifically, the text addresses that class of problems where the soils have a matric suction or where the pore–water pressure is negative.

Whether the soil is unsaturated or saturated, it is the negative pore–water pressure that gives rise to this unique class of soil mechanics problems. When the pore–water pressure is negative, it is advantageous, and generally necessary, to use two independent stress state variables to describe the behavior of the soil. This is in constrast to saturated soil mechanics problems where it is possible to relate the behavior of the soil to one stress state variable, namely, the effective stress variable.

The terms *saturated soil mechanics* and *unsaturated soil mechanics* are primarily used to designate conditions where the pore–water pressures are positive and negative, respectively. Soils situated above the groundwater table have negative pore–water pressures. The engineering problems involved may range from the expansion of a swelling clay to the loss of shear strength in a slope. Microclimatic conditions in an area produce a surface flux boundary condition which produces flow through the upper portion of the soil profile.

It would appear that most problems addressed in *saturated* soil mechanics have a counter problem of interest in *unsaturated* soils. In addition, the remolding and compacting of soils is an important part of many engineering projects. Compacted soils have negative pore–water pressures. The range of subjects of interest involving negative pore–water pressures are vast, and the problems are becoming of increasing relevance, particularly in arid regions.

An attempt has been made to write this textbook in an introductory manner. However, the subject matter is inherently complex. The need for such a book is clearly demonstrated by engineering needs associated with various projects around the world. The frustrations are expressed primarily by engineers who have received advanced training in conventional soil mechanics, only to discover difficult problems in practice involving unsaturated soils for which their knowledge is limited.

The textbook makes no attempt to redevelop concepts well known to saturated soil mechanics. Rather, the book is designed to be an extension of classical saturated soil mechanics. As far as is possible, the principles and concepts for unsaturated soils are developed as extensions of the principles and concepts for saturated soils. In this way, the reader should be able to readily grasp the formulations required for unsaturated soil mechanics.

The general format for the textbook is similar to that used in most classical soil mechanics textbooks. The book starts by introducing the breadth of unsaturated soil mechanics problems. It then presents material related to the: 1) volume-mass properties, 2) stress state variables, 3) flow behavior, and 4) pore pressure parameters for unsaturated soils. The book then goes on to present material on the: 5) shear strength and 6) volume change behavior of unsaturated soils. The latter part of the book concludes with material on the transient processes of interest to geotechnical engineering.

A brief summary of the chapters of the textbook is as follows. Chapter 1 presents a brief history of developments related to the behavior of unsaturated soils. The need for an understanding of unsaturated soil mechanics is presented, along with the scope and description of common geotechnical problems. The nature of an unsaturated soil element is described, concentrating on the difference between a saturated and an unsaturated soil. Chapter 2 presents the phase properties and the volume–mass relations of interest to unsaturated soils. This chapter provides some overlap with classical soil mechanics, but emphasizes extensions to the theory. The steps involved in all derivations are described in detail in order to assist the reader in this relatively new field.

Chapter 3 is devoted to describing the stress state variables of relevance in solving engineering problems associated with soils having negative pore–water pressures. The concept of the stress state is presented in detail because of its extreme importance in understanding the formulations presented later in the textbook. One needs only to examine the importance of the role of the effective stress concept in the development of saturated soil mechanics to realize the importance of an acceptable description of the stress state for unsaturated soils. The authors believe that a thorough understanding of the stress state provides the basis for developing a transferable science for unsaturated soil mechanics.

A knowledge of the stress state reveals that the measurement of the pore–water pressure is mandatory. The measurement of highly negative pore–water pressures and soil suction is difficult. Chapter 4 summarizes techniques and devices that have been developed and used to measure negative pore–water pressures and soil suction.

There are three fundamental soil properties that are commonly associated with soil mechanics problems. The properties are: 1) coefficient of permeability, 2) shear strength parameters, and 3) volume change coefficients. These properties are covered in the next nine chapters. Each of the properties is addressed from three standpoints. First, the theory related to the soil property is presented. Second, the measurement of pertinent soil properties is discussed, along with the presentation of typical values. Third, the application of the soil properties to specific soil mechanics problems is formulated and discussed. The logistics of these chapters is as follows:

**Chapters Presenting the Following Material**

| Soil Property | Theory | Measurement | Application |
| --- | --- | --- | --- |
| Permeability | 5 | 6 | 7 |
| Shear Strength | 9 | 10 | 11 |
| Volume Change | 12 | 13 | 14 |

Descriptions of the equipment required for the measurement of the soil properties are presented under each of the "Measurement" chapters. The main application problems presented pertaining to permeability are two-dimensional, earth dam seepage analyses. For shear strength, the applications are lateral earth pressure, bearing capacity, and slope stability problems, with most emphasis on the latter. The primary volume change problem is the prediction of the heave of light structures.

Chapter 8 presents the theory and typical test results associated with pore pressure parameters. Its location in the text is dictated by its importance in discussing undrained loading and the shear strength of soils.

The theory of consolidation, as well as unsteady-state flow analysis, require the combining of the volume change characteristics of a soil with its permeability characteristics. These analyses have formed an integral part of saturated soil mechanics and greatly assist the engineer in understanding soil behavior. Chapter 15 deals with the one-dimensional theory of consolidation, while Chapter 16 presents two- and three-dimensional, unsteady-state flow for unsaturated soils. The theory related to surface flux boundary conditions, as it relates to microclimatic conditions, is briefly presented in Chapter 16.

There is a great need for case histories to illustrate and substantiate the theories related to unsaturated soil behavior. One of the main objectives of this book is to synthesize the available research information and solidify an unsaturated soil theoretical context in order to form a basis for future studies in the form of case histories.

The book is the result of many years of study, research, and help from numerous persons. We thank the many authors and publishers for permission to reproduce figures and use information from research papers.

We want to acknowledge the support provided for the preparation of the manuscript. We thank Professor P. N. Nikiforuk, Dean of the College of Engineering, University of Saskatchewan, Saskatoon, Canada, and Professor Chen Charng Ning, Dean of the School of Civil and Structural Engineering, Nanyang Technological University, Singapore, for their encouragement and support. Thanks to Mr. A. W. Clifton, Clifton Associates Ltd., Regina, Canada, who was particularly instrumental in ensuring that the theoretical concepts and formulations for unsaturated soils were in a form which could readily be put into engineering practice. Several students and colleagues provided invaluable assistance in the review of the manuscript. Recognition is due to Dr. S. L. Barbour and Dr. G. W. Wilson for their review of several chapters. Miss E. Imre of Budapest, Hungary, provided helpful review of several chapters. Dr. D. E. Pufahl reviewed Chapter 2, and Dr. D. Y. F. Ho reviewed Chapters 9 and 10. We are particularly grateful to Professor N. R. Morgenstern who has continued to provide insight and encouragement into the study of the behavior of unsaturated soils.

We also wish to thank the typists, Mrs. Gladie Russell, Mr. Mark Vanstone, Miss Tracey Regier, Miss Kerri Fischer, and Mrs. Leslie Pavier for their endurance and meticulous typing of our many drafts. We are particularly grateful to Mrs. Pavier who organized the many persons involved in producing the final manuscript. Mr. J. L. Loi took a keen interest in the drafting of the figures, the replotting of figures to SI units, and the checking of data. Miss Kyla Fischer and Ms. Deidre S. Komarychka performed meticulous work in preparing the figures. The authors wish to acknowledge the excellent editing and proofreading of all the chapters by Mr. Sai K. Vanapalli, Mr. Julian Gan and Dr. A. Xing. Mr. L. Lam analyzed several of the example problems in Chapters 7 and 16. Mr. J. Lau and Mr. K. Fredlund organized the extensive list of references for the book. The work and efforts of other graduate students are deeply appreciated.

D. G. FREDLUND
H. RAHARDJO

*University of Saskatchewan*
*April 1993*

# CONTENTS

| CHAPTER 1 | Introduction to Unsaturated Soil Mechanics | 1 |
|---|---|---|
| | 1.1 Role of Climate | 1 |
| | 1.2 Types of Problems | 3 |
| |     1.2.1 Construction and Operation of a Dam | 3 |
| |     1.2.2 Natural Slopes Subjected to Environmental Changes | 5 |
| |     1.2.3 Mounding Below Waste Retention Ponds | 6 |
| |     1.2.4 Stability of Vertical or Near Vertical Excavations | 6 |
| |     1.2.5 Lateral Earth Pressures | 7 |
| |     1.2.6 Bearing Capacity for Shallow Foundations | 7 |
| |     1.2.7 Ground Movements Involving Expansive Soils | 8 |
| |     1.2.8 Collapsing Soils | 9 |
| |     1.2.9 Summary of Unsaturated Soils Examples | 9 |
| | 1.3 Typical Profiles of Unsaturated Soils | 9 |
| |     1.3.1 Typical Tropical Residual Soil Profile | 10 |
| |     1.3.2 Typical Expansive Soils Profile | 11 |
| | 1.4 Need for Unsaturated Soil Mechanics | 12 |
| | 1.5 Scope of the Book | 13 |
| | 1.6 Phases of an Unsaturated Soil | 14 |
| |     1.6.1 Definition of a Phase | 14 |
| |     1.6.2 Air–Water Interface or Contractile Skin | 14 |
| | 1.7 Terminology and Definitions | 15 |
| | 1.8 Historical Developments | 16 |
| CHAPTER 2 | Phase Properties and Relations | 20 |
| | 2.1 Properties of the Individual Phases | 20 |
| |     2.1.1 Density and Specific Volume | 21 |
| |         *Soil particles* | 21 |
| |         *Water phase* | 21 |
| |         *Air phase* | 21 |
| |     2.1.2 Viscosity | 23 |
| |     2.1.3 Surface Tension | 24 |
| | 2.2 Interaction of Air and Water | 25 |
| |     2.2.1 Solid, Liquid, and Vapor States of Water | 26 |
| |     2.2.2 Water Vapor | 26 |
| |     2.2.3 Air Dissolving in Water | 27 |
| |         *Solubility of Air in Water* | 28 |
| |         *Diffusion of Gases Through Water* | 28 |

|  |  |  |  |
|---|---|---|---|
|  | 2.3 | Volume–Mass Relations | 29 |
|  |  | 2.3.1 Porosity | 29 |
|  |  | 2.3.2 Void Ratio | 30 |
|  |  | 2.3.3 Degree of Saturation | 30 |
|  |  | 2.3.4 Water Content | 31 |
|  |  | 2.3.5 Soil Density | 32 |
|  |  | 2.3.6 Basic Volume–Mass Relationship | 32 |
|  |  | 2.3.7 Changes in Volume–Mass Properties | 33 |
|  |  | 2.3.8 Density of Mixtures Subjected to Compression of the Air Phase | 34 |
|  |  | *Piston-porous stone analogy* | 34 |
|  |  | *Conservation of mass applied to a mixture* | 36 |
|  |  | *Soil particles–water–air mixture* | 37 |
|  |  | *Air–water mixture* | 37 |
| CHAPTER 3 |  | Stress State Variables | 38 |
|  | 3.1 | History of the Description of the Stress State | 38 |
|  |  | 3.1.1 Effective Stress Concept for a Saturated Soil | 38 |
|  |  | 3.1.2 Proposed Effective Stress Equation for an Unsaturated Soil | 39 |
|  | 3.2 | Stress State Variables for Unsaturated Soils | 42 |
|  |  | 3.2.1 Equilibrium Analysis for Unsaturated Soils | 42 |
|  |  | *Normal and shear stresses on a soil element* | 42 |
|  |  | *Equilibrium equations* | 43 |
|  |  | 3.2.2 Stress State Variables | 43 |
|  |  | *Other combinations of stress state variables* | 44 |
|  |  | 3.2.3 Saturated Soils as a Special Case of Unsaturated Soils | 45 |
|  |  | 3.2.4 Dry Soils | 45 |
|  | 3.3 | Limiting Stress State Conditions | 46 |
|  | 3.4 | Experimental Testing of the Stress State Variables | 47 |
|  |  | 3.4.1 The Concept of Axis Translation | 47 |
|  |  | 3.4.2 Null Tests to Test Stress State Variables | 48 |
|  |  | 3.4.3 Other Experimental Evidence in Support of the Proposed Stress State Variables | 48 |
|  | 3.5 | Stress Analysis | 49 |
|  |  | 3.5.1 *In Situ* Stress State Component Profiles | 49 |
|  |  | *Coefficient of lateral earth pressure* | 52 |
|  |  | *Matric suction profile* | 53 |
|  |  | *Ground surface condition* | 53 |
|  |  | *Environmental conditions* | 53 |
|  |  | *Vegetation* | 53 |
|  |  | *Water table* | 54 |
|  |  | *Permeability of the soil profile* | 54 |
|  |  | 3.5.2 Extended Mohr Diagram | 54 |
|  |  | *Equation of Mohr circles* | 55 |
|  |  | *Construction of Mohr circles* | 56 |
|  |  | 3.5.3 Stress Invariants | 58 |
|  |  | 3.5.4 Stress Points | 59 |
|  |  | 3.5.5 Stress Paths | 59 |
|  | 3.6 | Role of Osmotic Suction | 63 |

| CHAPTER 4 | Measurements of Soil Suction | 64 |
|---|---|---|
| | 4.1 Theory of Soil Suction | 64 |
| |     4.1.1 Components of Soil Suction | 64 |
| |     4.1.2 Typical Suction Values and Their Measuring Devices | 66 |
| | 4.2 Capillarity | 67 |
| |     4.2.1 Capillary Height | 67 |
| |     4.2.2 Capillary Pressure | 68 |
| |     4.2.3 Height of Capillary Rise and Radius Effects | 69 |
| | 4.3 Measurements of Total Suction | 70 |
| |     4.3.1 Psychrometers | 70 |
| |         *Seebeck effects* | 70 |
| |         *Peltier effects* | 70 |
| |         *Peltier psychrometer* | 71 |
| |         *Psychrometer calibration* | 73 |
| |         *Psychrometer performance* | 74 |
| |     4.3.2 Filter paper | 77 |
| |         *Principle of measurement (filter paper method)* | 77 |
| |         *Measurement and calibration techniques (filter paper method)* | 77 |
| |         *The use of the filter paper method in practice* | 79 |
| | 4.4 Measurements of Matric Suction | 80 |
| |     4.4.1 High Air Entry Disks | 81 |
| |     4.4.2 Direct measurements | 82 |
| |         *Tensiometers* | 83 |
| |         *Servicing the tensiometer prior to installation* | 84 |
| |         *Servicing the tensiometer after installation* | 86 |
| |         *Jet fill tensiometers* | 86 |
| |         *Small tip tensiometer* | 86 |
| |         *Quick Draw tensiometers* | 88 |
| |         *Tensiometer performance for field measurements* | 88 |
| |         *Osmotic tensiometers* | 90 |
| |         *Axis-translation technique* | 91 |
| |     4.4.3 Indirect Measurements | 93 |
| |         *Thermal conductivity sensors* | 95 |
| |         *Theory of operation* | 97 |
| |         *Calibration of sensors* | 97 |
| |         *Typical results of matric suction measurements* | 99 |
| |         *The MCS 6000 sensors* | 99 |
| |         *The AGWA-II sensors* | 100 |
| | 4.5 Measurements of Osmotic Suction | 104 |
| |     4.5.1 Squeezing technique | 105 |
| CHAPTER 5 | Flow Laws | 107 |
| | 5.1 Flow of Water | 107 |
| |     5.1.1 Driving Potential for Water Phase | 108 |
| |     5.1.2 Darcy's Law for Unsaturated Soils | 110 |
| |     5.1.3 Coefficient of Permeability with Respect to the Water Phase | 110 |
| |         *Fluid and porous medium components* | 110 |

|  |  |  |
|---|---|---|
| | *Relationship between permeability and volume–mass properties* | 111 |
| | *Effect of variations in degree of saturation on permeability* | 111 |
| | *Relationship between coefficient of permeability and degree of saturation* | 111 |
| | *Relationship between water coefficient of permeability and matric suction* | 113 |
| | *Relationship between water coefficient of permeability and volumetric water content* | 113 |
| | *Hysteresis of the permeability function* | 116 |
| 5.2 | Flow of Air | 117 |
| | 5.2.1 Driving Potential for Air Phase | 117 |
| | 5.2.2 Fick's Law for Air Phase | 117 |
| | 5.2.3 Coefficient of Permeability with Respect to Air Phase | 119 |
| | *Relationship between air coefficient of permeability and degree of saturation* | 120 |
| | *Relationship between air coefficient of permeability and matric suction* | 120 |
| 5.3 | Diffusion | 121 |
| | 5.3.1 Air Diffusion Through Water | 121 |
| | 5.3.2 Chemical Diffusion Through Water | 123 |
| 5.4 | Summary of Flow Laws | 123 |

**CHAPTER 6   Measurement of Permeability — 124**

| | | |
|---|---|---|
| 6.1 | Measurement of Water Coefficient of Permeability | 124 |
| | 6.1.1 Direct Methods to Measure Water Coefficient of Permeability | 124 |
| | *Laboratory test methods* | 124 |
| | *Steady-state method* | 124 |
| | *Apparatus for steady-state method* | 125 |
| | *Computations using steady-state method* | 126 |
| | *Presentation of water coefficients of permeability* | 126 |
| | *Difficulties with the steady-state method* | 127 |
| | *Instantaneous profile method* | 127 |
| | *Instantaneous profile method proposed by Hamilton et al.* | 128 |
| | *Computations for the instantaneous profile method* | 129 |
| | *In situ field methods* | 130 |
| | *In situ instantaneous profile method* | 130 |
| | *Computations for the in situ instantaneous profile method* | 131 |
| | 6.1.2 Indirect Methods to Compute Water Coefficient of Permeability | 133 |
| | *Tempe pressure cell apparatus and test procedure* | 133 |
| | *Volumetric pressure plate extractor apparatus and test procedure* | 134 |
| | *Test procedure for the volumetric pressure plate extractor* | 135 |
| | *Drying portion of soil–water characteristic curve* | 136 |

|  |  |  | *Wetting portion of the soil–water characteristic curve* | 136 |
|---|---|---|---|---|
|  |  |  | *Computation of $k_w$ using the soil–water characteristic curve* | 136 |
|  | 6.2 | Measurement of Air Coefficient of Permeability | | 138 |
|  |  |  | *Triaxial permeameter cell for the measurement of air permeability* | 140 |
|  |  |  | *Triaxial permeameter cell for air and water permeability measurements* | 140 |
|  | 6.3 | Measurement of Diffusion | | 143 |
|  |  | 6.3.1 | Mechanism of Air Diffusion Through High Air Entry Disks | 144 |
|  |  | 6.3.2 | Measurements of the Coefficient of Diffusion | 144 |
|  |  |  | *Procedure for computing diffusion properties* | 145 |
|  |  | 6.3.3 | Diffused Air Volume Indicators | 146 |
|  |  |  | *Bubble pump to measure diffused air volume* | 146 |
|  |  |  | *Diffused air volume indicator (DAVI)* | 146 |
|  |  |  | *Procedure for measuring diffused air volume* | 148 |
|  |  |  | *Computation of diffused air volume* | 148 |
|  |  |  | *Accuracy of the diffused air volume indicator* | 149 |
| CHAPTER 7 | Steady-State Flow | | | 150 |
|  | 7.1 | Steady-State Water Flow | | 150 |
|  |  | 7.1.1 | Variation of Coefficient of Permeability with Space for an Unsaturated Soil | 151 |
|  |  |  | *Heterogeneous, isotropic steady-state seepage* | 151 |
|  |  |  | *Heterogeneous, anisotropic steady-state seepage* | 151 |
|  |  | 7.1.2 | One-Dimensional Flow | 152 |
|  |  |  | *Formulation for one-dimensional flow* | 153 |
|  |  |  | *Solution for one-dimensional flow* | 154 |
|  |  |  | *Finite difference method* | 155 |
|  |  |  | *Head boundary condition* | 155 |
|  |  |  | *Flux boundary condition* | 156 |
|  |  | 7.1.3 | Two-Dimensional Flow | 159 |
|  |  |  | *Formulation for two-dimensional flow* | 159 |
|  |  |  | *Solutions for two-dimensional flow* | 160 |
|  |  |  | *Seepage analysis using the finite element method* | 161 |
|  |  |  | *Examples of two-dimensional problems* | 164 |
|  |  |  | *Infinite slope* | 171 |
|  |  | 7.1.4 | Three-Dimensional Flow | 173 |
|  | 7.2 | Steady-State Air Flow | | 175 |
|  |  | 7.2.1 | One-Dimensional Flow | 175 |
|  |  | 7.2.2 | Two-Dimensional Flow | 176 |
|  | 7.3 | Steady-State Air Diffusion Through Water | | 177 |
| CHAPTER 8 | Pore Pressure Parameters | | | 178 |
|  | 8.1 | Compressibility of Pore Fluids | | 178 |
|  |  | 8.1.1 | Air Compressibility | 179 |
|  |  | 8.1.2 | Water Compressibility | 179 |
|  |  | 8.1.3 | Compressibility of Air–Water Mixtures | 179 |
|  |  |  | *The use of pore pressure parameters in the compressibility equation* | 181 |

|  |  |  |  |
|---|---|---|---|
|  | 8.1.4 | Components of Compressibility of an Air–Water Mixture | 181 |
|  |  | *Effects of free air on the compressibility of the mixture* | 182 |
|  |  | *Effects of dissolved air on the compressibility of the mixture* | 182 |
|  | 8.1.5 | Other Relations for Compressibility of Air–Water Mixture | 182 |
|  |  | *Limitation of Kelvin's equation in formulating the compressibility equation* | 183 |
| 8.2 | Derivations of Pore Pressure Parameters | | 184 |
|  | 8.2.1 | Tangent and Secant Pore Pressure Parameters | 185 |
|  | 8.2.2 | Summary of Necessary Constitutive Relations | 186 |
|  | 8.2.3 | Drained and Undrained Loading | 188 |
|  | 8.2.4 | Total Stress and Soil Anisotropy | 190 |
|  | 8.2.5 | $K_0$-Loading | 191 |
|  | 8.2.6 | Hilf's Analysis | 192 |
|  | 8.2.7 | Isotropic Loading | 194 |
|  | 8.2.8 | Uniaxial Loading | 196 |
|  | 8.2.9 | Triaxial Loading | 196 |
|  | 8.2.10 | Three-Dimensional Loading | 199 |
|  | 8.2.11 | $\alpha$ Parameters | 200 |
| 8.3 | Solutions of the Pore Pressure Equations and Comparisons with Experimental Results | | 201 |
|  | 8.3.1 | Secant $B'_h$ Pore Pressure Parameter Derived from Hilf's Analysis | 201 |
|  | 8.3.2 | Graphical Procedure for Hilf's Analysis | 202 |
|  | 8.3.3 | Experimental Results of Tangent $B$ Pore Pressure Parameters for Isotropic Loading | 204 |
|  | 8.3.4 | Theoretical Prediction of $B$ Pore Pressure Parameters for Isotropic Loading | 206 |
|  | 8.3.5 | Experimental Results of Tangent $B$ and $A$ Parameters for Triaxial Loading | 215 |
|  | 8.3.6 | Experimental Measurements of the $\alpha$ Parameter | 216 |

**CHAPTER 9  Shear Strength Theory** — 217

| | | | |
|---|---|---|---|
| 9.1 | History of Shear Strength | | 217 |
|  | 9.1.1 | Data Associated with Incomplete Stress Variable Measurements | 224 |
| 9.2 | Failure Envelope for Unsaturated Soils | | 225 |
|  | 9.2.1 | Failure Criteria | 225 |
|  | 9.2.2 | Shear Strength Equation | 227 |
|  | 9.2.3 | Extended Mohr–Coulomb Failure Envelope | 228 |
|  | 9.2.4 | Use of $(\sigma - u_w)$ and $(u_a - u_w)$ to Define Shear Strength | 230 |
|  | 9.2.5 | Mohr–Coulomb and Stress Point Envelopes | 231 |
| 9.3 | Triaxial Tests on Unsaturated Soils | | 236 |
|  | 9.3.1 | Consolidated Drained Test | 238 |
|  | 9.3.2 | Constant Water Content Test | 238 |
|  | 9.3.3 | Consolidated Undrained Test with Pore Pressure Measurements | 240 |
|  | 9.3.4 | Undrained Test | 243 |
|  | 9.3.5 | Unconfined Compression Test | 245 |

|  |  |  |  | |
|---|---|---|---|---|
|  | 9.4 | Direct Shear Tests on Unsatured Soils | | 247 |
|  | 9.5 | Selection of Strain Rate | | 248 |
|  |  | 9.5.1 | Background on Strain Rates for Triaxial Testing | 248 |
|  |  | 9.5.2 | Strain Rates for Triaxial Tests | 250 |
|  |  | 9.5.3 | Displacement Rate for Direct Shear Tests | 254 |
|  | 9.6 | Multistage Testing | | 255 |
|  | 9.7 | Nonlinearity of Failure Envelope | | 255 |
|  | 9.8 | Relationships Between $\phi^b$ and $\chi$ | | 258 |
| CHAPTER 10 | | Measurement of Shear Strength Parameters | | 260 |
|  | 10.1 | Special Design Considerations | | 260 |
|  |  | 10.1.1 | Axis-Translation Technique | 260 |
|  |  | 10.1.2 | Pore–Water Pressure Control or Measurement | 263 |
|  |  |  | *Saturation procedure for a high air entry disk* | 266 |
|  |  | 10.1.3 | Pressure Response Below the Ceramic Disk | 266 |
|  |  | 10.1.4 | Pore–Air Pressure Control or Measurement | 272 |
|  |  | 10.1.5 | Water Volume Change Measurement | 273 |
|  |  | 10.1.6 | Air Volume Change Measurement | 275 |
|  |  | 10.1.7 | Overall Volume Change Measurement | 275 |
|  |  | 10.1.8 | Specimen Preparation | 276 |
|  |  | 10.1.9 | Backpressuring to Produce Saturation | 277 |
|  | 10.2 | Test Procedures for Triaxial Tests | | 279 |
|  |  | 10.2.1 | Consolidated Drained Test | 280 |
|  |  | 10.2.2 | Constant Water Content Test | 281 |
|  |  | 10.2.3 | Consolidated Undrained Test with Pore Pressure Measurements | 281 |
|  |  | 10.2.4 | Undrained Test | 282 |
|  |  | 10.2.5 | Unconfined Compression Test | 282 |
|  | 10.3 | Test Procedures for Direct Shear Tests | | 282 |
|  | 10.4 | Typical Test Results | | 284 |
|  |  | 10.4.1 | Triaxial Test Results | 284 |
|  |  |  | *Consolidated drained triaxial tests* | 284 |
|  |  |  | *Constant water content triaxial tests* | 286 |
|  |  |  | *Nonlinear shear strength versus matric suction* | 286 |
|  |  |  | *Undrained and unconfined compression tests* | 288 |
|  |  | 10.4.2 | Direct Shear Test Results | 289 |
| CHAPTER 11 | | Plastic and Limit Equilibrium | | 297 |
|  | 11.1 | Earth Pressures | | 297 |
|  |  | 11.1.1 | At Rest Earth Pressure Conditions | 298 |
|  |  | 11.1.2 | Estimation of Depth of Cracking | 300 |
|  |  | 11.1.3 | Extended Rankine Theory of Earth Pressures | 301 |
|  |  |  | *Active earth pressure* | 303 |
|  |  |  | *Coefficient of active earth pressure* | 304 |
|  |  |  | *Active earth pressure distribution (constant matric suction with depth)* | 304 |
|  |  |  | *Tension zone depth* | 304 |
|  |  |  | *Active earth pressure distribution (linear decrease in matric suction to the water table)* | 304 |

|  |  |  | |
|---|---|---|---|
|  |  | *Active earth pressure distribution when the soil has tension cracks* | 305 |
|  |  | *Passive earth pressure* | 307 |
|  |  | *Coefficient of passive earth pressure* | 307 |
|  |  | *Passive earth pressure distribution (constant matric suction with depth)* | 307 |
|  |  | *Passive earth pressure distribution (linear decrease in matric suction to the water table)* | 308 |
|  |  | *Deformations with active and passive states* | 308 |
|  | 11.1.4 | Total Lateral Earth Force | 309 |
|  |  | *Active earth force* | 310 |
|  |  | *Passive earth force* | 311 |
|  | 11.1.5 | Effect of Changes in Matric Suction on the Active and Passive Earth Pressure | 312 |
|  |  | *Relationship between swelling pressures and the earth pressures* | 313 |
|  | 11.1.6 | Unsupported Excavations | 313 |
|  |  | *Effect of tension cracks on the unsupported height* | 314 |
| 11.2 | Bearing Capacity | | 315 |
|  | 11.2.1 | Terzaghi Bearing Capacity Theory | 315 |
|  | 11.2.2 | Assessment of Shear Strength Parameters and a Design Matric Suction | 317 |
|  |  | *Stress state variable approach* | 317 |
|  |  | *Total stress approach* | 318 |
|  | 11.2.3 | Bearing Capacity of Layered Systems | 319 |
| 11.3 | Slope Stability | | 320 |
|  | 11.3.1 | Location of the Critical Slip Surface | 320 |
|  | 11.3.2 | General Limit Equilibrium (GLE) Method | 321 |
|  |  | *Shear force mobilized equation* | 323 |
|  |  | *Normal force equation* | 324 |
|  |  | *Factor of safety with respect to moment equilibrium* | 324 |
|  |  | *Factor of safety with respect to force equilibrium* | 325 |
|  |  | *Interslice force function* | 325 |
|  |  | *Procedures for solving the factors of safety equation* | 327 |
|  |  | *Pore-water pressure designation* | 328 |
|  | 11.3.3 | Other Limit Equilibrium Methods | 330 |
|  | 11.3.4 | Numerical Difficulties Associated with the Limit Equilibrium Method of Slices | 332 |
|  | 11.3.5 | Effects of Negative Pore-Water Pressure on Slope Stability | 333 |
|  |  | *The "total cohesion" method* | 333 |
|  |  | *Two examples using the "total cohesion" method* | 334 |
|  |  | *Example no. 1* | 334 |
|  |  | *Example no. 2* | 338 |
|  |  | *The "extended shear strength" method* | 340 |
|  |  | *General layout of problems and soil properties* | 340 |
|  |  | *Initial conditions for the seepage analysis* | 342 |
|  |  | *Seepage and slope stability results under high-intensity rainfall conditions* | 344 |

| | | | |
|---|---|---|---|
| CHAPTER 12 | | Volume Change Theory | 346 |
| | 12.1 | Literature Review | 346 |
| | 12.2 | Concepts of Volume Change and Deformation | 349 |
| | | 12.2.1 Continuity Requirements | 349 |
| | | 12.2.2 Overall Volume Change | 350 |
| | | 12.2.3 Water and Air Volume Changes | 351 |
| | 12.3 | Constitutive Relations | 351 |
| | | 12.3.1 Elasticity Form | 351 |
| | |     *Water phase constitutive relation* | 353 |
| | |     *Change in the volume of air* | 353 |
| | |     *Isotropic loading* | 354 |
| | |     *Uniaxial loading* | 354 |
| | |     *Triaxial loading* | 354 |
| | |     *$K_0$-loading* | 356 |
| | |     *Plane strain loading* | 357 |
| | |     *Plane stress loading* | 357 |
| | | 12.3.2 Compressibility Form | 357 |
| | | 12.3.3 Volume–Mass Form (Soil Mechanics Terminology) | 358 |
| | | 12.3.4 Use of $(\sigma - u_w)$ and $(u_a - u_w)$ to Formulate Constitutive Relations | 358 |
| | 12.4 | Experimental Verifications for Uniqueness of Constitutive Surfaces | 360 |
| | | 12.4.1 Sign Convention for Volumetric Deformation Properties | 361 |
| | | 12.4.2 Verification of Uniqueness of the Constitutive Surfaces Using Small Stress Changes | 361 |
| | | 12.4.3 Verification of the Constitutive Surfaces Using Large Stress State Variable Changes | 363 |
| | 12.5 | Relationship Among Volumetric Deformation Coefficients | 365 |
| | | 12.5.1 Relationship of Volumetric Deformation Coefficients for the Void Ratio and Water Content Surfaces | 366 |
| | | 12.5.2 Relationship of Volumetric Deformation Coefficients for the Volume–Mass Form of the Constitutive Surfaces | 367 |
| | | 12.5.3 Laboratory Tests Used for Obtaining Volumetric Deformation Coefficients | 367 |
| | | 12.5.4 Relationship of Volumetric Deformation Coefficients for Unloading Surfaces | 369 |
| | | 12.5.5 Relationship of Volumetric Deformation Coefficients for Loading and Unloading Surfaces | 370 |
| | | 12.5.6 Constitutive Surfaces on a Semi-Logarithmic Plot | 370 |
| CHAPTER 13 | | Measurements of Volume Change Indices | 374 |
| | 13.1 | Literature Review | 374 |
| | 13.2 | Test Procedures and Equipments | 376 |
| | | 13.2.1 Loading Constitutive Surfaces | 377 |
| | |     *Oedometer tests* | 378 |
| | |     *Pressure plate drying tests* | 379 |
| | |     *Shrinkage tests* | 380 |

| | | *Determination of volume change indices* | 380 |
| | | *Determination of volume change indices associated with the transition plane* | 382 |
| | | *Typical results from pressure plate tests* | 386 |
| | | *Determination of in situ stress state using oedometer test results* | 388 |
| | | *"Constant volume" test* | 388 |
| | | *"Free-swell" test* | 389 |
| | | *Correction for the compressibility of the apparatus* | 389 |
| | | *Correction for sampling disturbance* | 390 |
| | 13.2.2 | Unloading Constitutive Surfaces | 392 |
| | | *Unloading tests after compression* | 392 |
| | | *Pressure plate wetting tests* | 393 |
| | | *Free-swell tests* | 394 |
| | | *Determination of volume change indices* | 395 |

# CHAPTER 14  Volume Change Predictions  397

14.1 Literature Review  397
    14.1.1 Factors Affecting Total Heave  401

14.2 Past, Present, and Future States of Stress  403
    14.2.1 Stress State History  404
    14.2.2 *In Situ* Stress State  405
    14.2.3 Future Stress State and Ground Movements  406

14.3 Theory of Heave Predictions  406
    14.3.1 Total Heave Formulations  407
    14.3.2 Prediction of Final Pore-Water Pressures  408
    14.3.3 Example of Heave Calculations  408
    14.3.4 Case Histories  410
        *Slab-on-grade floor, Regina, Saskatchewan*  410
        *Eston school, Eston, Saskatchewan*  411

14.4 Control Factors in Heave Prediction and Reduction  411
    14.4.1 Closed-Form Heave Equation when Swelling Pressure is Constant  412
    14.4.2 Effect of Correcting the Swelling Pressure on the Prediction of Total Heave  413
    14.4.3 Example with Wetting from the Top to a Specified Depth  414
    14.4.4 Example with a Portion of the Profile Removed by Excavation and Backfilled with a Nonexpansive Soil  415

14.5 Notes on Collapsible Soils  417

# CHAPTER 15  One-Dimensional Consolidation and Swelling  419

15.1 Literature Review  419

15.2 Physical Relations Required for the Formulation  420

15.3 Derivation of Consolidation Equations  422
    15.3.1 Water Phase Partial Differential Equation  423
        *Saturated condition*  424
        *Dry soil condition*  424
        *Special case of an unsaturated soil condition*  424
    15.3.2 Air Phase Partial Differential Equation  425
        *Saturated soil condition*  426

|  |  |  |  | CONTENTS | xxiii |
|---|---|---|---|---|---|

|  |  | *Dry soil condition* | 426 |
|  |  | *Special case of an unsaturated soil* | 426 |
| 15.4 | Solution of Consolidation Equations Using Finite Difference Technique | | 427 |
| 15.5 | Typical Consolidation Test Results for Unsaturated Soils | | 429 |
|  | 15.5.1 | Tests on Compacted Kaolin | 429 |
|  |  | *Presentation of results* | 429 |
|  |  | *Theoretical analyses* | 430 |
|  | 15.5.2 | Tests on Silty Sand | 433 |
|  |  | *Presentation of results* | 433 |
|  |  | *Theoretical analysis* | 435 |
| 15.6 | Dimensionless Consolidation Parameters | | 437 |

**CHAPTER 16** **Two- and Three-Dimensional Unsteady-State Flow and Nonisothermal Analyses** — 440

| 16.1 | Uncoupled Two-Dimensional Formulations | | 440 |
|---|---|---|---|
|  | 16.1.1 | Unsteady-State Seepage in Isotropic Soil | 440 |
|  |  | *Water phase partial differential equation* | 441 |
|  |  | *Air phase partial differential equation* | 441 |
|  | 16.1.2 | Unsteady-State Seepage in an Anisotropic Soil | 441 |
|  |  | *Anisotropy in permeability* | 442 |
|  |  | *Water phase partial differential equation* | 443 |
|  |  | *Seepage analysis using the finite element method* | 444 |
|  |  | *Examples of two-dimensional problems and their solutions* | 447 |
|  |  | *Example of water flow through an earth dam* | 447 |
|  |  | *Example of groundwater seepage below a lagoon* | 447 |
|  |  | *Example of seepage within a layered hill slope* | 449 |
| 16.2 | Coupled Formulations of Three-Dimensional Consolidation | | 456 |
|  | 16.2.1 | Constitutive Relations | 456 |
|  |  | *Soil structure* | 461 |
|  |  | *Water phase* | 463 |
|  |  | *Air phase* | 472 |
|  | 16.2.2 | Coupled Consolidation Equations | 472 |
|  |  | *Equilibrium equations* | 472 |
|  |  | *Water phase continuity* | 473 |
|  |  | *Air phase continuity* | 473 |
| 16.3 | Nonisothermal Flow | | 473 |
|  | 16.3.1 | Air Phase Partial Differential Equation | 473 |
|  | 16.3.2 | Fluid and Vapor Flow Equation for the Water Phase | 474 |
|  | 16.3.3 | Heat Flow Equation | 474 |
|  | 16.3.4 | Atmospheric Boundary Conditions | 475 |
|  |  | *Surface boundary conditions for air and fluid water flow* | 475 |
|  |  | *Surface boundary conditions for water vapor flow* | 475 |
|  |  | *Surface boundary conditions for heat flow* | 477 |

| | | | |
|---|---|---|---|
| **APPENDIX A** | | Units and Symbols | 479 |
| **APPENDIX B** | | Theoretical Justification for Stress State Variables | 483 |
| | B.1 | Equilibrium Equations for Unsaturated Soils | 483 |
| | B.2 | Total or Overall Equilibrium | 483 |
| | B.3 | Independent Phase Equilibrium | 484 |
| | | B.3.1 Water Phase Equilibrium | 485 |
| | | B.3.2 Air Phase Equilibrium | 485 |
| | | B.3.3 Contractile Skin Equilibrium | 485 |
| | B.4 | Equilibrium of the Soil Structure (i.e., Arrangement of Soil Particles) | 488 |
| | B.5 | Other Combinations of Stress State Variables | 489 |
| | | References | 490 |
| | | About the Authors | 508 |
| | | Index | 510 |

# CHAPTER 1

## *Introduction to Unsaturated Soil Mechanics*

Soil mechanics involves a combination of engineering mechanics and the properties of soils. This description is broad and can encompass a wide range of soil types. These soils could either be saturated with water or have other fluids in the voids (e.g., air). The development of classical soil mechanics has led to an emphasis on particular types of soils. The common soil types are saturated sands, silts and clays, and dry sands. These materials have been the emphasis in soil mechanics textbooks. More and more, it is realized that attention must be given to a broader spectrum of materials. This can be illustrated by the increasing number of research conferences directed towards special classes of soil types and problems.

There are numerous materials encountered in engineering practice whose behavior is not consistent with the principles and concepts of classical, saturated soil mechanics. Commonly, it is the presence of more than two phases that results in a material that is difficult to deal with in engineering practice. Soils that are unsaturated form the largest category of materials which do not adhere in behavior to classical, saturated soil mechanics.

The general field of soil mechanics can be subdivided into that portion dealing with saturated soils and that portion dealing with unsaturated soils (Fig. 1.1). The differentiation between saturated and unsaturated soils becomes necessary due to basic differences in their nature and engineering behavior. An unsaturated soil has more than two phases, and the pore-water pressure is negative relative to the pore-air pressure. Any soil near the ground surface, present in a relatively dry environment, will be subjected to negative pore-water pressures and possible desaturation.

The process of excavating, remolding, and recompacting a soil also results in an unsaturated material. These materials form a large category of soils that have been difficult to consider within the framework of classical soil mechanics.

Natural surficial deposits of soil are at relatively low water contents over a large area of the earth. Highly plastic clays subjected to a changing environment have produced the category of materials known as swelling soils. The shrinkage of soils may pose an equally severe situation. Loose silty soils often undergo collapse when subjected to wetting, and possibly a loading environment. The pore-water pressure in both of the above cases is initially negative, and volume changes occur as a result of increases in the pore-water pressure.

Residual soils have been of particular concern in recent years. Once again, the primary factor contributing to their unusual behavior is their negative pore-water pressures. Attempts have been made to use saturated soil mechanics design procedures on these soils with limited success.

An unsaturated soil is commonly defined as having three phases, namely, 1) solids, 2) water, and 3) air. However, it may be more correct to recognize the existence of a fourth phase, namely, that of the air-water interface or contractile skin (Fredlund and Morgenstern, 1977). The justification and need for a fourth phase is discussed later in this chapter. The presence of even the smallest amount of air renders a soil unsaturated. A small amount of air, likely occurring as occluded air bubbles, renders the pore fluid compressible. Generally, it is a larger amount of air which makes the air phase continuous throughout the soil. At the same time, the pore-air and pore-water pressures begin to differ significantly, with the result that the principles and concepts involved differ from those of classical, saturated soil mechanics. These differing conditions are addressed throughout this book.

## 1.1 ROLE OF CLIMATE

Climate plays an important role in whether a soil is saturated or unsaturated. Water is removed from the soil either by evaporation from the ground surface or by evapotranspiration from a vegetative cover (Fig. 1.2). These processes produce an upward flux of water out of the soil. On the other hand, rainfall and other forms of precipitation provide a downward flux into the soil. The difference be-

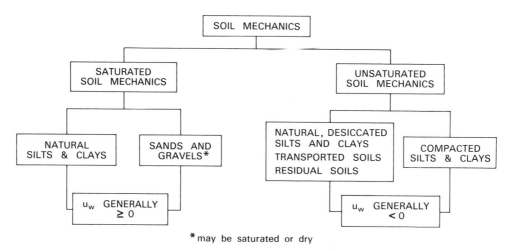

**Figure 1.1** Categorization of soil mechanics.

tween these two flux conditions on a local scale largely dictates the pore-water pressure conditions in the soil.

A net upward flux produces a gradual drying, cracking, and desiccation of the soil mass, whereas a net downward flux eventually saturates a soil mass. The depth of the water table is influenced, amongst other things, by the net surface flux. A hydrostatic line relative to the groundwater table represents an equilibrium condition where there is no flux at ground surface. During dry periods, the pore-water pressures become more negative than those represented by the hydrostatic line. The opposite condition occurs during wet periods.

Grasses, trees, and other plants growing on the ground surface dry the soil by applying a tension to the pore-water through evapotranspiration (Dorsey, 1940). Most plants are capable of applying 1-2 MPa (10-20 atm) of tension to the pore-water prior to reaching their wilting point (Taylor and Ashcroft, 1972). Evapotranspiration also results in the consolidation and desaturation of the soil mass.

The tension applied to the pore-water acts in all directions, and can readily exceed the lateral confining pressure in the soil. When this happens, a secondary mode of desaturation commences (i.e., cracking).

Year after year, the deposit is subjected to varying and changing environmental conditions. These produce changes in the pore-water pressure distribution, which in turn result in shrinking and swelling of the soil deposit. The pore-water pressure distribution with depth can take on a wide variety of shapes as a result of environmental changes (Fig. 1.2).

Significant areas of the earth's surface are classified as arid zones. The annual evaporation from the ground surface in these regions exceeds the annual precipitation. Figure 1.3 shows the climatic classification of the extremely arid, arid, and semi-arid areas of the world. Meigs (1953) used the Thornthwaite moisture index (Thornthwaite, 1948) to map these zones. He excluded the cold deserts. Regions with a Thornthwaite moisture index less than −40 indicate arid areas. About 33% of the earth's surface is considered arid and semi-arid (Dregne, 1976). The distribution of extremely arid, arid, and semi-arid areas in North America is shown in Fig. 1.4. These areas cover much of the region bounded by the Gulf of Mexico in the south, up into Canada in the north, over to the west coast.

Arid and semi-arid areas usually have a deep groundwater table. Soils located above the water table have neg-

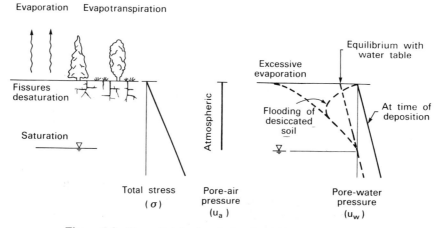

**Figure 1.2** Stress distribution during the desiccation of a soil.

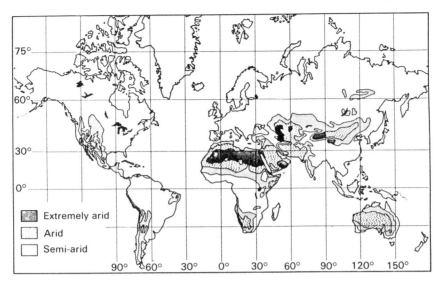

**Figure 1.3** Extremely arid, arid, and semi-arid areas of the world (from Meigs, 1953 and Dregne, 1976).

ative pore-water pressures. The soils are desaturated due to the excessive evaporation and evapotranspiration. Climatic changes highly influence the water content of the soil in the proximity of the ground surface. Upon wetting, the pore-water pressures increase, tending toward positive values. As a result, changes occur in the volume and shear strength of the soil. Many soils exhibit extreme swelling or expansion when wetted. Other soils are known for their significant loss of shear strength upon wetting. Changes in the negative pore-water pressures associated with heavy rainfalls are the cause of numerous slope failures. Reductions in the bearing capacity and resilient modulus of soils are also associated with increases in the pore-water pressures. These phenomena indicate the important role that negative pore-water pressures play in controlling the mechanical behavior of unsaturated oils.

## 1.2 TYPES OF PROBLEMS

The types of problems of interest in unsaturated soil mechanics are similar to those of interest in saturated soil mechanics. Common to all unsaturated soil situations are the

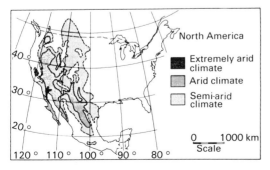

**Figure 1.4** Extremely arid, arid, and semi-arid areas of North America (from Meigs, 1953).

negative pressures in the pore-water. The type of problem involving negative pore-water pressures that has received the most attention is that of swelling or expansive clays. However, an attempt is made in this book to broaden the scope of problems to which the principles and concepts of unsaturated soil mechanics can be applied.

Several typical problems are described to illustrate relevant questions which might be asked by the geotechnical engineer. Throughout this book, an attempt is made to respond to these questions, mainly from a theoretical standpoint.

### 1.2.1 Construction and Operation of a Dam

Let us consider the construction of a homogeneous rolled earth dam. The construction involves compacting soil in approximately 150 mm (6 in) lifts from its base to the full height of the dam. The compacted soil would have an initial degree of saturation of about 80%. Figure 1.5 shows a dam at approximately one half of its design height, with a lift of soil having just been placed. The pore-air pressure in the layer of soil being compacted is approximately equal to the atmospheric pressure. The pore-water pressure is negative, often considerably lower than zero absolute pressure.

The soil at lower elevations in the fill is compressed by the placement of the overlying fill. Each layer of fill constitutes an increase in total stress to the embankment. Compression results in a change in the pore-air and pore-water pressures. The construction of the fill is generally rapid enough that the soil undergoes volume change under essentially undrained conditions. At any time during construction, the pore-air and pore-water pressures can be contoured as shown in Fig. 1.6.

In reality, some dissipation of the pore pressures will occur as the fill is being placed. The pore-air pressure will

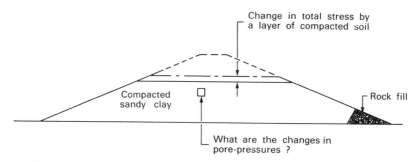

**Figure 1.5** Changes in pore pressure due to the placement of fill on a partly constructed dam.

dissipate to the atmosphere. The pore-water pressure may also be influenced by evaporation and infiltration at the surface of the dam. All pore pressure changes produce volume changes since the stress state is being changed.

There are many questions that can be asked, and there are many analyses that would be useful to the geotechnical engineer. During the early stages of construction, some relevant questions are:

- What is the magnitude of the pore-air and pore-water pressure induced as each layer of fill is placed?
- Is pore-air pressure of significance?
- Does the engineer only need to be concerned with the pore-water pressures?
- Does an induced pore-air pressure result in an increase or a decrease in the stability of the dam? Or would the computed factor of safety be conservative if the pore-air pressures are assumed to be zero?
- What is the effect of air going into solution and subsequently coming out of solution?
- Will the pore-air pressure dissipate to atmospheric conditions much faster than the pore-water pressures can come to equilibrium?

- What deformations would be anticipated as a result of changes in the total stress and the dissipation of the induced pore-air and pore-water pressures?
- What are the boundary conditions for the air and water phases during the placement of the fill?

Once the construction of the dam is complete, the filling of the reservoir will change the pore pressures in a manner similar to that shown in Fig. 1.7. This indicates a transient process with new boundary conditions. Some questions that might be asked are:

- What are the boundary conditions associated with the equalization processes once the filling of the reservoir is underway?
- How will the pore-air and pore-water pressures change with time, and what are the new equilibrium conditions?
- Will further deformation take place as the pore-air and pore-water pressures change in the absence of a change in total stress? If so, how much deformation can be anticipated as steady-state conditions are established?

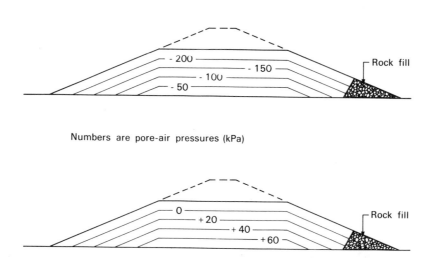

**Figure 1.6** Typical pore-water and pore-air pressures after partial construction of the dam.

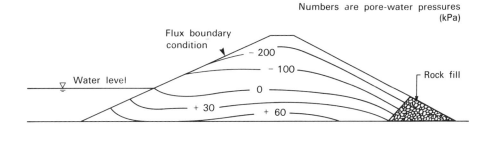

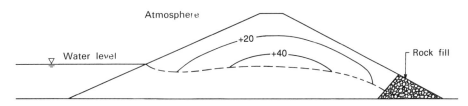

**Figure 1.7** Typical pore–water and pore–air pressures after some dissipation of pore pressures and partial filling of the reservoir.

- What changes take place in the limit equilibrium factor of safety of the dam as the reservoir is being filled and pore-water pressures tend to a steady-state condition?

After steady-state conditions are established, changes in the environment may give rise to further questions (Fig. 1.8).

- Does water flow across the phreatic surface under steady-state conditions?
- What effect will a prolonged dry or wet period have on the pore pressures in the dam?
- Could a prolonged dry period produce cracking of the dam? If so, to what depth might the cracks extend?
- Could a prolonged wet period result in the local or overall instability of the dam?

Answers to all of the above questions involve an understanding of the behavior of unsaturated soils. The questions involve analyses associated with saturated/unsaturated seepage, the change in volume of the soil mass, and the change in shear strength. The change in the shear strength could be expressed as a change in the factor of safety. These questions are similar to those asked when dealing with saturated soils; however, there is one primary difference. In the case of unsaturated soils problems, the flux boundary conditions produced by changes in the environment play a more important role.

### 1.2.2 Natural Slopes Subjected to Environmental Changes

Natural slopes are subjected to a continuously changing environment (Fig. 1.9). An engineer may be asked to investigate the present stability of a slope, and predict what would happen if the geometry of the slope were changed or if the environmental conditions should happen to change. In this case, boreholes may be drilled and undisturbed samples obtained for laboratory tests. Most or all of the potential slip surfaces may lie above the groundwater table. In other words, the potential slip surface may pass through unsaturated soils with negative pore-water pressures. Typical questions that might need to be addressed are:

- What effect could changes in the geometry have on the pore pressure conditions?

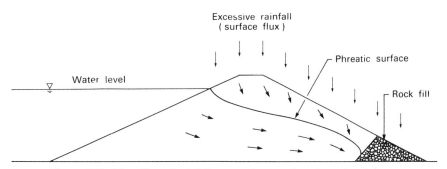

**Figure 1.8** The effect of rainfall on steady-state flow through a dam.

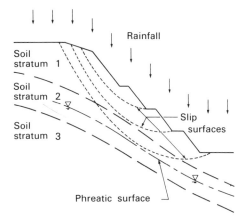

**Figure 1.9** An example of the effect of excavations on a natural slope subjected to environmental changes.

- What changes in pore pressures would result from a prolonged period of precipitation? How could reasonable pore pressures be predicted?
- Could the location of a potential slip surface change as a result of precipitation?
- How significantly would a slope stability analysis be affected if negative pore-water pressures were ignored?
- What would be the limit equilibrium factor of safety of the slope as a function of time?
- What lateral deformations might be anticipated as a result of changes in pore pressures?

Similar questions might be of concern with respect to relatively flat slopes. Surface sloughing commonly occurs on slopes following prolonged periods of precipitation. These failures have received little attention from an analytical standpoint. One of the main difficulties appears to have been associated with the assessment of pore-water pressures in the zone above the groundwater table.

The slow, gradual, downslope creep of soil is another aspect which has not received much attention in the literature. It has been observed, however, that the movements occur in response to seasonal, environment changes. Wetting and drying, freezing and thawing are known to be important factors. It would appear that an understanding of unsaturated soil behavior is imperative in formulating an analytical solution to these problems.

### 1.2.3 Mounding Below Waste Retention Ponds

Waste materials from mining and industry operations are often stored as a liquid or slurry retained by low-level dikes (Fig. 1.10). Soil profiles with a deep water table are considered to be ideal locations for these waste ponds. The soils above the water table have negative pore-water pressures and may be unsaturated. It has often been assumed that as long as the pore-water pressure remained negative, there is little or no movement of fluids downward from the waste pond. However, in recent years, it has been observed

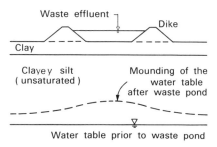

**Figure 1.10** An example of mounding below a waste pond due to seepage through an unsaturated soil.

that a mounding of the water table may occur below the waste pond, even though the intermediate soil may remain unsaturated. Now, engineers realize that significant volumes of water and contaminants can move through the soil profile, even though negative pore-water pressures are retained.

Questions of importance with respect to this type of problem would be:
- How should seepage be modeled for this situation? What are the boundary conditions?
- How should the coefficient of permeability of the unsaturated soil be characterized? The coefficient of permeability is a function of the negative pore-water pressure, and thereby becomes a variable in a seepage analysis.
- What equipment and procedures should be used to characterize the coefficient of permeability in the laboratory?
- How do the contaminant transport numerical models interface with unsaturated flow modeling?
- What would be the effect on the water table mounding if a clay liner were placed at the base of the retention pond?

### 1.2.4 Stability of Vertical or Near Vertical Excavations

Vertical or near vertical excavations are often used for the installation of a foundation or a pipeline (Fig. 1.11). It is well known that the backslope in a moist silty or clayey soil will stand at a near vertical slope for some time before failing. Failure of the backslope is a function of the soil type, the depth of the excavation, the depth of tension cracks, the amount of precipitation, as well other factors. In the event that the contractor should leave the excavation open longer than planned or, should a high precipitation period be encountered, the backslope may fail, causing damage and possible loss of life.

The excavations being referred to are in soils above the groundwater table where the pore-water pressures are negative. The excavation of soil also produces a further decrease in the pore-water pressures. This results in an increase in the shear strength of the soil. With time, there

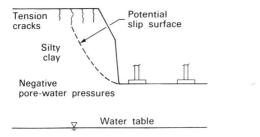

**Figure 1.11** An example of potential instability of a near vertical excavation during the construction of a foundation.

will generally be a gradual increase in the pore-water pressures in the backslope, and correspondingly, a loss in strength. The increase in the pore-water pressure is the primary factor contributing to the instability of the excavation. Engineers often place the responsibility for ensuring backslope stability onto the contractor. Predictions associated with this problem require an understanding of unsaturated soil behavior.

Some relevant questions that might be asked are:
- How long will the excavation backslope stand prior to failing?
- How could the excavation backslope be analytically modeled, and what would be the boundary conditions?
- What soil parameters are required for the above modeling?
- What *in situ* measurements could be taken to indicate incipient instability? Also, could soil suction measurements be of value?
- What effect would a ground surface covering (e.g., plastic sheeting) have on the stability of the backslope?
- What would be the effect of temporary bracing, and how much bracing would be required to ensure stability?

### 1.2.5 Lateral Earth Pressures

Figure 1.12 shows two situations where an understanding of lateral earth pressures is necessary. Another situation might involve lateral pressure against a grade beam placed on piles. Let us assume that in each situation, a relatively dry clayey soil has been placed and compacted. With time, water may seep into the soil, causing it to expand in both a vertical and horizontal direction. Although these situations may illustrate the development of high lateral earth pressures, they are not necessarily good design procedures.

Some questions that might be asked are:
- How high might the lateral pressures be against a vertical wall upon wetting of the backfill?
- What are the magnitudes of the active and passive earth pressures for an unsaturated soil?
- Are the lateral pressures related to the "swelling pressure" of the soil?

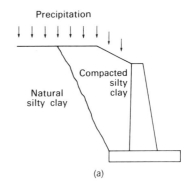

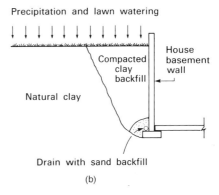

**Figure 1.12** Examples of lateral earth pressures generated subsequent to backfilling with dry soils. (a) Lateral earth pressures against a retaining wall as water infiltrates the compacted backfill; (b) lateral earth pressure against a house basement wall.

- Is there a relationship between the "swelling pressure" of a soil and the passive earth pressure?
- How much lateral movement might be anticipated as a result of the backfill becoming saturated?

### 1.2.6 Bearing Capacity for Shallow Foundations

The foundations for light structures are generally shallow spread footings (Fig. 1.13). The bearing capacity of the underlying (clayey) soils is computed based on the unconfined compressive strength of the soil. Shallow footings can easily be constructed when the water table is below the elevation of the footings. In most cases, the water table is at a considerable depth, and the soil below the footing has a negative pore-water pressure. Undisturbed samples, held intact by negative pore-water pressures, are routinely tested in the laboratory. The assumption is made that the pore-water pressure conditions in the field will remain relatively constant with time, and therefore, the unconfined compressive strength will also remain essentially unchanged. Based on this assumption, and a relatively high design factor of safety, the bearing capacity of the soil is computed.

The above design procedure has involved soils with negative pore-water pressures. It appears that the engineer has almost been oblivious to the problems related to the long-term retention of negative pore-water pressure when dealing with bearing capacity problems. Almost the opposite

# 1 INTRODUCTION TO UNSATURATED SOIL MECHANICS

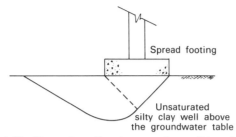

**Figure 1.13** Illustration of bearing capacity conditions for a light structure placed on soils with negative pore-water pressure.

attitude has been taken towards negative pore-water pressures when dealing with slope stability problems. That is, the attitude of the engineer has generally been that negative pore-water pressures cannot be relied upon to contribute to the shear strength of the soil on a long-term basis when dealing with slope stability problems. The two, seemingly opposite attitudes or perceptions, give rise to the question, "How constant are the negative pore-water pressures with respect to time?" Or, a more probing question might be, "Has the engineer's attitude towards negative pore-water pressures been strongly influenced by expediency?" This is a crucial question which requires further research and debate.

Other questions related to the design of shallow footings that might be asked are:
- What changes in pore-water pressures might occur as a result of sampling soils from above the water table?
- What effect does the *in situ* negative pore-water pressure and a reduced degree of saturation have on the measured, unconfined compressive strength? How should the laboratory results be interpreted?
- Would *confined* compression tests more accurately simulate the strength of an unsaturated soil for bearing capacity design?
- How much loss in strength could occur as a result of watering the lawn surrounding the building?

## 1.2.7 Ground Movements Involving Expansive Soils

There is no problem involving soils with negative pore-water pressures that has received more attention than the prediction of heave associated with the wetting of an expansive soil. Light structures such as a roadway or a small building are often subjected to severe distress subsequent to construction, as a result of changes in the surrounding environment (Figs. 1.14 and 1.15). Changes in the environment may occur as a result of the removal of trees, grass, and the excessive watering of a lawn around a new structure. The zone of soil undergoing volume change on an annual basis has been referred to as the "active zone." The higher the swelling properties of the soil, the greater will be the amount of heave to the structure.

It has been common practice to obtain undisturbed soil samples from the upper portion of the profile for one-dimensional oedometer testing in the laboratory. The laboratory results are used to provide quantitative estimates of potential heave. Numerous laboratory testing techniques and analytical procedures have been used in practice. Questions related to these procedures are discussed in detail in Chapter 13. However, other relevant questions that can be asked are:

- How much heave can be anticipated if the soil is flooded?
- How much heave can be anticipated if the negative pore-water pressures go to zero or remain at a slightly negative value?
- What is a reasonable final stress condition to assume for the pore-water pressures?
- What is the effect of prewetting or flooding the soil prior to constructing the foundation? How long must the prewetting continue? And what ground movements might continue subsequent to discontinuing the flooding?
- What is the effect of surcharging a swelling soil? How

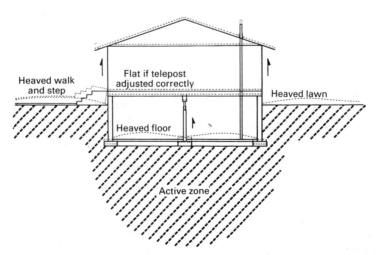

**Figure 1.14** Ground movements associated with the construction of shallow footings on an expansive soil (Hamilton, 1977).

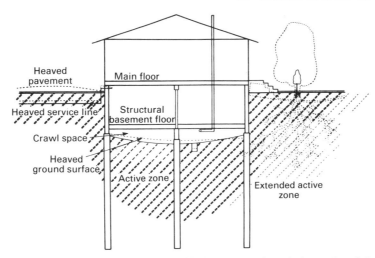

**Figure 1.15** Ground movements associated with the construction of a house founded on piles in expansive soils (Canadian Geotechnical Society, 1985).

much might the potential heave be reduced by surcharging?
- What is the effect of placing an impervious membrane around the perimeter of the footings?
- How much differential movement might one or more large trees produce on the foundation?
- What are the satisfactory laboratory testing procedures for measuring the swelling properties of an expansive soil?

### 1.2.8 Collapsing Soils

In many respects, collapsing soils can be thought of as behaving in an opposite manner to expansive soils. In both the expansive and collapsing cases, the initial pore-water pressures are negative. In both cases, movements are the result of an increase in the negative pore-water pressure. The wetting of a collapsing soil, however, results in a volume decrease. In this case, the soil is described as having a metastable soil structure.

The collapse of the soil structure may occur within a man-made or natural earth slope or in soil underlying a foundation. Research associated with the behavior of collapsing soils has been limited, and many questions remain to be answered from both a research and practice standpoint:

- How should collapsing soils be tested in the laboratory?
- How should the laboratory data be interpreted and applied to practical problems?

### 1.2.9 Summary of Unsaturated Soils Examples

The above examples show that there are many practical situations involving unsaturated soils that require an understanding of the seepage, volume change, and shear strength characteristics. In fact, there is often an interaction among, and a simultaneous interest in, all three of the aspects of unsaturated soil mechanics. Typically, a flux boundary condition produces an unsteady-state saturated/unsaturated flow situation which results in a volume change and a change in the shear strength of the soil. The change in shear strength is generally translated into a change in factor of safety. There may also be an interest in quantifying the change of other volume-mass soil properties (i.e., water content and degree of saturation).

The classical one-dimensional theory of consolidation is of central importance in *saturated* soil mechanics. The theory of consolidation predicts the change in pore-water pressure with respect to depth and time in response to a change in total stress. The changes in pore-water pressure are used to predict the volume change. The theory of consolidation does not play as important a role for *unsaturated* soils as it does for saturated soils. The application of a total stress to an unsaturated soil produces larger instantaneous volume changes, but smaller volume changes with respect to time. The induced pore-water pressures are considerably smaller than the applied total stress. The more common boundary condition for unsaturated soils is a change in flux as opposed to a change in total stress for a saturated soil. Nevertheless, the theory of consolidation for *unsaturated* soils plays an important phenomenological role. It assists the engineer in visualizing complex mechanisms, providing a qualitative "feel" for the behavior of an unsaturated soil.

## 1.3 TYPICAL PROFILES OF UNSATURATED SOILS

The microclimatic conditions in an area are the main factor causing a soil deposit to be unsaturated. Therefore, unsaturated soils or soils with negative pore-water pressures can occur in essentially any geological deposit. An unsaturated soil could be a residual soil, a lacustrine deposit, a bedrock formation, and so on. However, there are certain geologi-

10   1 INTRODUCTION TO UNSATURATED SOIL MECHANICS

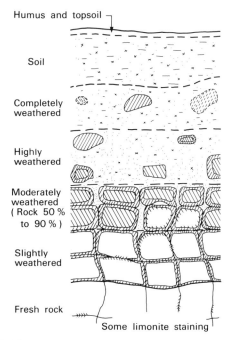

**Figure 1.16** Schematic diagram showing a typical tropical residual soil profile (from Little, 1969).

cal categories of soils with negative pore-water pressures that have received considerable attention in the research literature. A few examples will illustrate some of the features common to these deposits.

### 1.3.1 Typical Tropical Residual Soil Profile

Tropical residual soils have some unique characteristics related to their composition and the environment under which they develop. Most distinctive is the microstructure which changes in a gradational manner with depth (Vargas, 1985; Brand, 1985). The *in situ* water content of residual soils is generally greater than its optimum water content for compaction. Their density, plasticity index, and compressibility are likely to be less than corresponding values for temperate zone soils with comparable liquid limits. Their strength and permeability are likely to be greater than those of temperate zone soils with comparable liquid limits (Mitchell and Sitar, 1982).

Most classical concepts related to soil properties and soil behavior have been developed for temperate zone soils, and there has been difficulty in accurately modeling procedures and conditions to which residual soils will be subjected. Engineers appear to be slowly recognizing that residual soils are generally soils with negative *in situ* pore-water pressures, and that much of the unusual behavior exhibited during laboratory testing is related to a matric suction change in the soil (Fredlund and Rahardjo, 1985).

A typical deep, tropical weathering profile is shown in Fig. 1.16 (Little, 1969). Boundaries between layers are generally not clearly defined. Numerous systems of classification have been proposed based primarily on the degree of weathering and engineering properties (Deere and Patton, 1971; Tuncer and Lohnes, 1977; Brand, 1982).

Zones of completely weathered or highly weathered rock that contain particulate soil but retain the original rock structure are termed saprolite. Once the deposit has essentially no resemblance of the parent rock, it is termed a lateritic or residual soil.

Figure 1.17 shows the profile and soil properties for a porous, saprolite soil from basalt in Brazil (Vargas, 1985).

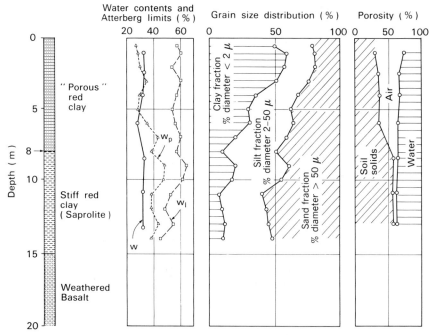

**Figure 1.17** Porous saprolite soil from basalt near Londrina, Brazil (from Vargas, 1985).

1.3 TYPICAL PROFILES OF UNSATURATED SOILS   11

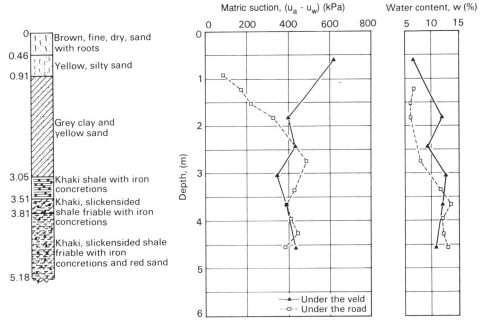

**Figure 1.18** Description of the soil profile at Welkom, South Africa, along with suction and water content profiles (from de Bruijn, 1965).

The region has a hot, humid summer and a mild, dry winter climate, with an annual rainfall of less than 1500 mm. The structure is highly porous, and in some cases may be unstable, resulting in collapse upon saturation. The soil deposit is unsaturated, and the *in situ* pore-water pressure is negative.

### 1.3.2 Typical Expansive Soils Profile

Expansive soils deposits may range from lacustrine deposits to bedrock shale deposits. In general, expansive soils have a high plasticity (i.e., high liquid limit) and are relatively stiff or dense. The above description is typical, but not exclusive. The expansive nature of the soil is most obvious near ground surface where the profile is subjected to seasonal, environmental changes. The pore-water pressure is initially negative and the deposit is generally unsaturated. These soils often have some montmorillonite clay mineral present. The higher the amount of monovalent cations absorbed to the clay mineral (e.g., sodium), the more severe the expansive soils problem.

Expansive soils deposits and their related engineering problems have been reported in many countries. One of the first countries to embark on research into expansive soils was South Africa. Figure 1.18 shows a typical expansive soil profile, along with water content and soil suction values from near Welkom, South Africa (de Bruijn, 1965). The area is well known for its foundation problems, giving rise to extensive damage to buildings and roads.

The soil profile (Fig. 1.18) shows a clay and sand deposit overlying a shale. The low water contents and high soil suctions give rise to high swelling upon wetting. The data show that the soil is wetter and the suctions are lower in the zone below a covered roadway than below an open field.

Figure 1.19 shows a soil profile of an expansive soil in Tel Aviv, Israel (Katzir, 1974). The upper portion of the excavation consisted of a highly plastic, slickensided, fissured clay to a depth of 8-10 m. The water table was at 14 m. The upper portion had a liquid limit of 60%, a plastic limit of 25%, and a shrinkage limit of 11%. These properties are typical of swelling clay profiles.

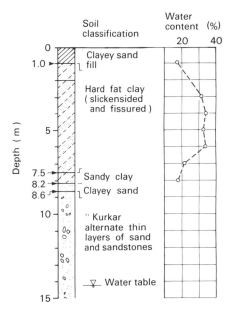

**Figure 1.19** Profile of expansive soil from Tel Aviv, Israel (from Katzir, 1974).

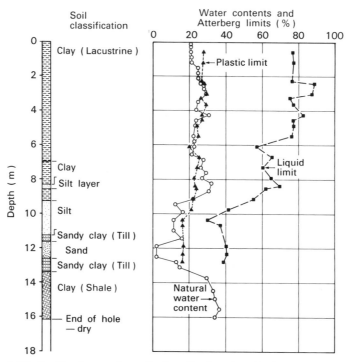

**Figure 1.20** Profile of expansive soils from Regina, Canada (from Fredlund, 1973).

Extensive areas of western Canada are covered by preglacial, lacustrine clay sediments that are known for their expansive nature. Regina, Saskatchewan is located in a semi-arid area where the annual precipitation is approximately 350 mm. A typical soil profile is shown in Fig. 1.20 (Fredlund, 1973). The average liquid limit is 75% and the average plastic limit is 25%. The shrinkage limit is typically 15%. The lacustrine clay is classified as a calcium montmorillonite. Buildings founded on shallow footings often experience 50-150 mm of movement subsequent to construction.

The above soil profiles are typical of conditions which are found in many parts of the world. In each case, the natural water contents are low and the pore-water pressures are negative. Throughout each season, and from season to season, the soil expands and contracts in response to changes in the environment.

## 1.4 NEED FOR UNSATURATED SOIL MECHANICS

The success of the practice of soil mechanics can be traced largely to the ability of engineers to relate observed soil behavior to stress conditions. This ability has led to the transferability of the science and a relatively consistent engineering practice. Although this has been true for saturated soils, it has not been the case for unsaturated soils. Difficulty has been experienced in extending classical soil mechanics to embrace unsaturated soils. This can be borne out by the empirical nature of much of the research associated with unsaturated soils.

The question can be asked: "Why hasn't a practical science developed and flourished for unsaturated soils?" A cursory examination may suggest that there is no need for such a science. However, this is certainly not the case when the problems associated with expansive soils are considered. Jones and Holtz (1973) reported that in the United States alone: "Each year, shrinking and swelling soils inflict at least $2.3 billion in damages to houses, buildings, roads, and pipelines—more than twice the damage from floods, hurricanes, tornadoes, and earthquakes!" They also reported that 60% of the new houses built in the United States will experience minor damage during their useful lives and 10% will experience significant damage—some beyond repair.

In 1980, Krohn and Slosson estimated that $7 billion are spent each year in the United States as a result of damage to all types of structures built on expansive soils. Snethen (1986) stated that, "While few people have ever heard of expansive soils and even fewer realize the magnitude of the damage they cause, more than one fifth of American families live on such soils and no state is immune from the problem they cause. Expansive soils have been called the "hidden disaster": while they do not cause loss of life, economically, they are one of the United States costliest natural hazards." The problem extends to many other countries of the world. In Canada, Hamilton (1977) stated that: "Volume changing clay subsoils constitute the most costly natural hazard to buildings on shallow foundations in Canada and the United States. In the Prairie Provinces alone, a million or more Canadians live in communities built on subsoils of very high potential expansion.''

It would appear that, internationally, most countries in the world have problems with expansive soils. Many countries have reported their problems at research conferences. Some countries reporting problems with expansive soils are: Australia, Argentina, Burma, China, Cuba, Ethiopia, Ghana, Great Britain, India, Iran, Israel, Kenya, Mexico, Morocco, South Africa, Spain, Turkey, and Venezuela. In general, the more arid the climate, the more severe is the problem. A series of conferences have been held to help cope with the expansive soil problem. These are:

1) First Symposium on Expansive Clays, South Africa (1957).
2) Symposium on Expansive Clays at the Colorado School of Mines in Golden, Colorado (1959).
3) Conference on Pore Pressure and Suction in Soils, London (1960).
4) First International Conference on Expansive Soils, Texas A & M, College Station, Texas (1965).
5) A Symposium in Print, Moisture Equilibria and Moisture Changes in Soils, Australia (1965).
6) Second International Conference on Expansive Soils, Texas A & M, College Station, Texas (1969).
7) Third International Conference on Expansive Soils, Haifa, Israel (1973).
8) Fourth International Conference on Expansive Soils, Denver, Colorado (1980).
9) Fifth International Conference on Expansive Soils, Adelaide, Australia (1984).
10) Sixth International Conference on Expansive Soils, New Delhi, India (1987).
11) Seventh International Conference on Expansive Soils, Dallas, Texas (1992).

There is also the need for reliable engineering design associated with compacted soils, collapsing soils, and residual soils. Soils that collapse usually have an initially low density. The soils may or may not be subjected to additional load, but they are usually given access to water. The water causes an increase in the pore-water pressures, with the result that the soil volume decreases. The process is similar to that occurring in an expansive soil, but the direction of volume change is the opposite. Examples of soil collapse have recently been reported in numerous countries. In the United States, for example, Johnpeer (1986) reported examples of collapse in New Mexico. Leaky septic tanks were a common initiating factor in the soil collapse.

Inhabited areas with steep slopes consisting of residual soils are sometimes the site of catastrophic landslides which claim many lives. A widely publicized case is the landslide at Po-Shan in Hong Kong which claimed 67 lives (The Commission of Inquiry, 1972). Similar problems have been reported in South American countries and other parts of the world.

The soils involved are often residual in genesis and have deep water tables. The surface soils have negative pore-water pressures which play a significant role in the stability of the slope. However, heavy, continuous rainfalls can result in increased pore-water pressures to a significant depth, resulting in the instability of the slope. The pore-water pressures along the slip surface at the time of failure may be negative or positive.

There appear to be two main reasons why a practical science has not developed for unsaturated soils (Fredlund, 1979). First, there has been the lack of an appropriate science with a theoretical base. This commences with a lack of appreciation of the engineering problems and an inability to place the solution within a theoretical context. The stress conditions and mechanisms involved, as well as the soil properties that must be measured, do not appear to have been fully understood. The boundary conditions for an analysis are generally related to the environment and are difficult to predict. Research work has largely remained empirical in nature, with little coherence and synthesis. There has been poor communication among engineers, and design procedures are not widely accepted and adhered to.

Second, there appears to be the lack of a system for financial recovery for services rendered by the engineer. In the case of expansive soil problems, the possible liability to the engineer is often too great relative to the financial remuneration. Other areas of practice are more profitable to consultants. The owner often reasons that the cost outweighs the risk. The hazard to life and injury is largely absent, and for this reason little attention has been given to the problem by government agencies. Although the problem basically remains with the owner, it is the engineer who has the greatest potential for circumventing possible problems.

Certainly, there is a need for an appropriate technology for unsaturated soil behavior. Such a technology must: 1) be practical, 2) not be too costly to employ, 3) have a sound theoretical basis, and 4) run parallel in concept to conventional saturated soil mechanics.

## 1.5 SCOPE OF THE BOOK

This book addresses the field of unsaturated soil mechanics. An attempt is made to cover all of the aspects normally associated with soil mechanics. When the term ''unsaturated soil mechanics'' is used, the authors are referring to soils which have negative pore-water pressures.

The aspects of interest to geotechnical engineering fall into three main categories. These can be listed as problems related to: 1) the flow of water through porous media, 2) the shear strength, and 3) the volume change behavior of unsaturated soils. No attempt is made to duplicate or redevelop information already available in classic saturated soil mechanics books. This book should be used to assist the geotechnical engineer in understanding soil mechanics concepts unique to unsaturated soils. At the same time, these concepts have been developed and organized to appear as logical and simple extensions of classical saturated soil mechanics. Subjects such as clay mineralogy and physico-chemical properties of soils are vitally important to understanding why soils behave in a certain manner. However, the readers are referred to excellent references for coverage of these subjects (Yong and Warkentin, 1966; Mitchell, 1976).

Most soil mechanics problems can be related to one (or more) of the three main soil properties. These categories are 1) the coefficient of permeability, 2) the shear strength parameters, and 3) the volume change indices. There are three chapters written to cover each category of soil properties. The chapters can be described as: 1) the theory related to the soil property, 2) the measurement of the soil property, and 3) the application of the soil property to one or more soil mechanics problems. Chapters 5, 6, and 7 present the theory, measurement, and application, respectively, relative to the air and water flow. Chapters 9, 10, and 11 present the theory, measurement, and application, respectively, relative to shear strength. Chapters 12, 13, and 14 present the theory, measurement, and application, respectively, relative to volume change. The equipment required and details on the testing procedures are presented under the "measurement" chapters in each case.

Considerable attention is devoted to the description of the stress state variables required to describe unsaturated soil behavior (Chapter 3). The space devoted to this topic is in keeping with the importance of this fundamental concept. Chapter 4 describes the techniques and devices available to measure soil suction. This is an area of ongoing research, and an attempt is made to present the most recent information on the measurement of total, matric, and osmotic suctions.

The concept, theory, and application of pore pressure parameters is presented in Chapter 8. Pore pressure parameters involve the combination of the constitutive relations for the soil and the phase relations (Chapter 2) to predict the change in pore-air and pore-water pressures as a result of applying a total stress and not allowing drainage.

The theory of consolidation, as well as unsteady-state flow analyses, combine the volume change properties and the flow laws of a soil. The one-dimensional consolidation theory is presented in Chapter 15, while two- and three-dimensional, unsteady-state flow is presented in Chapter 16.

The main objective of this book is to synthesize theories associated with the behavior of unsaturated soils. The theoretical derivations are presented in considerable detail because unsaturated soil behavior is a relatively new area of study and many of the derivations are not readily available to engineers. There is a need for case histories, and it is anticipated that these will become more common in future decades. Hopefully, as the analyses are related to case histories, the engineer can benefit from the consistent theoretical context provided by this book.

## 1.6 PHASES OF AN UNSATURATED SOIL

An unsaturated soil is a mixture of several phases. It is important to establish the number of phases comprising the soil since it has an influence on how the stress state of the mixture is defined. First, it is important to define what is meant by a phase. On the basis of the definition of a phase, it is proposed that an unsaturated soil actually consists of four phases rather than the commonly referred to three phases. It is postulated that in addition to the solid, air, and water phases, there is the air-water interface that can be referred to as the contractile skin. Let us pursue the justification for reference to the contractile skin as an independent phase.

### 1.6.1 Definition of a Phase

In order for a portion of a mixture to qualify as an independent phase, it must have: 1) differing properties from the contiguous materials, and 2) definite bounding surfaces. These two conditions must be met in order to identify an independent phase. It is easy to understand how a saturated soil consists of two phases (i.e., soil solids and water). It is also quite understandable that the air becomes another independent phase for an unsaturated soil. Each of these phases (i.e., soil solids, water, and air) obviously meet the requirements for designation as a phase.

It is also possible for a phase to change state, as is the case when water freezes. The ice becomes an independent phase from the water. However, more important is the examination of the properties and extent of the air-water interface.

### 1.6.2 Air-Water Interface or Contractile Skin

The most distinctive property of the contractile skin is its ability to exert a tensile pull. It behaves like an elastic membrane under tension interwoven throughout the soil structure. It appears that most properties of the contractile skin are different from those of the contiguous water phase (Davies and Rideal, 1963). For example, its density is reduced, its heat conductance is increased, and its birefringence data are similar to those of ice. The transition from the liquid water to the contractile skin has been shown to be distinct or jumpwise (Derjaguin, 1965). It is interesting to note that insects such as the "water spider" walk on top of the contractile skin, and those such as the "backswimmer" walk on the bottom of the contractile skin

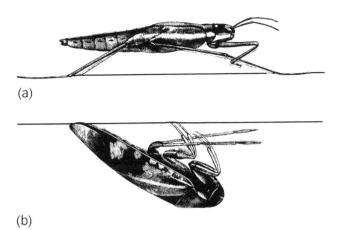

**Figure 1.21** Insects that live above and below the contractile skin. (a) Water strider; (b) Backswimmer. (From Milne and Milne, 1978).

(Milne and Milne, 1978). The water strider would sink into the water were it not for the contractile skin, whereas the backswimmer would pop out of the water (Fig. 1.21).

It is advantageous to recognize an unsaturated soil as a four-phase system when performing a stress analysis on an element (Fredlund and Morgenstern, 1977). From a behavioral standpoint, an unsaturated soil can be visualized as a mixture with two phases that come to equilibrium under applied stress gradients (i.e, soil particles and contractile skin) and two phases that flow under applied stress gradients (i.e., air and water).

From the standpoint of the volume-mass relations for an unsaturated soil, it is possible to consider the soil as a three-phase system since the volume of the contractile skin is small and its mass can be considered as part of the mass of water. However, when considering the stress analysis of a multiphase continuum, it is necessary to realize that the air-water interphase behaves as an independent phase.

Geotechnical engineers are familiar with a shrinkage-type experiment involving the drying of a small soil specimen as it is exposed to the atmosphere. The total stresses on the specimen remain unchanged at zero while the specimen undergoes a decrease in volume. The pore-water pressure goes increasingly negative during the experiment, but it is the contractile skin (or air-water interface) which acts like a thin rubber membrane, pulling the particles together.

## 1.7 TERMINOLOGY AND DEFINITIONS

Numerous approaches could be taken in the development of the discipline of unsaturated soil mechanics. The approaches may range from an empirical approach based strictly on experience to a particulate mechanics or quantum mechanics approach. The approach used throughout this book can be referred to as a macroscopic, phenomenological approach to unsaturated soil behavior. In other words, the science is developed around observable phenomena, while adhering to continuum mechanics principles. This approach has proven to be most successful in *saturated* soil mechanics and should be retained in *unsaturated* soil mechanics. An attempt is made to ensure a smooth transition in rationale between the saturated and unsaturated cases.

The authors are reticent to introduce new variables and terminology to unsaturated soil mechanics. However, the terminology associated with saturated soil mechanics is not sufficient when applied to unsaturated soil mechanics. As a result, a few more universally acceptable terms are proposed. These terms are common to continuum mechanics and are defined within a thermodynamic context. The following definitions, put in a succinct form, are based on numerous continuum mechanics and thermodynamic references:

1) *State*: Nonmaterial variables required for the characterization of a system.
2) *Stress state variable*: The nonmaterial variables required for the characterization of the stress condition.
3) *Deformation state variables*: The nonmaterial variables required for the characterization of deformation conditions or deviations from an initial state.
4) *Constitutive relations*: Single-valued equations expressing the relationship between state variables.

*The International Dictionary of Physics and Electronics* (Michels, 1961) defines *state* variables as:

"a limited set of dynamical variables of the system, such as pressure, temperature, volume, etc., which are sufficient to describe or specify the state of the system completely for the considerations at hand."

Fung (1965) describes the *state* of a system as that "information required for a complete characterization of the system for the purpose at hand." Typical state variables for an elastic body are given as those variables describing the strain field, the stress field, and its geometry. The state variables must be independent of the physical properties of the material.

Constitutive relations, on the other hand, are single-valued expressions which relate one state variable to one or more other state variables (Fung, 1969). A stress versus strain relationship is a constitutive relation which describes the mechanical behavior of a material. The material properties involved may be an elastic modulus and a Poisson's ratio. The ideal gas equation relates pressure to density, and temperature and is called a constitutive equation. The gas constant is the material property. Simple, idealized constitutive equations are well established for a nonviscous

# 16  1 INTRODUCTION TO UNSATURATED SOIL MECHANICS

fluid, a Newtonian viscous fluid, and a perfectly elastic solid (Fung, 1969).

Other examples of constitutive equations relating stress state variables are the shear strength equation and the pore pressure parameter equations. Examples of constitutive equations relating stress state variables to deformation state variables are the stress versus strain equations, stress versus volumetric change equations, and the soil-water characteristic curve. From the above definitions, it is clear that the physical properties of a system are part of the constitutive relations for the system and are not to be a part of the description of the stress state. The use of the above concepts was advocated in a general manner by Matyas and Radhakrishna in 1968.

## 1.8 HISTORICAL DEVELOPMENTS

The first ISSMFE conference (International Society for Soil Mechanics and Foundation Engineering) in 1936 provided a forum for the establishment of principles and equations relevant to saturated soil mechanics. These principles and equations have remained pivotal throughout subsequent decades of research. This same conference was also a forum for numerous research papers on unsaturated soil behavior. Unfortunately, a parallel set of principles and equations did not immediately emerge for unsaturated soils. In subsequent years, a science and technology for unsaturated soils has been slow to develop (Fredlund, 1979). Not until the research at Imperial College in the late 1950's did the concepts for understanding unsaturated soils behavior begin to be established (Bishop, 1959). The research of Lytton (1967) in the United States did much to ensure that the understanding of unsaturated soil behavior was founded upon a sound theoretical basis. Namely, that all theories were consistent with the principles put forth in continuum mechanics. The following is a brief history, highlighting the early developments in our understanding of the behavior of unsaturated soils. Most of the early research on unsaturated soils was related to the flow of water in the zone of negative pore-water pressure (i.e., capillary flow).

Associated with the urban development in the 1930's was the construction of numerous engineering works such as irrigation and transportation projects. One of the first problems that appeared to perplex civil engineers was that of the movement of water above the groundwater table. The term "capillarity" was adopted to describe the phenomenon of water flow upward from the static groundwater table. This term was selected because of the similarity to the operation of a capillarity tube. Example problems were illustrated by Hogentogler and Barber in 1941. In their first example [Fig. 1.22(a)], water was shown to move up and over the impervious core of a dam, even when the core extended above the reservoir elevation. Water was reported to move over the impervious wall and result in seepage problems along the downstream face of the dam.

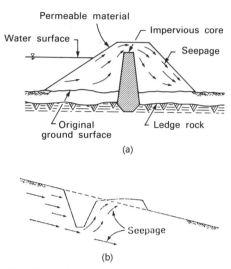

**Figure 1.22** Illustrations of unsaturated flow phenomena in the field. (a) Earth dam with an impervious core illustrating the syphon effect; (b) interceptor ditch for a highway and a side-hill location. (From Hogentogler and Barber 1941).

A second example, Fig. 1.22(b), illustrated the ineffectiveness of cutoff ditches in intercepting groundwater flow. The freezing and thawing of capillarity water lead to embankment instability and subgrade failures. These problems initiated research on soil capillarity for at least two decades.

Two research papers on soil capillarity were presented by Ostashev at the ISSMFE conference in 1936. Numerous factors, including pore-water pressure and capillary force, were described as affecting the mechanism of capillary flow. The Fourier heat flow equation was proposed for modeling the moisture flow process. At the same conference, two apparatuses were proposed for measuring the capillary pressure and capillary rise of water in soils (Boulichev, 1936).

Hogentogler and Barber (1941) attempted to give a comprehensive review of the nature of the capillary water. It was suggested that capillary water responded in accordance with the capillary rise equation for a fine bore tube. A capillarimeter was built to study the capillary phenomenon. The apparatus was a modification of the Bartell cell originally proposed in 1896 for use in studying lime and cement mortars. The capillarimeter measured the air entry value of the soil. It became the primary apparatus used for studying soil capillarity for several decades.

The phenomenon of capillary flow was illustrated using a simple beaker and sand-filled tube as shown in Fig. 1.23. Hogentogler and Barber (1941) offered the following comment on capillary flow, stating that, ". . . it is considered that capillary moisture of this type conforms strictly to recognized concepts of surface tension, the force of gravity and the principles of hydraulics as applied to free water. . . ." It was also suggested that the strength of an

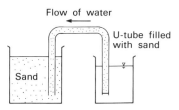

**Figure 1.23** Beaker and sand-filled glass tube for demonstrating the phenomenon of capillary flow through a sand (from Hogentogler and Barber).

unsaturated soil was greatly influenced by the state of stress in the capillary water. The state of stress in the capillary water was also related to evapotranspiration and the relative humidity of the air above the soil. It was suggested that the "stabilizing effect of capillary saturation" could be used in a practical way to stabilize the face of an excavation. Possible, practical concepts and applications associated with negative pore-water pressures appear to have been well developed by these researchers.

Terzaghi (1943) in his book, *Theoretical Soil Mechanics*, summarized the work of Hogentogler and Barber (1941) and endorsed the concepts related to the capillary tube model. The importance of the air–water interface was emphasized with respect to its effect on soil behavior. An equation was derived for the time required for the rise of water in the capillary zone. The assumption was made that the porosity, $n$, and the coefficient of permeability, $k$, were constants.

$$t = \frac{nh_c}{k}\left[\log\left(\frac{h_c}{h_c - z}\right) - \frac{z}{h_c}\right] \quad (1.1)$$

where:

- $t$ = time
- $n$ = porosity
- $h_c$ = maximum capillary rise
- $z$ = vertical distance above the groundwater table, corresponding to the elapsed time, $t$.

Valle-Rodas (1944) performed open tube and capillarimeter tests on uniform sands. In the open tests, sand was placed in a glass tube which had one end immersed in water. Experimental results showed an uneven water content in the sand in the capillary zone (Fig. 1.24). Moreover, it was found that there was a hysteresis in water content with respect to wetting and drying.

Further open tube and capillarimeter tests were performed by Lane and Washburn (1946) on cohesionless soils ranging from silts to gravels. The results indicated reasonable agreement between the measured capillary rise and the values predicted by Terzaghi's equation for the height of capillary rise. Terzaghi's equation for the prediction of the rate of capillary rise [Eq. (1.1)] appeared to overestimate the rate of capillary rise (Fig. 1.25). It was postulated that

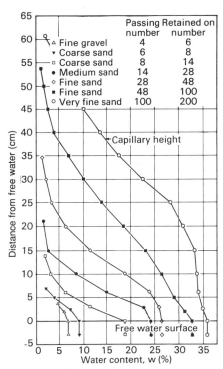

**Figure 1.24** Distribution of capillary water in a sand of varying gradation (from Valle-Rodas, 1944).

the discrepancy was due primarily to changing permeability in the capillary zone. It was shown that reduced values for the coefficient of permeability gave better correlations. Krynine (1948) reanalyzed the results of many capillary tests run between 1934 and 1946 and came to a similar conclusion. His results are summarized in Fig. 1.26.

Sitz (1948) noted that capillary water would rise to more than 10 m above a static groundwater table. He suggested that capillary water be subdivided into gravitational and molecular water. It was postulated that gravitational capillary water had properties similar to ordinary water, while molecular capillary water was assumed to have unique

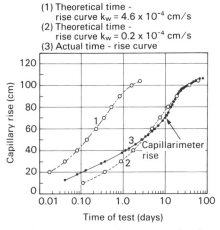

**Figure 1.25** Capillary rise versus time curves showing the actual and theoretical rise (from Lane and Washburn, 1946).

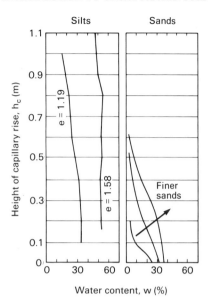

**Figure 1.26** Capillary rise tests on silts and fine sands (from Krynine, 1948).

properties. The property of water which was of primary interest was its ability to withstand high tensile stresses without cavitating or boiling.

Bernatzik (1948) observed an increase in the strength of a soil as a result of the air–water menisci. He suggested the use of an unconfined compression type of test on a soil specimen in order to study capillary tension.

Lambe (1951a) performed open tube capillary rise and drainage tests on graded sands and silts with various initial degrees of saturation. Negative pore-water pressures were measured at various locations along the specimens (Figs. 1.27 and 1.28). He again noted that the degree of saturation in the capillary zone was not 100%, and that the soil property controlling flow was noted to be different from the saturated coefficient of permeability. It was concluded that the hydraulic gradient was not uniform across the capillary zone. A large part of the available head appeared to be lost in the zone immediately behind the wetting front.

The historical review of the period up to the 1950's shows that most of the attention given to unsaturated soils was related to capillary flow. An attempt was made to use the capillary tube rise model (see Chapter 4) to explain the observed phenomenon. Although this model was of some value, it had limitations which became increasingly obvious. In fact, attempts to totally rely on the capillary tube rise model appear to be a significant factor in the slow development of unsaturated soil mechanics.

Research into the volume change and shear strength of unsaturated soils commenced with new impetus in the late 1950's. Some of the researchers were Black and Croney (1957), Bishop et al., (1960), Aitchison (1967), and Williams (1957). The research resulted in the proposal of several so-called effective stress equations for unsaturated soils. During the next decade, reservations were expressed regarding the use of a single-valued effective stress equation. There has subsequently been a slow change towards the acceptance of two independent stress state variables (Coleman, 1962; Matyas and Radhakrishna, 1968; Fredlund and Morgenstern, 1977).

One of the pioneers to strongly advocate a sound theo-

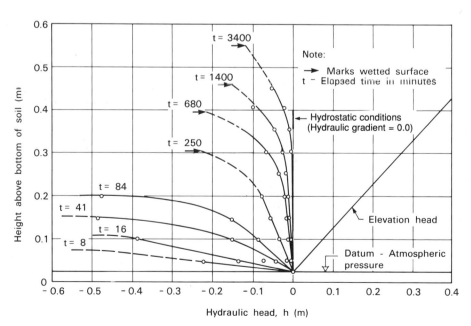

**Figure 1.27** Typical data from a capillary rise test (from Lambe, 1951a).

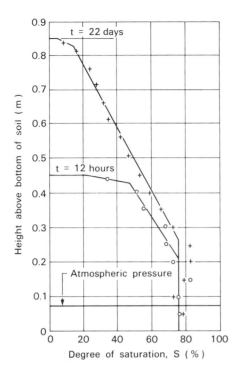

**Figure 1.28** Degree of saturation versus distance above the bottom of a soil column (from Lambe, 1951a).

## 1.8 HISTORICAL DEVELOPMENTS

retical basis for unsaturated soils' theories was Lytton (1967). His research drew upon the mixtures theories in continuum mechanics. These principles were applied to the multiphase, unsaturated soil system. Excellent direction was given for future research.

There are separate sections in this book devoted to a review of the historical developments related to each of the fundamental properties common to unsaturated soil mechanics. Details concerning the historical development towards the use of independent stress state variables are given in Chapter 3. The development and application of saturated–unsaturated flow modeling, subsequent to emphasis on the capillary model, are presented in Chapter 5. Developments in the area of the shear strength of unsaturated soils are presented in Chapter 9. Similar developments in the area of volume change in unsaturated soils are given in Chapter 12. In each of the above areas, much of the original work resulted in a somewhat semi-empirical approach towards understanding the behavior of unsaturated soils. With time, the principles of continuum mechanics that were found to be successful in saturated soil mechanics have also become the basis for unsaturated soil mechanics. It is on this infant, but sound, theoretical basis that the theory in this book has been assembled.

# CHAPTER 2

# *Phase Properties and Relations*

This chapter deals with three major topics. The first topic is the basic properties of each phase of an unsaturated soil. This information is used in later chapters when describing the behavior of the soil as a system of phases. The second topic deals with the understanding of the interaction between air and water. The third topic deals with establishing useful volume–mass relations for solving engineering problems.

An unsaturated soil has commonly been referred to as a three-phase system. However, more recently, the realization of the important role of the air–water interface (i.e., the contractile skin) has warranted its inclusion as an additional phase when considering certain physical mechanisms.

When the air phase is continuous, the contractile skin interacts with the soil particles and provides an influence on the mechanical behavior of the soil. An element of unsaturated soil with a continuous air phase is idealized in Fig. 2.1. When the air phase consists of occluded air bubbles, the fluid becomes significantly compressible.

The mass and volume of each phase can be schematically represented by a phase diagram. Figure 2.2(a) shows a rigorous four-phase diagram for an unsaturated soil. The thickness of the contractile skin is in the order of only a few molecular layers. Therefore, the physical subdivision of the contractile skin is unnecessary when establishing volume–mass relations for an unsaturated soil. The contractile skin is considered as part of the water phase without significant error. A simplified three-phase diagram, depicted in Fig. 2.2(b), can be used in writing the volume–mass relationships. The term "soil solids" is used when referring to the summation of masses and volumes of all the soil particles.

## 2.1 PROPERTIES OF THE INDIVIDUAL PHASES

An understanding of the basic properties of the soil particles, water, air, and contractile skin should precede the consideration of the behavior of the soil system. This sec-

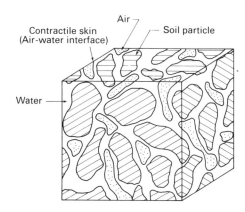

**Figure 2.1** An element of unsaturated soil with a continuous air phase.

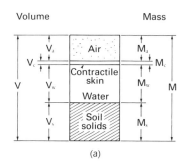

(a)

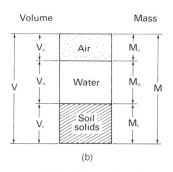

(b)

**Figure 2.2** Rigorous and simplified phase diagrams for an unsaturated soil. (a) Rigorous four phase unsaturated soil system; (b) simplified three phase diagram.

tion discusses the gravimetric and volumetric properties pertinent to each phase. The most important property of the contractile skin is its ability to exert a tensile pull. This property is called "surface tension" and is discussed later in this chapter.

### 2.1.1 Density and Specific Volume

Density, $\rho$, is defined as the ratio of mass to volume. Each phase of a soil has its own density. The density of each phase can be formulated from the phase diagram shown in Fig. 2.3.

Specific volume, $v_0$, is generally defined as the inverse of density; therefore, specific volume is the ratio of volume to mass.

Unit weight, $\gamma$, is a useful term in soil mechanics. It is the product of density, $\rho$, and gravitational acceleration, $g$ (i.e., 9.81 m/s$^2$).

*Soil Particles*

The density of the soil particles, $\rho_s$, is defined as follows (Fig. 2.3):

$$\rho_s = \frac{M_s}{V_s}. \qquad (2.1)$$

The density of the soil particles is commonly expressed as a dimensionless variable called specific gravity, $G_s$. The specific gravity of the soil particles is defined as the ratio of the density of the soil particles to the density of water at a temperature of 4°C under atmospheric pressure conditions (i.e., 101.3 kPa). In the SI system of units, this variable is now referred to as the relative density of the soil particles.

$$G_s = \frac{\rho_s}{\rho_w}. \qquad (2.2)$$

The density of water at 4°C and 101.3 kPa is 1000 kg/m$^3$. Table 2.1 presents typical values of specific gravity for several common minerals.

*Water Phase*

The density of water, $\rho_w$, is defined as follows:

$$\rho_w = \frac{M_w}{V_w}. \qquad (2.3)$$

Water is essentially a homogeneous substance the world over, except for variations produced by salts and isotopes of hydrogen and oxygen (Dorsey, 1940). Distilled water under the pressure of its saturated vapor is called pure, saturated water. The density of pure, saturated water must be measured experimentally. Figure 2.4 shows the density of pure water under various applied pressures and temperatures.

For soil mechanics problems, the variation in the density of water due to temperature differences is more significant

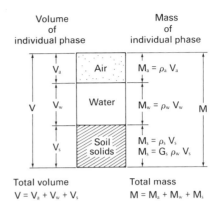

**Figure 2.3** Phase diagram for an unsaturated soil.

**Table 2.1 Specific Gravity of Several Minerals (from Lambe and Whitman, 1979)**

| Mineral | Specific Gravity, $G_s$ |
|---|---|
| Quartz | 2.65 |
| K–Feldspars | 2.54–2.57 |
| Na–Ca–Feldspars | 2.62–2.76 |
| Calcite | 2.72 |
| Dolomite | 2.85 |
| Muscovite | 2.7–3.1 |
| Biotite | 2.8–3.2 |
| Chlorite | 2.6–2.9 |
| Pyrophyllite | 2.84 |
| Serpentine | 2.2–2.7 |
| Kaolinite | 2.61[a]; 2.64 ± 0.02 |
| Halloysite (2H$_2$O) | 2.55 |
| Illite | 2.84[a]; 2.60–2.86 |
| Montmorillonite | 2.74[a]; 2.75–2.78 |
| Attapulgite | 2.30 |

[a]Calculated from crystal structure.

than its variation due to an applied pressure. For isothermal conditions, the density of water is commonly taken as 1000 kg/m$^3$.

*Air Phase*

The density of air, $\rho_a$, can be expressed as

$$\rho_a = \frac{M_a}{V_a}. \qquad (2.4)$$

The specific volume of air, $v_{a0}$, is

$$v_{a0} = \frac{V_a}{M_a}. \qquad (2.5)$$

Air behaves as a mixture of several gases (Table 2.2) and also varying amounts of water vapor. The mixture is called

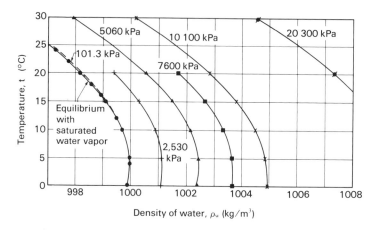

**Figure 2.4** Density of pure water for various applied pressures and temperatures (from Dorsey, 1940).

**Table 2.2 Composition of Dry Air**[a]

|  | Percentage by Volume | Density (kg/m³) | Molecular Mass (on Basis of Natural Scale, $O = 16$) (kg/kmol) |
|---|---|---|---|
| Nitrogen ($N_2$) | 78.08 | 1.25055 | 28.016 |
| Oxygen ($O_2$) | 20.95 | 1.42904 | 32.000 |
| Argon (Ar) | 0.934 | 1.7837 | 39.944 |
| Carbon dioxide ($CO_2$) | 0.031 | 1.9769 | 44.010 |
| Neon (Ne) | $1.82 \times 10^{-3}$ | 0.90035 | 20.183 |
| Helium (He) | $5.24 \times 10^{-4}$ | 0.17847 | 4.003 |
| Krypton (Kr) | $1.14 \times 10^{-4}$ | 3.708 | 83.800 |
| Hydrogen ($H_2$) | $\approx 5.0 \times 10^{-5}$ | 0.08988 | 2.016 |
| Xenon (Xe) | $8.70 \times 10^{-6}$ | 5.851 | 131.300 |
| Ozone ($O_3$) | $1 \times 10^{-6}$ to $1 \times 10^{-5}$ | 2.144 | 48.000 |
| Air | 100.0 | 1.2929 | 28.966 |

[a]Under standard conditions (i.e., 101.3 kPa and 0°C) with no water vapor.

dry air when no water vapor is present, and is called moist air when water vapor is present.

Dry and moist air can be considered to behave as an "ideal" gas under pressures and temperatures commonly encountered in geotechnical engineering. The ideal gas law can be written

$$\bar{u}_a V_a = \frac{M_a}{\omega_a} RT \quad (2.6)$$

where

- $\bar{u}_a$ = absolute air pressure (note that a bar, ⁻, sign indicates an absolute pressure, i.e., $\bar{u}_a = u_a + \bar{u}_{\text{atm}}$) (kN/m² or kPa)
- $u_a$ = gauge air pressure (kN/m² or kPa)
- $\bar{u}_{\text{atm}}$ = atmospheric pressure (i.e., 101.3 kPa or 1 atm)
- $V_a$ = volume of air (m³)
- $M_a$ = mass of air (kg)
- $\omega_a$ = molecular mass of air (kg/kmol)
- $R$ = universal (molar) gas constant [i.e., 8.31432 J/(mol · K)]
- $T$ = absolute temperature (i.e., $T = t^0 + 273.16$) (K)
- $t^0$ = temperature (°C).

The right-hand side of Eq. (2.6) (i.e., $(M_a/\omega_a)RT$) is a constant for a gas in a closed system with a constant mass and temperature. Under these conditions, Eq. (2.6) can be rewritten as Boyle's law:

$$\bar{u}_{a1} V_{a1} = \bar{u}_{a2} V_{a2} \quad (2.7)$$

where

$\bar{u}_{a1}, V_{a1}$ = absolute pressure and volume of air, respectively, at condition 1

$\bar{u}_{a2}, V_{a2}$ = absolute pressure and volume of air, respectively, at condition 2.

Rearranging the "ideal" gas equation (i.e., Eq. 2.6) gives

$$\frac{M_a}{V_a} = \frac{\omega_a}{RT} \bar{u}_a. \quad (2.8)$$

Substituting Eq. (2.4) into Eq. (2.8) gives an equation for the density of air:

$$\rho_a = \frac{\omega_a}{RT} \bar{u}_a. \quad (2.9)$$

The molecular mass of air, $\omega_a$, depends on the composition of the mixture of dry air and water vapor. The dry air has a molecular mass of 28.966 kg/kmol, and the molecular mass of the water vapor ($H_2O$) is 18.016 kg/kmol. The composition of air, namely, nitrogen ($N_2$) and oxygen ($O_2$), are essentially constant in the atmosphere. The carbon dioxide ($CO_2$) content in air may vary, depending on environmental conditions, such as the rate of consumption of fossil fuels. However, the constituent of air that can vary most is water vapor. The volume percentage of water vapor in the air may range from as little as 0.000002% to as high as 4–5% (Harrison, 1965). The molecular mass of air is affected by the change in each constituent. This consequently affects the density of air.

The concentration of water vapor in the air is commonly expressed in terms of relative humidity:

$$RH = \frac{\bar{u}_v(100)}{\bar{u}_{v0}} \quad (2.10)$$

where

RH = relative humidity (%)
$\bar{u}_v$ = partial pressure of water vapor in the air (kPa)
$\bar{u}_{v0}$ = saturation pressure of water vapor at the same temperature (kPa).

Table 2.3 presents the values of air density for different absolute air pressures ($\bar{u}_a$) and temperatures, $t^0$. The figures in the top portion of Table 2.3 were computed for air with a relative humidity of 50% and 0.04% carbon dioxide by volume. For air having relative humidities other than 50%, a correction should be applied as shown in the bottom portion of Table 2.3. Although the corrections are small, it should be noted that the density of air decreases as the relative humidity increases. This indicates that the moist air is lighter than the dry air.

### 2.1.2 Viscosity

All fluids resist a change of form or the action of shearing. This resistance is expressed by the property called viscosity. The absolute (dynamic) viscosity, $\mu$, of a fluid is defined as the resistance of the fluid to a shearing force applied by sliding one plate over another with the fluid placed in between. The absolute viscosity depends on the pressure and temperature. However, the influence of pressure is negligible for the range of pressures commonly encountered in typical civil engineering applications.

The viscosities of water and air at atmospheric pressure (i.e., 101.3 kPa) and different temperatures are given in Tables 2.4 and 2.5, respectively. Figure 2.5 presents the absolute viscosities of water, air, and several other materials at different temperatures. The viscosities of liquids are shown to decrease with an increase in temperature, while the viscosity of air increases as the temperature increases.

**Table 2.3 Density of Air at Different Absolute Pressures, Temperatures, and Relative Humidities (from Kaye and Laby, 1973)**

| | Density of Air, $\rho_a$ (kg/m³) | | |
|---|---|---|---|
| | | Temperature, $t^0$ (°C) | |
| Absolute Air Pressure, $\bar{u}_a$ (kPa) | 10 | 20 | 30 |
| 80 | 0.982 | 0.946 | 0.910 |
| 85 | 1.043 | 1.005 | 0.968 |
| 90 | 1.105 | 1.065 | 1.025 |
| 95 | 1.167 | 1.124 | 1.083 |
| 100 | 1.228 | 1.184 | 1.140 |
| 101 | 1.240 | 1.196 | 1.152 |
| 105 | 1.290 | 1.243 | 1.198 |

| Density Adjustments for Humidity (kg/m³) | | |
|---|---|---|
| | Temperature, $t^0$ (°C) | |
| Relative Humidity, RH (%) | 10 | 20 |
| 20 | +0.003 | +0.006 |
| 25 | +0.003 | +0.005 |
| 30 | +0.002 | +0.004 |
| 35 | +0.002 | +0.003 |
| 40 | +0.001 | +0.002 |
| 45 | +0.001 | +0.001 |
| 50 | 0 | 0 |
| 55 | −0.001 | −0.001 |
| 60 | −0.001 | −0.002 |
| 65 | −0.002 | −0.003 |
| 70 | −0.002 | −0.004 |
| 75 | −0.003 | −0.005 |
| 80 | −0.003 | −0.006 |

**Table 2.4 Viscosity of Water at 101.3 kPa (Modified from Tuma, 1976)**

| Temperature, $t^0$ (°C) | Absolute (Dynamic) Viscosity, $\mu$ ($\times 10^{-3}$ N·s/m²) | Temperature, $t^0$ (°C) | Absolute (Dynamic) Viscosity, $\mu$ ($\times 10^{-3}$ N·s/m²) |
|---|---|---|---|
| 0 | 1.794 | 55 | 0.507 |
| 5 | 1.519 | 60 | 0.470 |
| 10 | 1.310 | 65 | 0.437 |
| 15 | 1.144 | 70 | 0.407 |
| 20 | 1.009 | 75 | 0.381 |
| 25 | 0.895 | 80 | 0.357 |
| 30 | 0.800 | 85 | 0.336 |
| 35 | 0.731 | 90 | 0.317 |
| 40 | 0.654 | 95 | 0.299 |
| 45 | 0.597 | 100 | 0.284 |
| 50 | 0.548 | | |

**Table 2.5 Viscosity of Air at 101.3 kPa (Modified from Tuma, 1976; and Kaye and Laby, 1973)**

| Temperature, $t^0$ (°C) | Absolute (dynamic) Viscosity, $\mu$ ($\times 10^{-5}$ N·s/m²) | Sources |
|---|---|---|
| −20 | 1.604 | Tuma, 1976 |
| −10 | 1.667 | Tuma, 1976 |
| 0 | 1.705 | Tuma, 1976 |
| 10 | 1.761 | Tuma, 1976 |
| 20 | 1.785 | Tuma, 1976 |
| 30 | 1.864 | Tuma, 1976 |
| 40 | 1.909 | Tuma, 1976 |
| 50 | 1.96 | Kaye and Laby, 1973 |
| 100 | 2.20 | Kaye and Laby, 1973 |
| 200 | 2.61 | Kaye and Laby, 1973 |

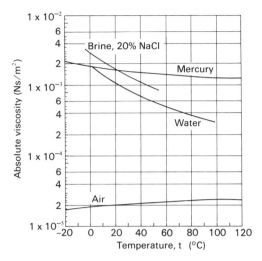

**Figure 2.5** Viscosity of fluids at different temperatures (from Streeter and Wylie, 1975).

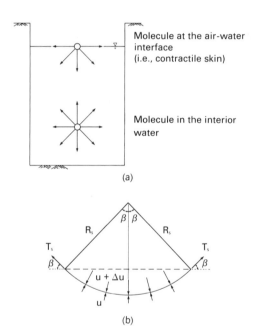

**Figure 2.6** Surface tension phenomenon at the air–water interface. (a) Intermolecular forces on contractile skin and water; (b) pressures and surface tension acting on a curved two-dimensional surface.

### 2.1.3 Surface Tension

The air–water interface (i.e., contractile skin) possesses a property called surface tension. The phenomenon of surface tension results from the intermolecular forces acting on molecules in the contractile skin. These forces are different from those that act on molecules in the interior of the water [Fig. 2.6(a)].

A molecule in the interior of the water experiences equal forces in all directions, which means there is no unbalanced force. A water molecule within the contractile skin experiences an unbalanced force towards the interior of the water. In order for the contractile skin to be in equilibrium, a tensile pull is generated along the contractile skin. The property of the contractile skin that allows it to exert a tensile pull is called its surface tension, $T_s$. Surface tension is measured as the tensile force per unit length of the contractile skin (i.e., units of N/m). Surface tension is tangential to the contractile skin surface. Its magnitude decreases as temperature increases. Table 2.6 gives surface tension values for the contractile skin at different temperatures.

**Table 2.6 Surface Tension of Contractile Skin (i.e., Air–Water Interface) (from Kaye and Laby, 1973).**

| Temperature, $t^0$ (°C) | Surface Tension, $T_s$ (mN/m) |
| --- | --- |
| 0 | 75.7 |
| 10 | 74.2 |
| 15 | 73.5 |
| 20 | 72.75 |
| 25 | 72.0 |
| 30 | 71.2 |
| 40 | 69.6 |
| 50 | 67.9 |
| 60 | 66.2 |
| 70 | 64.4 |
| 80 | 62.6 |
| 100 | 58.8 |

The surface tension causes the contractile skin to behave like an elastic membrane. This behavior is similar to an inflated balloon which has a greater pressure inside the balloon than outside. If a flexible two-dimensional membrane is subjected to different pressures on each side, the membrane must assume a concave curvature towards the larger pressure and exert a tension in the membrane in order to be in equilibrium. The pressure difference across the curved surface can be related to the surface tension and the radius of curvature of the surface by considering equilibrium across the membrane [Fig. 2.6(b)].

The pressures acting on the membrane are $u$ and $(u + \Delta u)$. The membrane has a radius of curvature of, $R_s$, and a surface tension, $T_s$. The horizontal forces along the membrane balance each other. Force equilibrium in the vertical direction requires that

$$2T_s \sin \beta = 2 \Delta u R_s \sin \beta \qquad (2.11)$$

where

$2R_s \sin \beta$ = length of the membrane projected onto the horizontal plane.

Rearranging Eq. (2.11) gives

$$\Delta u = \frac{T_s}{R_s}. \qquad (2.12)$$

Equation (2.12) gives the pressure difference across a two-dimensional curved surface with a radius, $R_s$, and a surface tension, $T_s$. For a warped or saddle-shaped surface (i.e., three-dimensional membrane), Eq. (2.12) can be extended using the Laplace equation (Fig. 2.7)

$$\Delta u = T_s \left( \frac{1}{R_1} + \frac{1}{R_2} \right) \qquad (2.13)$$

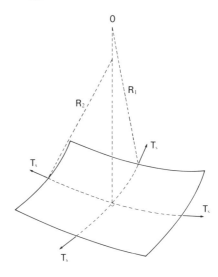

**Figure 2.7** Surface tension on a warped membrane.

where

$R_1$ and $R_2$ = radii of curvature of a warped membrane in two orthogonal principal planes.

If the radius of curvature is the same in all directions (i.e., $R_1$ and $R_2$ are equal to $R_s$), Eq. (2.13) becomes

$$\Delta u = \frac{2T_s}{R_s}. \qquad (2.14)$$

In an unsaturated soil, the contractile skin would be subjected to an air pressure, $u_a$, which is greater than the water pressure, $u_w$. The pressure difference, $(u_a - u_w)$, is referred to as matric suction. The pressure difference causes the contractile skin to curve in accordance with Eq. (2.14):

$$(u_a - u_w) = \frac{2T_s}{R_s} \qquad (2.15)$$

where

$(u_a - u_w)$ = matric suction or the difference between pore-air and pore-water pressures acting on the contractile skin.

Equation (2.15) is referred to as Kelvin's capillary model equation. As the matric suction of a soil increases, the radius of curvature of the contractile skin decreases. The curved contractile skin is often called a meniscus. When the pressure difference between the pore-air and pore-water goes to zero, the radius of curvature, $R_s$, goes to infinity. Therefore, a flat air–water interface exists when the matric suction goes to zero.

## 2.2 INTERACTION OF AIR AND WATER

Air and water can be combined as immiscible and/or miscible mixtures. The *immiscible* mixture is a combination of free air and water without any interaction. The immiscible

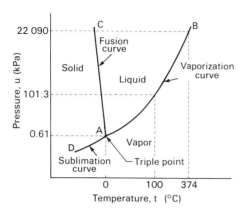

**Figure 2.8** State diagram for water (not to scale; from Van Haveren and Brown, 1972).

mixture is characterized by the separation produced by the contractile skin. A *miscible* air–water mixture can have two forms. First, air dissolves in water and can occupy approximately 2% by volume of the water (Dorsey, 1940). Second, water vapor can be present in the air. All of the above types of mixtures are dealt with in the following sections. Consideration is also given to all possible states for the water.

### 2.2.1 Solid, Liquid, and Vapor States of Water

Water can be found in one of three states: the solid state as ice, the liquid state as water, and the gaseous state as water vapor (Fig. 2.8). Throughout the text, the words "water" or "water phase" refer to the liquid state of water. The state of water depends on the pressure and temperature environment.

Three lines are drawn on the water state diagram (Fig. 2.8). These are the vaporization curve, *AB*, the fusion curve, *AC*, and the sublimation curve, *AD*. The vaporization curve, *AB*, is also called the vapor pressure curve of water. It gives combination values of temperature and pressure for which the liquid and vapor states of water can coexist in equilibrium. The fusion curve, *AC*, separates the solid and the liquid states of water, and the sublimation curve, *AD*, separates the solid and the vapor states of water. The solid state can coexist in equilibrium with the liquid state along the fusion curve, and with the vapor state along the sublimation curve.

The vaporization, fusion, and sublimation curves meet at point *A*. This point is called the triple point of water where the solid, liquid, and vapor states of water can coexist in equilibrium. The triple point of water is achieved at a temperature of 0°C and a pressure of 0.61 kPa.

### 2.2.2 Water Vapor

The vaporization curve, *AB*, in Fig. 2.8 represents an equilibrium condition between the liquid and vapor states of water. In this state of equilibrium, evaporation and condensation processes occur simultaneously at the same rate. The rate of condensation depends on the pressure in the water vapor which reaches its saturation value on the vaporization line. On the other hand, the evaporation rate depends only on temperature. Therefore, a unique relationship exists between the saturation water vapor pressure and temperature, which is described by the vaporization curve. Saturation water vapor pressures, $\bar{u}_{v0}$, are presented in Table 2.7.

In the atmosphere, water vapor is mixed with air. However, the presence of the air has no effect on the behavior of the water vapor. This phenomenon is expressed by Dalton's law of partial pressures. Dalton's law states that the pressure of a mixture of gases is equal to the sum of the

**Table 2.7 Saturation Pressures of Water Vapor at Different Temperatures[a] (from Kaye and Laby, 1973)**

| $t^0$ (°C) | 0 | 1 | 2 | 3 | 4 | 5 | 6 | 7 | 8 | 9 |
|---|---|---|---|---|---|---|---|---|---|---|
| 0 | 0.6107 | 0.6566 | 0.7055 | 0.7576 | 0.8130 | 0.8720 | 0.9348 | 1.0015 | 1.0724 | 1.1477 |
| 10 | 1.2276 | 1.3123 | 1.4022 | 1.4974 | 1.5983 | 1.7051 | 1.8180 | 1.9375 | 2.0639 | 2.1974 |
| 20 | 2.3384 | 2.4872 | 2.6443 | 2.8099 | 2.9846 | 3.1686 | 3.3625 | 3.5666 | 3.7814 | 4.0074 |
| 30 | 4.2451 | 4.4949 | 4.7574 | 5.0332 | 5.3226 | 5.6264 | 5.9451 | 6.2793 | 6.6296 | 6.9967 |

| $t^0$ (°C) | 0 | 2 | 4 | 6 | 8 | 10 | 12 | 14 | 16 | 18 |
|---|---|---|---|---|---|---|---|---|---|---|
| 40 | 7.3812 | 8.2053 | 9.1075 | 10.094 | 11.171 | 12.345 | 13.623 | 15.013 | 16.522 | 18.160 |
| 60 | 19.933 | 21.852 | 23.926 | 26.164 | 28.578 | 31.177 | 33.974 | 36.980 | 40.206 | 43.667 |
| 80 | 47.375 | 51.344 | 55.857 | 60.121 | 64.960 | 70.120 | 75.617 | 81.468 | 87.691 | 94.304 |
| 100 | 101.325 | 108.77 | 116.67 | 125.03 | 133.88 | 143.25 | 153.14 | 163.59 | 174.61 | 186.24 |
| 120 | 198.49 | 211.39 | 224.97 | 239.26 | 254.27 | 270.03 | 286.58 | 303.95 | 322.16 | 341.23 |

[a]Saturation water vapor pressures, $\bar{u}_{v0}$, are in kPa.

partial pressures that each individual gas would exert if it alone filled the entire volume. In other words, the behavior of a particular gas of a mixture of gases is independent of the other gases. Therefore, the partial pressure of water vapor in the atmosphere which is in equilibrium with water is the saturation pressure given in Table 2.7. Similarly the presence of air above water does not change the state equilibrium of water (Fig. 2.8).

In nature, the water vapor in air is usually not in equilibrium with adjacent bodies of water. This means that the partial pressure of the water vapor in air, $\bar{u}_v$, is usually not the same as the saturation pressure of the water vapor, $\bar{u}_{v0}$, at the corresponding temperature. The water vapor in air at a given temperature is therefore said to be undersaturated, saturated, or supersaturated when the partial pressure of water vapor, $\bar{u}_v$, is less than, equal to, or greater than the saturation water vapor pressure. The saturated condition indicates an equilibrium between the water vapor and the water where evaporation and condensation take place at the same rate. On the other hand, the undersaturated and supersaturated states of water vapor are not equilibrium conditions. The supersaturated state indicates an excess of water vapor which eventually condenses. In this case, the rate of condensation exceeds the evaporation rate until the partial pressure of water vapor, $\bar{u}_v$, has been reduced to the saturation pressure, $\bar{u}_{v0}$. In the undersaturated state, there is a lack of water vapor relative to the equilibrium condition. Therefore, the rate of evaporation exceeds the rate of condensation until the partial pressure of the water vapor, $\bar{u}_v$, has reached the saturation water vapor pressure, $\bar{u}_{v0}$.

The partial pressure of the water vapor in air defines the degree to which the air is saturated with water vapor at a specific temperature. The degree of saturation with respect to water vapor is referred to as the relative humidity, RH [Eq. (2.10)].

### 2.2.3 Air Dissolving in Water

Water molecules form a lattice structure with openings referred to as a "cage" that can be occupied by a gas (Rodebush and Busswell, 1958), as illustrated in Fig. 2.9. Air dissolves into the water and fills the "cages" which have a volume of approximately 2% by total volume.

The water lattice is relatively rigid and stable (Dorsey, 1940), and the density of water changes little as a consequence of the presence of the dissolved air. Figure 2.10 shows the effect of dissolved air on the density of water for various temperatures.

An analogy using a cylinder with a piston and porous stone is useful in analyzing the behavior of air–water mixtures. Consider a cylinder with a porous stone at its base and a frictionless piston at the top (Fig. 2.11). The porous stone has pores equaling 2% of its volume. The porous stone is used to simulate the behavior of water. In this model, the cylinder contains free air above the porous stone. An imaginary valve is situated at the boundary be-

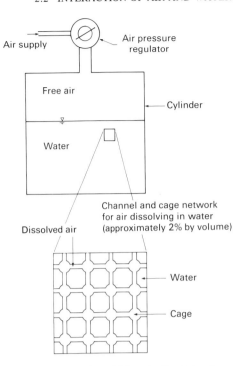

**Figure 2.9** Visualization aid for air dissolving in water.

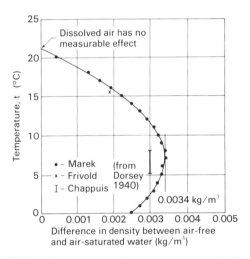

**Figure 2.10** Effect of dissolved air on the density of water (from Dorsey, 1940).

tween the free air and the porous stone to control the movement of air into the porous stone. The movement of air into the porous stone represents the movement of air into water.

Let us suppose there is an initial pressure applied equally to the free air and to the air in the porous stone in the cylinder. The imaginary valve is then closed. If the load on the piston is increased, the free air above the porous stone will be compressed following Boyle's law [Eq. (2.7)]. The imaginary valve is then opened, and some air will move into the porous stone in accordance with Henry's law. Henry's law states that the mass of gas dissolved in a fixed

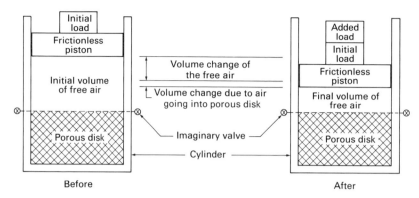

**Figure 2.11** Piston and porous stone analogy.

quantity of liquid, at constant temperature, is directly proportional to the absolute pressure of the gas above the solution (Sisler et al., 1953). This process will continue as the piston load is increased. Eventually, all the free air will move into the porous stone, and any additional applied load will be carried by the porous stone.

The above analogy cannot totally simulate the situation in an unsaturated soil. In the presence of a solid, such as soil particles, the air and water pressures can have different magnitudes. The air and water pressures in a soil can also change at differing rates during a process. In the analogy, the free air and the water (i.e., porous stone) have the same pressure. The possibility of a difference between the air and water pressures is later shown to be of significance in the compressibility formulation.

The mass of air going into or coming out of water is time dependent. This time dependency can either be ignored or taken into consideration, depending upon the engineering problem under consideration. The amount of air that can be dissolved in water is referred to as its solubility, and the rate of solution is referred to as its diffusivity.

### Solubility of Air in Water

The volume of dissolved air in water is essentially independent of air or water pressures. This can be demonstrated using the ideal gas law and Henry's law. The ideal gas law [Eq. (2.6)] can be rearranged and applied to a gas dissolving in water at a certain temperature and pressure:

$$V_d = \frac{M_d}{\bar{u}_a} \frac{RT}{\omega_a} \qquad (2.16)$$

where

$V_d$ = volume of dissolved air in water
$M_d$ = mass of dissolved air in water
$\bar{u}_a$ = absolute pressure of the dissolved air.

The absolute pressure of the dissolved air is equal to the absolute pressure of the free air under equilibrium conditions. Referring to the piston and porous stone analogy, an increase in the piston load will increase the pressure in the free air, and therefore more free air will go into the porous stone (i.e., water). After some time, an equilibrium condition will be reached where the pressure in the free air and the dissolved air are equal. If the piston load is then increased, the process will be repeated.

The mass of dissolved air at equilibrium is dependent upon the corresponding absolute air pressure as stated in Henry's law. If the temperature remains constant throughout the process, the ratio of the mass and the absolute pressure of the dissolved air is constant:

$$\frac{M_{d1}}{\bar{u}_{a1}} = \frac{M_{d2}}{\bar{u}_{a2}} = \text{constant} \qquad (2.17)$$

where

$M_{d1}, \bar{u}_{a1}$ = mass and absolute pressure of the dissolved air, respectively, at condition 1
$M_{d2}, \bar{u}_{a2}$ = mass and absolute pressure of the dissolved air, respectively, at condition 2.

The volume of dissolved air in water, $V_d$, is computed using Eq. (2.16) and by considering the relationship shown in (Eq. 2.17). At a constant temperature, the volume of dissolved air in water is a constant for different pressures.

The ratio between the *mass* of each gas that can be dissolved in a liquid and the mass of the liquid is called the coefficient of solubility, $H$. Table 2.8 presents the coefficients of solubility of oxygen, nitrogen, and air in water, over a temperature range. All coefficients of solubility are referenced to a standard pressure of 101.3 kPa.

The ratio of the *volume* of dissolved gas, $V_d$, in a liquid to the volume of the liquid is called the volumetric coefficient of solubility, $h$, which varies slightly with temperature. Values for the volumetric coefficient of solubility for air in water under different temperatures are given in Table 2.8.

### Diffusion of Gases Through Water

The rate at which air can pass through water is described by Fick's law of diffusion. The rate at which mass is transferred across a unit area is equal to the product of the coefficient of diffusion, $D$, and the concentration gradient. In the diffusion of air through water, the concentration differ-

Table 2.8 Solubility of Gases in Water (Under a Pressure of 101.3 kPa) (from Dorsey, 1940)

| Temperature, $t^0$ (°C) | Coefficient of Solubility, $H^a$ | | | Volumetric Coefficient of Solubility, $h^b$ |
|---|---|---|---|---|
| | Oxygen | Nitrogen, Argon, etc. | Air | Air |
| 0  | $14.56 \times 10^{-6}$ | $23.87 \times 10^{-6}$ | $38.43 \times 10^{-6}$ | 0.02918 |
| 4  | $13.06 \times 10^{-6}$ | $21.59 \times 10^{-6}$ | $34.65 \times 10^{-6}$ | 0.02632 |
| 10 | $11.25 \times 10^{-6}$ | $18.82 \times 10^{-6}$ | $30.07 \times 10^{-6}$ | 0.02284 |
| 15 | $10.07 \times 10^{-6}$ | $17.00 \times 10^{-6}$ | $27.07 \times 10^{-6}$ | 0.02055 |
| 20 | $9.11 \times 10^{-6}$  | $15.51 \times 10^{-6}$ | $24.62 \times 10^{-6}$ | 0.01868 |
| 25 | $8.28 \times 10^{-6}$  | $14.24 \times 10^{-6}$ | $22.52 \times 10^{-6}$ | 0.01708 |
| 30 | $7.55 \times 10^{-6}$  | $13.10 \times 10^{-6}$ | $20.65 \times 10^{-6}$ | 0.01564 |

[a] At standard atmospheric pressure.
[b] $h = (\rho_w/\rho_a)H$.

ence is equal to the difference in density between the free air and the dissolved air in the water.

Under constant temperature conditions, the density of air is a function of the air pressure [Eq. (2.9)]. An increase in pressure in the free air will develop a pressure difference between the free and dissolved air. This pressure difference becomes the driving potential for the free air to diffuse into the water.

The gases composing air individually diffuse into water. The coefficients of diffusion, $D$, for each component of air through water are tabulated in Table 2.9. Combined gases forming air dissolve in water at a rate of approximately $2.0 \times 10^{-9}$ m²/s (U.S. Research Council, 1933).

Barden and Sides (1967) measured the coefficient of diffusion for air through the water phase of both saturated and compacted clays. Their results are presented in Table 2.10. The study concluded that the coefficient of diffusion appears to decrease with decreasing water content of the soil. The coefficient of diffusion for air through the water in a soil appears to differ by several orders of magnitude from the coefficient of diffusion for air through free water.

## 2.3 VOLUME–MASS RELATIONS

The volume–mass relations of the soil particles, water, and air phases are useful properties in engineering practice. The derivations combine the gravimetric and volumetric properties of a soil.

### 2.3.1 Porosity

Porosity, $n$, in percent is defined as the ratio of the volume of voids, $V_v$, to the total volume, $V$ (Fig. 2.12):

$$n = \frac{V_v(100)}{V}. \qquad (2.18)$$

Table 2.9 Coefficients of Diffusion for Certain Gases in Water (from Kohn, 1965)

| Gas | Temperature, $t^0$ (°C) | Coefficient of Diffusion, $D$ (m²/s) |
|---|---|---|
| $CO_2$ | 20 | $1.7 \times 10^{-9}$ |
| $N_2$  | 22 | $2.0 \times 10^{-9}$ |
| $H_2$  | 21 | $5.2 \times 10^{-9}$ |
| $O_2$  | 25 | $2.92 \times 10^{-9}$ |

Similarly, porosity type terms can be defined with respect to each of the phases of a soil:

$$n_s = \frac{V_s(100)}{V} \qquad (2.19)$$

$$n_w = \frac{V_w(100)}{V} \qquad (2.20)$$

$$n_a = \frac{V_a(100)}{V} \qquad (2.21)$$

$$n_c = \frac{V_c(100)}{V} \qquad (2.22)$$

where

$n_s$ = soil particle porosity (%)
$n_w$ = water porosity (%)
$n_a$ = air porosity (%)
$n_c$ = contractile skin porosity (%).

The volume associated with the contractile skin can be

Table 2.10 Coefficient of Diffusion for Air Through Different Materials (from Barden and Sides, 1967)

| Material | Water Content, $w$ (%) | Coefficient of Diffusion, $D$ (m$^2$/s) |
|---|---|---|
| Free water | — | $2.2 \times 10^{-9}$ |
| Natural rubber | — | $1.1 \times 10^{-10}$ |
| Kaolin consolidated at 414 kPa (oriented parallel to flow) | 49 | $4.5 \times 10^{-10}$ |
| Kaolin consolidated at 414 kPa (oriented perpendicular to flow) | 49 | $3.2 \times 10^{-10}$ |
| Kaolin consolidated at 483 kPa | 47 | $3.0 \times 10^{-10}$ |
| Kaolin consolidated at 34.5 kPa | 75 | $6.2 \times 10^{-10}$ |
| Derwent clay (illite) consolidated at 34.5 kPa | 53 | $4.7 \times 10^{-10}$ |
| Jackson clay and 4% bentonite consolidated at 34.5 kPa | 39 | $< 1.0 \times 10^{-11}$ |
| Compacted Westwater clay | 16 | $1.0 \times 10^{-11}$ |
| Saturated ceramic | 49 | $1.6 \times 10^{-10}$ |
| Saturated coarse stone | 21 | $2.5 \times 10^{-5}$ |

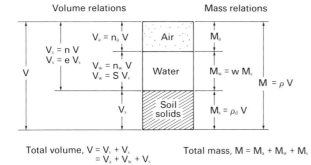

Figure 2.12 Volume–mass relations.

assumed to be negligible or part of the water phase. The water and air porosities represent their volumetric percentages in the soil. The soil particle porosity can be visualized as the percentage of the overall volume comprised of soil particles. The sum of the porosities of all phases must equal 100%. Therefore, the following soil porosity equation can be written as

$$n_s + n = n_s + n_a + n_w = 100(\%). \quad (2.23)$$

The water porosity, $n_w$, expressed in decimal form, is commonly referred to as the volumetric water content, $\theta_w$, in soil science and soil physics literature. The volumetric water content notation is also used throughout this book. Typical values of porosity for some soils are given in Table 2.11.

### 2.3.2 Void Ratio

Void ratio, $e$, is defined as the ratio of the volume of voids, $V_v$, to the volume of soil solids, $V_s$ (Fig. 2.12):

$$e = \frac{V_v}{V_s}. \quad (2.24)$$

The relationship between porosity and void ratio is obtained by equating the volume of voids, $V_v$, from the two equations [i.e., Eqs. (2.18) and (2.24)]:

$$n = \frac{e}{1 + e}. \quad (2.25)$$

Typical values for void ratio are shown in Table 2.11.

### 2.3.3 Degree of Saturation

The percentage of the void space which contains water is expressed as the degree of saturation, $S$ (%):

$$S = \frac{V_w(100)}{V_v}. \quad (2.26)$$

The degree of saturation, $S$, can be used to subdivide soils into three groups.

1) Dry soils (i.e., $S = 0\%$): Dry soil consists of soil particles and air. No water is present.
2) Saturated soils (i.e., $S = 100\%$): All of the voids in the soil are filled with water.
3) Unsaturated soils (i.e., $0\% < S < 100\%$): An unsaturated soil can be further subdivided, depending upon whether the air phase is continuous or occluded.

**Table 2.11 Typical Values of Porosity, Void Ratio, and Dry Density (Modified from Hough, 1969)**

| Soil Type | Void Ratio, $e$ maximum | Void Ratio, $e$ minimum | Porosity, $n$ (%) maximum | Porosity, $n$ (%) minimum | Density, $\rho$ (kg/m$^3$) maximum | Density, $\rho$ (kg/m$^3$) minimum |
|---|---|---|---|---|---|---|
| Granular Materials: 1) Uniform Materials | | | | | | |
| a) Equal spheres (theoretical values) | 0.92 | 0.35 | 47.6 | 26.0 | — | — |
| b) Standard Ottawa sand | 0.80 | 0.50 | 44.0 | 33.0 | 1762 | 1474 |
| c) Clean, uniform sand (fine or medium) | 1.0 | 0.40 | 50.0 | 29.0 | 1890 | 1330 |
| d) Uniform, inorganic silt | 1.1 | 0.40 | 52.0 | 29.0 | 1890 | 1281 |
| Granular Materials: 2) Well-Graded Materials | | | | | | |
| a) Silty sand | 0.90 | 0.30 | 47.0 | 23.0 | 2034 | 1394 |
| b) Clean, fine to coarse sand | 0.95 | 0.20 | 49.0 | 17.0 | 2210 | 1362 |
| c) Micaceous sand | 1.20 | 0.40 | 55.0 | 29.0 | 1922 | 1217 |
| d) Silty sand and gravel | 0.85 | 0.14 | 46.0 | 12.0 | 2239 | 1426 |
| Mixed Soils | | | | | | |
| a) Sandy or silty clay | 1.8 | 0.25 | 64.0 | 20.0 | 2162 | 961 |
| b) Skip-graded silty clay with stones or rock fragments | 1.0 | 0.20 | 50.0 | 17.0 | 2243 | 1346 |
| c) Well-graded gravel, sand, silt, and clay mixture | 0.70 | 0.13 | 41.0 | 11.0 | 2371 | 1602 |
| Clay Soils | | | | | | |
| a) Clay (30–50% clay sizes) | 2.4 | 0.50 | 71.0 | 33.0 | 1794 | 801 |
| b) Colloidal clay (−0.002 mm ≥ 50%) | 12.0 | 0.60 | 92.0 | 37.0 | 1698 | 308 |
| Organic Soils | | | | | | |
| a) Organic silt | 3.0 | 0.55 | 75.0 | 35.0 | 1762 | 641 |
| b) Organic clay (30–50% clay sizes) | 4.4 | 0.70 | 81.0 | 41.0 | 1602 | 481 |

*General Note:* Tabulation is based on $G_s = 2.65$ for granular soils, $G_s = 2.70$ for clays, and $G_s = 2.60$ for organic soils.

This subdivision is primarily a function of the degree of saturation. An unsaturated soil with a continuous air phase generally has a degree of saturation less than approximately 80% (i.e., $S < 80\%$). Occluded air bubbles commonly occur in unsaturated soils having a degree of saturation greater than approximately 90% (i.e., $S > 90\%$). The transition zone between continuous air phase and occluded air bubbles occurs when the degree of saturation is between approximately 80–90% (i.e., $80\% < S < 90\%$).

### 2.3.4 Water Content

Water content, $w$, is defined as the ratio of the mass of water, $M_w$, to the mass of soil solids, $M_s$ (Fig. 2.12). It is presented as a percentage [i.e., $w$ (%)]:

$$w = \frac{M_w (100)}{M_s}. \quad (2.27)$$

Water content, $w$, is also referred to as the gravimetric water content.

The volumetric water content, $\theta_w$, is defined as the ratio of the volume of water, $V_w$, to the total volume of the soil, $V$:

$$\theta_w = \frac{V_w}{V}. \quad (2.28)$$

The volumetric water content can also be expressed in terms of porosity, degree of saturation, and void ratio (Fig. 2.12). The volumetric water content can be written as

$$\theta_w = \frac{S V_v}{V}. \quad (2.29)$$

Since $V_v/V$ is equal to the porosity of the soil, Eq. (2.29) becomes

$$\theta_w = Sn. \quad (2.30)$$

Substituting Eq. (2.25) into Eq. (2.30) yields another form for the volumetric water content equation:

$$\theta_w = \frac{Se}{1 + e}. \quad (2.31)$$

### 2.3.5 Soil Density

Two commonly used soil density definitions are the total density and the dry density. The total density of a soil, $\rho$, is the ratio of the total mass, $M$, to the total volume of the soil, $V$ (Fig. 2.12):

$$\rho = \frac{M}{V}. \quad (2.32)$$

The total density is also called the bulk density. The dry density of a soil, $\rho_d$, is defined as the ratio of the mass of the soil solids, $M_s$, to the total volume of the soil, $V$ (Fig. 2.12):

$$\rho_d = \frac{M_s}{V}. \quad (2.33)$$

Typical minimum and maximum dry densities for various soils are presented in Table 2.11.

Other soil density definitions are the saturated density and the buoyant density. The saturated density of a soil is the total density of the soil for the case where the voids are filled with water (i.e., $V_a = 0$ and $S = 100\%$). The buoyant density of a soil is the difference between the saturated density of the soil and the density of water.

### 2.3.6 Basic Volume–Mass Relationship

The volume and mass for each phase can be related to one another using basic relations from the phase diagram (Fig. 2.3) and the volume–mass relations shown in Fig. 2.12.

The mass of the water phase of a soil, $M_w$, is the product of the volume and density of water (Fig. 2.3):

$$M_w = \rho_w V_w. \quad (2.34)$$

The volume of water, $V_w$, can also be computed from the volume relation given in Fig. 2.12 (i.e., left-hand side):

$$V_w = SeV_s. \quad (2.35)$$

The relationship given in Eq. (2.35) is shown in Fig. 2.13 (i.e., left-hand side). Equation (2.35) can then be rewritten as

$$M_w = \rho_w SeV_s. \quad (2.36)$$

The mass of the water, $M_w$, can also be related to the mass of the soil solids, $M_s$:

$$M_w = wM_s. \quad (2.37)$$

The mass of the soil solids, $M_s$, is obtained from the phase diagram in Fig. 2.3:

$$M_s = G_s \rho_w V_s. \quad (2.38)$$

Substituting Eq. (2.38) into Eq. (2.37) yields

$$M_w = wG_s \rho_w V_s. \quad (2.39)$$

Equating Eqs. (2.39) and (2.36) results in a *basic volume–mass relationship* for soils:

$$Se = wG_s. \quad (2.40)$$

The total and dry densities of a soil defined in Eqs. (2.32) and (2.33), respectively, can also be expressed in terms of the volume–mass properties of the soil (i.e., $S$, $e$, $w$, and $G_s$). Assuming that the mass of the air phase, $M_a$, is negligible, the total mass of the soil is the sum of the mass of the water, $M_w$, and the mass of the soil solids, $M_s$. The total volume of the soil, $V$, is given by the volume of the soil solids, $V_s$, and the volume of the voids, $V_v$. Therefore, the equation for the total density of a soil, $\rho$, can be rewritten using Eqs. (2.24), (2.38), and (2.39):

$$\rho = \frac{M_s + M_w}{V_s + V_v} \quad (2.41)$$

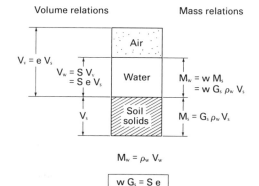

**Figure 2.13** Derivation of the basic volume–mass relationship.

$$\rho = \frac{G_s \rho_w V_s + w G_s \rho_w V_s}{V_s + e V_s} \quad (2.42)$$

$$\rho = \frac{G_s(1 + w)}{1 + e} \rho_w. \quad (2.43)$$

Substituting the basic volume–mass relationship in Eq. (2.40) into Eq. (2.43) gives the following equation for the total density:

$$\rho = \frac{G_s + Se}{1 + e} \rho_w. \quad (2.44)$$

The dry density of a soil, $\rho_d$, is obtained by eliminating the mass of the water, $M_w$, from Eq. (2.41):

$$\rho_d = \frac{G_s}{1 + e} \rho_w. \quad (2.45)$$

The relationship between total density, $\rho$, and dry density, $\rho_d$, for different water contents is presented graphically in Fig. 2.14. The specific gravity of the soil solids, $G_s$, and the density of the water, $\rho_w$, are properties described in Section 2.1. If any two of the volume–mass properties of a soil (e.g., $e$, $w$, or $S$) are known, the total density of the soil, $\rho$, can be computed in accordance with Eq. (2.43) or Eq. (2.44). The dry density of the soil, $\rho_d$, is computed using Eq. (2.45) provided the void ratio, $e$, or the porosity, $n$, of the soil are known.

The dry density curve corresponding to a degree of saturation of 100% is called the "zero air voids" curve. The dry density curves for different degrees of saturation are commonly presented in connection with soil compaction data (Fig. 2.15). Compaction is a mechanical process used to increase the dry density of soils (i.e., densification). The compaction process is not discussed in this book, since numerous soil mechanics textbooks deal with this subject.

The relationship between the gravimetric water contents, $w$, and the volumetric water content, $\theta_w$, can be established by substituting the basic volume–mass relationship [i.e., Eq. (2.40)] into Eq. (2.31):

$$\theta_w = \frac{SwG_s}{S + wG_s}. \quad (2.46)$$

### 2.3.7 Changes in Volume–Mass Properties

The basic volume–mass relationship [Eq. (2.40)] applies to any combination of $S$, $e$, and $w$. Any change in one of these volume–mass properties (i.e., $S$, $e$, and $w$) may produce changes in the other two properties. Changes in two of the volume–mass quantities must be determined or measured in order to compute a change in the third quantity. If

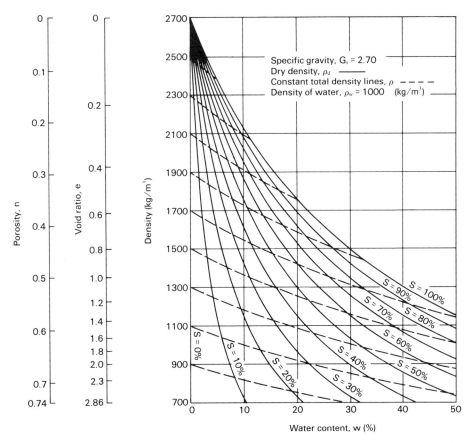

**Figure 2.14** Volume–mass relations for unsaturated soils.

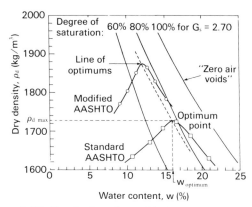

**Figure 2.15** Standard and modified AASHTO compaction curves.

changes in the void ratio, $e$, and the water content, $w$, are known, the change in the degree of saturation, $S$, can be computed. Similarly, if the changes in $S$ and $e$ or in $S$ and $w$ are known, then the change in $w$ or $e$, respectively, can be computed.

The relationship between the changes in the volume–mass properties can be derived from the *basic volume–mass relationship* expressed in Eq. (2.40). Consider a soil that undergoes a process such that changes occur in the volume–mass properties of the soil. Prior to the process, the volume–mass properties of the soil have the following relationship:

$$S_i e_i = w_i G_s \qquad (2.47)$$

where

$S_i$ = initial degree of saturation
$e_i$ = initial void ratio
$w_i$ = initial water content.

At the end of the process, the soil has final volume–mass properties which are also related by the basic volume–mass relation:

$$S_f e_f = w_f G_s \qquad (2.48)$$

where

$S_f$ = final degree of saturation
$e_f$ = final void ratio
$w_f$ = final water content.

The following relationships between initial and final conditions can be written:

$$S_f = S_i + \Delta S \qquad (2.49)$$

$$e_f = e_i + \Delta e \qquad (2.50)$$

$$w_f = w_i + \Delta w \qquad (2.51)$$

where

$\Delta S$ = change in the degree of saturation
$\Delta e$ = change in the void ratio
$\Delta w$ = change in the water content.

Substituting Eqs. (2.49), (2.50), (2.51), and (2.47) into Eq. (2.48) gives

$$S_i \Delta e + \Delta S e_i + \Delta S \Delta e = \Delta w G_s. \qquad (2.52)$$

The change in the degree of saturation, $\Delta S$, can be written in terms of the change in void ratio, $\Delta e$, and the change in water content, $\Delta w$:

$$\Delta S = \frac{(\Delta w G_s - S_i \Delta e)}{e_f}. \qquad (2.53)$$

Similarly, the change in the void ratio, $\Delta e$, is obtained by substituting Eq. (2.49) into Eq. (2.52) and solving for $\Delta e$:

$$\Delta e = \frac{(\Delta w G_s - \Delta S e_i)}{S_f}. \qquad (2.54)$$

The change in water content, $\Delta w$, can be written as follows:

$$\Delta w = \frac{(S_f \Delta e + \Delta S e_i)}{G_s}. \qquad (2.55)$$

### 2.3.8 Density of Mixtures Subjected to Compression of the Air Phase

Soil mixtures can occur in various forms in nature. A mixture of soil particles and air constitutes a dry soil, while a mixture of soil particles and water constitutes a saturated soil. Between these extremes lies the category of unsaturated soils which consist of soil particles, water, and air in differing volumetric percentages.

A mixture of soil particles, water, and air has a total density which has been defined in Eqs. (2.43) and (2.44). The dry density, $\rho_d$, of the mixture is expressed in Eq. (2.45) by considering only the mass of the soil particles.

The density of a soil mixture can also be formulated for situations where there is a change in the volume of air due to compression. The formulation can be visualized with the assistance of the piston–porous stone analogy described in Fig. 2.11. The density can be derived in a general form for a mixture of soil particles, water, and air. The derivations can then be extended to solve for the density of mixtures of soil particles and air, soil particles and water, and air and water, as well as the density of each phase.

*Piston–Porous Stone Analogy*

Consider a cubic element of soil that consists of soil particles at the bottom, water in the middle, and free air at the top. All sides are impervious and fixed, with the exception of the top which is a sealed frictionless piston. Initial volume–mass properties for each phase are shown in Table 2.12. The water phase initially has an amount of dissolved air which is in equilibrium with the free air. The air pressure is then increased by placing an additional load on the piston.

As a result of the pressure difference between the free

**Table 2.12 Symbols and Notations for the Initial and Final Stages of Each Phase in the Mixture of Soil Solids, Water, and Air**

| Volume–Mass Properties of Each Phase | Initial Stage | Final Stage |
|---|---|---|
| **Soil Solids** | | |
| Volume | $V_s$ | $V_s$ |
| Mass | $M_s$ | $M_s$ |
| Density | $\rho_s = M_s/V_s$ | $\rho_s = M_s/V_s$ |
| **Water Phase (Water and dissolved air)** | | |
| Volume of water and dissolved air | $V_w$ | $V_w$ |
| Mass of water | $M_w$ | $M_w$ |
| Density of water | $\rho_w = M_w/V_w$ | $\rho_w = M_w/V_w$ |
| Mass of dissolved air | $M_{di}$ | $M_d$ |
| Mass of water and dissolved air | $M_w + M_{di}$ | $M_w + M_d$ |
| Density of water and dissolved air | $\rho_{wi} = (M_w + M_{di})/V_w$ | $\rho_{wf} = (M_w + M_d)/V_w$ |
| Degree of saturation | $S_i = V_w/V_v$ | $S$ |
| Water content | $w_i = M_w/M_s$ | $w$ |
| **Air Phase (free and dissolved air)** | | |
| Volume of free air | $V_{ai}$ | $V_a$ |
| Volume of dissolved air | $V_d$ | $V_d$ |
| Absolute pressure of air | $\bar{u}_{ai}$ | $\bar{u}_a$ |
| Density of air | $\rho_{ai}$ | $\rho_a$ |
| Volumetric coefficient of solubility | $h = V_d/V_w$ | $h$ |
| **Voids (water and air)** | | |
| Void ratio | $e_i = V_v/V_s$ | $e$ |

and dissolved air, some of the free air dissolves into the water in accordance with Henry's law. This process is time dependent. At the end of the process, the pressure in the free and dissolved air are the same. The final volume–mass symbols and notations for water and air are summarized in Table 2.12. The soil particles are assumed to be incompressible (i.e., $V_s$ constant), and the mass of soil particles, $M_s$, is also constant throughout the process. Similarly, the mass and volume of water (i.e., $M_w$ and $V_w$, respectively) are constant during the process.

After an additional load is applied to the piston, the volume and pressure of the free air change. However, the total mass of air (i.e., free air and dissolved air) remains constant. At the end of the process, the pressure in the free air and dissolved air has changed from $\bar{u}_{ai}$ to $\bar{u}_a$, and the volume of free air has changed from $V_{ai}$ to $V_a$. The volume of dissolved air, $V_d$, was shown in Eq. (2.17) to be essentially constant for different air pressures.

Boyle's law states that the product of the volume and absolute pressure of a fixed amount of gas is constant under constant temperature conditions [Eq. (2.7)]. Applying Boyle's law to the free air and dissolved air at a constant temperature gives the final volume of free air, $V_a$:

$$(V_{ai} + V_d)\bar{u}_{ai} = (V_a + V_d)\bar{u}_a \quad (2.56)$$

$$V_a = (V_{ai} + V_d)\frac{\bar{u}_{ai}}{\bar{u}_a} - V_d \quad (2.57)$$

where

$V_{ai}$ = initial volume of free air corresponding to the initial absolute air pressure, $\bar{u}_{ai}$
$\bar{u}_{ai}$ = initial absolute air pressure
$V_d$ = volume of dissolved air
$V_a$ = final volume of free air corresponding to the final absolute air pressure, $\bar{u}_a$
$\bar{u}_a$ = final absolute air pressure.

The density of the free and dissolved air also changes

due to the change in pressure [see Eq. (2.9)]. At a constant temperature, the initial and final densities of air are related as

$$\rho_a = \frac{\bar{u}_a}{\bar{u}_{ai}} \rho_{ai} \tag{2.58}$$

where

$\rho_{ai}$ = initial density of air (free and dissolved) corresponding to the initial air pressure
$\rho_a$ = final density of air (free and dissolved) corresponding to the final air pressure

The water phase can be assumed to be incompressible (i.e., $V_w$ is constant) during the process. However, the mass of water and dissolved air is changed at the end of the process due to a change in the mass of dissolved air from $M_{di}$ to $M_d$. This corresponds to the change in absolute air pressure from $\bar{u}_{ai}$ to $\bar{u}_a$ as described by Henry's law. The final density of the water phase (i.e., water plus dissolved air), $\rho_{wf}$, is computed as the ratio of the total mass of water and dissolved air to the total volume of water and dissolved air under final conditions:

$$\rho_{wf} = \frac{M_w + M_d}{V_w} \tag{2.59}$$

where

$\rho_{wf}$ = final density of the water phase (i.e., water plus dissolved air)
$M_w$ = mass of water before and after the change in air pressure
$M_d$ = final mass of dissolved air
$V_w$ = total volume of water and dissolved air before and after the change in air pressure

The mass of dissolved air, $M_d$, is given as the product of the dissolved air volume, $V_d$, and the density of air, $\rho_a$ [Eq. (2.58)]:

$$M_d = V_d \left( \frac{\bar{u}_a}{\bar{u}_{ai}} \rho_{ai} \right). \tag{2.60}$$

The final density of water, $\rho_{wf}$, can be obtained by substituting Eq. (2.60) into Eq. (2.59).

$$\rho_{wf} = \frac{M_w}{V_w} + \left( \frac{\bar{u}_a}{\bar{u}_{ai}} \rho_{ai} \right) \frac{V_d}{V_w}. \tag{2.61}$$

The first term in Eq. (2.61) (i.e., $M_w/V_w$) is the density of water without dissolved air, $\rho_w$.

### Conservation of Mass Applied to a Mixture

The final density of a mixture can be written by satisfying the conservation of mass of the element:

$$\rho V = \rho_s V_s + \rho_{wf} V_w + \rho_a V_a \tag{2.62}$$

where

$\rho$ = total density of the mixture (i.e., soil particles, water, and air) after the change in air pressure
$V$ = total volume of the mixture after the change in absolute air pressure (i.e., $V_s + V_w + V_a$)
$\rho_s$ = density of soil solids
$V_s$ = volume of soil solids.

The density of the mixture can then be derived by substituting Eqs. (2.57), (2.58), and (2.61) into Eq. (2.62)

$$\rho = \frac{\rho_s V_s + \left\{ \rho_w + \left( \frac{\bar{u}_a}{\bar{u}_{ai}} \rho_{ai} \right) \frac{V_d}{V_w} \right\} V_w + \left( \frac{\bar{u}_a}{\bar{u}_{ai}} \rho_{ai} \right) \left\{ (V_{ai} + V_d) \frac{\bar{u}_{ai}}{\bar{u}_a} - V_d \right\}}{V_s + V_w + (V_{ai} + V_d) \frac{\bar{u}_{ai}}{\bar{u}_a} - V_d}. \tag{2.63}$$

This equation can be reduced to the following form:

$$\rho = \frac{\rho_s V_s + \rho_w V_w + \rho_{ai} V_{ai} + \rho_{ai} V_d}{V_s + V_w + V_{ai} \frac{\bar{u}_{ai}}{\bar{u}_a} + V_d \left( \frac{\bar{u}_{ai}}{\bar{u}_a} - 1 \right)}. \tag{2.64}$$

The volume of dissolved air, $V_d$, is related to the volume of water, $V_w$, by the volumetric coefficient of solubility, $h$ (i.e., $V_d = hV_w$). Equation (2.64) can therefore be rewritten as follows:

$$\rho = \frac{\rho_s V_s + \rho_w V_w + \rho_{ai} V_{ai} + \rho_{ai} h V_w}{V_s + V_w + V_{ai} \frac{\bar{u}_{ai}}{\bar{u}_a} + h V_w \left( \frac{\bar{u}_{ai}}{\bar{u}_a} - 1 \right)}. \tag{2.65}$$

The volume of each phase can be expressed in terms of the volume of soil particles, $V_s$, using the initial void ratio, $e_i$, and the initial degree of saturation, $S_i$. Therefore, the volume of water, $V_w$, can be written as

$$V_w = S_i e_i V_s \tag{2.66}$$

and the volume of the initial free air, $V_{ai}$, can be written as

$$V_{ai} = (1 - S_i) e_i V_s. \tag{2.67}$$

Substituting Eqs. (2.66) and (2.67) into Eq. (2.65) and dividing the top and bottom of the equation by $V_s$ gives

$$\rho = \frac{\rho_s + \rho_w S_i e_i + \rho_{ai}(1 - S_i)e_i + \rho_{ai} h S_i e_i}{1 + S_i e_i + (1 - S_i)e_i \frac{\bar{u}_{ai}}{\bar{u}_a} + h S_i e_i \left( \frac{\bar{u}_{ai}}{\bar{u}_a} - 1 \right)}. \tag{2.68}$$

Substituting the basic volume–mass relationship (i.e., $S_i e_i = w_i G_s$) into Eq. (2.68) yields another form for the density of the soil (i.e., the soil particles–water–air mixture):

$$\rho = \frac{\rho_s + \rho_w w_i G_s + \rho_{ai}(e_i - w_i G_s) + \rho_{ai} h w_i G_s}{1 + w_i G_s + (e_i - w_i G_s)\dfrac{\bar{u}_{ai}}{\bar{u}_a} + h w_i G_s \left(\dfrac{\bar{u}_{ai}}{\bar{u}_a} - 1\right)}. \quad (2.69)$$

Equations (2.68) and (2.69) are general mixture equations for the density after the soil has been subjected to a change in air pressure. The equations incorporate the effect of air going into solution due to the change in the air pressure. Where the air does not have time to dissolve in the water, the volumetric coefficient of solubility, $h$, can be set to zero.

The equations use the initial degree of saturation and the initial void ratio as a reference. The density of the mixture before a change in air pressure can also be obtained from these equations by setting $\bar{u}_a$ equal to $\bar{u}_{ai}$. This also applies to a mixture which does not experience any change in the air pressure (i.e., $\bar{u}_a$ equals $\bar{u}_{ai}$).

*Soil Particles–Water–Air Mixture*

The general equation for the density of mixtures can be specialized to the density of particular mixtures. Consider the case of an unsaturated soil not subjected to a pressure change.

Let us assume that the mass of the air phase is negligible (i.e., $\rho_{ai} = 0$), and that there is no change in the pore-air pressure (i.e., $\bar{u}_a$ equals $\bar{u}_{ai}$). Under this condition, the water content, the degree of saturation, and the void ratio remain constant (i.e., $w_i = w$, $S_i = S$, and $e_i = e$). Equation (2.69) then specializes to the following form:

$$\rho = \frac{\rho_s + \rho_w w G_s}{1 + e}. \quad (2.70)$$

Noting that the density of the solids can be written in terms of the specific gravity of the solids (i.e., $\rho_s = G_s \rho_w$), Eq. (2.70) becomes

$$\rho = \frac{G_s + Se}{1 + e}\rho_w. \quad (2.71)$$

Equation (2.71) gives the total density of a soil at a constant air pressure, and is the same as the expressions derived previously [i.e., Eq. (2.44)].

*Air–Water Mixture*

The general mixture equation can also be shown to specialize to the case of an air and water mixture. The volume of the soil particles, $V_s$, is set to zero, and the initial void ratio, $e_i$, becomes infinity. Rearranging Eq. (2.68) by dividing the top and bottom portions by the initial void ratio, $e_i$, yields the following form:

$$\rho = \frac{\dfrac{\rho_s}{e_i} + \rho_w S_i + \rho_{ai}(1 - S_i) + \rho_{ai} h S_i}{\dfrac{1}{e_i} + S_i + (1 - S_i)\dfrac{\bar{u}_{ai}}{\bar{u}_a} + h S_i \left(\dfrac{\bar{u}_{ai}}{\bar{u}_a} - 1\right)}. \quad (2.72)$$

Setting the initial void ratio to infinity gives the density of an air-water mixture after a change in air pressure from $\bar{u}_{ai}$ to $\bar{u}_a$:

$$\rho_m = \frac{\rho_w S_i + \rho_{ai}(1 - S_i) + \rho_{ai} h S_i}{S_i + (1 - S_i)\dfrac{\bar{u}_{ai}}{\bar{u}_a} + h S_i \left(\dfrac{\bar{u}_{ai}}{\bar{u}_a} - 1\right)} \quad (2.73)$$

where

$\rho_m$ = density of an air-water mixture after a change in air pressure from $\bar{u}_{ai}$ to $\bar{u}_a$.

The air-water mixture always has a density between that of air and water. As the final air pressure, $\bar{u}_a$, increases, the final density of the air-water mixture also increases. If the air pressure is continuously increased, the density of the air-water mixture will reach the density of the water phase (i.e., water plus dissolved air). If the density of air is assumed to be negligible, the density of the air-water mixture reaches the density of water (i.e., $\rho_m$ approaches $\rho_w$). At this stage, the free air has been dissolved into the water and the degree of saturation approaches 100%.

The magnitude of the air pressure required to bring the air-water mixture to saturation can be found from Eq. (2.73) by substituting ($\rho_m = \rho_w$ and $\rho_{ai} = 0$), and dividing the top and bottom portions of the equation by the initial degree of saturation, $S_i$:

$$\rho_w = \frac{\rho_w}{1 + \left(\dfrac{1}{S_i} - 1\right)\dfrac{\bar{u}_{ai}}{\bar{u}_a} + h\left(\dfrac{\bar{u}_{ai}}{\bar{u}_a} - 1\right)}. \quad (2.74)$$

Solving Eq. (2.74) for the absolute pore-air pressure gives

$$\bar{u}_a = \left(1 + \frac{1 - S_i}{S_i h}\right)\bar{u}_{ai}. \quad (2.75)$$

Equation (2.75) shows that the final absolute air pressure required to dissolve all of the free air depends upon the initial degree of saturation of the mixture, $S_i$, and the initial absolute air pressure, $\bar{u}_{ai}$. The lower the initial degree of saturation, the larger the air pressure required to saturate the air-water mixture. The absolute air pressure, $\bar{u}_a$, obtained from Eq. (2.75) is essentially the same as the maximum air pressure required to saturate a soil, as suggested by Bishop and Eldin (1950), and Schuurman (1966).

# CHAPTER 3

# Stress State Variables

The mechanical behavior of a soil (i.e., the volume change and shear strength behavior) can be described in terms of the state of stress in the soil. The state of stress in a soil consists of certain combinations of stress variables that can be referred to as stress state variables. A more complete definition of the term "stress state variable" is given in Chapter 1. These variables should be independent of the physical properties of the soil. The number of stress state variables required for the description of the stress state of a soil depends primarily upon the number of phases involved.

The effective stress, $(\sigma - u_w)$, for saturated soils has often been regarded as a physical law. More correctly, the effective stress is simply a stress state variable that can be used to describe the behavior of a saturated soil. The effective stress variable is applicable to sands, silts, or clays because it is independent of the soil properties. The volume change process and the shear strength characteristics of a saturated soil are both controlled by the effective stress.

Satisfactory stress state variables for an unsaturated soil have been considerably more difficult to establish. Only recently has there been some agreement on the most acceptable stress state variables to use in practice. An attempt is made in this chapter to review: 1) the historical development of an acceptable stress state description, 2) the mathematical form and experimental evidence pertinent to stress state descriptions, and 3) their application to practical problems. The magnitudes of the stress state variables in the field are of importance to practicing engineers for analyzing engineering problems. A smooth transition in the stress state description when going from the unsaturated to the saturated soil state is demonstrated.

The stress analysis for an unsaturated soil is presented as an extension of the saturated soil theory. The principles used in the stress analysis of a saturated soil (e.g., Mohr diagrams) are equally applicable to unsaturated soils.

## 3.1 HISTORY OF THE DESCRIPTION OF THE STRESS STATE

The effective stress concept has been well accepted and studied for saturated soils. Numerous attempts have been made to develop a similar concept of effective stress for unsaturated soils. However, unsaturated soils are more complex, and it has been more difficult to arrive at a consensus regarding the description of the stress state. The use of a single-valued effective stress for unsaturated soils has encountered many difficulties, and has led numerous researchers to the realization that two independent stress state variables should be used for unsaturated soils.

### 3.1.1 Effective Stress Concept for a Saturated Soil

Soil mechanics as a science has been successfully applied to many geotechnical problems involving saturated soils. This success is due to the ability of engineers to relate observed soil behavior to stress conditions in the soil. Terzaghi (1936) described the stress state variable controlling the behavior of a saturated soils as follows:

> The stresses in any point of a section through a mass of soil can be computed from the total principal stresses $\sigma_1$, $\sigma_2$, $\sigma_3$ which act at this point. If the voids of the soil are filled with water under a stress, $u_w$, the total principal stresses consist of two parts. One part, $u_w$, acts in the water and in the solid in every direction with equal intensity. It is called the neutral (or the pore-water) pressure. The balance $\sigma'_1 = \sigma_1 - u_w$, $\sigma'_2 = \sigma_2 - u_w$, and $\sigma'_3 = \sigma_3 - u_w$ represents an excess over the neutral stress, $u_w$, and it has its seat exclusively in the solid phase of the soil. All the measurable effects of a change in stress, such as compression, distortion, and a change in shearing resistance, are exclusively due to changes in the effective stress $\sigma'_1$, $\sigma'_2$, and $\sigma'_3$.

The stress state variable for a saturated soil has been called the effective stress and is commonly expressed in the form of an equation:

$$\sigma' = \sigma - u_w \quad (3.1)$$

where

$\sigma'$ = effective normal stress
$\sigma$ = total normal stress
$u_w$ = pore-water pressure.

Equation (3.1) is referred to as the effective stress equation, and is a definition of the stress state variable for saturated soils. Evidence has shown that only a single-valued effective stress or one stress state variable [i.e., $(\sigma - u_w)$] is required to describe the mechanical behavior of a saturated soil. A more complete description of the stress state involves writing the effective stress for each of three orthogonal directions and including shear components. The validity of the effective stress as a stress state variable for saturated soils has been well accepted and experimentally verified (Rendulic, 1936; Bishop and Eldin, 1950; Laughton, 1955; Skempton, 1961).

The effective stress concept forms the fundamental basis for studying saturated soil mechanics. All mechanical aspects of a saturated soil are governed by the effective stress. The change in volume and shear strength is controlled by a change in the effective stress. In other words, an effective stress change will alter the equilibrium state of a saturated soil. The success of the effective stress concept in describing saturated soil behavior has often resulted in effective stress being considered as a law. Although the effective stress equation in Eq. (3.1) is not a physical law, the effective stress has proven to be the only stress state variable controlling the behavior of a saturated soil.

There have been attempts to express the effective stress in different forms. However, the proposed, modified effective stress equations have involved the incorporation of a soil property (e.g., contact area between soil particles) into the description of the stress state. The incorporation of a soil property produces a constitutive relation which becomes questionable as a stress state variable from a continuum mechanics standpoint.

### 3.1.2 Proposed Effective Stress Equation for an Unsaturated Soil

Unsaturated soil behavior is more complex than saturated soil behavior. Unsaturated soils have commonly been viewed as a three-phase system (Lambe and Whitman, 1979). More recently, the contractile skin (i.e., the air-water interface) has been introduced as a fourth and independent phase (Fredlund and Morgenstern, 1977). Consideration of the contractile skin as the fourth phase is later used in the theoretical stress analysis for an unsaturated soil.

It is desirable that the concept of effective stress for saturated soils be extended to unsaturated soils. All proposed, so-called "effective stress" equations have attempted to provide a single-valued effective stress or one stress state variable for an unsaturated soil. Soil parameters are used in all proposed equations. The incorporation of a soil property in the description of the stress state leads to difficulties. Variables used for the description of a stress state should be independent of soil properties (Fung, 1977). Experimental results have shown that the soil properties measured do not yield a single-valued relationship to the proposed effective stress. In other words, the soil property in the proposed effective stress equation has different magnitudes for different problems (i.e., volume change and shear strength), different stress paths, and different types of soil (Jennings and Burland, 1962; Coleman, 1962; Bishop and Blight, 1963; Burland, 1964; Burland, 1965; Blight, 1965).

More recently, there has been a tendency to use the stress state variables for an unsaturated soil in an independent manner. In other words, the effective stress equation has been separated into two independent stress state variables, and the need for the incorporation of soil properties in the stress state description no longer exists. Consideration of the contractile skin as a phase lends support to the theoretical justification for two independent stress state variables for an unsaturated soil (Fredlund and Morgenstern, 1977). The use of two independent stress state variables is advocated throughout this book.

In 1941, Biot proposed a general theory of consolidation for an unsaturated soil with occluded air bubbles. The constitutive equations relating stress and strain were formulated in terms of the effective stress, $(\sigma - u_w)$, and the pore-water pressure, $u_w$. In other words, the need for separating the effects of total stress and pore-water pressure was recognized. Personnel at the Roads Research Laboratory were probably the first group to recognize the importance of soil suction in relation to road and air field design (Croney, 1952). Croney et al., (1958) proposed the following form of an effective stress equation for an unsaturated soil:

$$\sigma' = \sigma - \beta' u_w \quad (3.2)$$

where

$\sigma'$ = effective normal stress
$\sigma$ = total normal stress
$\beta'$ = the holding or bonding factor which is a measure of the number of bonds under tension, effective in contributing to the shear strength of the soil
$u_w$ = pore-water pressure.

Bishop (1959) suggested a tentative expression for effective stress which has gained widespread reference (i.e., Lecture in Oslo, Norway, in 1955):

$$\sigma' = (\sigma - u_a) + \chi(u_a - u_w) \quad (3.3)$$

where

$u_a$ = pore-air pressure
$\chi$ = a parameter related to the degree of saturation of the soil.

The magnitude of the $\chi$ parameter is unity for a saturated soil and zero for a dry soil. The relationship between $\chi$ and the degree of saturation, $S$, was obtained experimentally. Experiments were performed on cohesionless silt (Donald, 1961) and compacted soils (Blight 1961), as shown in Fig. 3.1(a) and 3.1(b), respectively. Figure 3.1 demonstrates the influence of the soil type on the $\chi$ parameter (Bishop and Henkel, 1962). Bishop et al. (1960) presented the results of triaxial tests performed on saturated and unsaturated soils in an attempt to substantiate the use of Bishop's equation [i.e., Eq. (3.3)].

Bishop and Donald (1961) published the results of triaxial tests on an unsaturated silt in which the total, pore-air, and pore-water pressures were controlled independently. During the tests, the confining, pore-air, and pore-water pressures (i.e., $\sigma_3$, $u_a$, and $u_w$) were varied in such a way that the $(\sigma_3 - u_a)$ and $(u_a - u_w)$ variables remained constant. The results showed that the stress–strain curve remained monotonic during these changes. This lent credibility to the use of Eq. (3.3); however, the test results equally justify the use of independent stress state variables.

Aitchison (1961) proposed the following effective stress equation at the Conference on Pore Pressure and Suction in Soils, London, in 1960:

$$\sigma' = \sigma + \psi p'' \qquad (3.4)$$

where

$p''$ = pore-water pressure deficiency
$\psi$ = a parameter with values ranging from zero to one.

Jennings (1961) also proposed an effective stress equation at the same conference:

$$\sigma' = \sigma + \beta p'' \qquad (3.5)$$

where

$p''$ = negative pore-water pressure taken as a positive value
$\beta$ = a statistical factor of the same type as the contact area. This factor should be measured experimentally.

Equations (3.2), (3.3), (3.4), and (3.5) are equivalent when the pore-air pressure used in all four equations is the same (i.e., $\beta' = \chi = \psi = \beta$). Only Bishop's form [i.e., Eq. (3.3)] references the total and pore-water pressures to the pore-air pressure. The other equations simply use gauge pressures which are referenced to the external air pressure.

Jennings and Burland (1962) appear to be the first to suggest that Bishop's equation did not provide an adequate relationship between volume change and effective stress for most soils, particularly those below a critical degree of saturation. The critical degree of saturation was estimated to be approximately 20% for silts and sands, and as high as 85-90% for clays.

Coleman (1962) suggested the use of "reduced" stress variables, $(\sigma_1 - u_a)$, $(\sigma_3 - u_a)$, and $(u_w - u_a)$, to represent the axial, confining, and pore-water pressures, respectively, in triaxial tests. The constitutive relations for volume change in unsaturated soils were then formulated in terms of the above stress variables.

In 1963, Bishop and Blight reevaluated the proposed effective stress equation [i.e., Eq. (3.3)] for unsaturated soils. It was noted that a variation in matric suction, $(u_a - u_w)$, did not result in the same change in effective stress as did a change in the net normal stress, $(\sigma - u_a)$. A graphical presentation was suggested for volume change (or void ratio change, $\Delta e$) versus the $(\sigma - u_a)$ and $(u_a - u_w)$ stress variables. This further reinforced the use of the stress state variables in an independent manner. Blight (1965) concluded that the proposed effective stress equation depends

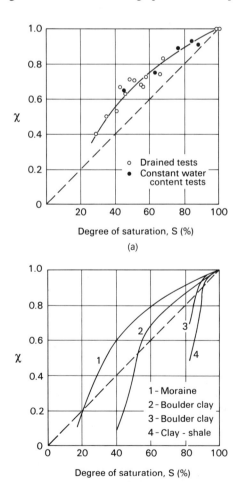

**Figure 3.1** The relationship between the $\chi$ parameter and the degree of saturation, $S$. (a) $\chi$ values for a cohesionless silt (after Donald, 1961); (b) $\chi$ values for compacted soils (after Blight, 1961).

on the type of process to which the soil was subjected. Burland (1964, 1965) further questioned the validity of the proposed effective stress equation, and suggested that the mechanical behavior of unsaturated soils should be independently related to the stress variables, $(\sigma - u_a)$ and $(u_a - u_w)$, whenever possible.

Richards (1966) incorporated a solute suction component into the effective stress equation:

$$\sigma' = \sigma - u_a + \chi_m(h_m + u_a) + \chi_s(h_s + u_a) \quad (3.6)$$

where

$\chi_m$ = effective stress parameter for matric suction
$h_m$ = matric suction
$\chi_s$ = effective stress parameter for solute suction
$h_s$ = solute suction.

Little reference has subsequently been made to this equation. Aitchison (1967) pointed out the complexity associated with the $\chi$ parameter. He stated that a specific value of $\chi$ may only relate to a single combination of $(\sigma)$ and $(u_a - u_w)$ for a particular stress path. It was suggested that the terms $(\sigma)$ and $(u_a - u_w)$ be separated in analyzing the behavior of unsaturated soils. Later, constitutive relationship data (Aitchison and Woodburn, 1969) were presented in accordance with the proposed independent stress variables.

Matyas and Radhakrishna (1968) introduced the concept of "state parameters" in describing the volumetric behavior of unsaturated soils. Volume change was presented as a three-dimensional surface with respect to the state parameters, $(\sigma - u_a)$ and $(u_a - u_w)$. Barden et al. (1969a) also suggested that the volume change of unsaturated soils be analyzed in terms of the separate components of applied stress, $(\sigma - u_a)$, and suction, $(u_a - u_w)$.

Brackley (1971) examined the application of the effective stress principle to the volume change behavior of unsaturated soils. He concluded from his test results that there was a limit to the use of a single-valued effective stress equation.

Aitchison (1965a, 1973) presented an effective stress equation slightly modified from that of Richards (1966):

$$\sigma' = \sigma + \chi_m p_m'' + \chi_s p_s'' \quad (3.7)$$

where

$p_m''$ = matric suction, $(u_a - u_w)$
$p_s''$ = solute suction
$\chi_m$ and $\chi_s$ = soil parameters which are normally within the range of 0–1, which are dependent upon the stress path.

The above history shows that considerable effort has been extended in the search for a single-valued effective stress equation for unsaturated soils. Numerous effective stress equations have been proposed. All equations incorporate a soil parameter in order to form a single-valued effective stress variable. Experiments have demonstrated that the effective stress equation is not single-valued. Rather, there is a dependence on the stress path followed. The soil parameter used in the effective stress equation appears to be difficult to evaluate. In general, the proposed effective stress equations have not received much recent attention in describing the mechanical behavior of unsaturated soils. In referring to the application of Bishop's effective stress equation, Morgenstern (1979) stated that the equation has "—proved to have little impact on practice. The parameter $\chi$ when determined for volume change behaviour was found to differ when determined for shear strength."

Probably more important than the above experimental difficulties is the philosophical difficulty in justifying the use of soil properties in the description of a stress state. Morgenstern (1979) stated, "The effective stress is a stress variable and hence related to equilibrium considerations alone while [Equation 3.3] contains a parameter, $\chi$, that bears on constitutive behavior. This parameter is found by assuming that the behavior of a soil can be expressed uniquely in terms of a single effective stress variable and by matching unsaturated behaviour with saturated behavior in order to calculate $\chi$. Normally, we link equilibrium considerations to deformations through constitutive behavior and do not introduce constitutive behavior directly into the stress variable." Reexamination of the proposed effective stress equations has led many researchers to suggest the use of independent stress state variables [e.g., $(\sigma - u_a)$ and $(u_a - u_w)$] to describe the mechanical behavior of unsaturated soils.

Fredlund and Morgenstern (1977) presented a theoretical stress analysis of an unsaturated soil on the basis of multiphase continuum mechanics. The unsaturated soil was considered as a four-phase system. The soil particles were assumed to be incompressible and the soil was treated as though it were chemically inert. These assumptions are consistent with those used in saturated soil mechanics.

The analysis concluded that any two of three possible normal stress variables can be used to describe the stress state of an unsaturated soil. In other words, there are three possible combinations which can be used as stress state variables for an unsaturated soil. These are: 1) $(\sigma - u_a)$ and $(u_a - u_w)$, 2) $(\sigma - u_w)$ and $(u_a - u_w)$, and 3) $(\sigma - u_a)$ and $(\sigma - u_w)$. In a three-dimensional stress analysis, the stress state variables of an unsaturated soil form two independent stress tensors. These are discussed in the following sections. The proposed stress state variables for unsaturated soils have also been experimentally tested (Fredlund, 1973).

The stress state variables can then be used to formulate constitutive equations to describe the shear strength behavior and the volume change behavior of unsaturated soils. This eliminates the need to find a single-valued effective stress equation that is applicable to both shear strength and volume change problems. The use of independent stress state variables has produced a more meaningful description of unsaturated soil behavior, and forms the basis for formulations in this book.

## 3.2 STRESS STATE VARIABLES FOR UNSATURATED SOILS

The mechanical behavior of soils is controlled by the same stress variables which control the equilibrium of the soil structure. Therefore, the stress variables required to describe the equilibrium of the soil structure can be taken as the stress state variables for the soil. The stress state variables must be expressed in terms of the measurable stresses, such as the total stress, $\sigma$, the pore-water pressure, $u_w$, and the pore-air pressure, $u_a$. An equilibrium stress analysis can be performed for an unsaturated soil after considering the state of stress at a point in the soil.

### 3.2.1 Equilibrium Analysis for Unsaturated Soils

There are two types of forces that can act on an element of soil. These are body forces and surface forces. Body forces act through the centroid of the soil element, and are expressed as a force per unit volume. Gravitational and interaction forces between phases are examples of body forces. Surface forces, such as external loads, act only on the boundary surface of the soil element. The average value of a surface force per unit area tends to a limiting value as the surface area approaches zero. This limiting value is called the stress vector or the surface traction on a given surface. The component of the stress vector perpendicular to a plane is defined as a normal stress, $\sigma$. The stress components parallel to a plane are referred to as shear stresses, $\tau$.

There are an infinite number of planes (or surfaces) that can be passed through a *point* in a soil mass. The stress state at a point can be analyzed by considering all the stresses acting on the planes that form a cubical element of infinitesimal dimensions. In addition, body forces acting through the centroid of the soil element should be considered. A cubical element that is completely enclosed by imaginary, unbiased boundaries yields the conventional free body used for a stress equilibrium analysis (Fung, 1969; Biot, 1955; Hubbert and Rubey, 1959). Figure 3.2 shows a cubical soil element with infinitesimal dimensions of $dx$, $dy$, and $dz$ in the Cartesian coordinate system. The normal and shear stresses on each plane of the element are illustrated in Fig. 3.2. The body forces are not shown.

### Normal and Shear Stresses on a Soil Element

Normal and shear stresses act on every plane in the $x$-, $y$-, and $z$-directions. The normal stress, $\sigma$, has one subscript to denote the plane on which it acts. Soils are most commonly subjected to compressive normal stresses. In soil mechanics, a positive normal stress is used to indicate a compression in the soil. All the normal stresses shown in Fig. 3.2 are positive or compressive. Opposite directions would indicate negative normal stresses or tensions.

The shear stress, $\tau$, has two subscripts. The first subscript denotes the plane on which the shear stress acts, and the second subscript refers to the direction of the shear stress. As an example, the shear stress, $\tau_{yz}$, acts on the $y$-plane and in the $z$-direction. All of the shear stresses

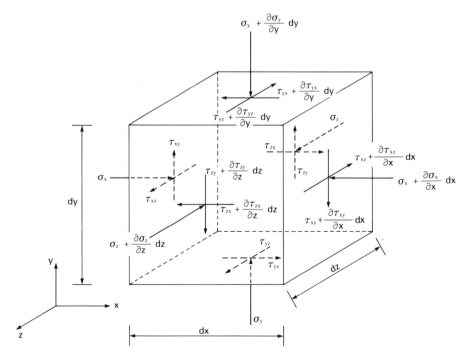

**Figure 3.2** Normal and shear stresses on a cubical soil element of infinitesimal dimensions.

shown in Fig. 3.2 have positive signs. Opposite directions would indicate negative shear stresses.

Equating the summation of moments about the $x$-, $y$-, and $z$-axes to zero results in the following shear stress relationships:

$$\tau_{xy} = \tau_{yx} \qquad (3.8)$$

$$\tau_{xz} = \tau_{zx} \qquad (3.9)$$

$$\tau_{yz} = \tau_{zy}. \qquad (3.10)$$

The stress components can vary from plane to plane across an element. The spatial variation of a stress component can be expressed as its derivative with respect to space. The stress variations in the $x$-, $y$-, and $z$-directions are expressed as stress fields (Fig. 3.2).

*Equilibrium Equations*

The stress equilibrium conditions for an unsaturated soil are presented in Appendix B. A cubical element of an unsaturated soil (Fig. 3.2) is used in the equilibrium analysis. Newton's second law is applied to the soil element by summing the forces in each direction (i.e., $x$-, $y$-, and $z$-directions). An equilibrium condition for an unsaturated soil element implies that the four phases (i.e., air, water, contractile skin, and soil particles) of the soil are in equilibrium. Each phase is assumed to behave as an independent, linear, continuous, and coincident stress field in each direction. An independent equilibrium equation can be written for each phase and superimposed using the principle of superposition. However, this may not give rise to equilibrium equations with stresses that can be measured. For example, the interparticle stresses cannot be measured directly. Therefore, it is necessary to combine the independent phases in such a way that measurable stresses appear in the equilibrium equation for the soil structure (i.e., the arrangement of soil particles).

The force equilibrium equations for the air phase, the water phase, and contractile skin, together with the total equilibrium equation for the soil element are used in formulating the equilibrium equation for the soil structure. In the $y$-direction, the equilibrium equation for the soil structure has the following form:

$$\frac{\partial \tau_{xy}}{\partial x} + \frac{\partial (\sigma_y - u_a)}{\partial y} + (n_w + n_c f^*) \frac{\partial (u_a - u_w)}{\partial y}$$

$$+ \frac{\partial \tau_{zy}}{\partial z} + (n_c + n_s) \frac{\partial u_a}{\partial y}$$

$$+ n_s \rho_s g - F_{sy}^w - F_{sy}^a$$

$$+ n_c (u_a - u_w) \frac{\partial f^*}{\partial y} = 0 \qquad (3.11)$$

where

- $\tau_{xy}$ = shear stress on the $x$-plane in the $y$-direction
- $\sigma_y$ = total normal stress in the $y$-direction (or on the $y$-plane)
- $u_a$ = pore–air pressure
- $f^*$ = interaction function between the equilibrium of the soil structure and the equilibrium of the contractile skin
- $(\sigma_y - u_a)$ = net normal stress in the $y$-direction
- $n_w$ = porosity relative to the water phase
- $n_c$ = porosity relative to the contractile skin
- $u_w$ = pore–water pressure
- $(u_a - u_w)$ = matric suction
- $\tau_{zy}$ = shear stress on the $z$-plane in the $y$-direction
- $n_s$ = porosity relative to the soil particles
- $g$ = gravitational acceleration
- $\rho_s$ = soil particle density
- $F_{sy}^w$ = interaction force (i.e., body force) between the water phase and the soil particles in the $y$-direction
- $F_{sy}^a$ = interaction force (i.e., body force) between the air phase and the soil particles in the $y$-direction.

Similar equilibrium equations can be written for the $x$- and $z$-directions. The stress variables that control the equilibrium of the soil structure [i.e., Eq. (3.11)] also control the equilibrium of the contractile skin through the interaction function, $f^*$.

### 3.2.2 Stress State Variables

Three independent sets of normal stresses (i.e., surface tractions) can be extracted from the equilibrium equation for the soil structure [Eq. (3.11)]. These are $(\sigma_y - u_a)$, $(u_a - u_w)$, and $(u_a)$, which govern the equilibrium of the soil structure and the contractile skin. The components of these variables are physically measurable quantities. The stress variable, $u_a$, can be eliminated when the soil particles and the water are assumed to be incompressible. The $(\sigma - u_a)$ and $(u_a - u_w)$ are referred to as the stress state variables for an unsaturated soil. More specifically, these are the surface tractions controlling the equilibrium of the soil structure and the contractile skin.

Similar stress state variables can also be extracted from the soil structure equilibrium equations for the $x$- and $z$-directions. The complete form of the stress state for an unsaturated soil can therefore be written as two independent stress tensors:

$$\begin{bmatrix} (\sigma_x - u_a) & \tau_{yx} & \tau_{zx} \\ \tau_{xy} & (\sigma_y - u_a) & \tau_{zy} \\ \tau_{xz} & \tau_{yz} & (\sigma_z - u_a) \end{bmatrix} \qquad (3.12)$$

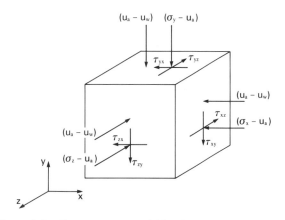

**Figure 3.3** The stress state variables for an unsaturated soil.

and

$$\begin{bmatrix} (u_a - u_w) & 0 & 0 \\ 0 & (u_a - u_w) & 0 \\ 0 & 0 & (u_a - u_w) \end{bmatrix}. \quad (3.13)$$

The above tensors cannot be combined into one matrix since the stress variables have different soil properties (i.e., porosities) outside the partial differential terms [see Eq. (3.11)]. The porosity terms are soil properties that should not be included in the description of the stress state of a soil. Figure 3.3 illustrates the two independent tensors acting at a point in an unsaturated soil.

In the case of compressible soil particles or pore fluid, an additional stress tensor, $u_a$, must be used to describe the stress state:

$$\begin{bmatrix} u_a & 0 & 0 \\ 0 & u_a & 0 \\ 0 & 0 & u_a \end{bmatrix} \quad (3.14)$$

The pore-air and pore-water pressures are usually expressed in terms of gauge pressure. This is a common practice in engineering. Under certain circumstances, such as when dealing with the gas law, the absolute air pressure must be used. Figure 3.4 illustrates the relationship between absolute and gauge pressures.

### Other Combinations of Stress State Variables

The equilibrium equation for the soil structure [i.e., Eq. (3.11)] can be formulated in a slightly different manner by using the pore-water pressure, $u_w$, or the total normal stress, $\sigma$, as a reference (see Appendix B). If the pore-water pressure, $u_w$, is used as a reference, the following combination of stress state variables, $(\sigma - u_w)$, $(u_a - u_w)$, and $(u_w)$, can be extracted from the equilibrium equations for the soil structure. The stress variable, $u_w$, is only of relevance for soils with compressible soil particles. If the total normal stress, $\sigma$, is used as a reference, the following combination of stress state variables, $(\sigma - u_a)$, $(\sigma - u_w)$, and $(\sigma)$, can be extracted from the equilibrium equations for the soil structure. The stress variable, $\sigma$, can be ignored when the soil particles are assumed to be incompressible.

In summary, there are three possible combinations of stress state variables that can be used to describe the stress state relevant to the soil structure and contractile skin in an unsaturated soil. These are tabulated in Table 3.1. The three combinations of stress state variables are obtained from equilibrium equations for the soil structure which are derived with respect to three different references (i.e., $u_a$, $u_w$, and $\sigma$). However, the $(\sigma - u_a)$ and $(u_a - u_w)$ combination appears to be the most satisfactory for use in engineering practice (Fredlund, 1979; Fredlund and Rahardjo, 1987). This combination is advantageous because the effects of a change in total normal stress can be separated from the effects caused by a change in the pore-water pressure. In addition, the pore-air pressure is atmospheric (i.e., zero gauge pressure) for most practical engineering problems.

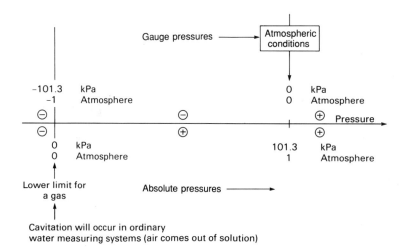

**Figure 3.4** Relationship between absolute and gauge pressures.

**Table 3.1 Possible Combinations of Stress State Variables for an Unsaturated Soil**

| Reference Pressure | Stress State Variables |
| --- | --- |
| Air, $u_a$ | $(\sigma - u_a)$ and $(u_a - u_w)$ |
| Water, $u_w$ | $(\sigma - u_w)$ and $(u_a - u_w)$ |
| Total, $\sigma$ | $(\sigma - u_a)$ and $(\sigma - u_w)$ |

Referencing the stress state to the pore-air pressure would appear to produce the most reasonable and simple combination of stress state variables. The $(\sigma - u_a)$ and $(u_a - u_w)$ combination is used throughout this book, and these stress variables are referred to as the net normal stress and the matric suction, respectively.

### 3.2.3 Saturated Soils as a Special Case of Unsaturated Soils

A saturated soil can be viewed as a special case of an unsaturated soil. The four phases in an unsaturated soil reduce to two phases for a saturated soil (i.e., soil particles and water). The phase equilibrium equations for a saturated soil can be derived using the same theory used for unsaturated soils (Appendix B). There is also a smooth transition between the stress state for a saturated soil and that of an unsaturated soil.

As an unsaturated soil approaches saturation, the degree of saturation, $S$, approaches 100%. The pore-water pressure, $u_w$, approaches the pore-air pressure, $u_a$, and the matric suction term, $(u_a - u_w)$, goes towards zero. Only the first stress tensor is retained for a saturated soil when considering this special case:

$$\begin{bmatrix} (\sigma_x - u_w) & \tau_{yx} & \tau_{zx} \\ \tau_{xy} & (\sigma_y - u_w) & \tau_{zy} \\ \tau_{xz} & \tau_{yz} & (\sigma_z - u_w) \end{bmatrix}. \quad (3.15)$$

The second stress tensors [i.e., Eq. (3.13)] disappears because the matric suction, $(u_a - u_w)$, goes towards zero. The pore-air pressure term in the first stress tensor [i.e., Eq. (3.12)] becomes the pore-water pressure, $u_w$, in the stress tensor for a saturated soil [i.e., Eq. (3.15)]. The stress state variables for saturated soils are shown diagrammatically in Fig. 3.5. The above rationale demonstrates the smooth transition in stress state description when going from an unsaturated soil to a saturated soil, and vice versa.

The stress tensor for a saturated soil indicates that the difference between the total stress and the pore-water pressure forms a stress state variable that can be used to describe the equilibrium. This stress state variable, $(\sigma - u_w)$, is commonly referred to as effective stress (Terzaghi, 1936). The so-called effective stress law is essentially a stress state variable which is required to describe the mechanical behavior of a saturated soil. For the case of compressible soil particles, an additional stress tensor (i.e., $u_w$) should be used to describe the complete stress state for a saturated soil (Skempton, 1961).

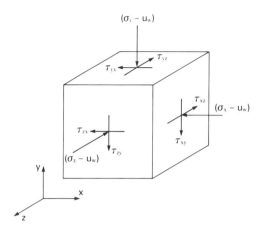

**Figure 3.5** The stress state variables for a saturated soil.

### 3.2.4 Dry Soils

Evaporation from a soil or air-drying a soil will bring the soil to a dry condition. As the soil dries, the matric suction increases. Numerous experiments have shown that the matric suction tends to a common limiting value in the range of 620–980 MPa as the water content approaches 0% (Fredlund, 1964). The relationship between the water content and the suction of a soil is commonly referred to as the soil-water characteristic curve. Figure 3.6 presents the soil-water characteristic curve for Regina clay. The gra-

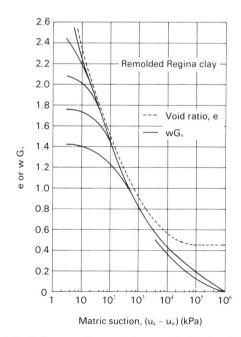

**Figure 3.6** Soil-water characteristic curve for Regina clay (from Fredlund, 1964).

46   3   STRESS STATE VARIABLES

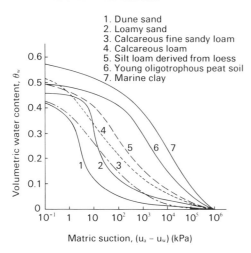

Figure 3.7  Soil–water characteristic curve for some Dutch soils (from Koorevaar et al., 1983).

vimetric water content, expressed in terms of $(wG_s)$, is plotted against matric suction. The void ratio, $e$, is also plotted against matric suction. The plot shows a decreasing void ratio and water content as the matric suction increases. Further results are shown in Fig. 3.7 where the volumetric water content, $\theta_w$, is plotted versus matric suction for various soils. The suction approaches a value of approximately 980 MPa (i.e., $9.8 \times 10^5$ kPa) at 0% water content, as shown in both figures. The above plots illustrate the continuous nature of the water content versus suction relationship. In other words, there does not appear to be any discontinuity in this relationship as the soil desaturates. In addition, the void ratio approaches the void ratio at the shrinkage limit of the soil as the water content approaches 0%, as shown in Fig. 3.8. Even for a sandy soil,

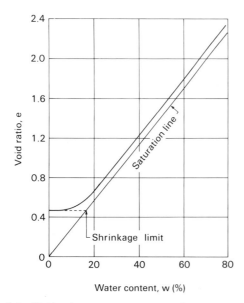

Figure 3.8  Void ratio versus water content for Regina clay (from Fredlund, 1964).

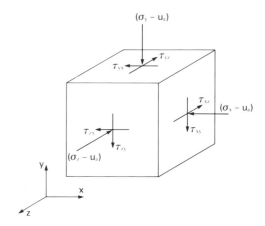

Figure 3.9  The stress state variables for a dry soil.

the soil suction continues to increase with drying to 0% water content.

The effects of a change in matric suction on the mechanical behavior of a soil may become negligible as the soil approaches a completely dry condition. In other words, a change in matric suction on a dry soil may not produce any significant change in the volume or shear strength of the soil. For these dry soils, the net normal stress, $(\sigma - u_a)$, may become the only stress state variable controlling their behavior.

The effect of a matric suction change on the volume change of Regina clay is demonstrated in Fig. 3.8. As the matric suction of the soil is increased, the water content is reduced and the volume of the soil decreases. However, prior to the soil becoming completely dry, the volume of the soil remains essentially constant regardless of the increase in matric suction.

As a soil becomes extremely dry, a matric suction change may no longer produce any significant changes in mechanical properties. Although matric suction remains a stress state variable, it may not be required in describing the behavior of the soil. Only the first stress tensor with $(\sigma - u_a)$ may be required for describing the volume decrease of a dry soil (Fig. 3.9):

$$\begin{bmatrix} (\sigma_x - u_a) & \tau_{yx} & \tau_{zx} \\ \tau_{xy} & (\sigma_y - u_a) & \tau_{zy} \\ \tau_{xz} & \tau_{yz} & (\sigma_z - u_a) \end{bmatrix}. \quad (3.16)$$

On the other hand, it may be necessary to consider matric suction as a stress state variable when examining the volume increase or swelling of a dry soil.

## 3.3  LIMITING STRESS STATE CONDITIONS

There is a hierarchy with respect to the magnitude of the individual stress components in an unsaturated soil:

$$\sigma > u_a > u_w. \quad (3.17)$$

## 3.4 EXPERIMENTAL TESTING OF THE STRESS STATE VARIABLES

The hierarchy in Eq. (3.17) must be maintained in order to ensure stable equilibrium conditions. Limiting stress state conditions occur when one of the stress state variables becomes zero. For example, if the pore-air pressure, $u_a$, is momentarily increased in excess of the total stress, $\sigma$, an "explosion" of the sample may occur. In other words, once the $(\sigma - u_a)$ variable goes to zero, a limiting stress state condition is reached. This limiting stress condition is utilized in the pressure plate test [Fig. 3.10(a)]. Let us suppose that an external air pressure greater than the pore-water pressure is applied to an unsaturated soil. The sample could be visualized as being surrounded with a rubber membrane which is subjected to a total stress equal to the external air pressure. The pore-air pressure is also equal to the external air pressure. In this case, the difference between the total stress, $\sigma$, and the pore-air pressure, $u_a$, is zero and the stress state variable $(\sigma - u_a)$ vanishes. The stress state variable, $(u_a - u_w)$, can be used to describe the behavior of the unsaturated soil under this limiting condition.

Another limiting stress state condition occurs when matric suction, $(u_a - u_w)$, vanishes. If the pore-water pressure is increased in excess of the pore-air pressure, the degree of saturation of the soil approaches 100%. The backpressure oedometer test [Fig. 3.10(b)] is an example involving the limiting condition where matric suction vanishes. As the backpressure is applied to the water phase of an initially unsaturated soil, the degree of saturation approaches 100%. The pore-water pressure approaches the pore-air pressure and the matric suction goes to zero. The behavior of the soil can now be described in terms of one stress state variable [i.e., $(\sigma - u_w)$]. A smooth transition from the unsaturated case to the saturated case takes place under the limiting stress state condition of pore-air pressure being equal to pore-water pressure.

A limiting condition occurs in saturated soils when the stress state variable $(\sigma - u_w)$ (i.e., the effective stress) reaches zero. At this point, the saturated soil becomes unstable. The soil is said to "quick." A further increase in the pore-water pressure results in a "boil" being formed.

## 3.4 EXPERIMENTAL TESTING OF THE STRESS STATE VARIABLES

The validity of the theoretical stress state variables should be experimentally tested. A suggested criterion was proposed by Fredlund and Morgenstern (1977):

> "A suitable set of independent stress state variables are those that produce no distortion or volume change of an element when the individual components of the stress state variables are modified but the stress state variables themselves are kept constant. Thus the stress state variables for each phase should produce equilibrium in that phase when a stress point in space is considered."

The experiments used by Fredlund and Morgenstern (1977) to test the stress state variables are called "null" tests. The working principle for the "null" tests is based upon the above criterion for testing stress state variables. The "null" tests consider the overall and water volume change (or equilibrium conditions) of an unsaturated soil. An axis-translation technique (Hilf, 1956) was used in testing the unsaturated soil. Similar null-type tests related to the shear strength of an unsaturated silt were performed by Bishop and Donald (1961).

### 3.4.1 The Concept of Axis Translation

Difficulties arise in testing unsaturated soils with negative pore-water pressures approaching $-1$ atm (i.e., zero absolute pressure). Water in the measuring system may start to cavitate when the water pressure approaches $-1$ atm (i.e., $-101.3$ kPa gauge). As cavitation occurs, the measuring system becomes filled with air. Then, water from the measuring system is forced into the soil.

The axis-translation technique is commonly used in the laboratory testing of unsaturated soils in order to prevent

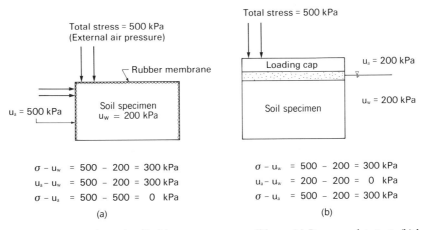

**Figure 3.10** Tests performed at limiting stress state conditions. (a) Pressure plate test; (b) backpressure oedometer test.

having to measure pore-water pressures less than zero absolute. The procedure involves a translation of the reference or pore-air pressure. The pore-water pressure can then be referenced to a positive air pressure (Hilf, 1956). Figure 3.11 presents results from null-type, pressure plate tests which demonstrate the use of the axis-translation technique in the measurement of matric suctions. This measuring technique is described in detail in Chapter 4. Unsaturated soil specimens were subjected to various external air pressures. The pore-air pressure, $u_a$, becomes equal to the externally applied air pressure. As a result, the pore-water pressure, $u_w$, undergoes the same pressure change as the change in the applied air pressure. In this way, the matric suction of the soil remains constant regardless of the translation of both the pore-air and pore-water pressures. Therefore, the pore-water pressure can be raised to a positive value that can be measured without cavitation occurring. The axis-translation technique has been successfully applied by numerous researchers to the volume change and shear strength testing of unsaturated soils (Bishop and Donald, 1961; Gibbs and Coffey, 1969b; Fredlund, 1973; Ho and Fredlund, 1982a; Gan et al. 1988).

The use of the axis-translation technique requires the control of the pore-air pressure and the control or measurement of the pore-water pressure. In a triaxial cell, the pore-air pressure is usually controlled through a coarse corundum disk placed on top of the soil sample. The pore-water pressure is controlled through a saturated high air entry ceramic disk sealed to the pedestal of the triaxial cell. The high air entry disk is a porous, ceramic disk which allows the passage of water, but prevents the flow of free air. Continuity between the water in the soil and the water in the ceramic disk is necessary in order to correctly establish the matric suction. The matric suction in the soil specimen must not exceed the air entry value of the ceramic disk. Air entry values for the ceramic disks generally range from about 50.5 kPa ($\frac{1}{2}$ bar) up to 1515 kPa (15 bars).

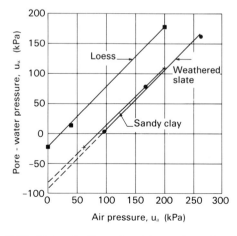

**Figure 3.11** Determination of matric suction using the axis-translation technique (from Hilf, 1956).

### 3.4.2 Null Tests to Test Stress State Variables

Null-type test data to "test" the stress state variables for unsaturated soils were published by Fredlund and Morgenstern in 1977. The components (i.e., $\sigma$, $u_a$, and $u_w$) of the proposed stress state variables were varied equally in order to maintain constant values for the stress state variables [i.e., $(\sigma - u_a)$, $(u_a - u_w)$, and $(\sigma - u_w)$]. In other words, the components of the stress state variables were increased or decreased by an equal amount while volume changes were monitored:

$$\Delta\sigma_x = \Delta\sigma_y = \Delta\sigma_z = \Delta u_w = \Delta u_a. \quad (3.18)$$

If the proposed stress state variables are valid, there should not be any change in the overall volume of the soil sample, and the degree of saturation of the soil should remain constant throughout the "null" test. In other words, positive results from the "null" test should show zero overall and water volume changes.

It is difficult to measure zero volume change over an extended period of testing. Slight volume changes may still occur due to one or more of the following reasons: 1) an imperfect testing procedure, 2) air diffusion through the high air entry disk, 3) water loss from the soil specimen through evaporation or diffusion, and 4) secondary consolidation.

A total of 19 "null" tests were performed on compacted kaolin. The soil was compacted according to the standard AASHTO procedure. Two types of equipment were used in performing the "null" tests. For the first apparatus, one-dimensional loading was applied using an enclosed, modified oedometer. The second apparatus involved isotropic loading using a modified triaxial cell. The axis-translation technique was used in both cases.

The pressure changes associated with the "null" tests on unsaturated soil samples are summarized in Table 3.2. The individual stress variables were varied in accordance with Eq. (3.18), while the stress state variables were kept constant. The measured volume changes of the overall sample and water inflow or outflow are given in Table 3.3. The results from one test are presented in Fig. 3.12. The results show essentially no volume change in the overall specimen and little water flow during the "null" tests. The stress state variables are therefore "tested" in the sense that they define equilibrium conditions for the unsaturated soil. In turn, the stress state variables are qualified for describing the mechanical behavior of unsaturated soils.

### 3.4.3 Other Experimental Evidence in Support of the Proposed Stress State Variables

Other data have been presented in the research literature which lend support to the use of the proposed stress state variables. Bishop and Donald (1961) performed a triaxial strength test on an unsaturated Braehead silt. The total (i.e., confining) pressure, $\sigma_3$, the pore-air pressure, and the pore-

Table 3.2 Pressure Changes for Null Tests on Unsaturated Soils (From Fredlund, 1973)

| Test Number | Initial Pressures (kPa) | | | Change in Pressures (kPa) | | |
|---|---|---|---|---|---|---|
| | Total, $\sigma$ | Air, $u_a$ | Water, $u_w$ | $\Delta\sigma$ | $\Delta u_a$ | $\Delta u_w$ |
| N-23 | 420.7 | 278.7 | 109.6 | +71.4 | +70.3 | +70.7 |
| N-24 | 359.4 | 270.9 | 3.0 | +135.9 | +135.9 | +140.5 |
| N-25 | 495.3 | 406.8 | 143.5 | +68.6 | +68.3 | +66.9 |
| N-26 | 701.7 | 613.2 | 498.3 | −204.3 | −204.3 | −204.9 |
| N-27 | 234.2 | 138.3 | 100.3 | +68.8 | +68.5 | +80.8 |
| N-28 | 474.8 | 394.6 | 32.3 | +136.6 | +137.4 | +137.9 |
| N-29 | 274.6 | 202.2 | 22.4 | +68.5 | +68.3 | +68.8 |
| N-30 | 343.1 | 270.5 | 91.2 | +68.8 | +68.5 | +68.8 |
| N-31 | 411.4 | 338.3 | 160.2 | +68.1 | +68.0 | +67.3 |
| N-32 | 479.5 | 406.3 | 227.5 | +69.5 | +70.1 | +69.7 |
| N-33 | 549.0 | 476.4 | 297.2 | +69.0 | +68.0 | +68.4 |
| N-34 | 272.8 | 202.2 | 73.1 | +66.9 | +65.9 | +66.1 |
| N-35 | 410.9 | 338.5 | 208.3 | +69.5 | +69.3 | +69.7 |
| N-36 | 480.4 | 407.8 | 278.0 | +67.1 | +65.9 | +65.9 |
| N-37 | 547.5 | 473.7 | 343.9 | +67.9 | +67.5 | +67.4 |
| N-38 | 615.4 | 541.2 | 411.3 | −66.0 | −64.1 | −63.7 |
| N-39 | 549.4 | 477.1 | 347.6 | −70.2 | −69.5 | −69.8 |
| N-40 | 479.2 | 407.6 | 277.8 | −66.6 | −66.9 | −66.4 |
| N-41 | 412.6 | 340.7 | 211.4 | −140.5 | −140.3 | −139.8 |

water pressure were varied by equal amounts in order to keep ($\sigma_3 - u_a$) and ($u_a - u_w$) constant. The pressure changes for individual stress components are given in Table 3.4. The values of ($\sigma_3 - u_a$) and ($u_a - u_w$) throughout the test are given in Table 3.5 (i.e., Combination 1). If ($\sigma_3 - u_a$) and ($u_a - u_w$) are valid stress state variables, it would be anticipated that the pressure variations should not produce any significant change in the shear strength of the soil. In other words, the stress versus strain curve of the soil should remain monotonic. The test results are plotted in Fig. 3.13. The results show that the stress versus strain relationship remains monotonic, substantiating the use of ($\sigma - u_a$) and ($u_a - u_w$) as valid stress state variables. As the matric suction variable was changed, towards the end of the test (i.e., portion 5), the behavior of the stress versus strain relationship was altered. Other small fluctuations in the stress versus strain curve were not believed to be of consequence. Bishop and Donald (1961) stated that:

> "The small temporary fluctuations in the stress strain curve are probably the result of a variation in rate of strain due to the change in end thrust on the loading ram as the cell pressure is changed."

Other combinations of stress components are equally justified, as shown in Table 3.5.

## 3.5 STRESS ANALYSIS

The proposed and tested stress state variables for unsaturated soils can be used in engineering practice in a manner similar to which the effective stress variable is used for saturated soils. *In situ* profiles can be drawn for each of the stress components. Their variation with depth and time is required for analyzing shear strength or volume change problems (i.e., slope instability and heave). Factors affecting the *in situ* stress profiles are described in order to better understand possible profile variations that may be observed in practice.

Most geotechnical engineering problems can be simplified from their three-dimensional form to either a two- or one-dimensional problem. This also applies for unsaturated soils, but the presentation of the stress state must be extended. An extended form of the Mohr diagram can be used to illustrate the role of matric suction. The extended Mohr diagram also helps illustrate the smooth transition to the conventional saturated soil case. The concepts of stress invariants, stress points, and stress paths are also applicable to unsaturated soil mechanics.

### 3.5.1 *In situ* Stress State Component Profiles

The magnitude and distribution of the stress components in the field are required prior to performing most geotech-

**Table 3.3  Volume Changes of the Specimen and Water Flow during Null Tests (From Fredlund, 1973)**

| Test Number | Specimen Volume Change (%) | | Water Volume Change (%) | Elapsed Time (min.) |
|---|---|---|---|---|
| | Immediate (%) | At Elapsed Time (%) | | |
| N-23 | 0.0 | −0.03 | −0.05 | 5800 |
| N-24 | +0.04 | +0.4 | −0.07 | 1500 |
| N-25 | +0.01 | 0.0 | −0.02 | 1650 |
| N-26 | −0.25 | −0.20 | — | 4300 |
| N-27 | 0.0 | −0.10 | −0.50 | 1880 |
| N-28 | −0.15 | −0.15 | −0.11 | 1900 |
| N-29 | −0.015 | +0.012 | −0.642 | 8700 |
| N-30 | −0.005 | +0.012 | −0.072 | 1350 |
| N-31 | — | +0.12 | −0.060 | 1380 |
| N-32 | — | +0.17 | −0.045 | 1390 |
| N-33 | — | +0.15 | −0.020 | 410 |
| N-34 | +0.055 | +0.060 | −0.105 | 4350 |
| N-35 | +0.015 | +0.033 | −0.060 | 5800 |
| N-36 | +0.010 | −0.020 | −0.035 | 2800 |
| N-37 | 0.0 | −0.005 | −0.050 | 5800 |
| N-38 | −0.015 | +0.002 | +0.010 | 2700 |
| N-39 | −0.010 | +0.005 | −0.005 | 1500 |
| N-40 | −0.007 | −0.005 | +0.015 | 5800 |
| N-41 | −0.030 | +0.007 | −0.040 | 2900 |

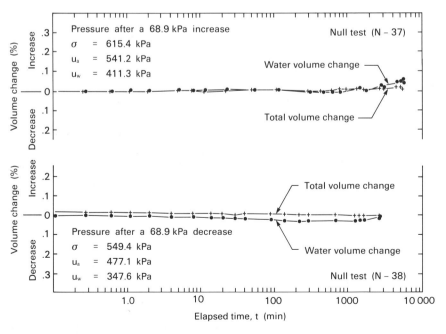

**Figure 3.12**  Results of null tests N-37 and N-38 on compacted kaolin (from Fredlund, 1973).

**Table 3.4 Pressure Changes in Bishop and Donald's (1961) Triaxial Strength Test Experiment on Braehead Silt**

| Portion of Stress-Strain Curve[a] | Confining Pressure, $\sigma_3$ (kPa) | Pore-Air Pressure, $u_a$ (kPa) | Pore-Water Pressure, $u_w$ (kPa) | Pressure Change (kPa) |
|---|---|---|---|---|
| 1 | 44.8 | 31.0 | −27.6 | 0.0 |
| 2 | 77.2 | 63.4 | +4.8 | +32.4 |
| 3 | 13.8 | 0.0 | −58.6 | −63.4 |
| 4 | 110.3 | 96.5 | +37.9 | +96.2 |
| 5 | 110.3 | 96.5 | +66.9 | varies |

[a]Portions 1, 2, 3, and 4 produced monotonic behavior with constant stress state variables, while matric suction was varied in portion 5.

**Table 3.5 Independent Stress State Variables Showing Monotonic Behavior (From Bishop and Donald's Data, 1961)**

| Portion of Stress Versus Strain Curve[a] | Combination 1 | | Combination 2 | | Combination 3 | |
|---|---|---|---|---|---|---|
| | $\sigma_3 - u_a$ (kPa) | $u_a - u_w$ (kPa) | $\sigma_3 - u_w$ (kPa) | $u_a - u_w$ (kPa) | $\sigma_3 - u_a$ (kPa) | $\sigma_3 - u_w$ (kPa) |
| 1 | 44.8 − 31.0 = 13.8 | 31.0 − (−27.6) = 58.6 | 72.4 | 58.6 | 13.8 | 72.4 |
| 2 | 77.2 − 63.4 = 13.8 | 63.4 − (+4.8) = 58.6 | 72.4 | 58.6 | 13.8 | 72.4 |
| 3 | 13.8 − 0.0 = 13.8 | 0 − (−58.6) = 58.6 | 72.4 | 58.6 | 13.8 | 72.4 |
| 4 | 110.3 − 96.5 = 13.8 | 96.5 − (+37.9) = 58.6 | 72.4 | 58.6 | 13.8 | 72.4 |
| 5 | 110.3 − 96.5 = 13.8 | 96.5 − (+66.9) = 29.6 | 72.4 | 29.6 | 13.8 | 43.4 |

[a]Portions 1, 2, 3, and 4 produced monotonic behavior.

nical analyses. The distribution of the stress components allows the computation of *in situ* profiles for the net normal stress, $(\sigma - u_a)$, and matric suction, $(u_a - u_w)$. As the soil becomes saturated, the two profiles revert to the classic effective stress, $(\sigma - u_w)$, profile. The present *in situ* profiles are generally based on field and/or laboratory measurements, while the final profiles are assumed or computed based on theoretical considerations.

The total normal stress in a soil is a function of the density or the total unit weight of the soil. The magnitude and distribution of the total normal stress is also affected by the application of external loads such as buildings or the removal of soil through excavation.

Let us consider a geostatic condition where the ground surface is horizontal and the vertical and horizontal planes do not have shear stress (Lambe and Whitman, 1979). The net normal stresses in the vertical and horizontal directions are related to the density of soil. The net normal stress in the vertical direction is called the overburden pressure, and can be computed as follows (see Fig. 3.14):

$$(\sigma_v - u_a) = \int_{z_2}^{z_1} \rho(z) \, g \, dz - u_a \quad (3.19)$$

where

$(\sigma_v - u_a)$ = vertical net normal stress
$u_a$ = pore-air pressure
$\rho(z)$ = density of the soil as a function of depth
$z_1$ = ground surface elevation
$z_2$ = elevation under consideration
$g$ = gravitational acceleration.

The vertical net normal stress distribution with respect to depth will be a straight line for the case where the density is constant. The pore-air pressure is generally assumed to be in equilibrium with atmospheric pressure (i.e., zero

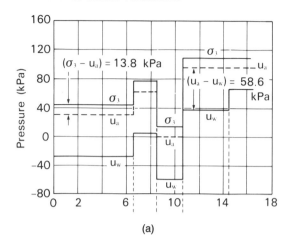

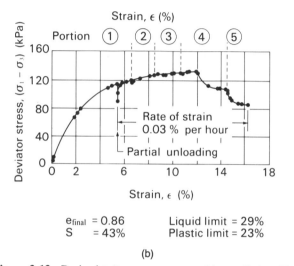

**Figure 3.13** Drained test on an unsaturated loose silt in which $\sigma_3$, $u_a$, and $u_w$ were varied, while keeping $(\sigma_3-u_a)$ and $(u_a-u_w)$ constant. (a) Pressure changes versus strain; (b) deviator stress versus strain (from Bishop and Donald, 1961).

gauge pressure). Fig. 3.14(a) shows a typical profile of the vertical net normal stress for geostatic conditions. When soil strata with distinctly different densities are encountered, the integration of Eq. (3.19) should be performed for each layer. In this case, the vertical net normal stress profile will not be a straight line.

*Coefficient of Lateral Earth Pressure*

The coefficient of lateral earth pressure, $K$, can be defined as the ratio of horizontal net normal stress to vertical net normal stress. This is a slight variation from saturated soil mechanics where horizontal and vertical stresses are not referenced to the pore-air pressure.

$$K = \frac{(\sigma_h - u_a)}{(\sigma_v - u_a)} \quad (3.20)$$

where

$(\sigma_h - u_a)$ = horizontal net normal stress.

For geostatic stress conditions where there is no horizontal strain, $K$ is defined as the coefficient of lateral earth pressure *at rest*, $K_0$ (Terzaghi, 1925). The coefficient of lateral earth pressure *at rest* depends on several factors, such as the type of soil, its stress history, and density (see Chapter 11). Saturated soils commonly have $K_0$ values ranging from as low as 0.4 to values in excess of 1.0. Unsaturated soils are commonly overconsolidated, and can have coefficients of earth pressure *at rest* greater than 1.0 (Brooker and Ireland, 1965). On the other hand, the coefficients can go to zero for the case where the soil becomes desiccated and cracked. A profile of the horizontal net normal stress at rest condition is shown in Fig. 3.14(b).

The effect of external loads and excavations on the net normal stress is presented in Chapter 11. The theory of elasticity, commonly used to compute the change in total stress, applies similarly for saturated and unsaturated soils.

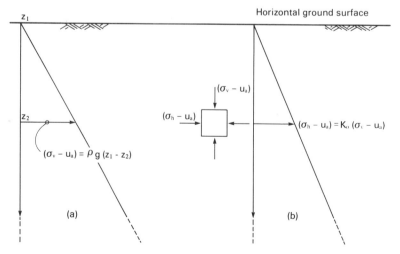

**Figure 3.14** *In situ* net normal stress profile under geostatic conditions. (a) Vertical net normal stress; (b) horizontal net normal stress.

## Matric Suction Profile

Matric suction is closely related to the surrounding environment and is of interest in analyzing geotechnical engineering problems. The *in situ* profile of pore–water pressures (and thus matric suction) may vary from time to time, as illustrated in Fig. 3.15. The variation in the soil suction profile is generally greater than variations commonly occurring in the net normal stress profile. Variations in the suction profile depend upon several factors, as illustrated by Blight (1980).

**Ground surface condition.** The matric suction profile below an uncovered ground surface is affected significantly by environmental changes, as shown in Fig. 3.16. Dry and wet seasons cause variations in the suction profile, particularly close to the ground surface. The suction profile beneath a covered ground surface is more constant with respect to time than is a profile below an uncovered surface. For example, the suction profile below a house or a pavement is less influenced by seasonal variations than the suction profile below an open field. However, moisture may slowly accumulate below the covered area on a long-term basis, causing a reduction in the soil suction. Figure 3.17 shows several matric suction profiles below a slope in Hong Kong. The sloping portion of the slope is covered by a layer of soil cement and lime plaster (i.e., locally referred to as Chunam) to prevent water infiltration into the slope. The top portion of the slope was exposed to the environment. In this particular case, the soil suction profile remains relatively constant throughout dry and wet (i.e., rainy) seasons.

**Environmental conditions.** The matric suction in the soil increases during dry seasons and decreases during wet seasons. Maximum changes in suction occur near ground surface. During a dry season, the evaporation rate is high, and it results in a net loss of water from the soil. The opposite condition may occur during a wet season.

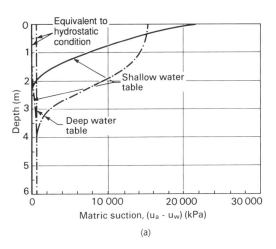

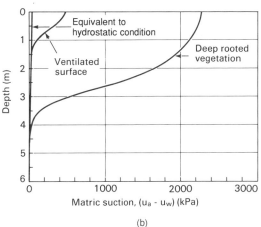

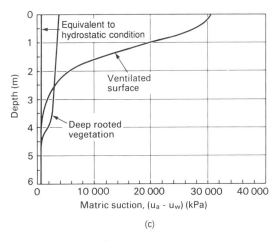

**Figure 3.16** Typical suction profiles below an uncovered ground surface. (a) Seasonal fluctuation; (b) drying influence on shallow water table condition; (c) drying influence on deep water table condition. (Modified from Blight, 1980).

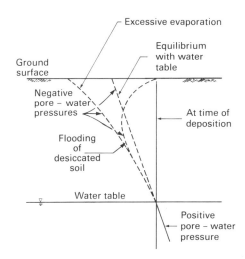

**Figure 3.15** Typical pore–water pressure profiles.

**Vegetation.** Vegetation on the ground surface has the ability to apply a tension to the pore–water of up to 1–2 MPa through the evapotranspiration process. Evapotranspiration results in the removal of water from the soil and an increase in the matric suction. The rate of evapotran-

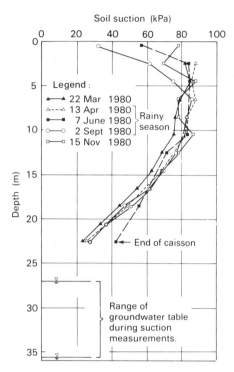

**Figure 3.17** *In situ* suction profiles in a steep slope in Hong Kong (from Sweeney, 1982).

spiration is a function of the microclimate, the type of vegetation, and the depth of the root zone.

**Water table.** The depth of the water table influences the magnitude of the matric suction. The deeper the water table, the higher the possible matric suction. The effect of the water table on the matric suction becomes particularly significant near ground surface (Blight, 1980).

**Permeability of the soil profile.** The permeability of a soil represents its ability to transmit and drain water. This, in turn, indicates the ability of the soil to change matric suction as a result of environmental changes. The permeability of an unsaturated soil varies widely with its degree of saturation. The permeability also depends on the type of soil. Different soil strata which have varying abilities to transmit water in turn affect the *in situ* matric suction profile. The relative effects of the environment, the water table, and the vegetation on the matric suction profiles are illustrated in Fig. 3.16.

Matric suction is a hydrostatic or isotropic pressure in that it has equal magnitude in all directions. The magnitude of the matric suction is often considerably higher than the magnitude of the net normal stress. Typical relative magnitudes between net normal stress and matric suction are shown in Fig. 3.18. This figure illustrates the importance of knowing the magnitude of the soil suction when studying the behavior of unsaturated soils.

### 3.5.2 Extended Mohr Diagram

The state of stress at a point in the soil is three-dimensional, but the concepts involved are more easily represented in a two-dimensional form. In two dimensions, there always exists a set of two mutually orthogonal principal planes with real-valued principal stresses. The principal planes are the planes on which there are no shear stresses. The direction of the principal planes depends on the general stress state at a point. The largest principal stress is called the major principal stress, and is given the symbol, $\sigma_1$. The smallest principal stress is called the minor principal stress, and is given the symbol, $\sigma_3$. In the case of a horizontal ground surface, the horizontal and vertical planes are the principal planes. The vertical net normal stress is generally the net major principal stress, $(\sigma_1 - u_a)$, and the horizontal net normal stress is the net minor principal stress, $(\sigma_3 - u_a)$.

If the magnitude and the direction of the stresses acting on any two mutually orthogonal planes (e.g., the principal planes) are known, the stress condition on any inclined

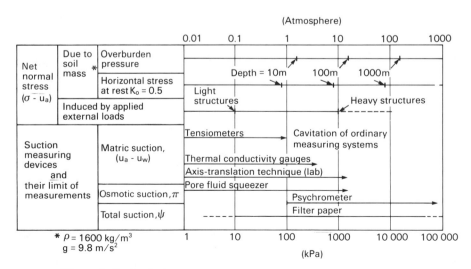

**Figure 3.18** Typical magnitudes of total normal stress and soil suction.

plane can be determined. In other words, the net normal stress and shear stress on any inclined plane can be computed from the known net principal stresses. The matric suction, $(u_a - u_w)$, on every inclined plane at a point is constant since it is an isotropic tensor. Therefore, only the net normal stress and shear stress on an inclined plane need to be considered.

*Equation of Mohr Circles*

Consider an unsaturated soil *at rest* with a horizontal ground surface. The net normal stress and shear stress on a plane with an inclination angle, $\alpha$, from the horizontal are illustrated in Fig. 3.19. The inclined plane has an infinitesimal length, $ds$, and results in a triangular free body element with horizontal and vertical planes. The horizontal plane has an infinitesimal length of $dx$. Its length can be written in terms of the sloping length, $ds$, and the angle, $\alpha$:

$$dx = ds \cos \alpha. \quad (3.21)$$

The vertical plane has an infinitesimal length of $dy$:

$$dy = ds \sin \alpha. \quad (3.22)$$

All the planes have a unit thickness in the perpendicular direction. The equilibrium of the triangular element requires that the summation of forces in the horizontal and vertical directions be equal to zero. Summing forces horizontally gives

$$-(\sigma_\alpha - u_a) \, ds \sin \alpha + \tau_\alpha \, ds \cos \alpha$$
$$+ (\sigma_3 - u_a) \, dy = 0. \quad (3.23)$$

Summing forces vertically gives

$$-(\sigma_\alpha - u_a) \, ds \cos \alpha - \tau_\alpha \, ds \sin \alpha$$
$$+ (\sigma_1 - u_a) \, dx = 0. \quad (3.24)$$

Substituting $dx$ and $dy$ [i.e., Eqs. (3.21) and (3.22)] into Eqs. (3.24) and (3.23), respectively, and multiplying Eq.

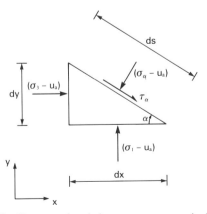

**Figure 3.19** Net normal and shear stresses on an inclined plane at a point in the soil mass below a horizontal ground surface.

(3.23) by $\sin \alpha$ and Eq. (3.24) by $\cos \alpha$, gives

$$-(\sigma_\alpha - u_a) \, ds \sin^2 \alpha + \tau_\alpha \, ds \sin \alpha \cos \alpha$$
$$+ (\sigma_3 - u_a) \, ds \sin^2 \alpha = 0 \quad (3.25)$$

and

$$-(\sigma_\alpha - u_a) \, ds \cos^2 \alpha - \tau_\alpha \, ds \sin \alpha \cos \alpha$$
$$+ (\sigma_1 - u_a) \, ds \cos^2 \alpha = 0. \quad (3.26)$$

Summing Eqs. (3.25) and (3.26) gives

$$-(\sigma_\alpha - u_a) \, ds \, (\sin^2 \alpha + \cos^2 \alpha)$$
$$+ (\sigma_3 - u_a) \, ds \sin^2 \alpha + (\sigma_1 - u_a) \, ds \cos^2 \alpha = 0. \quad (3.27)$$

Using trigonometric relations to solve for $(\sigma_\alpha - u_a)$ gives

$$(\sigma_\alpha - u_a) = (\sigma_1 - u_a) \left( \frac{1 + \cos 2\alpha}{2} \right)$$
$$+ (\sigma_3 - u_a) \left( \frac{1 - \cos 2\alpha}{2} \right). \quad (3.28)$$

Rearranging Eq. (3.28) gives

$$(\sigma_\alpha - u_a) = \left( \frac{\sigma_1 + \sigma_3}{2} - u_a \right)$$
$$+ \left( \frac{\sigma_1 - \sigma_3}{2} \right) \cos 2\alpha. \quad (3.29)$$

The shear stress, $\tau_\alpha$, is obtained by substituting $dx$ and $dy$ [i.e., Eqs. (3.21) and (3.22)] into Eq. (3.24) and (3.23), respectively, and multiplying Eq. (3.23) by $\cos \alpha$ and Eq. (3.24) by $\sin \alpha$:

$$-(\sigma_\alpha - u_a) \, ds \sin \alpha \cos \alpha + \tau_\alpha \, ds \cos^2 \alpha$$
$$+ (\sigma_3 - u_a) \, ds \sin \alpha \cos \alpha = 0 \quad (3.30)$$

$$-(\sigma_\alpha - u_a) \, ds \sin \alpha \cos \alpha - \tau_\alpha \, ds \sin^2 \alpha$$
$$+ (\sigma_1 - u_a) \, ds \sin \alpha \cos \alpha = 0. \quad (3.31)$$

Subtracting Eq. (3.31) from Eq. (3.30) gives

$$\tau_\alpha \, ds \, (\sin^2 \alpha + \cos^2 \alpha) + (\sigma_3 - u_a) \, ds \sin \alpha \cos \alpha$$
$$- (\sigma_1 - u_a) \, ds \sin \alpha \cos \alpha = 0. \quad (3.32)$$

Using trigonometric relations, it is possible to solve for $\tau_\alpha$:

$$\tau_\alpha = \left( \frac{\sigma_1 - \sigma_3}{2} \right) \sin 2\alpha. \quad (3.33)$$

Equations (3.29) and (3.33) give the net normal stress and the shear stress on an inclined plane through a point. The term $(\sigma_1 - \sigma_3)$ is called the deviator stress, and is an indication of the shear stress. For a given stress state, the largest shear stress, $[(\sigma_1 - \sigma_3)/2]$, occurs on a plane with an inclination angle, $\alpha$, such that $(\sin 2\alpha)$ will be equal to unity.

The net normal stress and shear stress at a point can also be determined using a graphical method. If Eqs. (3.29) and (3.33) are squared and added, the result is the equation of a circle:

$$\left[(\sigma_\alpha - u_a) - \left(\frac{\sigma_1 + \sigma_3}{2} - u_a\right)\right]^2 + \tau_\alpha^2 = \left(\frac{\sigma_1 - \sigma_3}{2}\right)^2.$$

(3.34)

The circle is known as the Mohr diagram, and represents the stress state at a point. In saturated soils, the Mohr diagram is often plotted with the principal effective normal stress as the abscissa and the shear stress as the ordinate. For unsaturated soils, an extended form of the Mohr diagram can be used as shown in Fig. 3.20. The extended Mohr diagram uses a third orthogonal axis to represent matric suction. The circle described in Eq. (3.34) is drawn on a plane with the net normal stress, $(\sigma - u_a)$, as the abscissa and the shear stress, $\tau$, as the ordinate. The center of the circle has an abscissa of $[(\sigma_1 + \sigma_3)/2 - u_a]$ and a radius of $[(\sigma_1 - \sigma_3)/2]$.

The matric suction must also be included as part of the description of the stress state. The matric suction determines the position of the Mohr diagram along the third axis. As the soil becomes saturated, the matric suction goes to zero, and the Mohr diagram moves to a single $[(\sigma - u_w)$ versus $\tau]$ plane.

### Construction of Mohr Circles

The construction of the Mohr diagram on the $[(\sigma - u_a)$ versus $\tau]$ plane is shown in Fig. 3.21. A compressive net normal stress is plotted as a positive net normal stress in accordance with the sign convention for the Mohr diagram. A shear stress that produces a counterclockwise moment about a point within the element is plotted as a positive shear stress. This shear stress sign convention is different from the convention used in continuum mechanics (Desai and Christian, 1977). Therefore, this convention should only be used for plotting the Mohr diagram. The major and minor net principal stresses $[(\sigma_1 - u_a)$ and $(\sigma_3 - u_a)]$ are plotted on the abscissa, and the center of the Mohr circle is located at $[(\sigma_1 + \sigma_3)/2 - u_a]$. The radius of the circle is $[(\sigma_1 - \sigma_3)/2]$. The Mohr circle represents the net normal stress and shear stress on any plane through a point in an unsaturated soil.

The net normal stress and shear stress on any plane can be determined if the pole point or the origin of planes is known. Any plane drawn through the pole point will intersect the Mohr diagram and give the net normal stress and shear stress acting on that plane. On the other hand, if the net normal stress and shear stress on a plane are known and plotted as a stress point on the Mohr circle, the direction of the plane under consideration is given by the orientation of a line joining the stress point and the pole point.

The pole point for the condition shown in Fig. 3.21 is determined from the known net normal stress and shear stress on a particular plane. Consider, for example, the case where the major principal stress acts on a horizontal plane. The stress condition on the horizontal plane is represented by the stress point $(\sigma_1 - u_a)$ on the Mohr circle. If a horizontal line is drawn through the stress point $(\sigma_1 - u_a)$, the line will intersect the Mohr circle at the stress point $(\sigma_3 - u_a)$. This is the pole point. The net normal stress and

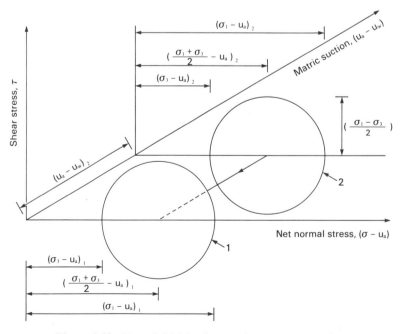

**Figure 3.20** Extended Mohr diagram for unsaturated soils.

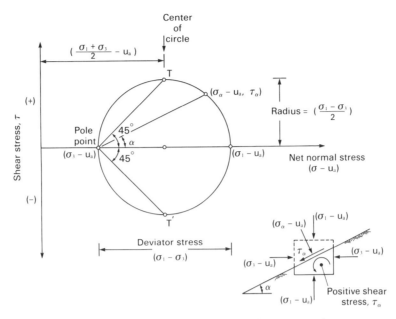

**Figure 3.21** Construction of a Mohr circle using net normal stresses.

shear stress on the inclined plane shown in Fig. 3.21 can then be determined using the same pole point. A line at an orientation, $\alpha$, can be drawn through the pole point to intersect the Mohr circle at the stress point $[(\sigma_\alpha - u_a), \tau_\alpha]$. The horizontal coordinate of the intersection point is the net normal stress, $(\sigma_\alpha - u_a)$, acting on the inclined plane (Fig. 3.21). The shear stress, $\tau_\alpha$, on the plane is positive and is given by the ordinate of the intersection point.

The plane with the maximum shear stress, $[+ (\sigma_1 - \sigma_3)/2]$, goes through the top point of the Mohr circle (i.e., stress point $T$ in Fig. 3.21). The maximum negative shear stress, $[- (\sigma_1 - \sigma_3)/2]$, occurs at the bottom point, $T'$, on the Mohr circle. The planes with the maximum positive and negative shear stresses are oriented at an angle of 45° from the principal planes or from the horizontal and vertical planes in this case.

The principal planes are not always the vertical and horizontal planes. A more general case is shown in Fig. 3.22 where shear stresses may be present on the vertical and horizontal planes. The principal stresses and principal planes can be found graphically using the known stresses on the vertical and horizontal planes. The vertical net normal stress, $(\sigma_y - u_a)$, is a compressive stress, and the horizontal net normal stress, $(\sigma_x - u_a)$, is negative because it is in tension. The matric suction, $(u_a - u_w)$, acts on every plane with equal magnitude. The shear stresses, $\tau_{xy}$ and $\tau_{yx}$, are always equal in magnitude and opposite in sign.

The extended Mohr circle for the stress state shown in Fig. 3.22 is presented in Fig. 3.23. The Mohr circle is drawn on the $[\tau$ and $(\sigma - u_a)]$ plane. Its position along the $(u_a - u_w)$ axis is determined by the magnitude of the matric suction. The first step in plotting the Mohr diagram is to plot the stress points which represent the stresses corresponding to the vertical and horizontal planes (i.e., $[(\sigma_x - u_a), \tau_{xy}]$ and $[(\sigma_y - u_a), \tau_{yx}]$, respectively). A line joining the two stress points intersects the $(\sigma - u_a)$ axis at a point $[(\sigma_x + \sigma_y)/2 - u_a]$. The intersection point is the center for the Mohr circle. The Mohr circle can then be drawn with the two stress points forming the diameter of the circle. The intersection points between the Mohr circle and the $(\sigma - u_a)$ axis (i.e., where the shear stress is zero) are the net major and net minor principal stresses [i.e., $(\sigma_1 - u_a)$ and $(\sigma_3 - u_a)$] (Fig. 3.23). The net minor principal stress is negative, which indicates that it is in tension.

The second step is to locate the pole point by drawing a horizontal plane through the stress point, $[(\sigma_y - u_a), \tau_{yx}]$. The intersection of the horizontal line and the Mohr circle is the pole point. The pole point can also be obtained by drawing a vertical line from the stress point corresponding

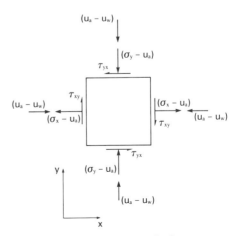

**Figure 3.22** General stress state at a point in an unsaturated soil.

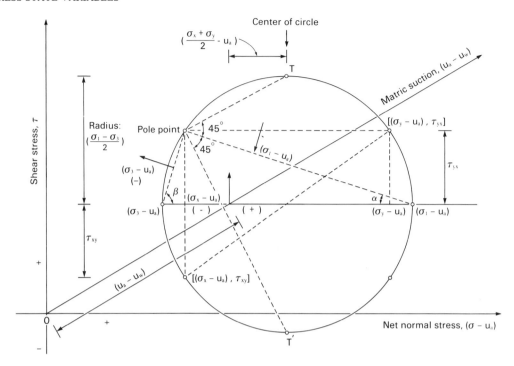

**Figure 3.23** Extended Mohr diagram showing the general stress state for an unsaturated soil element.

to the vertical plane (i.e., $[(\sigma_x - u_a), \tau_{xy}]$). A line joining the pole point and the net major or net minor principal stress point gives the orientation of the major or minor principal plane (Fig. 3.23). The major and minor principal planes are at an angle of $\alpha$ and $\beta$ with respect to the horizontal, respectively.

The top and the bottom stress points on the Mohr circle correspond to the planes on which the maximum and minimum shear stresses occur. The maximum and minimum shear stress planes are oriented at an angle of 45° from the principal planes (Fig. 3.23).

### 3.5.3 Stress Invariants

For a three-dimensional analysis, there are three principal stresses on three mutually orthogonal principal planes. The three principal stresses are named according to their magnitudes. These are the net major, net intermediate, and net minor principal stresses. The symbols used for the net major, net intermediate, and net minor principal stresses are $(\sigma_1 - u_a)$, $(\sigma_2 - u_a)$, and $(\sigma_3 - u_a)$, respectively. A corresponding Mohr circle is shown in Fig. 3.24. The matric suction acts equally on all three principal planes.

The principal stresses at a point can be visualized as the characterization of the physical state of stress. These principal stresses are independent of the selected coordinate system. The independent properties of principal stresses are expressed in terms of constants called stress invariants.

There are three stress invariants that can be derived from each of the two independent stress tensors for an unsaturated soil [refer to stress tensors (3.12) and (3.13)]. The first stress invariants of the first and second stress tensors, respectively, are

$$I_{11} = \sigma_1 + \sigma_2 + \sigma_3 - 3u_a \quad (3.35)$$

and

$$I_{12} = 3(u_a - u_w) \quad (3.36)$$

where

$I_{11}$ = first stress invariant of the first tensor
$I_{12}$ = first stress invariant of the second tensor.

The second stress invariants of the first and second stress tensors, respectively, are

$$I_{21} = (\sigma_1 - u_a)(\sigma_2 - u_a) + (\sigma_2 - u_a)(\sigma_3 - u_a)$$
$$+ (\sigma_3 - u_a)(\sigma_1 - u_a) \quad (3.37)$$

and

$$I_{22} = 3(u_a - u_w)^2 \quad (3.38)$$

where

$I_{21}$ = second stress invariant of the first tensor
$I_{22}$ = second stress invariant of the second tensor.

The third stress invariants of the first and second stress tensors, respectively, are

$$I_{31} = (\sigma_1 - u_a)(\sigma_2 - u_a)(\sigma_3 - u_a) \quad (3.39)$$

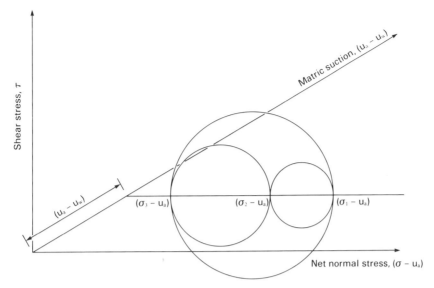

**Figure 3.24** The Mohr diagram for a three-dimensional stress analysis.

and

$$I_{32} = (u_a - u_w)^3 \quad (3.40)$$

where

$I_{31}$ = third stress invariant of the first tensor
$I_{32}$ = third stress invariant of the second tensor.

The stress invariants of the second tensor, $I_{12}$, $I_{22}$, and $I_{32}$, are related as follows:

$$I_{32} = (I_{12}/3)^3 = I_{12}I_{22}/9. \quad (3.41)$$

Therefore, only one stress invariant is required to represent the second tensor. In other words, a total of four stress invariants are required to characterize the stress state of an unsaturated soil as opposed to three stress invariants for a saturated soil.

### 3.5.4 Stress Points

Geotechnical analyses often require an understanding of the development or change in the stress state resulting from various loading patterns. These changes could be visualized by drawing a series of Mohr circles which follow the loading process. However, the pattern of the stress state change may become confusing when the loading pattern is complex. Therefore, it is better to use only one stress point on a Mohr circle to represent the stress state in the soil. A selected stress point can be used to define the stress path followed.

Figure 3.25 shows a Mohr circle for a two-dimensional case where the vertical and horizontal planes are principal planes. The stress point selected to represent the Mohr circle has the coordinates of $(p, q, r)$, where

$$p = \left(\frac{\sigma_v + \sigma_h}{2} - u_a\right) \quad \text{or} \quad \left(\frac{\sigma_1 + \sigma_3}{2} - u_a\right) \quad (3.42)$$

$$q = \left(\frac{\sigma_v - \sigma_h}{2}\right) \quad \text{or} \quad \left(\frac{\sigma_1 - \sigma_3}{2}\right) \quad (3.43)$$

$$r = (u_a - u_w) \quad (3.44)$$

and

$(\sigma_v - u_a)$ = vertical net normal stress
$(\sigma_h - u_a)$ = horizontal net normal stress
$(u_a - u_w)$ = matric suction.

The $q$-coordinate is one half the deviator stress $(\sigma_v - \sigma_h)$. The selected stress point represents the state of stress on a plane with an orientation of 45° from the principal planes, as shown in Fig. 3.25.

The vertical net normal stress for the condition shown in Fig. 3.25 is greater than the horizontal net normal stress [i.e., $(\sigma_v - u_a) > (\sigma_h - u_a)$]. This results in a positive $q$-coordinate. A negative $q$-coordinate would indicate the condition where $(\sigma_h - u_a)$ is greater than $(\sigma_v - u_a)$. For the hydrostatic or isotropic stress state [i.e., $(\sigma_h - u_a)$ equal to $(\sigma_v - u_a)$], the $q$-coordinate is equal to zero. A zero $q$-coordinate means the absence of shear stresses.

### 3.5.5 Stress Paths

A change in the stress state of a soil can be described using stress paths. A stress path is a curve drawn through the stress points for successive stress states (Lambe, 1967). As an example, consider a soil element where the initial condition has $(\sigma_h - u_a)$ equal to $(\sigma_v - u_a)$ at a particular matric suction value. This stress state is represented by point 0 in Fig. 3.26. The soil is then subjected to an increase in the vertical net normal stress, $\Delta(\sigma_v - u_a)$, while maintaining $(\sigma_h - u_a)$ and $(u_a - u_w)$ constant. As the vertical net normal stress is increased, the Mohr circle expands, as illustrated in Fig. 3.26. The stress point moves from point 0 to

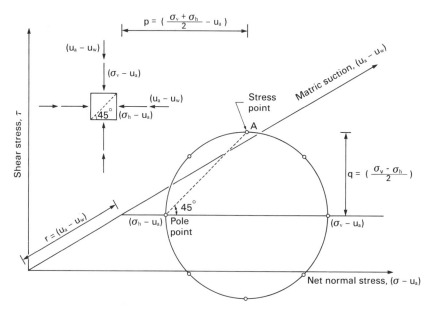

**Figure 3.25** Representative stress point for an extended Mohr circle.

points 1, 2, 3, 4, etc. These stress points represent a continuous series of Mohr circles or stress states. The stress path for this loading condition is shown in Fig. 3.27. The stress points are plotted on the $p$–$q$–$r$ diagram where $p$ is the abscissa, $q$ is the ordinate axis, and $r$ is the third orthogonal axis. The coordinates of the stress points, ($p$, $q$, $r$), are computed using Eqs. (3.42), (3.43), and (3.44). The $p$-, $q$-, $r$-coordinates represent the net normal stress, the shear stress, and the matric suction at each stage of loading. The stress path is established by joining the stress points. The stress path can be linear or curved, depending on the loading pattern.

The stress path shown in Fig. 3.27 illustrates a loading condition where the matric suction is maintained constant. Similar loading conditions can also be performed at other matric suction values. The stress paths are plotted on different planes, depending upon the matric suction value or the $r$-coordinate, as demonstrated in Fig. 3.28.

Figure 3.29 presents the stress paths for various loading patterns while maintaining a constant matric suction. The initial stress condition in the soil has $(\sigma_h - u_a)$ equal to $(\sigma_v - u_a)$. The magnitude and direction of the net normal stress changes determine the direction of the stress path on the $p$–$q$ plane.

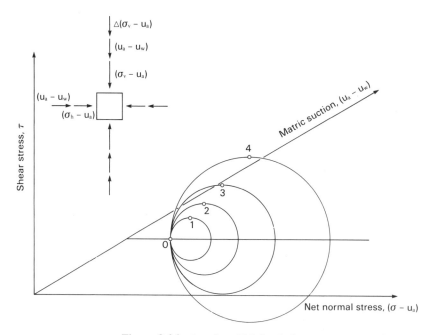

**Figure 3.26** A series of Mohr circles.

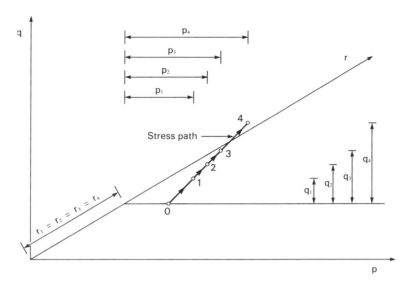

**Figure 3.27** A stress path for a series of stress states.

Stress states occurring in the field during deposition, desaturation, and soil sampling can be described using the stress path method, as illustrated in Fig. 3.30. The accumulation of soil sediments increases the vertical and horizontal effective normal stresses in accordance with the $K_0$-loading line, as indicated by the stress path $OA$. The shear stress in the soil increases during $K_0$-loading.

The $r$-coordinate can generally be considered equal to the pore-water pressure since the pore-air pressure in the field is usually atmospheric (i.e., zero gauge pressure). Therefore, matric suction, $(u_a - u_w)$, can be plotted as being equivalent in magnitude to the pore-water pressure. The accumulation of water in the soil due to rainfall can cause a soil to become saturated. As the soil becomes saturated, the stress state moves laterally on the saturation plane (i.e., $AB$) due to an increase in the positive pore-water pressure. Upon excessive evaporation, there will be a lowering of the groundwater table or a reduction in the pore-water pressure below atmospheric pressure. The drying process can be represented by the stress path $AC$ as the soil goes to an unsaturated condition. The wetting and drying processes occur repeatedly, and induce what is referred to as the stress history of the soil. Environmental changes cause a soil mass to repeatedly follow the stress paths $AB$, $BA$, $AC$, $CA$, and $AB$. The loadings of the soil due to drying and wetting are hydrostatic stress changes.

The drying process of a soil generally causes tension cracks to develop downward from the ground surface. The

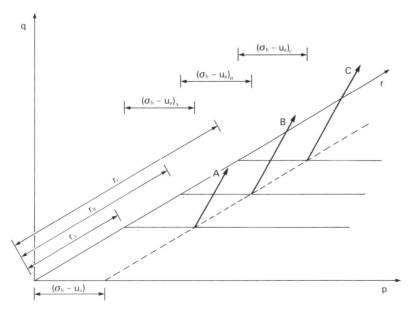

**Figure 3.28** Stress paths for different matric suction values.

**62**  3 STRESS STATE VARIABLES

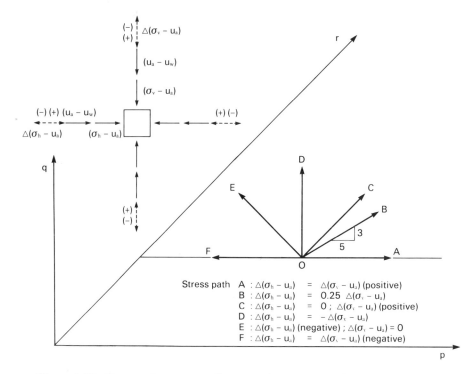

**Figure 3.29** Stress paths corresponding to various net normal stress loadings (modified after Lambe and Whitman, 1979).

tension cracks reduce the horizontal net normal stress. Upon subsequent wetting, the stress paths can become more complicated than those shown in Fig. 3.30.

When a soil sample is removed from the ground, the overburden pressure and the horizontal normal stress are removed. The removal of these stresses results in a tendency for the sample to expand. The expansion is resisted by an increase in matric suction or a further decrease in pore-water pressure. The changes in pore-water pressure due to changes in the total stress field can be defined in terms of the pore pressure parameters (see Chapter 8). The stress path followed during the sampling process is illustrated by the stress path $CD$. At point $D$, the net vertical and net horizontal stresses are zero, but the matric suction is slightly higher than the *in situ* matric suction. The soil sample now has a hydrostatic stress state (i.e., equal matric suction in all directions). The stress path method is later used to describe the shear strength and volume change behavior of unsaturated soils in Chapters 9 and 12, respectively.

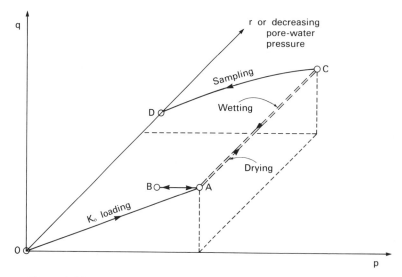

**Figure 3.30** Stress paths for $K_0$-loading, wetting, drying, and sampling.

## 3.6 ROLE OF OSMOTIC SUCTION

The total suction, $\psi$, of a soil is made up of two components, namely, the matric suction, $(u_a - u_w)$, and the osmotic suction, $\pi$:

$$\psi = (u_a - u_w) + \pi. \qquad (3.45)$$

Matric suction is known to vary with time due to environmental changes. Any change in suction affects the overall equilibrium of the soil mass. Changes in suction may be caused by a change in either one or both components of soil suction.

The role of osmotic suction has commonly been associated more with unsaturated soils than with saturated soils. However, osmotic suction is related to the salt content in the pore-water which is present in both saturated and unsaturated soils. The role of osmotic suction is therefore equally applicable to both unsaturated and saturated soils. Osmotic suction changes have an effect on the mechanical behavior of a soil. If the salt content in a soil changes, there will be a change in the overall volume and shear strength of the soil.

Most engineering problems involving unsaturated soils are commonly the result of environmental changes. The accumulation of water below a house may result in a reduction in matric suction and subsequent heaving of the structure. Similarly, the stability of an unsaturated soil slope may be endangered by excessive rainfall that reduces the suction in the soil. These changes primarily affect the matric suction component. Osmotic suction changes are generally less significant.

Figure 3.31 shows the relative importance of changes in osmotic suction as compared to matric suction when water content is varied. The total and matric suction curves are almost congruent one to another, particularly in the higher water content range. In other words, a change in total suction is essentially equivalent to a change in the matric suction [i.e., $\Delta \psi \approx \Delta (u_a - u_w)$]. For most geotechnical problems involving unsaturated soils, matric suction changes can be substituted for total suction changes, and vice versa.

There is a second reason why it is generally not necessary to take osmotic suction into account. The reason is related to the procedures commonly used in solving geotechnical problems. Generally, changes in osmotic suction that occur in the field are simulated during the laboratory testing for pertinent soil properties. For example, let us consider the swelling process of a soil as a result of rainfall. The rainfall, which is distilled water, dilutes the pore-water and changes the osmotic suction. In the laboratory, the soil specimen is generally immersed in distilled water prior to performing the test (e.g., volume change test in an oedometer). The matric suction is released to zero by immersing the soil specimen. The osmotic suction in the sample may also be changed in the process. It is not necessary

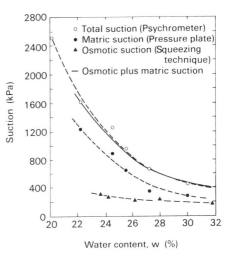

**Figure 3.31** Total, matric, and osmotic suction measurements on compacted Regina clay (from Krahn and Fredlund, 1972).

to know the change in osmotic suction provided the changes occurring in the field are simulated in the laboratory test.

In the case where the salt content of the soil is altered by chemical contamination, the effect of the osmotic suction change on the soil behavior may be significant. In this case, it is necessary to consider osmotic suction as part of the stress state. This applies equally for saturated and unsaturated soils. The role played by osmotic suction in influencing the mechanical behavior of a soil may or may not be of the same quantitative value as the role played by matric suction. The osmotic suction is more closely related to the diffuse double layer around the clay particles, whereas the matric suction is mainly associated with the air-water interface (i.e., contractile skin). It is possible to consider the osmotic suction, $\pi$, as an independent, isotropic stress state variable:

$$\begin{bmatrix} \pi & 0 & 0 \\ 0 & \pi & 0 \\ 0 & 0 & \pi \end{bmatrix}. \qquad (3.46)$$

In the case where both matric and osmotic suctions have the same quantitative influence on the behavior of a soil, the stress tensor (3.46) can be combined with the second stress tensor, (3.13):

$$\begin{bmatrix} \{(u_a - u_w) + \pi\} & 0 & 0 \\ 0 & \{(u_a - u_w) + \pi\} & 0 \\ 0 & 0 & \{(u_a - u_w) + \pi\} \end{bmatrix}.$$

$$(3.47)$$

Some research would indicate that it may be possible to algebraically combine the matric and osmotic components of suction when analyzing some geotechnical problems (Bailey, 1965; Chattopadhyay, 1972).

# CHAPTER 4

# *Measurements of Soil Suction*

The role of matric suction as one of the stress state variables for an unsaturated soil was illustrated in Chapter 3. The theory and components of soil suction will be presented first in this chapter, followed by a discussion of the capillary phenomena. Various devices and techniques for measuring soil suction and its components are described in detail in this chapter. Each device or technique is introduced with a history of its development, followed by its working principle, calibration technique, and performance.

## 4.1 THEORY OF SOIL SUCTION

The theoretical concept of soil suction was developed in soil physics in the early 1900's (Buckingham, 1907; Gardner and Widtsoe, 1921; Richards, 1928; Schofield, 1935; Edlefsen and Anderson, 1943; Childs and Collis-George, 1948; Bolt and Miller, 1958; Corey and Kemper, 1961; Corey et al., 1967). The soil suction theory was mainly developed in relation to the soil–water–plant system. The importance of soil suction in explaining the mechanical behavior of unsaturated soils relative to engineering problems was introduced at the Road Research Laboratory in England (Croney and Coleman, 1948; Croney et al., 1950). In 1965, the review panel for the soil mechanics symposium, "Moisture Equilibria and Moisture Changes in Soils" (Aitchison, 1965a), provided quantitative definitions of soil suction and its components from a thermodynamic context. These definitions have become accepted concepts in geotechnical engineering (Krahn and Fredlund, 1972; Wray, 1984; Fredlund and Rahardjo, 1988).

Soil suction is commonly referred to as the free energy state of soil water (Edlefsen and Anderson, 1943). The free energy of the soil water can be measured in terms of the partial vapor pressure of the soil water (Richards, 1965). The thermodynamic relationship between soil suction (or the free energy of the soil water) and the partial pressure of the pore–water vapor can be written as follows:

$$\psi = -\frac{RT}{v_{w0}\omega_v} \ln\left(\frac{\bar{u}_v}{\bar{u}_{v0}}\right) \quad (4.1)$$

where

- $\psi$ = soil suction or total suction (kPa)
- $R$ = universal (molar) gas constant [i.e., 8.31432 J/(mol K)]
- $T$ = absolute temperature [i.e., $T = (273.16 + t°)$ (K)]
- $t°$ = temperature (°C)
- $v_{w0}$ = specific volume of water or the inverse of the density of water [i.e., $1/\rho_w$] (m³/kg)]
- $\rho_w$ = density of water (i.e., 998 kg/m³ at $t° = 20°C$)
- $\omega_v$ = molecular mass of water vapor (i.e., 18.016 kg/kmol)
- $\bar{u}_v$ = partial pressure of pore–water vapor (kPa)
- $\bar{u}_{v0}$ = saturation pressure of water vapor over a flat surface of pure water at the same temperature (kPa).

Equation (4.1) shows that the reference state for quantifying the components of suction is the vapor pressure above a flat surface of pure water (i.e., water with no salts or impurities). The term $\bar{u}_v/\bar{u}_{v0}$ is called relative humidity, RH (%). If we select a reference temperature of 20°C, the constants in Eq. (4.1) give a value of 135 022 kPa. Equation (4.1) can now be written to give a fixed relationship between total suction in kilopascals and relative vapor pressure:

$$\psi = -135\ 022 \ln(\bar{u}_v/\bar{u}_{v0}). \quad (4.2)$$

Figure 4.1 shows a plot of Eq. (4.1) for three different temperatures. The soil suction, $\psi$, is equal to 0.0 when the relative humidity, RH (i.e., $\bar{u}_v/\bar{u}_{v0}$), is equal to 100% [Eq. (4.1)]. A relative humidity value less than 100% in a soil would indicate the presence of suction in the soil. Figure 4.1 also shows that suction can be extremely high. For example, a relative humidity of 94.24% at a temperature of 20°C corresponds to a soil suction of 8000 kPa. The range of suctions of interest in geotechnical engineering will correspond to high relative humidities.

### 4.1.1 Components of Soil Suction

The soil suction as quantified in terms of the relative humidity [Eq. (4.1)] is commonly called "total suction." It

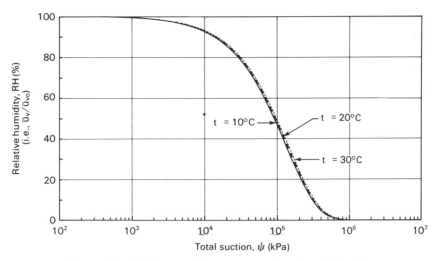

**Figure 4.1** Relative humidity versus total suction relationship.

has two components, namely, matric and osmotic suction. The total, matric, and osmotic suctions can be defined as follows (Aitchison, 1965a):

"Matric or capillary component of free energy—In suction terms, it is the equivalent suction derived from the measurement of the partial pressure of the water vapor in equilibrium with the soil water, relative to the partial pressure of the water vapor in equilibrium with a solution identical in composition with the soil water.

Osmotic (or solute) component of free energy—In suction terms, it is the equivalent suction derived from the measurement of the partial pressure of the water vapor in equilibrium with a solution identical in composition with the soil water, relative to the partial pressure of water vapor in equilibrium with free pure water.

Total suction or free energy of the soil water—In suction terms, it is the equivalent suction derived from the measurement of the partial pressure of the water vapor in equilibrium with a solution identical in composition with the soil water, relative to the partial pressure of the water vapor in equilibrium with free pure water."

The above definitions clearly state that the total suction corresponds to the free energy of the soil water, while the matric and osmotic suctions are the components of the free energy. In an equation form, this can be written as follows:

$$\psi = (u_a - u_w) + \pi \tag{4.3}$$

where

$(u_a - u_w)$ = matric suction
$u_a$ = pore-air pressure
$u_w$ = pore-water pressure
$\pi$ = osmotic suction.

The spelling of the term "matric" is in accordance with the recommendation of the Committee on Terminology of the Society of Soil Science of America. The definition is from their *Glossary of Soil Science Terminology* (1963, 1970, and 1979).

Figure 4.2 illustrates the concept of total suction and its component as related to the free energy of the soil water. The matric suction component is commonly associated with the capillary phenomenon arising from the surface tension of water. Surface tension has been described in Chapter 2, and is the result of the intermolecular forces acting on molecules in the contractile skin. The capillary phenomenon is usually illustrated by the rise of a water surface in a capillary tube (Fig. 4.2).

In soils, the pores with small radii act as capillary tubes that cause the soil water to rise above the water table (Fig. 4.3). The capillary water has a negative pressure with re-

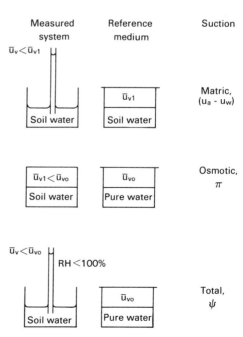

**Figure 4.2** Total suction and its components: matric and osmotic suction.

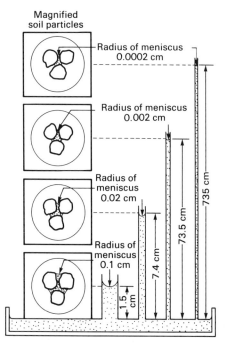

**Figure 4.3** Capillary tubes showing the air-water interfaces at different radii of curvatuve (from Janssen and Dempsey, 1980).

spect to the air pressure, which is generally atmospheric (i.e., $u_a = 0$) in the field. At low degrees of saturation, the pore-water pressures can be highly negative, with values as low as minus 7000 kPa (Olson and Langfelder, 1965). In this case, the adsorptive forces between soil particles are believed to play an important role in sustaining the highly negative pore-water pressures in soils.

Consider a capillary tube filled with a soil water. The surface of the water in the capillary tube is curved and is called a meniscus. On the other hand, the same soil water will have a flat surface when placed in a large container. The partial pressure of the water vapor above the *curved* surface of soil water, $\bar{u}_v$, is less than the partial pressure of the water vapor above a *flat* surface of the same soil water, $\bar{u}_{v1}$, (i.e., $\bar{u}_v < \bar{u}_{v1}$ in Fig. 4.2). In other words, the rela-

tive humidity in a soil will decrease due to the presence of curved water surfaces produced by the capillary phenomenon. The water vapor pressure or the relative humidity decreases as the radius of curvature of the water surface decreases. At the same time, the radius of curvature is inversely proportional to the difference between the air and water pressures across the surface [i.e., $(u_a - u_w)$] and is called matric suction. This means that one component of the total suction is matric suction, and it contributes to a reduction in the relative humidity.

The pore-water in a soil generally contains dissolved salts. The water vapor pressure over a flat surface of solvent, $\bar{u}_{v1}$, is less than the water vapor pressure over a flat surface of pure water, $\bar{u}_{v0}$. In other words, the relative humidity decreases with increasing dissolved salts in the pore-water of the soil. The decrease in relative humidity due to the presence of dissolved salts in the pore-water is referred to as the osmotic suction, $\pi$.

### 4.1.2 Typical Suction Values and Their Measuring Devices

Table 4.1 shows typical matric, osmotic, and total suction values for two soils which often form the subgrade for roads built in the province of Saskatchewan, Canada (Krahn and Fredlund, 1972). The Regina Clay is a highly plastic, inorganic clay with a liquid limit of 78% and a plastic limit of 31%. The glacial till has a liquid limit of 34% and a plastic limit of 17%. Suction values are given in Table 4.1 for soils compacted to standard AASHTO conditions, with the water contents at optimum and 2% dry of optimum.

Figure 4.4 shows experimental data illustrating that the matric plus the osmotic components of suction are equal to the total suction of the soil. The presented data are for glacial till specimens compacted under modified AASHTO conditions at various initial water contents. Each component of soil suction, and the total suction, were measured independently.

Several devices commonly used for measuring total, matric, and osmotic suctions are listed in Table 4.2. The ex-

**Table 4.1 Typical Suction Values for Compacted Soils**

| Soil Type | Water Content (%) | Matric Suction, $(u_a - u_w)$ (kPa) | Osmotic Suction, $\pi$ (kPa) | Total Suction, $\psi$ (kPa) |
|---|---|---|---|---|
| Regina clay: $\gamma_{max} = 13.81$ kN/m³ | 30.6 (optimum) | 273 | 187 | 460 |
|  | 28.6 | 354 | 202 | 556 |
| Glacial till: $\gamma_{max} = 19.24$ kN/m³ | 15.6 (optimum) | 310 | 290 | 600 |
|  | 13.6 | 556 | 293 | 849 |

## 4.2 CAPILLARITY

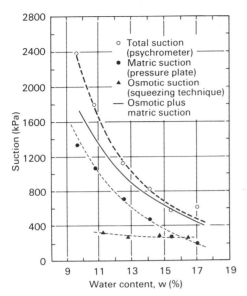

**Figure 4.4** Total, matric, and osmotic suctions for glacial till (from Krahn and Fredlund, 1972).

planation of each device is given later. The measurement range and comments related to each device are shown in Table 4.2.

## 4.2 CAPILLARITY

The capillary phenomenon is associated with the matric suction component of total suction. The height of water rise and the radius of curvature have direct implications on the water content versus matric suction relationship in soils (i.e., the soil–water characteristic curve). This relationship is different for the wetting and drying portions of the curve, and these differences can also be explained in terms of the capillary model.

### 4.2.1 Capillary Height

Consider a small glass tube that is inserted into water under atmospheric conditions (Fig. 4.5). The water rises up in the tube as a result of the surface tension in the contractile skin and the tendency of water to wet the surface of the glass tube (i.e., hygroscopic properties). This capillary behavior can be analyzed by considering the surface tension, $T_s$, acting around the circumference of the meniscus. The surface tension, $T_s$, acts at an angle, $\alpha$, from the vertical. The angle is known as the contact angle, and its magnitude depends on the adhesion between the molecules in the contractile skin and the material comprising the tube (i.e., glass).

Let us consider the vertical force equilibrium of the capillary water in the tube shown in Fig. 4.5. The vertical resultant of the surface tension (i.e., $2\pi r T_s \cos \alpha$) is responsible for holding the weight of the water column, which has a height of $h_c$ (i.e., $\pi r^2 h_c \rho_w g$):

$$2\pi r T_s \cos \alpha = \pi r^2 h_c \rho_w g \qquad (4.4)$$

where

$r$ = radius of the capillary tube
$T_s$ = surface tension of water
$\alpha$ = contact angle
$h_c$ = capillary height
$g$ = gravitational acceleration.

**Table 4.2 Devices for Measuring Soil Suction and Its Components**

| Name of Device | Suction Component Measured | Range (kPa) | Comments |
|---|---|---|---|
| Psychrometers | Total | $100^a$–~8000 | Constant temperature environment required |
| Filter paper | Total | (Entire range) | May measure matric suction when in good contact with moist soil |
| Tensiometers | Negative pore-water pressures or matric suction when pore-air pressure is atmospheric | 0–90 | Difficulties with cavitation and air diffusion through ceramic cup |
| Null-type pressure plate (axis translation) | Matric | 0–1500 | Range of measurement is a function of the air entry value of the ceramic disk |
| Thermal conductivity sensors | Matric | 0–~400+ | Indirect measurement using a variable pore size ceramic sensor |
| Pore fluid squeezer | Osmotic | (Entire range) | Used in conjuction with a psychrometer or electrical conductivity measurement |

$^a$Controlled temperature environment to ± 0.001°C.

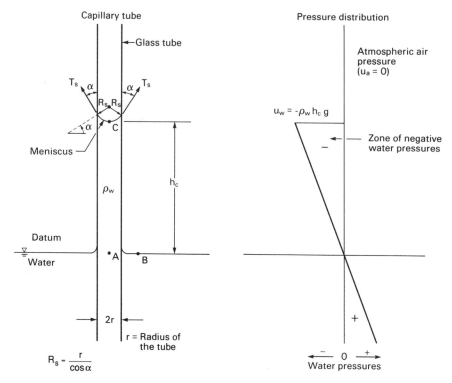

**Figure 4.5** Physical model and phenomenon related to capillarity.

Equation (4.4) can be rearranged to give the maximum height of water in the capillary tube, $h_c$:

$$h_c = \frac{2T_s}{\rho_w g R_s} \quad (4.5)$$

where

$R_s$ = radius of curvature of the meniscus (i.e., $r/\cos \alpha$).

The contact angle between the contractile skin for pure water and clean glass is zero (i.e., $\alpha = 0$). When the $\alpha$ angle is zero, the radius of curvature, $R_s$, is equal to the radius of the tube, $r$ (Fig. 4.5). Therefore, the capillary height of pure water in a clean glass is

$$h_c = \frac{2T_s}{\rho_w g r}. \quad (4.6)$$

The radius of the tube is analogous to the pore radius in soils. Equation (4.6) shows that the smaller the pore radius in the soil, the higher will be the capillary height. The capillary height can be plotted against the pore radius using Eq. (4.6) where the contact angle is assumed to be zero (Fig. 4.6).

### 4.2.2 Capillary Pressure

Points $A$, $B$, and $C$ in the capillary system shown in Fig. 4.5 are in hydrostatic equilibrium. The water pressures at points $A$ and $B$ are atmospheric (i.e., $u_w$ at $A$ = $u_w$ at $B$, which is equal to 0.0). The elevation of points $A$ and $B$ on the water surface is considered as the datum for the system (i.e., zero elevation). As a result, the hydraulic heads (i.e., elevation head plus pressure head) at points $A$ and $B$ are equal to zero.

Point $C$ is located at a height of $h_c$ from the datum (i.e., elevation head is equal to $h_c$). The hydrostatic equilibrium among points $C$, $B$, and $A$ requires that the hydraulic heads at all three points be equal. In other words, the hydraulic head at point $C$ is also equal to zero. This means that the

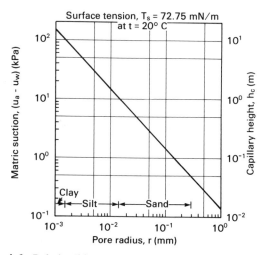

**Figure 4.6** Relationship among pore radius, matric suction, and capillary height.

pressure head at point $C$ is equal to the negative value of the elevation head at point $C$. The water pressure at point $C$ can be calculated as

$$u_w = -\rho_w g h_c \quad (4.7)$$

where

$u_w$ = water pressure.

The water pressures above point $A$ in the capillary tube are negative, as shown in Fig. 4.5. The water in the capillary tube is said to be under tension. On the other hand, the water pressures below point $A$ (i.e., water table) are positive due to hydrostatic pressure conditions. At point $C$, the air pressure is atmospheric (i.e., $u_a = 0$) and the water pressure is negative (i.e., $u_w = -\rho_w g h_c$). The matric suction, $(u_a - u_w)$, at point $C$ can then be expressed as follows:

$$(u_a - u_w) = \rho_w g h_c. \quad (4.8)$$

Substituting Eq. (4.5) into Eq. (4.8) gives rise to matric suction being written in terms of surface tension:

$$(u_a - u_w) = \frac{2T_s}{R_s}. \quad (4.9)$$

Equation (4.9) is the same as the equation for the pressure difference across a contractile skin as presented in Chapter 2. The radius of curvature, $R_s$, can be considered analogous to the pore radius, $r$, in a soil by assuming a zero contact angle (i.e., $\alpha = 0$). As a result, the smaller the pore radius of a soil, the higher the soil matric suction can be, as shown in Fig. 4.6.

The above explanation has demonstrated the ability of the surface tension to support a column of water, $h_c$, in a capillary tube. The surface tension associated with the contractile skin results in a reaction force on the wall of the capillary tube (Fig. 4.7). The vertical component of this reaction force produces compressive stresses on the wall of the tube. In other words, the weight of the water column is transferred to the tube through the contractile skin. In the case of a soil having a capillary zone, the contractile skin results in an increased compression of the soil structure. As a result, the presence of matric suction in an unsaturated soil increases the shear strength of the soil.

### 4.2.3 Height of the Capillary Rise and Radius Effects

The effects of the height of capillary rise and the radius of curvature on capillarity are illustrated in Fig. 4.8, as presented by Taylor (1948). A clean capillary tube of radius, $r$, allows pure water to rise to a maximum capillary height, $h_c$, as shown in Fig. 4.8(a). However, the water rise in a capillary tube may be restricted by the limited length of the tube [Fig. 4.8(b)]. A decrease in the capillary height results in an increase in the radius of curvature, $R_s$, as indicated by Eq. (4.5) (i.e., $h_c = 2T_s/(\rho_w g R_s)$). For a constant radius of the tube, the increase in $R_s$ causes an increase in the contact angle since $R_s$ is equal to $(r/\cos \alpha)$.

The radius or opening of the tube is a significant factor in the development of capillary rise, as illustrated in Fig. 4.8(c) and (d). In both cases, the tube has a bulb with a radius of $r_1$, which is larger than the radius of the tube, $r$. The presence of the bulb at the midheight of the capillary height, $h_c$, prevents the water from rising up beyond the base of the bulb [Fig. 4.8(c)]. In other words, the nonuniform opening along the capillary tube can prevent the full development of capillary height. On the other hand, the capillary height, $h_c$, can be fully developed if the bulb is filled by submerging it below the water surface and then raising it above the surface [Fig. 4.8(d)].

The development of capillary rise in a soil is also affected by the pore size distribution in the soil, as shown in Fig. 4.8(e). The water surface in the soil can rise to the capillary height, $h_c$, through continuous soil pores with radii that are smaller than or equal to $r$. Capillary heights greater than $h_c$ may also develop if the height of the soil column is extended. The higher capillary rise corresponds to the pore radii that are smaller than $r$. However, the water surface cannot rise within the large openings at the center of the soil column [Fig. 4.8(e)].

The above capillary tube analogy also applies to soil conditions in nature. The nonuniform pore size distribution in a soil can result in hysteresis in the soil–water characteristics curve. At a given matric suction, the soil water content during the wetting and drying processes are different, as illustrated by the examples shown in Fig. 4.8(c) and (d), respectively. In addition, the contact angle at an advancing interface during the wetting process is different from that at a receding interface during the drying process (Bear, 1979). The above factors, as well as the presence of entrapped air in the soil, are considered to be the main causes for hysteresis in the soil–water characteristic curve.

In spite of its simplicity, the capillary model has some limitations in its application to describing the mechanical

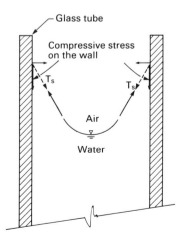

**Figure 4.7** Forces acting on a capillary tube.

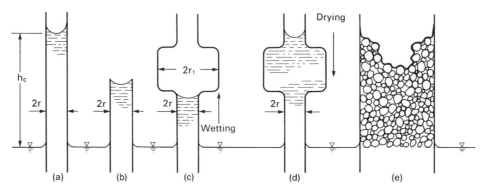

**Figure 4.8** Height and radius effects on capillarity (from Taylor, 1948).

behavior of unsaturated soils. An apparent anomaly will occur when the capillary model is incorporated into the formulation of pore fluid compressibility, as later explained in Chapter 8. The use of pore radius in the capillary equation [i.e., Eq. (4.9)] causes the model to be impractical for engineering practice. In addition, there are other factors that contribute to being able to sustain highly negative pore-water pressures in soils, such as the adsorptive forces between clay particles.

## 4.3 MEASUREMENTS OF TOTAL SUCTION

Environmental changes and changes in applied loads produce a change in the water content of the soil. The initial water content of compacted soils appears to have a direct relationship with the matric suction component (Fig. 4.4). On the other hand, the osmotic suction does not seem to be sensitive to the changes in the soil water content. As a result, a change in the total suction is quite representative of a change in the matric suction. Therefore, total suction measurements are of importance, particularly in the high suction ranges where the matric suction measurements are difficult to obtain.

The following sections discuss the direct and indirect measurements of total suction. The free energy of the soil water (i.e., total suction) can be determined by measuring the vapor pressure of the soil water or the relative humidity in the soil. The direct measurement of relative humidity in a soil can be conducted using a device called a psychrometer. The relative humidity in a soil can be indirectly measured using a filter paper as a measuring sensor. The filter paper is equilibrated with the suction in the soil.

### 4.3.1 Psychrometers

Thermocouple psychrometers can be used to measure the total suction of a soil by measuring the relative humidity in the air phase of the soil pores or the region near the soil. The relative humidity is related to the total suction in accordance with Eq. (4.1) where $(\overline{u}_v/\overline{u}_{v0})$ is equal to the relative humidity, RH.

There are two basic types of thermocouple psychrometers, namely, the wet-loop type (Richards and Ogata, 1958) and the Peltier type (Spanner, 1951). Both types of psychrometers operate on the basis of temperature difference measurements between a nonevaporating surface (i.e., dry bulb) and an evaporating surface (i.e., wet bulb). The difference in the temperatures between these surfaces is related to the relative humidity.

The wet-loop and the Peltier-type psychrometers differ in the manner by which the evaporating junction is wetted in order to induce evaporation. In the wet-loop psychrometer, the evaporating junction is wetted by placing a drop of water into a small silver ring. In the Peltier psychrometer, evaporation is induced by passing a Peltier current through the evaporating junction. The Peltier current causes the junction to cool below the dewpoint, resulting in the condensation of a minute quantity of water vapor on the junction. The Peltier psychrometer is most commonly used in geotechnical engineering and is described in the following sections. The Seebeck and the Peltier effects are the main principles behind the operation of the Peltier psychrometer.

*Seebeck Effects*

Seebeck (1821) discovered that an electromotive force (i.e., emf) was generated in a closed circuit of two dissimilar metals when the two junctions of the circuit have different temperatures [i.e., $T$ and $(T + \Delta T)$], as illustrated in Fig. 4.9. This phenomenon is referred to as the Seebeck effect, which allows the use of two dissimilar wires (i.e., a thermocouple) to measure temperature. One junction of the circuit is maintained at a constant temperature for a reference, while the other junction is used for sensing a difference in temperature. A microvoltmeter can be installed in the circuit to measure the Seebeck electromotive force, which is a function of the temperature difference between the two junctions.

*Peltier Effects*

Peltier (1834) discovered that when a current is passed through a circuit of two dissimilar metals, one of the junctions becomes warmer, while the other junction becomes

4.3 MEASUREMENTS OF TOTAL SUCTION    71

of maximum cooling. The maximum cooling results in a minimum dewpoint temperature that can be reached by the thermocouple. This, in turn, imposes a restriction on the lowest relative humidity (or the highest soil suction) that can be measured using a thermocouple psychrometer. The lower the relative humidity, the lower is the dewpoint temperature associated with its water vapor pressure.

*Peltier Psychrometer*

A typical Peltier psychrometer, often called a Spanner psychrometer, is shown in Fig. 4.11. The thermocouple consists of 0.025 mm diameter wires of constantan (i.e., copper–nickel) and chromel (i.e., chromium–nickel). The wires are welded together to form an evaporating or a measuring junction. The other ends of the wires are connected to 26 American Wire Gauge (AWG) copper lead wires to form a reference junction. The highly conductive copper wires have a large diameter (i.e., large thermal mass) in order to serve as heat sinks that can maintain a constant temperature at the reference junction. The heat sinks are required to adsorb the Joule heat generated near the reference junction as the measuring junction is being cooled.

The maximum degree of cooling generated by the chromel–constantan thermocouple is about 0.6°C below the ambient temperature (Brown and Bartos, 1982). This maximum cooling represents the lowest relative humidity (i.e., 94%) or the upper limit of the total suction (i.e., 8000 kPa) which can be measured using the thermocouple psychrometer. The lowest suction which can be measured using a thermocouple psychrometer is approximately 100 kPa under a controlled temperature environment of ±0.001°C (Krahn and Fredlund, 1972). This lower limit corresponds to a relative humidity approaching 100%. A slight lowering of the temperature as the 100% relative humidity is approached will immediately produce condensation on the thermocouple.

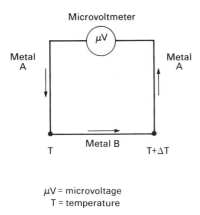

**Figure 4.9** Electrical circuit to illustrate the Seebeck effect.

cooler, as illustrated in Fig. 4.10. Passing the current in the opposite direction, as shown in Fig. 4.10, will produce a reverse thermal condition at the two junctions. This phenomenon is referred to as the Peltier effect, and it allows the use of thermocouples for the measurement of relative humidity.

The Peltier effect can be used to cool a thermocouple junction to reach the dewpoint temperature corresponding to the surrounding atmosphere. As a result, water vapor condenses on the junction. Upon termination of passing the current, the condensed water tends to evaporate to the surrounding atmosphere, causing a further reduction in the temperature at the junction. The temperature reduction is a function of the evaporation rate, which is in turn affected by the water vapor pressure in the atmosphere. If the ambient temperature and the temperature reduction due to evaporation are measured using the Seebeck effects, the relative humidity of the atmosphere can be computed.

There is a maximum degree of junction cooling that can be achieved using the Peltier current (Spanner, 1951). The electrical current that produces the Peltier cooling also produces the Joule heating effect. The Joule heating effect is the heat produced by the work done against friction along the thermocouple wires. Spanner (1951) showed that the net cooling effect is a quadratic function of the current, and that there is a maximum value beyond which the Joule heating will dominate.

Different types of thermocouples have different degrees

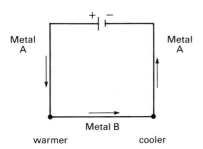

**Figure 4.10** Electrical circuit to illustrate the Peltier effect.

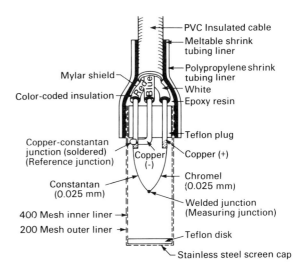

**Figure 4.11** Screen-caged single-junction Peltier thermocouple psychrometer (from Brown and Collins, 1980).

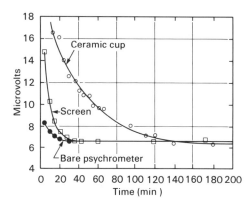

**Figure 4.12** Equilibration times for a thermocouple psychrometer with various protective coverings, placed over a 0.3 molal KC1 solution at 25°C (from Brown, 1970).

A protective housing is usually provided around the thermocouple wires. Protective covers take the form of a ceramic cup, a stainless steel screen, or a solid (stainless steel or teflon) tubing with a screen end window. The selection of the type of the protective cover depends on the application of the psychrometer. The time required for the water vapor equilibration is affected by the type of the protective cover, as demonstrated in Fig. 4.12. The ceramic cup appears to have a long equilibration time, which may not be practical in many situations.

The psychrometer device is connected to a control unit for applying the Peltier cooling current. The psychrometer is also connected to a microvoltmeter for measuring the generated electromotive force during the evaporation process.

Measurements of total suction are conducted by suspending a psychrometer in a closed system containing a soil specimen. The relative humidity is measured after equilibrium is attained between the air near the thermocouple and the pore-air in the soil specimen. Isothermal conditions among the temperature of the soil, the air, and the psychrometer must be achieved prior to conducting the measurements. A controlled temperature environment of $\pm\,0.001\,°C$ is required in order to measure total suctions to an accuracy of $\pm 10$ kPa (Krahn and Fredlund, 1972). Thermal equilibrium within the psychrometer is assured by obtaining a zero reading on the microvoltmeter.

The processes associated with the relative humidity measurement when using a Peltier psychrometer are best illustrated using Fig. 4.13 and the following explanation:

a) Isothermal equilibrium between the psychrometer and the surrounding atmosphere must be achieved prior to a measurement being taken. This is indicated by a zero voltage reading.

b) At an elapsed time of 15 s, a small electrical current (i.e., 5 mA) is passed through the psychrometer circuit from the constantan wire to the chromel wire for a period of 15 s. The passage of an electrical current in this direction causes the measuring junction to cool due to the Peltier effect. As the temperature at the measuring junction drops below the dewpoint corresponding to the surrounding atmosphere, water vapor condenses on the measuring junction. During the condensation process, the temperature at the measuring junction remains at the corresponding dewpoint temperature.

c) At the end of the 15 s period of cooling, the Peltier current is then terminated.

d) As soon as the cooling process is stopped, the condensed water on the measuring junction starts to

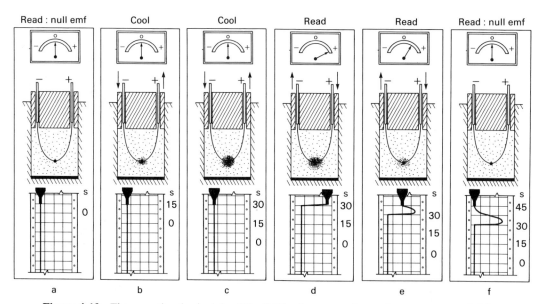

**Figure 4.13** The operational principle of the Peltier thermocouple psychrometer suspended in a sealed chamber over a soil specimen (from Van Havered and Brown, 1972).

evaporate to the surrounding atmosphere. The temperature at the measuring junction starts to drop below the dewpoint temperature as evaporation begins. As a result, the microvoltmeter records the electromotive force on a strip chart recorder. The generated electromotive force is a function of the temperature difference between the measuring junction and the reference junction (i.e., the Seebeck effect). The microvolt reading increases rapidly to a maximum value which is a function of the relative humidity in the surrounding atmosphere. The drier the atmosphere, the higher will be the microvolt output during the evaporation process.

e) Having reached the maximum output corresponding to the maximum evaporative cooling, the microvolt output decreases rapidly to a zero reading. The decreasing output indicates that the temperature at the measuring junction is increasing towards the ambient or the reference junction temperature.

f) The microvoltmeter gives a zero reading when the temperature at the measuring junction becomes equal to that of the reference junction.

*Psychrometer Calibration*

The calibration of a psychrometer consists of determining the relationship between microvolt outputs from the thermocouple and a known total suction value. The calibration is conducted by suspending the psychrometer over a salt solution with a known osmotic suction under isothermal conditions. The calibration is performed by mounting the psychrometer in a sealed chamber.

Filter papers are placed at the base of the chamber and generally saturated with a solution of NaCl or KCl. The osmotic suctions for NaCl and KCl solutions at different molalities and temperatures are summarized in Tables 4.3 and 4.4, respectively.

Under isothermal equilibrium conditions, the water vapor pressure or the relative humidity in the calibration chamber corresponds to the osmotic suction of the salt solution. Therefore, the psychrometer can be calibrated at various suction values by using different molalities (or osmotic suctions) for the salt solution. Isothermal conditions are maintained by placing the chamber in a constant temperature bath, as illustrated in Fig. 4.14.

The calibration process results in a set of calibration curves corresponding to various temperatures (Fig. 4.15). Each curve relates the psychrometer reading to a corresponding total suction. The maximum output from the microvoltmeter is taken as the psychrometer reading. The psychrometer can then be used to measure the total suction in a soil specimen by using the established calibration curves.

The calibration curves shown in Fig. 4.15 appear to increase from zero to a maximum microvolt value, and then decrease sharply to lower values. The maximum microvolt value indicates the maximum total suction that can be measured using psychrometers. Psychrometer readings beyond this point are highly variable, with the largest variability occurring at high temperatures. This characteristic occurs because there is a maximum degree of cooling that can be achieved. The curves in Fig. 4.15 indicate a range for the maximum measurable total suction from 7000 to 8000 kPa, corresponding to a temperature range between 0 and 35°C, respectively.

The response time of a psychrometer depends on its protective cover (Fig. 4.12) and the magnitude of total suction being measured (Fig. 4.16). The response time varies from

Table 4.3 Osmotic Suctions of NaCl solutions (from Lange, 1967)

| NaCl Molality | Temperature | | | | |
|---|---|---|---|---|---|
| | 0°C | 7.5°C | 15°C | 25°C | 35°C |
| | Osmotic Suction (kPa) | | | | |
| 0 | 0.0 | 0.0 | 0.0 | 0.0 | 0.0 |
| 0.2 | 836 | 860 | 884 | 915 | 946 |
| 0.5 | 2070 | 2136 | 2200 | 2281 | 2362 |
| 0.7 | 2901 | 2998 | 3091 | 3210 | 3328 |
| 1.0 | 4169 | 4318 | 4459 | 4640 | 4815 |
| 1.5 | 6359 | 6606 | 6837 | 7134 | 7411 |
| 1.7 | 7260 | 7550 | 7820 | 8170 | 8490 |
| 1.8 | 7730 | 8035 | 8330 | 8700 | 9040 |
| 1.9 | 8190 | 8530 | 8840 | 9240 | 9600 |
| 2.0 | 8670 | 9025 | 9360 | 9780 | 10 160 |

**Table 4.4 Osmotic Suctions for KCl Solutions (from Campbell and Gardner, 1971)**

| Molality | 0°C | 10°C | 15°C | 20°C | 25°C | 30°C | 35°C |
|---|---|---|---|---|---|---|---|
| 0 | 0.0 | 0.0 | 0.0 | 0.0 | 0.0 | 0.0 | 0.0 |
| 0.10 | 421 | 436 | 444 | 452 | 459 | 467 | 474 |
| 0.20 | 827 | 859 | 874 | 890 | 905 | 920 | 935 |
| 0.30 | 1229 | 1277 | 1300 | 1324 | 1347 | 1370 | 1392 |
| 0.40 | 1628 | 1693 | 1724 | 1757 | 1788 | 1819 | 1849 |
| 0.50 | 2025 | 2108 | 2148 | 2190 | 2230 | 2268 | 2306 |
| 0.60 | 2420 | 2523 | 2572 | 2623 | 2672 | 2719 | 2765 |
| 0.70 | 2814 | 2938 | 2996 | 3057 | 3116 | 3171 | 3226 |
| 0.80 | 3208 | 3353 | 3421 | 3492 | 3561 | 3625 | 3688 |
| 0.90 | 3601 | 3769 | 3846 | 3928 | 4007 | 4080 | 4153 |
| 1.00 | 3993 | 4185 | 4272 | 4366 | 4455 | 4538 | 4620 |

a few hours at several thousands kilopascals suction to about two weeks at 100 kPa suction (Richards, 1974). It appears that the psychrometer requires a considerably long time for equalization when used to measure low suction values.

Hamilton et al. (1981) reported serious problems associated with corrosion on the thermocouples. The response characteristics of a failing or dirty psychrometer is difficult to interpret, as illustrated in Fig. 4.17. The corrosion problem can be attributed to the acidic environment in the soil. It is important to clean the psychrometer thoroughly after each calibration or usage, in accordance with the instructions given by the manufacturer.

*Psychrometer Performance*

Psychrometers are useful for measuring high suctions in soils. *In situ* measurements of total suctions using psychrometers are generally not recommended because significant temperature fluctuations occur in the field. However, laboratory measurements can be conducted in a controlled temperature environment using undisturbed soil specimens from the field.

A small soil specimen is placed into a stainless steel or Lucite chamber, together with the thermocouple psychrometer, as illustrated in Fig. 4.18. The entire assembly is then placed in a constant temperature bath, as shown in Fig. 4.14. The temperature of the bath should be maintained at a constant temperature, within $\pm 0.001\,°C$ (Krahn and Fredlund, 1972). In other words, the thermoregulator must be able to respond to a fluctuation in temperature of $\pm 0.001\,°C$. The soil temperature is expected to be maintained within the same degree of accuracy or greater due to the buffering effect of the glass beaker.

Figures 4.19 and 4.20 present the relationships between total suction and initial water content for compacted glacial till and compacted Regina clay, respectively. The total suction measurements were conducted using thermocouple psychrometers. It should be noted that these relationships are different from the soil-water characteristic curves for the soils since the results were obtained from various soil

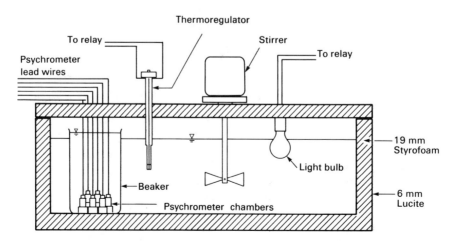

**Figure 4.14** Schematic diagram of a constant temperature bath (from Krahn and Fredlund, 1972).

4.3 MEASUREMENTS OF TOTAL SUCTION  75

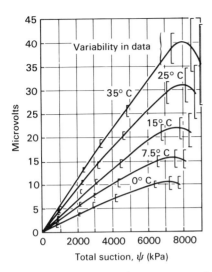

**Figure 4.15** Psychrometer calibration curves at various temperatures (from Brown and Cartos, 1982).

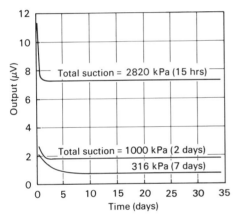

**Figure 4.16** Response times for laboratory psychrometers (from Richards, 1974).

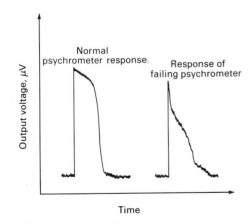

**Figure 4.17** Comparison of responses from a normal and a dirty or failing psychrometer, as obtained from the same suction (from Hamilton, 1979).

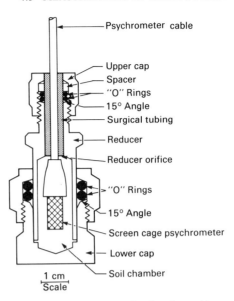

**Figure 4.18** Stainless steel sample chamber with a seal psychrometer in place (from Brown and Collins, 1980).

specimens compacted to different densities and at different water contents. A soil–water characteristic curve describes the water content versus suction relationship for a single soil specimen. Figures 4.19 and 4.20 clearly indicate a unique relationship between soil suction and initial water content for a particular compacted soil, regardless its dry densities. The *in situ* suction of the same *compacted* soil in the field can then be inferred from this type of relationship (Figs. 4.19 and 4.20) by measuring its water content. This applies only when the soil has just been compacted in the field.

Comparisons between suction measurements using ther-

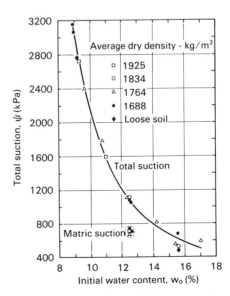

**Figure 4.19** Total suction versus initial water content relationship for glacial till (from Krahn and Fredlund, 1972).

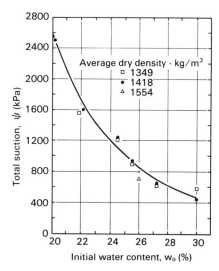

**Figure 4.20** Total suction versus initial water content relationship for Regina clay (from Krahn and Fredlund, 1972).

mocouple psychrometers and suction measurements using the filter paper technique are shown in Figs. 4.21 and 4.22. Figure 4.21 illustrates laboratory measurements of total suctions on soil samples from various depths at a location near Regina, Sask., Canada. The results indicate reasonably close agreement between both methods of total suction measurement as long as the filter paper is not in contact with the soil.

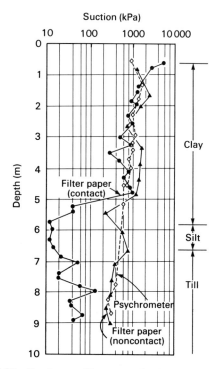

**Figure 4.21** Suction profile versus depth obtained using thermocouple psychrometers and the filter paper method (from van der Raadt et al., 1987).

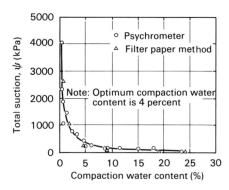

**Figure 4.22** Comparison of independent measurements of total suction on a compacted silty sand (from Daniel et al., 1981).

The basic Peltier psychrometer (Fig. 4.11) with a single measuring junction has been found to be extremely sensitive to slight temperature gradients. In addition, the single-junction psychrometer cannot be used to measure the ambient temperature around the measuring junction. Therefore, a double-junction Peltier psychrometer (Fig. 4.23) has been developed in order to eliminate the drawbacks associated with a single-junction psychrometer.

The double-junction psychrometer has two chromel–constantan thermocouples. The two constantan wires are attached to a constantan lead, while the two chromel wires are attached to different copper leads. The Peltier current is applied to the circuit $TN$ (Fig. 4.23), causing a cooling at the left measuring junction. The psychrometer output is measured between the two copper leads, $N$ and $P$, as a difference between the two junctions. The psychrometer output is relatively free from any extraneous electrical outputs associated with temperature fluctuations within the psychrometer chamber. Any electromotive force generated at one measuring junction due to a thermal gradient is compensated by an opposing electromotive force at the other

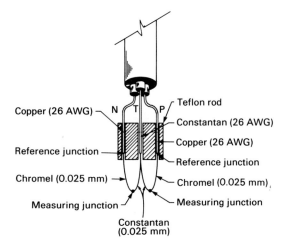

**Figure 4.23** Double-junction temperature-compensated Peltier thermocouple psychrometer (from Van Haveren and Brown, 1972 and Meeuwig, 1972).

measuring junction. Therefore, the double-junction psychrometer is called a temperature-compensated psychrometer. In addition, the output measured between the constantan lead wire, $T$, and the copper lead wire, $P$, gives the ambient temperature within the psychrometer chamber.

### 4.3.2 Filter Paper

The filter paper method for measuring soil suction was developed in the soil science discipline, and has since been used primarily in soil science and agronomy (Gardner, 1937; Fawcett and Collis-George, 1967; McQueen and Miller, 1968; Al-Khafaf and Hanks, 1974). Attempts have also been made to use the filter paper method in geotechnical engineering (Ho, 1979; Tang, 1979; McKeen, 1981; Khan, 1981; Ching and Fredlund, 1984; Gallen, 1985; McKeen, 1985; Chandler and Gutierrez, 1986). Recent experience with the use of the filter paper method on the studies of airport pavement subgrades and swelling potential profile of expansive soils by McKeen (1985), has indicated that this method deserves further consideration. At present, the filter paper method has not gained general acceptance in geotechnical engineering. There is a need for further research relative to the use of this technique in engineering.

*Principle of Measurement (Filter Paper Method)*

From a theoretical standpoint, it is possible to use the filter paper method to measure either the total or the matric suction of a soil. The filter paper is used as a sensor. The filter paper method is classified as an "indirect method" of measuring soil suction.

The filter paper method is based on the assumption that a filter paper will come to equilibrium (i.e., with respect to moisture flow) with a soil having a specific suction. Equilibrium can be reached by either liquid or vapor moisture exchange between the soil and the filter paper. When a dry filter paper is placed in direct contact with a soil specimen, it is assumed that water flows from the soil to the paper until equilibrium is achieved (Fig. 4.24). When a dry filter paper is suspended above a soil specimen (i.e., no direct contact with the soil), vapor flow of water will occur from the soil to the filter paper until equilibrium is achieved (Fig. 4.24). Having established equilibrium conditions, the water content of the filter paper is measured.

The water content of the filter paper corresponds to a suction value, as illustrated by the filter paper calibration curve shown in Fig. 4.25. Theoretically, the equilibrium water content of the filter paper corresponds to the *matric suction* of the soil when the paper is placed in *contact* with the water in the soil. On the other hand, the equilibrium water content of the filter paper corresponds to the *total suction* of the soil if the paper is *not in contact* with the soil. Therefore, the same calibration curve is used for both the matric and total suction measurements.

The filter paper method can be used to measure soil suc-

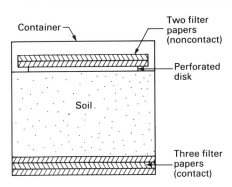

**Figure 4.24** Contact and noncontact filter paper methods for measuring matric and total suction, respectively (from Al-khafaf and Hanks, 1974).

tion over a wide range of values. The measurements are generally performed in the laboratory by equilibrating a filter paper with an undisturbed or disturbed soil specimen obtained from the field.

*Measurement and Calibration Techniques (Filter Paper Method)*

The following technique of measurement and calibration is written in accordance with a tentative ASTM standard on the filter paper method (ASTM Committee D18 on Soil and Rock). The filter paper sensor must be of the ash-free, quantitative Type II as specified by ASTM standard specification E832. Whatman No. 42 and Schleicher and Schuell No. 589 White Ribbon are two commonly used brands of filter paper. A typical filter paper has a disk size with a diameter of 55 mm. Filter papers from the same brand are considered to be "identical" in the sense that all filter paper disks have the same calibration curve.

The equipment associated with the filter paper method consists of large and small metal containers, an insulated box, a balance, and a drying oven. The large container must be able to contain a soil specimen of approximately 200 g. The container should be treated to prevent rusting. The

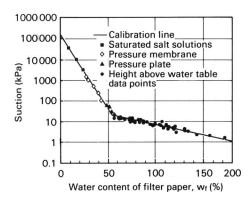

**Figure 4.25** A typical calibration curve showing measured filter paper water contents for applied suctions (from McQueen and Miller, 1968).

large container has an air-tight lid, and is used to equilibrate the soil specimen and the filter paper for a period of several days. The small container, with a volume of approximately 60 cm$^3$, is used to contain the filter paper during oven drying for its water content measurement. The small container should be as light as possible, considering the small mass of the filter paper.

An insulated box can be used to store the large containers with the soil specimens and filter papers during the equilibration period. The box should be maintained at a constant temperature within ±1°C. An accurate balance with a minimum capacity of 20 g and a readability of 0.0001 g should be used when weighing the filter paper during its water content measurement.

Some researchers pretreat the filter papers prior to its use in order to prevent fungal and bacterial growth during the equilibration period (Fawcett and Collis-George, 1967; and McQueen and Miller, 1968). Solutions of 3% Pentachlorophenol, $C_6Cl_5OH$, or 0.005% HgCl have been used to pretreat the filter papers. However, recent studies do not indicate any difference in the results obtained from the pretreated and untreated filter papers (Hamblin, 1984; Chandler and Gutierrez, 1986).

The most common practice is to have the filter paper initially dry, and then allow it to adsorb water from the soil specimen during equilibration. All calibration curves appears to have been established using initially dry filter papers. Therefore, if initially wet filter papers are to be used in the suction measurements, it may be necessary to establish new calibration curves using initially wet filter papers. There appears to be some hysteresis in the water content versus suction relationship for filter paper upon wetting and drying (Lykov, 1961).

The filter paper is initially oven dried for several hours. The dry filter paper is then cooled and stored in a desiccant container. Meanwhile, a soil specimen is placed in a large container. The soil specimen should almost fill the container (Fig. 4.24) in order to reduce the equilibration time. The "noncontact" procedure can be used by placing two dry filter papers on a perforated brass disk that is seated on top of the soil specimen, as shown in Fig. 4.24. The "contact" procedure can be used by placing three stacked filter papers in contact with the soil specimen, as illustrated in Fig. 4.24. For the contact procedure, the center filter paper is generally used for the suction measurement, while the outer filter papers are primarily used to protect the center paper from soil contamination.

Once the filter paper and the soil specimen are in the large container, the container is sealed with plastic electrical tape. The sealed container is then stored in the insulated box for equilibration. It appears that the ambient temperature does not affect the filter paper results provided the temperature variations during equilibration are minimized (Al-Khafaf and Hanks, 1974). Suctions in the filter paper should be allowed to equilibrate for a minimum period of seven days. Figure 4.26 illustrates the increasing water content of initially dry filter paper as equilibration occurs with the soil specimen. The results indicate that seven days is sufficient for the equilibration purpose.

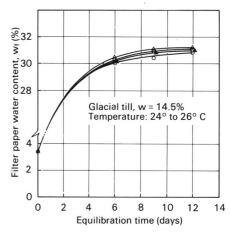

**Figure 4.26** Increasing water content of the initially dry filter paper during the equilibration period (from Tang, 1978).

At the end of the equilibration period, the filter papers are removed from the large container using a pair of tweezers, and their water contents are determined using the small metal containers. The water contents of both filter papers used in the "noncontact" measurement can be measured independently. In the "contact" measurement, the water content of the center filter paper is of primary importance. The other filter papers are primarily for protective purposes.

Extreme care must be exercised when measuring the small masses associated with the filter papers. The filter paper should be transferred from the large container to the small container within a short period of time (e.g., 3-5 s). This short period of transferring time will minimize water loss or gain between the filter paper and the surrounding atmosphere. The small container containing the filter paper must be closed and weighed immediately in order to determine the mass of the filter paper and the adsorbed water.

The container along with the filter paper is then placed in an oven at a temperature of 110 ± 5°C. In the oven, the lid of the container should be removed to allow water to escape from the filter paper. Having removed all of the water from the filter paper, the container, along with the dry filter paper, are weighed with the lid in place in order to determine the dry mass of the filter paper. The difference between the dry mass and the wet mass of the filter paper is used to compute the equilibrium water content of the filter paper.

The equilibrium suction is obtained from the calibration curve (Fig. 4.27) by using the measured equilibrium water content of the filter paper. The equilibrium suction is assumed to be equal to the suction in the soil specimen. Using the "noncontact" procedure, the suction values deter-

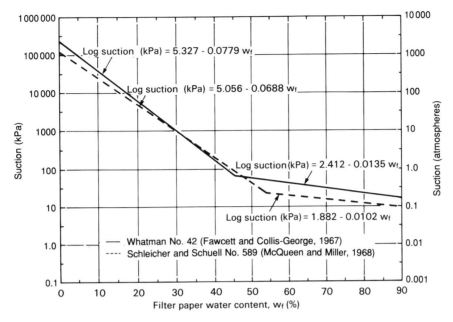

**Figure 4.27** Calibration curves for two types of filter paper.

mined from the two filter papers can be averaged when determining the soil suction, provided the two filter papers give similar water contents.

The calibration curve for a specific filter paper can be established by measuring the water content of filter paper in equilibrium with a salt solution having a known osmotic suction. In principle, the filter paper calibration is similar to the calibration of a psychrometer. The filter paper should be suspended above at least 50 cm$^3$ of a salt solution. The procedure for ensuring equilibration and measuring the water content is the same as those used during the measurements of soil suction. Various filter paper water contents can then be plotted against the differing osmotic suctions to give the calibration curve.

The calibration curve for filter papers always exhibits bi-linearity, as shown in Fig. 4.27. The lower part of the curve represents the high range of filter paper water contents where the water is believed to be held by the influence of capillary forces. On the other hand, the upper part of the calibration curve represents lower water contents where the water is believed to be held in an adsorbed water film within the filter paper (Miller and McQueen, 1978).

It should be emphasized that the filter paper technique is highly user-dependent, and great care must be taken when measuring the water content of the filter paper. The balance must be able to weigh to the nearest 0.0001 g. Each dry filter paper has a mass of about 0.52 g, and at a water content of 30%, the mass of water in the filter paper is about 0.16 g.

### The Use of the Filter Paper Method in Practice

A question of immediate concern to the practicing engineer is, "What is the accuracy of the suction measurement when using the filter paper method?" The question is still being answered, but the following typical results are presented in order to provide some indication of the accuracy of the filter paper method.

A comparison between the results of suction measurements using the filter paper method and psychrometers is shown in Fig. 4.21. The results from the noncontact filter paper agree fairly closely with the psychrometer results, indicating that total suction was being measured. However, the contact filter papers did not exhibit as consistent results with respect to depth. This is believed to be due to poor contact between the filter paper and the soil specimen, which resulted in the total suction being measured in many instances, instead of the matric suction over the depth range of 0–5 m (Fig. 4.21).

It appears to be difficult to ensure good contact between the soil specimen and the filter paper. For this reason, total suction will generally be measured when using the filter paper method. Figure 4.28 demonstrates a close agreement between total suction measurements obtained when using the filter paper method and thermocouple psychrometers. Total suction profiles in a montmorillonite clay in Texas (Fig. 4.29) have been predicted using the filter paper method (McKeen, 1981). The results appear to agree closely with the psychrometer measurements.

Figure 4.30 shows the results of filter paper measurements of total suction on a highly plastic clay from Eston, Canada (Ching and Fredlund, 1984). While augering several boreholes, water content samples were taken at about 0.3 m intervals. Three filter papers were included with each water content samples. These were allowed to equilibrate for one week. The measured suction on the highly swelling clay ranged from 2000 to 6000 kPa. Although there is no

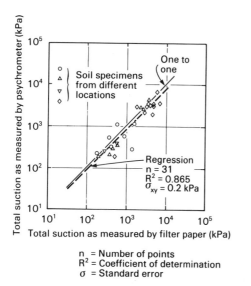

**Figure 4.28** Comparisons of total suction measurements obtained using the filter paper method and psychrometers (from McKeen, 1981).

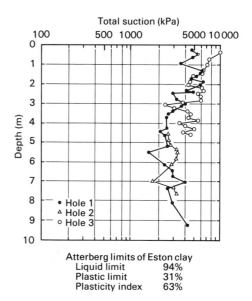

**Figure 4.30** Total suction profiles for Eston clay using the filter paper method.

direct confirmation of these measurements, the results appear to be reasonable for this deposit.

Another total suction profile (Fig. 4.31) was obtained from an excavation in shale using the filter paper method (McKeen, 1985). The drying effects near the surface of the excavation (i.e., down to a depth of 0.6 m) result in high soil suctions [Fig. 4.31(a)]. Correspondingly, this portion of the profile also has a low water content [Fig. 4.31(b)] in the soil. The filter paper method has also been used to estimate the *in situ* stress state of London clay with reasonable success (Chandler and Gutierrez, 1986).

It may also be possible to use the filter paper technique for *in situ* measurements of (total) suction (Fredlund, 1989). A proposed scheme for measuring suctions in subgrade soils is shown in Fig. 4.32. The filter papers would be left in place for about one week, and then removed for measurement of their water contents. New filter papers could then be installed and allowed to equalize for another week. Although this scheme has not been used to date, it appears to have possibilities as a low-cost, approximate technique to estimate total suction.

The filter paper method appears to have a wide range of measuring capability corresponding to soil suctions from a few kilopascals to several hundred thousand kilopascals (Fawcett and Collis-George, 1967; McQueen and Miller, 1968). However, the measurements must be performed with great care. In addition, only the "noncontact" filter paper procedure can be assured of measuring total suction. The "contact" filter paper procedure may measure either the total or the matric suction, depending on the degree of contact between the soil and the filter paper.

## 4.4 MEASUREMENTS OF MATRIC SUCTION

Matric suction can be measured either in a direct or indirect manner. The negative pore-water pressure is measured using direct methods. The pore-air pressure, which is generally atmospheric in the field, minus the negative pore-water pressure gives the matric suction.

High air entry ceramic disks are used for *direct* measurements of negative pore-water pressures. Therefore, the properties of high air entry ceramic disks are presented prior to describing different ways to perform a direct measurement.

Several types of porous sensors are used for performing *indirect* measurements of matric suction. The electrical and thermal properties of a standard ceramic are a function of

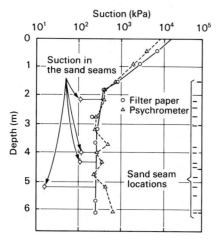

**Figure 4.29** Total suction profiles as determined using the filter paper method and thermocouple psychrometers (from McKeen, 1981).

## 4.4 MEASUREMENTS OF MATRIC SUCTION

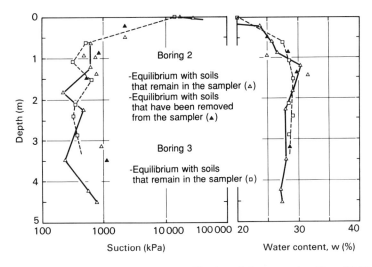

**Figure 4.31** Total suction and water content profiles in a shale excavation (from McKeen, 1985).

its water content, which in turn is a function of the matric suction. A measurement of the electrical or thermal properties of the sensor indicates the matric suction both in the sensor and in the surrounding soil. Indirect measurements of matric suction based on the thermal properties of the sensor are described in this chapter.

### 4.4.1 High Air Entry Disks

A high air entry disk has small pores of relatively uniform size. The disk acts as a membrane between air and water (Fig. 4.33). The disk is generally ceramic,[1] being made of sintered kaolin. Once the disk is saturated with water, air cannot pass through the disk due to the ability of the contractile skin to resist the flow of air.

The ability of the ceramic disk to withstand the flow of air results from the surface tension, $T_s$, developed by the contractile skin. The contractile skin acts like a thin membrane joining the small pores of radius, $R_s$, on the surface of the ceramic disk. The difference between the air pressure above the contractile skin and the water pressure below the contractile skin is defined as matric suction. The maximum matric suction that can be maintained across the surface of the disk is called its air entry value, $(u_a - u_w)_d$. The air entry value of the disk can be illustrated using Kelvin's equation:

$$(u_a - u_w)_d = \frac{2T_s}{R_s} \quad (4.10)$$

where

- $(u_a - u_w)_d$ = air entry value of the high air entry disk
- $T_s$ = surface tension of the contractile skin or the air-water interface (e.g., $T_s$ = 72.75 mN/m at 20°C)
- $R_s$ = radius of curvature of the contractile skin or the radius of the maximum pore size.

Surface tension, $T_s$, changes slightly with temperature. The air entry value of a disk is largely controlled by the

---

[1]Manufactured by Soilmoisture Equipment Corporation, Santa Barbara, CA.

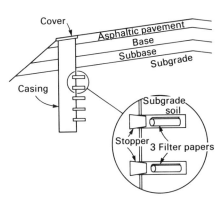

**Figure 4.32** Scheme for using filter papers to measure total suction.

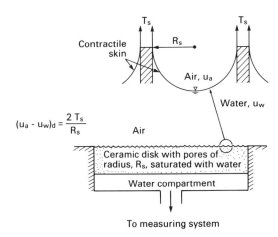

**Figure 4.33** Operating principle of a high air entry disk as described by Kelvin's capillary model.

**Table 4.5 High Air Entry Disks Used at Imperial College (From Blight, 1966)**

| Type of Disks | Porosity, $n$ (%) | Coefficient of Permeability with Respect to Water, $k_d$ (m/s) | Air Entry Value, $(u_a - u_w)_d$ (kPa) |
|---|---|---|---|
| Doulton Grade P6A | 23 | $2.1 \times 10^{-9}$ | 152 |
| Aerox "Celloton" Grade VI | 46 | $2.9 \times 10^{-8}$ | 214 |
| Kaolin-consolidated from a slurry and fired | 45 | $6.2 \times 10^{-10}$ | 317 |
| Kaolin-dust pressed and fired | 39 | $4.5 \times 10^{-10}$ | 524 |

radius of curvature, $R_s$, of the largest pore in the disk. The size of the pores is controlled by the preparation and sintering process used to manufacture the ceramic disk. The smaller the pore size in a disk, the greater will be its air entry value [Eq. (4.10)]. The properties of several types of high air entry disks used for unsaturated soils research at Imperial College, London, are listed in Table 4.5.

The ability of the high air entry disk to withstand a difference between air and water pressures makes the disk suitable for the direct measurement of negative pore–water pressures in an unsaturated soil. The disk is used as an interface between the unsaturated soil and the pore–water pressure measuring system. Water in the disk acts as a link between the pore–water in the soil and the water in the measuring system. At the same time, air cannot pass through the high air entry disk into the measuring system.

The separation of the air and water across a high air entry disk can be achieved only as long as the matric suction of the soil does not exceed the air entry value of the disk. Once the air entry value of the disk is exceeded, air will pass through the disk and enter the measuring system. The presence of air in the measuring system causes erroneous measurements of the pore–water pressure in a closed system.

Figure 4.34 shows the air passage characteristics of three disks mentioned in Table 4.5. The plots indicate the air entry value or the maximum matric suction sustainable across the disk.

The properties of several high air entry disks manufactured by Soilmoisture Equipment Corporation are tabulated in Table 4.6. The disks are identified according to their air entry values, which are expressed in bars (i.e., one bar is equal to 100 kPa). The water coefficient of permeability of a disk was measured by mounting the disk in a triaxial apparatus and placing water above the disk. An air pressure can then be applied to the water, producing a gradient across the high air entry disk. The volume of water flowing through the disk is measured using a water volume change indicator. Details on the equipment are presented in Chapter 10.

The flow of water through a high air entry disk is plotted against elapsed time, as shown in Fig. 4.35. The plot shows a straight line indicating steady-state seepage through the disk. The volume of water divided by the cross-sectional area of the disk and the elapsed time gives the coefficient of permeability of the disk. In general, the coefficient of permeability of the disk decreases with an increasing air entry value.

The air entry value and the permeability of a high entry disk should be measured prior to its use in unsaturated soil testings. Figure 4.36 presents the air passage characteristics of high air entry disks from Soilmoisture Equipment Corporation. The measured air entry values appear to be higher than the nominal values specified by the manufacturer. The results of measurements of the air entry values and the coefficients of permeability for various high air entry disks are summarized in Table 4.7.

### 4.4.2 Direct Measurements

There are two devices commonly used for the direct measurement of negative pore–water pressures. These are the tensiometer and the axis-translation apparatus. Tensiometers utilize a high air entry ceramic cup as an interface between the measuring system and the negative pore–water pressure in the soil. Tensiometers can be used in the lab-

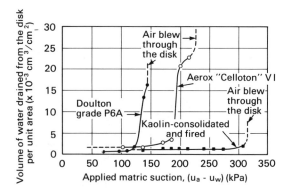

**Figure 4.34** Air passage characteristics of three high air entry disks (from Bishop and Henkel, 1962).

## 4.4 MEASUREMENTS OF MATRIC SUCTION

**Table 4.6 Properties of High Air Entry Disks Manufactured by Soilmoisture Equipment Corporation (Manufacturer's Results)**

| Type of Disks | Approximate Pore Diameter ($\times 10^{-3}$ mm) | Coefficient of Permeability with Respect to Water, $k_d$ (m/s) | Air Entry Value, $(u_a - u_w)_d$ (kPa) |
|---|---|---|---|
| 1/2 bar high flow | 6.0 | $3.11 \times 10^{-7}$ | 48–62 |
| 1 bar | 2.1 | $3.46 \times 10^{-9}$ | 138–207 |
| 1 bar high flow | 2.5 | $8.60 \times 10^{-8}$ | 131–193 |
| 2 bar | 1.2 | $1.73 \times 10^{-9}$ | 241–310 |
| 3 bar | 0.8 | $1.73 \times 10^{-9}$ | 317–483 |
| 5 bar | 0.5 | $1.21 \times 10^{-9}$ | >550 |
| 15 bar | 0.16 | $2.59 \times 10^{-11}$ | >1520 |

oratory and in the field. On the other hand, the axis-translation apparatus can be used only in the laboratory. The use of the axis-translation concept was described in Chapter 3.

### Tensiometers

A tensiometer measures the *negative pore-water pressure* in a soil. The tensiometer consists of a high air entry, porous ceramic cup connected to a pressure measuring device through a small bore tube. The tube is usually made from plastic due to its low heat conduction and noncorrosive nature. The tube and the cup are filled with deaired water. The cup can be inserted into a precored hole until there is good contact with the soil.

Once equilibrium is achieved between the soil and the measuring system, the water in the tensiometer will have the same negative pressure as the pore-water in the soil.

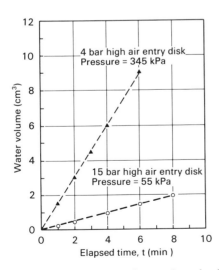

**Figure 4.35** Steady-state seepage of water through a high entry disk (from Fredlund, 1973).

The pore-water pressure that can be measured in a tensiometer is limited to approximately negative 90 kPa due to the possibility of cavitation of the water in the tensiometer. The measured negative pore-water pressure is numerically equal to the matric suction when the pore-air pressure is atmospheric (i.e., $u_a$ = zero gauge pressure). When the pore-air pressure is greater than atmospheric pressure (i.e., during axis translation), the tensiometer reading can be added to the ambient pore-air pressure reading to give the matric suction of the soil. The measured matric suction must not exceed the air entry value of the ceramic cup. The osmotic component of soil suction is not measured with tensiometers since soluble salts are free to move through the porous cup.

There are several types of tensiometers available from Soilmoisture Equipment Corporation. Figure 4.37 shows a regular tensiometer with a Bourdon-vacuum gauge to measure the negative pore-water pressure. The negative pore-water pressure in the tensiometer tube can also be measured using a water-mercury manometer or an electrical pressure transducer, as indicated in Fig. 4.38. Cassel and Klute (1986) discussed the sensitivity of various measuring devices on the response time of a tensiometer. In general, an increase in the gauge sensitivity will decrease the response time of the tensiometer. The increased gauge sensitivity also results in less water movement between the soil and the tensiometer, and subsequently a more accurate measurement of suction. A higher permeability of the ceramic cup will also result in a lower response time for the tensiometer.

The tensiometer tube in Fig. 4.37 has a diameter of approximately 20 mm and various lengths up to 1.5 m. In other words, tensiometer cups can be installed in the field to a depth of 1.5 m below the ground surface. However, the negative water pressure recorded at the ground surface must be corrected for the elevation head corresponding to the water column in the tensiometer. The longer the ten-

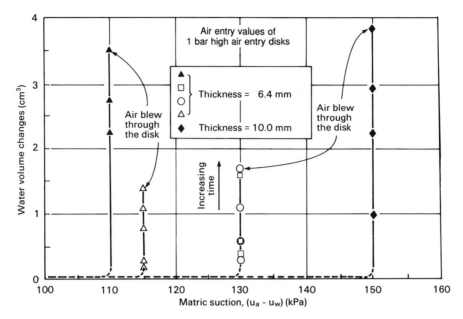

**Figure 4.36** Air passage characteristics of one bar, high air entry disks (from Rahardjo, 1990).

siometer tube, the greater will be the correction. This correction results in a more negative water pressure being measured than that recorded by the measuring device. A length of 1.5 m corresponds to a pressure correction of 15.2 kPa.

*Servicing the Tensiometer Prior to Installation*

The service cap at the top of the tensiometer tube (Fig. 4.37) is used to facilitate the filling of the tube with deaerated water, the sealing of the tube during measurements, and the servicing of the tensiometer. The tensiometer must be serviced properly prior to its use in order to obtain reliable results. Details on the preparation, installation, and usage of a tensiometer are presented by Cassel and Klute (1986). During preparation for installation, the ceramic cup should be checked for signs of plugging; air bubbles should be removed from the tensiometer, and the response time of the tensiometer should be checked. The ceramic cup can be checked by placing the empty tensiometer upright in a pail of water and allowing the cup to soak in the water overnight. An unplugged cup will allow water to fill the tensiometer tube.

The removal of air bubbles is performed by applying a

**Table 4.7 Permeability and Air Entry Value Measurements on High Air Entry Disks from Soilmoisture Equipment Corporation (from Fredlund, 1973; Rahardjo, 1990)**

| Type of Disks | Diameter of the Disk (mm) | Thickness of the disk (mm) | Air Entry Value of the Disk, $(u_a - u_w)_d$ (m/s) | Coefficient of Permeability of the Disk, $k_d$ (m/s) |
|---|---|---|---|---|
| 1 bar | 19.0 | 6.4 | 115 | $5.12 \times 10^{-8}$ |
| high flow | 19.0 | 6.4 | 130 | $3.92 \times 10^{-8}$ |
|  | 19.0 | 6.4 | 110 | $3.98 \times 10^{-8}$ |
|  | 19.0 | 6.4 | 130 | $5.09 \times 10^{-8}$ |
|  | 19.0 | 6.4 | 150 | $5.60 \times 10^{-8}$ |
|  | 101.6 | 10.0 | >200 | $4.20 \times 10^{-8}$ |
| 5 bar | 56.8 | 6.2 |  | $1.30 \times 10^{-9}$ |
| 15 bar | 56.8 | 3.1 |  | $8.41 \times 10^{-9}$ |
|  | 57.0 | 3.1 |  | $6.82 \times 10^{-10}$ |

4.4 MEASUREMENTS OF MATRIC SUCTION    85

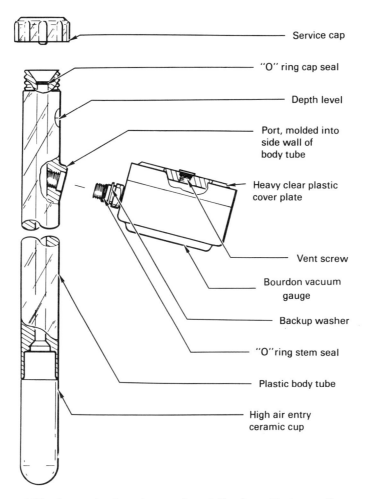

**Figure 4.37** Conventional tensiometer from Soilmoisture Equipment Corporation.

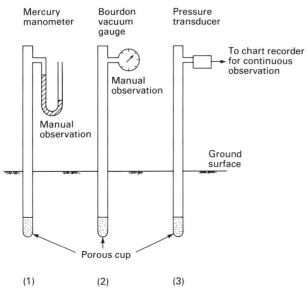

**Figure 4.38** Several measuring systems for a tensiometer (from Morrison, 1983).

vacuum of approximately 80 kPa to the top of the tensiometer tube for a period of 30–60 s. The vacuum can be applied using a hand-held vacuum pump, as shown in Fig. 4.39. This process will remove air bubbles from the ceramic cup, the Bourdon gauge, and from the imperfections on the wall of the tube. Having released the applied vacuum, deaerated water is added to refill the tube and the service cap is tightened in place. The water in the tube is then subjected to a negative pressure of approximately 80 kPa by allowing water to evaporate from the ceramic cup. Under this negative pressure, air bubbles may reappear in the tube, and the above procedure should be repeated until the tube is essentially free of air bubbles. It is important to have an air-free tensiometer tube in order to ensure correct readings and rapid responses.

The response time of a tensiometer can be checked by developing a negative water pressure of approximately 80 kPa by evaporation from the ceramic cup and then immersing the ceramic cup into water. The negative water pressure in the tensiometer should increase towards atmo-

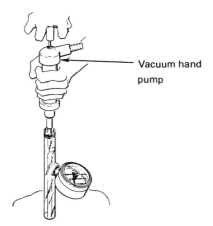

**Figure 4.39** Deairing the tensiometer using a hand-held vacuum pump (from Soilmoisture Equipment Corporation).

spheric pressure within 5 min after the immersion in water, as illustrated in Fig. 4.40. A sluggish response may indicate a plugged porous cup, the presence of entrapped air in the system, or a faulty gauge that requires rezeroing.

The ceramic cup must be soaked in water prior to its installation in order to avoid desaturation due to evaporation from the cup. The prepared tensiometer can then be installed in a predrilled hole in the field or in a soil specimen (e.g., a compacted specimen) in the laboratory. It is important to ensure good contact between the ceramic cup and the soil in order to establish continuity between the pore–water in the soil and the water in the tensiometer tube.

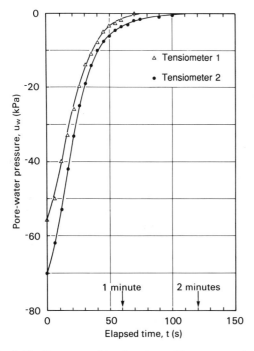

**Figure 4.40** Response of tensiometers with respect to time during immersion in water (from Tadepalli, 1990).

### Servicing the Tensiometer after Installation

After installation of the tensiometer, air bubbles may develop within the tensiometer due to several possible reasons. Dissolved air may come out of the solution as the water pressure decreases to a negative value. Air in the soil may diffuse through the water in the ceramic cup and come out of the solution inside the tube. When the water pressure approaches the vapor pressure of water at the ambient temperature, water molecules can move freely from the liquid to the vapor form (i.e., cavitation occurs). In other words, the measured absolute pressure is equal to the saturated water vapor pressure, $\bar{u}_{v0}$.

The minimum vacuum gauge pressure that can be theoretically developed in the tensiometer is ($-101.3$ kPa + $\bar{u}_{v0}$). In practice, however, the minimum gauge pressure that can be measured in a tensiometer is approximately $-90$ kPa due to the rapid accumulation of air bubbles as the saturated water vapor pressure is approached. In addition, pore–air passes through the high air entry cup if the matric suction of the soil exceeds the air entry value of the ceramic cup. The use of a ceramic cup with an air entry value greater than 100 kPa will not improve the measuring range of tensiometers since water in the tube will always cavitate when water pressure approaches approximately $-90$ kPa. It is necessary to check the tensiometer regularly in order to observe the development of air bubbles in the tube. This is particularly crucial when performing measurements on soils with high matric suctions. If air bubbles are allowed to accumulate in the tube, the pressure being read on the gauge will slowly increase towards zero (i.e., atmospheric pressure).

Air bubbles can be removed from the tensiometer using the vacuum pump (Fig. 4.39) and a procedure similar to that followed during the servicing of the tensiometer prior to installation. Having removed the air bubbles, deaerated water is added to refill the tube and the service cap is tightened in place. The tensiometer reading is then allowed to equilibrate.

### Jet Fill Tensiometers

A jet fill-type tensiometer is shown in Fig. 4.41. The jet fill type is an improved model of the regular tensiometer. A water reservoir is provided at the top of the tensiometer tube for the purpose of removing the air bubbles. The jet fill mechanism is similar to the action of a vacuum pump. The accumulated air bubbles are removed by pressing the button at the top to activate the jet fill action. The jet fill action causes water to be injected from the water reservoir to the tube of the tensiometer, and air bubbles move upward to the reservoir.

### Small tip Tensiometer

A small tip tensiometer with a flexible coaxial tubing is shown in Fig. 4.42. The tensiometer is prepared for installation using a similar procedure to that described for the

4.4 MEASUREMENTS OF MATRIC SUCTION    87

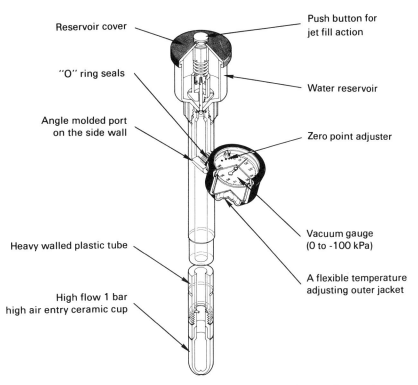

**Figure 4.41** Jet fill tensiometer from Soilmoisture Equipment Corporation.

regular tensiometer tube. A vacuum pump can be initially used to remove air bubbles from the top of the tensiometer tube. Subsequent removal of air bubbles can be performed by flushing through the coaxial tube.

Flushing is conducted by circulating water from the top of the tube and opening the water vent screw to the atmosphere. Water containing air bubbles will be forced into the inner nylon tube and released to the atmosphere through the opened water vent. Usually, this procedure is required on a daily basis. It is sometimes difficult to get an accurate

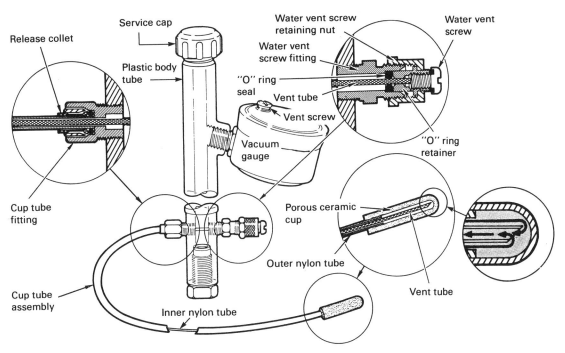

**Figure 4.42** Small tip tensiometer with flexible coaxial tubing (from Soilmoisture Equipment Corporation).

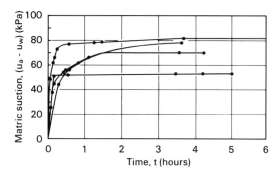

**Figure 4.43** Typical responses of the small tip tensiometer in decomposed volcanics (from Sweeney, 1982).

suction measurement because water moves from the tensiometer into the soil each time the tensiometer is serviced.

The small tip with the flexible tube allows the installation of the tensiometer into a relatively small soil specimen during laboratory experiments. As an example, the small tip tensiometer has been installed in a consolidation specimen to measure changes in matric suction during the collapse of a compacted silt (Tadepalli, 1990).

Figure 4.43 illustrates typical responses of the small tip tensiometer inserted into a decomposed volcanic soil. The response is plotted in terms of matric suctions by referencing measurements to atmospheric air pressures. Sweeney (1982) found that this type of tensiometer could only maintain matric suction equilibrium for about one or two days before the suction readings began to drop. However, the time that equilibrium can be maintained is a function of the matric suction value being measured.

### Quick Draw Tensiometers

Figure 4.44 shows a Quick Draw tensiometer equipped with a coring tool and a carrying case. The Quick Draw tensiometer has proven to be a particularly useful portable tensiometer to rapidly measure negative pore-water pressures. The water in the tensiometer is subjected to tension for only a short period of time during each measurement. Therefore, air diffusion through the ceramic cup with time is minimized.

The Quick Draw tensiometer can repeatedly measure pore-water pressures approaching $-1$ atm when it has been properly serviced. When it is not in use, the probe is maintained saturated in a carrying case which has water-saturated cotton surrounding the ceramic cup. The rapid response of the Quick Draw tensiometer is illustrated in Fig. 4.45.

### Tensiometer Performance for Field Measurements

Tensiometers have been used to measure negative pore-water pressures for numerous geotechnical engineering applications. One example is a cut slope consisting of 5–6 m of colluvium overlying a deep weathered granite (Fig. 4.46). Two observation shafts, made of concrete caisson rings, were constructed along the slope (i.e., shafts $A$ and $B$ in Fig. 4.46). Figure 4.47 illustrates the construction of the shaft, along with four circular openings along each caisson ring. The openings functioned as access holes for installing tensiometers into the soil at various locations along the shaft. The shaft was equipped with wood platforms at various elevations, a ladder, and lighting facilities, as shown in Fig. 4.47.

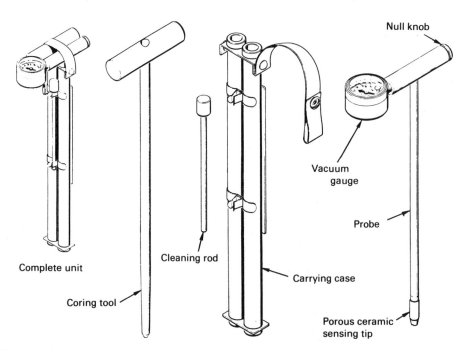

**Figure 4.44** Quick Draw tensiometer with coring tool and carrying case (from Soilmoisture Equipment Corporation).

4.4 MEASUREMENTS OF MATRIC SUCTION   89

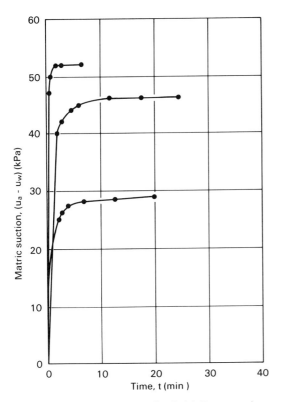

**Figure 4.45** Typical responses of a Quick Draw tensiometer in decomposed volcanics (from Sweeney, 1982).

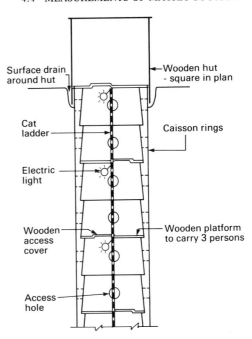

**Figure 4.47** Observation shaft for installing tensiometers in the field (from Sweeney, 1982).

Measurements of negative pore-water pressures along shafts A and B are presented in Figs. 4.48 and 4.49, respectively. The results are plotted in terms of matric suction and hydraulic head profiles throughout the depth of the shaft. The results indicate the fluctuation in matric suction values with respect to the time of the year. The largest variations occur near the ground surface (Fig. 4.48). The hydraulic heads along the depth of the shaft are plotted by adding the negative pore-water pressure head to the elevation head. The hydraulic head plots indicate a net downward water flow towards the water table through the unsaturated zone.

Quick Draw tensiometers have been used in the measurement of negative pore-water pressures along the sidewalls of two trenches excavated perpendicular to a railway embankment in western Canada. The soil consisted predominantly of an unsaturated silt. The results are plotted in Fig. 4.50 as contours of matric suction across the embankment. It appears that the matric suctions decrease to-

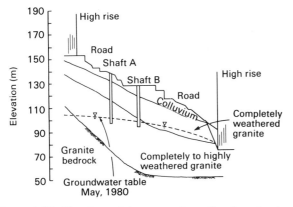

**Figure 4.46** Cut slope of decomposed granite where tensiometers were used to measure negative pore-water pressures (from Sweeney, 1982).

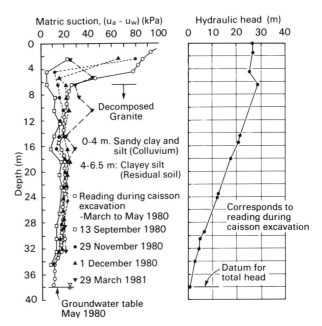

**Figure 4.48** Matric suction profile along shaft A (from Sweeney, 1982).

90  4  MEASUREMENTS OF SOIL SUCTION

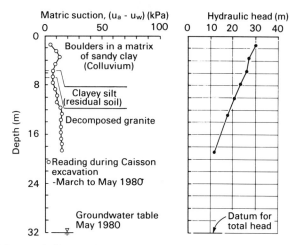

**Figure 4.49** Matric suction profile along shaft B (from Sweeney, 1982).

wards the ground surface due to the influence of the microclimatic conditions (i.e., excessive rainfall).

A scanning valve tensiometer system has also been used for recording several tensiometer readings through one central pressure transducer (Fig. 4.51). The hydraulic scanning valve rotates automatically from one tensiometer to another in order for the transducer to record the readings.

*Osmotic Tensiometers*

Attempts have been made to overcome the problem of water cavitation in a conventional tensiometer through the use of

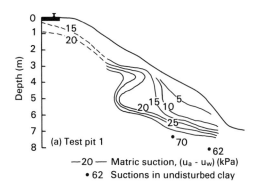
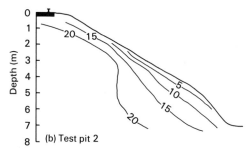

**Figure 4.50** Matric suction contours along the sidewalls of two test trenches in a silt embankment (from Krahn et al., 1989).

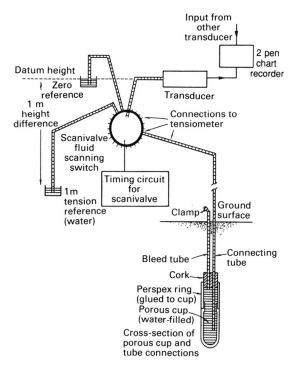

**Figure 4.51** A scanning valve tensiometer system (from Anderson and Burt, 1977).

an osmotic tensiometer (Peck and Rabbidge, 1966). The osmotic tensiometer uses an aqueous solution that has been internally prestressed to produce a positive gauge pressure. The positive pressure of the aqueous solution is then reduced by the negative pore-water pressure in an unsaturated soil when the osmotic tensiometer comes to equilibrium. The reduction of the pressure inside the osmotic tensiometer is measured using a pressure transducer to give the measured negative pore-water pressure in the soil.

Using the above concept, highly negative pore-water pressures can be measured without causing the tensiometer solution to go into tension. However, major difficulties associated with osmotic tensiometers have restricted their use (Peck and Rabbidge, 1969; Bocking and Fredlund, 1979).

The configuration and components of an osmotic tensiometer are shown in Fig. 4.52. The device consists of a closed chamber containing an aqueous solution of polyethylene glycol (PEG) with a molecular mass of 20 000. A filling port is provided on the sidewall for inserting the solution into the chamber. The upper end of the chamber is sealed with a pressure transducer for measuring the internal pressure of the solution. The lower end of the chamber is sealed with a semi-permeable membrane attached to a 15 bar, high air entry ceramic disk.

The initial internal chamber pressure of the solution is developed through an osmotic process. This chamber pressure is highly positive, and can be in the range of 1400–2000 kPa. This pressure is generated when the sensor is placed in water, and is regarded as the reference pressure

4.4 MEASUREMENTS OF MATRIC SUCTION   91

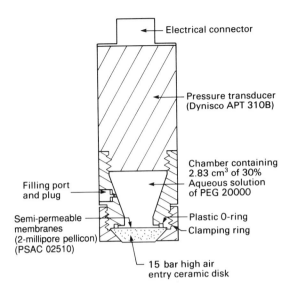

**Figure 4.52** Osmotic tensiometer constructed at the University of Saskatchewan (from Bocking and Fredlund, 1979).

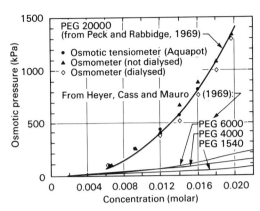

**Figure 4.53** Osmotic pressure of polyethylene glycol (PEG) solutions (from Bocking and Fredlund, 1979).

of the osmotic tensiometer. The reference pressure corresponds to equilibrium conditions at a standard temperature of 20°C with the device immersed in distilled water at zero gauge pressure. Therefore, the osmotic process is generated by immersing the osmotic tensiometer into water.

Water flows into the chamber as a result of the large difference between the solution concentration across the semi-permeable membrane. On the other hand, the solution molecules cannot pass through the semi-permeable membrane. However, water flows into the closed chamber, causing the solution pressure to increase as recorded by the pressure transducer. The flow of water will diminish as the internal pressure in the chamber increases. Eventually, equilibrium is reached between the internal pressure of the solution and the osmotic suction across the semi-permeable membrane. The reference pressure under equilibrium conditions is a function of the concentration and the molecular mass of the solution, as illustrated in Fig. 4.53 for the PEG (polyethylene glycol) solution.

The osmotic tensiometer can now be placed in contact with an unsaturated soil where the pore-water pressures are negative. During the equilibration process, a small amount of water will flow out of the chamber through the semi-permeable membrane. The solution pressure is reduced by an amount equal to the negative pore-water pressure in the soil. The solution pressure reduction is measured on the pressure transducer. Pore-air cannot enter the chamber because of the high air entry ceramic disk and the high positive pressure within the chamber.

The first major difficulty associated with the osmotic tensiometer is its inability to maintain a constant reference pressure with time. Figure 4.54 demonstrates the decreasing reference pressures of two osmotic tensiometers in equilibrium with distilled water at 20°C over a period of 260 days. There are several possible causes for the reduction in the reference pressure with time (Peck and Rabbidge, 1969). However, the most likely explanation for the pressure reduction is the minute leakage of the confined solute through the semi-permeable membrane. It was also found that the membrane is susceptible to physical deterioration with time (Bocking and Fredlund, 1979).

The second major difficulty associated with the osmotic tensiometer concerns changes in the internal reference pressure as a result of changes in the ambient temperature. Figure 4.55 shows the decreasing reference pressure of two osmotic tensiometers as the temperature increases. The pressures shown in Fig. 4.55 have been corrected for the pressure drift with time (Fig. 4.54). The effect of temperature on the reference pressure makes it extremely difficult to measure negative pore-water pressures other than in a controlled temperature environment. Even under controlled temperatures, there is significant drift of the reference pressure with time. To date, a solution has not been found to overcome these problems.

*Axis-Translation Technique*

Measurements of negative pore-water pressure can be made using the axis-translation technique. The measurement is

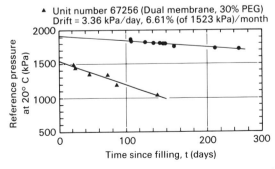

**Figure 4.54** Reference pressure drifts of two osmotic tensiometers with time (from Bocking and Fredlund, 1979).

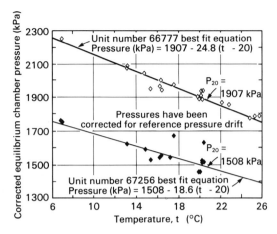

**Figure 4.55** Decreasing reference pressures of two osmotic tensiometers with increasing temperatures (from Bocking and Fredlund, 1979).

performed on either undisturbed or compacted specimens. This technique was originally proposed by Hilf (1956), as illustrated in Fig. 4.56.

An unsaturated soil specimen was placed in a closed pressure chamber. The pore-water pressure measuring probe consisted of a needle with a saturated high air entry ceramic tip. The probe was connected to a null-type pressure measuring system through a tube filled with deaired water, with a mercury plug in the middle. As soon as the probe was inserted into the specimen, the water in the tube tended to go into tension and the Bourdon gauge began registering a negative pressure. The tendency of the water in the measuring system to go further into tension was countered by increasing the air pressure in the chamber. Eventually, an equilibrium condition was achieved when the mercury plug (i.e., the null indicator) remained stationary. The difference between the air pressure in the chamber and the measured negative water pressure at equilibrium was taken to be the matric suction of the soil, $(u_a - u_w)$.

When the air pressure is atmospheric (i.e., $u_a = 0$), the matric suction value is numerically equal to the negative pore-water pressure. The axis-translation technique simply translates the origin of reference for the pore-water pressure from standard atmospheric conditions to the final air pressure in the chamber (i.e., axis translation; Hilf, 1956). As a result, the water pressure in the measuring system does not become highly negative, and the problem of cavitation is prevented.

Hilf (1956) demonstrated that the pore-water pressure increased by an amount equal to the increase in the ambient chamber air pressure. In other words, the soil matric suction remained constant when measured at various ambient air pressures. The condition of no flow maintained during the measurement of matric suction is the justification for the axis-translation technique.

The axis-translation technique has also been used by other researchers. Olson and Langfelder (1965), used the modified pressure plate apparatus shown in Fig. 4.57. The procedure used was as follows. A soil specimen is placed on top of a saturated high air entry disk in an air pressure chamber. The air entry value of the disk must be higher than the matric suction to be measured. A 1 kg mass is placed on top of the specimen to ensure a good contact between the soil specimen and the high air entry disk. The placement of the specimen onto the ceramic disk and the assemblage of the cell are performed as rapidly as possible (i.e., within approximately 30 s).

The water pressure in the compartment below the high air entry disk is maintained as close as possible to zero pressure by increasing the air pressure in the chamber. The pressure transducer connected to the water compartment is used as a null indicator.

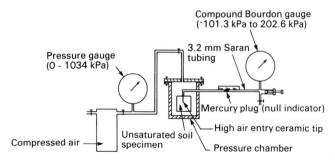

**Figure 4.56** Original setup for the null-type, axis-translation device for measuring negative pore-water pressures (from Hilf, 1956).

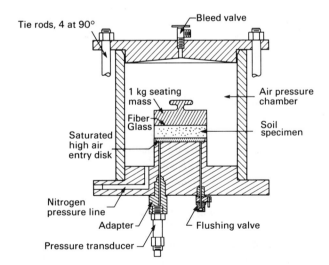

**Figure 4.57** Pressure plate apparatus for measuring negative pore-water pressures using the axis-translation technique (from Olson and Langfelder, 1965).

A similar pressure plate design was used in several research studies at the University of Saskatchewan. The device is shown in Fig. 4.58. It is important for the apparatus to have a system to flush air bubbles from below the high air entry disk in order to keep the compartment above the transducer saturated with water.

Pressure values are shown in Fig. 4.59 to illustrate how a highly negative pore-water pressure can be measured using this apparatus. Let us suppose that a soil specimen has an initial pore-water pressure of $-250$ kPa when placed onto the saturated, high air entry disk. The specimen will immediately tend to draw water up through the ceramic disk, causing the pressure transducer to commence registering a negative value. The cover must be quickly placed on top of the device, and the air pressure in the chamber is increased until there is no further tendency for the movement of water through the high air entry disk. At equilibrium, the chamber air pressure could be 255 kPa, while the water compartment would register 5 kPa. Therefore, the matric suction of the soil is 250 kPa.

Typical response time curves for the measurement of matric suction, using the axis-translation technique, are illustrated in Figs. 4.60 and 4.61 (Widger, 1976; Filson, 1980). The response curves generally exhibit an "S" shape, with a relatively fast equilibration time. Pressure response versus time is a function of the permeability characteristic of the high air entry disk and the soil. The results shown in Figs. 4.60 and 4.61 were obtained during matric suction measurements on a highly plastic clay (i.e., Regina clay).

Figures 4.62 and 4.63 present water content versus matric suction relationships for compacted specimens of Regina clay and glacial till, respectively. The matric suctions were measured using the axis-translation technique. Reasonably good agreement in the data has been obtained by

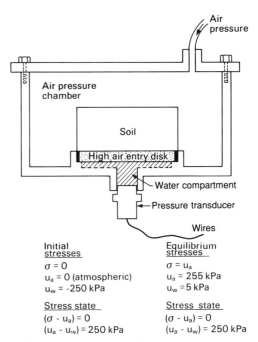

**Figure 4.59** Schematic showing the pressure changes associated with the measurement of matric suction using a null-type pressure plate apparatus (from Fredlund, 1989).

several researchers, indicating the reliability of this technique.

Similar water content versus matric suction relationships were obtained for several compacted soil types by Olson and Langfelder (1965); and Mou and Chu (1981). The results are shown in Figs. 4.64 and 4.65. Again, there is a distinct relationship between decreasing water contents and increasing matric suctions. There is, however, a difference in the water content versus matric suction relationships obtained from static and kneading compaction. This difference appears to be caused by different soil structures resulting from the different methods of compaction (Mou and Chu, 1981).

In general, the null-type axis-translation technique can be used to measure negative pore-water pressures in the laboratory with reasonable success and accuracy. High air entry disks with a maximum air entry value of up to 1500 kPa are commercially available (Table 4.6). Theoretical studies on the axis-translation technique suggested that the technique is best suited to soils with a continuous air phase (Bocking and Fredlund, 1980). The presence of occluded air bubbles in the soil specimen can result in an overestimation of the measured matric suction. In addition, air diffusing through the high air entry disk can cause an underestimating of the measured matric suction.

### 4.4.3 Indirect Measurements

The indirect measurement of soil matric suction can be made using a standard porous block as a measuring sensor.

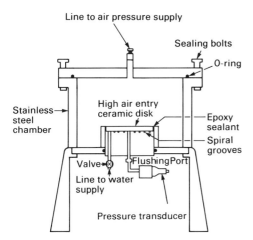

**Figure 4.58** Pressure plate apparatus for measuring negative pore-water pressures using the axis-translation technique (the design at the University of Saskatchewan).

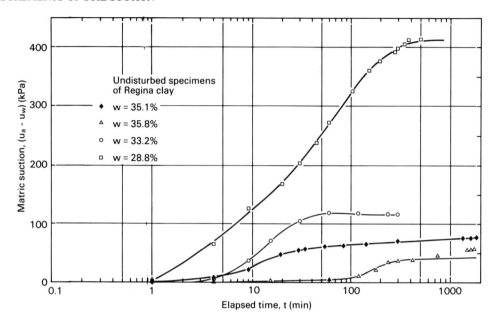

**Figure 4.60** Response versus time for matric suction measurements on Regina clay using the axis-translation technique (from Widger, 1976).

A wide range of porous materials has been examined for their soil–water characteristic relationship in order to select the most appropriate material for making the sensor. These materials include nylon, fiberglass, gypsum plaster, clay ceramics, sintered glass, and metal.

The porous block sensor must be brought into equilibrium with the matric suction in the soil. At equilibrium, the matric suction in the porous block and the soil are equal.

The matric suction is inferred from the water content of the porous block. The water content of the porous block can be determined by measuring the electrical or thermal properties of the porous block. These properties are a function of the water content, and can be established through calibrations. In the calibration process, the porous block is subjected to various matric suctions, and its electrical or thermal properties are measured. As a result, the measured

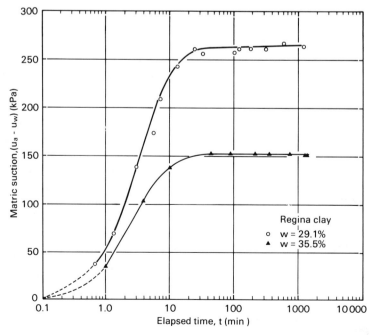

**Figure 4.61** Response versus time for matric suction measurements on Regina clay using the axis-translation technique (from Filson, 1980).

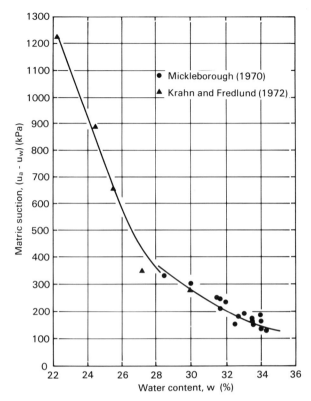

**Figure 4.62** Water content versus suction for specimens of compacted Regina clay.

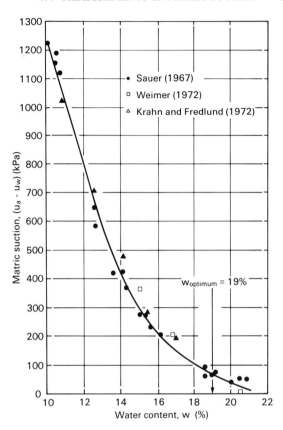

**Figure 4.63** Water content versus matric suction for specimens of compacted glacial till.

electrical or thermal property of the porous block in equilibrium with a soil can be used to determine the matric suction in the soil through use of the calibration curve.

The indirect measurements based on the electrical properties of a porous block have been found to be sensitive to the presence of dissolved salts in the pore–water (Richards, 1974). On the other hand, the indirect measurements based on the thermal properties of the porous block show little effect from the dissolved salts in the pore–water or variations in the ambient temperature. As a result, the thermal sensor is the most promising device for the indirect measurements of matric suction (Richards, 1974). Its working principle and application are explained in the following sections.

*Thermal Conductivity Sensors*

Thermal properties of a soil have been found to be indicative of the water content of a soil. Water is a better thermal conductor than air. The thermal conductivity of a soil increases with an increasing water content. This is particularly true where the change in water content is associated with a change in the degree of saturation of the soil.

Shaw and Baver (1939) developed a device consisting of a temperature sensor and heater which could be installed directly into the soil for thermal conductivity measurements. It was found that the presence of salts did not significantly affect the thermal conductivity of the soil. However, different soils required different calibrations in order to relate the thermal conductivity measurements to the water contents of the soil. Johnston (1942) suggested that the thermal conductivity sensor be enclosed in a porous medium that had a calibration curve. The porous cover could then be brought into equilibrium with the soil under consideration. Johnston (1942) used plaster of paris to encase the heating element.

In 1955, L.A. Richards patented an electrothermal element for measuring moisture in porous media (U.S. Patent 2 718 141). The element consisted of a resistance thermometer which was wrapped with a small heating coil. The electrothermal element was then mounted in a porous cup and sealed with ceramic cement. Richards proposed the use of a sandy silt material for the porous block. It was suggested that the porous cup should have an air entry value less than 10 kPa.

Bloodworth and Page (1957) studied three materials for use as a porous cup for the thermal conductivity sensors. Plaster of paris, fired clay, or ceramic and castone (i.e., a commercially available dental stone powder) were used in the study. The castone was found to be the best material for the porous cup.

Phene et al. (1971) developed a thermal conductivity sensor using a Germanium p–n diode as a temperature sen-

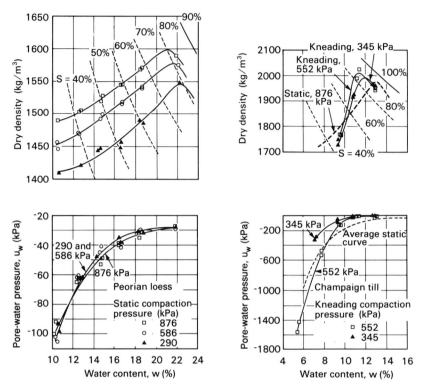

**Figure 4.64** Negative pore-water pressure measurements on compacted specimens using the axis-translation technique (from Olson and Langfelder, 1965).

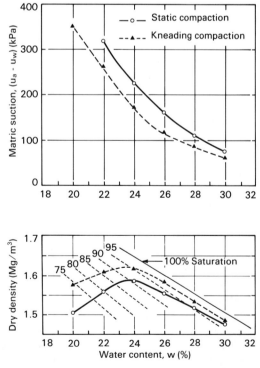

**Figure 4.65** Compaction curves and corresponding water content versus matric suction for an expansive clay in Texas (from Mou and Chu, 1981).

sor. The sensor was wrapped with 40-gauge Teflon-coated copper wire that served as the heating coil. The sensing unit was embedded in a porous block. The optimum dimensions of the porous block were calculated based on a theoretical analysis. The block must be large enough to contain the heat pulse (particularly for the saturated sensor) without interference from the surrounding soil. Also, it was found that the higher the ratio of the thermal conductivity to the diffusivity, the higher the precision with which the water content could be measured. The distribution of the pore sizes was also important.

Gypsum, ceramics, and mixtures of ceramics and castone were examined as potential porous block materials by Phene et al., (1971). It was found that the ceramic block exhibited a linear response and provided a stable solid matrix.

In the mid-1970's, Moisture Control System Inc. of Findlay, OH, manufactured the MCS 6000 thermal conductivity sensor. The sensor was built using the same design and construction procedures used by Phene et al., (1971). The manufactured sensors were subjected to a two-point calibration. The suggested calibration curves were assumed to be linear from zero suction to a suction of 300 kPa. Above 300 kPa, the calibration curves were empirically extrapolated. In the region above 300 kPa, the calibration curves became highly nonlinear and less accurate.

The MCS 6000 sensors have been used for matric suction measurements in the laboratory and in the field (Picornell et al., 1983; Lee and Fredlund, 1984). The sensors appeared to be quite suitable for field usage, being insensitive to temperature and salinity changes. Relatively accurate measurements of matric suction were obtained in the range of 0–300 kPa. Curtis and Johnston (1987) used the MCS 6000 sensors in a groundwater recharge study. The sensors were found to be quite responsive and sensitive. The results were in good agreement with piezometer and neutron probe data. However, Moisture Control System Inc. discontinued production in early 1980, and the MCS 6000 sensor is no longer commercially available.

In 1981, Agwatronics Inc. in Merced, CA, commenced production of the AGWA thermal conductivity sensors. The design of the sensor was changed from previous designs, but was based on the research by Phene et al., (1971). There were several difficulties associated with the AGWA sensor that resulted in their replacement by a new design, the AGWA-II sensor in 1984.

A thorough calibration study on the AGWA-II sensors was undertaken at the University of Saskatchewan, Canada (Wong et al., 1989; Fredlund and Wong, 1989). Several other difficulties were reported with the use of the AGWA-II sensors. These include the deterioration of the electronics and the porous block with time. The AGWA-II sensors have been used for laboratory and field measurements of matric suctions on several research studies (van der Raadt et al., 1987; Sattler and Fredlund, 1989; Rahardjo et al., 1989).

### Theory of Operation

A thermal conductivity sensor consists of a porous ceramic block containing a temperature sensing element and a miniature heater (Fig. 4.66). The thermal conductivity of the porous block varies in accordance with the water content of the block. The water content of the porous block is dependent upon the matric suctions applied to the block by the surrounding soil. Therefore, the thermal conductivity of the porous block can be calibrated with respect to an applied matric suction.

A calibrated sensor can then be used to measure the matric suction by placing the sensor in the soil and allowing it to come to equilibrium with the state of stress in the pore-water (i.e., the matric suction of the soil). Thermal conductivity measurements at equilibrium are related to the matric suction of the soil.

Thermal conductivity measurements are performed by measuring heat dissipation within the porous block. A controlled amount of heat is generated by the heater at the center of the block. A portion of the generated heat will be dissipated throughout the block. The amount of heat dissipation is controlled by the presence of water within the porous block. The *change* in the thermal conductivity of the sensor is directly related to the *change* in water content of the block. In other words, more heat will be dissipated as the water content in the block increases.

The undissipated heat will result in a temperature rise at the center of the block. The temperature rise is measured by the sensing element after a specified time interval, and its magnitude is inversely proportional to the water content of the porous block. The measured temperature rise is expressed in terms of a voltage output.

### Calibration of Sensors

AGWA-II sensors are usually subjected to a two-point calibration prior to shipment from the factory. One calibration reading is taken with the sensors placed in water (i.e., zero matric suction). A second calibration reading is taken with the sensors subjected to a suction of approximately 1 atm. This calibration procedure may be adequate for some applications. However, it has been suggested that a more rigorous calibration procedure is necessary when the sensors are used for geotechnical engineering applications (Fredlund and Wong, 1989).

A more thorough calibration of thermal conductivity sensors can be performed by applying a range of matric suction values to the sensors which are mounted in a soil. Readings of the change in voltage output is a measure of the thermal conductivity (or the water content) of the porous block under the applied matric suction. The matric suction can be applied to the sensor using a modified pressure plate apparatus (Wong et al., 1989; Fredlund and Wong, 1989).

The sensor is embedded in a soil which is placed on the pressure plate (Fig. 4.67). The soil on the pressure plate provides continuity between the water phase in the porous block and in the high air entry plate. In addition, the soil used in the calibration must be able to change its water content at a low matric suction (i.e., low air entry value), as shown in Fig. 4.68. The matric suction is applied by

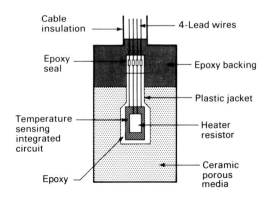

**Figure 4.66** A cross-sectional diagram of the AGWA-II thermal conductivity sensor (from Phene et al., 1971).

**98**  4  MEASUREMENTS OF SOIL SUCTION

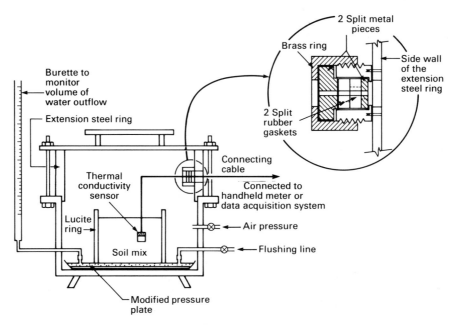

**Figure 4.67** Pressure plate calibration setup for thermal conductivity sensors (from Fredlund and Wong, 1989).

increasing the air pressure in the pressure plate apparatus, but maintaining the water pressure below the pressure plate at atmospheric conditions.

The change in voltage output from the sensor can be monitored periodically until matric suction equilibrium is achieved. The above procedure is repeated for various applied matric suctions in order to obtain a calibration curve. A number of thermal conductivity sensors can be calibrated simultaneously on the pressure plate. During calibration,

the pressure plate setup should be contained within a temperature-controlled box.

Figure 4.69 shows a typical response curve for the AGWA-II sensor resulting from the application of different air pressures during the calibration process. The curve indicates an increasing equalization time as the applied matric suctions increase. For the calibration soil indicated in Fig. 4.68, the sensor has an equalization time in the order of 50 h for an applied matric suction below 150 kPa. The

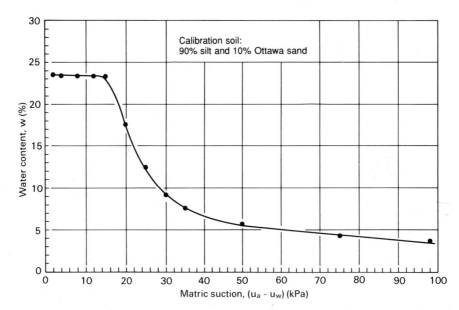

**Figure 4.68** Soil–water characteristic curve for the soil used in the calibration procedure (from Fredlund and Wong, 1989).

## 4.4 MEASUREMENTS OF MATRIC SUCTION

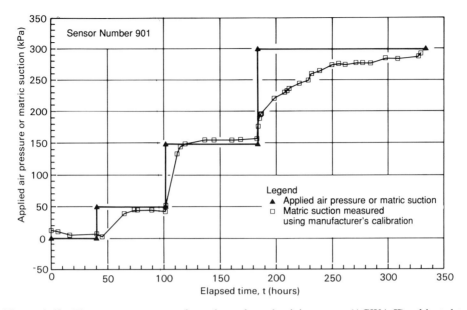

**Figure 4.69** Time response curves for a thermal conductivity sensor (AGWA-II) subjected to changes in applied air pressure (or matric suction).

equalization time for a sensor is affected by the permeability and thickness of the calibration soil. In addition, the permeability and the thickness of the high air entry disks also affect the equalization times.

More than 100 AGWA-II sensors have been calibrated and used at the University of Saskatchewan, Canada. Typical nonlinear calibration curves for the AGWA-II sensors are shown in Fig. 4.70. The nonlinear response of the sensors is likely related to the pore size distribution of the ceramic porous block. Similar nonlinearities were also observed on the calibration curves for the MCS 6000 sensor.

The nonlinear behavior of the AGWA-II sensors may be approximated by a bilinear curve, as illustrated in Fig. 4.70. The breaking points on the calibration curves are generally around 175 kPa. Relatively accurate measurements of matric suction can be made using the AGWA-II sensors, particularly within the range of 0–175 kPa. Matric suction measurements above 175 kPa correspond to the steeper portion of the calibration curve, which has a lower sensitivity to changes in thermal conductivity.

AGWA-II sensors have shown consistent, reproducible, and stable output readings with time (Fredlund and Wong, 1989). The sensors have been found to be responsive to both the wetting and drying processes. However, some failures have been experienced with the sensors, particularly when subjected to a positive water pressure. The failures are attributed to moisture coming into contact with the electronics sealed within the porous ceramic (Wong et al., 1989). Also, there have been continual problems with the porous blocks being too fragile. Therefore, the sensor must be handled with great care. Even so, there is a percentage of the sensors which crack or crumble during calibration or installation.

### Typical Results of Matric Suction Measurements

Laboratory and field measurements of matric suctions using the MCS 6000 and the AGWA-II thermal conductivity sensors have been made involving several types of soils. The soils have ranged from highly plastic clays to essentially nonplastic sands. The sensors have been installed either in an initially wet or an initially dry state. The results from the MCS 6000 sensors are presented first, followed by the results from the AGWA-II sensors.

### The MCS 6000 Sensors

Lee (1983) studied the performance of the MCS 6000 thermal conductivity sensor. The laboratory and field measurements of matric suctions in glacial till are shown in Figs. 4.71 and 4.72, respectively. The laboratory measurements

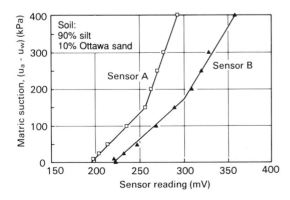

**Figure 4.70** Calibration curves for two AGWA-II thermal conductivity sensors.

100   4  MEASUREMENTS OF SOIL SUCTION

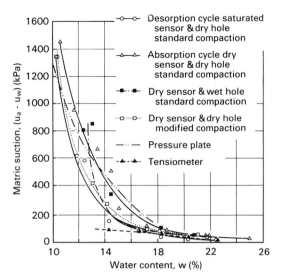

**Figure 4.71** Laboratory measurements of matric suction in glacial till using thermal conductivity sensors (MCS 6000).

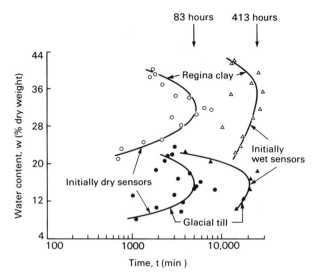

**Figure 4.73** Equalization times for the MCS 6000 sensors for glacial till and Regina clay compacted at various water contents.

were performed on compacted specimens. Figure 4.71 indicates that the initially wet sensor gives a lower matric suction than the initially dry sensor for the same water content in the soil.

The equalization times required for the MCS sensor are shown in Figure 4.73 for measurements in glacial till and Regina clay. The initially wet sensors have longer equalization times (i.e., maximum 413 h) than the initially dry sensors (i.e., 83 h). This pattern was consistent in both soils.

Unreliable suction measurements using thermal conductivity sensors have been attributed to poor contact between the porous block and the soil, the entrapment of air during installation (Nagpal and Boersma, 1973), and temperature and hysteretic effects. Poor contact between the porous block and the soil will cause the sensor to read a high suction value (Richard, 1974). The temperature effects on the MCS 6000 sensor readings in Regina clay are illustrated in Fig. 4.74.

### The AGWA-II Sensors

Results of laboratory measurements using the AGWA-II sensors on highly plastic clays from Sceptre and Regina, Saskatchewan are shown in Figs. 4.75, 4.76, and 4.77. The soils were sampled in the field using Shelby tubes. Matric suction measurements on compacted soils have also been performed on a silt from Brazil (Fig. 4.78). The results indicate that a considerably longer equalization time was required for the sensor to equilibrate when the water content of the specimen was low (Fig. 4.78) than when the water content of the specimen was high (Figs. 4.75, 4.76, and 4.77). The longer equalization time is attributed to the

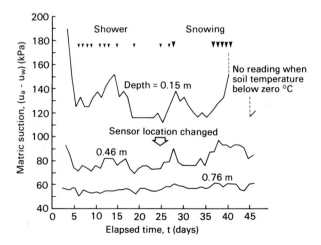

**Figure 4.72** Field measurements of matric suction in glacial till using thermal conductivity sensors (MCS-6000).

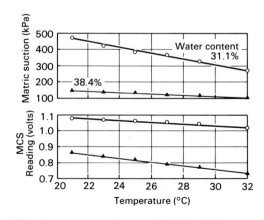

**Figure 4.74** Temperature effect on the MCS 6000 sensor readings in Regina clay.

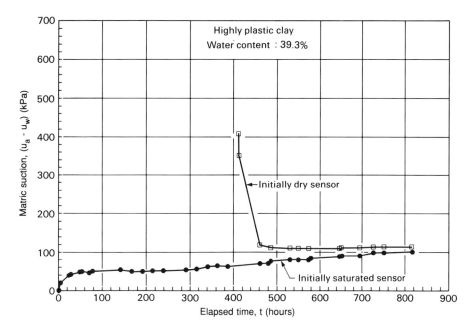

**Figure 4.75** Laboratory measurements of matric suction on a highly plastic clay from Sceptre, Sask., Canada ($w = 39.3\%$).

lower coefficient of permeability of the soil specimen as its water content decreases.

Several laboratory measurements were conducted using two sensors inserted into each soil specimen. One sensor was initially air-dried, and the other was initially saturated. The initially saturated sensor was submerged in water for about two days prior to being installed in the soil. The sensors were inserted into predrilled holes in either end of the soil specimen. The specimen with the installed sensors was wrapped in aluminum foil to prevent moisture loss during the measurement. The responses of both sensors were monitored immediately and at various elapsed times after their installation. The results indicate that the time required for the initially dry sensor to come to equilibrium with the

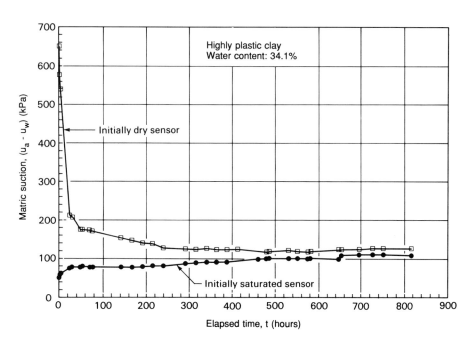

**Figure 4.76** Laboratory measurements of matric suctions on a highly plastic clay from Sceptre, Sask., Canada ($w = 34.1\%$).

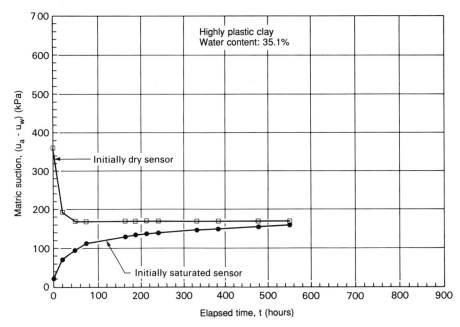

**Figure 4.77** Laboratory measurements of matric suction on a highly plastic clay from Darke Hall, Regina, Sask., Canada ($w = 35.1\%$).

soil specimen is less than the equilibrium time required for the initially saturated sensor to come to equilibrium.

On the basis of numerous laboratory experiments, it would appear that the AGWA-II sensors that were initially dry yielded a matric suction value which was close to the correct value. In general, the initially dry sensor should yield a value which was slightly high. On the other hand, the initially wet sensor yields a value which was too low.

Table 4.8 gives the interpretation of the results presented in Figs. 4.75–4.78 inclusive.

On the basis of many laboratory tests, it is recommended that if only one sensor is installed in an undisturbed sample, the sensor should be initially dry. If the sensors have been calibrated using at lest seven data points, the readings obtained in the laboratory should be accurate to within at least 15 kPa of the correct value, provided the matric suc-

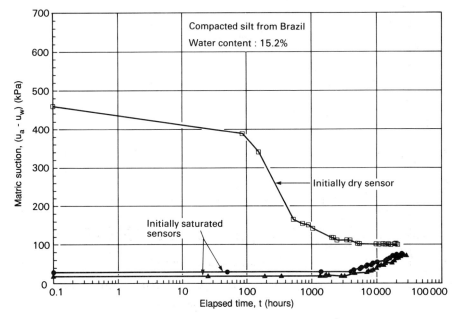

**Figure 4.78** Laboratory measurements of matric suctions on a compacted silt from Brazil ($w = 15.2\%$).

## 4.4 MEASUREMENTS OF MATRIC SUCTION

**Table 4.8 Interpretation of Laboratory Matric Suction Measurements**

| Soil Type | Figure No. | Water Content (%) | Initially Dry Sensor (kPa) | Initially Wet Sensor (kPa) | Best Estimate (kPa) |
|---|---|---|---|---|---|
| Sceptre clay | 4.75 | 39.3 | 120 | 100 | 114 |
| Sceptre clay | 4.76 | 34.1 | 136 | 108 | 126 |
| Regina clay | 4.77 | 35.1 | 160 | 150 | 157 |
| Brazil silt | 4.78 | 15.2 | 100 | 68 | 90 |

tion reading is in the range of 0–300 kPa. It may take four-seven days before equilibrium is achieved. If the sensors are left *in situ* for a long period of time, the measurements should be even more accurate.

Results from laboratory measurements of matric suction have been used to establish the negative pore-water pressures in undisturbed samples of Winnipeg clay taken from various depths within a railway embankment (Sattler et al., 1990). The samples were brought to the laboratory for matric suction measurements using the AGWA-II sensors. The measured matric suctions were corrected for the removal of the overburden stress, and plotted as a negative pore-water pressure profile (Fig. 4.79). The results indicated that the negative pore-water pressures approached zero at the average water table, and were, in general, more negative than the hydrostatic line above the water table.

Field measurements of matric suction under a controlled environment have been conducted in the subgrade soils of a Department of Highways indoor test track at Regina, Saskatchewan (Loi et al., 1989). The temperature and the relative humidity within the test track facility were controlled. Twenty-two AGWA-II sensors were installed in the subgrade of the test track. The subgrade consisted of a highly plastic clay and a glacial till. The sensors were initially air-dried and installed into predrilled holes at various depths in the subgrade. The sensor outputs were recorded twice a day.

Typical matric suction measurements on the compacted Regina clay and glacial till subgrade are presented in Figs. 4.80 and 4.81. Consistent readings of matric suction ranging from 50 to 400 kPa were monitored over a period of more than five months prior to flooding the test track. The

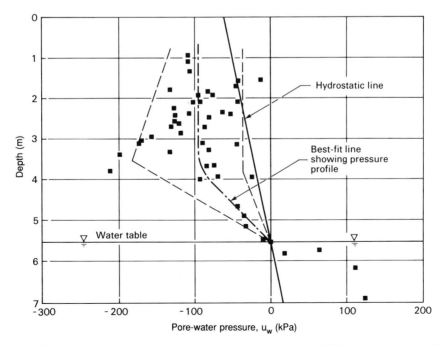

**Figure 4.79** Negative pore–water pressures measured using the AGWA-II thermal conductivity sensors on undisturbed samples.

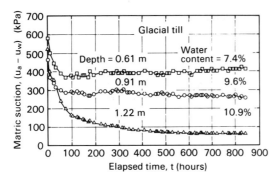

**Figure 4.80** Measurements of matric suction using the AGWA-II thermal conductivity sensors under a controlled environment in the test track facility (Department of Highways, Regina, Canada).

sensor responded quickly upon flooding (Fig. 4.82). The results demonstrated that the AGWA-II sensors can provide stable measurements of matric suction over a relatively long period of time.

Matric suction variations in the field can be related to environmental changes. Several AGWA-II sensors have been installed at various depths in the subgrade below a railroad. The soil was a highly plastic Regina clay that exhibited high swelling potentials. Matric suctions in the soil were monitored at various times of the year. The results clearly indicate seasonal variations of matric suctions in the field, with the greatest variation occurring near ground surface (Fig. 4.83).

Thermal conductivity sensors appear to be a promising device for measuring matric suction either in the laboratory or in the field. However, proper calibration should be performed on each sensor prior to its use. The calibration study on the AGWA-II sensors revealed that the sensors are quite sensitive for measuring matric suctions up to 175 kPa.

It is possible that further improvements on thermal conductivity type sensors will further enhance their performance. For example, a better seal around the electronics within the sensor could reduce the influence of soil water. Also, a stronger, more durable porous block would produce a better sensor for geotechnical engineering applications. These improvements would reduce the mortality rate of the sensor.

## 4.5 MEASUREMENTS OF OSMOTIC SUCTION

Several procedures can be used to measure the osmotic suction of a soil. For example, it is possible to add distilled water to a soil until the soil is in a near fluid condition, and then drain off some effluent and measure its electrical conductivity. The conductivity measurement can then be linearly extrapolated to the osmotic suction corresponding to the natural water content. This is known as the saturation extract procedure. Although the procedure is simple, it does not yield an accurate measurement of the *in situ* osmotic suction (Krahn and Fredlund, 1972).

A psychrometer can also be placed over the fluid extract to measure the osmotic suction, but this procedure, like-

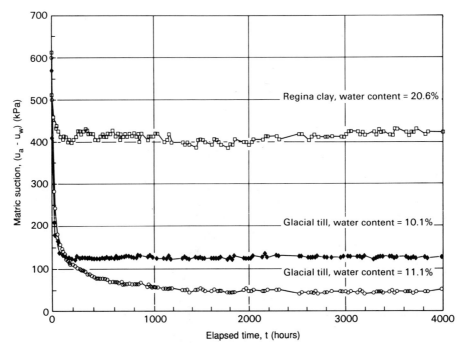

**Figure 4.81** Measurements of matric suction using the AGWA-II thermal conductivity sensors under a controlled environment (Test track facility, Department of Highways, Regina, Canada).

4.5 MEASUREMENTS OF OSMOTIC SUCTION    105

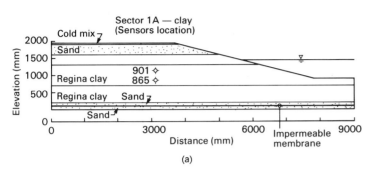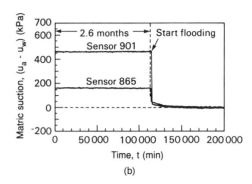

**Figure 4.82** Cross-section and location of measurements of matric suction using the AGWA-II thermal conductivity sensors under a controlled environment (a) sensor locations; (b) sensor responses, (Test track facility, Department of Highways, Regina, Canada).

wise, gives poor results. It is the use of the pore fluid squeezer technique that has proven to give the most reasonable measurements of osmotic suction.

### 4.5.1 Squeezing Technique

The *osmotic suction* of a soil can be indirectly estimated by measuring the electrical conductivity of the pore-water from the soil. Pure water has a low electrical conductivity in comparison to pore-water which contains dissolved salts. The electrical conductivity of the pore-water from the soil can be used to indicate the total concentration of dissolved salts which is related to the osmotic suction of the soil.

The pore-water in the soil can be extracted using a pore fluid squeezer which consists of a heavy-walled cylinder and piston squeezer (Fig. 4.84). The electrical resistivity (or electrical conductivity) of the pore-water is then measured. A calibration curve (Fig. 4.85) can be used to relate the electrical conductivity to the osmotic pressure of the soil. The results of squeezing technique measurements appear to be affected by the magnitude of the extraction pressure applied. Krahn and Fredlund (1972) used an extraction pressure of 34.5 MPa in the osmotic suction measurements on the glacial till and Regina clay.

Figures 4.86 and 4.87 present the results of osmotic suction measurements on glacial till and Regina clay, respectively. The measurements were conducted using the squeezing technique. The measured osmotic suctions are shown to agree closely with the total minus the matric suction measurements. In this case, the total and the matric suctions were measured independently. The discrepancies

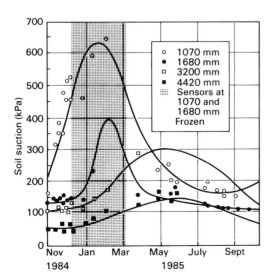

**Figure 4.83** Summary plot of matric suction measurements versus time of year for various depths in Regina clay in Saskatchewan (from van der Raadt, 1988).

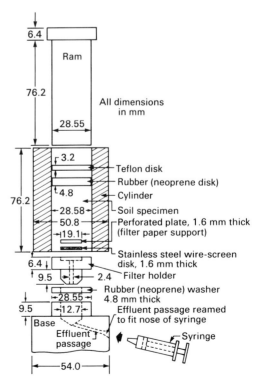

**Figure 4.84** The design of the pore fluid squeezer (from Manheim, 1966).

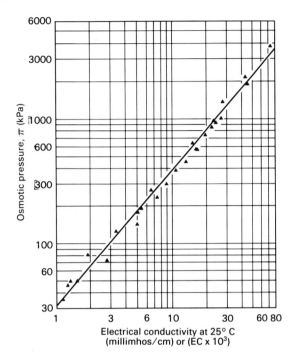

**Figure 4.85** Osmotic pressure versus electrical conductivity relationship for pore-water containing mixtures of dissolved salts (from *U.S.D.A. Agricultural Handbook No. 60*, 1950).

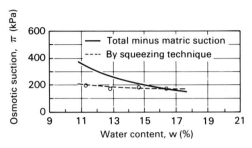

**Figure 4.86** Osmotic suction versus water content for glacial till (from Krahn and Fredlund, 1972).

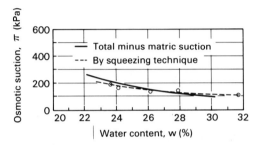

**Figure 4.87** Osmotic suction versus water content for Regina clay (from Krahn and Fredlund, 1972).

shown at low water contents for the glacial till (Fig. 4.85) are believed to be attributable to inaccurate measurements of matric suction (Krahn and Fredlund, 1972).

The close agreement exhibited in Figs. 4.86 and 4.87 indicates the reliability of the squeezing technique for osmotic suction measurements. The results also support the validity of the matric and osmotic suctions being components of the total suction [i.e., Eq. (4.3)].

It appears that the osmotic suction is relatively constant at various water contents (Figs. 4.86 and 4.87). Therefore, it is possible to use the osmotic suction as a relatively fixed value that can be subtracted from the total suction measurements in order to give the matric suction values.

# CHAPTER 5

# *Flow Laws*

Two phases of an unsaturated soil can be classified as fluids (i.e., water and air). The analysis of fluid flow requires a law to relate the flow rate with a driving potential using appropriate coefficients. The air in an unsaturated soil may be in an occluded form when the degree of saturation is relatively high. At lower degrees of saturation, the air phase is predominantly continuous. The form of the flow laws may vary for each of these cases. In addition, there may be the movement of air through the water phase, which is referred to as air diffusion through the pore–water (Fig. 5.1).

A knowledge of the driving potentials that cause air and water to flow or to diffuse is necessary for understanding the flow mechanisms. Throughout this chapter, the driving potentials of the water phase are given in terms of "heads." Water flow is caused by a hydraulic head gradient, where the hydraulic head consists of an elevation head plus a pressure head. A diffusion process is usually considered to occur under the influence of a chemical concentration or a thermal gradient. Water can also flow in response to an electrical gradient (Casagrande, 1952).

The concept of hydraulic head and the flow of air and water through unsaturated soils are presented in this chapter. A brief discussion on the diffusion process is also presented, together with its associated driving potential. Flows due to chemical, thermal, and electrical gradients are not discussed.

## 5.1 FLOW OF WATER

Several concepts have been used to explain the flow of water through an unsaturated soil. For example, a water content gradient, or a matric suction gradient, or a hydraulic head gradient have all been considered as driving potentials. However, it is important to use the form of the flow law that most fundamentally governs the movement of water.

A gradient in water content has sometimes been used to describe the flow of water through unsaturated soils. It is assumed that water flows from a point of high water content to a point of lower water content. This type of flow law, however, does not have a fundamental basis since water can also flow from a region of low water content to a region of high water content when there are variations in the soil types involved, hysteretic effects, or stress history variations are encountered. Therefore, a water content gradient should not be used as a fundamental driving potential for the flow of water (Fredlund, 1981).

In an unsaturated soil, a matric suction gradient has sometimes been considered to be the driving potential for water flow. However, the flow of water does not fundamentally and exclusively depend upon the matric suction gradient. Figure 5.2 demonstrates three hypothetical cases where the air and water pressure gradients are controlled across an unsaturated soil element at a constant elevation. In all cases, the air and water pressures on the left-hand side are greater than the pressures on the right-hand side.

The matric suction on the left-hand side may be smaller than on the right-hand side (Case 1), equal to the right-hand side (Case 2), or larger than on the right-hand side (Case 3). However, air and water will flow from left to right in response to the pressure gradient in the individual phases, regardless of the matric suction gradient. Even in Case 2, where the matric suction gradient is zero, air and water will still flow.

Flow can be defined more appropriately in terms of a hydraulic head gradient (i.e., a pressure head gradient in this case) for each of the phases. Therefore, the matric suction gradient is not the fundamental driving potential for the flow of water in an unsaturated soil. In the special case where the air pressure gradient is zero, the matric suction gradient is numerically equal to the pressure gradient in the water. This is the common situation in nature, and is probably the reason for the proposal of the matric suction form for water flow. However, the elevation head component has then been omitted.

The flow of water through a soil is not only governed by the pressure gradient, but also by the gradient due to ele-

108   5   FLOW LAWS

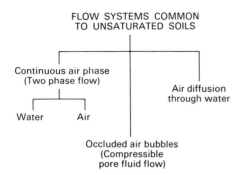

**Figure 5.1** Flow systems common to unsaturated soils.

vation differences. The pressure and elevation gradients are combined to give a hydraulic head gradient as the fundamental driving potential. The hydraulic head gradient in a specific fluid phase is the driving potential for flow in that phase. This is equally true for saturated and unsaturated soils.

### 5.1.1   Driving Potential for Water Phase

The driving potential for the flow of water defines the energy or capacity to do work. The energy at a point is computed relative to a datum. The datum is chosen arbitrarily because only the gradient in the energy between two points is of importance in describing flow.

A point in the water phase has three primary components of energy, namely, gravitational, pressure, and velocity. Figure 5.3 shows point $A$ in the water phase which is located at an elevation, $y$, above an arbitrary datum. Let us consider the energy state of point $A$. Point $A$ has a gravitational energy, $E_g$, which can be written

$$E_g = M_w g y \tag{5.1}$$

where

$E_g$ = gravitational energy
$M_w$ = mass of water at point $A$
$g$ = gravitational acceleration
$y$ = elevation of point $A$ above the datum.

The component of energy due to the pressure at point $A$ is given as follows (Freeze and Cherry, 1979):

$$E_p = M_w \int_0^{u_w} \frac{V_w}{M_w} du_w \tag{5.2}$$

where

$E_p$ = pressure energy
$u_w$ = pore-water pressure at point $A$
$V_w$ = volume of water at point $A$.

Equation (5.2) can also be written

$$E_p = M_w \int_0^{u_w} \frac{du_w}{\rho_w} \tag{5.3}$$

where

$\rho_w$ = density of water.

When the water density, $\rho_w$, is constant, Eq. (5.3) takes

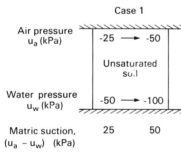

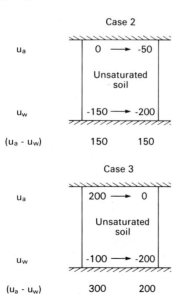

**Figure 5.2** Pressure and matric suction gradients across an unsaturated soil element.

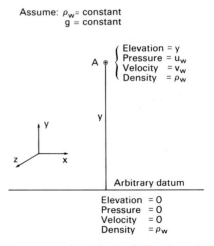

**Figure 5.3** Energy at point $A$ in the $y$-direction relative to an arbitrary datum.

the following form:

$$E_p = \frac{M_w u_w}{\rho_w}. \qquad (5.4)$$

The flow rate of water at point $A$ gives rise to a velocity energy, $E_v$:

$$E_v = \frac{M_w v_w^2}{2} \qquad (5.5)$$

where

$E_v$ = velocity energy
$v_w$ = flow rate of water at point $A$ (i.e., in the $y$-direction).

In total, the potential energy at point $A$ is the summation of the gravitational, pressure, and velocity components:

$$E = M_w g y + \frac{M_w u_w}{\rho_w} + \frac{M_w v_w^2}{2} \qquad (5.6)$$

where

$E$ = total energy.

The total energy at point $A$ can be expressed as energy per unit weight, which is called a potential or a hydraulic head. The hydraulic head, $h_w$, at point $A$ is obtained by dividing Eq. (5.6) by the *weight* of water at point $A$ (i.e., $M_w g$):

$$h_w = y + \frac{u_w}{\rho_w g} + \frac{v_w^2}{2g} \qquad (5.7)$$

where

$h_w$ = hydraulic head or total head.

The hydraulic head consists of three components, namely, the gravitational head, $y$, the pressure head, $(u_w/\rho_w g)$, and the velocity head, $(v_w^2/2g)$. The velocity head in a soil is negligible in comparison with the gravitational and the pressure heads. Equation (5.7) can therefore be simplified to yield an expression for the hydraulic head at any point in the soil mass:

$$h_w = y + \frac{u_w}{\rho_w g}. \qquad (5.8)$$

The heads expressed in Eq. (5.8) have the dimension of length. Hydraulic head is a measurable quantity, the gradient of which causes flow in saturated and unsaturated soils. Devices such as piezometers and tensiometers can be used to measure the *in situ* pore-water pressure at a point (Fig. 5.4). The distance between the elevation of the point under consideration and the datum indicates the elevation head (i.e., $y_A$ and $y_B$).

A piezometer can be used to measure the pore-water pressure at a point when the pore-water pressure is positive (e.g., point $B$ in Fig. 5.4). A tensiometer can be used to measure the pore-water pressure when the pressure is negative (e.g., point $A$ in Fig. 5.4).

The water level in the measuring device will rise or drop, depending upon the pore-water pressure at the point under consideration. For example, the water level in the piezometer rises above the elevation of point $B$ at a distance equal to the positive pore-water pressure head at point $B$. Alternately, the water level in the tensiometer drops below the elevation of point $A$ to a distance equal to the negative pore-water pressure head at point $A$. The distance between the water level in the measuring device and the datum is the sum of the gravitational and pressure heads (i.e., the hydraulic head).

In Fig. 5.4, point $A$ has a higher total head than point $B$ [i.e., $h_w(A) > h_w(B)$]. Water will flow from point $A$ to point $B$ due to the total head gradient between these two points. The driving potential causing flow in the water phase has the same form for both saturated (i.e., point $B$) and unsaturated (i.e., point $A$) soils (Freeze and Cherry, 1979). Water will flow from a point of high total head to

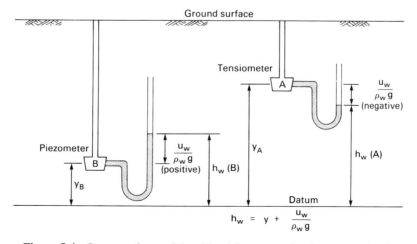

**Figure 5.4** Concept of potential and head for saturated and unsaturated soils.

a point of low total head, regardless of whether the pore-water pressures are positive or negative.

Osmotic suction has sometimes been included as a component in the total head equation for flow. However, it is better to visualize the osmotic suction gradient as the driving potential for the osmotic diffusion process (Corey and Kemper, 1961). Osmotic diffusion is a process where ionic or molecular constituents move as a result of their kinetic activity. For example, an osmotic gradient across a semipermeable membrane causes the movement of water through the membrane. On the other hand, the bulk flow of solutions (i.e., pure water and dissolved salts) in the absence of a semi-permeable membrane is governed by the hydraulic head gradient. Therefore, it would appear superior to analyze the bulk flow of water separately from the osmotic diffusion process since two independent mechanisms are involved (Corey, 1977). A brief explanation of the diffusion process is given in Section 5.3.

### 5.1.2 Darcy's Law for Unsaturated Soils

The flow of water in a saturated soil is commonly described using Darcy's law. Darcy (1856) postulated that the rate of water flow through a soil mass was proportional to the hydraulic head gradient:

$$v_w = -k_w \frac{\partial h_w}{\partial y} \qquad (5.9)$$

where

$v_w$ = flow rate of water
$k_w$ = coefficient of permeability with respect to the water phase
$\partial h_w/\partial y$ = hydraulic head gradient in the $y$-direction, which can be designated as $i_{wy}$.

The coefficient of proportionality between the flow rate of water and the hydraulic head gradient is called the coefficient of permeability, $k_w$. The coefficient of permeability is relatively constant for a specific saturated soil. Eq. (5.9) can also be written for the $x$- and $z$-directions. The negative sign in Eq. (5.9) indicates that water flows in the direction of a decreasing hydraulic head.

Darcy's law also applies for the flow of water through an unsaturated soil (Buckingham, 1907; Richard, 1931; Childs and Collis-George, 1950). However, the coefficient of permeability in an unsaturated soil cannot generally be assumed to be constant. Rather, the coefficient of permeability is a variable which is predominantly a function of the water content or the matric suction of the unsaturated soil.

Water can be visualized as flowing only through the pore space filled with water. The air-filled pores are nonconductive channels to the flow of water. Therefore, the air-filled pores in an unsaturated soil can be considered as behaving similarly to the solid phase, and the soil can be treated as a saturated soil having a reduced water content (Childs, 1969). Subsequently, the validity of Darcy's law can be verified in the unsaturated soil in a similar manner to its verification for a saturated soil. However, the volume of water (or water content) should be constant while the hydraulic head gradient is varied.

Experiments to verify Darcy's law for unsaturated soils have been performed, and the results are presented in Fig. 5.5 (Childs and Collis-George, 1950). A column of unsaturated soil with a uniform water content and a constant water pressure head was subjected to various gradients of gravitational head. The results indicate that at a specific water content, the coefficient of permeability, $k_w$, is constant for various hydraulic head gradients (i.e., in this case, only the gravitational head was varied) applied to the unsaturated soil. In other words, the rate of water flow through an unsaturated soil is linearly proportional to the hydraulic head gradient, with the coefficient of permeability being a constant, similar to the situation for a saturated soil. This confirms that Darcy's law [i.e., Eq. (5.9)] can also be applied to unsaturated soils. In an unsaturated soil, however, the magnitude of the coefficient of permeability will differ for different volumetric water contents, $\theta_w$, as depicted in Fig. 5.5.

### 5.1.3 Coefficient of Permeability with Respect to the Water Phase

The coefficient of permeability with respect to the water phase, $k_w$, is a measure of the space available for water to flow through the soil. The coefficient of permeability depends upon the properties of the fluid and the properties of the porous medium. Different types of fluid (e.g., water and oil) or different types of soil (e.g., sand and clay) produce different values for the coefficient of permeability, $k_w$.

#### Fluid and Porous Medium Components

The coefficients of permeability with respect to water, $k_w$, can be expressed in terms of the intrinsic permeability, $K$:

$$k_w = \frac{\rho_w g}{\mu_w} K \qquad (5.10)$$

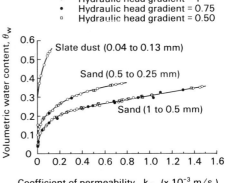

**Figure 5.5** Experimental verification of Darcy's law for water flow through an unsaturated soil (from Childs and Collis-George, 1950).

where

$\mu_w$ = absolute (dynamic) viscosity of water
$K$ = intrinsic permeability of the soil.

Equation (5.10) shows the influence of the fluid density, $\rho_w$, and the fluid viscosity, $\mu_w$, on the coefficient of permeability, $k_w$. The intrinsic permeability of a soil, $K$, represents the characteristics of the porous medium and is independent of the fluid properties.

The fluid properties are commonly considered to be constant during the flow process. The characteristics of the porous medium are a function of the volume–mass properties of the soil. The intrinsic permeability is used in numerous disciplines. However, in geotechnical engineering, the coefficient of permeability, $k_w$, is the most commonly used term, and will therefore be used throughout this book.

### Relationship Between Permeability and Volume–Mass Properties

The coefficient of permeability, $k_w$, is a function of any two of three possible volume–mass properties (Lloret and Alonso, 1980; Fredlund, 1981):

$$k_w = k_w(S, e) \quad (5.11)$$

or

$$k_w = k_w(e, w) \quad (5.12)$$

or

$$k_w = k_w(w, S) \quad (5.13)$$

where

$S$ = degree of saturation
$e$ = void ratio
$w$ = water content.

In a *saturated* soil, the coefficient of permeability is a function of the void ratio (Lambe and Whitman, 1979). However, the coefficient of permeability of a saturated soil is generally assumed to be a constant when analyzing problems such as transient flow.

In an *unsaturated* soil, the coefficient of permeability is significantly affected by combined changes in the void ratio and the degree of saturation (or water content) of the soil. Water flows through the pore space filled with water; therefore, the percentage of the voids filled with water is an important factor. As a soil becomes unsaturated, air first replaces some of the water in the large pores, and this causes the water to flow through the smaller pores with an increased tortuosity to the flow path. A further increase in the matric suction of the soil leads to a further decrease in the pore volume occupied by water. In other words, the air–water interface is drawn closer and closer to the soil particles (Fig. 5.6). As a result, the coefficient of permeability with respect to the water phase decreases rapidly as the space available for water flow reduces.

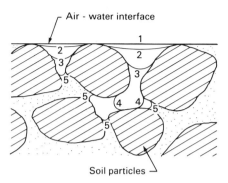

**Figure 5.6** Development of an unsaturated soil by the withdrawal of the air–water interface at different stages of matric suction or degree of saturation (i.e., stages 1–5) (from Childs, 1969).

### Effect of Variations in Degree of Saturation on Permeability

The coefficient of permeability of an unsaturated soil can vary considerably during a transient process as a result of changes in the volume–mass properties. The change in void ratio in an unsaturated soil may be small, and its effect on the coefficient of permeability may be secondary. However, the effect of a change in degree of saturation may be highly significant. As a result, the coefficient of permeability is often described as a singular function of the degree of saturation, $S$, or the volumetric water content, $\theta_w$.

A change in matric suction can produce a more significant change in the degree of saturation or water content than can be produced by a change in net normal stress. The degree of saturation has been commonly described as a function of matric suction. The relationship is called the matric suction versus degree of saturation curve [Fig. 5.7(a)].

Numerous semi-empirical equations for the coefficient of permeability have been derived using either the matric suction versus degree of saturation curve or the soil–water characteristic curve. In either case, the soil pore size distribution forms the basis for predicting the coefficient of permeability. The pore size distribution concept is somewhat new to geotechnical engineering. The pore size distribution has been used in other disciplines to give reasonable estimates of the permeability characteristics of a soil.

The prediction of the coefficient of permeability from the matric suction versus degree of saturation curve is discussed first, followed by the coefficient of permeability prediction using the soil–water characteristic curve.

### Relationship Between Coefficient of Permeability and Degree of Saturation

Coefficient of permeability functions obtained from the matric suction versus degree of saturation curve have been proposed by Burdine (1952) and Brooks and Corey (1964). The matric suction versus degree of saturation curve ex-

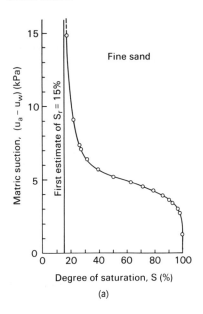

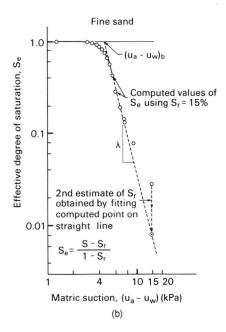

**Figure 5.7** Determination of the air entry value, $(u_a - u_w)_b$, residual degree of saturation, $S_r$, and pore size distribution index, $\lambda$. (a) Matric suction versus degree of saturation curve; (b) effective degree of saturation versus matric suction curve. (From Brooks and Corey, 1964).

hibits hysteresis. Only the drainage curve is used in their derivations. In addition, the soil structure is assumed to be incompressible.

There are three soil parameters that can be identified from the matric suction versus degree of saturation curve. These are the air entry value of the soil, $(u_a - u_w)_b$, the residual degree of saturation, $S_r$, and the pore size distribution index, $\lambda$. These parameters can readily be visualized if the saturation condition is expressed in terms of an effective degree of saturation, $S_e$ (Corey, 1954) [Fig. 5.7(b)]:

$$S_e = \frac{S - S_r}{1 - S_r} \quad (5.14)$$

where

$S_e$ = effective degree of saturation
$S_r$ = residual degree of saturation.

The residual degree of saturation, $S_r$, is defined as the degree of saturation at which an increase in matric suction does not produce a significant change in the degree of saturation [see Fig. 5.7(a)]. The values for all degree of saturation variables used in Eq. (5.14) are in decimal form.

The effective degree of saturation can be computed by first estimating the residual degree of saturation [see Fig. 5.7(a)]. The effective degree of saturation is then plotted against the matric suction, as illustrated in Fig. 5.7(b). A horizontal and a sloping line can be drawn through the points. However, points at high matric suction values may not lie on the straight line used for the first estimate of the residual degree of saturation. Therefore, the point with the highest matric suction must be forced to lie on the straight line by estimating a new value of $S_r$ [see Fig. 5.7(b)]. A second estimate of the residual degree of saturation is then used to recompute the values for the effective degree of saturation. A new plot of matric suction versus effective degree of saturation curve can then be obtained. The above procedure is repeated until all of the points on the sloping line constitute a straight line. This usually occurs by the second estimate of the residual degree of saturation.

The air entry value of the soil, $(u_a - u_w)_b$, is the matric suction value that must be exceeded before air recedes into the soil pores. The air entry value is also referred to as the "displacement pressure" in petroleum engineering or the "bubbling pressure" in ceramics engineering (Corey, 1977). It is a measure of the maximum pore size in a soil. The intersection point between the straight sloping line and the saturation ordinate (i.e., $S_e = 1.0$) in Fig. 5.7(b) defines the air entry value of the soil. The sloping line for the points having matric suctions greater than the air entry value can be described by the following equation:

$$S_e = \left\{ \frac{(u_a - u_w)_b}{(u_a - u_w)} \right\}^\lambda \quad \text{for } (u_a - u_w) > (u_a - u_w)_b$$

(5.15)

where

$\lambda$ = pore size distribution index, which is defined as the negative slope of the effective degree of saturation, $S_e$, versus matric suction, $(u_a - u_w)$, curve.

Soils with a wide range of pore sizes have a small value for $\lambda$. The more uniform the distribution of the pore sizes in a soil, the larger is the value for $\lambda$. Figure 5.8 presents typical $\lambda$ values for various soils which have been obtained from matric suction versus degree of saturation curves.

5.1 FLOW OF WATER   113

The empirical constant, $\delta$, is related to the pore size distribution index:

$$\delta = \frac{2 + 3\lambda}{\lambda}. \quad (5.17)$$

Table 5.1 presents several $\delta$ values and their corresponding pore size distribution indices, $\lambda$, for various soil types.

Water coefficients of permeability, $k_w$, corresponding to various degrees of saturation can be computed using Eq. (5.16), and can be expressed as the relative water phase coefficient of permeability, $k_{rw}$ (%):

$$k_{rw} = \frac{k_w (100)}{k_s}. \quad (5.18)$$

Experimental data for a sandstone expressed in terms of the relative permeability are shown in Fig. 5.9. In the experiments, a hydrocarbon liquid was used instead of water in order to produce a more stable soil structure and consistent fluid properties. The results are essentially the same as for water flow since the relative permeability is not a function of the fluid properties. However, the interactions between the water and the soil particles may produce some differences from the results obtained using water and those obtained using hydrocarbons.

### Relationship Between Water Coefficient of Permeability and Matric Suction

The coefficient of permeability with respect to the water phase, $k_w$, can also be expressed as a function of the matric suction by substituting the effective degree of saturation, $S_e$ [i.e., Eq. (5.15)], into the permeability function [i.e., Eq. (5.16)] (Brooks and Corey, 1964). Several other relationships between the coefficient of permeability and matric suction have also been proposed (Gardner, 1958a; Arbhabhirama and Kridakorn, 1968) and these are summarized in Table 5.2.

The relationship between the coefficient of permeability and matric suction proposed by Gardner (1958a) [Eq. (5.20) in Table 5.2] is presented in Fig. 5.10. The equation provides a flexible permeability function which is defined by two constants, "$a$" and "$n$." The constant "$n$" defines the slope of the function, and the "$a$" constant is related to the breaking point of the function. Four typical functions with differing values of "$a$" and "$n$" are illustrated in Fig. 5.10. The permeability functions are written in terms of matric suction in Eq. (5.20); however, these equations could also be written in terms of total suction.

### Relationship Between Water Coefficient of Permeability and Volumetric Water Content

The water phase coefficient of permeability, $k_w$, can also be related to the volumetric water content, $\theta_w$ (Buckingham, 1907; Richards, 1931; Moore, 1939). A coefficient of permeability function, $k_w(\theta_w)$, has been proposed

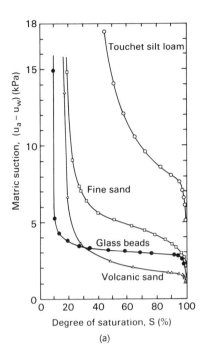

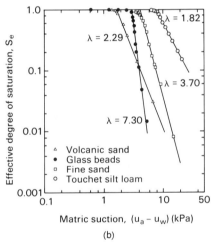

**Figure 5.8** Typical matric suction versus degree of saturation curves for various soils with their corresponding $\lambda$ values; (a) Matric suction versus degree of saturation curves, (b) effective degree of saturation versus matric suction. (From Brooks and Corey, 1964).

The coefficient of permeability with respect to the water phase, $k_w$, can be predicted from the matric suction versus degree of saturation curves as follows (Brooks and Corey, 1964):

$$k_w = k_s \quad \text{for } (u_a - u_w) \leq (u_a - u_w)_b$$

$$k_w = k_s S_e^\delta \quad \text{for } (u_a - u_w) > (u_a - u_w)_b \quad (5.16)$$

where

$k_s$ = coefficient of permeability with respect to the water phase for the soil at saturation (i.e., $S = 100\%$)

$\delta$ = an empirical constant.

**Table 5.1  Suggested Values of the Constant, $\delta$, and the Pore Size Distribution Index, $\lambda$, for Various Soils**

| Soils | $\delta$ Values | $\lambda$ Values | Source |
|---|---|---|---|
| Uniform sand | 3.0 | $\infty$ | Irmay [1954] |
| Soil and porous rocks | 4.0 | 2.0 | Corey [1954] |
| Natural sand deposits | 3.5 | 4.0 | Averjanov [1950] |

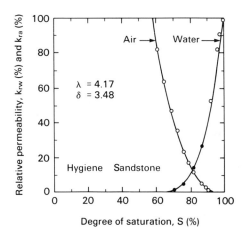

**Figure 5.9**  Relative permeability of water and air as a function of the degree of saturation during drainage (from Brooks and Corey, 1964).

using the configurations of the pore space filled with water (Childs and Collis-George, 1950). The soil is assumed to have a random distribution of pores of various sizes and an incompressible soil structure. The permeability function, $k_w(\theta_w)$, is written as the sum of a series of terms obtained from the statistical probability of interconnections between water-filled pores of varying sizes.

The volumetric water content, $\theta_w$, can be plotted as a function of matric suction, $(u_a - u_w)$, and the plot is called the soil–water characteristic cure. Therefore, the permeability function, $k_w(\theta_w)$, can also be expressed in terms of matric suction (Marshall, 1958; and Millington and Quirk, 1959, 1961). In other words, the soil–water characteristics curve can be visualized as an indication of the configuration of water-filled pores. The coefficient of permeability is obtained by dividing the soil–water characteristic curve into "$m$" equal intervals along the volumetric water con-

**Table 5.2  Relationships between Water Coefficient of Permeability and Matric Suction**

| Equations | Number | Source | Symbols |
|---|---|---|---|
| $k_w = k_s$ for $(u_a - u_w) \leq (u_a - u_w)_b$ <br><br> $k_w = k_s \left\{ \dfrac{(u_a - u_w)_b}{(u_a - u_w)} \right\}^\eta$ for $(u_a - u_w) > (u_a - u_w)_b$ | (5.19) | Brooks and Corey (1964) | $\eta$ = empirical constant $\eta = 2 + 3\lambda$ |
| $k_w = \dfrac{k_s}{1 + a \left\{ \dfrac{(u_a - u_w)}{\rho_w g} \right\}^n}$ | (5.20) | Gardner (1958a) | $a, n$ = constant |
| $k_w = \dfrac{k_s}{\left\{ \dfrac{(u_a - u_w)}{(u_a - u_w)_b} \right\}^{n'} + 1}$ | (5.21) | Arbhabhirama and Kridakorn (1968) | $n'$ = constant |

5.1 FLOW OF WATER   115

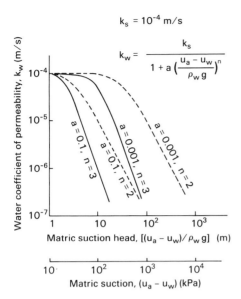

**Figure 5.10** Gardner's equation for the water coefficient of permeability as a function of the matric suction.

tent axis, as shown in Fig. 5.11. The matric suction corresponding to the midpoint of each interval is used to calculate the coefficient of permeability.

The permeability function, $k_w(\theta_w)$, is derived based on Poiseuille's equation. The following permeability function has a similar form to the function presented by Kunze *et al.*, (1968). The function is slightly modified in order to use SI units and matric suction instead of pore-water pressure head. Variables used in the equation are illustrated in Fig. 5.11.

$$k_w(\theta_w)_i = \frac{k_s}{k_{sc}} \frac{T_s^2 \rho_w g}{2\mu_w} \frac{\theta_s^p}{N^2} \sum_{j=i}^{m} \{(2j + 1 - 2i)(u_a - u_w)_j^{-2}\}$$

$$i = 1, 2, \cdots, m \qquad (5.22)$$

where

$k_w(\theta_w)_i$ = calculated water coefficient of permeability (m/s) for a specified volumetric water content, $(\theta_w)_i$, corresponding to the $i$th interval

$i$ = interval number which increases with the decreasing volumetric water content; for example, $i = 1$ identifies the first interval that closely corresponds to the saturated volumetric water content, $\theta_s$; $i = m$ identifies the last interval corresponding to the lowest volumetric water content, $\theta_L$, on the experimental soil–water characteristic curve

$j$ = a counter from "$i$" to "$m$"

$k_s$ = measured saturated coefficient of permeability (m/s)

$k_{sc}$ = calculated saturated coefficient of permeability (m/s)

$T_s$ = surface tension of water (kN/m)

$\rho_w$ = water density (kg/m$^3$)

$g$ = gravitational acceleration (m/s$^2$)

$\mu_w$ = absolute viscosity of water (N · s/m$^2$)

$\theta_s$ = volumetric water content at saturation (i.e., $S = 100\%$) (Green and Corey, 1971a)

$p$ = a constant which accounts for the interaction of pores of various sizes; the magnitude of "$p$" can be assumed to be equal to 2.0 (Green and Corey, 1971a)

$m$ = total number of intervals between the saturated volumetric water content, $\theta_s$, and the *lowest* volumetric water content, $\theta_L$, on the experimental soil–water characteristic curve

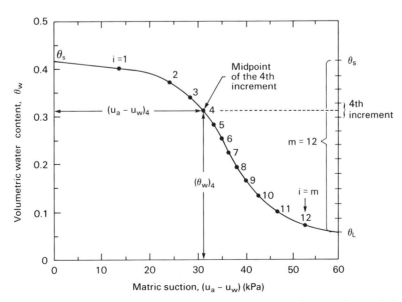

**Figure 5.11** Prediction of the coefficient of permeability from the soil–water characteristic curve.

$N$ = total number of intervals computed between the saturated volumetric water content, $\theta_s$, and *zero* volumetric water content (i.e., $\theta_w = 0$) (note: $N = m(\theta_s/(\theta_s - \theta_L))$, $m \leq N$; and $m = N$ when $\theta_L = 0$)

$(u_a - u_w)_j$ = matric suction (kPa) corresponding to the midpoint of the $j$th interval.

The calculation of the water coefficient of permeability, $k_w$, at a specific volumetric water content, $(\theta_w)_i$, involves the summation of the matric suction values that correspond to the volumetric water contents at and below $(\theta_w)_i$. Several procedures have been proposed for the calculation of the permeability function, $k_w(\theta_w)$, using Eq. (5.22). Basically, the difference between the various procedures lies in the interpretation of the pore interaction term [i.e., $\theta_s^p/N$ in Eq. (5.22)] (Green and Corey, 1971a). The matching factor, $(k_s/k_{sc})$, based on the saturated coefficient of permeability is necessary in order to provide a more accurate computation of the unsaturated coefficients of permeability.

The above computational procedure for obtaining the permeability function appears to be most successful for sandy soils having a relatively narrow pore size distribution (Nielsen et al., 1972). A comparison between the permeability function, $k_w(\theta_w)$, computed from Eq. (5.22) and experimental data is shown in Fig. 5.12 for a fine sand. The soil–water characteristic curve for the sand and the comparison of its permeability function are shown in Fig. 5.12(a) and (b), respectively.

The coefficient of permeability, $k_w$, at a specific volumetric water content, $\theta_w$, is *computed* directly from Eq. (5.22). The shape of the permeability function is determined by the terms inside the summation sign portion of the equation as obtained from the soil–water characteristic curve. However, the magnitude of the permeability function needs to be adjusted with reference to the measured saturated coefficient of permeability, $k_s$, by using the matching factor. Therefore, if the saturated coefficient of permeability is measured, the permeability function can be *predicted* directly from the soil–water characteristic curve because all of the terms in front of the summation sign in Eq. (5.22) can be considered as an adjusting factor. As a result, the permeability function, $k_w(\theta_w)$, can be written as follows:

$$k_w(\theta_w)_i = \frac{k_s}{k_{sc}} A_d \sum_{j=i}^{m} [(2j + 1 - 2i)(u_a - u_w)_j^{-2}]$$

$$i = 1, 2, \cdots, m \quad (5.23)$$

where

$A_d$ = adjusting constant [i.e., $(T_s^2 \rho_w g/2\mu_w)(\theta_s^p/N^2)$ (m $\cdot$ s$^{-1}$ $\cdot$ kPa$^2$)].

The technique for predicting the permeability function using Eq. (5.23) is explained in Chapter 6.

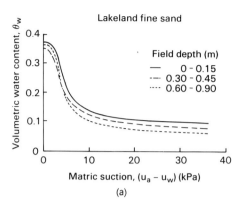

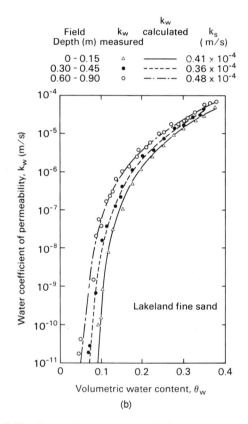

**Figure 5.12** Comparisons between calculated and measured unsaturated permeabilities for Lakeland fine sand. (a) Soil–water characteristic curves; (b) water coefficient of permeability as a function of volumetric water content (from Elzeftawy and Cartwright, 1981).

### Hysteresis of the Permeability Function

The coefficient of permeability, $k_w$, is generally assumed to be uniquely related to the degree of saturation, $S$, or the volumetric water content, $\theta_w$. This assumption is reasonable since the volume of water flow is a direct function of the volume of water in the soil. The relationships between the degree of saturation (or volumetric water content) and the coefficient of permeability appear to exhibit little hysteresis (Nielsen and Biggar, 1961; Topp and Miller, 1966; Corey, 1977; and Hillel, 1982). Nielsen *et al.*, (1972)

5.2 FLOW OF AIR    117

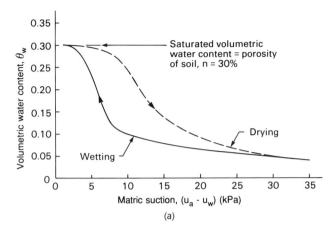

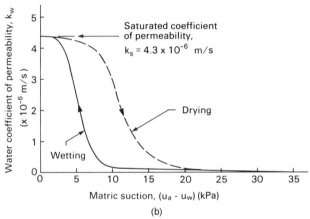

**Figure 5.13** Similar hysteresis forms in the volumeric water content and water coefficient of permeability when plotted as a function of $(u_a - u_w)$ for a naturally deposited sand. (a) Volumetric water content versus matric solution; (b) water coefficient of permeability versus matric suction (from Liakopoulos, 1965a).

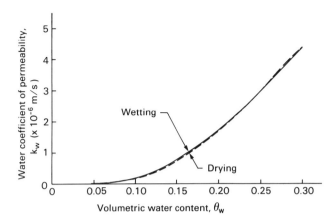

**Figure 5.14** Essentially no hysteresis is shown in the relationship between water coefficient of permeability versus volumetric water content.

## 5.2 FLOW OF AIR

The air phase of an unsaturated soil can be found in two forms. These are the continuous air phase form and the occluded air bubble form. The air phase generally becomes continuous as the degree of saturation reduces to around 85% or lower (Corey, 1957). The flow of air through an unsaturated soil commences at this point.

Under naturally occurring conditions, the flow of air through a soil may be caused by factors such as variations in barometric pressure, water infiltration by rain that compresses the air in the soil pores, and temperature changes. The flow of air in compacted fills may be due to applied loads.

When the degree of saturation is above about 90%, the air phase becomes occluded, and air flow is reduced to a diffusion process through the pore-water (Matyas, 1967).

### 5.2.1 Driving Potential for Air Phase

The flow of air in the continuous air phase form is governed by a concentration or pressure gradient. The elevation gradient has a negligible effect. The pressure gradient is most commonly considered as the only driving potential for the air phase. Both Fick's and Darcy's laws have been used to describe the flow of air through a porous media.

### 5.2.2 Fick's Law for Air Phase

Fick's law (1855) is often used to describe the diffusion of gases through liquids. A modified form of Fick's law can be applied to the air flow process. Fick's first law states that the rate of mass transfer of the diffusing substance across a unit area is proportional to the concentration gradient of the diffusing substance.

In the case of air flow through an unsaturated soil, the porous medium (i.e., soil) can be used as the reference in order to be consistent with the permeability concept for the

stated: "The function $k_w(\theta_w)$ is well behaved, inasmuch as for coarse-textured soils, it is approximately the same for both wetting and drying." However, this is not the case for the relationship between the water coefficient of permeability, $k_w$, and matric suction, $(u_a - u_w)$. Since there is hysteresis in the relationship between the volume of water in a soil and the stress state [i.e., namely, $(u_a - u_w)$], there will also be hysteresis in the relationship between the coefficient of permeability and matric suction.

The degree of saturation or volumetric water content shows significant hysteresis when plotted versus matric suction [Fig. 5.13(a)]. As a result, the coefficient of permeability, which is directly related to the volumetric water content or degree of saturation, will also show significant hysteresis when plotted versus matric suction. Figure 5.13(a) and (b) demonstrate a similar hysteresis form for both the volumetric water content, $\theta_w$, and the coefficient of permeability, $k_w$, when plotted against matric suction. However, if the coefficient of permeability is cross-plotted against volumetric water content, the resulting plot shows essentially no hysteresis, as demonstrated in Fig. 5.14.

water phase. This means that the mass rate of flow and the concentration gradient in the air are computed with respect to a unit area and a unit volume of the soil:

$$J_a = -D_a \frac{\partial C}{\partial y} \quad (5.24)$$

where

$J_a$ = mass rate of air flowing across a unit area of the soil
$D_a$ = transmission constant for air flow through a soil
$C$ = concentration of the air expressed in terms of the mass of air per unit volume of soil
$\partial C/\partial y$ = concentration gradient in the $y$-direction.

The negative sign in Eq. (5.24) indicates that air flows in the direction of a decreasing concentration gradient. Equation (5.24) can similarly be written for the $x$- and $z$-directions.

The concentration of air with respect to a unit volume of the soil can be written as

$$C = \frac{M_a}{V_a/(1-S)n} \quad (5.25)$$

where

$M_a$ = mass of air in the soil
$V_a$ = volume of air in the soil
$S$ = degree of saturation
$n$ = porosity of the soil.

Substituting the density of air, $\rho_a$, for $(M_a/V_a)$ in Eq. (5.25), gives

$$C = \rho_a(1-S)n. \quad (5.26)$$

Air density is related to the absolute air pressure in accordance with the gas law (i.e., $\rho_a = (\omega_a \bar{u}_a)/RT$), as explained in Chapter 2. Therefore, the concentration gradient in Eq. (5.24) can also be expressed with respect to a pressure gradient in the air. The gauge air pressure is used in reformulating Eq. (5.24), since only the gradient is of importance.

$$J_a = -D_a \frac{\partial C}{\partial u_a} \frac{\partial u_a}{\partial y} \quad (5.27)$$

where

$u_a$ = pore–air pressure
$\partial u_a/\partial y$ = pore–air pressure gradient in the $y$-direction (or similarly in the $x$- and $z$-directions).

A modified form of Fick's law is obtained from Eq. (5.27) by introducing a coefficient of transmission for air flow through soils, $D_a^*$:

$$D_a^* = D_a \frac{\partial C}{\partial u_a} \quad (5.28)$$

or

$$D_a^* = D_a \frac{\partial[\rho_a(1-S)n]}{\partial u_a}. \quad (5.29)$$

The coefficient of transmission, $D_a^*$, is a function of the volume–mass properties of the soil (i.e., $S$ and $n$) and the air density. Substituting $D_a^*$ [i.e., Eq. (5.28)] into Eq. (5.27) results in the following form:

$$J_a = -D_a^* \frac{\partial u_a}{\partial y}. \quad (5.30)$$

This modified form of Fick's law has been used in geotechnical engineering to describe air flow through soils (Blight, 1971).

The coefficient of transmission, $D_a^*$, can be related to the air coefficient of permeability which is given the symbol, $k_a$. The air coefficient of permeability, $k_a$, is the value measured in the laboratory.

A steady-state air flow can be established through an unsaturated soil specimen with respect to an average matric suction or an average degree of saturation. The soil specimen is treated as an element of soil having one value for its air coefficient of permeability that corresponds to the average matric suction or degree of saturation. This means that the air coefficient of permeability is assumed to be constant throughout the soil specimen. Steady-state air flow is produced by applying an air pressure gradient across the two ends of the soil specimen. The amount of air flowing through the soil specimen is measured at the exit point as a flow under constant pressure conditions (i.e., usually at 101.3 kPa absolute or zero gauge pressure) (Matyas, 1967). In other words, the mass rate of the air flow is measured at a constant air density, $\rho_{ma}$. Equation (5.30) can be rewritten for this particular case as follows:

$$\rho_{ma} \frac{\partial V_a}{\partial t} = -D_a^* \frac{\partial u_a}{\partial y} \quad (5.31)$$

or

$$v_a = -D_a^* \frac{1}{\rho_{ma}} \frac{\partial u_a}{\partial y} \quad (5.32)$$

where

$\rho_{ma}$ = constant air density corresponding to the pressure used in the measurement of the mass rate (i.e., at the exit point of flow)
$\partial V_a/\partial t$ = volume rate of the air flow across a unit area of the soil at the exit point of flow; designated as flow rate, $v_a$.

The pore–air pressure, $u_a$, in Eq. (5.32) can also be expressed in terms of the pore-air pressure head, $h_a$, using a constant air density, $\rho_{ma}$:

$$v_a = -D_a^* \, g \, \frac{\partial h_a}{\partial y} \quad (5.33)$$

where

$h_a$ = pore-air pressure head (i.e., $u_a/\rho_{ma} g$)
$\partial h_a/\partial y$ = pore-air pressure head gradient in the $y$-direction; designated as $i_{ay}$.

Equation (5.33) has the same form as Darcy's equation for the air phase:

$$v_a = -k_a \frac{\partial h_a}{\partial y} \quad (5.34)$$

where the relationship between the air coefficient of transmission, $D_a^*$, and the air coefficient of permeability, $k_a$, is defined as follows:

$$k_a = D_a^* \, g. \quad (5.35)$$

The hydraulic head gradient in Eq. (5.34) consists of the pore-air pressure head gradient as the driving potential. Equation (5.34) has been used in geotechnical engineering to compute the air coefficient of permeability, $k_a$ (Barden, 1965; Matyas, 1967; Langfelder et al., 1968; Barden et al., 1969b; Barden and Pavlakis, 1971).

Air permeability measurements can be performed at various matric suctions or different degrees of saturation in order to establish the functional relationship, $k_a(u_a - u_w)$, or $k_a(S)$. This relationship also applies to the air coefficient of transmission, $D_a^*$, since the two coefficients are related by a constant, "$g$" [Eq. (5.35)].

Experimental verifications have been performed for Fick's and Darcy's law (Blight, 1971), and some results are presented in Fig. 5.15. A series of permeability tests was performed by establishing steady-state air flows through dry soils. The soils were assumed to have a rigid structure because no measurable volume change occurred during the tests. The flow measurements were referenced to the air-filled pore space (Blight, 1971). In order to use the bulk soil as the reference, the mass rate of air flow must be multiplied by the air porosity, $n_a$, as shown in Fig. 5.15(a).

The applicability of Fick's law to air flow [i.e., Eq. (5.30)] is demonstrated in Fig. 5.15(a). For a small change in the pore-air pressure gradient, the mass flow rate, $J_a$, is almost linearly proportional to the pore-air pressure gradient, ($\partial u_a/\partial y$), with $D_a^*$ being the coefficient of proportionality. It should be noted that the air pressure gradients used in the above experiment were high. The magnitudes of $D_a^*$ and $k_a$ vary with the volume-mass properties of an unsaturated soil.

### 5.2.3 Coefficient of Permeability with Respect to Air Phase

Several relationships have been proposed between the air coefficient of permeability and the volume-mass properties of a soil. The coefficient of transmission, $D_a^*$, can either be computed in accordance with Eq. (5.35) or measured directly in experiments. The coefficient of permeability for

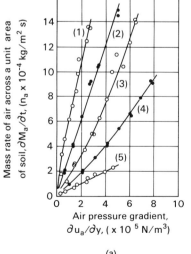
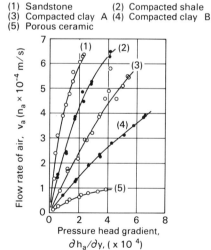

**Figure 5.15** Verifications of Fick's and Darcy's laws for air flow. (a) Mass rate of air flow versus air pressure gradient (Fick's law); (b) flow rate of air versus pressure gradient (Darcy's law).

the air phase, $k_a$, is a function of the fluid (i.e., air) and soil volume–mass properties, as described in Eq. (5.10). The fluid properties are generally considered to be constant during the flow process. Therefore, the air coefficient of permeability can be expressed as a function of the volume–mass properties of the soil. In this case, the volumetric percentage of air in the pores is an important factor since air flows through the pore space filled with air. As the matric suction increases or the degree of saturation decreases, the air coefficient of permeability increases.

*Relationship Between Air Coefficient of Permeability and Degree of Saturation*

The prediction of the air coefficient of permeability based on the pore size distribution and the matric suction versus degree of saturation curve has also been proposed for the air phase. The air coefficient permeability function, $k_a$, is essentially the inverse of the water coefficient of permeability function, $k_w$. The following equation has been used by Brooks and Corey (1964) to describe the $k_a(S_e)$ function:

$$k_a = 0.0 \quad \text{for } (u_a - u_w) \leq (u_a - u_w)_b$$

$$k_a = k_d(1 - S_e)^2(1 - S_e^{(2+\lambda)/\lambda})$$
$$\text{for } (u_a - u_w) > (u_a - u_w)_b \quad (5.36)$$

where

$k_d$ = coefficient of permeability with respect to the air phase for a soil at a degree of saturation of zero.

The values of $k_a$ at different degrees of saturation can be computed using Eq. (5.36) and expressed in terms of the relative coefficient of permeability of air, $k_{ra}$ (%):

$$k_{ra} = \frac{k_a(100)}{k_d}. \quad (5.37)$$

A comparison between Eq. (5.36) and experimental data is shown in Fig. 5.9 for Hygiene sandstone.

*Relationship Between Air Coefficient of Permeability and Matric Suction*

Another form of Eq. (5.36) is obtained when the effective degree of saturation, $S_e$, is expressed in terms of matric suction, as described in Eq. (5.15) (Brooks and Corey, 1964).

$$k_a = 0.0 \quad \text{for } (u_a - u_w) \leq (u_a - u_w)_b$$

$$k_a = k_d \left[1 - \left\{\frac{(u_a - u_w)_b}{(u_a - u_w)}\right\}^\lambda\right]^2 \left[1 - \left\{\frac{(u_a - u_w)_b}{(u_a - u_w)}\right\}^{2+\lambda}\right]$$
$$\text{for } (u_a - u_w) > (u_a - u_w)_b. \quad (5.38)$$

Figure 5.16 illustrates the agreement between measured data and the theoretical air coefficient of permeability function described using Eq. (5.38).

Several studies have been conducted on the air permea-

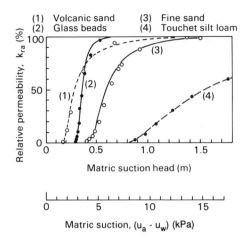

**Figure 5.16** Computed and measured data for the relative permeability of air as a function of matric suction (from Brooks and Corey, 1964).

bility of compacted soils. The coefficient of permeability with respect to air, $k_a$, decreases as the soil water content or degree of saturation increases (Ladd, 1960; Olson, 1963; Langfelder et al., 1968; Barden and Pavlakis, 1971). Figure 5.17 presents air and water coefficients of permeability for a soil compacted at different water contents or matric suction values. The air and water coefficients of permeability, $k_a$ and $k_w$, were measured on the same soil specimen during steady-state flow conditions induced by small pressure gradients (see Chapter 6). The air coefficient of permeability, $k_a$, decreases rapidly as the optimum water content is approached. At this point, the air phase becomes occluded, and the flow of air takes place as a diffusion of air through water. The occluded stage for soils with a high

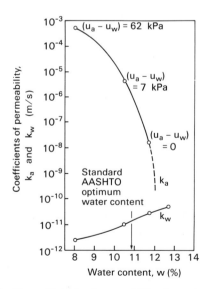

**Figure 5.17** Air coefficients of permeability, $k_a$, and water coefficients of permeability, $k_w$ as a function of the gravimetric water content for the Westwater soil (from Barden and Pavlakis, 1971).

clay content usually occurs at water contents higher than the optimum water content (Matyas, 1967; Barden and Pavlakis, 1971).

Although the air coefficient of permeability decreases and the water coefficient of permeability increases with increasing water content, the air permeability values remain significantly greater than the water permeability values for all water contents (Fig. 5.17). The difference in air and water viscosities is one of the reasons for the air coefficient of permeability being greater than the water coefficient of permeability.

The coefficient of permeability is inversely proportional to the absolute (i.e., dynamic) viscosity of the fluid, $\mu$, as shown in Eq. (5.10). The absolute viscosity of water, $\mu_w$, is approximately 56 times the absolute viscosity of air, $\mu_a$, at an absolute pressure of 101.3 kPa and a temperature of 20°C (see Chapter 2). Assuming that the volume–mass properties of a soil do not differ for completely saturated and completely dry conditions, the saturated water coefficient of permeability would be expected to be 56 times smaller than the air coefficient of permeability at the dry condition (Koorevar et al., 1983). It should be noted that this may not be the case for many soils.

Another factor affecting the measured air coefficient of permeability is the method of compaction. A dynamically compacted soil usually has a higher air coefficient of permeability than a statically compacted soil at the same density.

The air coefficient of transmission, $D_a^*$, can be obtained by dividing the air coefficient of permeability, $k_a$, by the gravitational acceleration, $g$. If the gravitational acceleration is assumed to be constant, $D_a^*$ functions are similar to the above air permeability, $k_a$, functions.

## 5.3 DIFFUSION

The diffusion process occurs in response to a concentration gradient. Ionic or molecular movement will take place from regions of higher concentration to regions of lower concentration. The air and water phase in a soil (i.e., soil voids) are the conducting media for diffusion processes. On the other hand, the soil structure determines the path length and cross-sectional area available for diffusion. The transport of gases (e.g., $O_2$ and $CO_2$), water vapor, and chemicals are examples of diffusion processes in soils.

There are two diffusion mechanisms common to unsaturated soil behavior. The first type of diffusion involves the flow of air through the pore–water in a saturated or unsaturated soil (Matyas, 1967; Barden and Sides, 1967). Another example of air diffusion involves the passage of air through the water in a high air entry ceramic disk. This type of diffusion involves gases dissolving into water and subsequently coming out of water, as explained in Chapter 2.

The second type of diffusion involves the movement of constituents through the water phase due to a chemical concentration gradient or an osmotic suction gradient. This type of diffusion process is not discussed in detail in this text. The following section considers the diffusion of air through water.

### 5.3.1 Air Diffusion Through Water

Fick's law can be used to describe the diffusion process (see Section 5.2.2). The concentration gradient which provides the driving potential for the diffusion process is expressed with respect to the soil voids (i.e., air and water phases). In other words, the mass rate of diffusion and the concentration gradient are expressed with respect to a unit area and a unit volume of the soil voids, respectively.

The formulation of Fick's law for diffusion in the $y$-direction is as follows:

$$\frac{\partial M}{\partial t} = -D \frac{\partial C}{\partial y} \qquad (5.39)$$

where

$\partial M/\partial t$ = mass rate of the air diffusing across a unit area of the soil voids
$D$ = coefficient of diffusion
$C$ = concentration of the diffusing air expressed in terms of mass per unit volume of the soil voids
$\partial C/\partial y$ = concentration gradient in the $y$-direction (or similarly in the $x$- or $z$-direction).

The diffusion equation can appear in several forms, similar to the forms presented for the flow of air through a porous medium. The concentration gradient for gases or water vapor (i.e., $\partial C/\partial y$) can be expressed in terms of their partial pressures. Consider a constituent diffusing through the pore–water in a soil. Equation (5.39) can be rewritten with respect to the partial pressure of the diffusing constituent:

$$\frac{\partial M}{\partial t} = -D \frac{\partial C}{\partial \bar{u}_i} \frac{\partial \bar{u}_i}{\partial y} \qquad (5.40)$$

where

$\bar{u}_i$ = partial pressure of the diffusing constituent
$\partial C/\partial \bar{u}_i$ = change in concentration with respect to a change in partial pressure
$\partial \bar{u}_i/\partial y$ = partial pressure gradient in the $y$-direction (or similarly in the $x$- or $z$-direction).

The mass rate of the constituent diffusing across a unit area of the soil voids (i.e., $\partial M/\partial t$) can also be determined by measuring the volume of the diffused constituent under constant pressure conditions. The ideal gas law is applied to the diffusing constituent in order to obtain the mass flow rate:

$$\frac{\partial M}{\partial t} = \bar{u}_{fi} \frac{\omega_i}{RT} \frac{\partial V_{fi}}{\partial t} \qquad (5.41)$$

where

$\bar{u}_{fi}$ = absolute constant pressure used in the volume measurement of the diffusing constituent
$\omega_i$ = molecular mass of the diffusing constituent
$R$ = universal (molar) gas constant
$T$ = absolute temperature
$\partial V_{fi}/\partial t$ = flow rate of the diffusing constituent across a unit area of the soil voids
$V_{fi}$ = volume of the diffusing constituent across a unit area of the soil voids.

The change in concentration of the diffusing constituent relative to a change in partial pressure (i.e., $\partial C/\partial \bar{u}_i$) is obtained by considering the change in density of the dissolved constituent in the pore-water. The density of the dissolved constituent in the pore-water is the ratio of the mass of dissolved constituent to the volume of water:

$$\frac{\partial C}{\partial \bar{u}_i} = \frac{\partial (M_{di}/V_w)}{\partial \bar{u}_i} \tag{5.42}$$

where

$M_{di}$ = mass of the dissolved constituent in the pore-water
$V_w$ = volume of water.

Applying the ideal gas law to Eq. (5.42) yields the following equation:

$$\frac{\partial C}{\partial \bar{u}_i} = \frac{\partial \left( \frac{V_{di}}{V_w} \bar{u}_i \frac{\omega_i}{RT} \right)}{\partial \bar{u}_i} \tag{5.43}$$

where

$V_{di}$ = volume of the dissolved constituent in the pore-water.

The ratio of the volume of dissolved constituent to the volume of water (i.e., $V_{di}/V_w$) is referred to as the volumetric coefficient of solubility, $h$. Under isothermal conditions, $h$ is essentially a constant (see Chapter 2).

$$\frac{\partial C}{\partial \bar{u}_i} = h \frac{\omega_i}{RT} \tag{5.44}$$

where

$h$ = volumetric coefficient of solubility for the constituent in water.

Substituting Eqs. (5.41) and (5.44) into Eq. (5.40) results in the following diffusion equation (van Amerongen, 1946):

$$v_{fi} = -\frac{Dh}{\bar{u}_{fi}} \frac{\partial \bar{u}_i}{\partial y} \tag{5.45}$$

where

$v_{fi}$ = flow rate of the diffusing constituent across a unit area of the soil voids (i.e., $\partial V_{fi}/\partial t$).

Equation (5.45) can be applied to air or gas diffusion through the pore-water in a soil or through free water or some other material such as a rubber membrane (Poulos, 1964). The partial pressure in Eq. (5.45) can be expressed in terms of the partial pressure head, $h_{fi}$, (i.e., $h_{fi} = \bar{u}_i/\rho_{fi} g$) with respect to the constituent density, $\rho_{fi}$. The density, $\rho_{fi}$, corresponds to the absolute constant pressure, $\bar{u}_{fi}$, used in the measurement of the diffusing constituent volume. The absolute constant pressure, $\bar{u}_{fi}$, is usually chosen to correspond to atmospheric conditions (i.e., 101.3 kPa), and $\rho_{fi}$ is the constituent density at the corresponding pressure.

$$v_{fi} = -Dh \frac{\rho_{fi} g}{\bar{u}_{fi}} \frac{\partial h_{fi}}{\partial y} \tag{5.46}$$

where

$\rho_{fi}$ = constituent density at the constant pressure, $\bar{u}_{fi}$, used in the volume measurement of the diffusing constituent
$h_{fi}$ = partial pressure head ($\bar{u}_{fi}/\rho_{fi} g$).

Equation (5.46) has a similar form to Darcy's law. Therefore, Eq. (5.46) can be considered as a modified form of Darcy's law for air flow through an unsaturated soil with occluded air bubbles where the air flow is of the diffusion form.

$$v_{fi} = -k_{fi} \frac{\partial h_{fi}}{\partial y} \tag{5.47}$$

where

$k_{fi}$ = diffusion coefficient of permeability for air through an unsaturated soil with occluded air bubbles.

The diffusion coefficient of permeability can then be written as follows:

$$k_{fi} = Dh \frac{\rho_{fi} g}{\bar{u}_{fi}}. \tag{5.48}$$

Substituting the ideal gas law into Eq. (5.48) results in another form for the diffusion coefficient of permeability:

$$k_{fi} = D \frac{h \omega_i g}{RT}. \tag{5.49}$$

Equation (5.49) indicates that under isothermal conditions, the coefficient of permeability (i.e., diffusion type) is directly proportional to the coefficient of diffusion since the term ($h\omega_i g/RT$) is a constant.

The coefficients of diffusion for several gases through water and for air through different materials are presented in Chapter 2. The diffusion coefficients, $D$, for air through

## Table 5.3 Summary of Flow Laws

| Type of Flow | Phase | Driving Potential | Flow Law (in y-Direction) | Comments |
|---|---|---|---|---|
| Bulk flow | Water | $h_w = y + \dfrac{u_w}{\rho_w g}$ | $v_w = -k_w \dfrac{\partial h_w}{\partial y}$ | Darcy's law |
| | Air | $u_a$ | $\dfrac{\partial M}{\partial t} = -D_a^* \dfrac{\partial u_a}{\partial y}$ | Fick's law |
| | | $h_a = \dfrac{u_a}{\rho_a g}$ | $v_a = -k_a \dfrac{\partial h_a}{\partial y}$ | Darcy's law $k_a = D_a^* g$ |
| Gas constituent diffusion | Gases, including water vapor and air through the pore-water in a soil | $C$ | $\dfrac{\partial M}{\partial t} = -D \dfrac{\partial C}{\partial y}$ | Fick's law $k_{fi} = D \dfrac{h \omega_i g}{RT}$ |
| | | $u_i$ | $v_{fi} = -k_{fi} \dfrac{\partial h_{fi}}{\partial y}$ | Darcy's law (obtained from Fick's, Henry's, and Darcy's laws) |
| Chemical diffusion | Pure water in osmotic diffusion | $\pi$ | $\dfrac{\partial M}{\partial t} = -D_o^* \dfrac{\partial \pi}{\partial y}$ | Fick's law |

water were computed in accordance with Eq. (5.47) (Barden and Sides, 1967). The diffusion values for porous media (e.g., soils) appear to be smaller than the diffusion values for free water. This has been attributed to factors such as the tortuosity within the soil and the higher viscosity of the adsorbed water close to the clay surface. As a result, the diffusion values decrease as the soil water content decreases.

### 5.3.2 Chemical Diffusion Through Water

The flow of water induced by an osmotic suction gradient (or a chemical concentration gradient) across a semi-permeable membrane can be expressed as follows:

$$\frac{\partial M}{\partial t} = -D_o \frac{\partial \pi}{\partial y} \quad (5.50)$$

where

$\partial M/\partial t$ = mass rate of pure water diffusion across a unit area of a semi-permeable membrane
$D_o$ = coefficient of diffusion with respect to osmotic suction (i.e., $D \partial C/\partial \pi$)
$C$ = concentration of the chemical
$\pi$ = osmotic suction

$\partial \pi/\partial y$ = osmotic suction gradient in the y-direction (or similarly in the x- or z-direction).

A semi-permeable membrane restricts the passage of the dissolved salts, but allows the passage of solvent molecules (e.g., water molecules). Clay soils may be considered as "leaky" semi-permeable membranes because of the negative charges on the clay surfaces (Barbour, 1987). Dissolved salts are not free to diffuse through clay particles because of the adsorption of the cations to the clay surfaces and the repulsion of the anions. This, however, may not completely restrict the passage of dissolved salts, as would be the case for a perfectly semi-permeable membrane. Therefore, pure water diffusion through a perfect, semi-permeable membrane [i.e., Eq. (5.50)] may not fully describe the flow mechanism related to the osmotic suction gradient in soils.

## 5.4 SUMMARY OF FLOW LAWS

Several flow laws related to the fluid phases of an unsaturated soil have been described in the preceding sections. A summary of the flow laws is given in Table 5.3.

# CHAPTER 6

# *Measurement of Permeability*

The application of flow laws to engineering problems requires the quantification of the hydraulic properties of a soil. The coefficient of permeability, $k$, in Darcy's law and the coefficient of diffusion, $D$, in Fick's law are examples of hydraulic properties. These properties must be determined using techniques which have been experimentally verified in order to obtain a reliable flow analysis for water and/or air movement in an unsaturated soil.

## 6.1 MEASUREMENT OF WATER COEFFICIENT OF PERMEABILITY

The water phase coefficient of permeability for a soil can be determined using either direct or indirect techniques. Direct measurements of permeability can be performed either in the laboratory or in the field. The direct measurements are commonly referred to as permeability tests, and the apparatus used in performing the test in the laboratory is called a permeameter.

Indirect methods can also be used to compute the coefficient of permeability. These methods use the volume-mass properties of the soil and the soil–water characteristic curve. The saturated coefficient of permeability is also required for the indirect prediction of the water phase permeability. The air permeability of the dry soil is required for the indirect prediction of the air phase permeability.

### 6.1.1 Direct Methods to Measure Water Coefficient of Permeability

The coefficient of permeability for a soil is best obtained from a direct measurement since there is no proven theoretical prediction (Hillel, 1982). The hydraulic head gradient and the flow rate are determined from pore–water pressure and water content measurements when using direct methods to measure permeability.

In some cases, either the pore–water pressure or the water content is measured, while the other variable is inferred from the soil–water characteristic curve. Measurements can be performed either in the laboratory or *in situ*. Laboratory tests are most economical, but *in situ* tests may better represent actual conditions. Unfortunately, the *in situ* field methods are not as advanced and standardized as the laboratory methods.

*Laboratory Test Methods*

Various laboratory methods can be used for measuring the coefficient of permeability with respect to the water phase, $k_w$, in unsaturated soils (Klute, 1972). All methods assume the validity of Darcy's law, which states that the coefficient of permeability is the ratio of the flow rate to the hydraulic head gradient. The flow rate and the hydraulic head gradient are the variables usually measured during a test. The flow rate and the hydraulic head gradient can either be kept constant with time (i.e., time independent) or varied with time during the test. Correspondingly, the various testing procedures can be categorized into two primary groups, namely, steady-state methods where the quantity of flow is time-independent, and unsteady-state methods where the quantity of flow is time-dependent.

The steady-state method is described first, followed by a description of the unsteady-state, instantaneous profile method. The measurement of the saturated coefficient of permeability has been described in numerous soil mechanics books, and is not repeated herein.

**Steady-state method.** The steady-state method for the measurement of the water coefficient of permeability is performed by maintaining a constant hydraulic head gradient across an unsaturated soil specimen. The matric suction and water content of the soil are also maintained constant. The constant hydraulic head gradient produces a steady-state water flow across the specimen. Steady-state conditions are achieved when the flow rate entering the soil is equal to the flow rate leaving the soil. The coefficient of permeability, $k_w$, which corresponds to the applied matric suction or water content, is computed. The experiment can be repeated for different magnitudes of matric suction or water content. The steady-state method can be used for both compacted and undisturbed specimens.

## 6.1 MEASUREMENT OF WATER COEFFICIENT OF PERMEABILITY

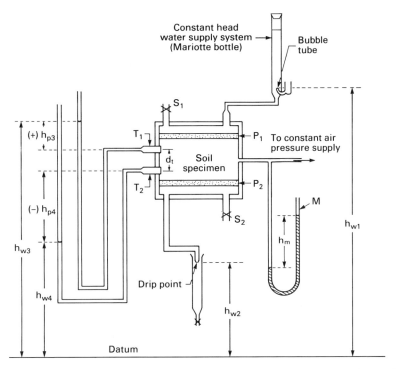

**Figure 6.1** Apparatus for the measurement of coefficient of permeability using the steady-state method (from Klute, 1965a).

**Apparatus for steady-state method.** An assemblage of the equipment used for the steady-state method is shown in Fig. 6.1 (Klute, 1965). A cylindrical soil specimen is placed in a permeameter between two high air entry ceramic plates: $P_1$ at the top of the specimen, and $P_2$ at the bottom of the specimen. Two tensiometers, $T_1$ and $T_2$, are installed along the height of the specimen for the measurement of pore-water pressures. An air supply maintains a constant pore-air pressure, which is measured using the manometer, $M$.

A water supply is applied to the top of the porous plate, $P_1$, to develop a constant hydraulic head gradient across the soil in the vertical direction. The water supply provides a constant hydraulic head, $h_{w1}$, by means of a Mariotte bottle, as shown in Fig. 6.2. It is also possible to use a simple overflow system. Water flows one-dimensionally through the ceramic plate, $P_1$, the soil specimen, and the ceramic plate, $P_2$. The outflow of water is maintained at a constant hydraulic head, $h_{w2}$, below the porous plate, $P_2$, by controlling the outflow elevation. Valves, $S_1$ and $S_2$, are used to flush air bubbles that may accumulate in the water compartment adjacent to the porous plate.

The permeability test is started using low matric suction values. The matric suctions are increased in steps as the permeability is measured. In other words, the test is commenced at a condition near saturation, and proceeds through a drying process in accordance with the following procedure. The matric suction, $(u_a - u_w)$, is set to a specified value by controlling the pore-air pressure, as indicated by the height in the manometer fluid, $h_m$. The pore-water pressure is measured using the tensiometers mounted along the specimen. The pore-water pressure head is determined from the elevation of the fluid interface in the attached manometer relative to the tensiometer elevation. The pressure

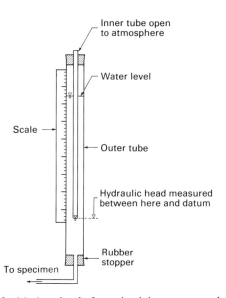

**Figure 6.2** Mariotte bottle for maintaining a constant hydraulic head and measuring the volume of water inflow (from U.S. Department of Interior, 1974).

head is positive when the fluid interface is above the elevation of the tensiometer, and is negative when the interface is below the tensiometer elevation. The tensiometer readings in Fig. 6.1 indicate pressure heads of (+) $h_{p3}$ and (−) $h_{p4}$ from tensiometers, $T_1$ and $T_2$, respectively.

A constant hydraulic head gradient is then applied to the specimen as depicted by hydraulic heads, $h_{w1}$ and $h_{w2}$. Steady-state flow conditions are achieved when the inflow and outflow rates are equal, and the tensiometer readings are constant with time. The volume of water, $Q$, flowing across the cross-sectional area of the soil, $A$, in a period of time, $t$, is measured and used to compute the flow rate. The hydraulic heads, $h_{w3}$ and $h_{w4}$, correspond to tensiometers, $T_1$ and $T_2$, which are placed a distance, $d_t$, apart. By knowing the distance, $d_t$, it is possible to compute the hydraulic head gradient in the soil. The measured hydraulic head gradient is used in calculating the coefficient of permeability owing to uncertainties associated with the hydraulic head changes across the porous plates. The contact planes between the specimen and the porous plates also produce uncertainties in predicting heads throughout the specimen (Klute, 1972).

**Computations using steady-state method.** The coefficient of permeability, $k_w$, is computed as follows:

$$k_w = \left(\frac{Q}{At}\right)\left(\frac{d_t}{h_{w3} - h_{w4}}\right). \quad (6.1)$$

The pore–air pressure is assumed to be uniform throughout the specimen:

$$u_a = \rho_m g h_m \quad (6.2)$$

where

$\rho_m$ = density of the manometer fluid
$g$ = gravitational acceleration
$h_m$ = height of the manometer fluid.

The applied hydraulic head gradient causes the pore-water pressures to differ at tensiometers, $T_1$ and $T_2$. The average pore–water pressure is computed as follows:

$$(u_w)_{\text{ave}} = \left(\frac{h_{p3} + h_{p4}}{2}\right)\rho_w g \quad (6.3)$$

where

$\rho_w$ = density of water
$(u_w)_{\text{ave}}$ = average of pore-water pressure.

The coefficient of permeability, $k_w$, computed using Eq. (6.1), corresponds to the average matric suction in the soil.

$$(u_a - u_w)_{\text{ave}} = \rho_m g h_m - \left(\frac{h_{p3} + h_{p4}}{2}\right)\rho_w g \quad (6.4)$$

where

$(u_a - u_w)_{\text{ave}}$ = average matric suction.

The water content of the soil specimen can be measured directly using either destructive or nondestructive techniques. Using a destructive technique, water contents are measured after each stage in the permeability test. The soil specimen can therefore be used for only one stage of the permeability test. Several "identical" specimens must be prepared in order to obtain permeabilities at different matric suctions or water contents.

Using a nondestructive technique, water contents at different matric suctions are measured using a single soil specimen. This can be done using a gamma attenuation technique, or by weighing the soil specimen (together with the apparatus) in order to obtain the change in water content after each stage of the test. The initial and final water contents are computed from the initial and final masses of the soil specimen. The mass of soil solids is obtained by oven-drying the specimen at the end of the test.

An indirect technique has also been used to infer the water content at a particular matric suction by referring to the soil–water characteristic curve. The curve would be determined from an independent specimen of the same soil (e.g., using a Tempe cell; see Section 6.1.2). Ingersoll (1981) showed that the soil–water characteristic curve and the coefficient of permeability could be obtained simultaneously using the apparatus shown in Fig. 6.1. In this case, the change in water content during each increment of matric suction was measured.

Having measured the coefficient of permeability, $k_w$, corresponding to a particular matric suction or water content, the permeability test is then repeated for higher values of matric suction. The matric suction of the soil specimen is increased by increasing the applied air pressure or by decreasing the applied water pressures. The average pore-water pressure can be decreased by reducing the values of $h_{w1}$ and $h_{w2}$. At no time should the matric suction exceed the air entry value of the ceramic plates, or the matric suction limit that can be read on the tensiometers (i.e., approximately 90 kPa). The use of higher air entry ceramic plates and psychrometers allows testing at higher values of suction. High air entry ceramic plates with the highest possible coefficient of permeability should be selected in the design of the apparatus (Klute, 1972).

**Presentation of water coefficients of permeability.** At the conclusion of the test, a series of coefficient of permeability values is obtained. A typical set of data for the coefficient of permeability, $k_w$, versus matric suction is shown in Fig. 6.3. The measured water coefficients of permeability correspond to the drying curve. There is hysteresis in the water coefficient of the permeability versus matric suction relationship. Hysteresis can be seen when the permeability test is performed for both the drying and the wetting processes. However, the steady-state method is commonly performed only when going from a saturated to an unsaturated condition.

Another apparatus which uses a steady-state flow system

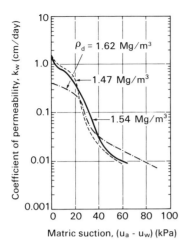

**Figure 6.3** Coefficient of water permeability as a function of matric suction obtained using the steady-state method (from Ingersoll, 1981).

is shown in Fig. 6.4 (Klute, 1965). A cylindrical soil specimen is placed between two high air entry ceramic plates located within an air pressure chamber. The soil specimen is subjected to an all-around controlled air pressure, $u_a$. The pore-water pressure, $u_w$, is measured by means of tensiometers placed along the length of the specimen. A constant hydraulic head gradient is applied across the specimen, as described in Fig. 6.1. The permeability test is performed using the procedure previously described. The matric suction in the specimen is increased by increasing the air pressure.

The specimen diameter is typically in the order of 25–100 mm, and its length ranges from 10 to 500 mm. The longer the length of the specimen, the longer is the time required to reach steady-state conditions. Therefore, the length of the specimen should be selected to be as short as possible. A specimen height in the order of 10–50 mm is suggested by Klute (1965). On the other hand, a short specimen may introduce inaccuracies in the pressure heads using the closely spaced tensiometers. A long specimen may require more than two tensiometers for accurate measure-

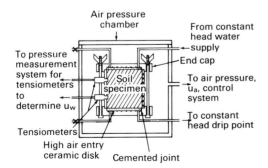

**Figure 6.4** Another apparatus used for the measurement of the coefficient of water permeability using the steady-state method (from Klute, 1965a).

ments of pore-water pressure. The steady-state method requires a long time for testing; however, the results are generally considered to be more accurate than those from the unsteady-state method (Hillel, 1982).

**Difficulties with the steady-state method.** There are several difficulties associated with the steady-state method (Klute, 1965, 1972, and Olson and Daniel, 1981). The main difficulty arises from the low coefficient of permeability of unsaturated soils, particularly at high matric suctions. As a result, the water flow rates during the test are extremely low, and a long time is required to complete a series of permeability measurements. The low flow rates necessitate extremely accurate volume of water measurements. The possibility of water loss within the apparatus must be minimized. Air diffusion through the water, and water loss from the apparatus (e.g., Lucite walls), can cause errors in the volume of water measurements. Air bubbles below the porous disks must be periodically removed by flushing. Diffused air inside the tensiometers should also be flushed from time to time. A diffused air volume indicator can be used to measure the amount of diffused air in order to apply the necessary corrections to the water volume change measurements (see Section 6.3.3).

Water is commonly used as a permeating fluid in the permeability test. However, in some cases, an osmotic suction gradient may develop between the pore-water within the soil and pure water. This will result in an osmotically induced flow across the specimen, in addition to the bulk flow related to the hydraulic head gradient. The significance of osmotic flow increases as the water content of the soil decreases. Osmotically induced flow is prevented when the permeating fluid has a similar chemical composition to that of the pore-water.

Another testing difficulty is related to the contact between the soil specimen and the permeameter. As matric suction increases, the soil specimen may shrink and separate from the permeameter wall and the high air entry ceramic plates. The air gap between the soil and the wall of the permeameter does not allow the seepage of water since air is nonconductive to the flow of water. The separation of soil from the permeameter wall generally does not create a significant problem in testing unsaturated soils (Daniel, 1983). However, a good contact between the soil specimen and the porous plates is required in order to ensure the continuity of water flow. The use of loaded porous plates has been suggested to overcome the problem of separation of the soil from the porous plates (Elrick and Bowman, 1964; Richards and Wilson, 1936). It is also important to maintain good contact between the soil and the tensiometers for accurate measurements of pore-water pressure.

**Instantaneous profile method.** The instantaneous profile method is an unsteady-state method that can be used either in the laboratory or *in situ*. The method uses a cylindrical specimen of soil that is subjected to a continuous water flow from one end of the specimen. The test method has several variations. These differ mainly in the flow process used and in the measurement of the hydraulic head gradient and the flow rate. The flow process can be a wetting process where water flows into the specimen, or a drying process where water flows out of the specimen.

The hydraulic head gradient and flow rate at various points along the specimen can be obtained using one of several procedures (Klute, 1972). Using the first procedure, the water content and pore–water pressure head distributions can be measured independently. The water content distribution can be used to compute the flow rates. The pore–water pressure head gradient can be calculated from the measured pore–water pressure head distribution. The gravitational head gradient is obtained from the elevation difference.

Using the second procedure, the water content distribution is measured while the pore–water pressure head is inferred from the soil–water characteristic curve. Using the third procedure, the pore–water pressure head distribution is measured, and the water content is inferred from the soil–water characteristic curve. Tensiometers and psychrometers have been used to measure the pore–water pressure distribution. Of the above procedures, the first appears to be most satisfactory. All variations of the instantaneous profile procedure are based on the same theoretical principles. The hydraulic head gradient and the flow rate are determined concurrently and instantaneously at various elapsed times, after the flow of water has commenced.

**Instantaneous profile method proposed by Hamilton et al.** Figure 6.5 shows the apparatus and procedure for the instantaneous profile method, proposed by Hamilton et al. (1981). A water flux is controlled at one end of the soil specimen, while the other end is vented to the atmosphere. The water flows in the horizontal direction as a result of a gradient in the pore–water pressure head. Gravitational head gradient effects are thereby negligible. Hamilton et al. 1981 elected to measure the pore–water pressure head distribution during the unsteady-state water flow, and obtained water contents from the soil–water characteristic curve. The hydraulic head gradient and the flow rate vary with time during the test. The pore–water pressures, and therefore the hydraulic head gradients, are measured at several points along the soil specimen.

The change in water content is related to the change in the negative pore–water pressure (or matric suction) through use of the soil–water characteristic curve. Flow rates are then computed from the change in volumetric water content. The ratio between the flow rate and the hydraulic head gradient gives the coefficient of permeability. Measurements at different locations along the specimen at different times during the unsteady-flow process produce a series of coefficients of permeability. Each permeability corresponds to a particular matric suction or water content. The method does not require the assumption of uniform hydraulic properties in the soil, as is the case in the steady-state method.

The following procedure is used to illustrate a permeability test performed using the wetting process. A compacted or undisturbed soil specimen is inserted into a cylindrical permeameter (Fig. 6.5). Both ends of the permeameter are covered by end plates with O-ring seals. Water is supplied to the left end plate using a hypodermic needle, and is distributed across the soil surface through the use of several sheets of filter paper. Air in the specimen is vented to the atmosphere using a hypodermic needle at the right end plate. Filter paper is also placed across the right end surface. Several ports are provided along the permeameter wall for the installation of tensiometers or psychrometers.

Tensiometers are used for a relatively moist soil having a matric suction less than approximately 90 kPa. Thermocouple psychrometers can be used for measuring suctions ranging from approximately 100 to 8000 kPa. The tensiometers or psychrometers are inserted through the ports in the permeameter and extended into small holes drilled into the soil specimen. The entire apparatus should be placed in a temperature-controlled chamber with a high relative humidity when psychrometers are used.

The test commences with an unsaturated specimen, and proceeds towards a saturated condition. The initial suction is first measured under equilibrium conditions. The equilibrium condition is then altered by slowly injecting water into the specimen using the hypodermic needle. The water inflow rate should be selected such that the pore–water pressure is always negative across the length of the specimen.

Figure 6.6 demonstrates three suction profiles across the specimen generated by three different inflow rates. A sharp wetting front or saturation in any part of the soil should be prevented. Flow rates in the range of 0.2–5 $cm^3$/day are commonly satisfactory (Daniel, 1983). Soil suctions are measured at various time intervals (e.g., every 24 h). Psychrometers can be replaced with tensiometers when the soil suction drops below 90 kPa.

The test is terminated when the pore–water pressure at the entrance of the permeameter (i.e., the left end) becomes positive. Positive pore–water pressures may pro-

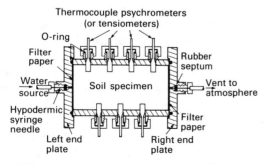

**Figure 6.5** Apparatus for measuring the coefficient of water permeability using the instantaneous profile method (from Hamilton et al. 1981).

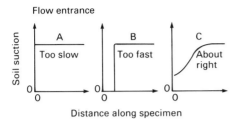

**Figure 6.6** Suction profiles associated with different water inflow rates (from Daniel, 1983).

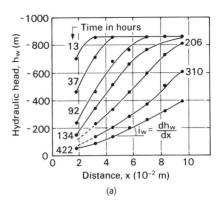

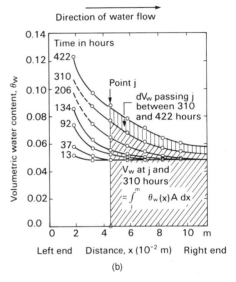

**Figure 6.7** Hydraulic head and water content profile during an unsteady-state flow test. (a) Profile of hydraulic head; (b) profile of volumetric water content (from Hamilton et al. 1981).

duce flow along the inside wall of permeameter or around the tensiometers. Upon completion of the test, the soil specimen is divided into several sections for final water content measurements. The final volumetric water contents along the specimen are plotted against the corresponding suctions to produce a soil–water characteristic curve. Results from independent measurements of the water content versus suction relationship can also be used in the determination of the soil–water characteristic curve. The entire test may take as long as two–three weeks.

**Computations for the instantaneous profile method.** The calculations for the coefficients of permeability are performed by plotting the pore-water pressure head versus the volumetric water content profiles from various points along the specimen, as illustrated in Fig. 6.7(a) and (b), respectively. The pore-water pressures are obtained from either the tensiometer or psychrometer readings. The total suction profile obtained from psychrometers can be taken as the negative pore-water pressure profile when the osmotic suction gradient is negligible and the air pressure is atmospheric. The pore-water pressure can be divided by the unit weight of water (i.e., $\rho_w g$) to give the pressure head. The hydraulic head gradient (i.e., pore-water pressure head gradient) at *a point* in the specimen *for a specific time* is equal to the slope of the hydraulic head profile [e.g., Fig. 6.7(a)]:

$$i_w = \frac{dh_w}{dx} \quad (6.5)$$

where

$i_w$ = hydraulic head gradient at a point for a specific time

$dh_w/dx$ = slope of the hydraulic head profile at the point under consideration.

The volumetric water content profile is obtained from the measured pore-water pressures and the soil–water characteristic curve. The absolute value of the negative pore-water pressure recorded by the tensiometer is equal to the matric suction. The flow rate, $v_w$, at a point is equal to the volume of water that flows across the cross-sectional area of the specimen, $A$, during a time interval, $dt$. Water flows from the left end of the permeameter to the right end, and there is no water flow out of the right end of the specimen. Therefore, the total volume of water passing through a point in the soil specimen during a period of time is equal to the water volume *change* that occurs between the point under consideration and the right end of the specimen during the specified time period.

The total volume of water present between point $j$ and the right end of the specimen (i.e., point $m$) at a *specific time* is obtained by integrating the volumetric water content profile over the specified time interval [Fig. 6.7(b)]:

$$V_w = \int_j^m \theta_w(x) A \, dx \quad (6.6)$$

where

$V_w$ = total volume of water in the soil between point $j$ and the right end of the specimen designated as point $m$

$\theta_w(x)$ = volumetric water content profile as a function of distance, $x$, for a specific time

$A$ = cross-sectional area of the specimen.

# 130  6 MEASUREMENT OF PERMEABILITY

The difference in volumes of water, $dV_w$, computed between two consecutive times (i.e., *an interval dt*) is the quantity of water flowing past point $j$ during the time interval under consideration [Fig. 6.7(b)]. The flow rate at the point is then computed as follows:

$$v_w = \frac{dV_w}{A dt}. \qquad (6.7)$$

The flow rate corresponds to an average value for the hydraulic head gradients obtained at two consecutive times. The coefficient of permeability, $k_w$, is computed by dividing the flow rate, $v_w$, by the average hydraulic head gradient, $i_{ave}$:

$$k_w = \frac{v_w}{i_{ave}}. \qquad (6.8)$$

Computations for the coefficient of permeability can be repeated for different points and different times. As a result, many coefficients of permeability can be computed corresponding to various water contents or suction values obtained from one test.

Figure 6.8 shows typical laboratory results of coefficient of permeability, $k_w$, for a clay as a function of suction. The permeability test was performed using the instantaneous profile method, and the data are presented in Fig. 6.7. Test procedural problems that may be encountered using the unsteady-state method are similar to those described previously for the steady-state method.

**In situ field method.** Nonhomogeneity and anisotropy of soils in the field make *in situ* measurements for the coefficient of permeability superior to laboratory measurements. Fissures, fractures, tension cracks, and root holes commonly encountered in unsaturated soils cannot be properly represented in small-scale laboratory specimens. Furthermore, laboratory specimens are subjected to sampling disturbance. On the other hand, *in situ* permeability tests have not been developed to the same extent as laboratory tests. Also, laboratory tests cost less than field tests. These are the main reasons why most testing has been performed in the laboratory.

The instantaneous profile method is generally considered to be the best method for permeability testing in the field (Klute, 1972). The procedure used in the field is similar in concept to the instantaneous profile method described for the laboratory. There are several variations to this method that have been investigated. These are summarized in Table 6.1.

The differences are mainly in the control of evaporation and in the manner used to determine the water content and the pore–water pressure profiles.

***In situ* instantaneous profile method.** Following is a description of an instantaneous profile, *in situ* test presented by Watson (1966) and Hillel *et al.* (1972). The method is also suitable for situations where the water table is deep and the soil is nonhomogeneous. The method is most suitable for soils of low plasticity. Numerous problems have been reported when using the test for highly plastic swelling soils. The test procedure by Watson (1966) and Hillel *et al.* (1972) considers a column of saturated soil that undergoes internal drainage while evaporation and infiltration are prevented. Water flows downward in response to the hydraulic head gradient. Water is assumed to flow in the vertical direction only, and the air phase is assumed to remain at atmospheric pressure conditions.

The pore–water pressure head and the volumetric water content profiles throughout a specified depth are measured during the unsteady-state flow process. Pore–water pressures are measured using a series of tensiometers at various depths. Water contents are measured using neutron moisture meter probes. Measurements are made at various elapsed times.

The pore–water pressure heads are added to the gravitational heads to give the hydraulic heads. The hydraulic head profiles are used to compute the hydraulic head gradients at various depths and at specified elapsed times. The flow rate at a point is computed from the change in the volumetric water content profile. A large number of coefficients or permeability can be computed for various volumetric water contents or matric suctions. These are calculated from the flow rates and the hydraulic head gradients [see Eq. (6.8)].

The field permeability test is commenced by selecting a representative plot, as shown in Fig. 6.9. The plot should be relatively large (e.g., 10 × 10 m) to eliminate boundary effects. All measurements should be performed close to the center of the plot. At least one neutron probe access tube should be installed near the center of the plot. It can be

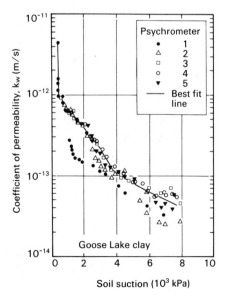

**Figure 6.8** Unsaturated coefficients of permeability obtained using the instantaneous profile method in the laboratory (from Hamilton *et al.* 1981).

**Table 6.1 Variations in the Instantaneous Profile Method for Field Permeability Tests (from Olson and Daniel, 1981)**

| Reference | Soil | Size of Plot, m | Evaporation Allowed? | Suction Probes | Water Content Probe | Maximum Depth of Probes, m |
|---|---|---|---|---|---|---|
| Richards et al. (1956) | Sandy loam | 2.6 × 2.6 | Yes | Tensiometers | Direct sampling | 0.8 |
| Ogata and Richards (1957) | Sandy loam | 2.6 × 2.6 | No | Tensiometers | Direct sampling | 0.5 |
| Nielsen et al. (1962) | Clay loam | 3.7 × 3.7 | No | Tensiometers | None | 1.5 |
| Rose et al. (1965) | Loam | — | Yes | None | Neutron method | 1.8 |
| Rose and Krishnan (1967) | Clayey sand | a | Yes | None | Neutron method | 1.6 |
| van Bavel et al. (1968) | Clay loam | 5 × 10 | No | Tensiometers | Neutron method | 1.6 |
| Davidson et al. (1969) | Loam and silty clay | 3 × 3 and 10 × 10 | No | Tensiometers | None | 1.8 |
| Hillel et al. (1972) | Sandy loam | 10 × 10 | No | Tensiometers | Neutron method | 1.5 |
| Nielsen et al. (1973) | Clay loam | 6.5 × 6.5 | No | Tensiometers | None | 1.8 |

*a* No special plot; the test was performed in the field after a heavy rain.

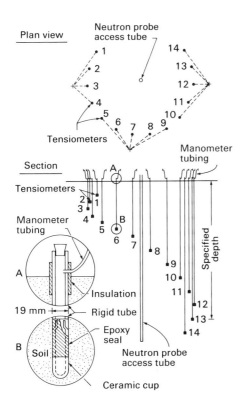

**Figure 6.9** Neutron probe and tensiometer arrangement for the instantaneous profile method in the field (from Wilson, 1971).

surrounded by several tensiometers located at various depths. The tensiometers should be installed in the proximity of the access tube without interfering with the neutron readings (e.g., at a distance of 0.50 m). In the vertical direction, the tensiometers should be embedded at depth intervals of about 0.30 m. The selected interval is primarily a function of the soil conditions.

The plot is first saturated by flooding with water until the soil profile becomes wet under steady-state infiltration conditions. Steady-state infiltration is indicated by constant readings on the tensiometers. The water flux across the ground surface is then stopped. The plot surface can be covered by plastic sheets to prevent evaporation or infiltration. The only subsequent process which occurs within the soil column is a downward seepage of water. During the subsequent unsteady-state process, the pore–water pressures and water contents are measured. The frequency of the measurements is highest near the beginning of the process. When the drainage process has slowed down, the readings can be less frequent.

**Computations for the *in situ* instantaneous profile method.** The calculation procedure for the coefficient of permeability, $k_w$, is similar to that described for the laboratory instantaneous profile test. The negative pore–water pressure heads at various depths are plotted for each elapsed time, as shown in Fig. 6.10(a). The pore–water pressure

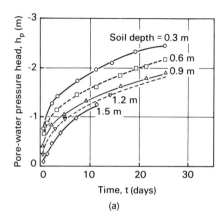

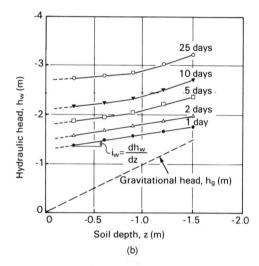

**Figure 6.10** Determination of the hydraulic head gradients during an unsteady-state flow test (a) Pore-water pressure heads at various times; (b) profile of hydraulic heads at various times. (from Hillel et al. 1972).

head, $h_p$, and the gravitational head, $h_g$, are added to give the hydraulic head, $h_w$. When water flows downward from the ground surface, it is reasonable to choose the ground surface as the datum for computing the gravitational head. The gravitational head is negative from the ground surface, as illustrated by the dashed line in Fig. 6.10(b). The summation of the pore-water pressure heads and the gravitational heads at each depth are used to profile the hydraulic head at various times:

$$h_w = h_g + h_p \quad (6.9)$$

where

$h_w$ = hydraulic head
$h_g$ = gravitational head
$h_p$ = pore-water pressure head.

The hydraulic head gradient at *a specific depth* for *a particular elapsed time* can be computed from the slope of the hydraulic head profile at that depth:

$$i_w = \frac{dh_w}{dz} \quad (6.10)$$

where

$i_w$ = hydraulic head gradient at a specific depth for a particular elapsed time
$dh_w/dz$ = slope of the hydraulic head profile at the depth under consideration.

The water flow rate is obtained from the volumetric water content profiles at various depths and times, as shown in Fig. 6.11. Water flows downward from the ground surface. There is no water flux at the ground surface. The total volume of water present between the ground surface (i.e., $z = 0$) and a depth $z$ below the ground surface at *a specified time* can be computed by integrating the volumetric water content profiles:

$$V_w = \int_o^z \theta_w(z) A dz \quad (6.11)$$

where

$V_w$ = volume of water in the soil between the ground surface and a depth, $z$
$\theta_w(z)$ = volumetric water content profile as a function of depth, $z$
$A$ = cross-sectional area of the plot.

The volume of water passing through a depth during a period of time is equal to the volumetric water content change in the region between the ground surface and the depth under consideration for a specific time period. The flow rate at the depth under consideration, $v_w$, is calculated in accordance with Eq. (6.7). The flow rate corresponds to an average hydraulic gradient computed at two consecutive times. The coefficient of permeability is the ratio of the flow rate, $v_w$, to the average hydraulic head gradient, $i_{ave}$, as given by Eq. (6.8). The computed coefficient of perme-

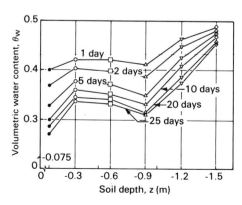

**Figure 6.11** Volumetric water content profiles during an *in situ* unsteady-state flow test (modified from Hillel et al. 1972).

ability, $k_w$, can also be related to the water content or the matric suction.

Coefficients of permeability can be computed for various depths and times. Figure 6.12 presents coefficients of permeability values as a function of volumetric water content. The permeability test was performed using the instantaneous profile method.

There is a special case where the hydraulic head gradient in Eq. (6.10) is unity during the unsteady-state flow process. This condition frequently occurs in a uniform, deeply wetted soil profile, in the absence of a shallow water table (Klute, 1972). A unity hydraulic head gradient indicates that the pore-water pressure head gradient is negligible, and only the gravitational head gradient is of consequence. In this case, the coefficient of permeability is equal to the flow rate.

The advantage of the *in situ* instantaneous profile method is that relatively simple equipment is required. The method is not applicable when there is significant water flow in the horizontal direction. Horizontal flow may occur as a result of either an impeding layer or a highly permeable layer. The method is not practical when the matric suctions exceed approximately 50 kPa because of the extremely slow drainage process.

### 6.1.2 Indirect Methods to Compute Water Coefficient of Permeability

Direct measurements of the water coefficient of permeability for an unsaturated soil are often difficult to perform. Attempts have been made to theoretically predict the coefficient of permeability on the basis of the soil pore size distribution. These predictions are commonly referred to as an indirect method to determine permeability.

The indirect method can be performed using either the matric suction versus degree of saturation curve or the soil-water characteristic curve. These curves are explained in Chapter 5. The prediction of coefficient of permeability from the matric suction versus degree of saturation curve has also been discussed in Chapter 5. Readers are referred to Brooks and Corey (1964, 1966) for further details.

The following discussion deals with the coefficient of permeability prediction using the soil-water characteristic curve. Equipment used in determining the soil-water characteristic curve is first presented, followed by a discussion of the analytical technique for computing the coefficients of permeability.

There are two commercially available pieces of equipment that are commonly used for the measurement of the soil-water characteristic curve. These are the Tempe pressure cell and the volumetric pressure plate extractor.[2] Both pieces of equipment operate on the same principle as the pressure plate apparatus (see Chapter 3 and ASTM D2325-68 (1974)).

A soil specimen is placed on a high air entry disk in the pressure chamber. The air pressure in the chamber can be raised to a prespecified value above atmospheric pressure (i.e., above zero gauge pressure). The air pressure applied to the soil causes the pore-water to drain. At equilibrium, the soil has a water content which corresponds to a specific matric suction. The matric suction of the soil is equal to the magnitude of the gauge air pressure in the chamber since the pore-water pressure is maintained at atmospheric conditions.

### Tempe Pressure Cell Apparatus and Test Procedure

A Tempe pressure cell assemblage is shown in Fig. 6.13, and its cross section is shown in Fig. 6.14. A soil specimen is placed on the high air entry disk inside the retaining cylinder of the Tempe pressure cell. An outlet tube located at the base plate underneath the high air entry disk allows the drainage of water from the soil specimen. Air pressure is supplied through the inlet tube on the top plate. The top and the bottom plates are fastened together during the test.

A test is started by saturating tne high air entry disk. Usually, the soil specimen is also saturated at the start of the test. After saturating the specimen, excess water is removed from the cell. The top plate is then mounted and tightened into place, and air pressure is applied to the specimen. The air pressure is set equal to the desired matric suction value.

Once the air pressure is applied, water starts draining from the specimen through the high air entry disk until equilibrium is reached. The matric suction in the soil is then equal to the applied air pressure. The time required to reach equilibrium depends upon the thickness and perme-

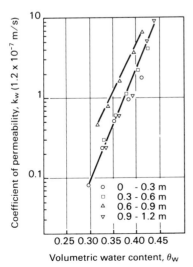

**Figure 6.12** Unsaturated coefficients of permeability measured using the field instantaneous profile method (after Hillel *et al.* 1972).

---

[2]Both pieces of equipment are manufactured by Soilmoisture Equipment Corporation, Santa Barbara, CA.

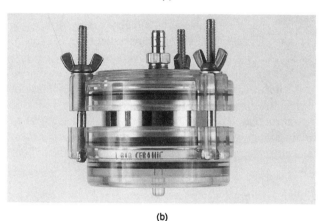

**Figure 6.13** Assemblage of the Tempe pressure cell. (a) Disassembled components of a Tempe cell; (b) assembled Tempe cell (from Soilmoisture Equipment Corporation, 1985).

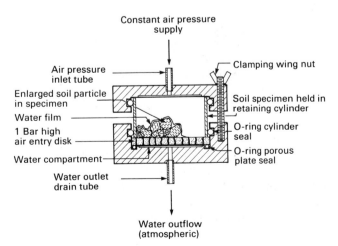

**Figure 6.14** Cross section of a Tempe pressure cell (from Soilmoisture Equipment Corporation, 1985).

ability of the specimen and the permeability of the high air entry disk. The change in water content is measured by weighing the specimen and the cell after equilibrium is reached.

The procedure is then repeated at higher applied air pressures (i.e., higher matric suctions). The Tempe pressure cell is commonly used for matric suctions up to 100 kPa. The cell containing the soil specimen must be weighed after equilibrium is attained for each applied pressure. Once the highest matric suction has been applied, the air pressure in the cell is released and the soil specimen is removed. The water content corresponding to the highest matric suction is measured by oven-drying the soil specimen. This water content, together with the previous changes in weight, are used to back-calculate the water contents corresponding to the other suction values. The matric suctions are then plotted against their corresponding water contents to give the soil–water characteristic curve.

### Volumetric Pressure Plate Extractor Apparatus and Test Procedure

Figure 6.15 shows an assemblage of a volumetric pressure plate extractor. The maximum matric suction that can be attained with the volumetric pressure plate extractor is 200 kPa. This apparatus can also be used to study the hysteresis of the soil–water characteristic curves associated with the drying and wetting of the soil. For this purpose, some hysteresis attachments are required, as outlined in Fig. 6.16.

The hysteresis attachments provide a more accurate volumetric measurement of water flow in or out of the soil specimen. The hysteresis attachments consist of a heater block, vapor saturator, air trap, ballast tube, and burette. The heater block is attached to the top plate to prevent water from condensing on the inside walls of the extractor. The heater block maintains the walls of the extractor at a slightly higher temperature than the soil temperature. Water condensation on the walls would introduce an error in determining the water content, particularly for long-term tests. The vapor saturator is used to completely saturate the inflow air to the volumetric pressure plate extractor. The saturated air surrounding the soil specimen will prevent the soil from drying by evaporation.

The air trap is provided to collect air that may diffuse through the high air entry disk. A "level mark" is provided on the stem of the air trap as a reference point in measuring the volume of water. A ballast tube is provided as a horizontal storage for water flowing in or out of the soil specimen under atmospheric conditions. A "level mark" on the ballast tub also serves as a reference point. A burette is used to store or supply water. The change in the volume of water in the burette during the equalization process is equal to the water volume change in the soil specimen.

The drying and wetting processes are performed on the

6.1 MEASUREMENT OF WATER COEFFICIENT OF PERMEABILITY 135

**Figure 6.15** Assemblage of the volumetric pressure plate extractor. (a) Disassembled components of a volumetric pressure plate extractor, (b) assembled volumetric pressure plate extractor (from Soilmoisture Equipment Corporation (1985)

same soil specimen when using the volumetric pressure plate extractor. During the drying process, the matric suction is increased, and pore-water drains from the specimen into the ballast tube. During the wetting process, the matric suction is decreased, and water in the ballast tube is absorbed by the specimen. The drying branch of the soil-water characteristic curve is first measured, followed by

measurements for the wetting branch. Drying and wetting curves can be further repeated subsequent to the first cycle.

### Test Procedure for the Volumetric Pressure Plate Extractor

The test procedure commences with the insertion of the soil specimen into the retaining ring, which is then placed on

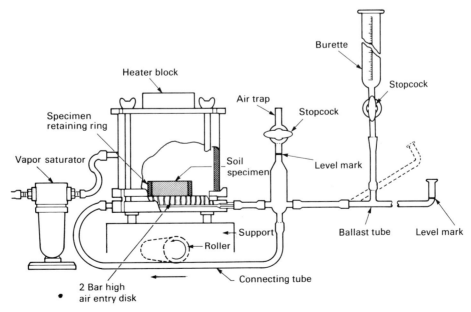

**Figure 6.16** Setup of the volumetric pressure plate extrator with its hysteresis attachments (from Soilmoisture Equipment Corporation, 1985).

top of the high air entry disk that has been saturated. The specimen is first saturated, and the hysteresis attachments are connected to the extractor. The hysteresis attachments are filled with water to the level marks. The air bubbles underneath the disk should be flushed prior to commencing the test by running the roller over the connecting tube. This action will pump water from the air trap through the grooves beneath the disk to force air bubbles into the air trap. The trapped air bubbles are then released by adjusting the water level in the air trap to the level mark. This is accomplished by opening the stopcock at the top of the air trap and applying a small vacuum to the outlet stem of the air trap. The vertical position of the ballast tube is made level with the top surface of the disk or with the center of the specimen by placing the extractor on a support.

### *Drying Portion of Soil–Water Characteristic Curve*

A starting point on the first drying curve should be established by applying a low matric suction to the soil specimen (i.e., raising the air pressure in the extractor to a low pressure). Water starts to drain from the specimen through the high air entry disk to the ballast tube. When the ballast tube is full of water, it should be drained to the burette. This is accomplished by applying a small vacuum to the top of the burette and opening the stopcock on the burette carefully. The outflow of water from the specimen stops when equilibrium is reached.

Diffused air is removed from the grooves underneath the disk using the pumping procedure described previously. During the test, air diffuses through the pore-water and the water in the high air entry disk, and comes out of the solution in the grooves beneath the disk (see detailed explanation in Section 6.3). The removal of diffused air should be performed each time equilibrium is reached. This produces a more accurate measurement of water flow from the specimen. Also, the accumulation of diffused air below the high air entry disk will prevent the uptake of water by the soil during the wetting process.

The above-described process is then repeated at increasing matric suctions (i.e., increasing air pressure in the extractor) until the drying curve is complete. Provision in the ballast tube should be made for the additional water outflow from the specimen at successively increasing matric suctions. The change in water volume reading in the burette for two successive equilibrium pressures provides the information necessary for the calculation of the water contents of the soil.

### *Wetting Portion of the Soil–Water Characteristic Curve*

Upon completion of the drying process, the test can be continued with the wetting process. The soil matric suction is reduced by decreasing the air pressure in the extractor. A decrease in the air pressure causes water to flow from the ballast tube back into the soil specimen. The water volume required for the backflow may be in excess of the water volume in the ballast tube. In this case, water should be added to the ballast tube by opening the burette stopcock. Equilibrium is reached when the water flow from the ballast tube into the specimen has stopped. Following equilibrium, the water levels in the air trap and the ballast tube are again adjusted to their level marks, and the burette reading is taken for computing the water volume change. The procedure is repeated at decreasing matric suctions until the desired range of the wetting curve is obtained. Subsequent cycles of drying and wetting can also be performed if desired.

The final water content of the specimen corresponding to the last matric suction is measured at the end of the test. The final water content and the water volume changes between two successive applied pressures are used to calculate the water content corresponding to each matric suction. A typical plot of matric suction versus water content for the drying and wetting processes is shown in Fig. 6.17. The plot shows the hysteresis effect between the drying and the wetting curves.

The volumetric pressure plate extractor has also been used to measure the coefficient of permeability in unsaturated soils using the outflow method described by Klute (1965b).

### *Computation of $k_w$ Using the Soil–Water Characteristic Curve*

The soil–water characteristic curve obtained from either the Tempe pressure cell or the volumetric pressure plate extractor can be used to compute the coefficient of permeability function [i.e., $k_w(\theta_w)$]. The following example is used to illustrate the technique by which the coefficient of permeability, $k_w(\theta_w)$, can be computed as a function of water content.

Let us consider the soil–water characteristic curve shown in Fig. 6.18. The computation of $k_w(\theta_w)$ from the drying curve is illustrated below. The drying curve is first divided

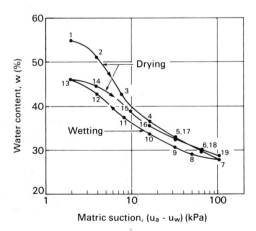

**Figure 6.17** Soil–water characteristic curves for the drying and wetting processes for Aiken clay loam (from Richards and Fireman, 1943).

6.1 MEASUREMENT OF WATER COEFFICIENT OF PERMEABILITY 137

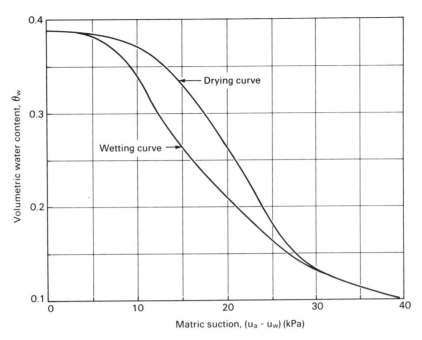

**Figure 6.18** Soil–water characteristic curve of find sand (i.e., tailings) (from Gonzalez and Adams, 1980).

into "$m$" equal intervals of volumetric water content, as shown in Fig. 6.19. In this case, the drying curve has a maximum and minimum volumetric water contents of 0.388 and 0.102, respectively. A division of the drying curve into 20 intervals with 20 midpoints is illustrated in Fig. 6.19. The first volumetric water content corresponds to saturated conditions (i.e., $(u_a - u_w)$ equal to zero). Each volumetric water content midpoint, $(\theta_w)_i$, corresponds to a particular matric suction, $(u_a - u_w)_i$. The midpoints are numbered starting from point 1 (i.e., "$i$" equal to 1) to point 20 (i.e., "$i$" equal to "$m$").

The permeability function, $k_w(\theta_w)$, is predicted in accor-

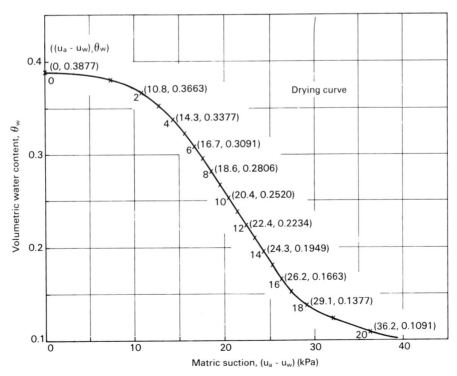

**Figure 6.19** Prediction for coefficient of permeability using a soil–water characteristic curve.

dance with the following equation (see Chapter 5):

$$k_w(\theta_w)_i = \frac{k_s}{k_{sc}} A_d \sum_{j=i}^{m} \{(2j + 1 - 2i)(u_a - u_w)_j^{-2}\}$$

$$i = 1, 2, \cdots, m \quad (6.12)$$

where

$k_w(\theta_w)_i$ = predicted water coefficient of permeability for a volumetric water content, $(\theta_w)_i$, corresponding to the $i$th interval (m/s)

$i$ = interval number which increases as the volumetric water content decreases. For example, $i = 1$ identifies the first interval which is close to the saturated volumetric water content, $\theta_s$; $i = m$ identifies the last interval corresponding to the lowest volumetric water content on the experimental soil-water characteristic curve, $\theta_L$

$j$ = a counter from "$i$" to "$m$"

$m$ = total number of intervals between the saturated volumetric water content, $\theta_s$, and the lowest volumetric water content on the experimental soil-water characteristic curve, $\theta_L$ (i.e., $m$ equal to 20)

$k_s$ = measured saturated coefficient of permeability (i.e., in this example, equal to $5.8 \times 10^{-8}$ m/s)

$k_{sc}$ = saturated coefficient of permeability (m/s)

$A_d$ = adjusting constant which is equal to

$$\frac{T_s^2 \rho_w g}{2\mu_w} \frac{\theta_s^p}{N^2} \text{ (m} \cdot \text{s}^{-1} \text{ kPa}^2)$$

$T_s$ = surface tension of water (kN/m)

$\rho_w$ = water density (kg/m$^3$)

$g$ = gravitational acceleration (m/s$^2$)

$\mu_w$ = absolute viscosity of water (N $\cdot$ s/m$^2$)

$\theta_s$ = volumetric water content at saturation or at a suction equal to zero

$p$ = a constant which accounts for the interaction of pores of various sizes; the magnitude of "$p$" can be set to 2.0 (Green and Corey, 1971a and 1971b)

$N$ = total number of intervals computed between the saturated volumetric water content, $\theta_s$, and zero volumetric water content (i.e., $\theta_w = 0.0$)

$(u_a - u_w)_j$ = matric suction corresponding to the $j$th interval (kPa).

The term $(\sum_{j=i}^{m} \{(2j + 1 - 2i)(u_a - u_w)^{-2}\})$ in Eq. (6.12) describes the shape of the permeability function. The $A_d$ term is used to factor or scale the coefficient of permeability function. The $A_d$ term is a constant when predicting the coefficients of permeability. However, the coefficient of permeability values, $k_w$, are adjusted in accordance with the saturated coefficient of permeability, $k_s$, by use of the term $(k_s/k_{sc})$. Therefore, any computed value of the $A_d$ term will not affect the final values of $k_w$. In this example, the term $A_d$ is not computed, but is assumed to be unity for simplicity in calculations.

The saturated coefficient of permeability, $k_s$, is independently measured in the laboratory, and in this example has a value of $5.83 \times 10^{-8}$ (m/s) (Gonzalez and Adams, 1980). The value of $k_{sc}$ is computed as follows:

$$k_{sc} = A_d \sum_{j=i}^{m} \{(2j + 1 - 2i)(u_a - u_w)_j^{-2}\}$$

$$i = 0, 1, 2, \cdots, m. \quad (6.13)$$

Substituting matric suctions corresponding to the midpoints along the drying curve into Eq. (6.13) and assuming that $A_d$ is equal to 1.0 results in a $k_{sc}$ value of 0.93 m/s. The computation of $k_{sc}$ includes the saturation volumetric water content, $\theta_s$, as point 0 (i.e., $i = 0$). The ratio of $k_s$ to $k_{sc}$ (i.e., $k_s/k_{sc}$ is equal to $6.28 \times 10^{-8}$) is used in all subsequent calculations for the unsaturated coefficients of permeability.

The coefficient of permeability corresponding to a specific midpoint, $(\theta_w)_i$, is computed in accordance with Eq. (6.12) using a $(k_s/k_{sc})$ ratio of $6.28 \times 10^{-8}$ and an $A_d$ value of 1.0. The permeability values, $k_w(\theta_w)_i$, are computed by substituting the matric suctions associated with the corresponding midpoints into Eq. (6.12). The computed coefficients of permeability along the drying curve are tabulated in Table 6.2 for 40 intervals (i.e., $m = 40$). A comparison between the computed and measured values, $k_w(\theta_w)_i$, is shown in Fig. 6.20. The measured permeability values, $k_w(\theta_w)i$, were obtained from a laboratory test using the steady-state method (Gonzalez and Adams, 1980). Similar computations of permeability, $k_w(\theta_w)_i$, can also be performed for the wetting curve. For both the wetting and drying curves, the $k_w(\theta_w)_i$ calculations should proceed from a low to a high matric suction.

## 6.2 MEASUREMENT OF AIR COEFFICIENT OF PERMEABILITY

There are direct and indirect methods available for obtaining the coefficient of permeability with respect to air, $k_a$. An indirect method for the calculation of the air coefficient of permeability has been explained in Chapter 5. This indirect method is based upon the matric suction versus degree of saturation curve (Brooks and Corey, 1964). The air coefficient of permeability can be computed as a function of the effective degree of saturation, $S_e$, or as a function of matric suction. This section presents direct methods for the measurement of the air coefficient of permeability using a triaxial cell permeameter.

The testing procedure used in the direct measurement for the water coefficient of permeability, $k_w$, can also be used in the measurement of the air coefficient of permeability,

**Table 6.2 Computed Coefficients of Permeability of Various Matric Suctions**

| Matric Suction, $(u_a - u_w)$ (kPa) | Coefficient of Permeability, $k_w(\theta_w)_i$ (m/s) | Matric Suction, $(u_a - u_w)$ (kPa) | Coefficient of Permeability, $k_w(\theta_w)_i$ (m/s) |
|---|---|---|---|
| 0.00  | $5.83 \times 10^{-8}$ | 21.42 | $9.44 \times 10^{-9}$ |
| 8.38  | $5.36 \times 10^{-8}$ | 21.80 | $8.36 \times 10^{-9}$ |
| 10.10 | $4.94 \times 10^{-8}$ | 22.22 | $7.35 \times 10^{-9}$ |
| 11.18 | $4.56 \times 10^{-8}$ | 22.62 | $6.42 \times 10^{-9}$ |
| 12.22 | $4.21 \times 10^{-8}$ | 23.10 | $5.57 \times 10^{-9}$ |
| 12.98 | $3.88 \times 10^{-8}$ | 23.52 | $4.79 \times 10^{-9}$ |
| 13.83 | $3.58 \times 10^{-8}$ | 23.90 | $4.08 \times 10^{-9}$ |
| 14.50 | $3.30 \times 10^{-8}$ | 24.34 | $3.43 \times 10^{-9}$ |
| 15.21 | $3.04 \times 10^{-8}$ | 24.59 | $2.85 \times 10^{-9}$ |
| 15.80 | $2.79 \times 10^{-8}$ | 25.30 | $2.33 \times 10^{-9}$ |
| 16.40 | $2.57 \times 10^{-8}$ | 25.83 | $1.87 \times 10^{-9}$ |
| 17.00 | $2.35 \times 10^{-8}$ | 26.40 | $1.47 \times 10^{-9}$ |
| 17.43 | $2.15 \times 10^{-8}$ | 27.10 | $1.13 \times 10^{-9}$ |
| 18.01 | $1.96 \times 10^{-8}$ | 27.79 | $8.34 \times 10^{-10}$ |
| 18.49 | $1.79 \times 10^{-8}$ | 28.61 | $5.90 \times 10^{-10}$ |
| 19.07 | $1.62 \times 10^{-8}$ | 29.68 | $3.93 \times 10^{-10}$ |
| 19.58 | $1.47 \times 10^{-8}$ | 31.02 | $2.41 \times 10^{-10}$ |
| 20.00 | $1.32 \times 10^{-8}$ | 32.78 | $1.30 \times 10^{-10}$ |
| 20.50 | $1.19 \times 10^{-8}$ | 34.80 | $5.49 \times 10^{-11}$ |
| 20.90 | $1.06 \times 10^{-8}$ | 37.78 | $1.31 \times 10^{-11}$ |

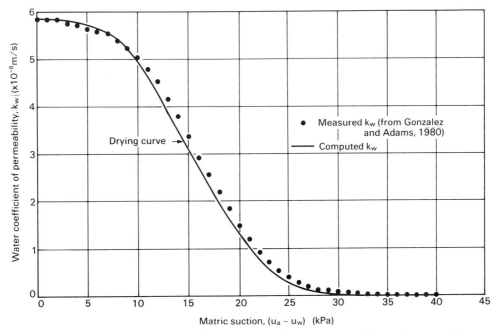

**Figure 6.20** Comparisons between the computed and measured coefficients of permeability.

$k_a$. In principle, permeabilities are measured at different water contents by changing the matric suction of the soil. A change in either the pore-air pressure, $u_a$, or the pore-water pressure, $u_w$, causes a change in matric suction. As a result, the volume of water in the soil changes. The total volume or the void ratio of the soil may or may not change significantly, depending upon the compressibility characteristics of the soil. These changes in the soil volume–mass properties (i.e., $S$, $e$, or $w$) alter the permeability coefficients for the soil.

An alternative permeability testing procedure involves altering the net normal stress, $(\sigma - u_a)$, and the matric suction, $(u_a - u_w)$, in order to change the volume–mass properties. This procedure has been used to obtain direct measurements of the air and water coefficients of permeability. An increase in the all-around stress, $\sigma$, on an unsaturated soil will result in a decrease in total volume, even when the water phase is undrained. The result is an increase in the degree of saturation and a decrease in matric suction. These changes affect the coefficients of permeability with respect to water, $k_w$, and air, $k_a$.

The net normal stress of an unsaturated soil can be altered using a one-dimensional oedometer (M.I.T., 1963; Barden et al., 1969) or a triaxial permeameter cell (Matyas, 1967; Barden et al., 1969; Barden and Pavlakis, 1971; Blight [57]). Permeability measurements performed using an oedometer sometimes encounter difficulties in obtaining a reliable seal between the soil specimen and the oedometer ring when the matric suction is increased (Barden and Pavlakis, 1971).

### Triaxial Permeameter Cell for the Measurement of Air Permeability

The triaxial permeameter cell shown in Fig. 6.21 was developed by Matyas (1967) to measure the air coefficient of permeability. A soil specimen is placed between two dry coarse porous stones. The soil specimen is subjected to an isotropic confining pressure, $\sigma$ (i.e., cell pressure). A constant air pressure is applied to the base of the specimen. The outflow of air from the top of the specimen is collected in a graduated burette having an air–oil interface. The air flow is measured under atmospheric pressure conditions by adjusting the two legs of the U-tube and maintaining the air–oil interfaces at the center elevation of the specimen.

The air coefficients of permeability for Sasumua clay were measured at various confining pressures using the above-described permeameter. A steady-state air flow was maintained during measurements of permeability.

Two specimens were compacted at an optimum water content of 60.5%. The standard AASHTO compaction curve for Sasumua clay, which contains a relatively high clay content, is shown in Fig. 6.22(a). Values for the air coefficient of permeability were obtained by applying Darcy's law. The results are plotted in Fig. 6.22(b) as a function of the confining pressure, $\sigma$. Decreasing air permeability values were observed as the confining pressure was increased or the matric suction was decreased.

### Triaxial Permeameter Cell for Air and Water Permeability Measurements

The triaxial permeameter cell developed by Barden et al. (1969) and Barden and Pavlakis (1971) can be used to measure both the water and air coefficients of permeability from one soil specimen (Fig. 6.23). The permeameter is designed to allow the independent control of the total stress, $\sigma$, the pore-air pressure, $u_a$, and the pore-water pressure, $u_w$. A soil specimen is usually subjected to an isotropic confining pressure, $\sigma$, although an anisotropic confining pressure could also be applied.

The stress state of the soil specimen is set to prescribed values prior to commencing the permeability measurements. After stress equilibrium conditions are obtained, the water and air coefficients of permeability are measured. Small pressure gradients in the water and air phases are applied to the specimen without changing the mean pore pressures. The induced flows of water and air are then measured under steady-state flow conditions. Measurements of the water and air coefficients of permeability, $k_w$ and $k_a$, can be repeated for various stress state variable conditions [i.e., $(\sigma - u_a)$ and $(u_a - u_w)$]. Changes in void ratio, $e$, and degree of saturation, $S$, of the soil are therefore varied during the measurement of permeability.

The triaxial permeameter has the advantage of being able to measure the water and air coefficients of permeability on the same soil specimen. A soil specimen is placed between the top and bottom high air entry disks (Fig. 6.23). Barden and Pavlakis (1971) used Aerox Celloton VI disks with an air entry value of 200 kPa and a coefficient of permeability with respect to water of $2.3 \times 10^{-3}$ m/s under saturated conditions. The water coefficients of permeability for the soil tested were on the order of $10^{-9}$ to $10^{-12}$ m/s for

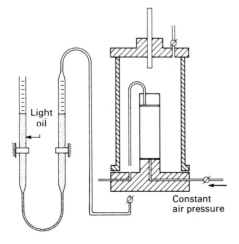

**Figure 6.21** Triaxial permeameter cell for measuring the air coefficient of permeability (from Matyas, 1967).

## 6.2 MEASUREMENT OF AIR COEFFICIENT OF PERMEABILITY

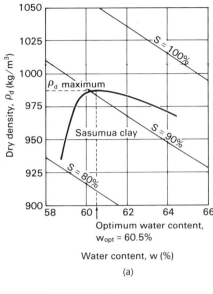

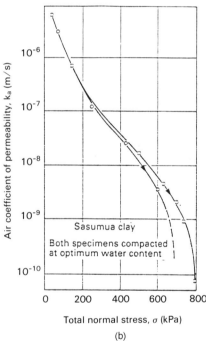

**Figure 6.22** Air coefficients of permeability for Sasumua clay obtained using a triaxial permeameter. (a) Standard AASHTO compaction curve; (b) measured air coefficient of permeability curves. (from Matyas, 1967).

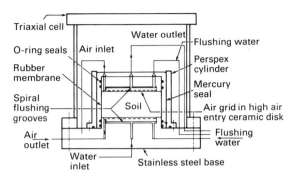

**Figure 6.23** Triaxial cell permeameter for measuring air and water coefficients of permeability (from Barden, *et al.* 1969b).

specimen. The arrangement for the water and air inlets and outlets is illustrated in Fig. 6.23, with more details given in Fig. 6.24. The inlet and outlet for water flow are connected directly to the top and bottom high air entry disks. Water must flow through both the high air entry disks and

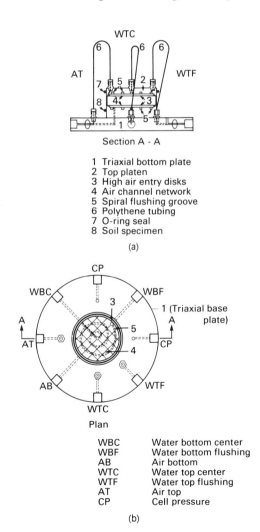

**Figure 6.24** Inlets and outlets for water and air flows in a triaxial cell permeameter. (a) Front view; (b) top view (from Barden and Pavlakis, 1971).

saturated to unsaturated conditions, respectively. It is important to use disks with a coefficient of permeability at least two orders of magnitude greater than the coefficient of permeability of the soil being tested. The high-permeability disks overcome the problem of impeded water flow.

The soil specimen is enclosed in a rubber membrane which is attached to the top and bottom plates by O-ring seals. Water flows from the bottom to the top of the specimen, while air flows from the top to the bottom of the

the soil specimen. The continuity of water flow is assured by periodically flushing air bubbles that may diffuse through the high air entry disks. The flushing mechanism is possible because of the spiral grooves cut in the top and bottom plates.

Air is applied directly to the soil specimen through a circular groove cut around the perimeter of the high air entry disks. These circular grooves are cross-connected by a network of channels on the surface of the disk. This allows a uniform distribution of the air to the soil surface. Air can flow through these channels without interfering with the water since the applied air pressure is always greater than the applied water pressure. On the other hand, air cannot flow through the high air entry disks as long as the matric suction of the soil is less than the air entry value of the disks. This arrangement maintains a separation in the flow of water and air.

Figure 6.25(a) shows a circuit drawing of the plumbing associated with the water phase. Constant water pressures can be applied to the top and the bottom of the specimen and measured using manometer, $M$. A small water pressure gradient in the upward direction can be applied to the soil specimen and measured using the differential manometer, $DM$. The inlet and outlet rates of water flow are measured using a horizontal constant bore glass capillary tube, GCT. The tube contains a small air bubble which is used as an indicator.

Figure 6.25(b) shows the circuit drawing for the plumbing associated with the air phase. The air pressure is maintained constant by use of the regulator, $R$. The air pressure is supplied to the top and bottom of the specimen through valves, $A$ and $B$. The air pressure can be measured on manometer, $M$.

A condensation chamber is placed between the regulator, $R$, and valves, $A$ and $B$, in order to collect condensing water vapor. The condensation chamber also maintains a saturated atmosphere in the air supply to the specimen. The air pressure can also be superimposed upon the confining pressure, $\sigma$, in order to maintain a constant net normal stress, $(\sigma - u_a)$. A small air pressure gradient can also be applied to the soil specimen and measured using the differential manometer, $DM$. A horizontal glass tube containing a dyed paraffin slug can be used to measure the rate of air flow during the constant head permeability test, as shown in Fig. 6.25(b).

The following example illustrates typical test results obtained using the triaxial permeameter cell. Water and air permeability measurements were performed on Backwater boulder clay (Barden and Pavlakis, 1971). After removing particles greater than the sand size, the soil was compacted in accordance with the standard AASHTO designation. The results of the compaction test for the Backwater soil are presented in Fig. 6.26. The soil specimen was 100 mm in diameter and 25 mm in thickness. The sample was compacted at a water content of 10.8% (i.e., drier than the optimum water content of 12.2%). Prior to placing the specimen in the triaxial permeameter, the water circuit was deaired by applying a pressure of 700 kPa and flushing for several days.

A soil specimen was installed, and an initial net normal stress, $(\sigma - u_a)$, was gradually built up in the soil specimen by adjusting the confining pressure, $\sigma$, and the air pressure. At equilibrium, the confining pressure, $\sigma$, was equal to 160 kPa and the pore-air pressure, $u_a$, was equal to 125 kPa. The pore-water pressure, $u_w$, was positive (i.e., 31 kPa) and could be measured using a pressure transducer below the base plate. The initial matric suction, $(u_a - u_w)$, was 94 kPa.

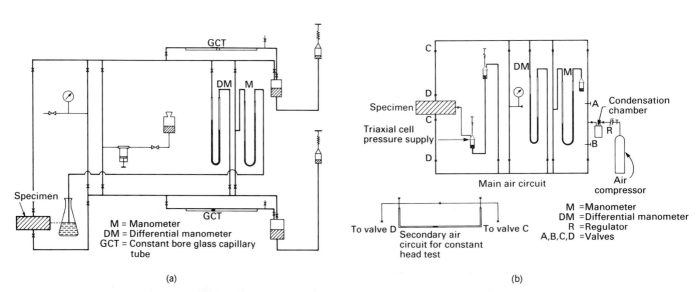

**Figure 6.25** Water and air circuit plumbing for the triaxial permeater cell. (a) Water circuit plumbing; (b) air circuit plumbing (from Barden and Pavlakis, 1971).

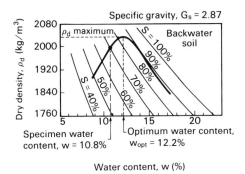

**Figure 6.26** Compaction test results for Backwater boulder clay (from Barden and Pavlakis, 1971).

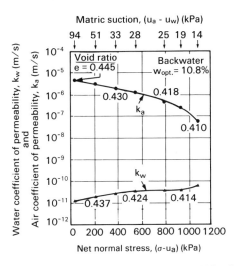

**Figure 6.28** Water and air coefficients of permeability for Backwater boulder clay obtained using a triaxial permeameter cell (from Barden and Pavlakis, 1971).

After equilibrium, permeability tests were performed by applying small pressure gradients to the water and air phases. The pressure gradients were applied such that the mean pore-water and pore-air pressure values remained constant at 31 and 125 kPa, respectively. Figure 6.27 illustrates an example of the pressure gradients in each phase. After establishing steady-state flow conditions, water and air flow rates were measured. This allowed the computation of the air and water coefficients of permeability for Backwater boulder clay corresponding to $(\sigma - u_a)$ and $(u_a - u_w)$ stress states of 35 and 94 kPa, respectively.

The test was repeated at higher confining pressure values, $(\sigma - u_a)$. The cell pressure, $\sigma$, was raised while maintaining an undrained condition for the water phase and a constant pore-air pressure of 125 kPa. Each increment of cell pressure caused the soil specimen to compress. However, the flow of water was not permitted. Only air was allowed to flow out of the specimen. After equilibrium was reached at each net confining pressure, $(\sigma - u_a)$, the pore-water pressure was measured to obtain a new value for matric suction. A higher value of pore-water pressure or a lower value of matric suction was measured at each increasing cell pressure. The permeability test was then repeated after the establishment of each new stress state.

The diffused air volume in the base plate should be measured in order to obtain the correct water volume change measurement (see Section 6.3).

The permeability test on Backwater boulder clay was performed at increasing values of net confining pressure, $(\sigma - u_a)$, from 35 to 1000 kPa, and at decreasing matric suction values from 94 to 14 kPa. Several sets of air and water permeability measurements are shown in Fig. 6.28. The results show decreasing air permeability values and increasing water permeability values as the soil specimen is compressed and the matric suction is decreased.

## 6.3 MEASUREMENT OF DIFFUSION

Long periods of time are often required for the laboratory testing of unsaturated soils. These long periods make it possible for air to diffuse through water. Although a high air entry disk is placed between the soil specimen and the measuring system to resist the passage of free air, air diffuses through the water in the disk (Bishop and Donald, 1961). The diffused air accumulates beneath the disk and introduces an error in either the pore-water pressure measurement associated with undrained tests or in the water volume change measurement associated with drained tests. Air diffusion is a common and important problem which must be addressed when testing unsaturated soils (Bishop, 1969).

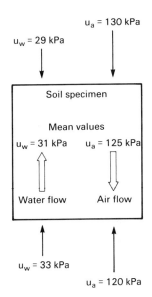

**Figure 6.27** Example showing the pressure gradient arrangement.

### 6.3.1 Mechanism of Air Diffusion Through High Air Entry Disks

*Undrained* shear strength tests with pore-water pressure measurements, can be performed by maintaining the water phase as a closed system. However, air can diffuse through the high air entry disk and come out of solution in the "compartment" below the disk, as illustrated in Fig. 6.29(a). Diffused air displaces the water in the "compartment" [Fig. 6.29(b)], and water is forced upward through the disk, back into the soil specimen. The water pressure measurement gradually changes from its original equilibrium value to the applied air pressure. In other words, the matric suction of the soil, which should tend towards a constant value, continuously decreases. This problem is common to null-type measurements of matric suction, as shown by the results in Fig. 6.30. The measured matric suction increases until the flow of diffused air forces water into the soil specimen, reducing the matric suction. Therefore, air diffusion imposes a limitation on this measuring technique.

*Drained* shear strength tests and volume change tests on unsaturated soils are performed by controlling the pore-water pressure, $u_w$, and the pore-air pressure, $u_a$. The measured water volume change consists of the flow of water to and from the soil specimen, plus the flow of diffused air through the water in the high air entry disk. Therefore, the diffused air volume should be measured independently and a correction applied to the measured water volume change.

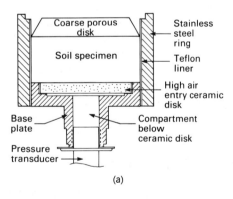

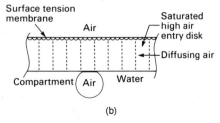

**Figure 6.29** Replacement of water from the "compartment" below the high air entry disk by diffused air. (a) Assembly showing the compartment below the high air entry disk; (b) air diffusion through a saturated high air entry disk (from Fredlund, 1975).

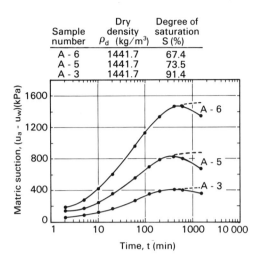

**Figure 6.30** Development of matric suction in statically compacted Regina clay (from Pufahl, 1970).

In addition, the diffused air beneath the disk can prevent water uptake should the soil specimen tend to expand or dilate during the test.

The volume of air diffusing through a high air entry disk over a period of several days can, in some cases, exceed the total volume of water in the soil specimen (Fredlund, 1973). A drained test extending more than one day should take into account the volume of diffused air if it is necessary to know the change in water content or degree of saturation (Fredlund, 1975).

The time required for air diffusion depends on the thickness of the high air entry disk and the matric suction being applied or measured. Attempts have been made to theoretically predict the volume of diffused air; however, there are many factors affecting the diffusion rate, making a theoretical prediction unreliable.

### 6.3.2 Measurements of the Coefficient of Diffusion

The mass rate of air diffusing across a unit area of a medium (i.e., soil, water, rubber membrane, etc.) is equal to the product of the coefficient of diffusion, $D$, and the air concentration gradient. The coefficient of diffusion for air through various media was measured by Barden and Sides (1967) using the apparatus shown in Fig. 6.31. A 0.5 mm thick latex rubber membrane was placed on the base of the specimen. The volume of diffused air was measured under atmospheric conditions using a 50 mm slug of water in a fine bore, horizontal glass tube.

Saturated and unsaturated clays were tested by Barden and Sides (1967). The saturated specimens were prepared by mixing dry powdered clay with distilled water to yield a water content equal to approximately twice the liquid limit. A vacuum was applied to the mixtures for 4 h. The mixture was then placed into the diffusion cell and consolidated to the desired pressure. The specimen was then trimmed to the required thickness.

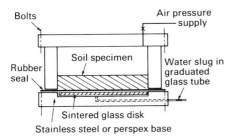

**Figure 6.31** Apparatus for measuring the coefficient of diffusion (from Barden and Sides, 1967).

Unsaturated specimens were prepared by compaction using one-half the standard AASHTO designation energy and a water content 5% wet of optimum. This procedure was assumed to yield an occluded air phase in the specimen. A mixture of clay and 4% bentonite was first consolidated in the diffusion cell. The center portion of the consolidated mixture was scribed out. This resulted in a thin-walled annulus which was used to seal the sides of the specimen against the cell wall. The soil specimen was then compacted inside this annulus.

An air pressure of varying magnitudes was applied to the top of the specimen. The applied air pressure established a pressure gradient across the specimen relative to the atmospheric pressure below the specimen. The pressure at each stage was held constant until steady-state diffused air flow was observed. The air pressure was maintained for an additional 48 h, and the resulting flow rate was measured.

The flow rates for air, $v_{fa}$, diffusing through kaolin are plotted against the applied air pressure in Fig. 6.32. The flow rate was computed as the diffused air volume flow rate divided by the area of soil voids. The area of soil voids (i.e., water and air) was computed as the cross-sectional area multiplied by the soil porosity.

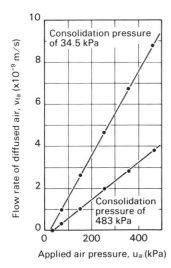

**Figure 6.32** Air diffusion test results for kaolin (from Barden and Sides, 1967).

### 6.3 MEASUREMENT OF DIFFUSION   145

*Procedure for Computing Diffusion Properties*

The coefficient of diffusion, $D$, of air through soil can be computed by considering a layered system consisting of the soil specimen and the rubber membrane, as illustrated in Fig. 6.33. The analysis is developed in terms of permeability, which is then converted to a coefficient of diffusion. An equivalent system can be used to replace the layered system. Continuity of one-dimensional flow requires that the flow rate be the same through each layer. The sum of the changes in hydraulic head for each layer must equal the total change in hydraulic head for the equivalent system:

$$\left(\frac{L_f + L_r}{k_{ea}}\right) v_f = \left(\frac{L_f}{k_{fa}} + \frac{L_r}{k_{ra}}\right) v_{fa}. \quad (6.14)$$

The equivalent coefficient of permeability for the layered system, $k_{ea}$, can be written as

$$k_{ea} = \frac{L_f + L_r}{\left(\dfrac{L_f}{k_{fa}} + \dfrac{L_r}{k_{ra}}\right)}. \quad (6.15)$$

The equivalent coefficient of permeability, $k_{ea}$, is related to the coefficient of permeability for each layer (i.e., $k_{fa}$ and $k_{ra}$) and the thickness of each layer (i.e., $L_f$ and $L_r$). The equivalent coefficient of permeability for the soil specimen and rubber membrane system, $k_{ea}$, can also be written in accordance with Darcy's law (see Chapter 5):

$$v_{fa} = -k_{ea}\left(\frac{\Delta h_{ea}}{L_f + L_r}\right) \quad (6.16)$$

where

$v_{fa}$ = diffused air flow rate measured at atmospheric conditions (i.e., 101.3 kPa absolute pressure)

$k_{ea}$ = equivalent coefficient of permeability for the soil specimen and rubber membrane system

$\Delta h_{ea}$ = total air pressure change in terms of hydraulic head across the system [i.e., $\Delta h_{ea} = \Delta u_a/(\rho_{fa} g)$]

$\Delta u_a$ = change in air pressure (i.e., gauge or absolute) across the system; in the above case, $\Delta u_a = u_a$ since the air pressure is atmospheric beneath the membrane

$\rho_{fa}$ = air density at an absolute pressure of 101.3 kPa; used in the flow rate measurement

$L_f, L_r$ = thickness of the soil and the rubber membrane, respectively.

Equation (6.16) can be used to compute the equivalent coefficient of permeability, $k_{ea}$, using the diffused air flow rate, $v_{fa}$, measurements. Substituting the value of $k_{ea}$ obtained from Eq. (6.16) into Eq. (6.15) gives the coefficient of permeability (i.e., a diffusion-type coefficient) for air

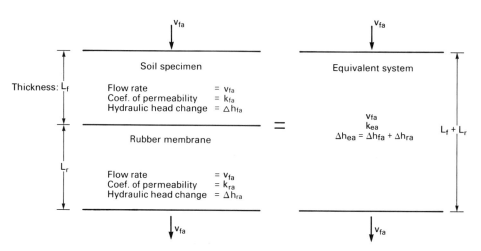

**Figure 6.33** Equivalent permeability for a layered system.

through soil, $k_{fa}$. The coefficient of permeability for the rubber membrane, $k_{ra}$, must be measured independently.

The coefficient of diffusion, $D$, for air through the soil pore–water can be computed from the coefficient of permeability, $k_{fa}$, as derived in Chapter 5:

$$k_{fa} = Dh \frac{\rho_{fa} g}{\bar{u}_{fa}} \quad (6.17)$$

where

$D$ = coefficient of diffusion for air through the soil pore–water

$h$ = volumetric coefficient of solubility for air in water, usually taken as 0.02

$\bar{u}_{fa}$ = absolute constant air pressure applied while measuring the flow rate of diffused air (i.e., 101.3 kPa).

Computed coefficients of diffusion for air through various soils are presented in Chapter 2. There is need for further studies which measure the coefficients of diffusion for air through the pore–water of other soils.

### 6.3.3 Diffused Air Volume Indicators

Several devices have been proposed to measure the volume of diffused air, and thereby apply a correction to measurements of overall water volume change. One example is the air trap and roller used in conjunction with the volumetric pressure plate apparatus (see Section 6.1.2). The measured inflow or outflow of water is then adjusted by releasing the trapped diffused air. This adjustment is a diffused air correction applied to the overall water volume change measurement.

Two other devices have been used to measure the volume of diffused air, namely, "the bubble pump" (Bishop and Donald, 1961) and "a diffused air volume indicator (DAVI)," consisting of an inverted burette (Fredlund, 1975).

#### Bubble Pump to Measure Diffused Air Volume

Figure 6.34 shows a perspex bubble pump connected to the compartment below a high air entry disk in a triaxial cell. A titled U-tube containing mercury is used to circulate water through the base plate and to collect the diffused air in a calibrated bubble trap. This is accomplished by rocking the U-tube to establish a differential pressure of about 1.0 kPa across the base plate. This action is continued for approximately 30 s.

Flexible connecting tubes and perspex valves with soft rubber washers enable the mercury in the U-tube to act as a piston in a double-acting pump during the rocking process. The volume of diffused air is measured in the 5 cm$^3$ bubble trap.

#### Diffused Air Volume Indicator (DAVI)

The diffused air volume indicator proposed by Fredlund (1975) differs from the bubble pump in that it is relatively simple to construct and can function under a substantial backpressure. A variable pressure gradient of 7–70 kPa can be established across the base plate. Figure 6.35 shows the layout of the diffused air volume indicator which can be connected to the base plate of a triaxial or oedometer apparatus.

The diffused air volume indicator consists of a 10 cm$^3$ graduated burette and a Lucite exit tube, placed inside an 80 mm diameter Lucite cylinder. A vent on the top of the burette can be opened when filling water into the burette. The vent remains closed during operation of the indicator. Two O-rings are placed around the top of the burette to form a seal with the Lucite cylinder. The base of the bu-

6.3 MEASUREMENT OF DIFFUSION    147

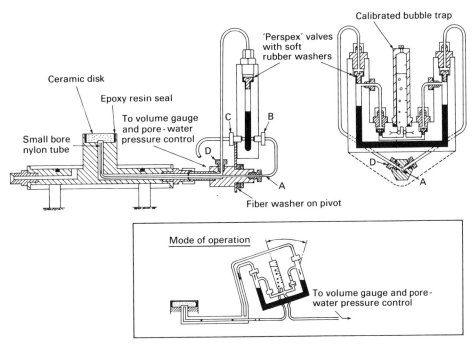

**Figure 6.34** Bubble pump and air trap (from Bishop and Donald, 1961).

rette is sealed against the Lucite cylinder using a threaded brass sleeve tightened against an O-ring. The exit tube is connected to the burette to form a U-tube. A constant air pressure, $u_{ab}$, is supplied into the Lucite cylinder to establish a pressure gradient relative to the controlled water pressure below the ceramic disk.

The burette is first filled with water to any desired level. The commercial cleaner called "Fantastik" should be added to the water in the burette. This substance has a low surface tension and promotes the upward movement of air bubbles in the burette.

The valve connecting the base plate and the diffused air

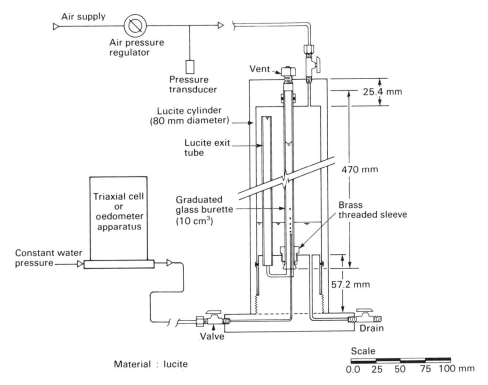

**Figure 6.35** Diffused air volume indicator (from Fredlund, 1975).

volume indicator is only opened when making a diffused air volume measurement. The water pressure below the disk is controlled at a constant pressure, $u_c$. A backpressure, $u_{ab}$ (i.e., $u_{ab} < u_c$), can be applied to the air in the Lucite cylinder in order to establish a pressure gradient of up to a maximum of 70 kPa across the base plate. The pressure gradient between the water pressure in the base plate and the air pressure in the diffused air volume indicator is the driving force for flushing water containing air bubbles from the base plate into the burette. The pressure gradient should not exceed 70 kPa in order to avoid cavitation of the water due to the spontaneous release of dissolved air from the water below the base plate.

### Procedure for Measuring Diffused Air Volume

The initial water level reading in the burette, $h_{a1}$ (see Fig. 6.36) can be taken when the air pressure in the Lucite cylinder is at atmospheric pressure or at the applied backpressure (i.e., at the air pressure, $u_{ab}$). The connecting valve is then opened momentarily, applying a pressure gradient across the base plate. This flushes the diffused air into the burette. The flushed diffused air displaces the water in the burette while water flows out of the exit tube. After removing all of the diffused air from the base plate, a new water level reading, $h_{a2}$, is recorded.

### Computation of Diffused Air Volume

The diffused air volume in the base plate, $V_{fa}$, is registered on the water volume change indicator as though it were water leaving the soil specimen. The measured water volume changes from the soil specimen must therefore be corrected for the volume of diffused air.

The diffused air pressure below the high air entry disk is assumed to be equal to the water pressure in the base plate compartment, $u_c$. The diffused air mass, $M_{fa}$, can be expressed as follows:

$$M_{fa} = \frac{\omega_a}{RT} \bar{u}_{fa} V_{fa} \quad (6.18)$$

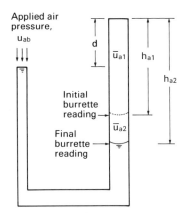

**Figure 6.36** Computations for the diffused air volume under isothermal conditions.

where

$M_{fa}$ = mass of diffused air
$\omega_a$ = molecular mass of diffused air
$R$ = universal (molar) gas constant
$T$ = absolute temperature (i.e., $T = 273.16 + t°$ where $t°$ = temperature in degrees Celsius)
$\bar{u}_{fa}$ = absolute diffused air pressure in the base plate (i.e., $\bar{u}_{fa} = \bar{u}_{atm} + u_c$)
$\bar{u}_{atm}$ = atmospheric pressure (i.e., 101.3 kPa)
$u_c$ = water pressure in the base plate
$V_{fa}$ = volume of diffused air in the base plate.

The diffused air volume in the base plate, $V_{fa}$, is computed by applying the ideal gas law to the air volume measured in the diffused air volume indicator. Let us consider the initial and final conditions in the diffused air volume indicator shown in the Fig. 6.36. The initial mass of air in the burette, $M_{a1}$, can be computed from its absolute pressure, $\bar{u}_{a1}$, and its volume, $V_{a1}$:

$$M_{a1} = \frac{\omega_a}{RT} \bar{u}_{a1} V_{a1} \quad (6.19)$$

where

$M_{a1}$ = initial mass of air in the burette
$\bar{u}_{a1}$ = absolute initial air pressure in the burette (i.e., $\bar{u}_{atm} + u_{ab} + (h_{a1} - d)\rho_w g$)
$u_{ab}$ = gauge applied air pressure in the Lucite cylinder
$h_{a1}$ = initial reading of the water level in the burette
$d$ = height difference between the burette and the exit tube
$V_{a1}$ = initial volume of air in the burette (i.e., $h_{a1}A$)
$A$ = cross-sectional area of the burette.

Similarly, the final mass of air in the burette can be expressed as

$$M_{a2} = \frac{\omega_a}{RT} \bar{u}_{a2} V_{a2} \quad (6.20)$$

where

$M_{a2}$ = final mass of air in the burette
$\bar{u}_{a2}$ = absolute final air pressure in the burette (i.e., $\bar{u}_{atm} + u_{ab} + (h_{a2} - d)\rho_w g$)
$h_{a2}$ = final reading of the water level in the burette
$V_{a2}$ = final volume of air in the burette (i.e., $h_{a2}A$).

When the initial and final burette readings are at atmospheric pressure, $u_{ab}$ is equal to zero in Eqs. (6.19) and (6.20). The difference between the initial and the final mass of air in the burette (i.e., $M_{a2} - M_{a1}$) is the amount of diffused air removed from the base plate, $M_{fa}$. The corresponding mathematical relationship is used to compute the diffused air volume, $V_{fa}$, under isothermal conditions:

$$\bar{u}_{a2}V_{a2} - \bar{u}_{a1}V_{a1} = \bar{u}_{fa}V_{fa}. \quad (6.21)$$

Eq. (6.21) can be solved for the diffused air volume

$$V_{fa} = \frac{\bar{u}_{a2}V_{a2} - \bar{u}_{a1}V_{a1}}{\bar{u}_{fa}}. \quad (6.22)$$

It is also possible to compute the amount of diffused air removed from the base plate, $M_{fa}$, by reading the initial and final water levels in the burette at the same applied air pressure (Fredlund, 1975). For example, if the initial burette reading was taken at atmospheric pressure, the air in the Lucite cylinder should be depressurized to atmospheric pressure conditions before the final reading is recorded. The diffused air volume is computed as,

$$V_{fa} = \frac{\bar{u}_{a1}(V_{a2} - V_{a1})}{\bar{u}_{fa}}. \quad (6.23)$$

The base plate is generally flushed once or twice a day. Corrections to intermediate water volume change readings can be made using a linear interpolation of diffused air flow with respect to time. There are no temperature corrections necessary, provided the temperatures of the diffused air volume indicator and the base plate of the triaxial or oedometer apparatus are the same.

### Accuracy of the Diffused Air Volume Indicator

The accuracy of the diffused air volume indicator has been tested by diffusing air through a saturated high air entry ceramic disk. An air pressure is applied to the top of a saturated ceramic disk. The saturated disk is connected to the water volume change indicator. A change in the water volume reading with time indicates the accumulation of diffused air below the high air entry disk. The ceramic disk

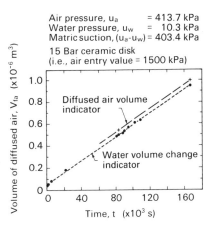

**Figure 6.37** Steady-state diffusion of air through a saturated ceramic disk measured using a water volume change indicator and a diffused air volume indicator (from Fredlund, 1973).

remains saturated since the air entry value of the disk is not exceeded.

The diffused air volume can be measured periodically using the diffused air volume indicator and continuously using the water volume change indicator. The two results should be essentially the same if the diffused air volume indicator performs satisfactorily.

Figure 6.37 compares volumes of diffused air measured on the water volume change indicator and on the diffused air volume indicator. The air diffusion test was performed on a saturated, 15 bar ceramic disk (i.e., 1500 kPa air entry value). Both the water volume change indicator and the diffused air volume indicators recorded essentially the same volumes of diffused air over the period of two days. The straight lines indicate steady-state diffusion of air through water in the high air entry disk.

# CHAPTER 7

# Steady-State Flow

Engineers are often interested in knowing the direction and quantity of flow through porous media. The pore pressure variations resulting from the flow process are also of interest. This information is required in predicting the volume change and shear strength change associated with the flow of water or air.

Chapter 5 outlined the driving potentials and the flow laws that govern the behavior of water flow, air flow, and air diffusion through water. The hydraulic properties of the soil with respect to each fluid phase were given in terms of coefficients of permeability. These properties are required in the application of the flow laws to engineering problems. In Chapter 6, various methods commonly used to obtain coefficients of permeability were described. This chapter presents the application of the flow laws and the associated coefficients of permeability to the analysis of practical seepage problems. The analysis is performed for air and water flow under isothermal conditions. The effect of air diffusion through water, air dissolving into water, and the movement of water vapor are not given consideration.

Seepage problems are usually categorized as steady-state or unsteady-state flow analyses. For steady-state analyses, the hydraulic head and the coefficient of permeability at any point in the soil mass remain constant with respect to time. For unsteady-state flow analyses, the hydraulic head (and possibly the coefficient of permeability) change with respect to time. Changes are usually in response to a change in the boundary conditions with respect to time. Steady-state flow analyses are considered in this chapter.

The quantity of flow of an incompressible fluid such as water is expressed in terms of a flux, $q$. Flux is equal to a flow rate, $v$, multiplied by a cross-sectional area, $A$. On the other hand, the quantity of flow of a compressible fluid such as air is usually expressed in terms of a mass rate. The governing partial differential seepage equations are derived in a manner consistent with the conservation of mass. The conservation of mass for steady-state seepage of an incompressible fluid dictates that the flux into an element must equal the flux out of an element. In other words, the net flux must be zero at any point in the soil mass. For a compressible fluid, the net mass rate through an element must be zero in order to satisfy the conservation of mass for steady-state seepage conditions.

## 7.1 STEADY-STATE WATER FLOW

The slow movement of water through soil is commonly referred to as seepage or percolation. Seepage analyses may form an important part of studies related to slope stability, groundwater contamination control, and earth dam design. Seepage analyses involve the computation of the rate and direction of water flow and the pore-water pressure distributions within the flow regime.

The flow of water in the saturated zone has been the primary concern in conventional seepage analyses. However, water flow in the unsaturated zone is of increasing interest to engineers. For example, the seepage through a dam has commonly been analyzed by considering only the zone below an empirically computed line of seepage (Casagrande, 1937). Recent studies have illustrated that there is a continuous flow of water between the saturated and unsaturated zones, as shown in Fig. 7.1(a) (Freeze, 1971; Papagiannakis and Fredlund, 1984). Another example is shown in Fig. 7.1(b), which illustrates the effect of infiltration and evaporation on the phreatic surface within a slope. A constant water flux across the surface boundary may develop a steady-state water flux through the unsaturated zone of the slope.

Water flow through unsaturated soils is governed by the same law as flow through saturated soils (i.e., Darcy's law). The main difference is that the water coefficient of permeability is assumed to be a constant for saturated soils, while it must be assumed to be a function of suction, water content, or some other variable for unsaturated soils. Also, the pore-water pressure generally has a positive gauge value in a saturated soil and a negative gauge value in an unsaturated soil. In spite of these differences, the formulation of the partial differential flow equation is similar in

## 7.1 STEADY-STATE WATER FLOW

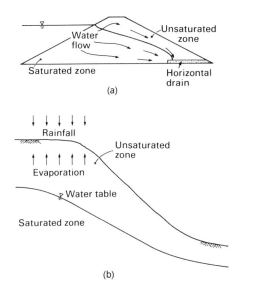

**Figure 7.1** Examples involving flow through unsaturated soils. (a) Water flow in the saturated and unsaturated zones of an earth dam; (b) water flow across the boundary of a slope.

both cases. There is also a smooth transition when going from the unsaturated to the saturated case (Fredlund, 1981).

### 7.1.1 Variation of Coefficient of Permeability with Space for an Unsaturated Soil

For steady-state seepage analyses, the coefficient of permeability is a constant with respect to time at each point in a soil. However, the coefficient of permeability usually varies from one point to another in an unsaturated soil. A spatial variation in permeability in a saturated soil can be attributed to a heterogeneous distribution of the *soil solids*. For unsaturated soils, it is more appropriate to consider the heterogeneous volume distribution of the *pore-fluid* (i.e., pore–water). This is the main reason for a spatial variation in the coefficient of permeability. Although the soil solid distribution may be homogeneous, the pore–fluid volume distribution can be heterogeneous due to spatial variations in matric suction. A point with a high matric suction (or a low water content) has a lower water coefficient of permeability than a point having a low matric suction.

Several functional relationships between the water coefficient of permeability and matric suction [i.e., $k_w(u_a - u_w)$] or volumetric water content [i.e., $k_w(\theta_w)$] have been described in Chapter 5. Coefficients of permeability for different points in a soil are obtained from the permeability function. The magnitude of the coefficient of permeability depends on the matric suction (or water content). In addition, the coefficient of permeability at a point may vary with respect to direction. This condition is referred to as anisotropy. The largest coefficient of permeability is called the major coefficient of permeability. The smallest coefficient of permeability is in a direction perpendicular to the largest permeability, and is called the minor coefficient of permeability.

### Heterogeneous, Isotropic Steady-State Seepage

Permeability conditions in unsaturated soils can be classified into three groups, as illustrated in Fig. 7.2. This classification is based on the pattern of permeability variation. A soil is called heterogeneous, isotropic if the coefficient of permeability in the $x$-direction, $k_x$, is equal to the coefficient of permeability in the $y$-direction at any point within the soil mass (i.e., $k_x = k_y$ at $A$ and $k_x = k_y$ at $B$) [see Fig. 7.2(a)]. However, the magnitude of the coefficient of permeability can vary from point $A$ to point $B$, depending upon the matric suction in the soil. The variation in the coefficient of permeability with respect to matric suction is often assumed to follow a single-valued functional relationship.

### Heterogeneous, Anisotropic Steady-State Seepage

Figure 7.2(b) illustrates the heterogeneous, anisotropic case. Here, the ratio of the coefficient of permeability in the $x$-direction, $k_x$, to the coefficient of permeability in the $y$-direction, $k_y$, is a constant at any point (i.e., $(k_x/k_y)$ at $A$

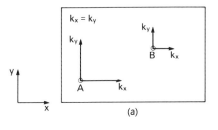

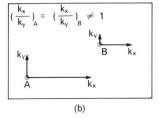

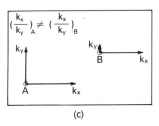

**Figure 7.2** Coefficient of permeability variations in an unsaturated soil. (a) Heterogeneous, isotropic conditions; (b) heterogeneous, anisotropic conditions; (c) continuous variation in permeability with space.

= ($k_x/k_y$) at $B$ = a constant not equal to unity). The magnitude of the coefficients of permeability, $k_x$ and $k_y$, can also vary with matric suction from one location to another, but their ratio is assumed to remain constant. Anisotropic conditions can also be oriented in any two perpendicular directions. The general case of any orientation for the major coefficient of permeability is not considered in the following formulations.

The third case is where there is a continuous variation in the coefficient of permeability [Fig. 7.2(c)]. The permeability ratio ($k_x/k_y$) may not be a constant from one location to another (i.e., ($k_x/k_y$) at $A \neq (k_x/k_y)$ at $B$), and different directions may have different permeability functions.

The following steady-state seepage formulations deal with the heterogeneous, isotropic and heterogeneous, anisotropic cases. The case where there is a continuous variation in permeability with space requires further study and is not presented in this text. All of the steady-state seepage analyses assume that the pore-air pressure has reached a constant equilibrium value. Where the equilibrium pore-air pressure is atmospheric, the water coefficient of permeability function with respect to matric suction, $k_w(u_a - u_w)$, has the same absolute value as the permeability function with respect to pore-water pressure, $k_w(-u_w)$.

### 7.1.2 One-Dimensional Flow

There are numerous situations where the water flow is predominantly in one direction. Let us consider a covered ground surface, with the water table located at a specified depth as shown in Fig. 7.3, in which the surface cover prevents any vertical flow of water from the ground surface.

The pore-water pressures are negative under static equilibrium conditions with respect to the water table. The negative pore-water pressure head has a linear distribution with depth (i.e., line 1). Its magnitude is equal to the gravitational head (i.e., elevation head) measured relative to the water table. In other words, the hydraulic head (i.e., the gravitational head plus the pore-water pressure head) is zero throughout the soil profile. This means that the change in head, and likewise the hydraulic gradient, is equal to zero. Therefore, there can be no flow of water in the vertical direction (i.e., $q_{wy} = 0$).

If the cover were removed from the ground surface, the soil surface would be exposed to the environment. Environmental changes could produce flow in a vertical direction, and subsequently alter the negative pore-water pressure head profile. Steady-state evaporation would cause the pore-water pressures to become more negative, as illustrated by line 2 in Fig. 7.3. The hydraulic head changes to a negative value since the gravitational head remains constant. The hydraulic head has a nonlinear distribution from a zero value at the water table to a more negative value at ground surface. An assumption is made that the water table remains at a constant elevation. The nonlinearity of the hydraulic head profile is caused by the spatial variation in the coefficient of permeability. Water flows in the direction of the decreasing hydraulic head. In other words, water flows from the water table upward to the ground surface. The upward constant flux of water is designated as positive for steady-state evaporation.

Steady-state infiltration causes a downward water flow. The negative pore-water pressure increases from the static equilibrium condition. This condition is indicated by line 3 in Fig. 7.3. The hydraulic head profile starts with a positive value at ground surface and decreases to zero at the water table. Therefore, water flows downward with a constant, negative flux for steady-state infiltration.

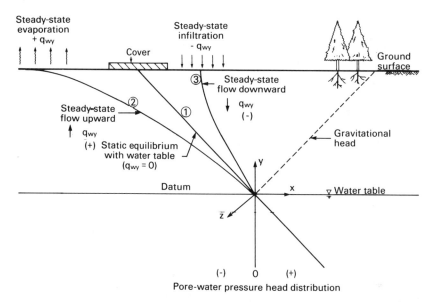

**Figure 7.3** Static equilibrium and steady-state flow conditions in the zone of negative pore-water pressures.

The above one-dimensional flow cases involve flux boundary conditions. A steady rate of evaporation or infiltration can be used as the boundary condition at ground surface. The water table acts as the lower boundary condition, giving a fixed zero pore–water pressure head.

The steady-state procedure for measuring the water coefficient of permeability in the laboratory is also a one-dimensional flow example, as illustrated in Chapter 6. In the laboratory measurement of the coefficient of permeability, however, the hydraulic heads are controlled as boundary conditions at the top and bottom of the soil specimen. Techniques to analyze both head and flux boundary conditions are explained in the following sections.

### Formulation for One-Dimensional Flow

Consider an unsaturated soil element with one-dimensional water flow in the $y$-direction (Fig. 7.4). The element has infinitesimal $dx$, $dy$, and $dz$ dimensions. The flow rate, $v_{wy}$, is assumed to be positive when water flows upward in the $y$-direction. Continuity requires that the volume of water flowing in and out of the element must be equal for steady-state conditions:

$$\left(v_{wy} + \frac{dv_{wy}}{dy} dy\right) dx\, dz - v_{wy}\, dx\, dz = 0 \quad (7.1)$$

where

$v_{wy}$ = water flow rate across a unit area of the soil in the $y$-direction
$dx, dy, dz$ = dimensions in the $x$-, $y$-, and $z$-directions, respectively.

The net flow can be written as follows:

$$\frac{dv_{wy}}{dy} dx\, dy\, dz = 0. \quad (7.2)$$

Substituting Darcy's law into the above equation yields

$$\frac{d\{-k_{wy}(u_a - u_w)dh_w/dy\}}{dy} dx\, dy\, dz = 0 \quad (7.3)$$

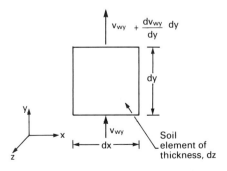

**Figure 7.4** One-dimensional water flow through an unsaturated soil element.

where

$k_{wy}(u_a - u_w)$ = water coefficient of permeability as a function of matric suction which varies with location in the $y$-direction
$dh_w/dy$ = hydraulic head gradient in the $y$-direction
$h_w$ = hydraulic head (i.e., gravitational head plus pore–water pressure head).

Equation (7.3) can be used to solve for the hydraulic head distribution in the $y$-direction through a soil mass. Since matric suction varies from one location to another, the coefficient of permeability also varies. However, for the remainder of the formulation, the $k_{wy}(u_a - u_w)$ term will be written as $k_{wy}$ for simplicity. Rewriting Eq. (7.3) and considering the nonzero dimensions for $dx$, $dy$, and $dz$ gives the following nonlinear differential equation:

$$k_{wy} \frac{d^2 h_w}{dy^2} + \frac{dk_{wy}}{dy} \frac{dh_w}{dy} = 0 \quad (7.4)$$

where

$dk_{wy}/dy$ = change in water coefficient of permeability in the $y$-direction due to a change in matric suction.

The nonlinearity of Eq. (7.4) is caused by its second term, which accounts for the variation in permeability with respect to space.

When the soil becomes *saturated*, the water coefficient of permeability, $k_{wy}$, can be taken as being equal to a single-valued, saturated coefficient of permeability, $k_s$. If the saturated soil is *heterogeneous* (e.g., layered soil), the coefficient of permeability, $k_s$, will again vary with respect to location. In a saturated soil, the heterogeneous distribution of the *soil solids* is the primary factor producing a varying coefficient of permeability. As a result, the flow equation can be written as follows:

$$k_s \frac{d^2 h_w}{dy^2} + \frac{dk_s}{dy} \frac{dh_w}{dy} = 0 \quad (7.5)$$

where

$k_s$ = saturated coefficient of permeability.

A comparison of Eqs. (7.4) and (7.5) reveals a similar form. In other words, the nonlinearity in the unsaturated soil flow equation produces the same form of equation as that required for a heterogeneous, saturated soil. In an *unsaturated* soil, the variation in the coefficient of permeability is caused by the *heterogeneous* distribution of the *pore–fluid* volume occurring as a result of different matric suction values.

If a saturated soil is *homogeneous*, the coefficient of permeability is constant for the soil mass. Substituting a nonzero, constant coefficient of permeability into Eq. (7.5)

**154**   **7 STEADY-STATE FLOW**

produces a linear differential equation:

$$\frac{d^2 h_w}{dy^2} = 0. \quad (7.6)$$

Equations similar to Eq. (7.6) can also be derived for one-dimensional flow in the $x$- and $z$-directions.

### Solution for One-Dimensional Flow

The differential equation for one-dimensional steady-state flow through a *homogeneous*, *saturated* soil [i.e., Eq. (7.6)] can be solved by integrating the equation twice. The result is a linear equation for the hydraulic head distribution in the $y$-direction:

$$h_w = C_1 y + C_2 \quad (7.7)$$

where

$C_1, C_2$ = constants of integration that can be determined for specified boundary conditions
$y$ = distance in the $y$-direction.

Figure 7.5 illustrates the case of one-dimensional steady-state flow through a homogeneous, saturated soil. A constant water pressure head is applied to the top of the soil column to establish a downward flow of water. The water coefficient of permeability is assumed to be constant throughout the column. The position of the water table at the base of the column is considered as the datum. The gravitational head distribution along the soil column is linear, equal to $h_{g1}$ at the base of the column (e.g., $h_{g1} = 0$), and $h_{gn}$ at the top of the column. The water pressure head distribution is also linear, equal to $h_{p1}$ at the base of the column (e.g., $h_{p1} = 0$), and $h_{pn}$ at the top of the column. These distributions can be used to compute the hydraulic heads.

The hydraulic heads at the top and the base of the column constitute the boundary conditions for this problem:

$h_{w1} = 0.0$          at $y$ equal to 0.0 (base)

$h_{wn} = h_{gn} + h_{pn}$    at $y$ equal to $h_{gn}$ (top)    (7.8)

where

$h_{w1}$ = hydraulic head at the base of the soil column
$h_{wn}$ = hydraulic head at the top of the soil column
$h_{gn}$ = gravitational head at the top of the soil column
$h_{pn}$ = pore–water pressure head at the top of the soil column.

Substituting the boundary conditions specified in Eq. (7.8) into Eq. (7.7) gives the constants of integration, $C_1$ and $C_2$:

$$C_1 = 1 + \frac{h_{pn}}{h_{gn}}$$

$$C_2 = 0.0. \quad (7.9)$$

The hydraulic head distribution can now be written as follows:

$$h_w = \left(1 + \frac{h_{pn}}{h_{gn}}\right) y. \quad (7.10)$$

The hydraulic head can be seen to vary linearly with depth. It has a value of $h_{wn}$ (i.e., $h_{gn}$ plus $h_{pn}$) at the top of the soil column, and a value of zero at the bottom of the soil column (Fig. 7.5). If the column is divided into ten depth intervals, each interval represents a change in hydraulic head of $0.1\, h_{wn}$. Therefore, points with equal hydraulic heads can be plotted as a horizontal line at each depth. These lines are called *equipotential lines*. The pore–

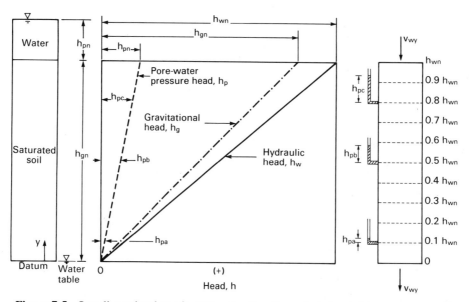

**Figure 7.5** One-dimensional steady-state water flow through a saturated homogeneous soil.

water pressure head, $h_p$, distribution is linear under steady-state seepage conditions. The linearity in the hydraulic head and the pore–water pressure head distributions is the result of the constant water coefficients of permeability.

The equation for one-dimensional steady-state flow through an *unsaturated* soil [i.e., Eq. (7.4)] requires a more complex solution than that for a saturated soil. A numerical solution can be used as an alternative to a closed-form solution. The finite difference method will be used to illustrate the solution to the flow equation for an unsaturated soil.

### Finite Difference Method

The seepage differential equation can be written in a finite difference form. Consider the situation where a function, $h(y)$, varies in space, as shown in Fig. 7.6. Values of the function at points along the curve can be computed using Taylor series:

$$h_{i+1} = h_i + \Delta y \left(\frac{dh}{dy}\right)_i + \frac{\Delta y^2}{2!} \left(\frac{d^2h}{dy^2}\right)_i$$
$$+ \frac{\Delta y^3}{3!} \left(\frac{d^3h}{dy^3}\right)_i + \cdots \quad (7.11)$$

$$h_{i-1} = h_i - \Delta y \left(\frac{dh}{dy}\right)_i + \frac{\Delta y^2}{2!} \left(\frac{d^2h}{dy^2}\right)_i$$
$$- \frac{\Delta y^3}{3!} \left(\frac{d^3h}{dy^3}\right)_i + \cdots \quad (7.12)$$

where

$i - 1, i, i + 1$ = three consecutive points spaced at increments, $\Delta y$.

Subtracting Eq. (7.12) from Eq. (7.11) and neglecting the higher order derivatives result in the first derivative of the function at point $(i)$:

$$\left(\frac{dh}{dy}\right)_i = \frac{h_{i+1} - h_{i-1}}{2\Delta y}. \quad (7.13)$$

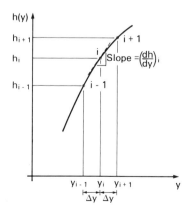

**Figure 7.6** Function $h(y)$ shown in a finite difference form.

Summing Eqs. (7.11) and (7.12) and again neglecting the higher order derivatives gives the second derivative of the function at point $(i)$:

$$\left(\frac{d^2h}{dy^2}\right)_i = \frac{h_{i+1} + h_{i-1} - 2h_i}{\Delta y^2}. \quad (7.14)$$

Equations (7.13) and (7.14) are called the central difference approximations for the first and second derivatives of the function, $h(y)$, at point $i$. These approximations can be used to solve the differential equation. Similar approximations can be derived for a function, $h(x)$, in the $x$-direction.

The use of an iterative finite difference technique in solving flow problems is illustrated in the following sections. One example involves the use of a head boundary condition, while another illustrates the use of a flux boundary condition.

### Head Boundary Condition

Steady-state evaporation from a column of unsaturated soil is illustrated in Fig. 7.7. A tensiometer is installed near the ground surface to measure the negative pore–water pressure. One-dimensional, steady-state flow is assumed when the tensiometer reading remains constant with respect to time. The pore–water pressure at the base of the column (i.e., the water table) is equal to zero.

The hydraulic head distribution along the length of the column is given by Eq. (7.4). This equation can be solved using the finite difference approximations in Eqs. (7.13) and (7.14). The column length is first discretized into $(n)$ equally spaced nodal points at a distance $\Delta y$ apart (Fig. 7.7). A central difference approximation is then applied to the hydraulic head and coefficient of permeability derivatives in Eq. (7.4). For example, Eq. (7.4) can be written in a finite difference form for point $(i)$:

$$k_{wy(i)} \left\{ \frac{h_{w(i+1)} + h_{w(i-1)} - 2h_{w(i)}}{(\Delta y)^2} \right\}$$
$$+ \left\{ \frac{k_{wy(i+1)} - k_{wy(i-1)}}{2\Delta y} \right\} \left\{ \frac{h_{w(i+1)} - h_{w(i+1)}}{2\Delta y} \right\} = 0$$
$$\quad (7.15)$$

where

$k_{wy(i)}, k_{wy(i-1)}, k_{wy(i+1)}$ = water coefficients of permeability in the $y$-direction at points $(i)$, $(i - 1)$, and $(i + 1)$, respectively

$h_{w(i)}, h_{w(i-1)}, h_{w(i+1)}$ = hydraulic heads at points $(i)$, $(i - 1)$, and $(i + 1)$, respectively.

Equation (7.15) can be rearranged after assuming equal

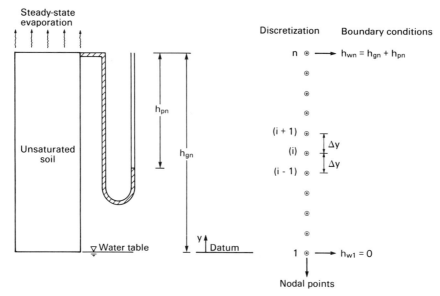

**Figure 7.7** One-dimensional, steady-state water flow through an unsaturated soil with a constant head boundary condition.

$\Delta y$ increments:

$$-\{8 \, k_{wy(i)}\} h_{w(i)} + \{4 \, k_{wy(i)} + k_{wy(i+1)} - k_{wy(i-1)}\}$$
$$\cdot h_{w(i+1)} + \{4 \, k_{wy(i)} + k_{wy(i-1)} - k_{wy(i+1)}\}$$
$$\cdot h_{w(i-1)} = 0. \qquad (7.16)$$

The hydraulic heads at the external points (i.e., points 1 and $n$) become the boundary conditions. The hydraulic head at point 1 is zero. The elevation of point ($n$) relative to the datum, $h_{gn}$, gives the gravitational head at point ($n$). The tensiometer reading near the ground surface indicates the negative pore–water pressure head at point ($n$) (i.e., $h_{pn}$). Therefore, the hydraulic head boundary condition at the top and the base of the soil column can be expressed mathematically:

$h_{w(1)} = 0.0$ at $y$ equal to $0.0$ (base)

$h_{w(n)} = h_{gn} + h_{pn}$ at $y$ equal to $h_{gn}$ (top). (7.17)

The finite difference seepage equation [i.e., Eq. (7.16)] can be written for the $(n - 2)$ internal points [i.e., points $2, 3, \cdots, (n - 1)$]. As a result, there are $(n - 2)$ equations that must be solved simultaneously for $(n - 2)$ hydraulic heads at intermediate points. The finite difference scheme illustrated by Eq. (7.16) is called an implicit form. The equation is also nonlinear because the coefficients of permeability, $k_{wy}$, are a function of matric suction, which in turn is related to hydraulic head, $h_w$. The nonlinear equations require several iterations to produce convergence. During each iteration, each equation is assumed to be linear by setting the water coefficients of permeability at each node to a constant value.

For the first iteration, the $k_{wy}$ values at all points can be set equal to the saturated coefficient of permeability, $k_s$. The $(n - 2)$ linearized equations can then be solved simultaneously using a procedure such as the Gaussian elimination technique. The computed hydraulic heads are used to calculate new values for the water coefficient of permeability. The coefficient of permeability values at each point must be in agreement with the coefficient of permeability versus matric suction function. The revised coefficient of permeability values, $k_{wy}$, are then used for the second iteration. New hydraulic heads are computed for all depths. The iterative procedure is repeated until there is no longer a significant change in the computed hydraulic heads and the computed coefficients of permeability.

Figure 7.8 illustrates typical distributions for the pore–water pressure and the hydraulic head along the unsaturated soil column. Flow is occurring under steady-state evaporation conditions. The nonlinearity of the flow equation [i.e., Eq. (7.16)] results in a nonlinear distribution of the hydraulic head and the pore–water pressure head. The equipotential lines are not equally spaced along the column. This is different from the uniformly spaced equipotential lines for the homogeneous, saturated soil column. The difference is the result of a varying coefficient of permeability throughout the unsaturated soil column. The above analysis can similarly be applied to the steady-state downward flow of water through an unsaturated soil. Once again, the hydraulic head boundary conditions at two points along the soil column must be known.

### Flux Boundary Condition

Infiltration into an unsaturated soil column is another example which can be used to illustrate the solution of the nonlinear differential flow equation (Fig. 7.9). Steady-state

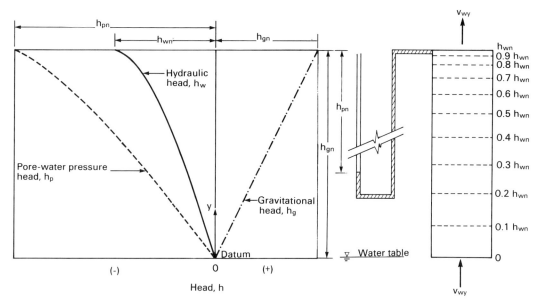

**Figure 7.8** Steady-state evaporation through an unsaturated soil column.

infiltration may be established as a result of sprinkling irrigation. Let us assume a constant downward water flux of $q_{wy}$. Steady-state flow can be described using Eq. (7.4). The hydraulic head distribution can be determined by solving the finite difference form of the steady-state flow equation [i.e., Eq. (7.16)]. The hydraulic head boundary condition at the ground surface is assumed to be unknown. However, the water flux, $q_{wy}$, is known, and is constant throughout the soil column for steady-state conditions.

The soil column is first discretized into $(n)$ nodal points with an equal spacing, $\Delta y$ (Fig. 7.9). The water flux at point $(i)$ can be expressed in terms of the hydraulic heads at points $(i + 1)$ and $(i - 1)$ using Darcy's law:

$$q_{wy} = -k_{wy(i)} \frac{h_{w(i+1)} - h_{w(i-1)}}{2\Delta y} A \quad (7.18)$$

where

$q_{wy}$ = water flux through the soil column during the steady-state flow; the flux is assumed to be positive in an upward direction and negative in a downward direction

$A$ = cross-sectional area of the soil column.

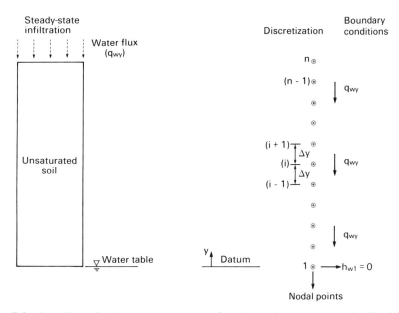

**Figure 7.9** One-dimensional steady-state water flow through an unsaturated soil with a flux boundary condition.

Equation (7.18) can be rearranged as follows:

$$h_{w(i+1)} = h_{w(i-1)} - \frac{2\Delta y}{A\, k_{wy(i)}} q_{wy}. \qquad (7.19)$$

Substituting Eq. (7.19) into the flow equation for point ($i$) [i.e., Eq. (7.16)] yields the following form:

$$-\{8 k_{wy(i)}\} h_{w(i)} + \{4 k_{wy(i)} + k_{wy(i+1)} - k_{wy(i-1)}\}$$
$$\cdot \left\{ h_{w(i-1)} - \frac{2\Delta y}{A\, k_{wy(i)}} q_{wy} \right\}$$
$$+ \{4 k_{wy(i)} + k_{wy(i-1)} - k_{wy(i+1)}\} h_{w(i-1)} = 0.$$
$$\qquad (7.20)$$

Equation (7.20) can now be solved for the hydraulic head at point ($i$):

$$h_{w(i)} = h_{w(i-1)}$$
$$- \left\{ \frac{4k_{wy(i)} + k_{wy(i+1)} - k_{wy(i-1)}}{8 k_{wy(i)}^2} \right\} \frac{2\Delta y}{A} q_{wy}.$$
$$\qquad (7.21)$$

The finite difference Eq. (7.21) is in an explicit form. Therefore, the hydraulic heads can be solved directly starting from a known boundary condition. Point 1 (Fig. 7.9) has a zero hydraulic head. Therefore, the base of the soil column is a suitable point to commence solving for the heads. Hydraulic heads can subsequently be solved point by point, up to the ground surface. Equation (7.21) is nonlinear since the coefficient of permeability, $k_{wy}$, is a function of the hydraulic head, $h_w$. The equation must be solved iteratively by setting the coefficients of permeability as constants for each iteration.

The coefficient of permeability at each node, $k_{wy}$, is assumed to be equal to the saturated coefficient of permeability, $k_s$, for the first iteration. The computed hydraulic heads, and subsequently the negative pore–water pressures, are used to revise the coefficients of permeability for the second iteration. This iterative procedure is repeated until there is convergence with respect to the hydraulic heads and the coefficients of permeability. When computing the hydraulic head at the ground surface, the $k_{wy(n+1)}$ value can be assumed to be equal to the $k_{wy(n)}$ value.

Typical distributions of pore–water pressure and hydraulic head during steady-state infiltration are illustrated in Fig. 7.10. The nonlinear distribution of the pore–water pressure and hydraulic head is produced by the nonlinearity of Eq. (7.21). As a result, the equipotential lines are not uniformly distributed along the soil column. The above analysis is also applicable to steady-state, upward flow (e.g., evaporation from ground surface) where the flux, $q_{wy}$, is known.

In the case of a *heterogeneous, saturated* soil, the coefficients of permeability can be replaced by $k_s$ in Eq. (7.21):

$$h_{w(i)} = h_{w(i-1)}$$
$$- \left\{ \frac{4k_{s(i)} + k_{s(i+1)} - k_{s(i-1)}}{8 k_{s(i)}^2} \right\} \frac{2\Delta y}{A} q_{wy}. \quad (7.22)$$

Equation (7.22) becomes linear when the soil is homogeneous:

$$h_{w(i)} = h_{w(i-1)} - \frac{\Delta y}{A k_s} q_{wy}. \qquad (7.23)$$

Equation (7.23) defines a linear distribution of the hydraulic head for a homogeneous, saturated soil column subjected to one-dimensional steady-state flow.

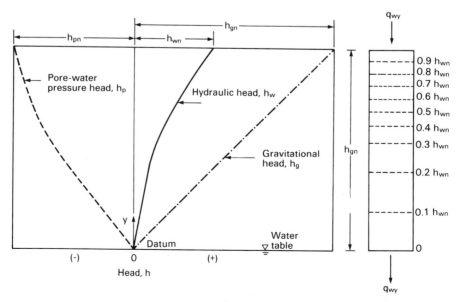

**Figure 7.10** Steady-state infiltration through an unsaturated soil.

### 7.1.3 Two-Dimensional Flow

Seepage through an earth dam is a classical example of two-dimensional flow. Water flow is in the cross-sectional plane of the dam, while flow perpendicular to the plane is assumed to be negligible. Until recently, it has been conventional practice to neglect the flow of water in the unsaturated zone of the dam. The analysis presented herein assumes that water flows in both the saturated and unsaturated zones in response to a hydraulic head driving potential.

The following two-dimensional formulation is an expanded form of the previous one-dimensional flow equation. The formulation is called an uncoupled solution since it only satisfies continuity. For a rigorous formulation of two-dimensional flow, continuity should be coupled with the force equilibrium equations.

*Formulation for Two-Dimensional Flow*

The following derivation is for the general case of a *heterogeneous, anisotropic*, unsaturated soil [Fig. 7.2(b)]. The coefficients of permeability in the $x$-direction, $k_{wx}$, and the $y$-direction, $k_{wy}$, are assumed to be related to the matric suction by the same permeability function, $k_w(u_a - u_w)$. The ratio of the coefficients of permeability in the $x$- and $y$-directions, $(k_{wx}/k_{wy})$, is assumed to be constant at any point within the soil mass.

A soil element with infinitesimal dimensions of $dx$, $dy$, and $dz$ is considered, but flow is assumed to be two-dimensional (Fig. 7.11). The flow rate, $v_{wx}$, is positive when water flows in the positive $x$-direction. The flow rate, $v_{wy}$, is positive for flow in the positive $y$-direction. Continuity for two-dimensional, steady-state flow can be expressed as follows:

$$\left(v_{wx} + \frac{\partial v_{wx}}{\partial x} dx - v_{wx}\right) dy\, dz$$
$$+ \left(v_{wy} + \frac{\partial v_{wy}}{\partial y} dy - v_{wy}\right) dx\, dz = 0 \quad (7.24)$$

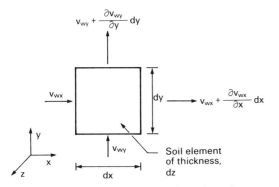

**Figure 7.11** Two-dimensional water flow through an unsaturated soil element.

where

$v_{wx}$ = water flow rate across a unit area of the soil in the $x$-direction.

Therefore, the net flux in the $x$- and $y$-directions is,

$$\left(\frac{\partial v_{wx}}{\partial x} + \frac{\partial v_{wy}}{\partial y}\right) dx\, dy\, dz = 0 \quad (7.25)$$

Substituting Darcy's laws into Eq. (7.25) results in a nonlinear partial differential equation:

$$\frac{\partial}{\partial x}\left\{k_{wx}(u_a - u_w)\frac{\partial h_w}{\partial x}\right\}$$
$$+ \frac{\partial}{\partial y}\left\{k_{wy}(u_a - u_w)\frac{\partial h_w}{\partial y}\right\} = 0 \quad (7.26)$$

where

$k_{wx}(u_a - u_w)$ = water coefficients of permeability as a function of matric suction; the permeability can vary with location in the $x$-direction

$\partial h_w / \partial_x$ = hydraulic head gradient in the $x$-direction.

For the remainder of the formulations, $k_{wx}(u_a - u_w)$ and $k_{wy}(u_a - u_w)$ are written as $k_{wx}$ and $k_{wy}$, respectively, for simplicity. Equation (7.26) describes the hydraulic head distribution in the $x$–$y$ plane for steady-state water flow. The nonlinearity of Eq. (7.26) becomes more obvious after an expansion of the equation:

$$k_{wx}\frac{\partial^2 h_w}{\partial x^2} + k_{wy}\frac{\partial^2 h_w}{\partial y^2} + \frac{\partial k_{wx}}{\partial x}\frac{\partial h_w}{\partial x} + \frac{\partial k_{wy}}{\partial y}\frac{\partial h_w}{\partial y} = 0$$
$$(7.27)$$

where

$\partial k_{wx}/\partial x$ = change in water coefficient of permeability in the $x$-direction.

The spatial variation of the coefficient of permeability given in the third and fourth terms in Eq. (7.27) produces nonlinearity in the governing flow equation.

For the *heterogeneous, isotropic* case, the coefficients of permeability in the $x$- and $y$-directions are equal (i.e., $k_{wx} = k_{wy} = k_w$). Therefore, Eq. (7.27) can be written as follows:

$$k_w\left(\frac{\partial^2 h_w}{\partial x^2} + \frac{\partial^2 h_w}{\partial y^2}\right) + \frac{\partial k_w}{\partial x}\frac{\partial h_w}{\partial x} + \frac{\partial k_w}{\partial y}\frac{\partial h_w}{\partial y} = 0 \quad (7.28)$$

where

$k_w$ = water coefficient of permeability in the $x$- and $y$-directions.

Table 7.1 summarizes the relevant equations for two-dimensional steady-state flow through unsaturated soils.

**Table 7.1 Two-Dimensional Steady-State Equations for *Unsaturated* Soils**

| Heterogeneous, Anisotropic | Heterogeneous, Isotropic |
|---|---|
| $k_{wx}\dfrac{\partial^2 h_w}{\partial x^2} + k_{wy}\dfrac{\partial^2 h_w}{\partial y^2}$ $+ \dfrac{\partial k_{wx}}{\partial x}\dfrac{\partial h_w}{\partial x}$ $+ \dfrac{\partial k_{wy}}{\partial y}\dfrac{\partial h_w}{\partial y} = 0$ | $k_w\left(\dfrac{\partial^2 h_w}{\partial x^2} + \dfrac{\partial^2 h_w}{\partial y^2}\right)$ $+ \dfrac{\partial k_w}{\partial x}\dfrac{\partial h_w}{\partial x}$ $+ \dfrac{\partial k_w}{\partial y}\dfrac{\partial h_w}{\partial y} = 0$ |
| Eq. (7.29) | Eq. (7.30) |

Seepage through a dam involves flow through the unsaturated and saturated zones. Flow through the saturated soil can be considered as a special case of flow through an unsaturated soil. For the *saturated* portion, the water coefficient of permeability becomes equal to the saturated coefficient of permeability, $k_s$. The saturated coefficients of permeability in the x- and y-directions, $k_{sx}$ and $k_{sy}$, respectively, may not be equal due to anisotropy. The saturated coefficients of permeability may vary with respect to location due to heterogeneity. A summary of steady-state equations for *saturated* soils under different conditions is presented in Table 7.2. Equations (7.31)–(7.34) are specialized forms that can be derived from the steady-state flow equation for *unsaturated* soils [i.e., Eq. (7.27)]. Therefore, steady-state seepage through saturated–unsaturated soils can be analyzed simultaneously using the same governing equation.

**Table 7.2 Two-Dimensional Steady-State Equations for *Saturated* Soils**

| Anisotropic | Isotropic |
|---|---|
| Heterogeneous | |
| $k_{sx}\dfrac{\partial^2 h_w}{\partial x^2} + k_{sy}\dfrac{\partial^2 h_w}{\partial y^2}$ $+ \dfrac{\partial k_{sx}}{\partial x}\dfrac{\partial h_w}{\partial x}$ $+ \dfrac{\partial k_{sy}}{\partial y}\dfrac{\partial h_w}{\partial y} = 0$ | $k_s\left(\dfrac{\partial^2 h_w}{\partial x^2} + \dfrac{\partial^2 h_w}{\partial y^2}\right)$ $+ \dfrac{\partial k_s}{\partial x}\dfrac{\partial h_w}{\partial x}$ $+ \dfrac{\partial k_s}{\partial y}\dfrac{\partial h_w}{\partial y} = 0$ |
| Eq. (7.31) | Eq. (7.32) |
| Homogeneous | |
| $k_{sx}\dfrac{\partial^2 h_w}{\partial x^2} + k_{sy}\dfrac{\partial^2 h_w}{\partial y^2} = 0$ | $\dfrac{\partial^2 h_w}{\partial x^2} + \dfrac{\partial^2 h_w}{\partial y^2} = 0$ |
| Eq. (7.33) | Eq. (7.34) |

### Solutions for Two-Dimensional Flow

The differential equation describing two-dimensional steady-state flow through a *homogeneous, isotropic saturated* soil [i.e., Eq. (7.34)] is called the Laplacian equation. It is a linear, partial differential equation. The solution of this equation describes the head at all points in a soil mass. The solution can be obtained using closed-form analytical methods, analog methods, or numerical methods. Often, a graphical method referred to as drawing a "flownet" has been used to solve the Laplacian equation (Casagrande, 1937).

The flownet solution results in two families of curves, referred to as flow lines and equipotential lines. The flownet solution has been used extensively to analyze problems involving seepage through saturated soils, and is explained in most soil mechanics textbooks. Boundary conditions for the soil domain must be known prior to the construction of the flownet. Either the head or the flux is prescribed along the boundary. A boundary condition exception is the case of a free surface. A network of flow lines and equipotential lines is sketched by trial and error in order to satisfy the boundary conditions and the requirement of right-angled, equidimensional elements.

A head boundary condition or an impermeable boundary condition can readily be imposed for most saturated soils problems. For example, steady-state seepage beneath a sheet pile wall has the boundary conditions shown in Fig. 7.12(a). However, the conditions are more difficult to assign when dealing with unsaturated soils.

Let us consider steady-state seepage through an earth dam [Fig. 7.12(b)]. In the past, the assumption has generally been made that the flow of water through the unsaturated zone is negligible due to its low permeability. In other words, the phreatic line is assumed to behave as an impervious, uppermost boundary when constructing the flownet. This uppermost boundary [i.e., line *BC* in Fig. 7.12(b)] is not only considered to be a phreatic line, but also an uppermost flow line. The uppermost boundary is referred to as a free surface under these special conditions (Freeze and Cherry, 1979). However, the position of the free surface is unknown, and it must be approximated prior to constructing the flownet.

The position of the free surface is usually determined using an empirical procedure (Casagrande, 1937). The assumption that the free surface is a phreatic line requires that the pore–water pressures be zero along this line. Equipotential lines must intersect the free surface at right angles since it is also an uppermost flow line. In other words, it is assumed that there is no flow across the free surface. The flownet can then be constructed.

The flownet technique has been developed primarily to analyze steady-state seepage through isotropic, homoge-

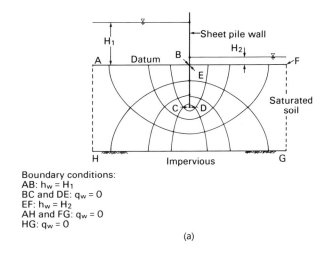

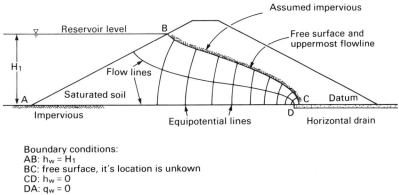

**Figure 7.12** Flownet constructions to solve the Laplacian equation. (a) Steady-state seepage throughout a homogeneous, isotropic *saturated* soil; (b) steady-state seepage throughout a homogeneous, isotropic earth dam.

neous, saturated soils. The flownet technique becomes complex and difficult to use when analyzing anisotropic, heterogeneous soil systems. There is an inherent problem associated with applying the flownet technique to saturated-unsaturated flow. Freeze (1971) stated that, "...the boundary conditions that are satisfied on the free surface specify that the pressure head must be atmospheric and the surface must be a streamline. Whereas the first of these conditions is true, the second is not."

Figure 7.13 compares two solutions of a saturated-unsaturated soil system. The flownet in Fig. 7.13(a) was drawn based on a numerical method solution for a saturated-unsaturated flow system. The flownet shown in Fig. 7.13(b) was constructed using an empirically defined free surface, thereby neglecting flow in the unsaturated zone. The free surface is a close approximation of the phreatic line from the saturated-unsaturated flow modeling. The incorrect assumption regarding the uppermost boundary condition can be avoided by realizing that there is flow between the saturated and unsaturated zones (Freeze, 1971; Papagiannakis and Fredlund, 1984).

Steady-state flow in the saturated and unsaturated zones can be analyzed simultaneously using the same governing equation [i.e., Eq. (7.26)]. Both zones are treated as a single domain. The water coefficient of permeability in the saturated zone is equal to $k_s$. The water coefficient of permeability, $k_w$, varies with respect to the matric suction in the unsaturated zone. The flownet technique is no longer applicable to saturated-unsaturated flow modeling when the governing flow equation is not of the Laplacian form. The general flow equation can be solved using a numerical technique such as the finite difference or the finite element method. Figure 7.14 shows several typical solutions by Freeze (1971) involving saturated-unsaturated flow modeling. The following section briefly describes the formulation of the finite element method in analyzing steady-state seepage through saturated-unsaturated soils.

### Seepage Analysis Using the Finite Element Method

The application of the finite element method requires the discretization of the soil mass into elements. Triangular and quadrilateral shapes of elements are commonly used for

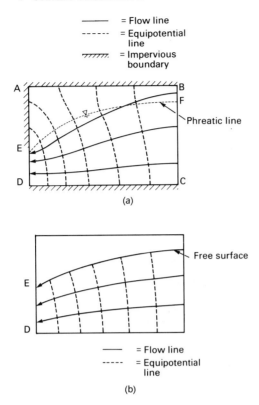

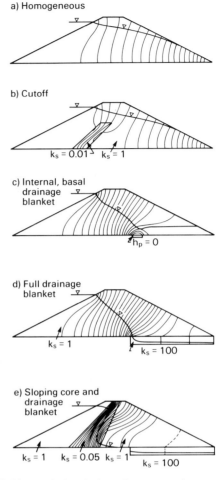

**Figure 7.13** Steady-state seepage in a saturated–unsaturated soil system. (a) Flownet constructed from *saturated–unsaturated* flow modeling; (b) flownet construction by considering flow in the *saturated* zone (after Freeze and Cherry, 1979).

**Figure 7.14** Typical solutions for saturated–unsaturated flow modeling of various dam sections (from Freeze, 1971).

two-dimensional problems. Figure 7.15 shows the cross section of a dam that has been discretized using triangular elements. The lines separating the elements intersect at nodal points. The hydraulic head at each nodal point is obtained by solving the governing flow equation and applying the boundary conditions.

The finite element formulation for steady-state seepage in two dimensions has been derived using the Galerkin principle of weighted residuals (Papagiannakis and Fredlund, 1984):

$$\int_A \begin{bmatrix} \frac{\partial}{\partial x}\{L\} \\ \frac{\partial}{\partial y}\{L\} \end{bmatrix}^T \begin{bmatrix} k_{wx} & 0 \\ 0 & k_{wy} \end{bmatrix} \begin{bmatrix} \frac{\partial}{\partial x}\{L\} \\ \frac{\partial}{\partial y}\{L\} \end{bmatrix} dA \{h_{wn}\}$$

$$\int_S \{L\}^T \bar{v}_w \, dS = 0 \qquad (7.35)$$

where

$\{L\}$ = matrix of the element area coordinates (i.e., $\{L_1 \, L_2 \, L_3\}$)

$L_1, L_2, L_3$ = area coordinates of points in the element that are related to the Cartesian coordinates of nodal points as follows (Fig. 7.16):

$L_1 = 1/2A\{(x_2 y_3 - x_3 y_2) + (y_2 - y_3)x + (x_3 - x_2)y\}$

$L_2 = 1/2A\{(x_3 y_1 - x_1 y_3) + (y_3 - y_1)x + (x_1 - x_3)y\}$

$L_3 = 1/2A\{(x_1 y_2 - x_2 y_1) + (y_1 - y_2)x + (x_2 - x_1)y\}$

$x_i, y_i (i = 1, 2, 3)$ = Cartesian coordinates of the three nodal points of an element

$x, y$ = Cartesian coordinates of a point within the element

$A$ = area of the element

$\begin{bmatrix} k_{wx} & 0 \\ 0 & k_{wy} \end{bmatrix}$ = matrix of the water coefficients of permeability (i.e., $[k_w]$)

$\{h_{wn}\}$ = matrix of hydraulic heads at the nodal points, that is,

$$\begin{Bmatrix} h_{w1} \\ h_{w2} \\ h_{w3} \end{Bmatrix}$$

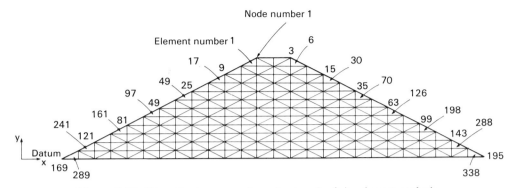

**Figure 7.15** Discretized cross section of a dam for finite element analysis.

$\bar{v}_w$ = external water flow rate in a direction perpendicular to the boundary of the element

$S$ = perimeter of the element.

Rearranging Eq. (7.35) yields a simplified form for the governing flow equation:

$$\int_A [B]^T [k_w][B] \, dA \, \{h_{wn}\} - \int_S [L]^T \bar{v}_w \, dS = 0 \quad (7.36)$$

where

$[B]$ = matrix of the derivatives of the area coordinates, which can be written as

$$\frac{1}{2A} \begin{Bmatrix} (y_2 - y_3) & (y_3 - y_1) & (y_1 - y_2) \\ (x_3 - x_2) & (x_1 - x_3) & (x_2 - x_1) \end{Bmatrix}.$$

Either the hydraulic head or the flow rate must be specified at boundary nodal points. Specified hydraulic heads at the boundary nodes are called Dirichlet boundary conditions. A specified flow rate across the boundary is referred to as a Neuman boundary condition. The second term in Eq. (7.36) accounts for the specified flow rate measured in a direction normal to the boundary. For example, a specified flow rate, $v_w$, in the vertical direction must be converted to a normal flow rate, $\bar{v}_w$, as illustrated in Fig. 7.17. The normal flow rate is in turn converted to a nodal flow, $Q_w$ (Segerlind 1984). Figure 7.17 shows the computation of the nodal flows, $Q_{wi}$ and $Q_{wj}$, at the boundary nodes ($i$) and ($j$), respectively. A positive nodal flow signifies that there is infiltration at the node or that the node acts as a "source." A negative nodal flow indicates evaporation or evapotranspiration at the node and that the node acts as a "sink." When the flow rate across a boundary is zero (e.g., impervious boundary), the second term in Eq. (7.36) disappears.

The finite element equation [Eq. (7.36)] can be written for each element and assembled to form a set of global flow equations. This is performed while satisfying nodal compatibility (Desai and Abel, 1972). Nodal compatibility requires that a particular node shared by the surrounding elements must have the same hydraulic head in all of the elements (Zienkiewicz, Desai 1975a).

Equation (7.36) is nonlinear because the coefficients of permeability are a function of matric suction, which is related to the hydraulic head at the nodal points, $\{h_{wn}\}$. The hydraulic heads are the unknown variables in Eq. (7.36). Equation (7.36) is solved by using an iterative method. For each iteration, the coefficient of permeability within an element is set to a value depending upon the average matric suction at the three nodal points. In this way, the global flow equations are linearized and can be solved simultaneously using a Gaussian elimination technique. The computed hydraulic head at each nodal point is again averaged to determine a new coefficient of permeability from the permeability function, $k_w(u_a - u_w)$. The above steps are repeated until the hydraulic heads and the coefficients of permeability no longer change by a significant amount.

The hydraulic head gradients in the $x$- and $y$-directions can be computed for an element by taking the derivative of the element hydraulic heads with respect to $x$ and $y$, respectively:

$$\begin{Bmatrix} i_x \\ i_y \end{Bmatrix} = [B] \{h_{wn}\} \quad (7.37)$$

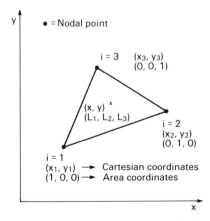

**Figure 7.16** Area coordinates in relation to the Cartesian coordinates for a triangular element.

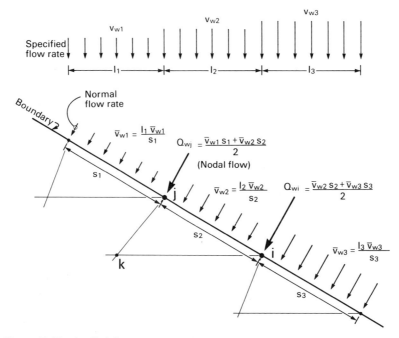

**Figure 7.17** Applied flow rate across the boundary expressed as nodal flows.

where

$i_x$, $i_y$ = hydraulic head gradient within an element in the $x$- and $y$-directions, respectively.

The element flow rates, $v_w$, can be calculated from the hydraulic head gradients and the coefficients of permeability in accordance with Darcy's law:

$$\begin{Bmatrix} v_{wx} \\ v_{wy} \end{Bmatrix} = [k_w][B]\{h_{wn}\} \quad (7.38)$$

where

$v_{wx}$, $v_{wy}$ = water flow rates within an element in the $x$- and $y$-directions, respectively.

The hydraulic head gradient and the flow rate at nodal points are computed by averaging the corresponding quantities from all elements surrounding the node. The weighted average is computed in proportion to the element areas.

### Examples of Two-Dimensional Problems

The following examples are presented to demonstrate the application of the finite element method to steady-state seepage through saturated–unsaturated soils. Lam (1984) has solved several classical problems of seepage through a dam using a saturated–unsaturated finite element seepage analysis. The cross section and discretization of the problem are illustrated in Fig. 7.15. A 10 m height of water is applied to the upstream of the dam. The permeability function used in the analysis is shown as function $A$ in Fig. 7.18. The saturated coefficient of permeability, $k_s$, is 1.0 × 10$^{-7}$ m/s. The pore–air pressure is assumed to be atmospheric. Therefore, the matric suction values in Fig. 7.18 are numerically equal to the pore–water pressures, and can be expressed as a pore–water pressure head, $h_p$. The base of the dam is chosen as the datum. The effects of anisotropy, infiltration, and the use of a core and a horizontal drain on seepage through the dam are illustrated later using additional examples.

The first example is an isotropic earth dam with a horizontal drain, as shown in Fig. 7.19. The 10 m height of water on the upstream of the dam gives a 10 m hydraulic head at each node along the upstream face. A zero hy-

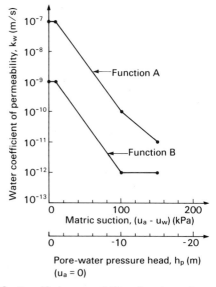

**Figure 7.18** Specified permeability functions for analyzing steady-state seepage through a dam.

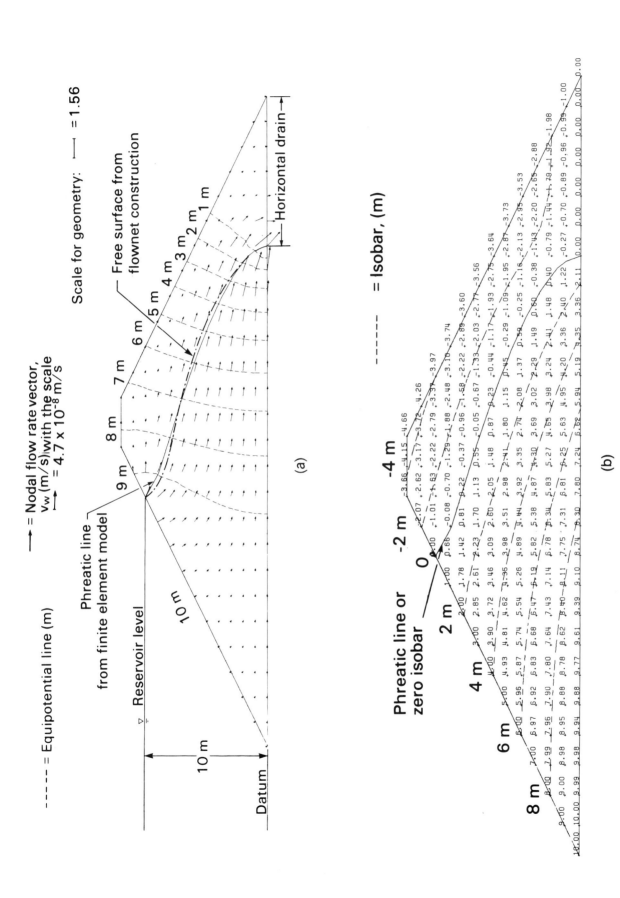

**Figure 7.19** Seepage through an isotropic earth dam with a horizontal drain. (a) Equipotential lines and nodal flow rate vectors through the dam; (b) contours of pore-water pressure head (isobars) through the dam.

draulic head is specified at nodes along the horizontal drain. Zero nodal flow is specified at nodes along the remaining boundaries. The results of the finite element analysis are presented in Fig. 7.19(a) and (b).

The phreatic line resulting from the saturated–unsaturated flow model is in close agreement with the empirical free surface from a conventional flownet construction. This observation supports the assumption that the empirical free surface is approximately equal to a phreatic line. However, water can flow across the phreatic line, as indicated by the nodal flow rate vectors. Water flow across the phreatic line into the unsaturated zone indicates that the phreatic line is not the uppermost flow line, as assumed in the flownet technique.

The difference between the phreatic line (from the finite element analysis) and the free surface (from the flownet technique) decreases as the permeability function for the unsaturated zone becomes steeper. A steep permeability function indicates a rapid reduction in the water coefficient of permeability for a small increase in matric suction. In this case, the quantity of water flow into the unsaturated zone is considerably reduced. This condition approaches the assumption associated with the conventional flownet technique. In other words, the phreatic line approaches the empirical free surface.

Equipotential lines extend from the saturated zone through the unsaturated zone, as shown in Fig. 7.19(a). Changes in hydraulic head between equipotential lines demonstrates that water flows in both the saturated and unsaturated zones. The amount of water flowing in the unsaturated zone depends upon the rate at which the coefficient of permeability changes with respect to matric suction.

The pore-water pressure heads at all of the nodes throughout the dam are shown in Fig. 7.19(b). Pore-water pressure heads are computed by subtracting the elevation head from the hydraulic head. Contour lines of equal pressure heads or isobars are also shown. The pressure heads range from positive to negative values, with the zero pressure head contoured as the phreatic line. The isobars are almost parallel to the phreatic line in the central section of the dam. Steady-state seepage in the central section of the dam tends towards an infinite slope situation. This case is further explained in the next section.

The flow of water in the saturated and unsaturated zones is approximately parallel to the phreatic line, as observed from the flow rate vectors in the central section of the dam [Fig. 7.19(a)]. This is not the situation for the sections close to the upstream face and the toe of the dam. Near the upstream face of the dam, water flows across the phreatic line from the saturated to the unsaturated zone and continues to flow in the unsaturated zone. The water in the saturated and unsaturated zones then flows essentially parallel to the phreatic line in the central section of the dam. The water in the saturated zone then flows across the phreatic line into the unsaturated zone at the toe of the dam.

Figures 7.20(a) and (b) show steady-state seepage for the above dam cross section when the soil is anisotropic. The water coefficient of permeability in the horizontal direction is assumed to be nine times larger than in the vertical direction (i.e., $k_{wx} = 9k_{wy}$). This ratio is assumed to be constant throughout the dam. One permeability function (i.e., function A in Fig. 7.18) is used for the x- and y-directions. The phreatic line is elongated in the direction of the major coefficient of permeability for the anisotropic case [Fig. 7.20(a)]. The saturated zone may reach the downstream face of the dam for higher ratios of the horizontal to vertical coefficients of permeability.

The third example shows an isotropic earth dam having a core with a lower coefficient of permeability and a horizontal drain, as illustrated in Fig. 7.21. The soil has a saturated coefficient of permeability, $k_s$, of $1.0 \times 10^{-7}$ m/s, and the permeability function is in accordance with function A in Fig. 7.18. The core has a saturated coefficient of permeability, $k_s$, of $1.0 \times 10^{-9}$ m/s and a permeability function in accordance with function B in Fig. 7.18. The boundary conditions used in the analysis are the same as those applied to the previous problems. The results show that most of the hydraulic head change occurs in the region around the core, as depicted by the concentrated distribution of equipotential lines in the core zone. As the difference in the coefficients of permeability between the soil and the core increases, greater hydraulic head changes take place in the core. The nodal flow rate vectors also indicate that a significant amount of water flows upward into the unsaturated zone and bypasses the relatively impermeable core (i.e., the siphon effect), as shown in Fig. 7.21(a).

The fourth example demonstrates the effect of a flux boundary (i.e., infiltration) on the isotropic earth dam shown in Fig. 7.19. The seepage analysis results are presented in Fig. 7.22. Steady-state infiltration is simulated by applying a positive nodal flow, $Q_w$, of $1.0 \times 10^{-8}$ m$^2$/s to each of the nodes along the upper boundary of the dam. The results can be compared to the case of zero flux across the upper boundary by comparing Figs. 7.19(a) and 7.22(a). Infiltration results in a rise in the phreatic line. Consequently, the pore-water pressures in the unsaturated zone increase [i.e., Fig. 7.22(b)] relative to the zero flux case [i.e., Fig. 7.19(b)].

The fifth example demonstrates the development of a seepage face on the downstream of the dam. In this case, there is no horizontal drain, and zero nodal flows are specified along the entire lower boundary. There is close agreement between the phreatic line obtained from the finite element analysis and the free surface obtained using the flownet technique (Fig. 7.23). The phreatic line extends to the downstream face of the dam. The phreatic line exits on the downstream face, and the portion below the exit point is called the seepage face. The seepage face has a zero pore-water pressure (i.e., atmospheric) boundary condition. In other words, the hydraulic head is equal to the gravitational head.

The location of the exit point is not known prior to per-

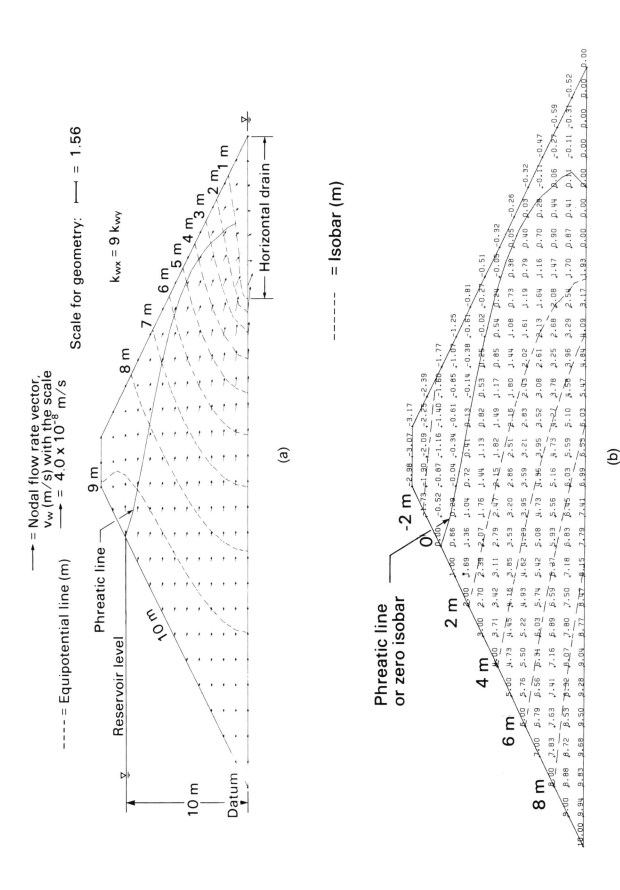

**Figure 7.20** Seepage through an anisotropic earth dam with a horizontal drain. (a) Equipotential lines and nodal flow rate vectors throughout the dam; (b) contours of pore-water pressure head (isobars) throughout the dam.

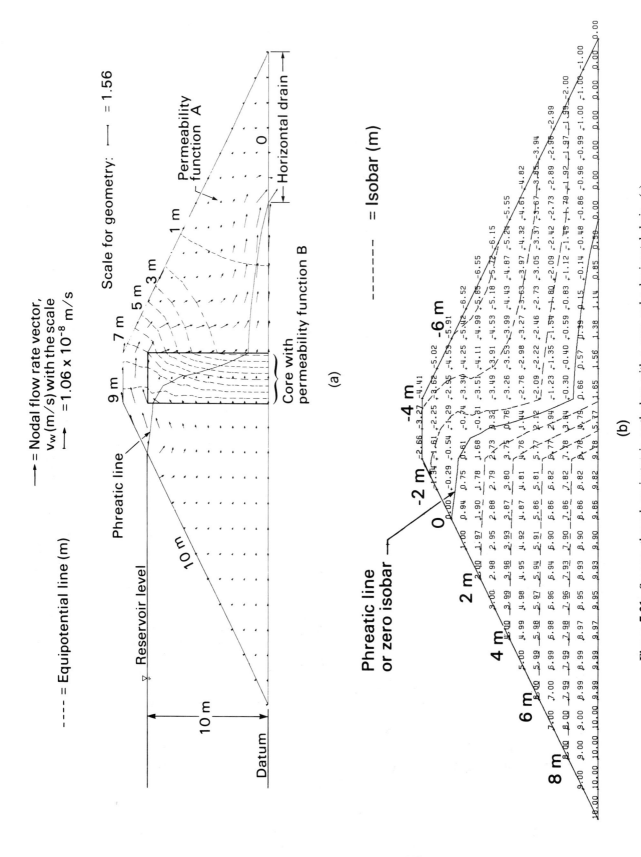

**Figure 7.21** Seepage through an isotropic earth dam with a core and a horizontal drain. (a) Equipotential lines and nodal flow rate vectors throughout the dam, (b) contours of pore-water pressure head (isobars) throughout the dam.

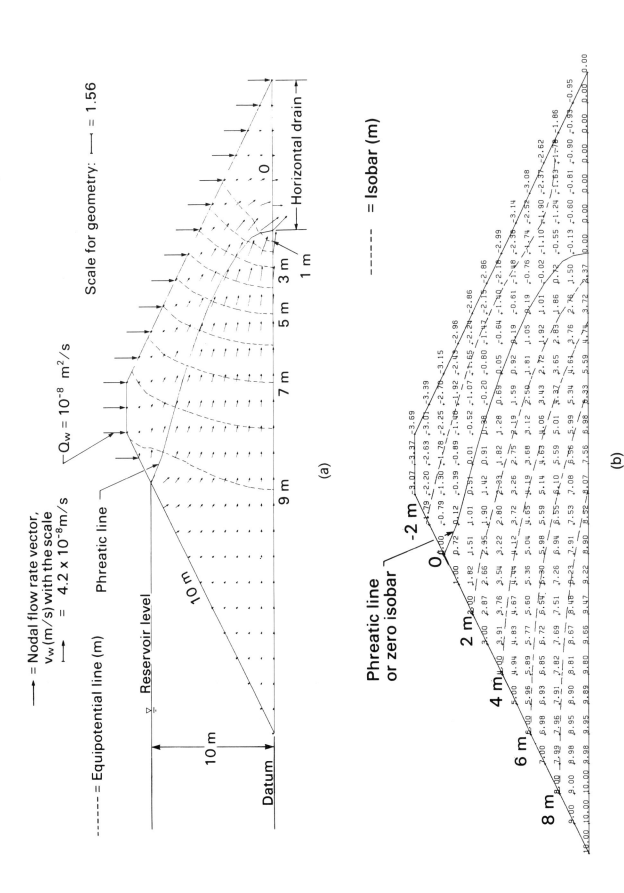

**Figure 7.22** Seepage through an isotropic earth dam with a horizontal drain under steady-state infiltration. (a) Equipotential lines and nodal flow rate vectors throughout the dam, (b) contours of pore-water pressure head (isobars) throughout the dam.

169

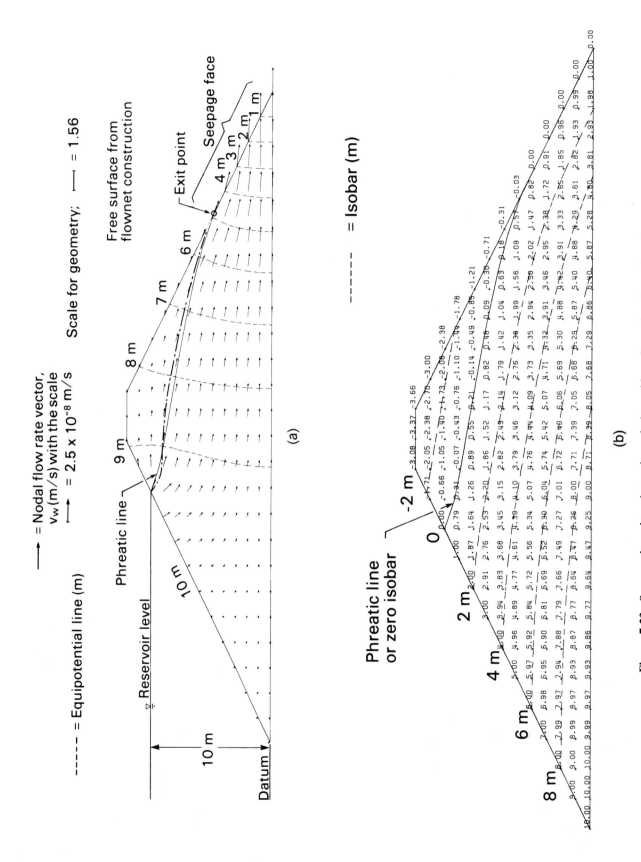

**Figure 7.23** Seepage through an isotropic earth dam with an impervious lower boundary. (a) Equipotential lines and nodal flow rate vectors throughout the dam, (b) contours of pore-water pressure head (isobars) throughout the dam.

170

forming the analysis. Therefore, the location of the exit point must be assumed in order to commence the analysis. The exit point can then be revised after each iteration by reevaluating the seepage face boundary condition. During the analysis, the nodal flows above the assumed exit point are set at zero. The hydraulic heads at or below the assumed exit point are specified as being equal to their gravitational heads. After convergence, the pore-water pressure head at the node directly above the assumed exit point is examined. A negative pore-water pressure head at this point indicates that the assumed exit point is correct. Otherwise, the seepage face boundary is revised by assuming a higher exit point for the phreatic line. The above procedure is repeated until the correct exit point is obtained.

The above examples deal with seepage through earth dams. However, the same type of finite element seepage analysis can be applied to other problems involving saturated-unsaturated flow.

*Infinite Slope*

A slope of infinite length is illustrated in Fig. 7.24. Let us consider the case where steady-state water flow is established within the slope and the phreatic line is parallel to the ground surface. Water flows through both the saturated and unsaturated zones, and is parallel to the phreatic line. The direction of the water flow indicates that there is no flow perpendicular to the phreatic line. In other words, the hydraulic head gradient is equal to zero in a direction perpendicular to the phreatic line. In this case, the lines drawn normal to the phreatic line are equipotential lines.

Isobars are parallel to the phreatic line. This is similar to the condition in the central section of a homogeneous dam, as shown earlier. The coefficient of permeability is essentially independent of the pore-water pressure in the saturated zone. Therefore, the saturated zone can be subdivided into several flow channels of equal size. An equal amount of water (i.e., water flux, $q_w$) flows through each channel. Lines separating the flow channels are referred to as flow lines.

The water coefficient of permeability depends on the negative pore-water pressure or the matric suction in the unsaturated zone. The pore-water pressure decreases from zero at the phreatic line to some negative value at ground surface. Similarly, the permeability decreases from the phreatic line to ground surface. As a result, increasingly larger flow channels are required in order to maintain the same quantity of water flow, $q_w$, as ground surface is approached.

The water flow in each channel is one-dimensional, in a direction parallel to the phreatic line. The coefficient of permeability varies in the direction perpendicular to flow. This condition can be compared to the previous case of water flow through a vertical column, as explained in Section 7.1.2. In the case of the vertical column, the coefficient of permeability varied in the flow direction, and the equipotential lines were not equally distributed throughout the soil column.

The above examples illustrate that equipotential lines and flow lines intersect at right angles for unsaturated flow problems, as long as the soil is isotropic. Heterogeneity with respect to the coefficient of permeability results in varying distances between either the flow lines or the equipotential lines; however, these lines cross at 90°.

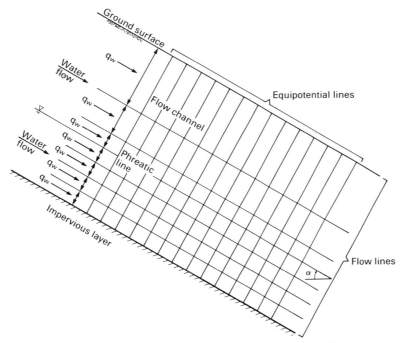

**Figure 7.24** Steady-state water flow through an infinite slope.

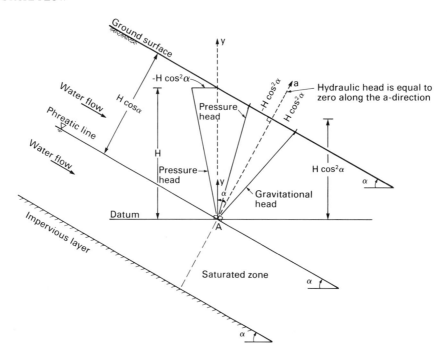

**Figure 7.25** Pore–water pressure distributions in the unsaturated zone of an infinite slope during steady-state seepage.

The pore–water pressure distribution in the unsaturated zone can be analyzed by considering a horizontal datum through an arbitrary point (e.g., point A in Fig. 7.25) on the phreatic line. The pore–water pressure distribution in a direction perpendicular to the phreatic line (i.e., in the $a$-direction) is first examined. The results are then used to analyze the pore–water pressure distribution in the $y$-direction (i.e., vertically). The gravitational head distribution in the $a$-direction is zero at point A (i.e., datum), and increases linearly to a gravitational head of ($H \cos^2\alpha$) at ground surface. The pore–water pressure head at a point in the $a$-direction must be negative and equal in magnitude to its gravitational head because the hydraulic heads are zero in the $a$-direction. Therefore, the pore–water pressure head distribution in the $a$-direction must start at zero at the datum (i.e., point A) and decrease linearly to ($-H \cos^2\alpha$) at ground surface. A pore–water pressure head of ($-H \cos^2\alpha$) applies to any point along the ground surface since every line parallel to the phreatic line is also an isobar.

The pore–water pressure head distribution in a vertical direction also commences with a zero value at point A, and decreases linearly to a head of ($-H \cos^2\alpha$) at ground surface. However, the pore–water pressure head is distributed along a length, ($H \cos \alpha$), in the $a$-direction, while the head is distributed along a length, $H$, in the vertical direction. The negative pore–water pressure head at a point on a vertical plane can therefore be expressed a follows:

$$h_{pi} = - y \cos^2\alpha \tag{7.39}$$

where

$h_{pi}$ = negative pore-water pressure head on a vertical plane (i.e., the $y$-direction) for an infinite slope

$y$ = vertical distance from the point under consideration to the datum (i.e., point A)

$\alpha$ = inclination angle of the slope and the phreatic line.

When the ground surface and the phreatic line are horizontal (i.e., $\alpha = 0$ or $\cos \alpha = 1$), the negative pore-water pressure head at a point along a vertical plane, $h_{ps}$, is equal to $-y$. This is the condition of static equilibrium above and below a horizontal water table, as shown in Fig. 7.3. The ratio between the pore-water pressure heads on a vertical plane through an infinite slope (i.e., $h_{pi} = -y \cos^2\alpha$) and the pore-water pressure heads associated with a horizontal ground surface (i.e., $h_{ps} = -y$) is plotted in Fig. 7.26. This ratio indicates the reduction in the pore-water pressures on a vertical plane as the slope, $\alpha$, becomes steeper (Fig. 7.26).

The gravitational head at a point along a vertical plane is equal to its elevation from the datum, $y$ (Fig. 7.25). The hydraulic head is computed as the sum of the gravitational and pore-water pressure heads:

$$h_{wi} = (1 - \cos^2\alpha)\, y. \tag{7.40}$$

Equation (7.40) indicates that there is a decrease in the hydraulic head as the datum is approached. In other words, there is a vertical downward component of water flow.

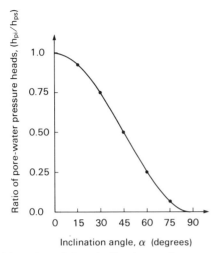

**Figure 7.26** Effect of slope inclination on the pore–water pressure distribution along a vertical plane.

The above analysis also applies to the pore–water pressure conditions below the phreatic line. Using the same horizontal line through point A, positive pore–water pressure heads along a vertical plane can be computed in accordance with Eq. (7.39). The hydraulic head [Eq. (7.40)] is zero at the phreatic line, and decreases linearly with depth along a vertical plane.

### 7.1.4 Three-Dimensional Flow

Sometimes it is necessary to use a three-dimensional flow analysis in order to simulate the flow system of interest. Three-dimensional flow can be formulated by expanding the two-dimensional flow equation to include the third direction. The three-dimensional equation is derived based on continuity, and the equation is referred to as the uncoupled equation of flow.

Let us consider an unsaturated soil having *heterogeneous, anisotropic* conditions [Fig. 7.2(b)]. The coefficient of permeability at a point varies in the $x$-, $y$-, and $z$-directions. However, the permeability variations in the three directions will be assumed to be governed by the same permeability function. Figure 7.27 shows a cubical soil element with water flow in the $x$-, $y$-, and $z$-directions. The soil element has infinitesimal dimensions of $dx$, $dy$, and $dz$. The flow rates, $v_{wx}$, $v_{wy}$, and $v_{wz}$, are assumed to be positive when water flows in the positive $x$-, $y$-, and $z$-directions. Continuity for three-dimensional, steady-state flow can be satisfied as follows:

$$\left( v_{wx} + \frac{\partial v_{wx}}{\partial x} dx - v_{wx} \right) dy\, dz$$
$$+ \left( v_{wy} + \frac{\partial v_{wy}}{\partial y} dy - v_{wy} \right) dx\, dz$$
$$+ \left( v_{wz} + \frac{\partial v_{wz}}{\partial z} dz - v_{wz} \right) dx\, dy = 0 \quad (7.41)$$

where

$v_{wz}$ = water flow rate across a unit area of the soil in the $z$-direction.

Equation (7.41) reduces to the following form:

$$\left( \frac{\partial v_{wx}}{\partial x} + \frac{\partial v_{wy}}{\partial y} + \frac{\partial v_{wz}}{\partial z} \right) dx\, dy\, dz = 0. \quad (7.42)$$

Substituting Darcy's law into Eq. (7.42) produces a nonlinear partial differential equation:

$$\frac{\partial}{\partial x} \left\{ k_{wx}(u_a - u_w) \frac{\partial h_w}{\partial x} \right\} + \frac{\partial}{\partial y} \left\{ k_{wy}(u_a - u_w) \frac{\partial h_w}{\partial y} \right\}$$
$$+ \frac{\partial}{\partial z} \left\{ k_{wz}(u_a - u_w) \frac{\partial h_w}{\partial z} \right\} = 0 \quad (7.43)$$

where

$k_{wz}(u_a - u_w)$ = water coefficient of permeability as a function of matric suction

$\partial h_w / \partial z$ = hydraulic head gradient in the $z$-direction.

For the remainder of the formulations, $k_{wx}(u_a - u_w)$, $k_{wy}(u_a - u_w)$, and $k_{wz}(u_a - u_w)$ are written as $k_{wx}$, $k_{wy}$, and

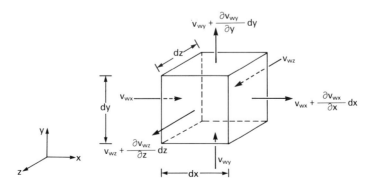

**Figure 7.27** Three-dimensional steady-state water flow through an unsaturated soil element.

**Table 7.3  Three-Dimensional Steady-State Equations for *Unsaturated* Soils**

| Heterogeneous, Anisotropic | Heterogeneous, Isotropic |
|---|---|
| $k_{wx}\dfrac{\partial^2 h_w}{\partial x^2} + k_{wy}\dfrac{\partial^2 h_w}{\partial y^2} + k_{wz}\dfrac{\partial^2 h_w}{\partial z^2}$ $+ \dfrac{\partial k_{wx}}{\partial x}\dfrac{\partial h_w}{\partial x} + \dfrac{\partial k_{wy}}{\partial y}\dfrac{\partial h_w}{\partial y} + \dfrac{\partial k_{wz}}{\partial z}\dfrac{\partial h_w}{\partial z} = 0$ | $k_w\left(\dfrac{\partial^2 h_w}{\partial x^2} + \dfrac{\partial^2 h_w}{\partial y^2} + \dfrac{\partial^2 h_w}{\partial z^2}\right)$ $+ \dfrac{\partial k_w}{\partial x}\dfrac{\partial h_w}{\partial x} + \dfrac{\partial k_w}{\partial y}\dfrac{\partial h_w}{\partial y} + \dfrac{\partial k_w}{\partial z}\dfrac{\partial h_w}{\partial z} = 0$ |
| Eq. (7.46) | Eq. (7.47) |

$k_{wz}$, respectively, for simplicity. The hydraulic head distribution in a soil mass during three-dimensional steady-state flow is described by Eq. (7.43). The nonlinearity of Eq. (7.43) is caused by permeability variations with respect to space. The nonlinearity can be illustrated by expanding the equation:

$$k_{wx}\frac{\partial^2 h_w}{\partial x^2} + k_{wy}\frac{\partial^2 h_w}{\partial y^2} + k_{wz}\frac{\partial^2 h_w}{\partial z^2} + \frac{\partial k_{wx}}{\partial x}\frac{\partial h_w}{\partial x}$$
$$+ \frac{\partial k_{wy}}{\partial y}\frac{\partial h_w}{\partial y} + \frac{\partial k_{wz}}{\partial z}\frac{\partial h_w}{\partial z} = 0 \qquad (7.44)$$

where

$\partial k_{wz}/\partial z$ = change in water coefficient of permeability in the $z$-direction.

The fourth, fifth, and sixth terms in Eq. (7.44) account for the spatial variation in the coefficient of permeability. In the case of two-dimensional flow, the hydraulic head gradient in the third direction is negligible (e.g., $\partial h_w/\partial z = 0$), and Eq. (7.44) reverts to Eq. (7.27).

For the *heterogeneous, isotropic* case, the coefficients of permeability in the $x$-, $y$-, and $z$-directions are equal, and Eq. (7.44) takes the following form:

$$k_w\left(\frac{\partial^2 h_w}{\partial x^2} + \frac{\partial^2 h_w}{\partial y^2} + \frac{\partial^2 h_w}{\partial z^2}\right) + \frac{\partial k_w}{\partial x}\frac{\partial h_w}{\partial x}$$
$$+ \frac{\partial k_w}{\partial y}\frac{\partial h_w}{\partial y} + \frac{\partial k_w}{\partial z}\frac{\partial h_w}{\partial z} = 0 \qquad (7.45)$$

where

$k_w$ = water coefficient of permeability in the $x$-, $y$-, and $z$-directions.

Table 7.3 summarizes the three-dimensional steady-state equations for unsaturated soils. For a saturated soil, the coefficient of permeability becomes equal to the saturated coefficient of permeability, $k_s$. A summary of three-dimensional, steady-state equations for *saturated* soils corresponding to various conditions is presented in Table 7.4.

The three dimensional steady-state flow equations can be

**Table 7.4  Three-Dimensional Steady-State Equations for *Saturated* Soils**

| Anisotropic | Isotropic |
|---|---|
| Heterogeneous | |
| $k_{sx}\dfrac{\partial^2 h_w}{\partial x^2} + k_{sy}\dfrac{\partial^2 h_w}{\partial y^2} + k_{sz}\dfrac{\partial^2 h_w}{\partial z^2}$ $+ \dfrac{\partial k_{sx}}{\partial x}\dfrac{\partial h_w}{\partial x} + \dfrac{\partial k_{sy}}{\partial y}\dfrac{\partial h_w}{\partial y} + \dfrac{\partial k_{sz}}{\partial z}\dfrac{\partial h_w}{\partial z} = 0$ | $k_s\left(\dfrac{\partial^2 h_w}{\partial x^2} + \dfrac{\partial^2 h_w}{\partial y^2} + \dfrac{\partial^2 h_w}{\partial z^2}\right)$ $+ \dfrac{\partial k_s}{\partial x}\dfrac{\partial h_w}{\partial x} + \dfrac{\partial k_s}{\partial y}\dfrac{\partial h_w}{\partial y} + \dfrac{\partial k_s}{\partial z}\dfrac{\partial h_w}{\partial z} = 0$ |
| Eq. (7.48) | Eq. (7.49) |
| Homogeneous | |
| $k_{sx}\dfrac{\partial^2 h_w}{\partial x^2} + k_{sy}\dfrac{\partial^2 h_w}{\partial y^2} + k_{sz}\dfrac{\partial^2 h_w}{\partial z^2} = 0$ | $\dfrac{\partial^2 h_w}{\partial x^2} + \dfrac{\partial^2 h_w}{\partial y^2} + \dfrac{\partial^2 h_w}{\partial z^2} = 0$ |
| Eq. (7.50) | Eq. (7.51) |

solved using numerical procedures such as the finite difference and finite element methods.

## 7.2 STEADY-STATE AIR FLOW

The bulk flow of air can occur through an unsaturated soil when the air phase is continuous. In many practical situations, the flow of air may not be of concern. However, it is of value to understand the formulations for compressible flow through porous media.

The air coefficient of transmission, $D_a^*$, or the air coefficient of permeability, $k_a$ (i.e., $D_a^* g$), is a function of the volume–mass properties or the stress state of the soil. The relationships between the air coefficient of permeability, $k_a$, and matric suction [i.e., $k_a(u_a - u_w)$] or degree of saturation [i.e., $k_a(S_e)$] are described in Chapter 5. The value of $k_a$ or $D_a^*$ may vary with location, depending upon the distribution of the pore–air volume in the soil. Possible variations in the air coefficient of permeability in an unsaturated soil are described using Fig. 7.2. The air coefficient of permeability at a point can be assumed to be constant with respect to time during steady-state air flow.

This section presents the steady-state formulations for one- and two-dimensional air flow using Fick's law. Heterogeneous, isotropic, and anisotropic situations are presented. Steady-state air flow is analyzed by assuming that the pore–water pressure has reached equilibrium. The following air flow equations can be solved using numerical methods such as the finite difference or the finite element methods. The manner of solving the equations is similar to that described in the previous sections.

### 7.2.1 One-Dimensional Flow

Consider an *unsaturated* (i.e., heterogeneous) soil element with one-dimensional air flow in the y-direction (Fig. 7.28). The air flow has a mass rate of flow, $J_{ay}$, under steady-state conditions. The mass rate is assumed to be positive for an upward air flow. The principle of continuity states that the mass of air flowing into the soil element must be equal to the mass of air flowing out of the element:

$$\left(J_{ay} + \frac{dJ_{ay}}{dy} dy\right) dx\, dz - J_{ay}\, dx\, dz = 0 \quad (7.52)$$

where

$J_{ay}$ = mass rate of air flow across a unit area of the soil in the y-direction.

Rearranging Eq. (7.52) gives the net mass rate of air flow

$$\left(\frac{dJ_{ay}}{dy}\right) dx\, dy\, dz = 0. \quad (7.53)$$

Substituting Fick's law (see Chapter 5) for the mass rate of flow into the above equation yields a nonlinear differential equation:

$$\frac{d\{-D_{ay}^*(u_a - u_w)\, du_a/dy\}}{dy} = 0 \quad (7.54)$$

where

$D_{ay}^*(u_a - u_w)$ = air coefficient of transmission as a function of matric suction
$du_a/dy$ = pore–air pressure gradient in the y-direction
$u_a$ = pore–air pressure.

The coefficient of transmission, $D_{ay}^*(u_a - u_w)$, will be written as $D_{ay}^*$ for simplicity. The spatial variation of $D_{ay}^*$ causes nonlinearity in Eq. (7.54):

$$D_{ay}^* \frac{d^2 u_a}{dy^2} + \frac{dD_{ay}^*}{dy} \frac{du_a}{dy} = 0 \quad (7.55)$$

where

$dD_{ay}^*/dy$ = change in the air coefficient of transmission in the y-direction.

Equations (7.54) and (7.55) describe the pore–air pressure distribution in the soil mass in the y-direction. The second term in Eq. (7.55) accounts for the spatial variation in the coefficient of transmission. The coefficient of transmission is obtained by dividing the air coefficient of permeability, $k_{ay}$, by the gravitational acceleration (i.e., $D_{ay}^* = k_{ay}/g$). In other words, the coefficients $D_{ay}^*$ and $k_{ay}$ have similar functional relationships to matric suction.

The measurement of the air coefficient of permeability using a triaxial permeameter cell (Chapter 6) is an application involving one-dimensional, steady-state air flow. In this case, however, the air coefficient of permeability is assumed to be constant throughout the soil specimen. Neglecting the change in the air coefficient of permeability with respect to location, Eq. (7.55) is reduced to a linear differential equation:

$$\frac{d^2 u_a}{dy^2} = 0. \quad (7.56)$$

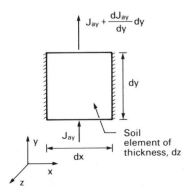

**Figure 7.28** One-dimensional steady-state air flow through an unsaturated soil element.

The pore–air pressure distribution in the y-direction is obtained by integrating Eq. (7.56) twice:

$$u_a = C_1 y + C_2 \qquad (7.57)$$

where

$C_1, C_2$ = constants of integration related to the boundary conditions
$y$ = distance in the y-direction.

Figure 7.29 illustrates the pore–air pressure distribution within a soil specimen during an air permeability test. The air pressures at both ends of the specimen (i.e., $u_a = u_{ab}$ at $y = 0.0$ and $u_a = u_{at} = 0.0$ at $y = h_s$) are the boundary conditions. Substituting the boundary conditions into Eq. (7.57) results in a linear equation for the pore–air pressure along the soil specimen (i.e., $u_a = (1 - y/h_s)u_{ab}$).

### 7.2.2 Two-Dimensional Flow

Two-dimensional, steady-state air flow is first formulated for the heterogeneous, anisotropic condition [Fig. 7.2(b)]. The air coefficients of transmission in the x- and y-directions, $D_{ax}^*$ and $D_{ay}^*$, are related to matric suction using the same transmission function, $D_a^*(u_a - u_w)$. The $(D_{ax}^*/D_{ay}^*)$ ratio will be assumed to be constant at any point within the soil mass. An element of soil subjected to two-dimensional air flow is shown in Fig. 7.30. Satisfying continuity for steady-state flow yields the following equation:

$$\left(J_{ax} + \frac{\partial J_{ax}}{\partial x} dx - J_{ax}\right) dy\, dz$$
$$+ \left(J_{ay} + \frac{\partial J_{ay}}{\partial y} dy - J_{ay}\right) dx\, dz = 0 \qquad (7.58)$$

where

$J_{ax}$ = mass rate of air flowing across a unit area of the soil in the x-direction.

Rearranging Eq. (7.58) results in the following equation:

$$\left(\frac{\partial J_{ax}}{\partial x} + \frac{\partial J_{ay}}{\partial y}\right) dx\, dy\, dz = 0. \qquad (7.59)$$

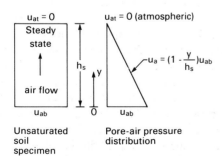

**Figure 7.29** Pore–air pressure distribution during the measurement of the air coefficient of permeability, $k_a$.

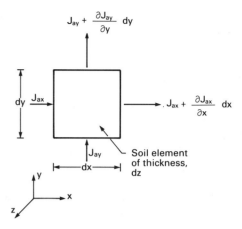

**Figure 7.30** An element subjected to two-dimensional air flow.

Substituting Fick's law for the mass rates, $J_{ax}$ and $J_{ay}$, into Eq. (7.59) gives the following nonlinear partial differential equation:

$$\frac{\partial}{\partial x}\left\{D_{ax}^*(u_a - u_w)\frac{\partial u_a}{\partial x}\right\}$$
$$+ \frac{\partial}{\partial y}\left\{D_{ay}^*(u_a - u_w)\frac{\partial u_a}{\partial y}\right\} = 0 \qquad (7.60)$$

where

$D_{ax}^*(u_a - u_w)$ = air coefficient of transmission as a function of matric suction
$du_a/dx$ = pore–air pressure gradient in the x-direction.

Let us write $D_{ax}^*(u_a - u_w)$ and $D_{ay}^*(u_a - u_w)$ simply as $D_{ax}^*$ and $D_{ay}^*$, respectively. The coefficient of transmission, $D_{ax}^*$, is related to the air coefficient of permeability, $k_{ax}^*$, by the gravitational acceleration (i.e., $D_{ax}^* = k_{ax}^*/g$). Expanding Eq. (7.60) results in the following flow equation:

$$D_{ax}^*\frac{\partial^2 u_a}{\partial x^2} + D_{ay}^*\frac{\partial^2 u_a}{\partial y^2} + \frac{\partial D_{ax}^*}{\partial x}\frac{\partial u_a}{\partial x} + \frac{\partial D_{ay}^*}{\partial y}\frac{\partial u_a}{\partial y} = 0$$

$$(7.61)$$

where

$\partial D_{ax}^*/\partial x$ = change in air coefficient of transmission in the x-direction.

Spatial variations in the coefficients of transmission are accounted for by the third and fourth terms in Eq. (7.61). These two terms produce the nonlinearity in the flow equation. When solved, Eq. (7.61) describes the pore–air pressure distribution in the $x$–$y$ plane of the soil mass during two-dimensional, steady-state air flow.

For the *heterogeneous*, *isotropic* case, the coefficients of transmission in the $x$- and $y$-directions are equal (i.e., $D_{ax}^* = D_{ay}^* = D_a^*$), and Eq. (7.61) becomes

$$D_a^* \left( \frac{\partial^2 u_a}{\partial x^2} + \frac{\partial^2 u_a}{\partial y^2} \right) + \frac{\partial D_a^*}{\partial x} \frac{\partial u_a}{\partial x} + \frac{\partial D_a^*}{\partial y} \frac{\partial u_a}{\partial y} = 0 \quad (7.62)$$

where

$D_a^*$ = air coefficient of transmission in the $x$- and $y$-directions.

These partial differential equations for air flow are similar in form to those previously presented for water flow.

## 7.3 STEADY-STATE AIR DIFFUSION THROUGH WATER

The diffusion of air through a saturated ceramic high air entry disk is one example of steady-state air diffusion through water. Another example is the diffusion of air through a saturated soil specimen. In each case, the diffused air pressure is dissipated across a region of water.

The measurement of the coefficient of diffusion can be used as an example of steady-state air diffusion through water. The coefficient is assumed to be a constant. The

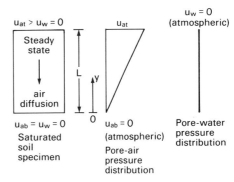

**Figure 7.31** Pore–air pressure distribution during the measurement of the coefficient of diffusion, $D$.

partial differential equation describing air diffusion takes the same form as that for air flow through an unsaturated soil [i.e., Eq. (7.56)]. The pore–air pressure distribution through the soil specimen is assumed to be linear.

An example showing the pore–air and pore–water pressure distributions across a saturated soil specimen during the measurement of the coefficient of diffusivity is shown in Fig. 7.31. The air pressure at each end of the specimen (i.e., $u_a = u_{ab} = 0.0$ at $y = 0.0$ and $u_a = u_{at}$ at $y = L$) are the boundary conditions. Substituting the boundary conditions into Eq. (7.57) yields a linear equation for the diffusing pore–air pressure distribution in the $y$-direction (i.e., $u_a = (y/L)u_{at}$).

# CHAPTER 8

# Pore Pressure Parameters

The mechanical behavior of unsaturated soils is directly affected by changes in the pore–air and pore–water pressures. Two classes of pore pressure conditions may develop in the field. First, there are the pore pressures associated with the flow or seepage through soils. This pore pressure condition was explained in Chapter 7. Second, there are the pore pressure conditions that are generated from the application of an external load, such as an engineered structure.

The pore pressures generated immediately after loading are commonly referred to as the undrained pore pressures. In the undrained condition, the applied total stress is carried by the soil structure, the pore–air and pore–water depending upon their relative compressibilities. The induced pore–air and pore–water pressures can be written as a function of the applied total stress. These excess pore pressures will be dissipated with time if the pore fluids are allowed to drain. The applied total stress is eventually carried by the soil structure.

This chapter presents the pore pressures generated from the application of total stress to the soil. The compressibilities of air, water, and air–water mixtures are first presented. The compressibility of the soil structure is summarized in the form of a constitutive relationship for an unsaturated soil. Equations which present the pore pressure as a function of the applied total stress require the use of the compressibility of the pore fluids. Isotropic and anisotropic soils under various undrained loading conditions are considered in the derivations.

The pore pressure response is expressed in terms of pore pressure parameters. Pore pressure parameters have proven to be useful in practice, particularly in earth dam construction. The pore pressures developed during construction can be estimated using the pore pressure parameters. The estimated pore pressures are required at the start of a transient analysis, such as consolidation (Chapter 15). Comparisons between the predictions and measurements of pore pressures generated by applied loads are presented and discussed.

This chapter mainly addresses the pore pressures generated under various loading conditions.

## 8.1 COMPRESSIBILITY OF PORE FLUIDS

During undrained compression of an unsaturated soil, the pore–air and pore–water are not allowed to flow out of the soil. Volume change occurs as a result of the compression of the air and, to lesser extent, the water. The compression of soil solids can be assumed negligible for the stress range commonly encountered in practice. The pore fluid volume change is related to the change in the pore–air and pore–water pressures. The pore–air and pore–water pressures increase as an unsaturated soil is compressed. The pore pressure increase is commonly referred to as an excess pore pressure. The volume change of a phase is related to a pressure change by use of its compressibility. Figure 8.1 defines the compressibility of a material at a point on the volume–pressure curve during undrained compression. Isothermal compressibility is defined as the volume change of a fixed mass with respect to a pressure change per unit volume at a constant temperature:

$$C = -\frac{1}{V}\frac{dV}{du} \quad (8.1)$$

where

$C$ = compressibility
$V$ = volume
$dV/du$ = volume change with respect to a pressure change
$du$ = a pressure change.

The term $(dV/du)$ in Eq. (8.1) has a negative sign because the volume decreases as the pressure increases. Therefore, a negative sign is used in Eq. (8.1) in order to give a positive compressibility.

In an unsaturated soil, the pore fluid consists of water,

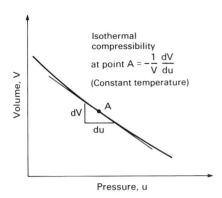

**Figure 8.1** Definition of isothermal compressibility.

free air, and air dissolved in water, as explained in Chapter 2. The individual compressibilities of air and water are required in formulating the compressibility of the mixture.

### 8.1.1 Air Compressibility

The isothermal compressibility of air can be expressed in accordance with Eq. (8.1):

$$C_a = \frac{-1}{V_a}\frac{dV_a}{du_a} \tag{8.2}$$

where

$C_a$ = isothermal compressibility of air
$V_a$ = volume of air
$dV_a/du_a$ = air volume change with respect to an air pressure change
$u_a$ = air pressure.

The volume versus pressure relation for air during isothermal, undrained compression can be expressed using Boyle's law:

$$V_a = \frac{\bar{u}_{ao}V_{ao}}{\bar{u}_a} \tag{8.3}$$

where

$\bar{u}_{ao}$ = initial absolute air pressure (i.e., $\bar{u}_{ao} = u_{ao} + \bar{u}_{atm}$)
$u_{ao}$ = initial gauge air pressure
$\bar{u}_{atm}$ = atmospheric pressure (i.e., 101.3 kPa)
$V_{ao}$ = initial volume of air
$\bar{u}_a$ = absolute air pressure (i.e., $\bar{u}_a = u_a + \bar{u}_{atm}$).

Differentiating the volume of air, $V_a$, with respect to the absolute air pressure, $\bar{u}_a$, gives

$$\frac{dV_a}{d\bar{u}_a} = -\frac{1}{\bar{u}_a^2}(\bar{u}_{ao}V_{ao}). \tag{8.4}$$

Equation (8.4) gives the volume change of air with respect to an infinitesimal change in the air pressure. Substituting Boyle's law into Eq. (8.4) gives

$$\frac{dV_a}{d\bar{u}_a} = -\frac{V_a}{\bar{u}_a}. \tag{8.5}$$

The volume derivative with respect to the absolute pressure, $(dV_a/d\bar{u}_a)$, is equal to the derivative with respect to the gauge pressure, $(dV_a/du_a)$, since the atmospheric pressure, $\bar{u}_{atm}$, is assumed to be constant. Therefore, Eq. (8.5) can be substituted into Eq. (8.2) to give the isothermal compressibility of air:

$$C_a = \frac{1}{\bar{u}_a}. \tag{8.6}$$

Equation (8.6) shows that the isothermal compressibility of air is inversely proportional to the absolute air pressure. In other words, the air compressibility decreases as the air pressure increases.

### 8.1.2 Water Compressibility

The compressibility of water is defined as follows:

$$C_w = -\frac{1}{V_w}\frac{dV_w}{du_w} \tag{8.7}$$

where

$C_w$ = water compressibility
$V_w$ = volume of water
$dV_w/du_w$ = water volume change with respect to water pressure change
$u_w$ = water pressure.

Figure 8.2 presents the results of water compressibility measurements (Dorsey, 1940). Dissolved air in water produces an insignificant difference between the compressibilities of air-free water and air-saturated water.

### 8.1.3 Compressibility of Air–Water Mixtures

The compressibility of an air–water mixture can be derived using the direct proportioning of the air and water com-

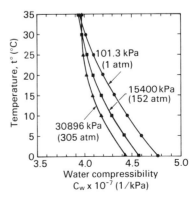

**Figure 8.2** Isothermal compressibility of air–saturated water (from Dorsey, 1940).

pressibilities. The conservation of mass and the compressibility definition in Eq. (8.1) must be adhered to. Let us consider the air, water, and solid volumetric relations, as shown in Fig. 8.3. Let us assume that the soil has a degree of saturation, $S$, and a porosity, $n$. The total volume of the air-water mixture is the sum of the volume of water, $V_w$, and the volume of air, $V_a$ (i.e., $V_w + V_a$). The volume of the dissolved air, $V_d$, is within the volume of water, $V_w$. The volumetric coefficient of solubility, $h$, gives the percentage of dissolved air with respect to the volume of water. The pore-air and pore-water pressures are $u_a$ and $u_w$, respectively, with $u_a$ always being greater than $u_w$. The soil is subjected to a compressive total stress, $\sigma$.

Let us apply an infinitesimal increase in total stress, $d\sigma$, to the undrained soil. The pore-air and pore-water pressures increase, while the volumes of air and water decrease. The compressibility of an air-water mixture for an infinitesimal increase in total stress can be written using the total stress as a reference:

$$C_{aw} = -\frac{1}{V_w + V_a}\left\{\frac{d(V_w - V_d)}{d\sigma} + \frac{d(V_a + V_d)}{d\sigma}\right\} \quad (8.8)$$

where

- $C_{aw}$ = compressibility of air-water mixture
- $(V_w + V_a)$ = volume of the air-water mixture
- $V_w$ = volume of water
- $V_a$ = volume of free air
- $d(V_w - V_d)/d\sigma$ = water volume change with respect to a total stress change
- $d(V_a + V_d)/d\sigma$ = air volume change with respect to a total stress change
- $V_d$ = volume of dissolved air.

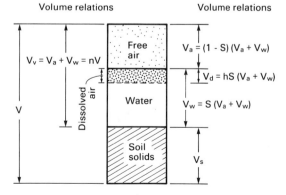

**Figure 8.3** Volumetric composition of the pore fluid in an unsaturated soil.

Equation (8.8) is a modification of the compressibility equation for air-water mixtures proposed by Fredlund (1976). The total stress change, $d\sigma$, is used as the reference pressure in Eq. (8.8), while the pore-water pressure change, $du_w$, was used as the reference pressure in Fredlund's (1976) compressibility equation. The term $(d(V_w - V_d)/d\sigma)$ in Eq. (8.8) is considered to be equal to $(dV_w/d\sigma)$ since the dissolved air is a fixed volume internal to the water. As such, its volume does not change. The total volume of water, $V_w$, is therefore used in computing the compressibility of water [i.e., $C_w = -(1/V_w)(dV_w/du_w)$].

The change in air volume occurs as a result of the compression of the free air in accordance with Boyle's law, and a further dissolving of free air into water in accordance with Henry's law. The total air volume change can be obtained directly using Boyle's law by considering the initial and final pressures and the volume conditions in the air phase. The free air compression and the air dissolving in water are assumed to be complete under final conditions. The free and dissolved air can be considered as one volume with a uniform pressure. Although the volume of dissolved air, $V_d$, is a fixed quantity, it is maintained in the formulation for clarity. Therefore, the dissolved air volume, $V_d$, also appears in Eq. (8.8).

Applying the chain rule of differentiation to Eq. (8.8) gives

$$C_{aw} = \frac{-1}{V_w + V_a}\left\{\frac{dV_w}{du_w}\frac{du_w}{d\sigma} + \frac{d(V_a + V_d)}{du_a}\frac{du_a}{d\sigma}\right\} \quad (8.9)$$

where

- $dV_w/du_w$ = water volume change with respect to a pore-water pressure change
- $du_w/d\sigma$ = water pressure change with respect to a total stress change
- $d(V_a + V_d)/du_a$ = air volume change with respect to a pore-air pressure change
- $du_a/d\sigma$ = air pressure change with respect to a total stress change.

Rearranging Eq. (8.9) gives

$$C_{aw} = -\left[\frac{V_w}{V_w + V_a}\frac{1}{V_w}\frac{dV_w}{du_w}\right]\frac{du_w}{d\sigma}$$
$$- \left[\frac{V_a + V_d}{V_w + V_a}\frac{1}{V_a + V_d}\frac{d(V_a + V_d)}{du_a}\right]\frac{du_a}{d\sigma}. \quad (8.10)$$

Substituting the volume relations in Fig. 8.3 and Eqs. (8.2) and (8.7) into Eq. (8.10) yields the compressibility of an air-water mixture:

$$C_{aw} = SC_w\left(\frac{du_w}{d\sigma}\right) + (1 - S + hS)C_a\left(\frac{du_a}{d\sigma}\right). \quad (8.11)$$

The isothermal compressibility of air, $C_a$, is equal to the inverse of the absolute air pressure:

$$C_{aw} = SC_w \left(\frac{du_w}{d\sigma}\right) + (1 - S + hS)\left(\frac{du_a}{d\sigma}\right)/\bar{u}_a. \quad (8.12)$$

### The Use of Pore Pressure Parameters in the Compressibility Equation

The ratio between the pore pressure change and the total stress change, $(du/d\sigma)$, in Eq. (8.12) is referred to as a pore pressure parameter. This parameter indicates the magnitude of the pore pressure change in response to a total stress change. The pore pressure parameter concept was first introduced by Skempton (1954) and Bishop (1954). The pore pressure parameters for the air and water phases are different (Bishop, 1961a; Bishop and Henkel, 1962) and depend primarily upon the degree of saturation of the soil. The parameters also vary depending on the loading conditions. These conditions are discussed in Section 8.2. The pore pressure parameters can also be directly measured in the laboratory. For isotropic loading conditions, the parameter is commonly called the $B$ pore pressure parameter, and it can be substituted into Eq. (8.12) as follows:

$$C_{aw} = SC_w B_w + \{(1 - S + hS)B_a/\bar{u}_a\} \quad (8.13)$$

where

- $B_w$ = pore-water pressure parameter for isotropic loading (i.e., $du_w/d\sigma_3$)
- $\sigma_3$ = isotropic (confining) total stress
- $B_a$ = pore-air pressure parameter for isotropic loading (i.e., $du_a/d\sigma_3$).

The compressibility of the pore fluid in an unsaturated soil [i.e., Eq. (8.13)] takes into account the matric suction of the soil through use of the $B_w$ and $B_a$ parameters. In the absence of soil solids, the $B_a$ and $B_w$ parameters are equal to one. In the presence of soil solids, however, the surface tension effects will result in the $B_a$ and $B_w$ values being less than 1.0, depending upon the matric suction of the soil. The pore-air and the pore-water pressures change at differing rates in response to the applied total stress. The $B_w$ value is greater than the $B_a$ value. The $B_a$ and $B_w$ parameters are low at low degrees of saturation, and both parameters approach an equal value of 1.0 at saturation. At this point, the matric suction of the soil goes to zero. The development of the $B_a$ and $B_w$ parameters during undrained compression is illustrated in Fig. 8.31.

### 8.1.4 Components of Compressibility of an Air-Water Mixture

The first term in the compressibility equation [i.e., Eq. (8.13)] accounts for the compressibility of the water portion of the mixture, while the second term accounts for the compressibility of the air portion. The compressibility of the air portion is due to the compression of free air (i.e., $(1 - S)B_a/\bar{u}_a$) and the air dissolving into water (i.e., $hSB_a/\bar{u}_a$). The contribution of each compressibility component to the overall compressibility of the air-water mixture is illustrated in Fig. 8.4 for various degrees of saturation. The case considered has an initial absolute air pressure, $\bar{u}_{ao}$, of 202.6 kPa (i.e., 2 atm). Values of $B_a$ and $B_w$ are assumed to be equal to 1.0 for all degrees of saturation. This assumption may be unrealistic for low degrees

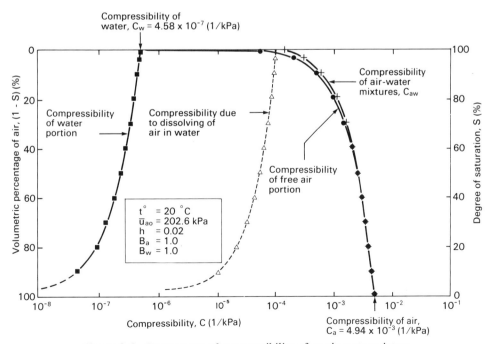

**Figure 8.4** Components of compressibility of an air-water mixture.

### Effects of Free Air on the Compressibility of the Mixture

Figure 8.4 shows that the compressibility of an air-water mixture is predominantly influenced by the compressibility of the free air portion (i.e., $(1 - S)B_a/\bar{u}_a$). When the soil voids are filled with air (i.e., $S$ is equal to 0.0), the compressibility of the pore fluid is equal to the isothermal compressibility of air (i.e., $4.94 \times 10^{-3}$ (1/kPa) at $\bar{u}_a$ equals 202.6 kPa). At saturation (i.e., $S$ is equal to 1.0), the soil voids are completely filled with water, and pore fluid compressibility becomes equal to that of water [i.e., $4.58 \times 10^{-7}$ (1/kPa)]. The inclusion of even 1% air in the soil is sufficient to significantly increase the pore fluid compressibility (see Fig. 8.4). The compressibility of the water is only of significance in computing the compressibility when the soil is fully saturated.

### Effects of Dissolved Air on the Compressibility of the Mixture

The solution of air in water gives the effect that the soil is compressible. The compressibility due to the solution of air in water (i.e., $hSB_a/\bar{u}_a$) is approximately two orders of magnitude greater than the compressibility of the water (i.e., $SC_wB_w$). Air dissolving in water significantly affects the compressibility of an air-water mixture when the free air volume becomes less than approximately 20% of the volume of voids. The effect of air solubility on the compressibility of an air-water mixture is shown in Fig. 8.5 for several initial absolute air pressures. The $B_a$ and $B_w$ parameters are assumed to be equal to 0.8 and 0.9, respectively.

The figure shows that the effect of air solubility on the compressibility of an air-water mixture is the same (i.e., on a logarithmic scale) for any initial air pressures. However, the effect of air dissolving in water does not result in a smooth transition for the compressibility of an air-water mixture as saturation is reached (i.e., $S$ is equal to 100%). There is a discontinuity at the point where there is no more free air to be dissolved in the water. As a result, the second term in Eq. (8.13) must be dropped, and the compressibility abruptly decreases to the compressibility of water. If the free air does not have time to be dissolved in the water, the term $(hSB_a/\bar{u}_a)$ in Eq. (8.13) must be set to zero. In this case, the compressibility of pore fluid has a smooth transition back to saturation, as shown in Fig. 8.5.

### 8.1.5 Other Relations for Compressibility of Air-Water Mixtures

The compressibility of an air-water mixture has been investigated by numerous researchers. For example, an equation for the compressibility of an air-water mixture has been proposed, neglecting the compressibility of the water and assuming zero matric suction (Bishop and Eldin, 1950; Skempton and Bishop, 1954). Therefore, the compressibility of the air is considered as representative of the compressibility of the pore fluid. This equation is suitable for conditions where the air phase constitutes a significant portion of the pore fluid. This proposed equation can also be obtained by ignoring the first term in Eq. (8.13). Also, the $B_a$ and $B_w$ values must be set to 1.0. The resulting equation

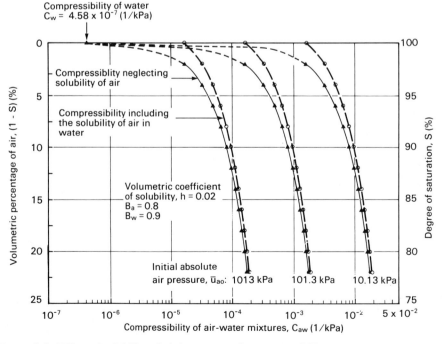

**Figure 8.5** Effect of solubility of air in water on the compressibility of an air-water mixture.

has the following form:

$$C_{aw} = (1 - S + hS)/\bar{u}_a. \quad (8.14)$$

Equation (8.14) is slightly different from the equation proposed by Bishop and Eldin (1950). The difference lies in the definition of air compressibility. Equation (8.14) is derived using the definition of air compressibility given in Eq. (8.2) [i.e., $C_a = -(1/V_a)/(dV_a/du_a)$]. The equation proposed by Bishop and Eldin (1950), was derived with the air compressibility defined with reference to the initial volume of air, $V_{ao}$:

$$C_a = -\frac{1}{V_{ao}} \frac{dV_a}{du_a}. \quad (8.15)$$

Substituting the term $(dV_a/du_a)$ from Eq. (8.4) into Eq. (8.15) yields a slightly different equation:

$$C_a = \frac{\bar{u}_{ao}}{\bar{u}_a^2}. \quad (8.16)$$

Equation 8.6 (i.e., $C_a = 1/\bar{u}_a$) describes the air compressibility at a point on the volume-pressure curve during undrained compression (see Fig. 8.1). Equation (8.16) gives the average air compressibility during an air pressure change from $\bar{u}_{ao}$ to $\bar{u}_a$ (Skempton and Bishop, 1954). The difference in the computed air compressibilities from both equations is negligible when the air pressure change is infinitesimal.

Replacing the air compressibility term in Eq. (8.14) (i.e., $1/\bar{u}_a$) with the average air compressibility (i.e., $\bar{u}_{ao}/\bar{u}_a^2$) yields the air-water mixture compressibility equation proposed by Bishop and Eldin (1950):

$$C_{aw} = (1 - S + hS)\bar{u}_{ao}/\bar{u}_a^2. \quad (8.17)$$

Koning (1963) suggested another equation for the compressibility of air-water mixtures. The equation was first derived by expressing the pore-air and pore-water pressure changes as a function of surface tension. However, changes in the pore-air and pore-water pressures were later assumed to be equal. This implies that $B_a$ and $B_w$ are equal to 1.0. Also, compressibility due to the solution of air in water was disregarded in the derivation (i.e., $hSB_a/\bar{u}_a = 0$). Applying the above assumptions to the compressibility of an air-water mixture gives the following equation (Koning, 1963):

$$C_{aw} = SC_w + (1 - S)/\bar{u}_a. \quad (8.18)$$

Equation (8.18) was also suggested by Verruijt (1969) in formulating the elastic storage of aquifers. The solubility of air in water and the effect of matric suction were neglected.

### Limitation of Kelvin's Equation in Formulating the Compressibility Equation

Kelvin's equation (i.e., $(u_a - u_w) = 2T_s/R_s$) relates matric suction, $(u_a - u_w)$, to surface tension, $T_s$, and the radius of curvature, $R_s$. Attempts have also been made to use Kelvin's equation in writing an equation for the compressibility of air-water mixtures (Schuurman, 1966; Barends, 1979). Equations were proposed for the case of occluded air bubbles in soils with a degree of saturation greater than approximately 85%. Kelvin's equation results in the incorporation of the radius of curvature, $R_s$, as a variable. However, $R_s$ is unmeasurable.

The proposed theory would suggest that the pore-air and pore-water pressures increase as the total stress increases during undrained compression. It has been postulated that there exist a critical pore-water pressure beyond which all of the free air would dissolve abruptly (i.e., collapse). This has become known as the air bubble collapse theory. Collapse occurs when a critical pore-water pressure is reached. This phenomenon would be reflected by an increase in compressibility at the critical pore-water pressure.

The difference between the pore-air and pore-water pressures (i.e., matric suction) is predicted to increase in the bubble collapse region, with an increase in total stress. However, there appears to be no experimental evidence to support this postulate. On the contrary, experimental evidence supports a continual increase of the pore-air and pore-water pressures, approaching a single value as the total stress is increased. This means that the matric suction decreases with an increasing total stress during undrained compression. There appear to be some conceptual difficulties when Kelvin's equation is incorporated into the equation for pore fluid compressibility.

In order to better understand the problem associated with incorporating Kelvin's equation into the equation for pore fluid compressibility, let us consider an element of soil with certain volume-mass properties. Let us assume there are numerous occluded air bubbles within the soil element, as depicted in Fig. 8.6(a). Kelvin's equation describes a microscopic phenomenon within the selected element. The radii of the occluded air bubbles are not measurable. Therefore, Kelvin's equation cannot be incorporated into a macroscopic-type formulation for compressibility.

It is of interest to draw a parallel comparison to the concept of effective stress and intergranular stresses in a saturated soil element [Fig. 8.6(b)]. From a microscopic point of view, there exist numerous intergranular stresses acting at the contacts between the soil particles in the element. From a macroscopic standpoint, there is an effective stress acting on the soil. The effective stress is the only stress state variable required to describe the behavior of the saturated soil. The intergranular stresses occur at a different level of definition, and are not measurable nor necessary when analyzing the behavior of the soil. Likewise, to incorporate Kelvin's equation into a macroscopic definition of pore fluid compressibility would be erroneous.

The net result of attempts to apply Kelvin's equation, together with Boyle's and Henry's laws, to the compressibility of an air-water mixture is an anomaly. The anomaly

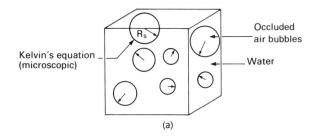

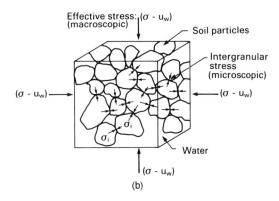

**Figure 8.6** Microscopic and macroscopic models for a soil. (a) Compressibility formulation for an air–water mixture; (b) effective stress and intergranular stresses in a saturated soil.

can be best understood by first considering an increase in total stress applied to an unsaturated soil specimen. The total stress increase causes the free air volume in the soil to compress and the pore-air pressure to increase, in accordance with Boyle's law. The increased air pressure drives more free air into solution, in accordance with Henry's law. As a result, the volume of free air further decreases. The decrease in the free air volume is accompanied by a decrease in the radius of curvature of the air–water menisci. Kelvin's equation would indicate that the matric suction, $(u_a - u_w)$, must increase.

The increase in matric suction implies that the pore-air pressure increase is larger than the increase in the pore-water pressure. The pore-air and pore-water pressure increases must be positive since the total stress increase is positive. The increase in air pressure in response to Kelvin's equation would further drive more air into solution, in accordance with Henry's law. Consequently, the volume of free air decreases, and the whole process is repeated over and over. In summary, it would appear that a slight increase in total stress could cause a chain reaction process which would reduce the free air volume to an infinitesimal size while the matric suction goes to infinity. This would suggest an air bubble instability problem as a result of a small increment in total stress. However, the above scenario is not in agreement with observed experimental results, as shown in Fig. 8.36.

Experimental results indicate that the pore-air and pore-water pressures gradually increase towards a single value as the matric suction approaches zero. This occurs as the total stress is increased under undrained loading. The process is gradual, caused by several increments of total stress.

The above apparent anomaly appears to be the result of incorporating Kelvin's equation (i.e., microscopic model) in a macroscopic formulation of pore fluid compressibility. This anomaly can also be partly understood by assuming that the decrease in the free air volume is not necessarily accompanied by a decrease in the *controlling* radius of curvature. Figure 8.7 shows that although the volume of the continuous air phase in zone 1 *decreases* due to an increase in the pore-air pressure from $u_{a1}$ to $u_{a2}$, the controlling radius may, in fact, *increase* from $R_{s1}$ to $R_{s2}$.

The above postulate is only possible for the nonspherical air bubbles such as those in zone 1. The nonspherical air bubbles have nonuniform radii of curvature over their surface, but the assumption is made that only the controlling radius is considered in Kelvin's equation. Nevertheless, the increase in total stress will eventually cause the air bubbles to take on a spherical form, as shown in zone 2 (Fig. 8.7). For spherical air bubbles, a decrease in volume must be followed by a decrease in the radius. In this case, the anomaly illustrated by the application of Kelvin's equation cannot be resolved.

On the basis of the above conceptual difficulties, it would appear best not to apply Kelvin's equation to the compressibility equation. It is recommended that the pore-air and pore-water pressures be measured or computed using $B_a$ and $B_w$ pore pressure parameters in the continuous pore-air zone (i.e., zone 1 in Fig. 8.7). Air pressure measurements are valid as long as the air phase is continuous. In the occluded zone (i.e., zone 2 in Fig. 8.7), the air bubbles do not interact with the soil structure. The presence of air bubbles merely renders the pore fluid compressible. Therefore, it is recommended that the pore-air and pore-water pressures be assumed to be equal in the occluded zone (i.e., zone 2).

## 8.2 DERIVATIONS OF PORE PRESSURE PARAMETERS

The pore pressure response for a change in total stress during undrained compression has been expressed in terms of pore pressure parameters in the previous section. In this section, derivations are presented for pore pressure parameters corresponding to various loading conditions. The equations take into account the relative compressibilities involved. One equation is derived for the water phase, and another for the air phase. These equations assist in analyzing the pore pressure response of an unsaturated soil during undrained loading. These equations are not meant, however, to replace direct measurements of the pore pressure parameters.

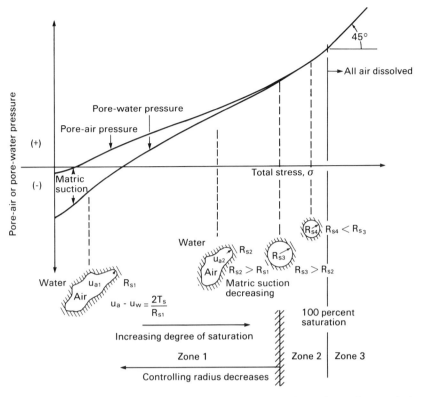

**Figure 8.7** Pore–air and pore–water pressure responses to a change in total stress during undrained compression.

### 8.2.1 Tangent and Secant Pore Pressure Parameters

The pore pressure parameters for an unsaturated soil can be defined either as a tangent- or secant-type parameter. These definitions are similar in concept to the tangent and secant modulus used in the theory of elasticity.

Consider the development of pore–air and pore–water pressures during isotropic, undrained compression, as shown in Fig. 8.8. The pore–water pressure increases faster than the pore–air pressure, in response to the increase in total confining stress. The pore pressure response to an increase in the confining stress is referred to as the $B$ pore pressure parameter. At a point during the loading of the soil (i.e., point 1), the tangent and the secant pore pressure parameters can be defined for both the air and water phases (Fig. 8.8). If the increase in the isotropic pressure is referenced to the initial condition, the secant $B'$ pore pressure parameter can be expressed as follows for the air phase:

$$B'_a = \frac{\Delta u_a}{\Delta \sigma_3} \quad (8.19)$$

where

$B'_a$ = secant pore pressure parameter for the air phase during isotropic, undrained compression

$\Delta u_a$ = increase in pore-air pressure due to an increase in isotropic pressure, $\Delta \sigma_3$; the change in pore–air pressure is measured from the initial condition (i.e., $u_a - u_{ao}$)

$u_{ao}$ = initial pore-air pressure

$\Delta \sigma_3$ = increase in isotropic pressure from the initial condition.

Similarly, for the water phase, the $B'$ pore pressure parameter can be expressed as

$$B'_w = \frac{\Delta u_w}{\Delta \sigma_3} \quad (8.20)$$

where

$B'_w$ = secant pore pressure parameter for the water phase during isotropic, undrained compression

$\Delta u_w$ = increase in pore-water pressure due to an increase in isotropic pressure, $\Delta \sigma_3$; the change in pore-water pressure is measured from the initial condition (i.e., $u_w - u_{wo}$)

$u_{wo}$ = initial pore-water pressure.

If an infinitesimal increase in the isotropic pressure is considered at point 1, the pore pressure response at point 1 can be expressed as the tangent $B$ pore pressure parameter (Fig. 8.8). The tangent $B$ pore pressure parameters for the

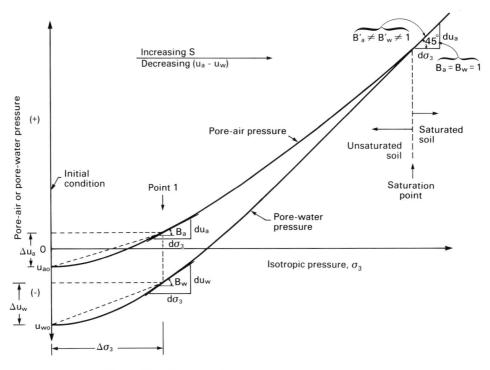

**Figure 8.8** Tangent and secant pore pressure parameters.

air and water phases are written as follows:

$$B_a = \frac{du_a}{d\sigma_3} \quad (8.21)$$

$$B_w = \frac{du_w}{d\sigma_3} \quad (8.22)$$

where

$B_a$ = tangent pore pressure parameter for the air phase during insotropic, undrained compression

$du_a$ = increase in pore-air pressure due to an infinitesimal increase in isotropic pressure, $d\sigma_3$

$d\sigma_3$ = infinitesimal increase in isotropic pressure

$B_w$ = tangent pore pressure parameter for the water pressure during isotropic, undrained compression

$du_w$ = increase in pore-water pressure due to an infinitesimal increase in isotropic pressure, $d\sigma_3$.

The concept of the tangent $B$ pore pressure parameter has been used in a previous section [i.e., Eq. (8.13)] where an infinitesimal increase in total stress was considered. The secant $B'$ pore pressure parameters are particularly useful in computing the final pore-air and pore-water pressures after a large change in total stress. The tangent $B$ pore pressure parameters can also be used to predict the final pore pressures under large changes in total stress by using a marching-forward technique with finite increments of total stress. The total stress is incrementally increased, commencing from an initial condition and proceeding to the final condition under investigation. As saturation is approached, the pore-water pressure approaches the pore-air pressure (i.e., $u_w \rightarrow u_a$ or $B_w \rightarrow B_a$), and the air bubbles dissolve in the water.

At saturation, a change in total stress causes an almost equal change in pore-water pressure. The small difference between the change in total stress and the change in pore-water pressure at saturation can be ignored due to the low compressibility of water relative to the compressibility of the soil structure (Skempton, 1954). In other words, the tangent $B$ parameter becomes equal to 1.0 (i.e., $B_a = B_w = 1.0$). A convention in laboratory testing has been to assume complete saturation of a soil when the $B$ pore pressure parameter reaches a value of 1.0 (Skempton, 1954). It is possible for the secant $B'$ pore pressure parameter to be less than 1.0 even though the soil has reached saturation. Also, the values of $B'_a$ and $B'_w$ are not equal at saturation since they are computed with respect to initial conditions where the pore-air pressure is greater than the pore-water pressure (Fig. 8.8).

In the following sections, the derivation for tangent pore pressure parameters for various loading conditions will be presented. The only exception is Hilf's analysis where a secant-type pore pressure parameter is derived. A prime sign will be assigned to the secant parameters (e.g., $B'$) in order to differentiate between the secant and the tangent pore pressure parameters.

### 8.2.2 Summary of Necessary Constitutive Relations

The derivation of pore pressure parameters requires volume change constitutive relations for an unsaturated soil. These constitutive relations describe the volume changes that take place under *drained* loading. The volume changes are expressed in terms of the stress state variable changes. These constitutive relations are formulated and explained in Chapter 12. This section provides a summary of the constitutive relations required for deriving the pore pressure parameter equations. The use of constitutive relations from *drained* loading conditions when deriving the pore pressure parameter equations for *undrained* loading is explained in Section 8.2.3.

Consider an unsaturated soil specimen that undergoes one-dimensional, drained compression. The stress state variables, $(\sigma - u_a)$ and $(u_a - u_w)$, change as the soil is compressed. As a result, the volume of the soil changes in accordance with the constitutive relation for the soil structure. Volume change is primarily the result of compression of the pore fluid since the soil solids are essentially incompressible. Figure 8.9(a) shows the volume of the voids, $V_v$, referenced to the initial total volume of the soil, $V_0$ (i.e., $V_v/V_0$) as a function of the stress state variables. The idealized three-dimensional surface presented in Fig. 8.9(a) is the constitutive relationship for the soil structure. For finite changes in the stress state, the constitutive relationship can be linearized, as depicted in Fig. 8.9(a). The linear equation for volume change can be written as proposed by Fredlund and Morgenstern (1976).

$$\frac{dV_v}{V_0} = m_1^s d(\sigma - u_a) + m_2^s d(u_a - u_w) \quad (8.23)$$

where

$dV_v/V_0$ = volume change referenced to the initial total volume of the soil
$V_v$ = volume of soil voids
$V_0$ = initial total volume of the soil
$m_1^s$ = coefficient of soil volume change with respect to a change in net normal stress
$d(\sigma - u_a)$ = change in net normal stress
$m_2^s$ = coefficient of soil volume change with respect to a change in matric suction
$d(u_a - u_w)$ = change in matric suction.

The continuity of a referential, unsaturated soil element requires that the volume change must be equal to the sum of the changes in volume associated with the air and water phases which fill the pore voids. Figure 8.9(b) and (c) illustrate idealized constitutive surfaces for the air and water phases, respectively. Changes in the volume of air and water can be regarded as a linear function of finite changes in the stress state variables (Fig. 8.9(b) and (c), respec-

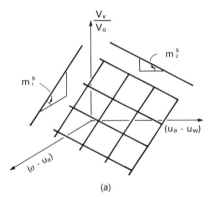

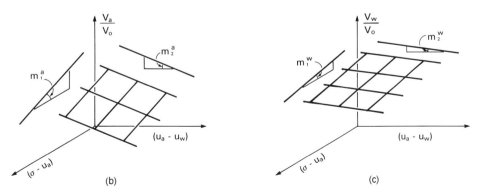

**Figure 8.9** Linearized portion of the constitutive relationships during one-dimensional, drained compression. (a) Soil structure constitutive surface; (b) air phase constitutive surface; (c) water phase constitutive surface.

tively). The change in the volume of air can be expressed as

$$\frac{dV_a}{V_0} = m_1^a d(\sigma - u_a) + m_2^a d(u_a - u_w) \quad (8.24)$$

where

$dV_a/V_0$ = change in the volume of air referenced to the initial total volume of the soil
$V_a$ = volume of air
$m_1^a$ = coefficient of air volume change with respect to a change in net normal stress
$m_2^a$ = coefficient of air volume change with respect to a change in matric suction.

The change in the volume of water can be written as

$$\frac{dV_w}{V_0} = m_1^w d(\sigma - u_a) + m_2^w d(u_a - u_w) \quad (8.25)$$

where

$dV_w/V_0$ = change in the volume of water referenced to the initial total volume of the soil
$V_w$ = volume of water
$m_1^w$ = coefficient of water volume change with respect to a change in net normal stress
$m_2^w$ = coefficient of water volume change with respect to a change in matric suction.

The continuity requirement for a referential element can be expressed as

$$\frac{dV_v}{V_0} = \frac{dV_a}{V_0} + \frac{dV_w}{V_0}. \quad (8.26)$$

The following conditions must be satisfied when considering volumetric continuity:

$$m_1^s = m_1^a + m_1^w \quad (8.27)$$

$$m_2^s = m_2^a + m_2^w. \quad (8.28)$$

The above constitutive relations can be used to compute volume changes that occur during undrained compression (Fig. 8.8). The changes in volume that occur between the initial and the final conditions can be calculated using a marching-forward technique with finite changes in the stress state variables [i.e., $d(\sigma - u_a)$ and $d(u_a - u_w)$]. The volume of the soil, water, and air must be accounted for after each increment of total stress.

Although there are three constitutive relationships (i.e., soil structure, air, and water) available, only two of the relationships are required in deriving the pore pressure parameter equations. The constitutive equations for the soil structure and the air phase will be used in the derivation of the pore pressure parameters.

### 8.2.3 Drained and Undrained Loading

This section explains how the constitutive relationships obtained from drained loading apply to undrained loading.

The application of an all-around, positive (i.e., compressive) total stress, $d\sigma$, either in drained or undrained loading, can cause a change in volume. In *drained* loading, air and water are allowed to drain from the soil subsequent to the application of the total stress increment. The stress state variables in the soil are altered and the volume of the soil changes. The volume change can be computed from the stress state variable changes in accordance with the constitutive relationship for the soil structure.

In *undrained* loading, the air and water are not allowed to drain from the soil. The total stress increase causes the pore-air and pore-water pressures to increase, and consequently the stress state variables also change. The increase in the pore fluid pressures occurs in response to a compression of the pore fluid. The soil volume change during undrained loading can be regarded as the volume change equivalent to the pore fluid compression, as illustrated in Fig. 8.10(a). The volume change equivalent to the pore fluid compression, $dV_v$, can be computed by multiplying the compressibility of the air-water mixture, $C_{aw}$ [i.e., Eq. (8.8)] by the pore fluid volume $V_v$ (i.e., $V_w + V_a = nV$) and the total stress increment, $d\sigma$:

$$\left(\frac{dV_v}{V_0}\right)_0 = C_{aw} n \, d\sigma \quad (8.29)$$

where

$(dV_v/V_0)_0$ = volume change during undrained compression, referenced to the initial total vol-

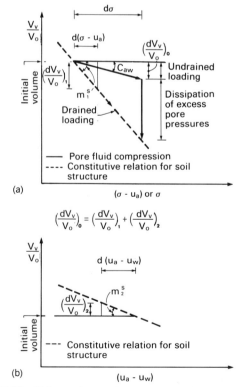

Figure 8.10 Volume changes in an unsaturated soil during loading. (a) Volume change with respect to a change in $(\sigma - u_a)$ or $\sigma$, (b) volume change with respect to a change in $(u_a - u_w)$.

ume of the soil; the volume change results from the compression of the pore fluids

$n$ = porosity (i.e., $V_v/V$ can be assumed to be equal to $V_v/V_0$ for small changes in stress state variables or volumes).

The volume change can also be written in terms of the stress state variable changes in accordance with the constitutive relationship for the soil structure. The increase in total stress results in an increase in net normal stress, $d(\sigma - u_a)$, and a decrease in matric suction, $d(u_a - u_w)$. The increase in net normal stress causes a decrease in volume [Fig. 8.10(a)]:

$$\left(\frac{dV_v}{V_0}\right)_1 = m_1^s \, d(\sigma - u_a) \qquad (8.30)$$

where

$(dV_v/V_0)_1$ = volume change due to the change in net normal stress, $d(\sigma - u_a)$, referenced to the initial total volume.

The decrease in matric suction generally causes an increase in soil volume [Fig. 8.10(b)]:

$$\left(\frac{dV_v}{V_0}\right)_2 = m_2^s \, d(u_a - u_w) \qquad (8.31)$$

where

$(dV_v/V_0)_2$ = volume change due to a change in matric suction, $d(u_a - u_w)$, referenced to the initial total volume.

The matric suction decrease in Eq. (8.31) yields a volume increase or swelling. The $m_2^s$ value for the unloading path must be used in Eq. (8.31) (see Chapter 13). The total volume change obtained from the constitutive relations for the soil structure can be written as the sum of Eqs. (8.30) and (8.31):

$$\left(\frac{dV_v}{V_0}\right)_1 + \left(\frac{dV_v}{V_0}\right)_2 = m_1^s \, d(\sigma - u_a) + m_2^s \, d(u_a - u_w). \qquad (8.32)$$

The total volume change obtained from the constitutive relation [i.e., Eq. (8.32)] can be equated to the volume change due to pore fluid compression [i.e., Eq. (8.29)]:

$$\left(\frac{dV_v}{V_0}\right)_1 + \left(\frac{dV_v}{V_0}\right)_2 = \left(\frac{dV_v}{V_0}\right)_0 \qquad (8.33)$$

or

$$m_1^s \, d(\sigma - u_a) + m_2^s \, d(u_a - u_w) = C_{aw} n \, d\sigma. \qquad (8.34)$$

In a *saturated* soil, the pore voids are filled with water. The pore fluid compressibility is equal to the compressibility of water, which is far less than the compressibility of the soil structure. A total stress increase, $d\sigma$, in undrained loading is almost entirely transferred to the water phase (i.e., $du_w \approx d\sigma$) or the pore-water pressure parameter approaches 1.0. The effective stress in undrained loading remains essentially unchanged [i.e., $d(\sigma - u_w) \approx 0.0$]. As a result, the volume change computed from the constitutive relation for the soil structure is extremely small [Fig. 8.11(a)]. The soil volume change obtained from the pore fluid compression is also extremely small because of the low compressibility of the water [Fig. 8.11(a)].

In a *dry* soil, the pore voids are primarily filled with air, which is much more compressible than the soil structure. This is particularly true when the pore-air pressure is low since the isothermal compressibility of air is equal to the inverse of the absolute air pressure. Therefore, a total stress increase, $d\sigma$, during undrained loading is almost entirely transferred to the soil structure. The pore pressure remains practically constant or the pore pressure parameter approaches zero. The volume change can be computed from the change in net normal stress, $(\sigma - u_a)$, in accordance with the constitutive relation for the soil structure [Fig. 8.11(b)]. The volume change due to a change in matric suction is negligible in a dry soil, as described in Chapter 3. The compression of the air during undrained loading follows the compression of the soil structure [Fig. 8.11(b)]. A substantial volume change is required for a significant pore-air pressure change due to the high compressibility of air.

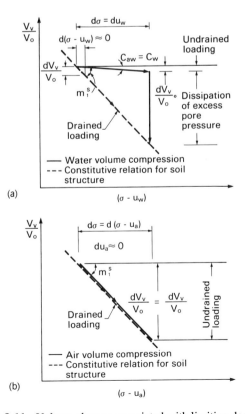

**Figure 8.11** Volume changes associated with limiting degree of saturation conditions. (a) Volume change in a saturated soil; (b) volume change in a dry soil.

## 8.2.4 Total Stress and Soil Anisotropy

The pore pressure parameters can be derived by considering various conditions of loading and soil anisotropy. The loading conditions are similar to those outlined by Lambe and Whitman (1979) for saturated soils. These conditions are summarized in Fig. 8.12, where increments in the major, intermediate, and minor principal stresses are denoted as $d\sigma_1$, $d\sigma_2$, and $d\sigma_3$, respectively. For $K_0$-loading, the soil can only change volume in *one direction* (i.e., the y-direction). For the other loading conditions, the soil can change volume *three-dimensionally* (i.e., x-, y-, and z-directions). These loadings differ in the magnitude and direction of the applied increments of total stress, as illustrated in Fig. 8.12.

An isotropic soil is defined as a soil with compressibilities which are constant with respect to the different directions. That is to say, the magnitudes of $m_1^s$ and $m_1^a$ are constant in the x-, y-, and z-directions. Only the constitutive equations for the soil structure and the air phase have been selected for use in the derivation of the pore pressure parameters.

Soil anisotropy is herein defined as the condition where the soil compressibility with respect to a change in each stress state variable varies with direction. The number of independent moduli required depends on whether the longitudinal strain in each direction is to be predicted independently or whether three-dimensional volume changes are to be predicted. The following statements assume that volume changes are being predicted. As a result, the compressibility $m_1^s$ varies in the x-, y-, and z-directions, and is designated as $m_{11}^s$, $m_{12}^s$, and $m_{13}^s$, respectively. Similarly, the air phase has the moduli, $m_{11}^a$, $m_{12}^a$, $m_{13}^a$, in the x-, y-, and z-directions, respectively. However, the compressibility with respect to a change in matric suction can be assumed to produce singular values (i.e., $m_2^s$ and $m_2^a$).

The following sections present the derivations of pore pressure parameters for an isotropic soil under various loading conditions. One-dimensional loading (i.e., $K_0$-

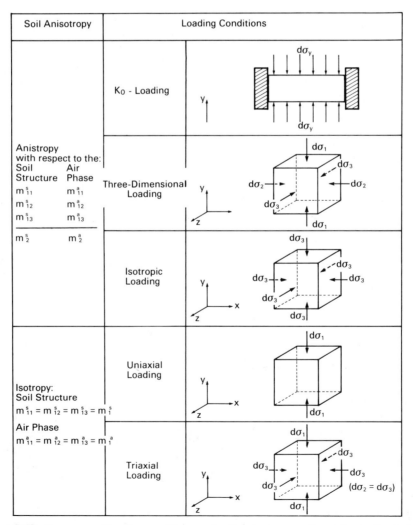

**Figure 8.12** Summary of loading conditions and soil anisotropy as it pertains to the derivation of pore pressure parameters.

loading) is a commonly used condition in soil mechanics. For this reason, the volume change moduli for $K_0$-loading will be given the subscript "k" (i.e., $m_{1k}^s$ and $m_{1k}^a$) in order to differentiate this loading condition from the other loading conditions.

### 8.2.5 $K_0$-Loading

The following pore pressure parameters are derived for $K_0$-undrained loading. This loading condition is illustrated in Fig. 8.13. Let us suppose that a total stress increment is applied in the vertical direction. The total stress increment is denoted as $d\sigma_y$, and the vertical direction is assumed to be the major principal stress direction. Equation (8.34) relates the soil volume change to the volume change due to the pore fluid compression. Substituting the $K_0$-loading condition (i.e., $d\sigma_y$) into Eq. (8.34) gives

$$m_{1k}^s \, d(\sigma_y - u_a) + m_2^s \, d(u_a - u_w) = C_{aw} n \, d\sigma_y \quad (8.35)$$

where

$m_{1k}^s$ = compressibility of the soil structure with respect to a change in net major principal stress for $K_0$-loading

$d\sigma_y$ = infinitesimal increase in the major principal stress.

The compressibility of an air-water mixture, $C_{aw}$, is calculated by substituting the change in total stress into Eq. (8.12):

$$C_{aw} = SC_w \left(\frac{du_w}{d\sigma_y}\right)$$
$$+ (1 - S + hS) \left(\frac{du_a}{d\sigma_y}\right) \bigg/ \bar{u}_a. \quad (8.36)$$

Combining Eqs. (8.35) and (8.36) results in the following relationship for $K_0$-undrained loading:

$$m_{1k}^s \, d(\sigma_y - u_a) + m_2^s \, d(u_a - u_w)$$
$$= SnC_w \, du_w + (1 - S + hS) n \, du_a/\bar{u}_a. \quad (8.37)$$

Rearranging Eq. (8.37) yields an expression for the change in the pore-water pressure, $du_w$, in response to a total stress change, $d\sigma_y$:

$$du_w = \left(\frac{m_2^s - m_{1k}^s - \{(1 - S + hS)n/\bar{u}_a\}}{m_2^s + SnC_w}\right) du_a$$
$$+ \left(\frac{m_{1k}^s}{m_2^s + SnC_w}\right) d\sigma_y. \quad (8.38)$$

The compressibility, $m_2^s$, can be written as a ratio of the compressibility with respect to a total stress change, $m_{1k}^s$:

$$R_{sk} = m_2^s/m_{1k}^s. \quad (8.39)$$

Substituting Eq. (8.39) into Eq. (8.38) gives

$$du_w = \left(\frac{R_{sk} - 1 - \{(1 - S + hS)n/(\bar{u}_a m_{1k}^s)\}}{R_{sk} + SnC_w/m_{1k}^s}\right) du_a$$
$$+ \left(\frac{1}{R_{sk} + SnC_w/m_{1k}^s}\right) d\sigma_y. \quad (8.40)$$

Equation (8.40) can be further simplified as follows:

$$du_w = R_{1k} \, du_a + R_{2k} \, d\sigma_y \quad (8.41)$$

where

$$R_{1k} = \frac{R_{sk} - 1 - \{(1 - S + hS)n/(\bar{u}_a m_{1k)}^s\}}{R_{sk} + SnC_w/m_{1k}^s}$$

$$R_{2k} = \frac{1}{R_{sk} + SnC_w/m_{1k}^s}.$$

There are two unknowns (i.e., $du_w$ and $du_a$) in Eq. (8.41). In order to compute the pore pressure changes, a second independent equation is required. The second equation is derived by considering the change in the volume of air. The change in volume is described by the constitutive relation for the air phase [i.e., Eq. (8.24)]. The volume change due to the compression of air, $dV_a$, is computed from the compressibility of the air [i.e., the second term in Eq. (8.36)] multiplied by the pore fluid volume (i.e., $nV$) and the total stress increment, $d\sigma_y$:

$$\frac{dV_a}{V_0} = \left\{\frac{(1 - S + hS)n}{\bar{u}_a}\right\} du_a. \quad (8.42)$$

The volume change given by the constitutive relation for the air phase [i.e., Eq. (8.24)] must be equal to the volume change due to compression of the air [i.e., Eq. (8.42)]:

$$m_{1k}^a \, d(\sigma_y - u_a) + m_2^a \, d(u_a - u_w) = \frac{(1 - S + hS)n}{\bar{u}_a} du_a \quad (8.43)$$

where

$m_{1k}^a$ = slope of the $(V_a/V_0)$ versus $(\sigma_y - u_a)$ plot for $K_0$-loading.

This second equation [i.e., Eq. (8.43)] can be written to give the change in the pore-air pressure, $du_a$, due to a total

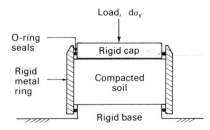

**Figure 8.13** Conditions simulated in the derivation of the pore pressure parameters for $K_0$-undrained loading.

stress increment, $d\sigma_y$:

$$du_a = \left[\frac{m_2^a}{m_2^a - m_{1k}^a - \{(1-S+hS)n/\bar{u}_a\}}\right] du_w$$
$$- \left[\frac{m_{1k}^a}{m_2^a - m_{1k}^a - \{(1-S+hS)n/\bar{u}_a\}}\right] d\sigma_y. \quad (8.44)$$

The relationship between the moduli, $m_2^a$ and $m_{1k}^a$, can be expressed as a ratio, $R_{ak}$:

$$R_{ak} = m_2^a/m_{1k}^a. \quad (8.45)$$

Equation (8.44) can be written in terms of the ratio, $R_{ak}$:

$$du_a = \left[\frac{R_{ak}}{R_{ak} - 1 - \{(1-S+hS)n/(\bar{u}_a m_{1k}^a)\}}\right] du_w$$
$$- \left[\frac{1}{R_{ak} - 1 - \{(1-S+hS)n/(\bar{u}_a m_{1k}^a)\}}\right] d\sigma_y. \quad (8.46)$$

In a simple form, Eq. (8.46) can be written

$$du_a = R_{3k} \, du_w - R_{4k} \, d\sigma_y \quad (8.47)$$

where

$$R_{3k} = \left[\frac{R_{ak}}{R_{ak} - 1 - \{(1-S+hS)n/(\bar{u}_a m_{1k}^s)\}}\right]$$

$$R_{4k} = \left[\frac{1}{R_{ak} - 1 - \{(1-S+hS)n/(\bar{u}_a m_{1k}^s)\}}\right].$$

The pore-air and pore-water pressure parameters for $K_0$-drained loading can be written as $B_{ak}$ and $B_{wk}$ parameters, respectively. These pore pressure parameters are defined as tangent-type parameters, referenced to a particular stress point:

$$B_{ak} = \frac{du_a}{d\sigma_y} \quad (8.48)$$

$$B_{wk} = \frac{du_w}{d\sigma_y}. \quad (8.49)$$

Equations (8.41) and (8.47) can be expressed in terms of the pore pressure parameters and written as follows:

$$B_{wk} = R_{1k} B_{ak} + R_{2k} \quad (8.50)$$

$$B_{ak} = R_{3k} B_{wk} - R_{4k}. \quad (8.51)$$

Substituting Eq. (8.50) into Eq. (8.51) results in an equation for the $B_{ak}$ pore-air pressure parameter:

$$B_{ak} = \frac{R_{2k} R_{3k} - R_{4k}}{1 - R_{1k} R_{3k}}. \quad (8.52)$$

Similarly, Eq. (8.51) can be substituted into Eq. (8.50) to give an equation for the $B_{wk}$ pore-water pressure param-

eter:

$$B_{wk} = \frac{R_{2k} - R_{1k} R_{4k}}{1 - R_{1k} R_{3k}}. \quad (8.53)$$

The pore-air and pore-water pressure responses at any point during $K_0$-undrained loading can be computed directly using the $B_{ak}$ and $B_{wk}$ pore pressure parameters.

### 8.2.6 Hilf's Analysis

Hilf [1948] outlined a procedure to compute the change in pore pressure in compacted earth fills as a result of an applied total stress. His information can be rearranged in the form of a pore pressure parameter equation. The derivation is based on the results of a one-dimensional oedometer test on a compacted soil, Boyle's law, and Henry's law. A relationship between total stress and pore pressure was established. A secant-type pore pressure parameter can be derived from this relationship. This method has been extensively used by the United States Bureau of Reclamation (i.e., U.S.B.R.), and has proven to be quite satisfactory in practice (Gibbs *et al*, 1960). This method is often referred to as the U.S.B.R. method for the prediction of pore pressures in compacted fills. The same formulation for the estimation of pore pressures in compacted soils has also been advanced by Bishop (1957).

Hilf (1948) stated: "To illustrate the role of air in the relation between consolidation and pore pressure, consider a sample of moist earth compacted in a laboratory cylinder (see Fig. 8.13). If a static load is applied by means of a tight-fitting piston, permitting neither air nor water to escape, it is found that there is a measurable reduction in volume of the soil mass." The reduction in volume was assumed to be the result of compression of free air and air dissolving into water. The soil solids and the water are assumed to be incompressible. Vapor pressure and temperature effects are also assumed to be negligible. The amount of dissolved air in the water is computed in accordance with Henry's law. The free and dissolved air are regarded as a single total volume of air at a particular pressure. The change in pore-air pressure between the initial and final loading conditions is computed using Boyle's law.

The initial and final conditions considered in Hilf's analysis are shown in Fig. 8.14. The total volume of air associated with the initial condition, $V_{ao}$, can be written as follows:

$$V_{ao} = \{(1 - S_0)n_0 + hS_0 n_0\} V_0 \quad (8.54)$$

where

$V_{ao}$ = initial volume of free and dissolved air
$S_0$ = initial degree of saturation
$n_0$ = initial porosity
$V_0$ = initial volume of the soil.

The first and second terms in Eq. (8.54) represent the free and dissolved air volumes, respectively. The initial

## 8.2 DERIVATIONS OF PORE PRESSURE PARAMETERS

absolute pore-air pressure is denoted as $\bar{u}_{ao}$, and can be assumed to be at atmospheric conditions (i.e., 101.3 kPa). An increment in the major principal stress, $\Delta \sigma_y$, is then applied to the soil specimen. The total volume of air decreases and the air pressure increases, in accordance with Boyle's law. The air volume change is equal to the void volume change, $\Delta V_v$, since the soil solids and water are assumed to be incompressible. Therefore, the air volume change can be written as a change in porosity (i.e., $\Delta n = \Delta V_v / V_0$) times the initial volume of soil, $V_0$, as illustrated in Fig. 8.14. The total volume of air under final conditions, $V_{af}$, can be expressed as

$$V_{af} = \{(1 - S_0)n_0 + hS_0n_0 - \Delta n\} V_0 \quad (8.55)$$

where

$V_{af}$ = final volume of free and dissolved air
$\Delta n$ = change (i.e., decrease) in porosity.

The final absolute air pressure, $\bar{u}_{af}$, can be written as the initial absolute pore-air pressure plus the change (i.e., increase) in pore-air pressure:

$$\bar{u}_{af} = \bar{u}_{ao} + \Delta \bar{u}_a \quad (8.56)$$

where

$\Delta \bar{u}_a$ = change (i.e., increase) in absolute pore-air pressure.

Boyle's law can be applied to the initial and final conditions of the free and dissolved air:

$$\bar{u}_{ao} V_{ao} = \bar{u}_{af} V_{af}. \quad (8.57)$$

Substituting the initial conditions [i.e., Eq. (8.54)] and the final conditions [i.e., Eqs. (8.55) and (8.56)] into Eq. (8.57) gives

$$\bar{u}_{ao}\{(1 - S_0)n_0 + hS_0n_0\} V_0$$
$$= (\bar{u}_{ao} + \Delta \bar{u}_a)\{(1 - S_0)n_0 + hS_0n_0 - \Delta n\} V_0. \quad (8.58)$$

Rearranging Eq. (8.58) yields an expression for the change in pore-air pressure, $\Delta \bar{u}_a$:

$$\Delta \bar{u}_a = \left[ \frac{\Delta n}{\{(1 - S_0)n_0 + hS_0n_0 - \Delta n\}} \right] \bar{u}_{ao}. \quad (8.59)$$

Equation (8.59) is commonly referred to as Hilf's equation. It provides a relationship between the change in pore-air pressure and the change in pore-air volume (i.e., $\Delta n$) during $K_0$-undrained loading. If the change in porosity is known, the change in pore-air pressure can be computed from Eq. (8.59).

In order to reach saturation, the soil volume change, $\Delta V_v$, must equal the volume of free air (i.e., $(1 - S_0)n_0 V_0$) (see Fig. 8.14). The change in porosity corresponding to this condition can be written as

$$\Delta n = (1 - S_0)n_0. \quad (8.60)$$

Substituting Eq. (8.60) into Eq. (8.59) gives the absolute pore-air pressure change (i.e., increase) required for saturating a soil:

$$\Delta \bar{u}_{as} = \left( \frac{1 - S_0}{S_0 h} \right) \bar{u}_{ao} \quad (8.61)$$

where

$\Delta \bar{u}_{as}$ = pore-air pressure change (i.e., increase) required for saturation.

Equation (8.61) is equivalent to the equation derived for computing the saturation pressure from the density of an

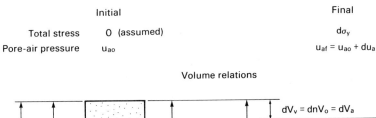
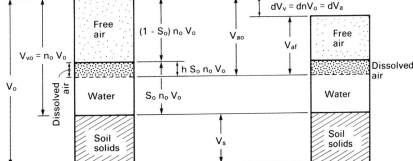

**Figure 8.14** Initial and final pressure and volume conditions considered in Hilf's analysis.

air–water mixture in Chapter 2. Equation (8.59) can be written in an alternate form by replacing $\Delta n$ with $(\Delta V_v/V_0)$:

$$\frac{\Delta V_v}{V_0} = \left(\frac{\Delta \bar{u}_a}{\bar{u}_{ao} + \Delta \bar{u}_a}\right)(1 - S_0 + hS_0)n_0 \quad (8.62)$$

where

$\Delta V_v/V_0$ = change in volume of voids, referenced to the initial volume of the soil (i.e., porosity change, $\Delta n$).

Equation (8.62) describes the volume change due to the compression of air. The soil volume change can also be predicted from the constitutive relations for the soil structure. Hilf (1948) assumed that the matric suction of the soil was negligible, and that the change in pore-air pressure was equal to the change in pore-water pressure (i.e., $\Delta u_a = \Delta u_w$). He also assumed that the constitutive relation for the soil structure could be obtained by saturating the soil in an oedometer test (i.e., under constant volume conditions) and measuring its volume change upon increased loading. The volume change of the soil structure can be expressed as

$$\frac{\Delta V_v}{V_0} = m_v \Delta(\sigma_y - u_a) \quad (8.63)$$

where

$m_v$ = coefficient of volume change measured on a saturated soil in a one-dimensional oedometer test.

The soil compressibility, $m_1^s$, is assumed to be equal to the coefficient of volume change, $m_v$, measured under saturated conditions. The pore-pressure parameter for $K_0$-undrained loading can be derived using Hilf's analysis by equating the volume change of the soil structure [i.e., Eq. (8.63)] to the volume change due to the compression of air [i.e., Eq. (8.62)]:

$$m_v \Delta \sigma_y - m_v \Delta \bar{u}_a = \left(\frac{\Delta \bar{u}_a}{\bar{u}_{ao} + \Delta \bar{u}_a}\right)(1 - S_0 + hS_0)n_0. \quad (8.64)$$

Equation (8.64) can be rearranged and solved for the change in pore-air pressure:

$$\Delta \bar{u}_a = \left[\frac{1}{1 + \frac{(1 - S_0 + hS_0)n_0}{(\bar{u}_{ao} + \Delta \bar{u}_a)m_v}}\right]\Delta \sigma_y. \quad (8.65)$$

The secant $B'_{ah}$ pore pressure derived from Hilf's analysis can be written as

$$B'_{ah} = \left[\frac{1}{1 + \frac{(1 - S_0 + hS_0)n_0}{(\bar{u}_{ao} + \Delta \bar{u}_a)m_v}}\right] \quad (8.66)$$

where

$B'_{ah}$ = secant pore-air pressure parameter (i.e., $\Delta u_a/\Delta \sigma_y$) for $K_0$-undrained loading, in accordance with Hilf's analysis. The prime signifies a secant-type parameter.

An increase in pore-air pressure due to a total stress increase can be predicted using Eq. (8.65). The final pore-air pressure can be computed from the initial pore-air pressure, $\bar{u}_{ao}$, and the change in pore-air pressure, $\Delta \bar{u}_a$. This procedure considers only the initial and final conditions, without using a marching-forward technique. The total stress increment does not have to be small, as required when using the tangent $B$ pore pressure parameter (see Section 8.2.5). However, Eq. (8.65) is nonlinear because the unknown term, $\Delta \bar{u}_a$, appears on both sides of the equation. An iterative technique is required in solving Eq. (8.65).

The coefficient of volume change, $m_v$, varies, depending upon the magnitude of total stress $\sigma_y$. Therefore, an average value of $m_v$ can be used in Eqs. (8.65) and (8.66) when considering a total stress change, $\Delta \sigma_y$. The secant pore-air pressure parameter, $B'_{ah}$, is obtained by dividing the pore-air pressure increase by the total stress increase [i.e., Eq. (8.66)]. The $B'_{ah}$ pore-air pressure parameter is assumed to be equal to the $B'_{wh}$ pore-water pressure parameter in Hilf's analysis. As a result, the difference between a change in pore-air pressure cannot be isolated when using Hilf's analysis. In the application of Hilf's analysis, changes in pore-water pressure have generally been assumed to be equal to the changes in pore-air pressure. The initial pore-water pressure is often taken as being equal to zero. Therefore, the computed pore pressures may be somewhat high or conservative for design purposes.

### 8.2.7 Isotropic Loading

Isotropic loading is a special case of three-dimensional loading, as illustrated in Fig. 8.12. Let us first consider the general case of an isotropic soil loaded three-dimensionally, and then specialize the equations to other cases. Three-dimensional loading consists of applying an increment of total stress in each of the three principal stress directions. The total stress increments in the $x$-, $y$-, and $z$-directions are denoted as $d\sigma_1$, $d\sigma_2$, and $d\sigma_3$, respectively. For an *isotropic soil* under *three-dimensional loading*, the constitutive relation for the soil structure can be expressed as

$$\frac{dV_v}{V_0} = m_1^s d\left(\frac{\sigma_1 + \sigma_2 + \sigma_3}{3} - u_a\right) + m_2^s d(u_a - u_w) \quad (8.67)$$

where

$\sigma_1$ = major principal stress
$\sigma_2$ = intermediate principal stress
$\sigma_3$ = minor principal stress.

In the case of isotropic loading, the total stress increments are equal in all directions (i.e., $d\sigma_1 = d\sigma_2 = d\sigma_3$). Therefore, the constitutive relation for an *isotropic soil* under *isotropic loading* can be written as follows:

$$\frac{dV_v}{V_0} = m_1^s d(\sigma_3 - u_a) + m_2^s d(u_a - u_w). \quad (8.68)$$

The compressibility of the air-water mixture, $C_{aw}$, is obtained from Eq. (8.12) by using the isotropic pressure increment, $d\sigma_3$, for the total stress increment, $d\sigma$:

$$C_{aw} = SC_w \left(\frac{du_w}{d\sigma_3}\right) + (1 - S + hS)\left(\frac{du_a}{d\sigma_3}\right)\bigg/\overline{u}_a. \quad (8.69)$$

Volume change due to pore fluid compression, $dV_v$, is computed from Eq. (8.69), multiplied by the pore fluid volume, $V_v$ (i.e., $nV$), and the isotropic pressure increment, $d\sigma_3$:

$$\frac{dV_v}{V_0} = SnC_w du_w + \{(1 - S + hS)n \, du_a/\overline{u}_a\}. \quad (8.70)$$

The change in volume of the soil structure, Eq. (8.68), must be equal to the change in volume due to pore fluid compression:

$$m_1^s d\sigma_3 - m_1^s du_a + m_2^s du_{ax} - m_2^s du_w$$
$$= SnC_w du_w + \{(1 - S + hS)n \, du_a/\overline{u}_a\}. \quad (8.71)$$

Rearranging Eq. (8.71) gives an equation for the change in pore-water pressure, $du_w$:

$$du_w = \left(\frac{m_2^s - m_1^s - \{(1 - S + hS)n/\overline{u}_a\}}{m_2^s + SnC_w}\right) du_a$$
$$+ \left(\frac{m_1^s}{m_2^s + SnC_w}\right) d\sigma_3. \quad (8.72)$$

A second equation is derived by considering the continuity of the air phase. The constitutive relation for the air phase of an *isotropic soil* under a *three-dimensional loading* can be written as

$$\frac{dV_a}{V_0} = m_1^a d(\sigma_{ave} - u_a) + m_2^a d(u_a - u_w). \quad (8.73)$$

For *isotropic loading*, the constitutive relation for the air phase of an *isotropic soil* is obtained by substituting the condition of ($d\sigma_1 = d\sigma_2 = d\sigma_3$) into Eq. (8.73):

$$\frac{dV_a}{V_0} = m_1^a d(\sigma_3 - u_a) + m_2^a d(u_a - u_w). \quad (8.74)$$

Volume change due to the compression of air, $dV_a$, is represented by the second term in Eq. (8.70):

$$\frac{dV_a}{V_0} = \frac{(1 - S + hS)n}{\overline{u}_a} du_a. \quad (8.75)$$

Volume change due to the compression of air must equal the volume change predicted by the constitutive relation:

$$m_1^a d\sigma_3 - m_1^a du_a + m_2^a du_a - m_2^a du_w$$
$$= (1 - S + hS)n \, du_a/\overline{u}_a. \quad (8.76)$$

Equation (8.76) can be solved for the change in pore-air pressure, $du_a$:

$$du_a = \left[\frac{m_2^a}{m_2^a - m_1^a - \{(1 - S + hS)n/\overline{u}_a\}}\right] du_w$$
$$- \left[\frac{m_1^a}{m_2^a - m_1^a - \{(1 - S + hS)n/\overline{u}_a\}}\right] d\sigma_3. \quad (8.77)$$

Changes in the pore-air and pore-water pressures, $du_a$ and $du_w$, due to a change in the isotropic total stress, $d\sigma_3$, can be computed by solving Eqs. (8.72) and (8.77). The tangent $B$ pore pressure parameter for isotropic, undrained loading can be defined as the ratio of the change in pore pressure to the change in isotropic stress.

The compressibility $m_2^s$ can be written as a ratio of $m_1^s$:

$$R_s = m_2^s/m_1^s. \quad (8.78)$$

Substituting Eq. (8.78) into Eq. (8.72) gives

$$du_w = \left\{\frac{R_s - 1 - \{(1 - S + hS)n/(\overline{u}_a m_1^s)\}}{R_s + (SnC_w/m_1^s)}\right\} du_a$$
$$+ \left\{\frac{1}{R_s + (SnC_w/m_1^s)}\right\} d\sigma_3. \quad (8.79)$$

Equation (8.79) can be written in a compressed form by collecting the soil properties into single terms:

$$du_w = R_1 du_a + R_2 d\sigma_3 \quad (8.80)$$

where

$$R_1 = \frac{R_s - 1 - \{(1 - S + hS)n/(\overline{u}_a m_1^s)\}}{R_s + (SnC_w/m_1^s)}$$

$$R_2 = \frac{1}{R_s + (SnC_w/m_1^s)}.$$

The $m_2^a$ coefficient can be related to the $m_1^a$ coefficient by the ratio, $R_a$:

$$R_a = m_2^a/m_1^a. \quad (8.81)$$

Equation (8.77) becomes

$$du_a = \left[\frac{R_a}{R_a - 1 - \{(1 - S + hS)n/(\overline{u}_a m_1^a)\}}\right] du_w$$
$$- \left[\frac{1}{R_a - 1 - \{(1 - S + hS)n/(\overline{u}_a m_1^a)\}}\right] d\sigma_3. \quad (8.82)$$

Equation (8.82) can be further simplified:

$$du_a = R_3 du_w - R_4 d\sigma_3 \quad (8.83)$$

where

$$R_3 = \left[\frac{R_a}{R_a - 1 - \{(1 - S + hS)n/(\overline{u}_a m_1^a)\}}\right]$$

$$R_4 = \left[\frac{1}{R_a - 1 - \{(1 - S + hS)n/(\overline{u}_a m_1^a)\}}\right].$$

Changes in pore-air and pore-water pressures for the isotropic soil, due to a finite change in the total isotropic pressure, $d\sigma_3$, can be computed from Eqs. (8.80) and (8.83). Similarly, the $B_a$ (i.e., $du_a/d\sigma_3$) and the $B_w$ (i.e., $du_w/d\sigma_3$) pore pressure parameters can be computed directly from Eqs. (8.80) and (8.83). These equations are similar in form to Eqs. (8.41) and (8.47) developed in Section 8.2.5. Following the same procedure presented in Section 8.2.5, the tangent $B$ pore pressure parameters for isotropic loading can be written as follows:

$$B_a = \frac{R_2 R_3 - R_4}{1 - R_1 R_3} \tag{8.84}$$

$$B_w = \frac{R_2 - R_1 R_4}{1 - R_1 R_3}. \tag{8.85}$$

### 8.2.8 Uniaxial Loading

A total stress is applied only in the major principal stress direction for uniaxial loading. For example, a finite increment of total stress, $d\sigma_1$, can be applied to a soil in the vertical or y-direction, as illustrated in Fig. 8.12. Uniaxial loading differs from $K_0$-loading (see Sections 8.2.5 and 8.2.6) in that the boundary conditions imposed on the volume change are different. For $K_0$-loading, the total stress increment, $d\sigma_1$, is applied vertically to a laterally confined soil. Movement occurs in the direction of the applied stress. For uniaxial loading, the soil is free to move in three directions, while the total stress increment is applied only in the major principal stress direction. The soil structure constitutive relation for an *isotropic soil* under *uniaxial loading* conditions can be derived from Eq. (8.67) using $d\sigma_1$ as the major principal stress increment, while setting $d\sigma_2$ and $d\sigma_3$ to zero:

$$\frac{dV_v}{V_0} = m_1^s d\left(\frac{1}{3}\sigma_1 - u_a\right) + m_2^s d(u_a - u_w). \tag{8.86}$$

The compressibility of the air-water mixture, $C_{aw}$, is obtained from Eq. (8.12) by substituting the average stress increment, $(d\sigma_1/3)$, for the total stress increment, $d\sigma$:

$$C_{aw} = SC_w \left(\frac{du_w}{\frac{1}{3}d\sigma_1}\right) + (1 - S + hS)\left(\frac{du_a}{\frac{1}{3}d\sigma_1}\right)/\overline{u}_a. \tag{8.87}$$

Volume change due to pore fluid compression, $dV_v$, is equal to the compressibility of the air-water mixture, $C_{aw}$, multiplied by the pore fluid volume, $V_v$ (i.e., $nV$), and the average stress increment, $(d\sigma_1/3)$. The final equation for the volume change due to pore fluid compression is the same as Eq. (8.70).

Equating the volume change of the soil structure [i.e., Eq. (8.87)] to the volume change due to the compression of pore fluid [i.e., Eq. (8.70)] results in the following relation:

$$\tfrac{1}{3}m_1^s d\sigma_1 - m_1^s du_a + m_2^s du_a - m_2^s du_w$$
$$= SnC_w du_w + (1 - S + hS)n\, du_a/\overline{u}_a. \tag{8.88}$$

Equation (8.88) can be rearranged to give the change in pore-water pressure, $du_w$, due to a finite increment in the major principal stress, $d\sigma_1$, under uniaxial, undrained loading:

$$du_w = \left[\frac{m_2^s - m_1^s - \{(1 - S + hS)n/\overline{u}_a\}}{m_2^s + SnC_w}\right] du_a$$
$$+ \left[\frac{(\tfrac{1}{3})m_1^s}{m_2^s + SnC_w}\right] d\sigma_1. \tag{8.89}$$

A second equation is formulated based on the continuity of the air phase. The constitutive relation for the air phase of an *isotropic soil* under *uniaxial loading* can be obtained by substituting $(d\sigma_1/3)$ into Eq. (8.73) and setting $d\sigma_2$ and $d\sigma_3$ equal to zero:

$$\frac{dV_a}{V_0} = m_1^a d\left(\frac{1}{3}\sigma_1 - u_a\right) + m_2^a d(u_a - u_w). \tag{8.90}$$

The volume change due to the overall compression of the air phase portion, $dV_a$, is given by the second term in Eq. (8.70) as presented in Eq. (8.75). Equating the volume change from the constitutive relation for the air phase [i.e., Eq. (8.90)] to the volume change due to pore-air compression [i.e., Eq. (8.75)] provides a second independent equation:

$$\tfrac{1}{3}m_1^a d\sigma_y - m_1^a du_a + m_2^a du_a - m_2^a du_w$$
$$= (1 - S + hS)n\, du_a/\overline{u}_a. \tag{8.91}$$

Equation (8.91) can be rearranged to give the change in pore-air pressure, $du_a$, due to a finite increment of the major principal stress, $d\sigma_1$, under uniaxial undrained loading:

$$du_a = \left[\frac{m_2^a}{m_2^a - m_1^a - \{(1 - S + hS)n/\overline{u}_a\}}\right] du_w$$

$$- \left[\frac{(\tfrac{1}{3})m_1^a}{m_2^a - m_1^a - \{(1 - S + hS)n/\overline{u}_a\}}\right] d\sigma_1. \tag{8.92}$$

Changes in pore-air and pore-water pressures, $du_a$ and $du_w$, caused by a finite increment of total stress, $d\sigma_1$, can be obtained by solving Eqs. (8.89) and (8.92). Let us define the tangent pore-air pressure parameter, $D_a$, as

$$D_a = \frac{du_a}{d\sigma_1} \quad (8.93)$$

where

$D_a$ = tangent pore-air pressure parameter for uniaxial, undrained loading.

Similarly, let us define the tangent pore-water pressure parameter, $D_w$, as

$$D_w = \frac{du_w}{d\sigma_1} \quad (8.94)$$

where

$D_w$ = tangent pore-water pressure parameter for uniaxial, undrained loading.

The form of the uniaxial, undrained loading equations [i.e., Eqs. (8.89) and (8.92)] and the isotropic, undrained loading equations [i.e., Eqs. (8.72) and (8.77)] are similar. The only difference is in the constant of the second term of the equation. The constant of the second term for uniaxial loading is one third of the similar constant for isotropic loading. Equations (8.89) and (8.92) can be further simplified using the ratios $R_s$ and $R_a$ (i.e., Eqs (8.78) and (8.81), respectively) in a similar manner to that used for isotropic loading conditions:

$$du_w = R_1 du_a + (\tfrac{1}{3})R_2 d\sigma_1 \quad (8.95)$$

$$du_a = R_3 du_w - (\tfrac{1}{3})R_4 d\sigma_1. \quad (8.96)$$

The tangent $D$ pore pressure parameters for an isotropic soil under uniaxial, undrained loading are solved directly from Eqs. (8.95) and (8.96). The equation for the tangent pore-air pressure parameter is

$$D_a = \frac{1}{3}\left(\frac{R_2 R_3 - R_4}{1 - R_1 R_3}\right). \quad (8.97)$$

The equation for the tangent pore-water pressure parameter is

$$D_w = \frac{1}{3}\left(\frac{R_2 - R_1 R_4}{1 - R_1 R_3}\right). \quad (8.98)$$

A comparison of Eqs. (8.97) and (8.84) reveals the following relationship between isotropic and uniaxial undrained loading:

$$D_a = \tfrac{1}{3}B_a. \quad (8.99)$$

Similarly, Eqs. (8.98) and (8.85) indicate that

$$D_w = \tfrac{1}{3}B_w. \quad (8.100)$$

Equations (8.99) and (8.100) show that for an elastic, isotropic soil, the tangent $D$ pore pressure parameters for uniaxial loading are one third of the tangent $B$ pore pressure parameters from isotropic loading. These formulations are similar to those derived for saturated soils (Lambe and Whitman, 1979).

### 8.2.9 Triaxial Loading

Triaxial loading conditions are illustrated in Fig. 8.12. Uniform increments of total stress are applied laterally (i.e., $d\sigma_2$ is equal to $d\sigma_3$) to the soil. The largest increment of total stress, $d\sigma_1$, is applied in the vertical direction. This triaxial loading condition can be viewed as comprising two loading conditions, as shown in Fig. 8.15. The first loading condition is a finite increment in the all-around pressure of $d\sigma_3$. The second loading condition is a finite, uniaxial loading in the major principal stress direction of a stress difference, $d(\sigma_1 - \sigma_3)$. The increment $d(\sigma_1 - \sigma_3)$ is commonly referred to as a deviator stress increment. The superposition of these two loading conditions closely simulates undrained triaxial testing in the laboratory. A soil specimen is first subjected to an isotropic pressure. The specimen is then loaded axially until the maximum deviator stress is reached.

The pore pressure development in the triaxial test can be viewed in terms of pore pressure generated from an isotropic load increment of $d\sigma_3$ and from a uniaxial load, $d(\sigma_1 - \sigma_3)$. For isotropic loading, the change in pore pressure is presented in terms of a $B$ pore pressure parameter (see Section 8.2.7). For uniaxial loading, the change in pore pressure is presented in terms of a $D$ pore pressure parameter (see Section 8.2.8). Therefore, the changes in pore-water pressure during triaxial loading can be written as follows:

$$du_w = B_w d\sigma_3 + D_w d(\sigma_1 - \sigma_3). \quad (8.101)$$

Equation (8.101) can be rearranged:

$$du_w = B_w\{d\sigma_3 + (D_w/B_w)\, d(\sigma_1 - \sigma_3)\}. \quad (8.102)$$

The term $(D_w/B_w)$ in Eq. (8.102) can be defined as another pore pressure parameter:

$$A_w = \frac{D_w}{B_w}. \quad (8.103)$$

Substituting Eq. (8.103) into Eq. (8.102) gives

$$du_w = B_w\{d\sigma_3 + A_w d(\sigma_1 - \sigma_3)\}. \quad (8.104)$$

Rearranging Eq. (8.104) to a more general form gives

$$du_w = B_w\left\{d\left(\frac{\sigma_1 + 2\sigma_3}{3}\right) + \frac{3A_w - 1}{3} d(\sigma_1 - \sigma_3)\right\}. \quad (8.105)$$

Equations (8.104) and (8.105) have the same form as the pore-water pressure generation equation for saturated soils under triaxial loading conditions (Skempton, 1954, Lambe and Whitman, 1979). The equations have also been pro-

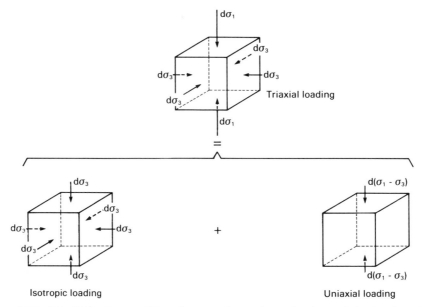

**Figure 8.15** Triaxial loading condition viewed as the resultant of an isotropic load and a uniaxial load.

posed by Bishop (1961a) and Bishop and Henkel (1962) for an unsaturated soil under triaxial loading conditions. In the special case of an *isotropic* and *elastic soil*, the $(D_w/B_w)$ ratio or the $A_w$ value is equal to $\frac{1}{3}$ [see Eq. (8.100)]:

$$du_w = B_w \, d\left(\frac{\sigma_1 + 2\sigma_3}{3}\right). \quad (8.106)$$

In similar manner, the change in pore-air pressure in an isotropic soil under the triaxial loading conditions can be written as

$$du_a = B_a \, d\sigma_3 + D_a \, d(\sigma_1 - \sigma_3) \quad (8.107)$$

or

$$du_a = B_a \{d\sigma_3 + (D_a/B_a) \, d(\sigma_1 - \sigma_3)\}. \quad (8.108)$$

Let us define the term $(D_a/B_a)$ as

$$A_a = \frac{D_a}{B_a}. \quad (8.109)$$

This results in the following equation:

$$du_a = B_a \{d\sigma_3 + A_a d(\sigma_1 - \sigma_3)\} \quad (8.110)$$

or

$$du_a = B_a \left\{ d\left(\frac{\sigma_1 + 2\sigma_3}{3}\right) + \frac{3A_a - 1}{3} d(\sigma_1 - \sigma_3) \right\}. \quad (8.111)$$

Again, the $A_a$ value is equal to $\frac{1}{3}$ for an *isotropic, elastic soil*:

$$du_a = B_a \, d\left(\frac{\sigma_1 + 2\sigma_3}{3}\right). \quad (8.112)$$

Equations (8.106) and (8.112) indicate that the pore pressure responses in an isotropic, elastic soil with an $A$ parameter of $\frac{1}{3}$ depend mainly on the mean principal stress increments (i.e., $d\{(\sigma_1 + 2\sigma_3)/3\}$). Pore pressure responses in soils with an $A$ parameter not equal to $\frac{1}{3}$ will be influenced by the shear stress [i.e., $d(\sigma_1 - \sigma_3)$] (Skempton, 1954).

The $A$ parameter is also equal to $\frac{1}{3}$ for saturated soils which are isotropic and elastic in their behavior. In general, however, the $A$ parameter for saturated soils can vary from a negative value to a value greater than 1.0. The $A$ parameter has been found to depend on factors such as the soil type, stress history, magnitude of strain, and time (Lambe and Whitman, 1979). Further studies are needed to fully understand the behavior of the $A$ parameters for unsaturated soils. However, the factors affecting the $A$ parameter in saturated soils will, in general, affect an unsaturated soil in a similar manner. For example, unsaturated soils are generally dense due to desiccation, and will tend to be slightly dilatant upon shearing. The net result is a low $A$ pore pressure parameter.

Equations (8.101) and (8.107) can be used to predict the changes in pore-water and pore-air pressures, respectively, due to finite increments of total stress under triaxial loading conditions. The pore pressure parameters (i.e., $B$, $D$, and $A$) must first be measured for the soil under investigation. For the special case of isotropic loading, the deviator stress increment is zero (i.e., $d(\sigma_1 - \sigma_3) = 0$), and the second terms in Eqs. (8.101) and (8.107) vanish. As a result, the $B$ pore pressure parameter can be used to predict changes in pore pressure.

In the case of uniaxial loading, the isotropic pressure increment is zero (i.e., $d\sigma_3 = 0$), and the first terms in Eqs.

(8.101) and (8.107) drop out. As a result, the $D$ pore pressure parameter can be used to predict the pore pressure change. The $D$ pore pressure parameter in Eqs. (8.101) and (8.107) is commonly referred to as the $\overline{A}$ pore pressure parameter in saturated soils (Skempton, 1954).

An alternative form for Eqs. (8.104) and (8.110) can be presented as follows:

$$du_w = B_w\{d\sigma_1 - (1 - A_w)(d\sigma_1 - d\sigma_3)\} \quad (8.113)$$

$$du_a = B_a\{d\sigma_1 - (1 - A_a)(d\sigma_1 - d\sigma_3)\} \quad (8.114)$$

Another tangent pore pressure parameter, called $\overline{B}$, can be defined as the ratio between the change in pore pressure and the increase in the major principal stress (Skempton, 1954):

$$\overline{B}_w = \frac{du_w}{d\sigma_1} \quad (8.115)$$

$$\overline{B}_a = \frac{du_a}{d\sigma_1}. \quad (8.116)$$

The $\overline{B}_w$ and $\overline{B}_a$ pore pressure parameters can be obtained from Eqs. (8.113) and (8.114):

$$\overline{B}_w = B_w\{1 - (1 - A_w)(1 - d\sigma_3/d\sigma_1)\} \quad (8.117)$$

$$\overline{B}_a = B_a\{1 - (1 - A_a)(1 - d\sigma_3/d\sigma_1)\}. \quad (8.118)$$

For $K_0$-loading, the $\overline{B}$ pore pressure parameters are equivalent to the $B_k$ parameters derived in Section 8.2.5.

### 8.2.10 Three-Dimensional Loading

Three-dimensional loading conditions are similar to triaxial loading conditions, except that the total stress increments in the lateral directions need not be equal. In the triaxial case, the total stress increments consist of an intermediate and a minor principal stress increment in the lateral directions (i.e., $d\sigma_2$ and $d\sigma_3$, respectively) and a major principal stress increment in the axial direction. Using the same principle of superposition, three-dimensional loading can be visualized as being composed of three loading conditions, as shown in Fig. 8.16. The first loading is an isotropic load of $d\sigma_3$. The second loading is a uniaxial load of $d(\sigma_1 - \sigma_3)$. The third loading is also a uniaxial load of $d(\sigma_2 - \sigma_3)$. The change in pore-water pressure under three-dimensional loading can be computed as the summation of the three components of loading:

$$du_w = B_w d\sigma_3 + D_w d(\sigma_1 - \sigma_3) + D_w d(\sigma_2 - \sigma_3). \quad (8.119)$$

Rewriting Eq. (8.119) gives

$$du_w = B_w\{d\sigma_3 + (D_w/B_w)\,d(\sigma_1 - \sigma_3) + (D_w/B_w)\,d(\sigma_2 - \sigma_3)\}. \quad (8.120)$$

Substituting the $A_w$ parameter into Eq. (8.120) results in the following relation:

$$du_w = B_w\{d\sigma_3 + A_w d(\sigma_1 - \sigma_3) + A_w d(\sigma_2 - \sigma_3)\}. \quad (8.121)$$

Rearranging Eq. (8.121) into a more general form gives

$$du_w = B_w\left\{d\left(\frac{\sigma_1 + \sigma_2 + \sigma_3}{3}\right) + \frac{3A_w - 1}{3}d(\sigma_1 - \sigma_3) + \frac{3A_w - 1}{3}d(\sigma_2 - \sigma_3)\right\}. \quad (8.122)$$

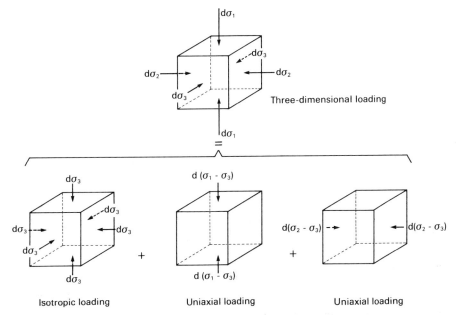

**Figure 8.16** Three-dimensional loading conditions viewed as the resultant of an isotropic load and two uniaxial loads.

The equation for the pore–air pressure change under the three-dimensional loading can be derived in the same manner:

$$du_a = B_a d\sigma_3 + D_a d(\sigma_1 - \sigma_3) + D_a d(\sigma_2 - \sigma_3). \quad (8.123)$$

Equation (8.123) can be rearranged as

$$du_a = B_a \{d\sigma_3 + (D_a/B_a) d(\sigma_1 - \sigma_3) + (D_a/B_a) d(\sigma_2 - \sigma_3)\}. \quad (8.124)$$

Equation (8.124) can be written using an $A$ pore-air pressure parameter:

$$du_a = B_a \{d\sigma_3 + A_a d(\sigma_1 - \sigma_3) + A_a d(\sigma_2 - \sigma_3)\}. \quad (8.125)$$

Rearranging Eq. (8.125) into a more general form gives

$$du_a = B_a \left\{ d\left(\frac{\sigma_1 + \sigma_2 + \sigma_3}{3}\right) + \frac{3A_a - 1}{3} d(\sigma_1 - \sigma_3) + \frac{3A_a - 1}{3} d(\sigma_2 - \sigma_3)\right\}. \quad (8.126)$$

Equations (8.121) and (8.125) can be used to predict the changes in pore-water and pore-air pressures, respectively, due to finite increments of total stress in three dimensions. The pore pressure parameters (i.e., $B$, $D$, and $A$) used in the equations are functions of the soil properties.

For an *isotropic, elastic* soil, the $A$ parameters will be equal to $\frac{1}{3}$. Substituting ($A_w = A_a = \frac{1}{3}$) into Eqs. (8.122) and (8.126) gives

$$du_w = B_w d\left(\frac{\sigma_1 + \sigma_2 + \sigma_3}{3}\right) \quad (8.127)$$

and

$$du_a = B_a d\left(\frac{\sigma_1 + \sigma_2 + \sigma_3}{3}\right). \quad (8.128)$$

Equations (8.127) and (8.128) indicate that the pore pressure responses in an isotropic, elastic soil (i.e., $A = \frac{1}{3}$) are mainly a function of the mean principal stress increments (i.e., $d\{(\sigma_1 + \sigma_2 + \sigma_3)/3\}$). The shear stresses [i.e., $d(\sigma_1 - \sigma_3)$ and $d(\sigma_2 - \sigma_3)$] will influence the pore pressure responses in a soil when the $A$ parameters are not equal to $\frac{1}{3}$.

### 8.2.11 α Parameters

Another important parameter proposed for unsaturated soils is the α parameter. The α parameter was first introduced by Croney (1952) and Croney and Coleman (1961) for loading under constant water content test conditions. In the constant water content test, the air phase is open to atmospheric pressure (i.e., drained), while the water phase is maintained as undrained (i.e., constant gravimetric water content). As total stress is applied, air leaves the soil and the pore–air pressure remains atmospheric. The negative pore-water pressure changes as the total stress changes since water is not allowed to drain. The ratio between the change in pore-water pressure and the change in total stress was originally defined as the α parameter.

The α parameter concept was later extended to a more general form of undrained loading (Bishop, 1961a, Bishop and Henkel, 1962). The pore–air pressure was now allowed to change as the total stress changed. In this general case, the α parameter is defined as the ratio of the change in matric suction, $d(u_a - u_w)$, to the change in net normal stress, $d(\sigma - u_a)$ (Fig. 8.17).

$$d(u_a - u_w) = -\alpha \, d(\sigma - u_a) \quad (8.129)$$

Equation (8.129) considers the α parameter to be constant for finite changes in the net normal stress and matric suction. Therefore, the α parameter concept is similar to the concept of the tangent pore pressure parameter. In general, an increase in net normal stress causes a decrease in matric suction. A negative sign is used in Eq. (8.129) in order to result in a positive α parameter for the general case. For the special case where an increase in net normal stress causes an increase in matric suction (e.g., soil dilates), the α parameter will have a negative value (Fig. 8.17).

In the case of *isotropic, undrained* loading, Eq. (8.129) reverts to the following form:

$$d(u_a - u_w) = -\alpha \, d(\sigma_3 - u_a). \quad (8.130)$$

Dividing Eq. (8.130) by the change in isotropic pressure, $d\sigma_3$, gives

$$\frac{du_a}{d\sigma_3} - \frac{du_w}{d\sigma_3} = -\alpha + \alpha \left(\frac{du_a}{d\sigma_3}\right). \quad (8.131)$$

Substituting the tangent $B$ pore pressure parameters for the air and water phases corresponding to isotropic, undrained loading [i.e., Eqs. (8.21) and (8.22)] into Eq.

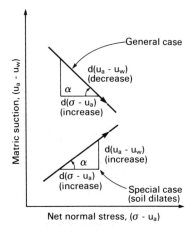

**Figure 8.17** Concept of the α parameter for the general and special cases.

(8.131) results in the following relationship:

$$B_w = \alpha + (1 - \alpha)B_a \quad (8.132)$$

or

$$\alpha = \frac{B_w - B_a}{1 - B_a}. \quad (8.133)$$

Equation (8.133) relates the $B_a$ and $B_w$ pore pressure parameters to the $\alpha$ parameter. The $\alpha$ parameters obtained from the undrained and the constant water content tests are equal for the same soil specimen (Bishop and Henkel, 1962) provided that both tests are performed under the same loading conditions. For the case of the constant water content test, the change in pore-air pressure or the value of $B_a$ is equal to zero. Setting $B_a$ equal to zero in Eq. (8.133) indicates that the $\alpha$ parameter is identical to the tangent pore-water pressure parameter, $B_w$, for a constant water content test.

## 8.3 SOLUTIONS OF THE PORE PRESSURE EQUATIONS AND COMPARISONS WITH EXPERIMENTAL RESULTS

Several pore pressure parameter equations are solved, and typical results are presented in this section. The numerical results are compared with data obtained from experimental results presented in the literature. The effects of changing some of the variables in the pore pressure equations are also illustrated. The theoretical pore pressure parameters can be used to estimate the pore pressures that can develop during construction.

### 8.3.1 Secant $B_h'$ Pore Pressure Parameter Derived from Hilf's Analysis

Measurements of pore-water pressures within compacted cores of several earth dams have been conducted and published by Hilf (1948). The measurements were made using piezometers with coarse porous tips. Figure 8.18 presents the pore-water pressure measurements from two piezometers installed in the Anderson Ranch dam. The pore-water pressures increased as the overburden pressure increased (i.e., $\Delta\sigma_y$) during construction.

The pore-water pressure development during construction can be simulated using Hilf's analysis [i.e., Eq. (8.65)]. Equation (8.65) is applicable to both pore-air and pore-water pressures in the sense that capillary effects are ignored in its derivation. The soil has an initial degree of saturation, $S_0$, and an initial porosity, $n_0$, of 87.4% and 28.1%, respectively. The volumetric coefficient of solubility, $h$, for air in water is assumed to be 0.02. The initial absolute pore-air pressure, $\bar{u}_{a0}$, is assumed to be atmospheric (i.e., 101.3 kPa). The coefficient of volume change, $m_v$, is measured in a conventional oedometer. Substituting the above data into Eq. (8.65) results in a relationship between $\Delta\bar{u}$ (or $\Delta u_a$) and $\Delta\sigma_y$. The relationship between $\Delta u_a$ and $\Delta\sigma_y$ is also assumed to be the relationship between $\Delta u_w$ and $\Delta\sigma_y$ since capillary effects are ignored. The pore-water pressure, $u_w$, can then be plotted against the major principal stress, $\sigma_y$, using the $\Delta u_w$ and $\Delta\sigma_y$ relationship, as shown in Fig. 8.18. The initial pore-water pressure is also assumed to be atmospheric (i.e., zero gauge pressure). An average value for the volume change coefficient, $m_v$, is often used over the range of total stress change.

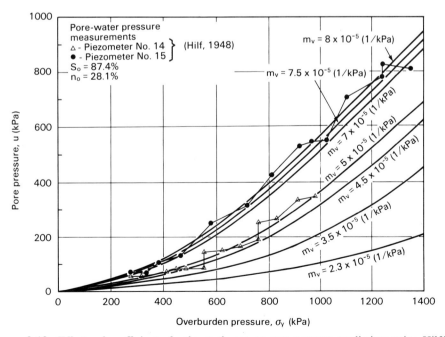

**Figure 8.18** Effects of coefficient of volume change on pore pressure predictions using Hilf's analysis.

Several curves relating the generated pore–water pressure to the applied total stress have been computed using several values for $m_v$. The results are presented in Fig. 8.18. The value of $m_v$ is assumed to be constant for each curve, regardless of the total stress change, $\Delta\sigma_y$. Theoretically, the value of $m_v$ should be varied for each curve of $u_w$ versus $\sigma_y$ depending upon the total stress level [Eq. (8.65)]. The shape of the relationship between $u_w$ and $\sigma_y$ varies significantly as the average value of $m_v$ is changed and the initial variables (i.e., $S_0$, $n_0$, and $\bar{u}_{a0}$) are kept constant. A higher soil compressibility results in a greater pore pressure response as a result of an increase in total stress. This is indicated by a steeper $u_w$ versus $\sigma_y$ curve for a more compressible soil (Fig. 8.18).

Reasonably close agreement between the piezometric measurements and the predicted pore-water pressures can be obtained using Hilf's analysis when appropriate soil compressibilities are used in the analysis. The close agreement is achieved in spite of using an average value for $m_v$. Therefore, the use of an average coefficient of volume change seems to be somewhat justified.

The secant $B'_h$ pore pressure parameter derived using Hilf's analysis can also be computed in accordance with Eq. (8.66). Figure 8.19 presents the secant $B'_h$ pore pressure parameters computed for two predicted curves shown in Fig. 8.18. The two predicted curves correspond to the average coefficients of volume change of $8 \times 10^{-5}$ and $5 \times 10^{-5}$ (1/kPa).

The shape of the relationship between pore pressure and total stress is also strongly influenced by the initial volume-mass properties of the soil. Figure 8.20 presents several curves relating pore-water pressure and total stress, while the initial degree of saturation, $S_0$, is varied. The average coefficient of volume change, $m_v$, is maintained at $8 \times 10^{-5}$ (1/kPa) for all curves. The results indicate that the pore pressure versus total stress relationship becomes steeper as the initial degree of saturation increases. In other words, the pore pressure response becomes more significant at higher degrees of saturation. A comparison between Figs. 8.18 and 8.20 suggests that the coefficient of volume change and the initial degree of saturation are both important variables affecting the pore pressure response.

### 8.3.2 Graphical Procedure for Hilf's Analysis

An analytical procedure for predicting pore-air pressures in accordance with Hilf's analysis was presented in Section 8.3.1. Hilf (1948) outlined a graphical procedure which used nomographs in the solution. This section describes the application of Hilf's graphical procedure. The final objective of the graphical procedure is to plot the relationship between pore pressure and total stress. This is achieved by combining oedometer test data and Hilf's equation.

The results of a conventional oedometer test on a soil under investigation are first plotted as shown in Fig. 8.21. The consolidation test results are plotted as the change in porosity, $\Delta n$ [i.e., $\Delta e/(1 + e)$], versus the change in effective stress, $\Delta(\sigma_y - u_w)$. Hilf's equation [i.e., Eq. (8.59)] can also be plotted on the same graph. The ordinate is the change in porosity, $\Delta n$, while the abscissa is the change in pore-water pressure $\Delta u_w$, (Fig. 8.21). Recall that Eq. (8.59) applies to both pore-air and pore-water pressures since capillary effects are ignored. The relationship between a change in porosity and a change in total stress can be established by combining the above two plots. For a

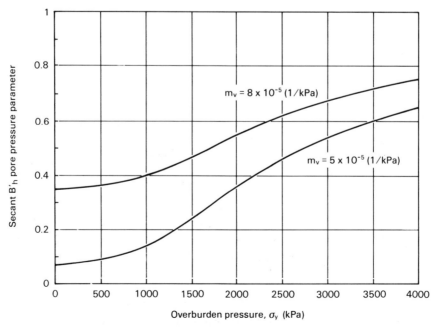

**Figure 8.19** Secant $B'_h$ pore pressure parameters obtained from Hilf's analysis.

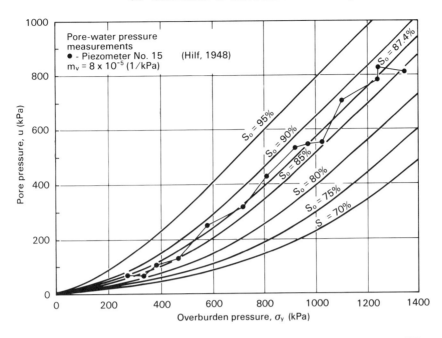

**Figure 8.20** Effect of initial degree of saturation on pore pressure predictions using Hilf's analysis.

given change in porosity, $\Delta n$, the change in total stress, $\Delta \sigma_y$, can be obtained by adding the magnitudes of $\Delta(\sigma_y - u_w)$ and $\Delta u_w$. This summing procedure is repeated for several changes in porosity until a curve of $\Delta n$ versus $\Delta \sigma_y$ is established.

The next step in the procedure is to cross-plot the pore-water pressure, $u_w$, against the major principal stress, $\sigma_y$, as shown in Fig. 8.22. The initial values of $u_w$ and $\sigma_y$ can be estimated. Hilf (1948) assumed that the initial pore-water pressure was atmospheric. The development of the pore-water pressure as the total stress is increased is obtained by cross-plotting the $\Delta n$ versus $\Delta u_w$ and the $\Delta n$ versus $\Delta \sigma_y$ graphs. The magnitudes of $\Delta u_w$ and the corresponding $\Delta \sigma_y$ are then added to the initial values of $u_w$ and $\sigma_y$.

The curve exhibits a nonlinear relationship between pore-water pressure and total pressure at low overburden pressures or at degrees of saturation less than 100%. At saturation pressure [i.e., computed from Eq. (8.61)], the relationship between pore pressure and total stress becomes linear, with a slope of 45°. This illustrates that at saturation, a change in total stress under undrained conditions produces an equal change in pore-water pressure. The pore-water pressure versus total stress plot can have different shapes, depending upon the initial degree of saturation of the soil (Fig. 8.20) and the compressibility of the soil (Fig. 8.18).

The analytical method outlined in Section 8.3.1 differs from the graphical method in the manner in which the compressibility of the soil is defined. In the analytical method, an average constant coefficient of volume change is assumed for the entire total stress change. This assumption

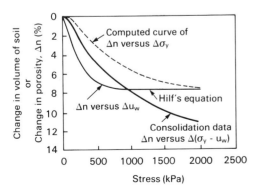

**Figure 8.21** Plots of the stress components versus volume change for Hilf's analysis.

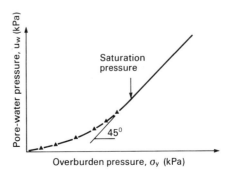

**Figure 8.22** Nonlinear pore-water pressure versus total stress relationship derived using Hilf's analysis.

**Table 8.1 Tangent $B_w$ Pore-Water Pressure Parameters for Various Unsaturated Soils (from Skempton and Bishop, 1954)**

| Soil | Maximum Dry Unit Weight, $\gamma_d$ (kN/m³) | Optimum Water Content, (%) | Dry Unit Weight, $\gamma_d$ (kN/m³) | Water Content $w$ (%) | Degree of Saturation, $S$ (%) | $B_w$ (for $\sigma_3 = 276$ kPa) |
|---|---|---|---|---|---|---|
| Boulder clay | 21.37 | 7.4 | 20.39 | 8.7 | 93 | 0.69 |
| (Liquid limit = 17%, | | | 21.37 | 7.4 | 87 | 0.33 |
| plastic limit = 15%) | | | 21.20 | 6.7 | 76 | 0.10 |
| Moraine | 20.42 | 10.4 | 19.79 | 11.4 | 89 | 0.89 |
| (Liquid limit = 22%, | | | 20.42 | 10.4 | 88 | 0.57 |
| plastic limit = 14%) | | | 20.26 | 9.6 | 80 | 0.35 |
| Volcanic clay | 10.99 | 49 | 10.99 | 52.8 | 99 | 0.47 |
| (Liquid limit = 85%, | | | 10.84 | 49.0 | 90 | 0.21 |
| plastic limit = 55%) | | | 10.68 | 51.0 | 90 | 0.12 |

is not required in the graphical method. The use of the nonlinear consolidation curve (i.e., Fig. 8.21) for obtaining the pore-water pressure versus the total stress relationship implies that varying coefficient of volume change values have been used in the graphical analysis.

### 8.3.3 Experimental Results of Tangent B Pore Pressure Parameters for Isotropic Loading

Tangent $B$ pore pressure parameters for isotropic loading are quite readily measured in the laboratory. A soil specimen is subjected to an equal all-around or isotropic pressure, $\sigma_3$. The pore-air and pore-water pressures are measured as the isotropic pressure is increased. The pore pressure response, $du$, due to a finite increment in total stress, $d\sigma_3$, is expressed as a tangent $B$ parameter. The $B$ pore pressure parameters vary as the isotropic pressure increases (see Fig. 8.8). The $B$ pore pressure parameters increase as the degree of saturation increases. Typical results for the water phase pore pressure parameters are shown in Table 8.1. Figure 8.23 shows the relationship between the $B_w$ pore pressure parameter and the degree of saturation for a clayey gravel.

An apparatus and test procedure were developed by Campbell (1973) to measure the independent generation of pore-air and pore-water pressures in an unsaturated soil specimen under isotropic loading. The specimens were statically compacted and subjected to isotropic pressures up to 6900 kPa. Two types of soil were investigated: Peorian loess (i.e., an inorganic silt), and Champaign till (i.e., a well-graded glacial till).

The development of pore-water pressure during the application of an isotropic pressure are shown in Figs. 8.24 and 8.25 for the Peorian loess and the Champaign till, respectively. The plots indicate a low pore pressure response (i.e., a flat curve) for the conditions of low initial water content or low initial degree of saturation. The pore pressure response increases as the initial water content or degree of saturation increases. As a result, the $B_w$ parameter also increases with an increase in the initial degree of saturation.

Specimens compacted near and above optimum water content exhibit a high, immediate pore-water pressure response to the application of an isotropic pressure increment (i.e., a steeper curve). At initial water contents well above optimum water content, the pore pressure response curves approximate a straight line, with a slope angle approaching 45°. In this case, the $B_w$ pore pressure parameter is constant and equal to unity. In other words, at saturation, an increment in total stress, $d\sigma_3$, is transferred entirely to pore

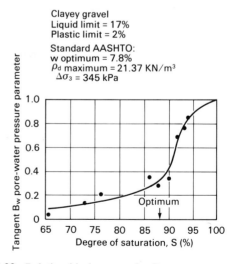

**Figure 8.23** Relationship between the $B_w$ pore-water pressure parameter and the degree of saturation for a clayey gravel (from Skempton, 1954).

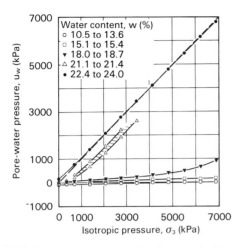

**Figure 8.24** Pore-water pressure development in Peorian loess under isotropic, undrained loading (from Campbell, 1973).

pressure (i.e., $du_w = d\sigma_3$). It is common practice in laboratory tests to ensure the saturation of the soil specimen by confirming that the $B_w$ parameter has reached approximately 1.0.

Figures 8.26 and 8.27 present the $B_w$ pore-water pressure parameter as a function of degree of saturation for Peorian loess and Champaign till, respectively. The plots show that the $B_w$ pore pressure parameter increases rapidly at a degree of saturation equal to 80% for Peorian loess and 90% for Champaign till. These degrees of saturation correspond to the conditions near optimum water content for each of the compacted soils. The $B_w$ parameter approaches 1.0 as the degree of saturation approaches 100%.

The $B_a$ pore-air pressure parameter also increases with an increase in degree of saturation of the Peorian loess and Champaign till specimens, as shown in Figs. 8.28 and 8.29. At saturation, the pore-water pressure approaches the pore-air pressure, and air may exist in the form of occluded air bubbles. Therefore, the pore-air pressure measuring de-

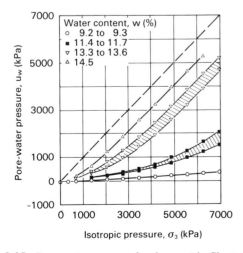

**Figure 8.25** Pore-water pressure development in Champaign till under isotropic, undrained loading (from Campbell, 1973).

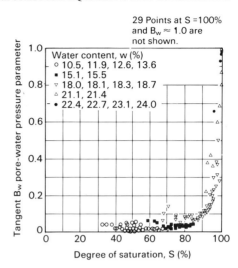

**Figure 8.26** $B_w$ pore-water pressure parameter as a function of degree of saturation for Peorian loess (from Campbell, 1973).

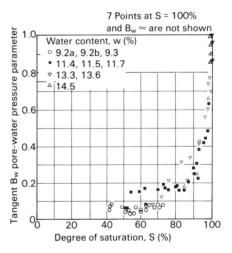

**Figure 8.27** $B_w$ pore-water pressure parameter as a function of degree of saturation for Champaign till (from Campbell, 1973).

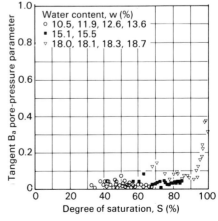

**Figure 8.28** $B_a$ pore-air pressure parameter as a function of degree of saturation for Peorian loess (from Campbell, 1973).

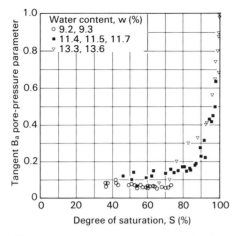

**Figure 8.29** $B_a$ pore pressure parameter as a function of degree of saturation for Champaign till (from Campbell, 1973).

vice likely begins to measure the pore-water pressure as saturation is approached. As a result, the $B_a$ parameter approaches a value of one at saturation, as exhibited in Fig. 8.29 for Champaign till. Theory would also suggest that this is reasonable.

The $B_w$ pore-water pressure parameter can also deviate from a value of 1.0 at saturation if the soil has a low compressibility (Lee et al. 1969). Table 8.2 shows theoretical predictions of $B_w$ parameters at conditions near saturation for four soils. The results show that the pore pressure response is dependent upon the soil compressibility. As the soil compressibility decreases, the $B_w$ pore pressure parameter at saturation becomes less than one.

### 8.3.4 Theoretical Prediction of B Pore Pressure Parameters for Isotropic Loading

Tangent $B$ pore pressure parameters for an isotropic soil subjected to isotropic loading can be predicted using Eqs. (8.84) and (8.85). The computations are performed by using a marching-forward technique with finite increments of isotropic pressure, $d\sigma_3$. The changes in total stress must start from a known initial condition, and extend to a desired final condition. The initial condition corresponds to a degree of saturation less than 100%. The initial degree of saturation, porosity, pore-air, and pore-water pressures are required to commence the calculations.

The difference between the pore-air pressure and the pore-water pressure gives the matric suction $(u_a - u_w)$. The volumetric coefficient of solubility, $h$, can be assumed to be 0.02, and the compressibility of water, $C_w$, is given in Fig. 8.2. The coefficients of volume change (i.e., $m_1^s$ and $m_1^a$) and their relationships (i.e., $R_s = m_2^s/m_1^s$ and $R_a = m_2^a/m_1^a$) must be measured or estimated. The coefficients of volume change and their relationships can either be assumed to remain constant throughout loading or to vary throughout loading. Both conditions can be accommodated using a marching-forward technique. The above-mentioned parameters can be substituted into Eqs. (8.84) and (8.85) to compute the tangent $B_a$ and $B_w$ pore pressure parameters, respectively.

The next step is to consider a finite increment of isotropic pressure, $d\sigma_3$. This causes an increase in the pore-air pressure, $du_a$, and in the pore-water pressure, $du_w$. The increase in the pore-air and pore-water pressures are obtained by multiplying the change in total stress, $d\sigma_3$, by $B_a$ and $B_w$, respectively. The isotropic pressure, the pore-air, and the pore-water pressures are then revised by adding the pressure increases to their respective initial values. These pressure changes result in a change in the stress state of the soil [i.e., $d(\sigma_3 - u_a)$ and $d(u_a - u_w)$]. Consequently, the volume of the soil also changes in accordance with the soil structure constitutive relation [i.e., Eq. (8.23)]. The volume change associated with the air phase is termed as $dn_a$ (i.e., $dV_a/V_0$) and is given in Eq. (8.24).

**Table 8.2** $B_w$ Pore–Water Pressure Parameters for Different Types of Soil at Complete and Nearly Complete Saturation (from Black and Lee, 1973)

| Class of Soil | Void Ratio,[a] $e$ | Soil Compressibility,[a] $m_1^s$ (1/kPa) | Pore–Water Pressure Parameter, $B_w$ | | |
|---|---|---|---|---|---|
| | | | $S = 100\%$ | $S = 99.5\%$ | $S = 99\%$ |
| Soft (i.e., normally consolidated clays) | 2.0 | $1.45 \times 10^{-3}$ | 0.9998 | 0.992 | 0.986 |
| Medium (i.e., compacted clays) | 0.6 | $1.45 \times 10^{-4}$ | 0.9988 | 0.963 | 0.930 |
| Stiff (i.e., stiff clays—sands) | 0.6 | $1.45 \times 10^{-5}$ | 0.9877 | 0.69 | 0.51 |
| Very stiff (i.e., very high consolidation pressures) | 0.4 | $1.45 \times 10^{-6}$ | 0.9130 | 0.20 | 0.10 |

[a] Approximate values.

The water volume change, $dn_w$ (i.e., $dV_w/V_0$), can be calculated as the difference between the total and the air volume changes.

The degree of saturation and the porosity of the soil are then revised using their initial values and the volume changes that have taken place, as illustrated in Fig. 8.30. The final volume of pore voids after the change in total stress can be written as

$$n'V' = n_0 V_0 - dn\, V_0 \qquad (8.134)$$

where

$n'$ = final porosity after the total stress increment
$V'$ = final volume of the soil after the total stress increment
$n_0$ = initial porosity before the total stress increment
$V_0$ = initial volume of the soil before the total stress increment
$dn$ = change in porosity as a result of the total stress increment (i.e., $dV_v/V_0$).

The final volume of the soil can be expressed in terms of its initial volume as

$$V' = (1 - dn)V_0. \qquad (8.135)$$

Substituting Eq. (8.135) into Eq. (8.134) gives

$$n'(1 - dn)V_0 = (n_0 - dn)V_0. \qquad (8.136)$$

The final porosity, $n'$, can then be written in terms of the initial porosity, $n_0$, and the porosity change, $dn$, as follows:

$$n' = \frac{n_0 - dn}{1 - dn}. \qquad (8.137)$$

The final volume of water in the soil after the total stress change can be derived from Fig. 8.30 as

$$S'n'V' = S_0 n_0 V_0 - (dn - dn_a)V_0 \qquad (8.138)$$

where

$S'$ = final degree of saturation after the total stress increment
$S$ = initial degree of saturation before the total stress increment
$dn_a$ = air volume change associated with the total stress increment (i.e., $dV_a/V_0$).

The final volume of the soil, $V'$, and the final porosity, $n'$, in Eq. (8.138) can be replaced by Eqs. (8.135) and (8.137), respectively.

$$S' \frac{n_0 - dn}{1 - dn}(1 - dn)V_0 = S_0 n_0 V_0 - (dn - dn_a)V_0. \qquad (8.139)$$

Rearranging Eq. (8.139) gives the final degree of saturation, $S'$, after the total stress increment:

$$S' = \frac{S_0 n_0 - (dn - dn_a)}{n_0 - dn}. \qquad (8.140)$$

The revised volume-mass properties are then considered as initial conditions for the next increment of isotropic pressure. Similarly, the revised isotropic pressure and the revised pore-air and pore-water pressures become the initial stress variables for the next increment. The coefficient of volume change can be modified if desired for the next increment.

The calculations proceed by computing the new values for $R_1$, $R_2$, $R_3$, $R_4$, and the tangent $B_a$ and $B_w$ pore pressure parameters for the next increment of isotropic pressure. The pore-air and pore-water pressure increases are computed using the new $B_a$ and $B_w$ pore pressure parameters, respectively. The volume properties (i.e., $S$ and $n$) and the stress variables (i.e., $\sigma_3$, $u_a$, and $u_w$) of the soil are again revised to give new values. The above procedure is repeated for subsequent finite increments of isotropic pressure until the

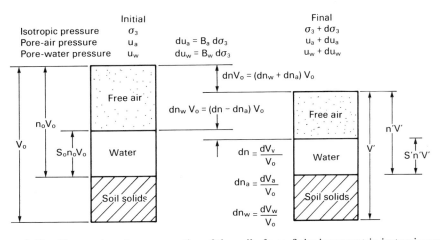

**Figure 8.30** Changes in volume properties of the soil after a finite increment in isotropic pressure.

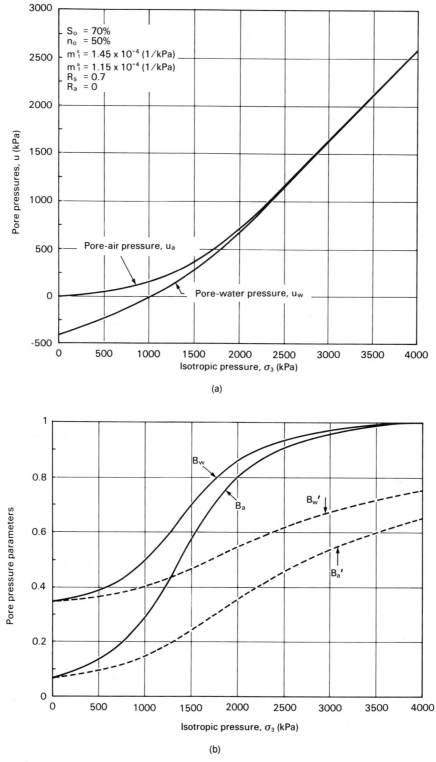

**Figure 8.31** Theoretical prediction of pore pressures and the associated parameters for a soil with an initial degree of saturation of 70%. (a) Development of pore-air and pore-water pressures; (b) development of the pore pressure parameters.

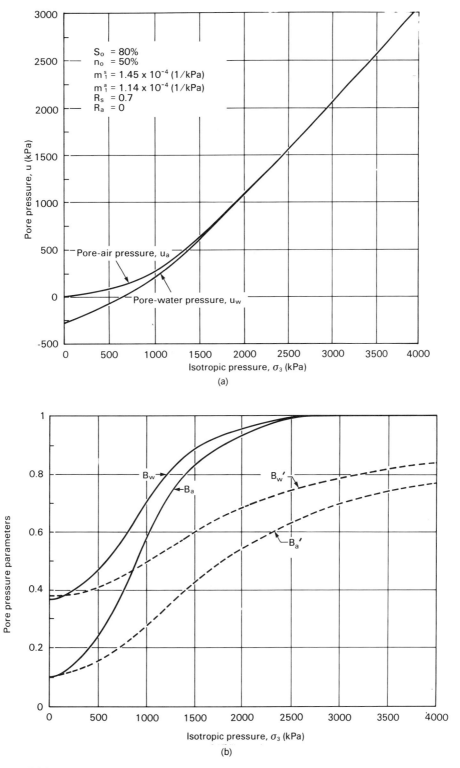

**Figure 8.32** Theoretical predictions of pore pressure parameters for a soil with an initial degree of saturation of 80%. (a) Development of pore–air and pore–water pressures; (b) development of pore pressure parameters.

desired final condition is reached. This procedure can be referred to as the marching-forward technique using finite increments of total stress.

A finite increment of total stress is used to compute a tangent type of $B$ pore pressure parameter. The soil volume change (i.e., $dV_v/V_0$) can be calculated using a linear constitutive relation. The secant $B'_a$ and $B'_w$ pore pressure parameters can also be calculated in accordance with Eqs. (8.19) and (8.20), respectively. The secant $B'$ pore pressure parameters at a point during loading are obtained by dividing the total increase in pore pressure by the total increment of isotropic pressure (see Fig. 8.8).

Several theoretical predictions of $B$ pore pressure parameters for an isotropic soil under an isotropic loading are presented. Equations (8.84) and (8.85) are solved simultaneously using the marching-forward procedure. Figure 8.31(a) shows the development of the pore-air and pore-water pressures in a soil specimen which is successively loaded under isotropic, undrained loading. The soil specimen has an initial degree of saturation of 70% and a corresponding initial matric suction of 414 kPa. The development of the tangent $B_a$ and $B_w$ pore pressure parameters during the loading process is presented in Fig. 8.31(b). The figure shows that the $B_w$ pore pressure parameter is greater than the $B_a$ pore pressure parameter, and that both parameters gradually increase to a common value of unity as saturation is approached. It can also be observed that the rate of increase for the $B_a$ pore pressure parameter is greater than that for the $B_w$ parameter. In other words, the pore-air pressure parameter increases more rapidly than the pore-water pressure parameter, particularly at low degrees of saturation [Fig. 8.31(a)]. There appears to be a discontinuity in the $B_a$ and $B_w$ curves at saturation. At this point, there is no free air to be driven into the water. Therefore, the term $(hSn)$ must be dropped from the equations for parameters $R_1$, $R_3$, and $R_4$ [i.e., Eqs. (8.80) and (8.83)]. As a result, the $B_a$ and $B_w$ parameters abruptly increase to one. This condition is similar to the abrupt change in the compressibility of air-water mixtures at saturation, as explained in Section 8.1.3. The secant $B'_a$ and $B'_w$ pore pressure parameters can easily be computed from the established pore pressure versus isotropic pressure curves. The results presented in Fig. 8.31(b) indicate that the secant $B'_a$ and $B'_w$ pore pressure parameters do not necessarily approach an equal value of unity as the saturation is approached. This situation was explained in Section 8.2.1.

Figure 8.32(a) presents theoretical calculations for the pore pressures in the soil described in Fig. 8.31(a), which has a higher initial degree of saturation, $S_0$. An initial degree of saturation of 80%, corresponding to an initial matric suction of 276 kPa, is used for Fig. 8.32(a). The pore pressure versus total stress curve becomes steeper as the initial degree of saturation increases. The tangent and secant pore pressure parameters for this condition are presented in Fig. 8.32(b). The discontinuity in the tangent $B_a$ and $B_w$ curves at saturation becomes more pronounced than those shown in Fig. 8.31(b). The general effect of initial degree of saturation on the pore pressures versus the isotropic pressure curves is illustrated in Fig. 8.33. The curve is significantly influenced by the initial degree of saturation, as previously illustrated in Hilf's analysis (see Section 8.3.1).

Figure 8.34 shows the development of pore pressures and pore pressure parameters for a compressible soil. The tangent $B_a$ and $B_w$ pore pressure parameters increase rapidly to an equal value of one.

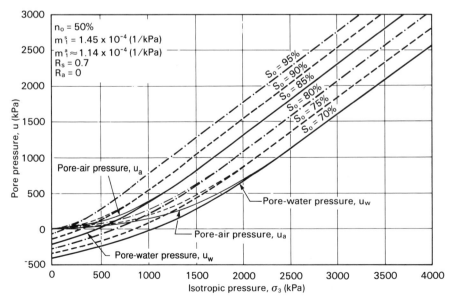

**Figure 8.33** Effect of initial degree of saturation on the pore-air and pore-water pressure versus isotropic pressure curves.

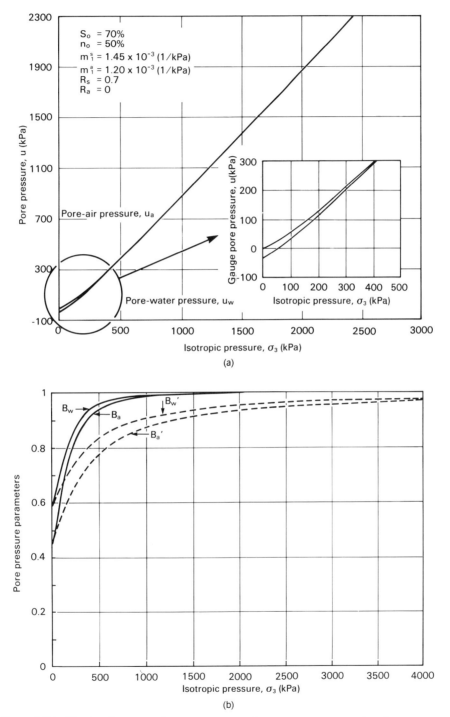

**Figure 8.34** The development of pore pressures and pore pressure parameters for a more compressible soil. (a) Development of pore–air and pore–water pressures; (b) pore pressure parameters.

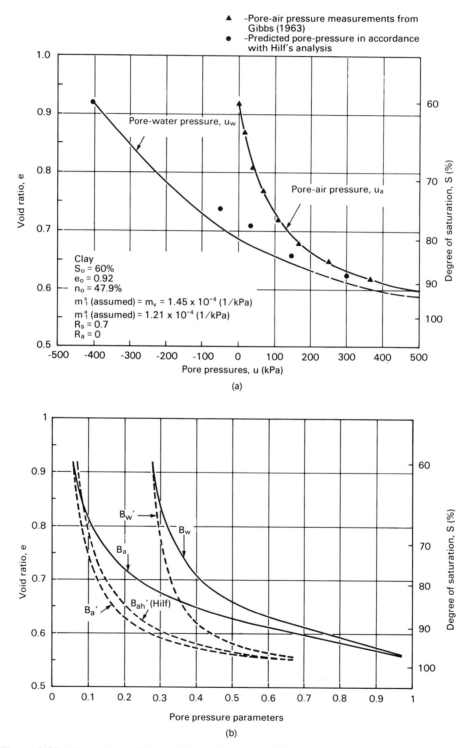

**Figure 8.35** Comparison of theoretical predictions and laboratory measurements of pore-water pressures under undrained loading conditions. (a) Development of pore–air and pore–water pressures; (b) pore pressure parameters.

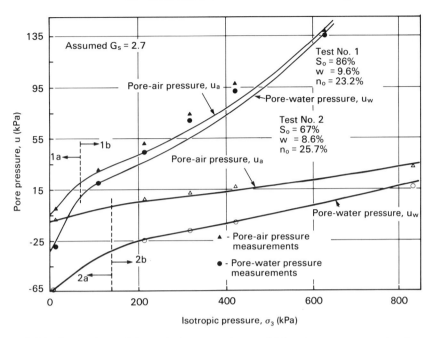

**Figure 8.36** Comparison of theoretical computations and laboratory measurements of pore–air and pore–water pressures (data from Bishop and Henkel, 1962).

A comparison of the theoretically computed pore pressures and pore-water pressure measurements (Gibbs, 1963) is presented in Fig. 8.35. Coefficients of volume change (i.e., $m_1^s$ and $m_1^a$) must be assumed in order to compute the tangent $B_a$ and $B_w$ pore pressure parameters. The coefficients are also assumed to be constant during the undrained loading process. The assumption of constant coefficients of volume change may contribute to deviations between the measured and predicted pore-water pressures. In general, the soil compressibility will decrease as the total stress increases. The predicted pore-air pressure from Eq. (8.83) is in good agreement with predictions using Hilf's analysis [i.e., Eq. (8.65)]. The agreement between Hilf's analysis and the more rigorous equations is the result of setting the parameter $R_a$ (i.e., $m_2^a/m_1^a$) to zero. This assumption means that the volume change associated with the air phase does not depend on the matric suction change, but only on the total stress change. This, in essence, is the assumption involved in Hilf's analysis. This agreement may not occur when the parameter $R_a$ is not zero.

The tangent $B_a$ and $B_w$ parameters increase to unity as saturation is approached, while the secant $B_a'$ and $B_w'$ parameters approach a value of 0.7 [Fig. 8.35(b)]. The secant $B_a'$ pore pressure parameter using the marching-forward

**Table 8.3 Coefficients of Volume Change used in the Theoretical Computations of Pore Pressures on Test Data Presented by Bishop and Henkel (1962)**

| Test No. | Coefficients of Volume Change | |
|---|---|---|
| No. 1 | $\sigma_3 < 70$ kPa (#1a) | $\sigma_3 > 70$ kPa (#1b) |
| Soil structure, $m_1^s$ | $4.0 \times 10^{-4}$ | $2.9 \times 10^{-5}$ |
| Air phase, $m_1^a$ | $2.6 \times 10^{-4}$ | $2.9 \times 10^{-5}$ |
| No. 2 | $\sigma_3 < 140$ kPa (#2a) | $\sigma_3 > 140$ kPa (#2b) |
| Soil structure, $m_1^s$ | $1.0 \times 10^{-4}$ | $2.6 \times 10^{-5}$ |
| Air phase, $m_1^a$ | $8.7 \times 10^{-5}$ | $2.6 \times 10^{-5}$ |

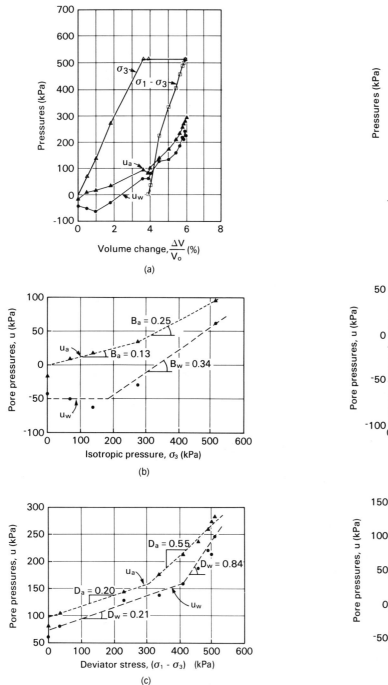

**Figure 8.37** Pore pressure development during undrained triaxial test no. 1. (a) Stress-strain behavior during an undrained, triaxial test (from Knodel and Coffey, 1966); (b) $B_a$ and $B_w$ pore pressure parameters; (c) $D_a$ and $D_w$ pore pressure parameters.

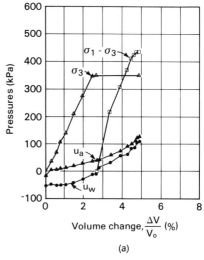

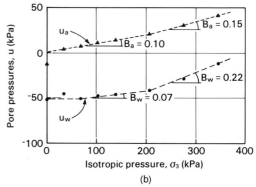

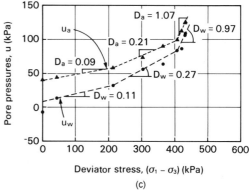

**Figure 8.38** Pore pressure development during undrained triaxial test no. 2. (a) Stress-strain behavior during an undrained, triaxial test (from Knodel and Coffey, 1966); (b) $B_a$ and $B_w$ pore pressure parameters; (c) $D_a$ and $D_w$ core pressure parameters.

technique is slightly different from the secant $B'_{ah}$ pore pressure parameter obtained from Hilf's analysis. The difference could be attributed to the assumption of zero matric suction in Hilf's analysis.

Measurements of pore-air and pore-water pressures for two unsaturated soils under isotropic, undrained loading have been presented by Bishop and Henkel (1962) and are shown in Fig. 8.36. The theoretical predictions of the pore-air and pore-water pressures can be made using varying coefficients of volume change, as outlined in Table 8.3. As

8.3 SOLUTIONS OF THE PORE PRESSURE EQUATIONS AND COMPARISONS 215

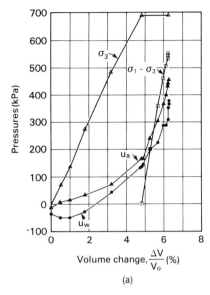

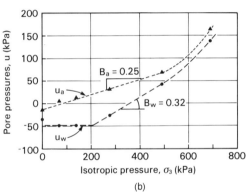

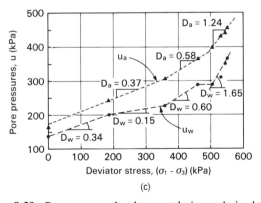

**Figure 8.39** Pore pressure development during undrained triaxial test no. 3. (a) Stress-strain behavior during an undrained, triaxial test (from Knodel and Coffey, 1966); (b) $B_a$ and $B_w$ pore pressure parameters; (c) $D_a$ and $D_w$ pore pressure parameters.

the isotropic pressure is increased, the soil compressibility is decreased. The results indicate that the theoretical computations better predict the measured pore pressures when the coefficients of volume change are varied during loading. Evidence indicates that the assessment of the coeffi-

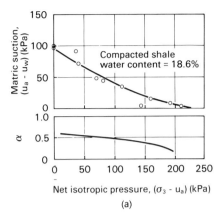

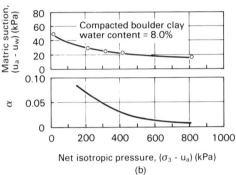

**Figure 8.40** Experimental results showing the $\alpha$ parameters of two soils. (a) Development of the $\alpha$ parameter for compacted shale; (b) development of the $\alpha$ parameter for a compacted boulder clay. (Bishop, 1961a).

cients of volume change during loading is an important factor in predicting the pore-water pressures.

### 8.3.5 Experimental Results of Tangent $B$ and $A$ Parameters for Triaxial Loading

Undrained, triaxial testing is commonly performed by first increasing the isotropic pressure of the soil specimen to a given minor principal stress, $\sigma_3$. The pore pressures developed during the isotropic pressure increase, $d\sigma_3$, can be written as tangent $B$ pore pressure parameters.

The second step in the triaxial test is to increase the vertical stress on the soil specimen to produce a maximum value for the major principal stress, $\sigma_1$. The minor principal stress, $\sigma_3$, remains constant. The change in pore pressures during an increment of deviator stress, $d(\sigma_1 - \sigma_3)$, gives the tangent $D$ pore pressure parameter. The resultant pore-air and pore-water pressures can be obtained by a superposition method, as expressed by Eqs. (8.107) and (8.101), respectively.

Figures 8.37–Fig. 8.39 present pore pressure measurements obtained from undrained, triaxial tests performed by the U.S. Department of the Interior Bureau of Reclamation (U.S.B.R. (1966)). The plotted volume changes are expressed in terms of the initial volume of the soil, $V_0$. The pore pressure parameters computed from the experimental results are the average tangent $B$ or $D$ parameters. The results indicate that the tangent $B$ and $D$ parameters are a function of the stress state in the soil and the degree of saturation of the soil. In general, the pore pressure parameters increase as the total stress on the soil increases.

### 8.3.6 Experimental Measurements of the $\alpha$ Parameter

Figure 8.40 presents two sets of experiments where the $\alpha$ parameter was measured on two compacted soils under isotropic loading (Bishop, 1961a). The $\alpha$ parameter is the ratio of the matric suction change, $d(u_a - u_w)$, to the net isotropic pressure change, $d(\sigma_3 - u_a)$. This is in accordance with the definition of the $\alpha$ parameter given in Eq. (8.129).

The first test is on a shale compacted at a water content slightly above optimum water content. The $\alpha$ parameter was initially about 0.6, and decreased as the net isotropic pressure increased, as shown in Fig. 8.40(a). The second test is on a boulder clay compacted slightly below optimum water content. The $\alpha$ parameter started with a value of about 0.1, and also decreased with increasing net isotropic pressures, as illustrated in Fig. 8.40(b). In other words, the change in matric suction due to a change in net isotropic pressure becomes insignificant at high total stresses or low matric suctions.

# CHAPTER 9

# *Shear Strength Theory*

Many geotechnical problems such as bearing capacity, lateral earth pressures, and slope stability are related to the shear strength of a soil. The shear strength of a soil can be related to the stress state in the soil. The stress state variables generally used for an unsaturated soil are the net normal stress, $(\sigma - u_a)$, and the matric suction, $(u_a - u_w)$, as explained in Chapter 3. This chapter describes how shear strength is formulated in terms of the stress state variables and the shear strength parameters. Techniques for measuring the shear strength parameters in the laboratory are outlined in Chapter 10. The application of the shear strength equation to different types of geotechnical problems is presented in Chapter 11.

A brief historical review of the shear strength theory and attempts to measure relevant soil properties is given in this chapter prior to formulating the shear strength equation. The shear strength test results discussed in the review are selected from the many references on this subject. The selection of research papers for reference is based primarily upon whether or not the researcher used proper procedures and techniques for the measurement or control of the pore pressures during the shearing process. The two commonly performed shear strength tests are the triaxial test and the direct shear test. The theory associated with various types of triaxial tests and direct shear tests for unsaturated soils are compared and discussed in this chapter. Measurement techniques and related equipment are described in Chapter 10. A theoretical model for predicting the strain rate required for testing unsaturated soils is also presented.

The shear strength equation for an unsaturated soil is presented, both in analytical and graphical forms. Both forms of presentations assist in visualizing the changes which occur when going from unsaturated to saturated conditions and vice versa. The possibility of nonlinearity in the shear strength failure envelope is discussed. Various possible methods for handling the nonlinearity are outlined.

Soil specimens which are "identical" in their initial conditions are required for the determination of the shear strength parameters in the laboratory. If the strength parameters of an undisturbed soil are to be measured, the tests should be performed on specimens with the same geological and stress history. On the other hand, if strength parameters for a compacted soil are being measured, the specimens should be compacted at the same initial water content and with the same compactive effort. The soil can then be allowed to equalize under a wide range of applied stress conditions. It is most important to realize that soils compacted at different water contents, to different densities, are "different" soils. In addition, the laboratory test should closely simulate the loading conditions that are likely to occur in the field. Various stress paths that can be simulated by the triaxial and the direct shear tests are described in Chapters 9 and 10.

## 9.1 HISTORY OF SHEAR STRENGTH

The shear strength of a *saturated* soil is described using the Mohr–Coulomb failure criterion and the effective stress concept (Terzaghi, 1936).

$$\tau_{ff} = c' + (\sigma_f - u_w)_f \tan \phi' \tag{9.1}$$

where

$\tau_{ff}$ = shear stress on the failure plane at failure

$c'$ = effective cohesion, which is the shear strength intercept when the effective normal stress is equal to zero

$(\sigma_f - u_w)_f$ = effective normal stress on the failure plane at failure

$\sigma_{ff}$ = total normal stress on the failure plane at failure

$u_{wf}$ = pore-water pressure at failure

$\phi'$ = effective angle of internal friction.

Equation (9.1) defines a line, as illustrated in Fig. 9.1 The line is commonly referred to as a failure envelope. This envelope represents possible combinations of shear stress and effective normal stress on the failure plane at failure. The shear and normal stresses in Eq. (9.1) are given

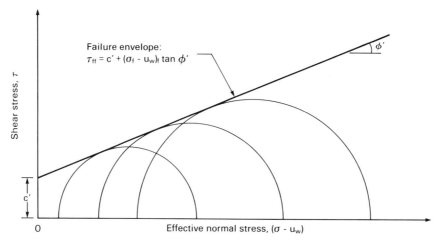

**Figure 9.1** Mohr-Coulomb failure envelope for a saturated soil.

the subscript "$f$." The "$f$" subscript within the brackets refers to the failure plane, and the "$f$" subscript outside of the brackets indicates the failure stress condition. One subscript "$f$" is given to the pore-water pressure to indicate the failure condition. The pore-water pressure acts equally on all planes (i.e., isotropic). The shear stress described by the failure envelope indicates the shear strength of the soil for each effective normal stress. The failure envelope is obtained by plotting a line tangent to a series of Mohr circles representing failure conditions. The slope of the line gives the effective angle of internal friction, $\phi'$, and its intercept on the ordinate is called the effective cohesion, $c'$. The direction of the failure plane in the soil is obtained by joining the pole point to the point of tangency between the Mohr circle and the failure envelope (see Chapter 3). The tangent point on the Mohr circle at failure represents the stress state on the failure plane at failure.

The use of effective stresses with the Mohr-Coulomb failure criterion has proven to be satisfactory in engineering practice associated with saturated soils. Similar attempts have been made to find a single-valued effective stress variable for unsaturated soils, as explained in Chapter 3. If this were possible, a similar shear strength equation could be proposed for unsaturated soils. However, increasing evidence supports the use of two independent stress state variables to define the stress state for an unsaturated soil, and consequently the shear strength (Matyas and Radhakrishna, 1968, Fredlund and Morgenstern, 1977).

Numerous shear strength tests and other related studies on unsaturated soils have been conducted during the past 30 years. This section presents a review of studies related to the shear strength of unsaturated soils. Similar to saturated soils, the shear strength testing of unsaturated soils can be viewed in two stages. The first stage is prior to shearing, where the soils can be consolidated to a specific set of stresses or left unconsolidated. The second stage involves the control of drainage during the shearing process. The pore-air and pore-water phases can be independently maintained as undrained or drained during shear.

In the drained condition, the pore fluid is allowed to completely drain from the specimen. The desire is that there be no excess pore pressure built up during shear. In other words, the pore pressure is externally *controlled* at a constant value during shear. In the undrained condition, no drainage of pore fluid is allowed, and changing pore pressures during shear may or may not be measured. It is important, however, to measure or control the pore-air and pore-water pressures when it is necessary to know the net normal stress and the matric suction at failure. The stress state variables at failure must be known in order to assess the shear strength of the soil in a fundamental manner.

Many shear strength tests on unsaturated soils have been performed without either controlling or measuring the pore-air and pore-water pressures during shear. In some cases, the matric suction of the soil has been measured at the beginning of the test. These results serve only as an indicator of the soil shear strength since the actual stresses at failure are unknown.

A high air entry disk with an appropriate air entry value should be used when measuring pore-water pressures in an unsaturated soil. The absence of a high air entry disk will limit the possible measurement of the difference between the pore-air and pore-water pressure to a fraction of an atmosphere. The interpretation of the results from shear strength tests on unsaturated soils becomes ambiguous when the stress state variables at failure are not known. The following literature review is grouped into two categories. The first category is a review of shear strength tests where there has been adequate control or measurement of the pore-air and pore-water pressures. The second category is a review of shear strength tests on unsaturated soils where there has been inadequate control or measurement of pore pressures during shear.

The concept of "strain" is used in presenting triaxial test results in the form of stress versus strain curves. Stress and strain concepts are discussed in detail in Chapters 3 and 12, respectively. Normal strain is defined as the ratio of the change in length to the original length. When a soil specimen is subjected to an axial normal stress, the normal strain in the axial direction can be defined as follows (Fig. 9.2):

$$\epsilon_y = \frac{L_o - L}{L_o} (100) \qquad (9.2)$$

where

$\epsilon_y$ = axial normal strain in the y-direction expressed as a percentage
$L_o$ = original length of the soil specimen
$L$ = final length of the soil specimen.

A series of direct shear tests on unsaturated fine sands and coarse silts were conducted by Donald (1956). The tests were performed in a modified direct shear box, as shown in Fig. 9.3(a). The pore-air and pore-water pressures were controlled during shear. The top of the direct shear box was exposed to the atmosphere in order to maintain the pore-air pressure, $u_a$, at atmospheric pressure, 101.3 kPa (i.e., zero gauge pressure). The pore-water pressure, $u_w$, was controlled at a negative value by applying a constant negative head to the water phase. The specimen was placed in contact with the water in the base of the shear box through use of a colloidon membrane. The water in the base of the shear box was then connected to a constant head overflow tube at a desired negative gauge pressure [Fig. 9.3(b)]. The pore-water pressure could be reduced to approximately zero absolute before cavitation occurred in the measuring system.

The soil specimens were consolidated under a total stress of approximately 48 kPa, with a uniform initial density. The desired negative pore-water pressure was applied for several hours in order for the specimens to reach equilib-

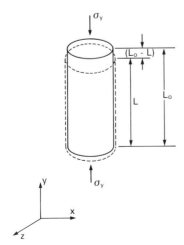

**Figure 9.2** Strain concept used in the triaxial test.

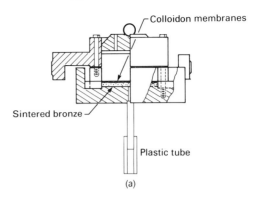

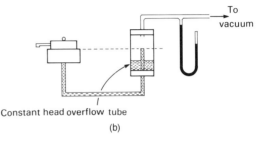

**Figure 9.3** Modified direct shear equipment for testing soils under low matric suction. (a) Modified direct shear box with a colloidon membrane; (b) system for applying a constant negative pore-water pressure (from Donald, 1956).

rium. The specimens were then sheared at a rate of 0.071 mm/s. The results from four types of sand are presented in Fig. 9.4. The shear strength at zero matric suction is the strength due to the applied total stress. As the matric suction is increased, the shear strength increases to a peak value and then decreases to a fairly constant value. As long as the specimens were saturated, the strengths of the sands appeared to increase at the same rate as for an increase in total stress. Once the sands desaturated, the rate of increase in strength decreased, and in fact, the strength decreased when the suction was increased beyond some limiting value.

The U.S. Bureau of Reclamation has performed a number of studies on the shear strength of unsaturated, compacted soils in conjunction with the construction of earth fill dams and embankments (Gibbs *et al.* 1960; Knodel and Coffey, 1966; Gibbs and Coffey, 1969). Undrained triaxial tests with pore-air and pore-water pressure measurements were performed. The pore-air pressure, $u_a$, was measured through the use of a coarse ceramic disk at one end of the specimen. The pore-water pressure, $u_w$, was measured at the other end of the specimen through the use of a high air entry disk. The pore-air and pore-water pressures were measured during the application of an isotropic pressure, $\sigma_3$, and subsequently during the application of the deviator stress, $(\sigma_1 - \sigma_3)$. The pore-air pressure measurements agreed closely with the pore-air pressure predictions using Hilf's analysis (Chapter 8).

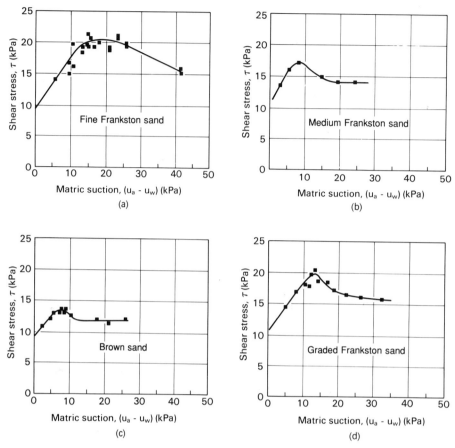

**Figure 9.4** Results of direct shear tests on sands under low matric suctions (modified from Donald, 1956).

No attempt was made to relate the measured shear strength to the matric suction, $(u_a - u_w)$. Rather, two sets of shear strength parameters were obtained by plotting two Mohr-Coulomb envelopes. The first envelope was tangent to Mohr circles plotted using the $(\sigma - u_w)$ stress variables [i.e., Eq. (9.1)]. The second envelope was tangent to Mohr circles plotted using the $(\sigma - u_a)$ stress variables. Figure 9.5 presents typical plots of two envelopes used to plot the shear strength data. The pore pressure measurements for undrained triaxial test no. 3 were presented in Chapter 8. The two failure envelopes indicated that there is a greater difference in their cohesion intercepts than in their friction angles.

An extensive research program on unsaturated soils was conducted at Imperial College, London, in the late 1950's and early 1960's. At the Research Conference on the Shear Strength of Cohesive Soils, Boulder, CO, Bishop et al. (1960) proposed testing techniques and presented the results of five types of shear strength tests on unsaturated soils. The types of tests were: 1) consolidated drained, 2) consolidated undrained, 3) constant water content, 4) undrained, and 5) unconfined compression tests. These are

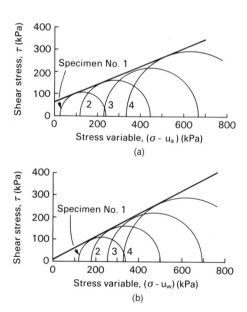

**Figure 9.5** Two procedures used by the U.S. Bureau of Reclamation to plot their shear strength data. (a) Failure envelope based on the $(\sigma - u_a)$ stress variables; (b) failure envelope based on the $(\sigma - u_w)$ stress variables (from Gibbs and Coffey, 1969).

explained in greater detail in Section 9.3. The tests were performed using a modified triaxial cell. The pore-air and pore-water pressures were either measured or controlled during the test.

Bishop (1961b) gave a discussion on the measurement of pore pressures in triaxial tests at the Conference on Pore Pressure and Suction in Soils in London. Tests confirmed that pore-water pressures could be measured directly through a saturated coarse porous ceramic disk sealed onto the base pedestal below a soil specimen. The pore-water pressure measurements were made by balancing the pressure in the measuring system, with the pore-water pressure measured using a null indicator to ensure a no-flow condition. This direct measurement, however, was limited to a gauge pressure range above negative 90 kPa. Bishop and Eldin (1950) successfully measured pore-water pressures down to negative 90 kPa in a saturated soil specimen during a consolidated undrained test with a carefully deaired measuring system. Pore-water pressures less than −1 atm can be measured using the axis-translation technique (Hilf, 1956; see Chapter 3).

The axis-translation technique translates the highly negative pore-water pressure to a pressure that can be measured without cavitation of the water in the measuring system. In addition, a high air entry disk with an air entry value greater than the matric suction being measured must be used in order to prevent the passage of pore-air into the measuring system. A single layer of glass fiber cloth with a low attraction for water was placed on the top of the specimen for pore-air pressure measurement or control.

The test results were presented in terms of stress points, as explained in Chapter 3, and were plotted with respect to the $\{(\sigma_1 + \sigma_3)/2 - u_w\}_f$ and $\{(\sigma_1 + \sigma_3)/2 - u_a\}_f$ stress variables at failure. Figure 9.6 shows a typical plot of constant water content test results on a compacted shale. The condition when the $[(\sigma_1 - \sigma_3)/(\sigma_3 - u_w)]_f$ ratio reached a maximum value was considered to be the failure condition.

In 1961, Bishop and Donald introduced a device called a "bubble pump" to remove and measure the air that diffused through the high air entry disk and that was released as free air in the triaxial cell base compartment. The working mechanism of the bubble pump was explained in Chapter 6.

Pore-air diffusion through the rubber membrane into the water in the triaxial cell was prevented by completely surrounding the membrane (i.e., specimen) with mercury rather than with water. The results of a consolidated drained test on an unsaturated loose silt were used to verify the significance and application of the $(\sigma - u_a)$ and $(u_a - u_w)$ stress variables. Laboratory testing techniques and details of various types of triaxial tests were explained and summarized by Bishop and Henkel in 1962.

The use of the axis-translation technique in the shear strength testing of unsaturated soils was examined by Bishop and Blight (1963). A compression test with the net confinement maintained at zero was conducted on a compacted Selset clay specimen using a stepwise series of axis-translation pressures. The results show a monotonic shear stress versus strain relationship as long as the matric suction remains constant during the test. A comparison between the shear strengths obtained from similar tests with and without axis translation was also performed on Talybont clay. The shear stress versus strain curves from the two types of tests agree closely. This experimentally confirms the applicability of the axis translation technique for the laboratory testing of unsaturated soils. In addition, the ability of the pore-water to withstand absolute tensions greater than 1 atm (i.e., 101.3 kPa) is confirmed since the

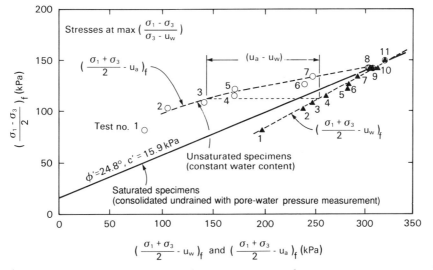

**Figure 9.6** Results of constant water content triaxial tests on a shale (clay fraction 22%) compacted at a water content of 18.6% (from Bishop et al., 1960).

test results without axis translation yielded essentially the same shear strength as those with axis translation.

The development of pore-air and pore-water pressures during undrained tests was also studied by Bishop and Blight (1963). Typical results of constant water content tests were presented and discussed. Donald (1963) presented further results of undrained tests on compacted Talybont clays with pore-air and pore-water pressure measurements. Pore-air and pore-water pressure changes during the compression were found to be a reflection of the volume change tendencies for the soil. The strain rate of testing affected the pore-air pressure response more than the pore-water pressure response. The matric suction of the soil specimen increased markedly with axial strain.

In 1963, a research program on the engineering behavior of unsaturated soils was undertaken by the Soil Engineering Division at the Massachusetts Institute of Technology (i.e., M.I.T.) in Boston. The triaxial apparatus was of the same design as that used by Bishop and Donald (1961), with the following exceptions (M.I.T., 1963). The null indicator for measuring pore-water pressure was replaced with an electrical pressure transducer. The glass fiber cloth at the top of the soil specimen, for measuring pore-air pressure, was substituted with a coarse porous disk. A series of consolidated undrained tests with pore pressure measurements and undrained tests with pore-air pressure control and pore-water pressure measurements were performed on compacted specimens. The specimens were a mixture of 80% ground quartz and 20% kaolin. Some difficulty was experienced in analyzing the test data using a single-valued stress variable. In particular, the data showed considerable scatter, and indicated that an increase in matric suction produced a slight decrease in shear strength. In general, the data appeared to be quite inconclusive.

Blight (1967) reported the results of several consolidated drained tests performed on unsaturated soil specimens. All specimens were compacted at a water content of 16.5% using the standard AASHTO compactive effort. The specimens were then brought to equilibrium at three matric suction values in a triaxial cell. Two specimens, subjected to a constant matric suction, were tested using two net confining pressures, $(\sigma_3 - u_a)$ (i.e., 13.8 and 27.6 kPa). The deviator stress versus strain curves obtained from these tests are shown in Fig. 9.7(a). The results indicate an increase in shear strength with increasing matric suction, and also with an increasing net confining pressure. The water volume changes and overall specimen volume changes during compression are presented in Fig. 9.7(b) and (c), respectively, for the specimens sheared under a constant matric suction of 137.9 kPa. Although pore-water was expelled from the specimen during shear, the overall volume of the specimen increased. In other words, the specimens dilated during compression.

The shear strength of two unsaturated, compacted soils from India, namely, Delhi silt and Dhanauri clay, were

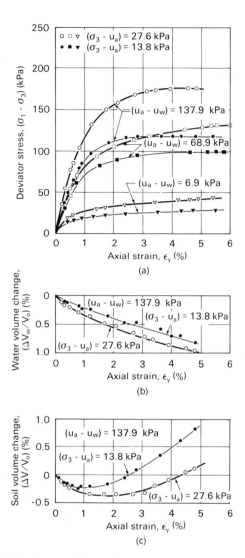

**Figure 9.7** Consolidated drained tests on an unsaturated silt. (a) Typical deviator stress versus strain curves; (b) water volume change versus strain relations; (c) specimen volume change versus strain relations (from Blight, 1967).

tested by Satija and Gulhati (1978 and 1979). Consolidated drained tests were performed with the pore pressures being maintained in a modified triaxial cell. Constant water content tests with pore-air pressure control and pore-water pressure measurement were also performed.

Research on the behavior of unsaturated soils was undertaken at the University of Saskatchewan, Canada, in the mid-1970's. In 1977, Fredlund and Morgenstern proposed the use of $(\sigma - u_a)$ and $(u_a - u_w)$ as independent stress state variables. In 1978, a shear strength equation for an unsaturated soil was proposed, making use of these independent stress state variable (Fredlund et al. 1978). The shear strength of an unsaturated soil was considered to consist of an effective cohesion, $c'$, and independent contributions from the net normal stress, $(\sigma - u_a)$, and a further contribution from the matric suction, $(u_a - u_w)$. The effective angle of internal friction, $\phi'$, was associated with the shear strength contribution from the net normal stress state variable. Another angle, namely, $\phi^b$, was introduced and related to the shear strength contribution from the matric suction stress state variable. Two sets of shear strength test results from Imperial College and one set of data from M.I.T. were used in the examination of the proposed shear strength equation. The test data indicated a failure surface which was essentially planar. The failure envelope was viewed as a three-dimensional surface. The three-dimensional plot with $(\sigma - u_a)$ and $(u_a - u_w)$ as abscissas can be visualized as an extension of the conventional Mohr–Coulomb failure envelope (Fredlund, 1979).

Satija (1978) conducted an experimental study on the shear strength behavior of unsaturated Dhanauri clay. Constant water content and consolidated drained tests were conducted on compacted specimens for various values of $(\sigma - u_a)$ and $(u_a - u_w)$ stresses. The triaxial apparatus was similar to that used in the M.I.T. research program (M.I.T., 1963). Pore pressures were either controlled or measured throughout the shear test. The appropriate strain rate was found to decrease with a decreasing degree of saturation of the soil (Satija and Gulhati, 1979). The results were presented as a three-dimensional surface where half of the deviator stress at failure, $\{(\sigma_1 - \sigma_3)/2\}_f$ was plotted with respect to the net minor principal stress at failure, $(\sigma_3 - u_a)_f$, and the matric suction at failure, $(u_a - u_w)_f$ (Gulhati and Satija, 1981). Some of the data from this program are reanalyzed and presented in Chapter 10.

A series of consolidated drained direct shear and triaxial tests on unsaturated Madrid grey clay were reported by Escario in 1980. The tests were performed under controlled matric suction conditions using the axis-translation technique. A modified shear box device, enclosed in a pressure chamber, was used to apply a controlled air pressure to the soil specimen. The specimen was placed on a high air entry disk in contact with water at atmospheric pressure. This arrangement is similar to the pressure plate technique, where the matric suction is controlled by varying the pore-air pressure, while the pore-water pressure is maintained constant. Prior to testing, the soil specimens were statically compacted and brought to the desired matric suction under an applied vertical normal stress. Typical results obtained from the direct shear tests are presented in Fig. 9.8. The failure envelopes exhibit almost a parallel upward translation, indicating an increase in the shear strength as the soil matric suction is increased.

The results of triaxial tests by Escario (1980) are shown in Fig. 9.9 The pore-water pressure was controlled at atmospheric conditions through a high air entry disk placed at the bottom of the soil specimen. An air pressure was applied to the soil specimen through a coarse porous disk placed on top of the soil specimen. The specimen was enclosed in a rubber membrane, and the confining pressure was applied using water as the medium in the triaxial cell.

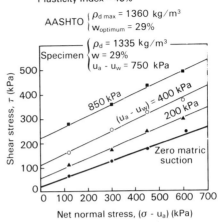

**Figure 9.8** Increase in shear strength for Madrid clay due to an increase in matric suction, obtained from direct shear tests (from Escario, 1980).

The results demonstrated an increase in shear strength with an increase in matric suction.

In 1982, a series of multistage triaxial tests was performed by Ho and Fredlund on unsaturated soils. Undisturbed specimens of two residual soils from Hong Kong were used in the testing program. The soils were a decomposed rhyolite and a decomposed granite. The program consisted of consolidated drained tests with the pore-air pressure was controlled from the top of the specimen through a coarse porous disk. The pore-water pressure was controlled from the bottom of the specimen using a high air entry disk sealed onto the base pedestal. The desired matric suction in the specimen was obtain by controlling the pore-air and pore-water pressures using the axis-translation technique. The strain rate required for shearing an unsaturated soil was discussed in detail using a theoretical formulation described by Ho and Fredlund (1982c).

The triaxial test results showed essentially a planar failure envelope when analyzed using the proposed shear strength equation. Typical two-dimensional projections of

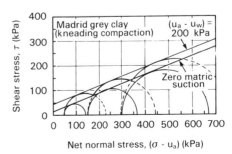

**Figure 9.9** Increase in shear strength due to matric suction for Madrid clay, obtained from triaxial tests (modified from Escario, 1980).

the failure envelope onto the shear stress, $\tau$, versus ($\sigma - u_a$) plane are presented in Fig. 9.10(a). The intersections between the failure envelope and the ordinate are plotted in Fig. 9.10(b). For a constant net confining pressure, the shear strength at failure increased with increasing matric suctions, as illustrated in Fig. 9.10(a). For a planar failure envelope, the internal friction angle, $\phi'$, remains essentially constant under saturated and unsaturated conditions. The effect of matric suction is clearly shown by the $\phi^b$ angle in Fig. 9.10(b).

Typical $\phi^b$ angles have been measured for various soils, and the results have been summarized by Fredlund (1985a). The experimental results showed that the angle $\phi^b$ is always smaller than or equal to the internal friction angle, $\phi'$.

Gan (1986) conducted a multistage direct shear testing program on an unsaturated glacial till. A modified direct shear box that allowed the control of the pore-air and pore-water pressures was used for testing. The shear box was enclosed in an air pressure chamber in order to control the pore-air pressure. The pore-water pressure was controlled through the base of the specimen using a high air entry disk. Consolidated drained direct shear tests were performed with matric suction being controlled during shear (i.e., axis-translation technique). Matric suctions ranged from 0 to 500 kPa, while the net normal stress was maintained at approximately 72 kPa. Typical test results are presented in Fig. 9.11(a), where the shear stress is plotted with respect to the matric suction axis (i.e., $\tau$ versus ($u_a - u_w$) plane) for a constant net normal stress at failure, ($\sigma_f - u_a)_f$. The results show some nonlinearity of the failure envelope on the shear stress versus matric suction plane. The $\phi^b$ angle commences at a value equal to $\phi'$ (i.e., 25.5° when measured under saturated conditions) for low matric suctions. The $\phi^b$ angle decreases to 7° at high matric suction values, as shown in Fig. 9.11(b).

The nonlinearity in the shear strength versus matric suction relationship was also observed by Escario and Sáez (1986). Direct shear tests were performed on three soils, namely, Madrid grey clay, red clay of Guadalix de la Sierra, and Madrid clayey sand. The tests were performed on a modified direct shear box using the procedure described by Escario (1980). A curved relationship between shear stress and matric suction was obtained as illustrated in Fig. 9.12(b) for Madrid grey clay. The nonlinearity of the shear stress versus matric suction relationship has become more noticeable as soils are being tested over a wider range of matric suctions.

### 9.1.1 Data Associated with Incomplete Stress Variable Measurements

Numerous shear strength tests on unsaturated soils have been conducted without a knowledge of the pore-air and/

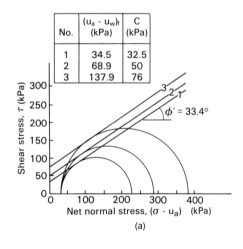

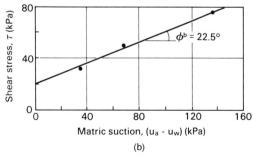

**Figure 9.10** Two-dimensional presentation of failure envelope for decomposed granite specimen No. 22. (a) Failure envelope projected onto the $\tau$ versus ($\sigma - u_a$) plane; (b) intersection line between the failure envelope and the $\tau$ versus ($u_a - u_w$) plane (from Ho and Fredlund, 1982a).

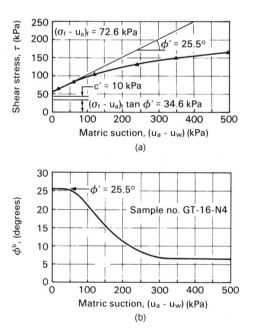

**Figure 9.11** Direct shear test results exhibiting a nonlinear relationship between $\tau$ versus ($u_a - u_w$). (a) Failure envelope projected onto the $\tau$ versus ($u_a - u_w$) plane; (b) varying $\phi^b$ with respect to matric suction (from Gan, 1986).

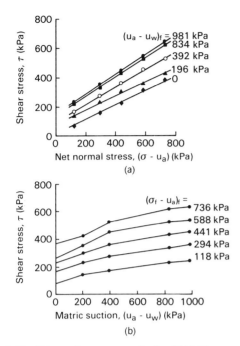

**Figure 9.12** Direct shear test results for Madrid grey clay, under controlled matric suctions. (a) Shear stress versus net confining pressure relationship for various matric suctions; (b) shear stress versus matric suction relationship (from Escario and Sáez, 1986).

or pore-water pressures at failure. Examples are unconfined compression tests where the initial matric suction of the specimens was established or measured (Aitchison, 1959; Blight, 1966; Williams and Shaykewich, 1970; Edil et al. 1981). Undrained triaxial tests with only pore-water pressure measurements during shear have also been performed (Kassiff, 1957).

Consolidated, undrained triaxial tests with only pore-water pressure measurements during shear have been performed by Neves (1971). Neves (1971) used a high air entry disk in making the pore-water pressure measurements. Komornik et al. (1980) carried out consolidated undrained tests where the initial matric suction of the specimens was established using osmotic suction equilibrium.

The interpretation of the above tests becomes more meaningful in view of the theory presented later in this chapter. The brevity of the presentation of data on tests where the pore pressures at failure were not measured should not be interpreted as a vote against these tests. Rather, these tests should be viewed as "total stress" type tests that can only be justified on the basis of a simulation of specific drainage conditions.

## 9.2 FAILURE ENVELOPE FOR UNSATURATED SOILS

The shear strength envelope is a measure of the ability of the soil to withstand applied shear stresses. The soil will fail when the applied shear stress exceeds the shear strength of the soil. The following discussions deal with several criteria for defining soil failure and present the related mathematical expressions.

### 9.2.1 Failure Criteria

There are numerous laboratory and field methods available for the measurement of shear strength. In the laboratory, soil specimens taken from the field can be tested under a range of stress state conditions that are likely to be encountered in the field. The results can be used to define the shear strength parameters of the soil. The initial conditions of the soil specimens must be essentially identical in order for the results to produce unique shear strength parameters for the soil. Only specimens with the same geological condition and stress history should be used to define a specific set of shear strength parameters.

Unsaturated soil specimens are sometimes prepared by compaction. In this case, the soil specimens must be compacted at the same initial water content to produce the same dry density in order to qualify as an "identical" soil. Specimens compacted at the same water content but at different dry densities, or vice versa, cannot be considered as "identical" soils, even though their classification properties are the same. Soils with differing density and water content conditions can yield different shear strength parameters, and should be considered as different soils (Fig. 9.13).

The shear strength test is performed by loading a soil specimen with increasing applied loads until a condition of failure is reached. There are several ways to perform the test, and there are several criteria for defining failure. Consider a consolidated drained triaxial compression test where the pore pressures in the soil specimen are maintained constant [Fig. 9.14(a)]. The soil specimen is subjected to a constant matric suction, and is surrounded by a constant net confining pressure (i.e., the net minor normal stress), $(\sigma_3 - u_a)$. The specimen is failed by increasing the net axial pressure (i.e., the net major normal stress), $(\sigma_1 - u_a)$. The difference between the major and minor normal

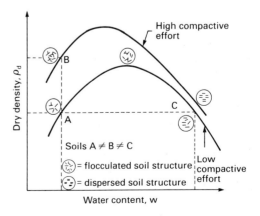

**Figure 9.13** The particle structure of clay specimens compacted at various dry densities and water contents (from Lambe, 1958).

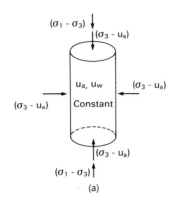

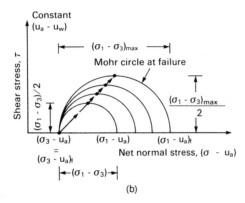

**Figure 9.14** Consolidated drained triaxial compression test data. (a) Applied stresses for a consolidated drained test; (b) Mohr circles illustrating changes in the stress states during shear.

stresses, commonly referred to as the deviator stress, $(\sigma_1 - \sigma_3)$, is a measure of the shear stress developed in the soil [see Fig. 9.14(b)]. As the soil is compressed, the deviator stress increases gradually until a maximum value is obtained, as illustrated in Fig. 9.14(b). The applied deviator stress is usually plotted with respect to the axial strain, $\epsilon_y$, and the plot is referred to as a "stress versus strain" curve. Figure 9.15(a) shows two stress versus strain curves for Dhanauri clay. The tests were performed as consolidated drained triaxial tests at two different net confining pressures.

The maximum deviator stress, $(\sigma_1 - \sigma_3)_{max}$, is an indicator of the shear strength of the soil, and has been used as a failure criterion. The net principal stresses corresponding to failure conditions are called the net major and net minor normal stresses at failure (i.e., $(\sigma_1 - u_a)_f$ and $(\sigma_3 - u_a)_f$, respectively), as indicated in Fig. 9.14(b).

An alternative failure criterion is the principal stress ratio defined as $(\sigma_1 - \sigma_3)_f/(\sigma_3 - u_w)_f$ (Bishop et al. 1960). A plot of the principal stress ratio versus the axial strain for an undrained triaxial test on a compacted shale is illustrated, along with the corresponding stress versus strain curve, in Fig. 9.16(a). In an undrained test, the maximum deviator stress, $(\sigma_1 - \sigma_3)_{max}$, and the maximum principal stress ratio, $(\sigma_1 - \sigma_3)_{max}/(\sigma_3 - u_w)_f$, may not occur at the

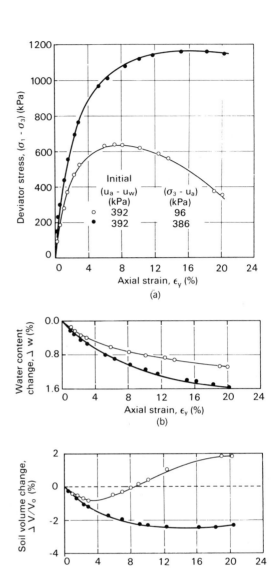

**Figure 9.15** Consolidated drained triaxial test results on Dhanauri clay. (a) Stress versus strain curve; (b) water content change versus strain curve; (c) soil volume change versus strain curve (from Satija, 1978).

same axial strain, as illustrated in Fig. 9.16(a). The maximum principal stress ratio is a function of the pore-water pressure measured during the undrained test [Fig. 9.16(b)]. On the other hand, the maximum deviator stress is not a direct function of the pore pressures. For the results presented in Fig. 9.16(a), the authors selected the maximum principal stress ratio as the failure criterion since it occurred prior to the maximum deviator stress.

In a drained test, the deviator stress curve has the same shape as the principal stress ratio curve since the pore pressures are maintained constant throughout the test. In other words, the denominator of the principal stress ratio, $(\sigma_3 - u_w)$, is a constant. It is possible that the use of the principal stress ratio as a failure criterion for unsaturated soils may

## 9.2 FAILURE ENVELOPE FOR UNSATURATED SOILS

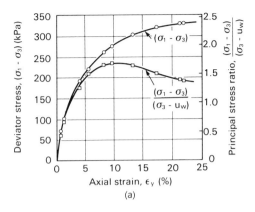

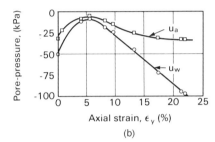

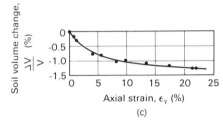

**Figure 9.16** Undrained triaxial tests on a compacted shale. (a) Stress versus strain curve; (b) pore pressures versus strain curve; (c) soil volume change versus strain curve (from Bishop et al., 1960).

require further study. It is not clear, for example, whether the pore-air pressure or the pore-water pressure should be used in calculating the principal stress ratio. In addition, the use of other ratios of the principal stresses may be possible. For example, the ratio $(\sigma_1 - u_a)/(\sigma_3 - u_a)$ or $(\sigma_1 - u_w)/(\sigma_3 - u_w)$ may also be possible as a failure criterion.

The above failure criteria depict some maximum combination of stresses that the soil can resist. However, sometimes the stress versus strain curve does not exhibit an obvious maximum point, even at large strains, as shown in Fig. 9.17. In this case, an arbitrary strain (e.g., 12%) is selected to represent the failure criterion. The limiting strain failure criterion is sometimes used when large deformations are required in order to mobilize the maximum shear stress. A limiting displacement definition of failure is sometimes used in direct shear testing.

The above-mentioned failure criteria have been proposed for the shear strength analysis of unsaturated soils with limited corroborating evidence. In general, the different failure criteria produce similar shear strength parameters. Fur-

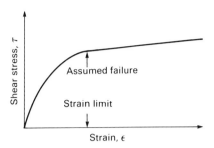

**Figure 9.17** Strain limit used as a failure criterion.

ther research is needed to establish the most appropriate failure criteria for unsaturated soils.

### 9.2.2 Shear Strength Equation

The shear strength of an *unsaturated* soil can be formulated in terms of independent stress state variables (Fredlund et al. 1978). Any two of the three possible stress state variables can be used for the shear strength equation. The stress state variables, $(\sigma - u_a)$ and $(u_a - u_w)$, have been shown to be the most advantageous combination for practice. Using these stress variables, the shear strength equation is written as follows:

$$\tau_{ff} = c' + (\sigma_f - u_a)_f \tan \phi' + (u_a - u_w)_f \tan \phi^b \quad (9.3)$$

where

- $c'$ = intercept of the "extended" Mohr-Coulomb failure envelope on the shear stress axis where the net normal stress and the matric suction at failure are equal to zero; it is also referred to as "effective cohesion"
- $(\sigma_f - u_a)_f$ = net normal stress state on the failure plane at failure
- $u_{af}$ = pore-air pressure on the failure plane at failure
- $\phi'$ = angle of internal friction associated with the net normal stress state variable, $(\sigma_f - u_a)_f$
- $(u_a - u_w)_f$ = matric suction on the failure plane at failure
- $\phi^b$ = angle indicating the rate of increase in shear strength relative to the matric suction, $(u_a - u_w)_f$.

A comparison of Eqs. (9.1) and (9.3) reveals that the shear strength equation for an unsaturated soil is an extension of the shear strength equation for a saturated soil. For an unsaturated soil, two stress state variables are used to describe its shear strength, while only one stress state variable [i.e., effective normal stress, $(\sigma_f - u_w)_f$] is required for a saturated soil.

The shear strength equation for an unsaturated soil exhibits a smooth transition to the shear strength equation for a saturated soil. As the soil approaches saturation, the pore-water pressure, $u_w$, approaches the pore–air pressure, $u_a$, and the matric suction, $(u_a - u_w)$, goes to zero. The matric suction component vanishes, and Eq. (9.3) reverts to the equation for a saturated soil.

### 9.2.3 Extended Mohr–Coulomb Failure Envelope

The failure envelope for a saturated soil is obtained by plotting a series of Mohr circles corresponding to failure conditions on a two-dimensional plot, as shown in Fig. 9.1. The line tangent to the Mohr circles is called the failure envelope, as described by Eq. (9.1). In the case of an unsaturated soil, the Mohr circles corresponding to failure conditions can be plotted in a three-dimensional manner, as illustrated in Fig. 9.18. The three-dimensional plot has the shear stress, $\tau$, as the ordinate and the two stress state variables, $(\sigma - u_a)$ and $(u_a - u_w)$, as abscissas. The frontal plane represents a saturated soil where the matric suction is zero. On the frontal plane, the $(\sigma - u_a)$ axis reverts to the $(\sigma - u_w)$ axis since the pore–air pressure becomes equal to the pore–water pressure at saturation.

The Mohr circles for an unsaturated soil are plotted with respect to the net normal stress axis, $(\sigma - u_a)$, in the same manner as the Mohr circles are plotted for saturated soils with respect to effective stress axis, $(\sigma - u_w)$. However, the location of the Mohr circle plot in the third dimension is a function of the matric suction (Fig. 9.18). The surface tangent to the Mohr circles at failure is referred to as the extended Mohr–Coulomb failure envelope for unsaturated soils. The extended Mohr–Coulomb failure envelope defines the shear strength of an unsaturated soil. The intersection line between the extended Mohr–Coulomb failure envelope and the frontal plane is the failure envelope for the saturated condition.

The inclination of the theoretical failure plane is defined by joining the tangent point on the Mohr circle to the pole point, as explained in Chapter 3. The tangent point on the Mohr circle at failure represents the stress state on the failure plane at failure.

The extended Mohr–Coulomb failure envelope may be a planar surface or it may be somewhat curved. The theory presented in this chapter assumes that the failure envelope is planar and can be described by Eq. (9.3). A curved failure envelope can also be described by Eq. (9.3) for limited changes in the stress state variables. Techniques for handling the non-linearity of the failure envelope are described in Section 9.7.

Figure 9.18 shows a planar failure envelope that intersects the shear stress axis, giving a cohesion intercept, $c'$. The envelope has slope angles of $\phi'$ and $\phi^b$ with respect to the $(\sigma - u_a)$ and $(u_a - u_w)$ axes, respectively. Both angles are assumed to be constants. The cohesion intercept, $c'$, and the slope angles, $\phi'$ and $\phi^b$, are the strength parameters used to relate the shear strength to the stress state variables. The shear strength parameters represent many factors which have been simulated in the test. Some of these factors are density, void ratio, degree of saturation, mineral composition, stress history, and strain rate. In other words, these factors have been combined and expressed mathematically in the strength parameters.

The mechanical behavior of an unsaturated soil is affected differently by changes in net normal stress than by changes in matric suction (Jennings and Burland, 1962). The increase in shear strength due to an increase in net normal stress is characterized by the friction angle, $\phi'$. On the other hand, the increase in shear strength caused by an increase in matric suction is described by the angle, $\phi^b$. The value of $\phi^b$ is consistently equal to or less than $\phi'$, as indicated in Table 9.1, for soils from various geographic locations.

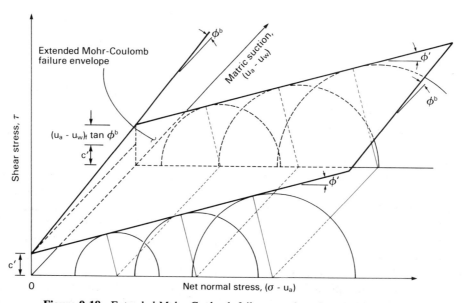

**Figure 9.18** Extended Mohr–Coulomb failure envelope for unsaturated soils.

## 9.2 FAILURE ENVELOPE FOR UNSATURATED SOILS

**Table 9.1 Experimental Values of $\phi^b$**

| Soil Type | $c'$ (kPa) | $\phi'$ (degrees) | $\phi^b$ (degrees) | Test Procedure | Reference |
|---|---|---|---|---|---|
| Compacted shale; $w = 18.6\%$ | 15.8 | 24.8 | 18.1 | Constant water content triaxial | Bishop et al. (1960) |
| Boulder clay; $w = 11.6\%$ | 9.6 | 27.3 | 21.7 | Constant water content triaxial | Bishop et al. (1960) |
| Dhanauri clay; $w = 22.2\%$, $\rho_d = 1580$ kg/m$^3$ | 37.3 | 28.5 | 16.2 | Consolidated drained triaxial | Satija, (1978) |
| Dhanauri clay; $w = 22.2\%$, $\rho_d = 1478$ kg/m$^3$ | 20.3 | 29.0 | 12.6 | Constant drained triaxial | Satija, (1978) |
| Dhanauri clay; $w = 22.2\%$, $\rho_d = 1580$ kg/m$^3$ | 15.5 | 28.5 | 22.6 | Consolidated water content triaxial | Satija, (1978) |
| Dhanauri clay; $w = 22.2\%$, $\rho_d = 1478$ kg/m$^3$ | 11.3 | 29.0 | 16.5 | Constant water content triaxial | Satija, (1978) |
| Madrid grey clay; $w = 29\%$, | 23.7 | 22.5[a] | 16.1 | Consolidated drained direct shear | Escario (1980) |
| Undisturbed decomposed granite; Hong Kong | 28.9 | 33.4 | 15.3 | Consolidated drained multistage triaxial | Ho and Fredlund (1982a) |
| Undisturbed decomposed rhyolite; Hong Kong | 7.4 | 35.3 | 13.8 | Consolidated drained multistage triaxial | Ho and Fredlund (1982a) |
| Tappen–Notch Hill silt; $w = 21.5\%$, $\rho_d = 1590$ kg/m$^3$ | 0.0 | 35.0 | 16.0 | Consolidated drained multistage triaxial | Krahn et al. (1989) |
| Compacted glacial till; $w = 12.2\%$, $\rho_d = 1810$ kg/m$^3$ | 10 | 25.3 | 7–25.5 | Consolidated drained multistage direct shear | Gan et al. (1988) |

[a]Average value.

The failure envelope intersects the shear stress versus matric suction plane along a line of intercepts, as illustrated in Fig. 9.19. The line of intercepts indicates an increase in strength as matric suction increases. In other words, the shear strength increase with respect to an increase in matric suction is defined by the angle, $\phi^b$. The equation for the line of intercepts is as follows:

$$c = c' + (u_a - u_w)_f \tan \phi^b \qquad (9.4)$$

where

$c$ = intercept of the extended Mohr–Coulomb failure envelope with the shear stress axis at a specific ma-

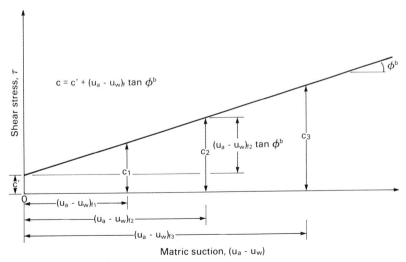

**Figure 9.19** Line of intercepts along the failure plane on the $\tau$ versus $(u_a - u_w)$ plane.

tric suction, $(u_a - u_w)_f$, and zero net normal stress; it can be referred to as the "total cohesion intercept."

The extended Mohr–Coulomb failure envelope can be presented as a horizontal projection onto the $\tau$ versus $(\sigma - u_a)$ plane. The horizontal projection can be made for various matric suction values, $(u_a - u_w)_f$. The horizontal projection of the failure envelope onto the $\tau$ versus $(\sigma - u_a)$ plane results in a series of contours shown in Fig. 9.20(a). The lines have different cohesion intercepts, depending upon their corresponding matric suctions. The cohesion intercept becomes the effective cohesion, $c'$, when the matric suction goes to zero. All lines of equal matric suction have the same slope angle, $\phi'$, as long as the failure plane is planar. The equation for these contour lines can be written as

$$\tau_{ff} = c + (\sigma_f - u_a)_f \tan \phi' \quad (9.5)$$

where

$c$ = total cohesion intercept.

Substituting Eq. (9.4) into Eq. (9.5) yields the equation for the extended Mohr–Coulomb failure envelope [i.e., Eq. (9.3)]. Equation (9.5) is the same as Eq. (9.3), and Fig. 9.20(b) is a two-dimensional representation of the extended Mohr–Coulomb failure envelope. The failure envelope projection illustrates the increase in shear strength as matric suction is increased at a specific net normal stress. The projected failure envelope is a simple, descriptive representation of the three-dimensional failure envelope. Equation (9.5) is also convenient to use when performing analytical studies involving unsaturated soils.

The inclusion of matric suction in the definition of the cohesion intercept does not necessarily suggest that matric suction is a cohesion component of shear strength. Rather, the matric suction component (i.e., $(u_a - u_w) \tan \phi^b$) is lumped with effective cohesion, $c'$, for the purpose of translating the three-dimensional failure envelope onto a two-dimensional representative plot. The suction component of shear strength has also been called the apparent or total cohesion (Taylor, 1948).

A smooth transition from the unsaturated to the saturated condition can be demonstrated using the extended Mohr–Coulomb failure envelope shown in Fig. 9.18. As the soil becomes saturated, the matric suction goes to zero and the pore–water pressure approaches the pore–air pressure. As a result, the three-dimensional failure envelope is reduced to the two-dimensional envelope of $\tau$ versus $(\sigma - u_w)$. The smooth transition can also be observed in Fig. 9.20(b). As the matric suction decreases, the failure envelope projection gradually lowers, approaching the failure envelope for the saturated condition. In this case, the cohesion intercept, $c$, approaches the effective cohesion, $c'$.

The extended Mohr–Coulomb failure envelope can also be projected horizontally onto the $\tau$ versus $(u_a - u_w)$ plane (Fig. 9.21). The horizontal projection is made for various net normal stresses at failure, $(\sigma_f - u_a)_f$ [Fig. 9.21(a)]. The resulting contour lines have an ordinate intercept of $\{c' + (\sigma_f - u_a)_f \tan \phi'\}$ and a slope angle of $\phi^b$ [Fig. 9.21(b)]. The horizontal projection shows that there is an increase in shear strength as the net normal stress is increased at a specific matric suction.

### 9.2.4 Use of $(\sigma - u_w)$ and $(u_a - u_w)$ to Define Shear Strength

The shear strength equation [i.e., Eq. (9.3)] has thus far been expressed using the $(\sigma - u_a)$ and $(u_a - u_w)$ stress state variables. The shear strength equation for an unsaturated soil can also be expressed in terms of other combinations of stress state variables, such as $(\sigma - u_w)$ and $(u_a - u_w)$:

$$\tau_{ff} = c' + (\sigma_f - u_w)_f \tan \phi' + (u_a - u_w)_f \tan \phi'' \quad (9.6)$$

where

$(\sigma_f - u_w)_f$ = net normal stress state with respect to the pore–water pressure on the failure plane at failure

$\phi''$ = friction angle associated with the matric suction stress state variable, $(u_a - u_w)_f$, when using the $(\sigma - u_w)$ and $(u_a - u_w)$ stress state variables in formulating the shear strength equation.

As the matric suction goes to zero, the third terms in Eqs. (9.6) and (9.3) disappear, and the pore–water pressure approaches the pore–air pressure. As a result, both equations revert to the shear strength equation for a saturated soil [i.e., Eq. (9.1)]. Therefore, the second term in both equations should have the same friction angle parameter, $\phi'$ (i.e., $(\sigma_f - u_a)_f \tan \phi'$ and $(\sigma_f - u_w)_f \tan \phi'$).

Equations (9.6) and (9.3) give the same shear strength for a soil at a specific stress state. As a result, Eq. (9.6) can be equated to Eq. (9.3):

$$\begin{aligned} -u_{af} \tan \phi' + (u_a - u_w)_f \tan \phi^b \\ = -u_{wf} \tan \phi' + (u_a - u_w)_f \tan \phi''. \end{aligned} \quad (9.7)$$

Rearranging Eq. (9.7) gives the relationship between the friction angles:

$$\tan \phi'' = \tan \phi^b - \tan \phi'. \quad (9.8)$$

Equation (9.8) shows that the friction angle $\phi''$ will generally be negative since the magnitude of $\phi^b$ is less than or equal to $\phi'$. Figure 9.22 displays the extended Mohr–Coulomb failure envelope when failure conditions are plotted with respect to the $(\sigma - u_w)$ and $(u_a - u_w)$ stress state variables [i.e., Eq. (9.6)] and with respect to the $(\sigma - u_a)$ and $(u_a - u_w)$ stress state variables [i.e., Eq. (9.3)].

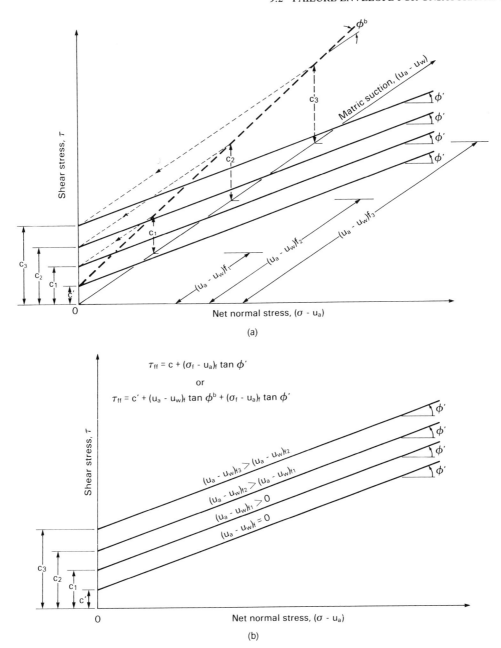

**Figure 9.20** Horizontal projection of the failure envelope onto the $\tau$ versus $(\sigma - u_a)$ plane, viewed parallel to the $(u_a - u_w)$ axis. (a) Failure envelope projections onto the $\tau$ versus $(\sigma - u_a)$ plane; (b) contour lines of the failure envelope onto the $\tau$ versus $(\sigma - u_a)$ plane.

### 9.2.5 Mohr–Coulomb and Stress Point Envelopes

The extended Mohr–Coulomb envelope has been defined as a surface tangent to the Mohr circles at failure. Each Mohr circle is constructed using the net minor and net major principal stresses at failure [i.e., $(\sigma_{3f} - u_{af})$ and $(\sigma_{1f} - u_{af})$], as shown in Fig. 9.23(b). The difference between the net minor and net major principal stress at failure is called the maximum deviator stress.

The top point of a Mohr circle with coordinates $(p_f, q_f,$ $r_f)$ can be used as to represent the stress conditions at failure. A detailed discussion on stress points and stress paths is given in Chapter 3. A stress point surface (i.e., stress point envelope) can be drawn through the stress points at failure [Fig. 9.23(b)]. The stress point envelope is another representation of the stress state of the soil under failure conditions. However, the stress point envelope and the extended Mohr–Coulomb failure envelope are different surfaces. Nevertheless, the stress point envelope can be used to represent the stress state at failure.

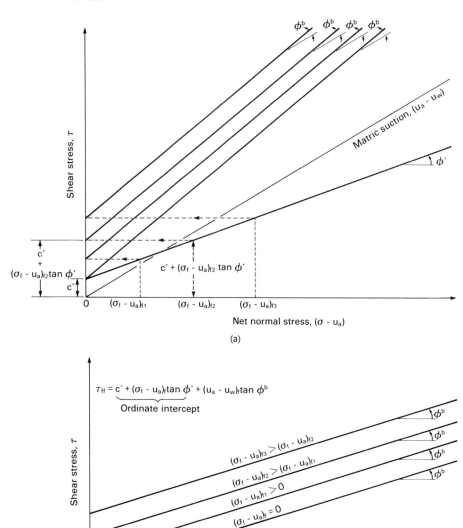

**Figure 9.21** Horizontal projections of the failure envelope onto the $\tau$ versus $(u_a - u_w)$ plane, viewed parallel to the $(\sigma - u_a)$ axis. (a) Failure envelope projections onto the $\tau$ versus $(u_a - u_w)$ plane; (b) contour lines of failure envelope on the $\tau$ versus $(u_a - u_w)$ plane.

The stress point envelope can be defined by the following equation:

$$q_f = d' + p_f \tan \psi' + r_f \tan \psi^b \quad (9.9)$$

where

- $q_f$ = half of the deviator stress at failure (i.e., $(\sigma_1 - \sigma_3)_f/2$)
- $\sigma_{1f}$ = major principal stress at failure
- $\sigma_{3f}$ = minor principal stress at failure
- $d'$ = intercept of the stress point envelope on the $q$ axis when $p_f$ and $r_f$ are equal to zero
- $p_f = ((\sigma_1 + \sigma_3)/2 - u_a)_f$; mean net normal stress at failure
- $\psi'$ = slope angle of the stress point envelope with respect to the stress variable, $p_f$
- $r_f$ = matric suction at failure [i.e., $(u_a - u_w)_f$]
- $\psi^b$ = slope angle of the stress point envelope with respect to the stress variable, $r_f$.

Figure 9.23(b) presents a planar stress point envelope corresponding to the planar extended Mohr–Coulomb failure envelope shown in Fig. 9.23(a). Equation (9.9) defines the stress point envelope. The frontal plane in Fig. 9.23(b)

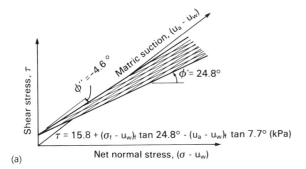

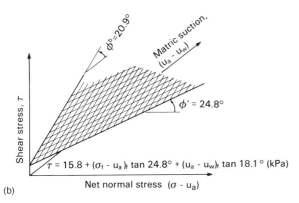

**Figure 9.22** Extended Mohr–Coulomb failure envelope plotted with respect to two possible combinations of stress state variables (a) Failure envelope defined in terms of the $(\sigma - u_w)$ and $(u_a - u_w)$ stress state variables; (b) failure envelope defined in terms of the $(\sigma - u_a)$ and $(u_a - u_w)$ stress state variables (data from Bishop et al., 1960).

represents the saturated condition where the matric suction is zero. As a result, the $((\sigma_1 + \sigma_3)/2 - u_a)$ axis reverts to the $((\sigma_1 + \sigma_3)/2 - u_w)$ axis on the frontal plane. The intersection line between the stress point envelope and the frontal plane is a line commonly referred to as the $K_f$-line in saturated soil mechanics (Lambe and Whitman, 1979). The $K_f$-line passes through the top points of the Mohr circles for saturated soils at failure. The $K_f$-line has a slope angle, $\psi'$, with respect to the $p$ axis and an ordinate intercept, $d'$, on the $q$ axis. Any line parallel to the $K_f$-line on the planar stress point envelope will have a slope angle, $\psi'$, with respect to the $p$ axis. The stress point envelope reverts to the $K_f$-line as the soil becomes saturated or when matric suction, $r_f$, is equal to zero.

The intersection line between the stress point envelope and the $q$ versus $r$ plane has a slope angle $\psi^b$, with respect to the $r$ axis [Fig. 9.23(b)]. The intersection line indicates that there is an increase in strength as the matric suction at failure, $r_f$, increases. The equation for the intersection line can be written as follows:

$$d = d' + r_f \tan \psi^b \quad (9.10)$$

where

$d$ = ordinate intercept of the stress point envelope on the $q$ axis at an $r_f$ and $p_f$ value equal to zero.

The ordinate intercept of the stress point envelope on the $q$ versus $r$ plane is equal to $d'$ when $r_f$ is zero. The ordinate intercept is equal to $d$ [i.e., Eq. (9.10)] when $r_f$ is not zero. The above variables, $d'$, $\psi'$, and $\psi^b$, are the required parameters for Eq. (9.9). The stress point envelope can also be represented by contour lines when the surface is projected onto the $q$ versus $p$ plane. The equation for the contour lines is obtained by substituting Eq. (9.10) into Eq. (9.9):

$$q_f = d + p_f \tan \psi'. \quad (9.11)$$

The stress point envelope can be related to the extended Mohr–Coulomb failure envelope by obtaining the relationships between the parameters used to define both envelopes (i.e., $c$, $\phi'$, $\phi^b$ and $d$, $\psi'$, $\psi^b$). Figure 9.24 presents Mohr circles on the $\tau$ versus $(\sigma - u_a)$ plane for a specific matric suction. The extended Mohr–Coulomb failure envelope is drawn tangent to the Mohr circles (e.g., at point $A$), whereas the stress point envelope passes through the top points of the Mohr circles (e.g., through points $B$). The extended Mohr–Coulomb failure envelope and the stress point envelope have slope angles of $\phi'$ and $\psi'$, respectively, with respect to the $(\sigma - u_a)$ axis. The distance between the tangent point $A$ and the top point $B$ (i.e., $\overline{AB}$) can be computed from triangle $ADC$ as being equal to ($q_f \sin \phi'$). As the Mohr circle moves to the left, the radius, $q_f$, decreases and eventually goes to zero. As a result, the distance between the tangent and the top points (i.e., $q_f \sin \phi'$) also decreases and eventually goes to zero. This means that the extended Mohr–Coulomb failure envelope and the stress point envelope converge to a point on the $(\sigma - u_a)$ axis (i.e., point $T$).

The relationship between the slope angles, $\phi'$ and $\psi'$, is obtained by equating the lengths, $\overline{TC}$, computed from triangles $TBC$ and $TAC$ as follows:

$$\frac{q_f}{\tan \psi'} = \frac{q_f}{\sin \phi'}. \quad (9.12)$$

The $q_f$ variable can be cancelled, giving

$$\tan \psi' = \sin \phi'. \quad (9.13)$$

The relationship between the cohesion intercept, $c$, and the ordinate intercept, $d$, can be computed by considering the distance between points $T$ and $O$ (i.e., $\overline{TO}$):

$$\frac{d}{\tan \psi'} = \frac{c}{\tan \phi'}. \quad (9.14)$$

Substituting Eq. (9.13) into Eq. (9.14) and rearranging Eq. (9.14) yields

$$d = c \cos \phi'. \quad (9.15)$$

When the matric suction at failure is equal to zero (i.e., the saturated condition), Eq. (9.15) becomes

$$d' = c' \cos \phi'. \quad (9.16)$$

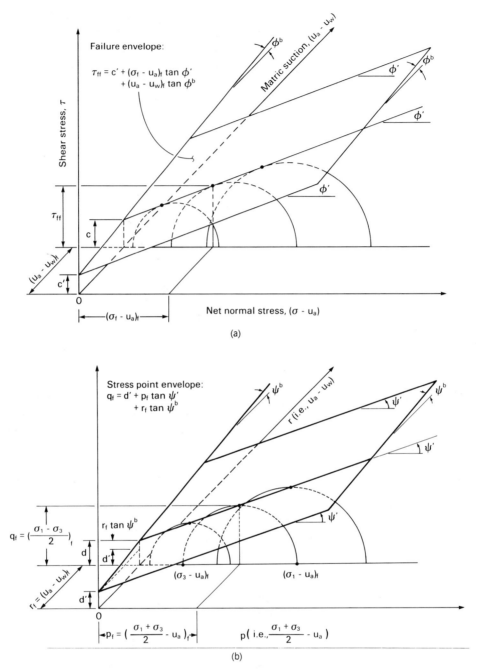

**Figure 9.23** Comparisons of the failure envelope and the corresponding stress point envelope. (a) Extended Mohr–Coulomb failure envelope; (b) stress point envelope.

Figure 9.25 shows the intersection lines of the extended Mohr–Coulomb failure envelope and the stress point envelope on the shear strength versus matric suction plane. The intersection lines associated with the extended Mohr–Coulomb failure envelope and the stress point envelope are defined by Eqs. (9.4) and (9.10), respectively. The ratio between the $d$ and $c$ values is always constant and equal to $\cos \phi'$ [i.e., Eq. (9.15)] at various matric suctions. As a result, the difference between the $d$ and $c$ values is not constant for different matric suctions. In other words, the intersection lines are not parallel, or put another way, $\phi^b$ is not equal to $\psi^b$. Substituting Eqs. (9.4) and (9.10) into Eq. (9.15) gives the following relationship:

$$d' + r_f \tan \psi^b = c' \cos \phi' + (u_a - u_w)_f \tan \phi^b \cos \phi'. \tag{9.17}$$

Equation (9.17) can be rearranged by substituting Eq.

## 9.2 FAILURE ENVELOPE FOR UNSATURATED SOILS

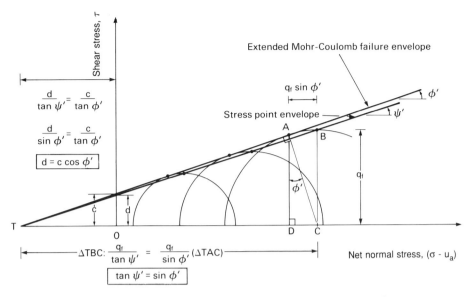

**Figure 9.24** Relationships among the variables $c$, $d$, $\phi'$, and $\psi'$.

(9.16) for $d'$ and substituting $(u_a - u_w)_f$ for $r_f$ in order to obtain the relationship among $\psi^b$, $\phi^b$, and $\phi'$:

$$\tan \psi^b = \tan \phi^b \cos \phi'. \qquad (9.18)$$

The above relationships [i.e., Eqs. (9.13), (9.15), (9.17), and (9.18)] can be used to define the stress point envelope corresponding to an extended Mohr–Coulomb failure envelope or vice versa. The extended Mohr–Coulomb failure envelope can be established by testing a soil in the saturated and unsaturated conditions. The Mohr–Coulomb failure envelope for the saturated condition gives the angle of internal friction, $\phi'$, and the effective cohesion, $c'$. Theoretically, the cohesion intercept, $c$, can be obtained from a single Mohr circle at a specific matric suction if a planar failure envelope is assumed. Figure 9.26 illustrates the construction of a Mohr circle at failure with its corresponding $p_f$ and $q_f$ values. A failure envelope with a slope angle of $\phi'$ is drawn tangent to the Mohr circle at point $A$. The envelope intersects the shear strength axis at point $B$ and the $(\sigma - u_a)$ axis at point $T$. The cohesion intercept, $c$, is computed from triangle $TBO$ (see Fig. 9.26):

$$c = \left[ \frac{q_f}{\sin \phi'} - p_f \right] \tan \phi'. \qquad (9.19)$$

Rearranging Eq. (9.19) gives

$$c = \frac{q_f}{\cos \phi'} - p_f \tan \phi'. \qquad (9.20)$$

The cohesion intercepts, $c$, at various matric suctions can be computed using Eq. (9.20) and plotted on the shear strength versus matric suction plane (see Fig. 9.25) in order to obtain the angle, $\phi^b$. Knowing the strength parameters, $c'$, $\phi'$, and $\phi^b$, the parameters for the stress point envelope (i.e., $d'$, $\phi'$, and $\psi^b$) can also be computed.

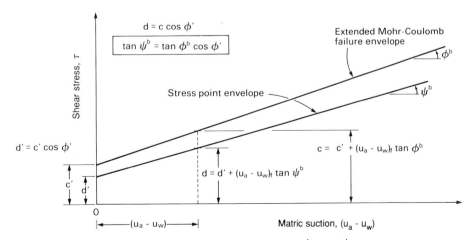

**Figure 9.25** Relationship among the $\phi'$, $\phi^b$, and $\psi^b$ angles.

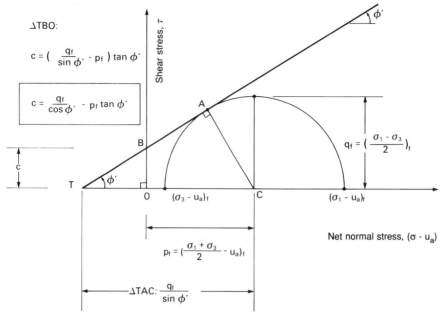

**Figure 9.26** Analytical procedure to obtain the cohesion intercept, $c$, from a single Mohr circle.

## 9.3 TRIAXIAL TESTS ON UNSATURATED SOILS

One of the most common tests used to measure the shear strength of a soil in the laboratory is the triaxial test. The theoretical concepts behind the measurement of shear strength are outlined in this section, while details on the equipment and measuring techniques, along with typical results, are presented in Chapter 10. There are various procedures available for triaxial testing, and these methods are explained and compared in this section. However, there are basic principles used in the triaxial test that are common to all test procedures. The triaxial test is usually performed on a cylindrical soil specimen enclosed in a rubber membrane, placed in the triaxial cell. The cell is filled with water and pressurized in order to apply a constant all-around pressure or confining pressure. The soil specimen can be subjected to an axial stress through a loading ram in contact with the top of the specimen.

The application of the confining pressure is considered as the first stage in a triaxial test. The soil specimen can either be allowed to drain (i.e., consolidate) during the application of the confining pressure or drainage can be prevented. The term consolidation is used to describe the process whereby excess pore pressures due to the applied stress are allowed to dissipate, resulting in volume change. This process is discussed in detail in Chapter 15. The consolidation process occurs subsequent to the application of the confining pressure if the pore fluids are allowed to drain. On the other hand, the consolidation process will *not* occur if the pore fluids are maintained in an undrained condition. The *consolidated* and *unconsolidated* conditions are used as the first criterion in categorizing triaxial tests.

The application of the axial stress is considered as the second stage or the shearing stage in the triaxial test. In a conventional triaxial test, the soil specimen is sheared by applying a compressive stress. The total confining pressure generally remains constant during shear. The axial stress is continuously increased until a failure condition is reached. The axial stress generally acts as the total major principal stress, $\sigma_1$, in the axial direction, while the isotropic confining pressure acts as the total minor principal stress, $\sigma_3$, in the lateral direction. The total intermediate principal stress, $\sigma_2$, is equal to the total minor principal stress, $\sigma_3$ (i.e., $\sigma_2 = \sigma_3$). Figure 9.27 illustrates the stress conditions associated with a consolidated drained triaxial test. The pore fluid drainage conditions during the shearing process are used as the second criterion in categorizing triaxial tests. When the pore fluid is allowed to flow in and out of the soil specimen during shear, the test is referred to as a *drained* test. On the other hand, a test is called an *undrained* test if the flow of pore fluid is prevented. The pore-air and pore-water phases can have different drainage conditions during shear.

Various triaxial test procedures are used for unsaturated soils based upon the drainage conditions adhered to during the first and second stages of the triaxial test. The triaxial test methods are usually given a two-word designation or abbreviated to a two-letter symbol. The designations are: 1) consolidated drained or CD test, 2) constant water content or CW test, 3) consolidated undrained or CU test with pore pressure measurements, 4) undrained test, and 5) unconfined compression or UC test. In the case of CD and CU tests, the first letter refers to the drainage condition prior to shear, while the second letter refers to the drainage condition during shear. The constant water content test is a special case where only the pore-air is kept in a drained

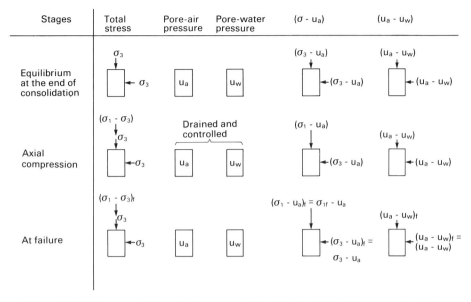

**Figure 9.27** Stress conditions during a consolidated drained triaxial compression test.

mode, while the pore-water phase is kept undrained during shear (i.e., constant water content). The pore-air and pore-water are not allowed to drain throughout the test for the undrained triaxial test. The unconfined compression test is a special loading condition of the undrained triaxial test. These five testing procedures are explained in the following sections. A summary of triaxial testing conditions used, together with the measurements performed, is given in Table 9.2. The air, water, or total volume changes may or may not be measured during shear.

The shear strength data obtained from triaxial tests can be analyzed using the stress state variables at failure or using the total stresses at failure when the pore pressures are not known. This concept is similar to the effective stress approach and the total stress approach used in saturated soil mechanics. In a drained test, the pore pressure is controlled at a desired value during shear. Any excess pore pressures caused by the applied load are dissipated by allowing the pore fluids to flow in or out of the soil specimen. The pore pressure at failure is known since it is controlled, and the stress state variables at failure can be used to analyze the shear strength data. In an undrained test, the excess pore pressure due to the applied load can build up because pore fluid flow is prevented during shear. If the changing pore pressures during shear are measured, the pore pressures at failure are known, and the stress state variables can be

**Table 9.2  Various Triaxial Tests for Unsaturated Soils**

| Test Methods | Consolidation Prior to Shearing Process | Drainage | | Shearing Process | | |
|---|---|---|---|---|---|---|
| | | Pore-Air | Pore-Water | Pore-Air Pressure, $u_a$ | Pore-Water Pressure, $u_w$ | Soil Volume Change, $\Delta V$ |
| Consolidated Drained (CD) | yes | yes | yes | C | C | M |
| Constant water content (CW) | yes | yes | no | C | M | M |
| Consolidated undrained (CU) | yes | no | no | M | M | — |
| Undrained | no | no | no | — | — | — |
| Unconfined compression (UC) | no | no | no | — | — | — |

$M$ = Measurement, $C$ = controlled.

**238** 9 SHEAR STRENGTH THEORY

computed. However, if pore pressure measurements are not made during undrained shear, the stress state variables are unknown. In this case, the shear strength can only be related to the total stress at failure.

The total stress approach should be applied in the field only for the case where it can be assumed that the strength measured in the laboratory has relevance to the drainage conditions being simulated in the field. In other words, the applied total stress that causes failure in the soil specimen is assumed to be the same as the applied total stress that will cause failure in the field. The above simulation basically assumes that the stress state variables control the shear strength of the soil; however, it is possible to perform the analysis using total stresses. It is difficult, however, to closely simulate field loading conditions with an undrained test in the laboratory. Rapid loading of a fine-grained soil may be assumed to be an undrained loading condition.

### 9.3.1 Consolidated Drained Test

The consolidated drained or CD test refers to a test condition where the soil specimen is consolidated first and then sheared under drained conditions for both the pore–air and pore–water phases, as illustrated in Fig. 9.27. The soil specimen is consolidated to a stress state representative of what is likely to be encountered in the field or in the design. The soil is generally consolidated under an isotropic confining pressure of $\sigma_3$, while the pore–air and pore–water pressures are controlled at pressures of $u_a$ and $u_w$, respectively. The pore–air and pore–water pressures can be controlled at positive values in order to establish a matric suction greater than 101.3 kPa (i.e., 1 atm) without cavitation in the pore–water pressure measuring system. This is referred to as the axis-translation technique. At the end of the consolidation process, the soil specimen has a net confining pressure of $(\sigma_3 - u_a)$ and a matric suction of $(u_a - u_w)$ (Fig. 9.27).

During the shearing process, the soil specimen is compressed in the axial direction by applying a deviator stress [i.e., $(\sigma_1 - \sigma_3)$]. During shear, the drainage valves for both pore–air and pore–water remain open (i.e., under drained conditions). The pore–air and pore–water pressures are controlled at constant pressures (i.e., their pressures at the end of consolidation). The deviator stress is applied slowly in order to prevent the development of excess pore–air or pore–water pressure in the soil. The net confining pressure, $(\sigma_3 - u_a)$, and the matric suction, $(u_a - u_w)$, remain constant throughout the test until failure conditions are reached, as indicated in Fig. 9.27 [i.e., $(\sigma_3 - u_a)_f = (\sigma_3 - u_a)$ and $(u_a - u_w)_f = (u_a - u_w)$]. Only the deviator stress, $(\sigma_1 - \sigma_3)$, keeps increasing during shear until the net major principal stress reaches a value of $(\sigma_1 - u_a)_f$ at failure.

Typical stress–strain curves for the consolidated drained triaxial test are shown in Fig. 9.15(a). The curves show an increase in the maximum deviator stress as the net confining pressure, $(\sigma_3 - u_a)$, is increased while the matric suction, $(u_a - u_w)$, remains constant. The volume change of the soil specimen during shear is usually measured with respect to the initial soil volume (i.e., $\Delta V/V_o$) and plotted versus strain [Fig. 9.15(c)]. Compression is conventionally given a negative sign while expansion has a positive sign. Figure 9.15(b) presents a plot of gravimetric water content change versus strain where an increase in water content is given a positive sign.

Typical stress paths followed during consolidated drained tests under a constant matric suction are illustrated in Fig. 9.28. The tests are performed on several specimens at various net confining pressures. For example, stress point $A$ represents the stress state at the end of consolidation when the soil specimen has a net confining pressure of $(\sigma_3 - u_a)$ and a matric suction of $(u_a - u_w)$. As the soil is compressed during shear, the stress point moves from point $A$ to point $B$ along the stress path $\overline{AB}$. Stress point $B$ represents the stress state at the condition of failure. When moving from stress point $A$ to stress point $B$, the Mohr circle diameter or the deviator stress increases until the failure condition is reached at stress point $B$. However, the net confining pressure and the matric suction remain constant throughout the stress path $\overline{AB}$. A line drawn tangent to the Mohr circles at failure (i.e., through stress points $C$, $C_1$, and $C_2$) represents the failure envelope corresponding to the matric suction used in the tests. The failure envelope has a slope angle of $\phi'$ with respect to the $(\sigma - u_a)$ axis. The friction angle appears to be essentially equal to the effective angle of internal friction obtained from shear strength tests on saturated soil specimens. The value of the $\phi'$ angle for compacted soils commonly ranges from 25° to 35°, as shown in Table 9.1. The effect of compactive effort on the strength parameters, $\phi'$ and $c'$, for a clayey sand is illustrated in Fig. 9.29.

Figure 9.30 presents the stress paths followed during consolidated drained tests under a constant net confining pressure and various matric suctions. The Mohr circle at failure increases in diameter as the matric suction at failure increases. The Mohr circle at failure is tangent to the failure envelope corresponding to the matric suction used in the test (e.g., at stress points, $C_1$, $C_2$, and $C_3$). However, stress points, $C_1$, $C_2$, and $C_3$, do not occur at the same net normal stress. Therefore, a line joining stress points, $C_1$, $C_2$, and $C_3$, will not give the angle, $\phi^b$. Rather, it is suggested that the failure envelope be extended to intersect the shear strength versus $(u_a - u_w)$ plane to give cohesion intercepts. A line joining the cohesion intercepts at various matric suctions gives the angle, $\phi^b$.

### 9.3.2 Constant Water Content Test

For the constant water content or CW triaxial test, the specimen is first consolidated and then sheared, with the pore–air phase allowed to drain while the pore–water phase

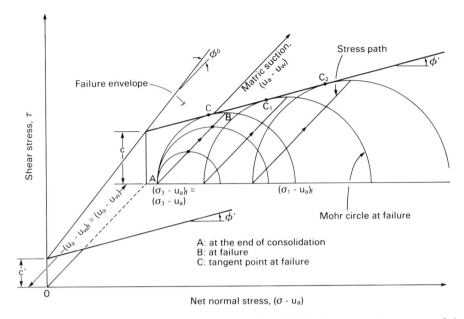

**Figure 9.28** Stress paths followed during a consolidated drained test at various net confining pressures under a constant matric suction.

is in an undrained mode. The consolidation procedure is similar to that of the consolidated drained test. The axis-translation technique can be used to impose matric suctions greater than 101.3 kPa. When equilibrium is reached at the end of consolidation, the soil specimen has a net confining pressure of $(\sigma_3 - u_a)$ and a matric suction of $(u_a - u_w)$. The specimen is sheared by increasing the deviator stress, $(\sigma_1 - \sigma_3)$, until failure is reached. During shear, the drainage valve for the pore–air remains open (i.e., under drained conditions), while the drainage valve for the pore–water is closed (i.e., under undrained conditions). The pore–air pressure, $u_a$, is maintained at the pressure applied during consolidation. The pore–water pressure, $u_w$, changes during shear under undrained loading conditions (see Chapter 8). The excess pore pressure is related to deviator stress by the $D$ pore pressure parameters. The net confining pressure, $(\sigma_3 - u_a)$, remains constant throughout the test, while the matric suction, $(u_a - u_w)$, changes, as illustrated in

Fig. 9.31 (i.e., $(\sigma_3 - u_a)_f = (\sigma_3 - u_a)$ and $(u_a - u_w)_f = (u_a - u_w) - \Delta u_{wf}$). The net major principal stress reaches a value of $(\sigma_1 - u_a)_f$ at failure.

Typical stress versus strain curves for the constant water content test are shown in Fig. 9.32(a). The stress versus strain curves obtained from the constant water content test are similar in shape to those obtained from the consolidated drained triaxial test. The maximum deviator stress increases with an increase in the net confining pressure, $(\sigma_3 - u_a)$, for soil specimens prepared at the same initial matric suction. The matric suction and the soil volume changes during shear are plotted in Fig. 9.32(b) and (c), respectively. The matric suction decreases during shear since the pore–air pressure is maintained at a constant value while the pore–water pressure increases. For the results presented, the degree of saturation of the soil specimen increases as the pore voids are compressed as a result of the pore–air being squeezed out of the soil. The water content remains constant. Figure 9.32(c) indicates that the soil specimens undergo compression until the maximum deviator stress is reached. The soil specimen with the lower net confining pressure dilates after reaching the maximum deviator stress. This is accompanied by a slight increase in matric suction.

A hypothetical stress path that may be followed by a soil specimen during a constant water content test is shown in Fig. 9.33. Stress point $A$ represents the stress state at the end of consolidation where the soil specimen has a net confining pressure of $(\sigma_3 - u_a)$ and a matric suction of $(u_a - u_w)$. As the soil is compressed during shear, the stress point is assumed to move from point $A$ to point $B$ along stress path $\overline{AB}$. Stress point $B$ represents the stress state at failure.

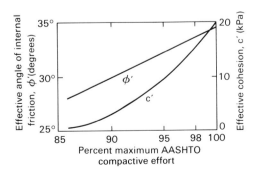

**Figure 9.29** Effect of compactive effort on $\phi'$ and $c'$ for a clayey sand (from Moretto *et al.*, 1963).

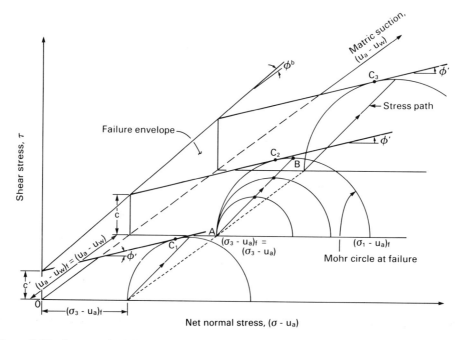

**Figure 9.30** Stress paths followed during consolidated drained tests at various matric suctions under a constant net confining pressure.

The net confining pressure remains constant at $(\sigma_3 - u_a)$ along stress path $\overline{AB}$ since the pore-air pressure is maintained at the pressure used during consolidation. The pore-water pressure is assumed to increase continuously during shear. This results in a reduction in the matric suction [i.e., $(u_a - u_w)_f < (u_a - u_w)$]. The failure envelope sloping at an angle of $\phi'$ can be drawn tangent to the Mohr circle at failure (e.g., at stress point $C$). The failure envelope intersects the shear strength versus matric suction plane at a cohesion intercept, $c$. The cohesion intercepts obtained at various matric suctions can be joined to give the $\phi^b$ angle.

### 9.3.3 Consolidated Undrained Test with Pore Pressure Measurements

The consolidated undrained or CU test uses the test condition where the soil specimen is consolidated first and then sheared, with both pore-air and pore-water under un-

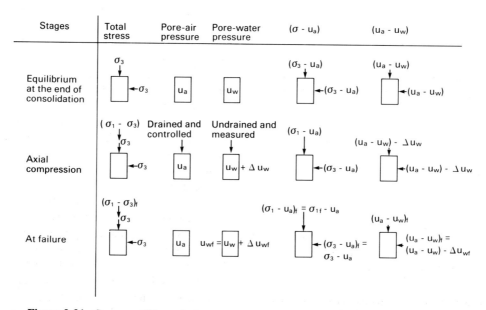

**Figure 9.31** Stress conditions during a constant water content triaxial compression test.

9.3 TRIAXIAL TESTS ON UNSATURATED SOILS  241

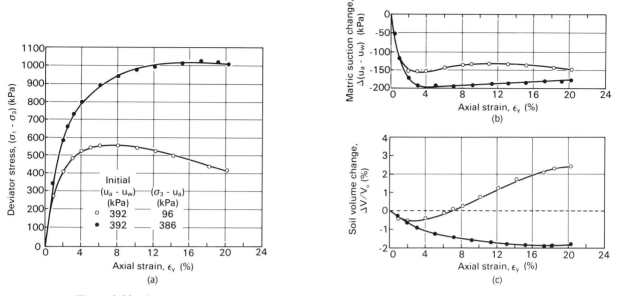

**Figure 9.32** Constant water content triaxial tests on Dhanauri clay. (a) Stress versus strain curve; (b) matric suction change versus strain; (c) soil volume change versus strain (from Satija, 1978).

drained conditions, as shown in Fig. 9.34. The consolidation process brings the soil specimen to the desired stress state [i.e., $(\sigma_3 - u_a)$ and $(u_a - u_w)$]. The axis-translation technique is used to establish matric suctions greater than 101.3 kPa. After equilibrium conditions have been reached, the soil specimen is sheared by increasing the axial load, $(\sigma_1 - \sigma_3)$, until failure is reached. The drainage valves for both the pore-air and pore-water pressures are closed (i.e., undrained conditions) during shear. Excess pore-air and pore-water pressures are developed during undrained loading. Under triaxial loading conditions, the excess pore pressures are related to the deviator stress by the $D$ pore pressure parameters (see Chapter 8). The pore-air and pore-water pressures should be measured during the shear process. The net confining pressure, $(\sigma_3 - u_a)$, and the matric suction, $(u_a - u_w)$, are altered throughout the test due to the changing pore-air and pore-water pressures. At failure, the magnitudes of the net major and minor principal stresses and the matric suction are a function of the pore pressures.

A typical stress path for a consolidated undrained test is illustrated in Fig. 9.35. The stress state at the end of con-

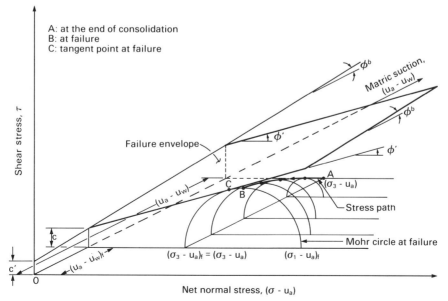

**Figure 9.33** Stress path followed during a constant water content test.

**242** 9 SHEAR STRENGTH THEORY

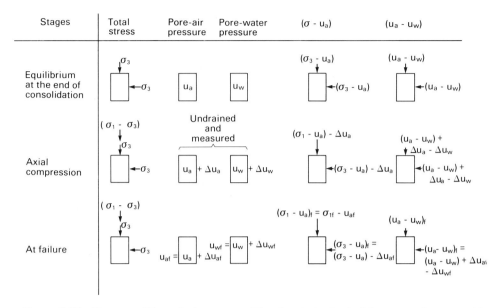

**Figure 9.34** Stress conditions during a consolidated undrained triaxial compression test with pore pressure measurements.

solidation is represented by stress point $A$ where the net confining pressure is $(\sigma_3 - u_a)$ and the matric suction is $(u_a - u_w)$. Shear causes the stress state to move from point $A$ to point $B$ along stress path $\overline{AB}$. The stress state at failure is represented by stress point $B$, corresponding to a different net confining pressure and matric suction from those associated with stress point $A$. In the example shown, the pore–air pressure is assumed to increase continuously during shear. This causes the net confining pressure to decrease [i.e., $(\sigma_3 - u_a)_f < (\sigma_3 - u_a)$]. The matric suction is also assumed to decrease continuously [i.e., $(u_a - u_w)_f$ $< (u_a - u_w)$]. The failure envelope is tangent to the Mohr circle at failure (e.g., at stress point $C$) and inclined at an angle of $\phi'$ with respect to the $(\sigma - u_a)$ axis. The failure envelope intersects the shear strength versus $(u_a - u_w)$ plane at a cohesion intercept, $c$. The intersection line joining the cohesion intercepts produced by tests at different matric suctions gives the angle, $\phi^b$.

It should be noted that it is difficult to maintain a fully undrained condition for the pore–air since air diffuses through the pore–water, the rubber membrane, and other parts of the triaxial apparatus.

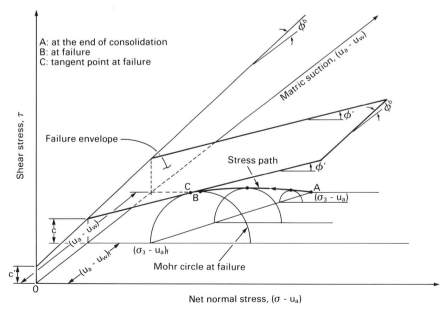

**Figure 9.35** Typical stress path followed during a consolidated undrained test.

## 9.3.4 Undrained Test

The pore-air and pore-water are not allowed to drain in the undrained test. This applies both when the confining pressure and the deviator stress are applied to the soil specimen (Fig. 9.36). The excess pore-air and pore-water pressures developed during the application of the confining pressure can be related to the isotropic confining pressure by use of the $B$ pore pressure parameters, as explained in Chapter 8. Although the excess pore pressures built up during the application of confining pressure are not allowed to dissipate, the volume of the soil specimen may change due to compression of pore-air. The soil has a net confining pressure, $(\sigma_3 - u_a)$, and a matric suction, $(u_a - u_w)$, after the application of the confining pressure.

The soil specimen is sheared by applying an axial stress, $(\sigma_1 - \sigma_3)$, until failure is reached. Undrained loading during shear causes a further development of excess pore-air and pore-water pressures. The excess pore pressures can be related to the deviator stress by the $D$ pore pressure parameters for triaxial loading conditions (see Chapter 8). Generally, the pore pressures are not measured during shear. Therefore, the undrained test results are commonly used in conjunction with a total stress formulation of a problem. Here, the shear strength is related to the total stress without a knowledge of the pore pressures at failure.

The net confining pressure, $(\sigma_3 - u_a)$, and the matric suction, $(u_a - u_w)$, vary throughout the shear process. However, the stress state variables during shear and at failure are unknown since the pore pressures are not measured. Figure 9.36 attempts to illustrate how the stress state variables would change during an undrained triaxial test. Only the total confining pressure, $\sigma_3$, and the deviator stress, $(\sigma_1 - \sigma_3)$, are measured values in this test. Typical stress-strain curves for the undrained test are shown in Fig. 9.16(a). These are the results of a special test in the sense that the pore-air and pore-water pressures were measured during shear [Fig. 9.16(b)]. The pore-water pressure measurements were limited to $-1$ atm (i.e., $-101.3$ kPa) (Bishop et al. 1960). The plot of soil volume change in Fig. 9.16(c) indicates that the soil specimen compressed during shear.

Hypothetical stress paths that may be followed by a soil specimen during an undrained test are illustrated in Fig. 9.37. Consider four identical specimens that are initially confined at four different total confining pressures. These are represented by stress points $A$, $A_1$, $A_2$, and $A_3$, where $\sigma_3$ at $A < \sigma_3$ at $A_1 < \sigma_3$ at $A_2 < \sigma_3$ at $A_3$. The application of the total confining pressure under undrained conditions results in the compression of the pore fluids and the development of excess pore-air and pore-water pressures. The pore pressure increases in an unsaturated soil are always less than the total stress increment applied. This is in keeping with a $B$ pore pressure parameter which must always be less than 1.0 for an unsaturated soil (see Chapter 8). Therefore, a higher total confining pressure results in a higher net confining pressure (i.e., $(\sigma_3 - u_a)$ at $A < (\sigma_3 - u_a)$ at $A_1 < (\sigma_3 - u_a)$ at $A_2 < (\sigma_3 - u_a)$ at $A_3$) and a lower matric suction (i.e., $(u_a - u_w)$ at $A > (u_a - u_w)$ at $A_1 > (u_a - u_w)$ at $A_2 > (u_a - u_w)$ at $A_3$). In other words, the four identical soil specimens are brought to four different initial stress states, represented by stress points $A$, $A_1$, $A_2$, and $A_3$ (Fig. 9.37). As the soil is sheared in undrained loading, the pore fluids are further compressed and the pore pressures may further increase. The stress point moves from point $A$ to point $B$ along the stress path $\overline{AB}$. The net confining pressure and the matric suction of the soil specimen decrease when going from stress point $A$ to stress point $B$.

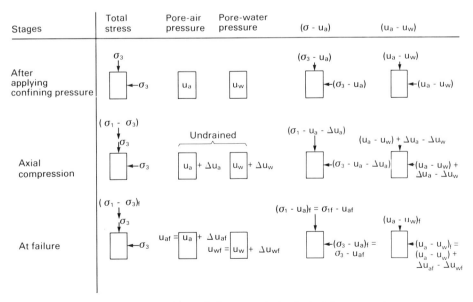

**Figure 9.36** Stress conditions during an undrained triaxial compression test.

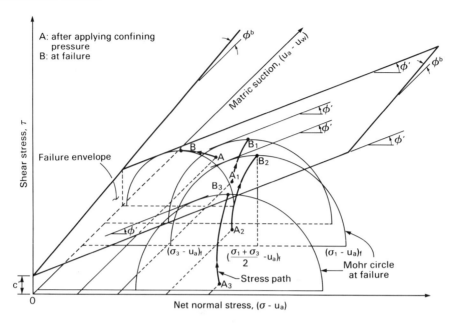

**Figure 9.37** Stress paths followed during an undrained test.

Stress point $B$ represents the stress state of the soil at failure. Similar stress paths are followed by the soil specimens at stress points $A_1$, $A_2$, and $A_3$ (i.e., $\overline{A_1B_1}$, $\overline{A_2B_2}$, and $\overline{A_3B_3}$).

Figure 9.37 indicates an increase in the diameter of the Mohr circle at failure as the initial total confining pressure increases (i.e., $(\sigma_1 - \sigma_3)_f$ at $B < (\sigma_1 - \sigma_3)_f$ at $B_1 < (\sigma_1 - \sigma_3)_f$ at $B_2 < (\sigma_1 - \sigma_3)_f$ at $B_3$). In other words, the shear strength of the soil increases with increasing initial total confining pressure, although the initial matric suction decreases. This occurs because the rate of shear strength increase caused by an increase in confining pressure is greater than the reduction in shear strength caused by a decrease in matric suction. This can also be visualized as occurring because the $\phi^b$ angle is always lower than the friction angle, $\phi'$.

The increase in shear strength due to the increase in the initial total confining pressure can also be demonstrated using a shear stress versus total normal stress plot (i.e., $\tau$ versus $\sigma$ plot), as shown in Fig. 9.38. The Mohr circles at failure are drawn using the total confining pressure at failure (i.e., $\sigma_{3f} = \sigma_3$) and the total major principal stress at failure, $\sigma_{1f}$. The diameter of the Mohr circle, $(\sigma_1 - \sigma_3)_f$, remains the same when the circle is plotted with respect to either the net normal stress, $(\sigma - u_a)$, or the total normal stress, $\sigma$. However, the position of the Mohr circles on the horizontal axis differs by a magnitude equal to the pore-air pressure at failure, $u_{af}$. This can be seen by comparing the Mohr circle associated with stress point $B_2$ in Figs. 9.37 and 9.38.

A typical curved envelope can be drawn tangent to the Mohr circles of the *unsaturated* soil specimens at failure (Fig. 9.38). The envelope defines the curved relationship between shear strength and total normal stress for unsaturated soils tested under undrained conditions. The curve indicates an increase in shear strength as the applied total stress increases. It may be practical in some cases to approximate the curved failure envelope with a straight line. Care should be taken not to exceed the limits to which the envelope is applicable. As the confining pressure increases, the matric suction decreases and the degree of saturation increases. Eventually, a point is reached where the soil approaches saturation. In the *saturated* condition, an increase in the confining pressure will be equally balanced by a pore-water pressure increase since the $B_w$ pore pressure parameter will equal 1.0. In other words, the effective confining stress $(\sigma_3 - u_w)$, remains constant regardless of the total confining pressure, $\sigma_3$. Therefore, the shear strength of the saturated soil should also remain constant for different total confining pressures. It is well known that undrained tests on saturated soil specimens at various confining pressures result in a series of Mohr circles at failure with equal diameters. A horizontal envelope with a slope of zero (i.e., $\phi_u = 0$; see Fig. 9.38) can be drawn tangent to the Mohr circles at failure. The ordinate intercept of the horizontal envelope on the shear strength axis is half the deviator stress at failure, and is referred to as the undrained shear strength, $c_u$. The application of the undrained shear strength to the analysis of saturated soil is referred to as the $\phi_u = 0$ concept.

The shear strength versus total normal stress relationship for an unsaturated soil does not produce a horizontal line. Rather, the undrained shear strength is a function of the applied total normal stress, as illustrated in Fig. 9.38.

9.3 TRIAXIAL TESTS ON UNSATURATED SOILS    245

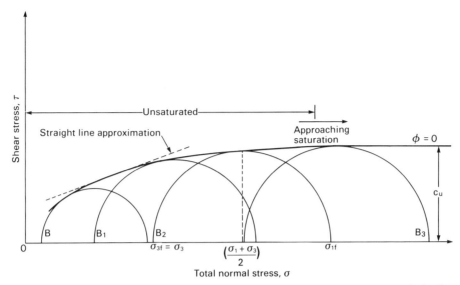

**Figure 9.38** Shear stress versus total normal stress relationship at failure for the undrained test.

### 9.3.5 Unconfined Compression Test

The unconfined compression or UC test is a special case of the undrained test. No confining pressure is applied to the soil specimen throughout the test (Fig. 9.39). The test can be performed by applying a load in a simple loading frame. At the start of the test, the unsaturated soil specimen has a negative pore-water pressure, and the pore-air pressure is assumed to be atmospheric. The soil matric suction, $(u_a - u_w)$, is therefore numerically equal to the pore-water pressure.

The soil specimen is sheared by applying an axial load until failure is reached. The deviator stress, $(\sigma_1 - \sigma_3)$, is equal to the major principal stress, $\sigma_1$, since the confining pressure, $\sigma_3$, is equal to zero. The compressive load is applied quickly in order to maintain undrained conditions. Theoretically, this should apply to both in the pore-air and pore-water phases. The pore-air and pore-water pressures are not measured during compression. The excess pore pressures developed during an unconfined compression test can be theoretically related to the major principal stress through use of the $D$ pore pressure parameter (see Chapter 8). Figure 9.39 illustrates typical changes in the stress state variables that would occur during unconfined compression if the pore pressures were measured.

Figure 9.40 illustrates two possible stress paths that may

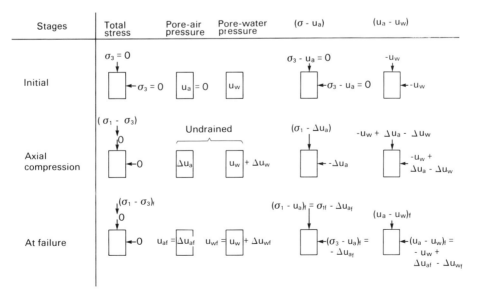

**Figure 9.39** Stress conditions during an unconfined compression test.

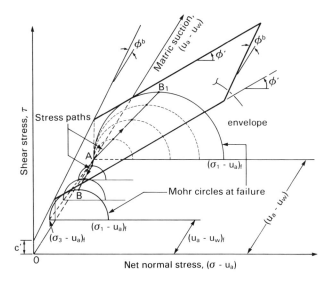

**Figure 9.40** Possible stress paths followed during an unconfined compression test.

be followed in an unsaturated soil specimen during the unconfined compression. The initial stress state is represented by stress point A where the soil has a zero net confining pressure (i.e., $(\sigma_3 - u_a) = 0$) and a matric suction, $(u_a - u_w)$. During an undrained compression, the matric suction can increase, decrease, or remain constant, depending upon the A parameter of the soil (see Chapter 9). Generally, the matric suction will decrease during an undrained compression test, and the stress state in the soil will move forward from point A to point B along the stress path $\overline{AB}$ (Fig. 9.40). The stress state of the soil at failure is represented by point B. The pore-air pressure is assumed to increase slightly during compression. This causes the net confining pressure to decrease along stress path $\overline{AB}$ to a negative value (i.e., $(\sigma_3 - u_a)_f = -\Delta u_{af}$). The matric suction at failure will be less than the initial matric suction (i.e., $(u_a - u_w)_f$ at $B <$ $(u_a - u_w)$ at A).

In the case of a constant matric suction during compression, the stress state of the soil could move from point A to point $B_1$ along the stress path $\overline{AB_1}$. In this case, the stress path lies in a plane of constant matric suction. Stress point $B_1$ represents the stress state of the soil at failure. The pore-air and pore-water pressures are assumed to remain constant during compression in order for the matric suction to be constant. As a result, the net confining pressure remains equal to zero until failure is reached (i.e., $(\sigma_3 - u_a) = (\sigma_3 - u_a)_f = 0$).

For the case of an increasing matric suction during compression (e.g., the soil dilates), the stress state in the soil will move backward from point A to a point (or plane) somewhere behind point A. This stress path is not shown in Fig. 9.40.

The stress paths illustrated in Fig. 9.40 are associated with unsaturated soil specimens. The soil may be unsatu-

rated in the field or become unsaturated during sampling due to the release of the overburden pressure. Unconfined compression test results on the *unsaturated* soil apply only to the condition where the total confining pressure, $\sigma_3$, is zero [Fig. 9.41(a)]. The deviator stress at failure, $(\sigma_1 - \sigma_3)_f$, is referred to as the unconfined compressive strength, $q_u$. The unconfined compressive strength is commonly taken as being equal to twice the undrained shear strength, $c_u$. However, as the confining pressure increases, the undrained shear strength for the unsaturated soil also increases. As a result, the compressive strength value may not be satisfactory in approximating the undrained shear strength, $c_u$, at a confining pressure greater than zero.

In some cases, the soil samples in the field remain saturated (i.e., S is equal to 100%), although its pore-water pressure is negative. The pore-water pressure can be negative because the soil is from some distance above the groundwater table, or it can be negative due to the release of overburden pressure, or both. The shear strength equation for saturated soils [i.e., Eq. (9.1)] can be used even though the pore-water pressures are negative. The undrained shear strength, $c_u$, for the soil remains constant at various confining pressures [Fig. 9.41(b)]. As a result, the

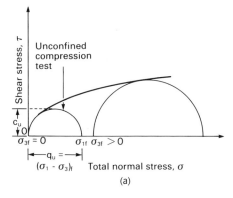

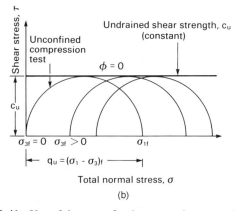

**Figure 9.41** Use of the unconfined compressive strength, $q_u$, to approximate the undrained shear strength, $c_u$, for an unsaturated soil and a saturated soil. (a) Relationship between $q_u$ and $c_u$ for an unsaturated soil; (b) relationship between $q_u$ and $c_u$ for a saturated soil (with negative pore-water pressures).

unconfined compressive strength, $q_u$, is twice the undrained shear strength, $c_u$, for the saturated soil, regardless of the magnitude of the confining pressure.

In summary, it is quite likely that unconfined compression tests on unsaturated soils will greatly underestimate the available shear strength. Unconfined compression tests are routinely performed for bearing capacity, slope stability, and other engineering studies. The engineer must understand that an approximate simulation of the field confining pressure is strongly recommended when performing laboratory tests. In other words, the undrained tests should be performed as confined compression tests when the soils are unsaturated.

## 9.4 DIRECT SHEAR TESTS ON UNSATURATED SOILS

A direct shear test apparatus basically consists of a split box, with a top and bottom portion. The test is generally performed using a consolidated drained procedure, as shown in Fig. 9.42. A soil specimen is placed in the direct shear box and consolidated under a vertical normal stress, $\sigma$. During consolidation, the pore-air and pore-water pressures must be controlled at selected pressures. The axis-translation technique can be used to impose a matric suction greater than 1 atm. The direct shear test can be conducted in an air-pressurized chamber in order to elevate the pore-air pressure to a magnitude above atmospheric pressure (i.e., 101.3 kPa). The pore-water pressure can be controlled below the soil specimen using a high air entry disk. At the end of the consolidation process, the soil specimen has a net vertical normal stress of $(\sigma_n - u_a)$ and a matric suction of $(u_a - u_w)$ (Fig. 9.42). Shearing is achieved by horizontally displacing the top half of the direct shear box relative to the bottom half. The soil specimen is sheared along a horizontal plane between the top and bottom halves of the direct shear box. The horizontal load required to shear the specimen, divided by the nominal area of the specimen, gives the shear stress on the shear plane. During shear, the pore-air and pore-water pressures are controlled at constant values. Shear stress is increased until the soil specimen fails. The failure plane has a shear stress designated as $\tau_{ff}$, corresponding to a net vertical normal stress of $(\sigma_f - u_a)_f$ [i.e., equal to $(\sigma - u_a)$] and a matric suction of $(u_a - u_w)_f$ [i.e., equal to $(u_a - u_w)$], as illustrated in Fig. 9.42. Figure 9.43 shows a typical plot of shear stress versus horizontal displacement for a direct shear test.

The failure envelope can be obtained from the results of direct shear tests without constructing the Mohr circles. The shear stress at failure, $\tau_{ff}$, is plotted as the ordinate, and $(\sigma_f - u_a)_f$ and $(u_a - u_w)_f$ are plotted as the abscissas to give a point on the failure envelope (Fig. 9.44). A line joining points of equal magnitude of $(\sigma_f - u_a)_f$ determines the $\phi^b$ angle (e.g., a line joining points $A$, $B$, and $C$ in Fig. 9.44). Similarly, a line can be drawn through the points of equal $(u_a - u_w)_f$ to give the angle of internal friction, $\phi'$ (e.g., a line drawn through points $A_1$, $A_2$, and $A$ in Fig. 9.44).

The direct shear test is particularly useful for testing unsaturated soils due to the short drainage path in the specimen. The low coefficient of permeability of unsaturated soils results in "times to failure" in triaxial tests which can be excessive. Other problems associated with testing unsaturated soils in a direct shear apparatus are similar to those common to saturated soils (e.g., stress concentrations, definition of the failure plane, and the rotation of principal stresses).

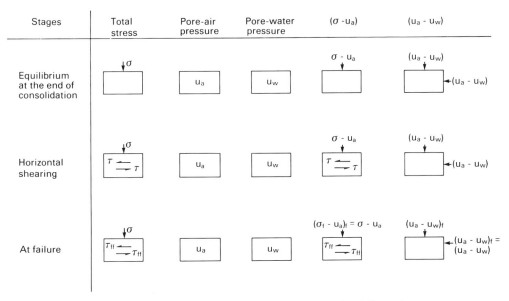

**Figure 9.42** Stress conditions during a consolidated drained direct shear test.

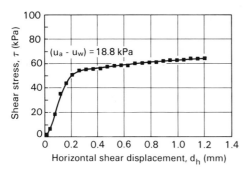

**Figure 9.43** A typical shear stress versus shear displacement curve from a direct shear test (from Gan, 1986).

## 9.5 SELECTION OF STRAIN RATE

The strength testing of unsaturated soils is commonly performed at a constant rate of strain. An appropriate strain rate must be selected prior to commencing a test. In *undrained* shear, the selected strain rate must ensure equalization of induced pore pressures throughout the specimen. In *drained* shear, the selected strain rate must ensure dissipation of induced pore pressures. The estimation of the strain rate for testing soils can be made partly on the basis of experimental evidence and partly on the basis of theory.

### 9.5.1 Background on Strain Rates for Triaxial Testing

Prior to the 1960's, unsaturated soils were tested in shear using similar procedures and equipment to those used for saturated soils. Coarse ceramic disks were used, and the strain rates were relatively high. In the early 1960's, the procedures and equipment used for testing unsaturated soils were modified. A high air entry ceramic disk was placed at the bottom of the soil specimen in order to measure or control the pore–water pressure independently from the pore–air pressure. Several empirical procedures were proposed for estimating the strain rate for testing unsaturated soils (Bishop *et al.* 1960; Lumb, 1966; Ruddock, 1966). The testing equipment was also modified to accommodate the independent control of the pore–air and pore–water pressures. The pore–air pressure was controlled or measured through the use of a coarse ceramic disk, commonly placed on the top of the soil specimen. The control or measurement of the pore–water pressure was made through the use of a high air entry ceramic disk. The disk was sealed onto the base pedestal. The modified equipment allowed the use of the axis-translation technique (Hilf, 1956). The above arrangement for the shear strength testing of unsaturated soils has generally been used up to the present.

Shear strength tests on unsaturated soils have been generally performed at relatively slow rates in order to ensure equalization or dissipation of induced pore pressures. Trial and error procedures have been used by several researchers to assess the appropriate strain rate. Donald (1961) recommended that the effect of strain rate on the maximum deviator stress be used as a criterion in assessing an appropriate strain rate. Gibson and Henkel (1954) and Bishop and Henkel (1962) presented test data showing the variation in strength (i.e., the deviator stress at failure) with strain rate. The shear strength levelled off below a particular strain rate, indicating an upper limit of strain rate for failing the soil specimen. No measurements were made to ensure that this strain rate limit was slow enough for complete pore pressure equalization or dissipation.

Satija and Gulhati (1979) concluded from their test data that deviator stress is not sensitive to the effect of varying strain rates [Figs. 9.45(a) and 9.46(a)]. It was suggested that changes in matric suction for CW tests [Fig. 9.45(b)]

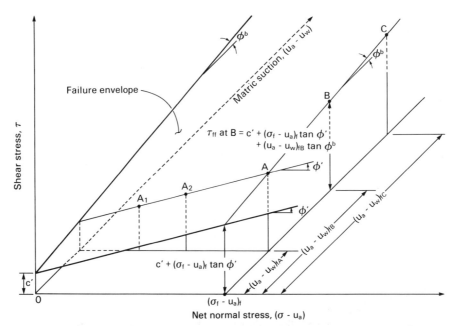

**Figure 9.44** Extended Mohr–Coulomb failure envelope established from direct shear test results.

## 9.5 SELECTION OF STRAIN RATE

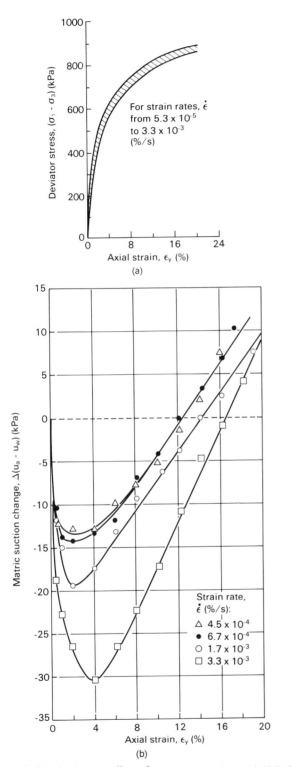

**Figure 9.45** Strain rate effects for constant water content tests on Dhanauri clay. (a) Effect of strain rate on deviator stress; (b) effect of strain rate on matric suction change (from Satija, 1978).

and changes in water content for CD tests [Fig. 9.46(b)] would be a more reasonable indicator when assessing an appropriate strain rate. Tests were performed on specimens of compacted Dhanauri clay using strain rates varying from $5.3 \times 10^{-5}$ to $3.3 \times 10^{-3}$ %/s. It was concluded that a

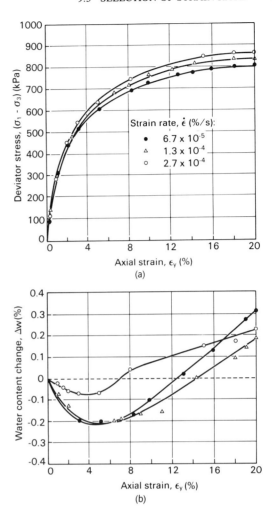

**Figure 9.46** Strain rate effects for consolidated drained tests on Dhanauri clay. (a) Effect of strain rate on deviator stress; (b) effect of strain rate on water content change (from Satija, 1978).

strain rate of $6.7 \times 10^{-4}$ (%/s) was adequate for CW tests on Dhanauri clay. The strain rate for CD tests was chosen to be one-fifth of the strain rate for CW tests (i.e., $1.3 \times 10^{-4}$ %/s).

The effect of strain rate on the equalization of pore-water pressures for undrained tests has been studied by Bishop *et al.* 1960. Two compacted clay shale specimens, 101.6 mm in diameter and 203.2 mm in height, were tested at two different strain rates (i.e., $6.9 \times 10^{-4}$ and $4.6 \times 10^{-5}$ %/s). The pore-water pressure was measured at the base of the specimen and at the midheight of the specimen. The results are presented in Figs. 9.47 and 9.48 for strain rates of $6.9 \times 10^{-4}$ and $4.6 \times 10^{-5}$ %/s, respectively. The higher strain rate resulted in significantly different pore-water pressures across the specimen [Fig. 9.47(b)]. The base measurement was higher than the midheight measurement. As a result, the principal stress ratios computed from the two pore-water pressure measurements were also different [Fig. 9.47(a)]. This difference affects the assessment of the shear strength. Closer agreement between the pore-

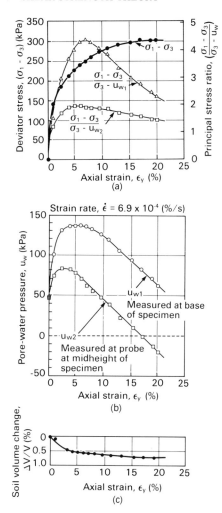

**Figure 9.47** An undrained test on a compacted clay shale specimen at a strain rate of $6.9 \times 10^{-4}$ %/s. (a) Shear stress versus strain curve; (b) pore-water pressure versus strain at two points of measurement; (c) total volume change versus strain (from Bishop et al., 1960).

water measurements at the base and at the midheight point was obtained when using the slower strain rate [Fig. 9.48(b)].

### 9.5.2 Strain Rates for Triaxial Tests

Strain rate can be defined as the rate at which a soil specimen is axially compressed:

$$\dot{\epsilon} = \frac{\epsilon_f}{t_f} \quad (9.21)$$

where

- $\dot{\epsilon}$ = strain rate for shearing a specimen in the triaxial test
- $\epsilon_f$ = strain of the soil specimen at failure
- $t_f$ = time required to fail the soil specimen or "time to failure."

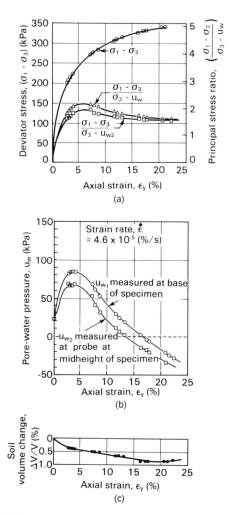

**Figure 9.48** An undrained test on a compacted clay shale specimen at a strain rate of $4.6 \times 10^{-5}$ %/s. (a) Shear stress versus strain curve; (b) pore-water pressure versus strain at two points of measurement; (c) total volume change versus strain (from Bishop et al., 1960).

The strain at failure, $\epsilon_f$, depends on the soil type and the stress history of the soil. Table 9.3 presents typical values of strain at failure, $\epsilon_f$, obtained from numerous triaxial testing programs on unsaturated soils. This information may be of value as a guide when attempting to establish a suitable strain rate, $\dot{\epsilon}$. The strain at failure can be revised as testing progresses.

Gibson and Henkel (1954) used the theory of consolidation in formulating a theoretical method for approximately determining the time required to failure for a specimen under drained conditions. The theory is applicable for both triaxial and direct shear tests. The strain rate used in the triaxial testing of saturated soils has commonly been estimated using this procedure (Bishop and Henkel, 1962). This theory has been extended to embrace unsaturated soils testing. For unsaturated soils, it is necessary to consider impeded flow through the high air entry disk at the base of

9.5 SELECTION OF STRAIN RATE    251

Table 9.3  Strain Rate and Strain at Failure for Triaxial Tests on Unsaturated Soils

| Soil Type | Triaxial Test | Strain Rate, $\dot{\epsilon}$ (%/s) | Approximate Strain at Failure, $\epsilon_f$ (%) | References |
|---|---|---|---|---|
| Boulder clay; $w = 11.6\%$ and % clay = 18% | CW | $3.5 \times 10^{-5}$ | 15 | Bishop et al. 1960 |
| Braehead silt | CW | $4.7 \times 10^{-5}$ | 11 | Bishop and Donald (1961) |
|  | CD | $8.3 \times 10^{-6}$ | 12 |  |
| Talybont boulder clay; $w = 9.75\%$ and % clay = 6% | Undrained with pore pressure measurements | $4.7 \times 10^{-7}$ | $\sigma_3 = 83$ kPa : 8.5  $\sigma_3 = 207$ kPa : 11 | Donald (1963) |
| Dhanauri clay; $w = 22.2\%$ and % clay = 25% | CW | $6.7 \times 10^{-4}$ | 20 | Satija and Gulhati (1979) |
|  | CD | $1.3 \times 10^{-4}$ | 20 |  |
| Undisturbed decomposed granite and rhyolite | CD | $1.7 \times 10^{-5}$ | Stage I: 3–5 | Ho and Fredlund (1982a) |
|  | Multistage | $6.7 \times 10^{-5}$ | Stage II: 1–3  Stage III: 1–3 |  |
| Clayey sand; $w = 14$–$17\%$ and % clay = 30% | Undrained and unconfined | $1.7 \times 10^{-3}$ | 15–20 | Chantawarangul (1983) |

the specimen and the physical properties of the soil (Ho and Fredlund, 1982c). The high air entry disk has a low coefficient of permeability, $k_d$. As a result, the disk not only prevents the passage of air, but also impedes the flow of water in and out of the specimen during shear. Another factor affecting the time to failure, $t_f$, is the low coefficient of permeability of the unsaturated soil (see Chapters 5 and 6).

Analytical considerations of the impeded flow problem from an unsaturated soil specimen can be solved in a manner similar to the problem of impeded flow caused by plugged porous stones above and below saturated soil specimens, as presented by Bishop and Gibson (1963). The solution gives the time required to fail the specimen when using a drained test. For an undrained test, a smaller value of time to failure, $t_f$, can be used. However, the high air entry disk is still controlling the pore pressure equalization. In other words, it is suggested that the time to failure value computed for a drained test can be considered as a conservative estimate of time to failure for an undrained test.

The ability of an unsaturated soil specimen and the measuring system to dissipate the excess pore pressures developed during drained shear are the main considerations in computing the time to failure. The excess pore pressure dissipation is essentially a one-dimensional consolidation process (see Chapter 15). Generally, the coefficient of permeability with respect to the air phase, $k_a$, is much larger than the coefficient of permeability with respect to the water phase, $k_w$. When the air phase is occluded, the air coeffi-

cient of permeability reduces to the diffusivity of air through water (see Chapter 6). It is therefore assumed that the dissipation of the excess pore-water pressure will, in general, require a longer time than the dissipation of the excess pore-air pressure. As a result, the time required to fail the unsaturated soil specimen can be computed by considering the dissipation of excess pore-water pressure. The above assumptions also apply when the air phase is occluded since the soil can then be treated as a saturated soil having a compressible pore-fluid.

The solution to the problem of excess pore-water pressure dissipation with impeded flow is mathematically equivalent to the heat conduction of a slab initially at a uniform temperature, being cooled at its surface in accordance with Newton's law (Carslaw and Jaeger, 1959). The solution can be written as a Fourier series (Gray, 1945). Figure 9.49 shows the dimensions and property variables of an unsaturated soil specimen and high air entry disk, required in the solution to the impeded dissipation of excess pore-water pressures. The high air entry disk is assumed to be incompressible, and its properties appear in the form of an impedance-type factor, $\lambda$:

$$\lambda = \frac{k_d d}{k_w L_d} \quad (9.22)$$

where

$\lambda$ = impedance factor  
$k_d$ = coefficient of permeability of the high air entry disk

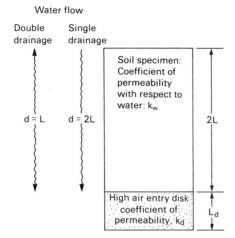

**Figure 9.49** Properties and variables required to model pore pressure equalization for an unsaturated soil specimen and a high air entry disk.

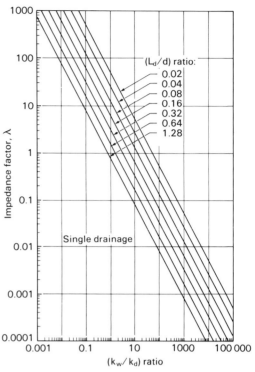

**Figure 9.50** Impedance factors, $\lambda$, for various ratios of $(k_w/k_d)$ and $(L_d/d)$ for single drainage.

$k_w$ = coefficient of permeability of the unsaturated soil with respect to the water phase
$d$ = length of drainage path in the soil
$L_d$ = thickness of the high air entry disk.

For single drainage, where the pore-water drains only to the bottom (or top) of the soil specimen, the length of drainage path, $d$, is equal to the height of the soil specimen, $2L$ (i.e., $d = 2L$; see Fig. 9.49). If the pore-water is allowed to drain from the top and bottom of the specimen (i.e., double drainage), the length of the drainage path becomes half of the specimen height (i.e., $d = L$). In unsaturated soil testing, the water is usually permitted to drain through the high air entry disk at the bottom of the specimen. Equation (9.22) is plotted in Fig. 9.50 in terms of dimensionless ratios of $(k_w/k_d)$ and $(L_d/d)$ for single drainage dissipation (i.e., $d = 2L$). The dimensionless form allows the assessment of the impedance factor, $\lambda$, for a wide range of soils and high air entry disks. The impedance factor, $\lambda$, can be obtained directly from Fig. 9.50 when the properties of the specimen and the high air entry disk are known. The coefficient of permeability with respect to the water phase, $k_w$, may vary depending upon the matric suction of the specimen. The coefficient of permeability of the soil is expressed as a function, $k_w(u_a - u_w)$, and is explained in Chapters 5 and 6. It may be necessary to reduce the rate of strain as matric suction is increased.

The time required to fail the soil specimen, $t_f$, can be expressed in terms of a desired degree of dissipation of excess pore-water pressure (Bishop and Gibson, 1963).

$$t_f = \frac{L^2}{\eta c_v^w (1 - \overline{U}_f)} \quad (9.23)$$

where

$L$ = half of the actual length of the soil specimen, which is the same for single or double drainage

$\eta = 0.75/(1 + 3/\lambda)$ for single drainage
$\eta = 3/(1 + 3/\lambda)$ for double drainage
$c_v^w = k_w/(\rho_w g m_2^w)$; the average coefficient of consolidation with respect to the water phase (see Chapter 15)
$\rho_w$ = density of water
$g$ = gravitational acceleration
$m_2^w$ = slope of the plot of water volume change, $(V_w/V)$, versus matric suction (see Chapters 12 and 13)
$\overline{U}_f$ = average degree of dissipation of the excess pore-water pressure at failure.

The effect of impeded flow on the time to failure, $t_f$, is taken into account through the impedance factor, $\lambda$, in the $\eta$ parameter in Eq. (9.23). The properties of the unsaturated soil specimen are taken into account in the $L$ and $c_v^w$ terms. Approximate properties for the unsaturated soil (i.e., $k_w$ and $m_2^w$) can be used to estimate the average coefficient of consolidation, $c_v^w$. Any desired value for the average degree of dissipation, $\overline{U}_f$, can be used, but a value of 95% is recommended (Gibson and Henkel, 1954, Bishop and Henkel, 1962). An average degree of dissipation of 95% means that 95% of the excess pore-water pressure developed during shear will be dissipated at the time of failure.

The time to failure, $t_f$ [i.e., Eq. (9.23)] can be plotted with respect to the impedance factor, $\lambda$, for various values of the coefficient of consolidation, $c_v^w$, as shown in Figs. 9.51 and 9.52. Both figures are generated for single drain-

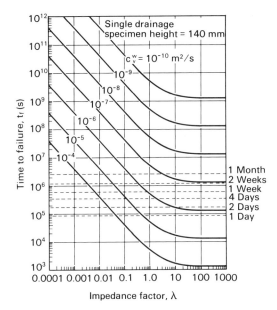

**Figure 9.51** Time to failure, $t_f$, for a drained triaxial test specimen, 140 mm in height.

age triaxial testing. Figures 9.51 and 9.52 correspond to specimen heights of 140 and 76 mm, respectively. These heights are common in triaxial testing. The plots indicate an increase in time to failure, with a decrease in the coefficient of consolidation or a decrease in the impedance factor, $\lambda$. The increase in time to failure, due to a decrease in the coefficient of consolidation, $c_v^w$, is uniform at various $\lambda$ values. However, the increase in time to failure due to a decrease in the impedance factor becomes more significant for impedance factors less than 10. When the impedance factor is greater than 10, the time to failure is almost constant for a specific coefficient of consolidation.

An impedance factor of 10 corresponds to a specific permeability ratio, $(k_w/k_d)$ (Fig. 9.50). As an example, an impedance factor of 10 corresponds to a $(k_w/k_d)$ ratio of 0.7 when the $(L_d/d)$ ratio is 0.16. Impedance values less than 10 would correspond to $(k_w/k_d)$ ratios greater than 0.7 (Fig. 9.50). A comparison between Fig. 9.50 and Figs. 9.51 or 9.52 indicates that the time to failure, $t_f$, is significantly affected by the impeded flow when the impedance factor is less than 10. In other words, the impeded flow problem exists even when the high air entry disk has a coefficient of permeability equal to the coefficient of permeability of the soil specimen. In many cases, the coefficient of permeability of the disk is lower than that of the soil. On the other hand, the effect of impeded flow on the time to failure, $t_f$, remains essentially unchanged when the impedance factor is greater than 10. This means that little improvement can be made on the time to failure, $t_f$, when the impedance factor has reached a value greater than 10.

It is evident from the above discussion that impeded flow due to the presence of the high air entry disk is often the governing factor influencing the time to failure. One way to reduce the time to failure is by using a thin high air entry disk. High air entry disks generally range in thickness from 0.32 to 0.95 cm. For a specific $(k_w/k_d)$ ratio and a specific specimen height, a decrease in the thickness of the high air entry disk causes an increase in the impedance factor (Fig. 9.50). This, in turn, reduces the time to failure, $t_f$ (Figs. 9.51 or 9.52). However, thicker disks are superior due to their ability to greatly reduce the diffusion rate of air through the disk.

The strain rate, $\dot{\epsilon}_f$, for triaxial testing of unsaturated soils can be computed in accordance with Eq. (9.21). The computed strain rate is only an estimation because of the assumptions involved in the theory and the difficulties in accurately assessing relevant soil properties. Nevertheless, the theory does provide a somewhat rational approach to obtaining the strain rate for testing. Typical strain rates that have been used for triaxial tests on unsaturated soils are tabulated in Table 9.3. The strain rate for testing should be relatively low, even when the soil appears to be quite pervious, since the strain rate is largely controlled by the high air entry disk. Equation (9.21) can also be used to obtain a conservative estimate of strain rate for undrained shear strength testing.

It appears that water can more readily flow out from a soil specimen through the high air entry disk than into the specimen through the high air entry disk. If a soil tends to dilate during shear, it may be difficult to ensure upward water flow through the low-permeability disk into the soil specimen. This problem can be somewhat alleviated by using the axis-translation technique and periodically flushing diffused air from below the high air entry disk. The axis translation ensures a positive water pressure below the high air entry disk.

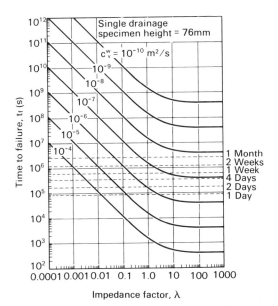

**Figure 9.52** Time to failure, $t_f$, for a drained triaxial test specimen, 76 mm in height.

**Table 9.4 Horizontal Displacement Rate and Horizontal Displacements at Failure for Several Direct Shear Tests on Unsaturated Soils**

| | Direct Shear Test | Displacement Rate, $\dot{d}_h$ (mm/s) | Displacement at Failure, $d_h$ (mm)[a] | References |
|---|---|---|---|---|
| Madrid grey clay | CD | $1.4 \times 10^{-4}$ | 3.5–5 | Escario (1980) |
| Madrid grey clay | CD | $2.8 \times 10^{-5}$ | 6.0–7.2 | Escario and Sáez (1986) |
| Red clay of Guadalix de la Sierra | CD | $2.8 \times 10^{-5}$ | 4.8–7.2 | Escario and Sáez (1986) |
| Madrid clayey sand | CD | $2.8 \times 10^{-5}$ | 2.4–4.8 | Escario and Sáez (1986) |
| Glacial till | CD (multistage) | $1.7 \times 10^{-4}$ | 1.2 | Gan (1986) |

[a] Square specimen of 50 × 50 mm.

### 9.5.3 Displacement Rate for Direct Shear Tests

The rate of horizontal shear displacement in a direct shear test is analogous to the strain rate in a triaxial test. The horizontal shear displacement rate can be defined as the relative rate at which the top and the bottom halves of the direct shear box are displaced:

$$\dot{d}_h = \frac{d_h}{t_f} \quad (9.24)$$

where

$\dot{d}$ = horizontal shear displacement rate for a direct shear test

$d_h$ = horizontal displacement of the soil specimen at failure.

The horizontal displacement at failure, $d_h$, depends on the soil type and the soil stress history. Typical horizontal displacements at failure, obtained from several testing programs, are shown in Table 9.4. An estimate of the horizontal displacement at failure is required in order to compute the horizontal shear displacement rate, $\dot{d}_h$.

The theoretical estimation for the time to failure, $t_f$, used in triaxial testing [i.e., Eqs. (9.22) and (9.23)] is also applicable to drained direct shear tests. The excess pore pressures developed during a drained direct shear test are also dissipated one-dimensionally in the vertical direction. The pore–water is commonly drained out of the bottom of the specimen through the high air entry disk. The impedance factor for direct shear tests can be computed using Eq. (9.22) (Fig. 9.50). The time to failure, $t_f$, for direct shear tests can be computed using Eq. (9.23) (Fig. 9.53). Figure 9.53 is a plot of the time to failure for various impedance factors and coefficients of consolidation. The plot is generated for a specimen height of 12.7 mm.

The horizontal shear displacement rate, $\dot{d}_h$, can be computed from Eq. (9.24) using estimated values of $d_h$ and $t_f$. Typical horizontal shear displacement rates that have been used in direct shear testing on unsaturated soils are listed in Table 9.4.

A comparison of Figs. 9.51, 9.52, and 9.53 reveals that the time required to fail the soil specimen is substantially reduced by using a thinner soil specimen. This can be observed from Eq. (9.23) where the time to failure is a function of the square of half the specimen height, $h$. The direct shear test has the advantage that a thin specimen can assist in expediting testing.

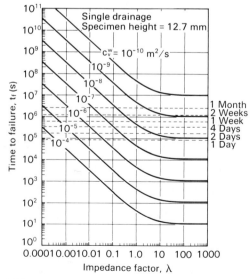

**Figure 9.53** Time to failure, $t_f$, for drained direct shear tests with a specimen height of 12.7 mm.

## 9.6 MULTISTAGE TESTING

Multistage testing can be performed using triaxial or direct shear equipment, and it consists of altering the stress path during the testing of a single soil specimen. Multistage testing is used to maximize the amount of shear strength information that can be obtained from one specimen. It also assists in eliminating the effect of soil variability on the results. Multistage triaxial testing of saturated soils usually consists of using several effective confining pressures, $(\sigma_3 - u_w)$, when testing one soil specimen. A Mohr–Coulomb failure envelope (Fig. 9.1) can be drawn tangent to the Mohr circles, corresponding to failure at the various effective confining pressures.

For unsaturated soils, multistage triaxial testing has been performed in conjunction with a consolidated drained type of test. However, the multistage triaxial testing is also applicable when using the constant water content or the consolidated undrained test procedure. For multistage, consolidated drained triaxial tests, the net confining pressure, $(\sigma_3 - u_a)$, is usually maintained constant, while the matric suction, $(u_a - u_w)$, is varied from one stage to another. Figure 9.54(a) illustrates a multistage triaxial test with three stages. Each stage corresponds to a different matric suction. A multistage consolidated drained triaxial test can also be performed by maintaining the matric suction constant and varying the net confining pressure.

Each stage of the multistage test should be terminated when failure is imminent. This can generally be determined by observing when the deviator stress tends to a maximum value. The soil specimen should not be subjected to excessive deformation, particularly during the early stages of loading. Difficulties related to excessive deformation in multistage triaxial testing have been reported by several researchers (Kenney and Watson, 1961; Wong, 1978; Ho and Fredlund, 1982b). When the specimen undergoes excessive deformation during the early stage of loading, the specimen will tend to develop distinct shear planes, and the strength may be reduced from its peak strength. The shear strength measured at successive stages may tend towards an ultimate or residual strength condition. The ultimate or residual shear strength is obtained when the deviator stress has levelled off after reaching its peak or maximum value. Excessive strain accumulation can be a problem in multistage testing, particularly for soils whose structure is sensitive to disturbance.

Two types of loading procedures were proposed for multistage testing by Ho and Fredlund (1982b). These were called the cyclic and the sustained loading procedures, and are shown in Fig. 9.54(a) and (b), respectively. For the cyclic loading procedure, the deviator stress, $(\sigma_1 - \sigma_3)$, is released to zero once a maximum value is reached. Each stage of loading commences from a condition of zero deviator stress [Fig. 9.54(a)]. For the sustained loading procedure, the deviator stress is sustained at its maximum value while the next stage of loading is applied. In other words, the maximum deviator stress is maintained while a new stress state is applied to the specimen [Fig. 9.54(b)]. The cyclic loading procedure appears to have an advantage over the sustained loading procedure in that it reduces the accumulated strain. The release of the deviator stress between stages in the cyclic loading procedure appears to minimize the accumulation of strain experienced during the sustained loading procedure. Part of the accumulated strain appears to be restored through elastic recovery when releasing the deviator stress between stages.

Typical stress versus displacement curves for a multistage direct shear test, using the cyclic loading procedure, are shown in Fig. 9.55.

## 9.7 NONLINEARITY OF FAILURE ENVELOPE

The shear strength theory for an unsaturated soil has been presented in Section 9.3 using an extended Mohr–Coulomb failure envelope. Recent shear strength test results on unsaturated soils indicate some nonlinearity in the shear strength versus matric suction failure envelope (Gan, 1986; Escario and Sáez, 1986). Figure 9.56 illustrates a typical nonlinear matric suction failure envelope. The $\phi^b$ angle ap-

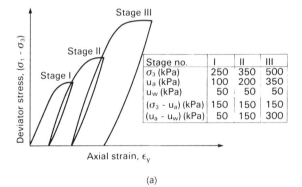

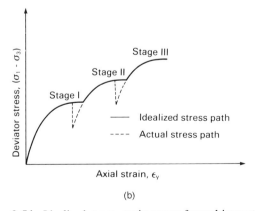

**Figure 9.54** Idealized stress–strain curves for multistage triaxial testing. (a) Cyclic loading procedure; (b) substained loading procedure.

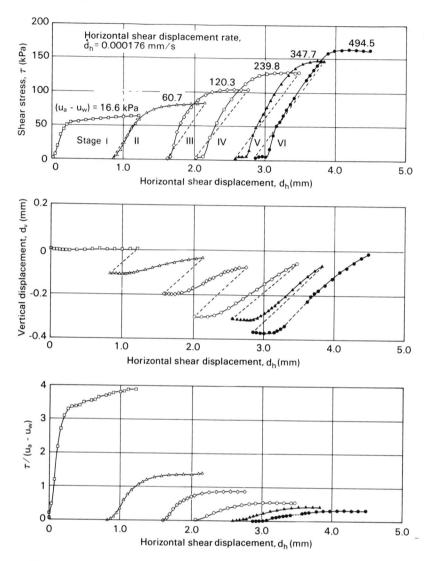

**Figure 9.55** Multistage direct shear test on a compacted glacial till specimen GT-16-N4 (from Gan, 1986).

pears to be equal to $\phi'$ at low matric suctions, and decreases to a lower value at high matric suctions. It appears that the $\phi^b$ angle is a function of matric suction, $(u_a - u_w)$. The variation in the $\phi^b$ angle with respect to matric suction can better be understood by considering the pore volume on which the pore-water pressure acts. Let us consider a direct shear test performed using the axis-translation technique for a wide range of matric suctions. Initially, the soil specimen is saturated and consolidated under a vertical normal stress, $\sigma$. The initial matric suction is maintained at zero by applying an external air pressure equal to the pore-water pressure.

The matric suction of the soil is equal to the difference between the gauge external air pressure and the gauge pore-water pressure. At the start of the test, the matric suction is the pore-water pressure referenced to the external air pressure. The total normal stress is referenced to the external air pressure, $(\sigma - u_a)$. Assuming that the $\phi'$ and $c'$ parameters are known, the soil shear strength under the initial conditions can be predicted from Eq. (9.3) as being equal to $(c' + (\sigma - u_a) \tan \phi')$. The initial condition is represented as point $A$ in Fig. 9.56.

Let us now apply a positive matric suction while keeping the net normal stress, $(\sigma - u_a)$, constant. At low matric suctions, the soil specimen remains saturated. When the soil remains saturated, the effects of pore-water pressure and total normal stress on the shear strength are characterized by the friction angle, $\phi'$. Both the pore-water pressure and the total normal stress are referenced to the same external air pressure, as expressed by the $(u_a - u_w)$ and $(\sigma - u_a)$ stress state variables, respectively. As a result, an increase in matric suction causes an increase in shear strength in accordance with the $\phi'$ angle. The shear stress versus matric suction envelope has a slope angle $\phi^b$ equal to $\phi'$.

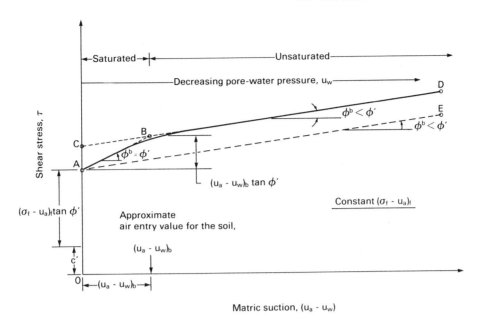

**Figure 9.56** Nonlinearity of the failure envelope on the $\tau$ versus $(u_a - u_w)$ plane.

This condition is maintained as long as the soil is saturated. This condition is indicated by line $AB$ in Fig. 9.56. Substituting $\phi'$ for $\phi^b$ into Eq. (9.3) results in the same shear strength equation as used for saturated soils [i.e., Eq. (9.1)]. Therefore, the shear strength equation for saturated soils, Eq. (9.1), is applicable when the pore–water pressures are negative, provided the soil remains saturated (Aitchison and Donald, 1956; Bishop, 1961b).

The pore–water can fill the entire volume of the pores of a soil, even when its pressure is negative. However, as matric suction is increased, water will be drawn out from the pores of the soil. When the air entry value of the soil specimen, $(u_a - u_w)_b$, is reached, air enters into the pores, resulting in a desaturation of the soil. For the unsaturated condition, the pore–water pressure is referenced to the pore–air pressure [i.e., $(u_a - u_w)$], which is now both external and internal to the soil. The total normal stress is similarly referenced to the pore–air pressure through the $(\sigma - u_a)$ stress state variable. The pore–water occupies only a portion of the soil pores, as indicated by a degree of saturation which is less than 100%. As a result, a further increase in matric suction proves to not be as effective as an increase in net normal stress in increasing the shear strength of the soil. Figure 9.56 indicates a decrease in the $\phi^b$ angle to a value lower than $\phi'$ as matric suction is increased beyond point $B$. The matric suction corresponding to point $B$ appears to be correlative with the air entry value of the soil specimen, $(u_a - u_w)_b$. Therefore, the effects of normal stress and pore–water pressure on the shear strength of an unsaturated soil are best independently evaluated in terms of the $(\sigma - u_a)$ and $(u_a - u_w)$ stress state variables, as shown in Eq. (9.3).

The air entry value of a soil largely depends on the grain size distribution of the soil. Coarse granular soils, such as sands, desaturate at lower matric suctions than fine-grained clayey soils (see Chapter 4). As a results, there is a relationship between the air entry value of the soil and the pore sizes in the soil. This relationship is typified by the water characteristic curve for a soil. A sand often has an air entry value less than 101.3 kPa, whereas a clayey soil can have an air entry value well beyond 101.3 kPa. The air entry value of a soil may also depend, to some extent, upon the net confining pressure of the soil. The air entry value may be used as an approximation of the breaking point of the nonlinear matric suction failure envelope (i.e., point $B$ in Fig. 9.56). Experimental results to date indicate a fairly constant value for the $\phi^b$ angle beyond the breaking point $B$ as shown in Fig. 9.56. Some results do not appear to exhibit any nonlinearity in the failure envelope with respect to matric suction.

The nonlinearity of the failure envelope illustrated in Fig. 9.56 can be handled in one of several ways for analysis purposes (Fredlund *et al.* 1987). Often, the failure envelope can be divided into two linear portions. As an example, the failure envelope shown in Fig. 9.56 can be divided into two linear sections, as represented by lines $\overline{AB}$ and $\overline{BD}$. When the matric suction is less than the matric suction at point $B$, $(u_a - u_w)_b$, the failure envelope will have a slope angle of $\phi'$. Line $\overline{AB}$ intersects the shear stress ordinate at point $A$. When the matric suction in the soil is greater than $(u_a - u_w)_b$, line $\overline{BD}$ should be used to represent the failure envelope. The $\phi^b$ angle will be less than $\phi'$. Line $\overline{BD}$ intersects the shear stress ordinate at point $C$.

An alternative procedure to simplifying the nonlinearity of the envelope involves the use of one linear envelope with a slope angle $\phi^b$ less than $\phi'$, starting from zero ma-

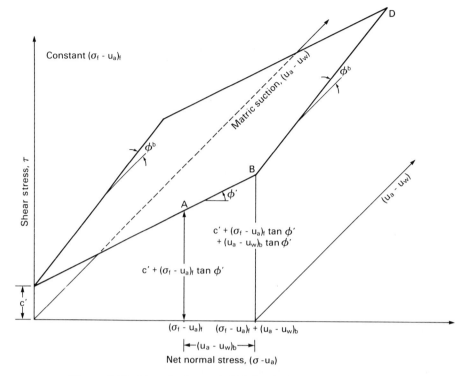

**Figure 9.57** Linearized extended Mohr–Coulomb failure envelope.

tric suction, as illustrated by line $\overline{AE}$. In this case, the shear strength predicted from the failure envelope $\overline{AE}$ will be conservative.

The failure envelope can also be discretized into several linear segments with varying $\phi^b$ angles and different ordinate intercepts. This would be of interest when the failure envelope is highly nonlinear. Each segment would correspond to a specific range of matric suction.

The nonlinear failure envelope shown in Fig. 9.56 can also be modified to a linear failure envelope using the method illustrated in Fig. 9.57. As pointed out previously, the matric suction along line $\overline{AB}$ in Fig. 9.56 produces the same increase in shear strength as the net normal stress. In other words, the effect of their changes can be characterized by using the same friction angle, $\phi'$. Therefore, line $\overline{AB}$ can be drawn on the shear strength versus $(\sigma - u_a)$ plane instead of on the shear strength versus $(u_a - u_w)$ plane. As the matric suction increases beyond the air entry value of the soil, $(u_a - u_w)_b$, the soil begins to desaturate. In this case, a third axis, $(u_a - u_w)$, can be used to separate the different effects of normal stress and matric suction on the shear strength. This causes line $\overline{BD}$ in Fig. 9.56 to be translated onto the shear strength versus matric suction plane, as shown in Fig. 9.58. In other words, the curved failure envelope may be linearized if the matric suction axis is drawn starting from the air entry value, $(u_a - u_w)_b$ (Fig. 9.57), as opposed to using zero matric suction. The air entry value can generally be assumed to remain constant for various net confining pressures.

## 9.8 RELATIONSHIPS BETWEEN $\varphi^b$ AND $\chi$

Bishop [39] proposed a shear strength equation for unsaturated soils which had the following form:

$$\tau_{ff} = c' + \{(\sigma_f - u_a)_f + \chi(u_a - u_w)_f\} \tan \phi' \quad (9.25)$$

where

$\chi$ = a parameter related to the degree of saturation of the soil.

Let us assume that the shear strength computed using Eq. (9.3) can be made to be equal to the shear strength given by Eq. (9.25). Then it is possible to illustrate the relation-

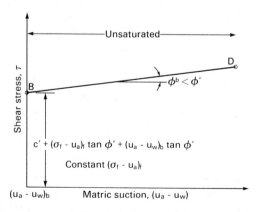

**Figure 9.58** Linearized failure envelope on the shear stress versus matric suction plane.

ship between $(\tan \phi^b)$ and $\chi$:

$$(u_a - u_w)_f \tan \phi^b = \chi(u_a - u_w)_f \tan \phi'. \quad (9.26)$$

It is then possible to solve for the parameter, $\chi$:

$$\chi = \frac{\tan \phi^b}{\tan \phi'}. \quad (9.27)$$

A graphical comparison between the $\phi^b$ representation of strength [i.e., Eq. (9.3)] and the $\chi$ representation of strength [i.e., Eq. (9.25)] is shown in Fig. 9.59. Using the $\phi^b$ method, the increase in shear strength due to matric suction is represented as an upward translation from the saturated failure envelope. The magnitude of the upward translation is equal to $\{(u_a - u_w) \tan \phi^b\}_f$ (i.e., point $A$ in Fig. 9.59). In this case, the failure envelope for the unsaturated soil is viewed as a third-dimension extension of the failure envelope for the saturated soil. On the other hand, the $\chi$ parameter method uses the same failure envelope for saturated and unsaturated conditions. Matric suction is assumed to produce an increase in the net normal stress. This increase is a fraction of the matric suction at failure (i.e., $\chi(u_a - u_w)_f$). The shear strength at point $A$ using the $\phi^b$ method is equivalent to the shear strength at point $A'$ in the $\chi$ parameter method, as depicted in Fig. 9.59.

Theoretically, only one $\chi$ value is obtained from Eq. (9.27) for a particular soil when the failure envelope is planar. A planar failure envelope uses one value of $\phi'$ and one value of $\phi^b$. If the failure envelope is bilinear with respect to the $\phi^b$, there will be two values for $\chi$. A $\chi$ value equal to 1.0 corresponds to the condition where $\phi^b$ is equal to $\phi'$. A $\chi$ value less than 1.0 corresponds to the condition when $\phi^b$ is less than $\phi'$. For envelopes which are highly curved with respect to matric suction, there will be various $\chi$ values corresponding to different matric suctions.

The $\chi$ parameter has commonly been correlated with the

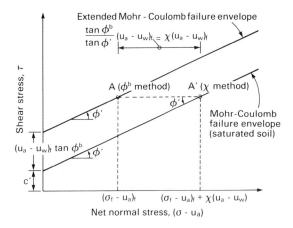

**Figure 9.59** Comparison of the $\phi^b$ and $\chi$ methods of designating shear strength.

degree of saturation of the soil. Unfortunately, the $\chi$ value has sometimes been obtained from shear strength tests on soil specimens compacted at different water contents. The different initial compacted water contents may have been used to give varying initial matric suctions. However, the soil specimens compacted at different water contents do not represent an "identical" soil. As a result, the $\chi$ values obtained from specimens compacted at different water contents are essentially obtained from different soils. The $\phi^b$ and $\chi$ relationship in Eq. (9.27) applies only to initially "identical" soils. These may be soils compacted at the same water content to the same dry density. The $\chi$ parameter need not be unique for both shear strength and volume change problems. The $\chi$ relationship given in Eq. (9.27) is only applicable to the evaluation of the shear strength of the soil. From a practical engineering standpoint, it would appear to be better to use the $(\sigma - u_a)$ and $(u_a - u_w)$ stress state variables in an independent manner for designating the shear strength of an unsaturated soil.

# CHAPTER 10

# *Measurement of Shear Strength Parameters*

The extended Mohr–Coulomb shear strength envelope for unsaturated soils requires that three shear strength parameters be defined, namely, $c'$, $\phi'$, and $\phi^b$, as explained in Chapter 9. These parameters can be measured in the laboratory.

Conventional triaxial and direct shear apparatuses require modifications prior to their use for testing unsaturated soils. The required modifications are outlined in the first part of this chapter. The $c'$ and $\phi'$ parameters can be measured using saturated soil specimens. The equipment and techniques used for testing saturated soils are well documented by Bishop and Henkel (1962) and Head (1986), and will not be repeated in this book. Only a brief section on backpressure is included as it relates to the behavior of soils with a pore fluid consisting of an air–water mixture. The tests on unsaturated specimens are performed to obtain the $\phi^b$ soil parameter.

The second part of the chapter provides general procedures for testing unsaturated soils using the modified triaxial and direct shear equipment. Computations and necessary corrections for analyzing the test results are similar to those required for saturated soils. Typical test results and their interpretation are given in the last section of the chapter.

## 10.1 SPECIAL-DESIGN CONSIDERATIONS

Conventional triaxial and direct shear equipment require modifications prior to their use for testing unsaturated soils. Several factors related to the nature of an unsaturated soil must be considered in modifying the equipment. The presence of air and water in the pores of the soil causes the testing procedures and techniques to be more complex than those required when testing saturated soils. The modifications must accommodate the independent measurement or control of the pore-air and pore-water pressures. In addition, the pore-water pressure is usually negative (gauge), and can result in water cavitation problems in the measuring system.

The theoretical concepts associated with triaxial testing on unsaturated soils are given in Chapter 9. In general, the various triaxial test methods are defined on the basis of the drainage conditions during the application of the confining pressure, $\sigma_3$, and the drainage conditions during the application of the deviator stress, $(\sigma_1 - \sigma_3)$. For undrained testing conditions, the pore fluid (i.e., pore-air or pore-water) is not allowed to drain. The excess pore-air and pore-water pressures developed during undrained loading conditions must be independently measured.

The pore-air and pore-water pressures are allowed to drain to prescribed pressures for drained testing. The volume changes associated with the pore fluid flow during a drained test can be measured. Conventional triaxial or direct shear equipment requires modification for the independent control of pore-air and pore-water pressures.

### 10.1.1 Axis-Translation Technique

The axis-translation technique forms the basis for the laboratory testing of unsaturated soils with high matric suctions. The pore-water pressure, $u_w$, in an unsaturated soil can be measured or controlled directly using a ceramic disk with fine pores (i.e., a high air entry disk).

Figure 10.1 illustrates the direct measurement of pore-water pressure relative to atmospheric air pressure conditions (i.e., $u_a = 0$). The soil specimen is placed on top of a saturated porous disk, and a U-tube is connected to the water compartment below the disk. The U-tube is moved up and down until there is no tendency for flow in or out of the specimen. However, this direct measurement has limitations.

The porous disk in contact with the soil must have an air entry value greater than the soil matric suction, as illustrated in Fig. 10.1(a). The air entry value of a porous disk is the matric suction at which air commences passing through the saturated porous disk. At this point, air enters the water compartment, which no longer maintains continuity between the pore-water and the water in the measuring system. The measuring system then becomes filled with air bubbles.

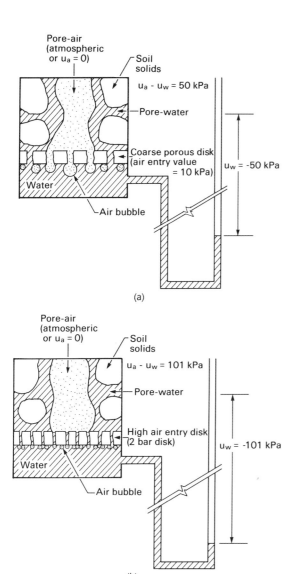

**Figure 10.1** Direct measurement of pore-water pressure in an unsaturated soil specimen (not to scale). (a) Air movement through the porous disk when its air entry value is exceeded; (b) air diffusion through the high air entry disk and water cavitation in the measuring system.

Another limitation of the direct measurement arises from the fact that water cavitates as a gauge pressure of $-1$ atm is approached (i.e., $-101$ kPa). As an example, consider an unsaturated soil specimen with a matric suction of 101 kPa. The pore-water pressure in the specimen is being measured relative to atmospheric air pressure (i.e., $u_a = 0$), as shown in Fig. 10.1(b). The magnitude of the gauge pore-water pressure that is to be measured is $-101$ kPa. A saturated porous disk with an air entry value of 202 kPa is suitable since it is higher than the soil matric suction (i.e., 101 kPa). Nevertheless, the water in the compartment below the disk starts to cavitate when the gauge water pressure approaches $-1$ atm. As a result, occluded air bubbles accumulate below the disk in the water compartment. This causes an error in the pore-water pressure measurement. The direct measurement of pore-water pressure is limited to a value greater than $-1$ atm. This is true regardless of the air entry value of the porous disk. This lower limit of $-101$ kPa always applies when attempting to measure the pore-water pressure.

The limitations of direct control or measurement of pore-water pressure in an unsaturated soil can be overcome by applying an axis-translation to the specimen. The axis-translation technique is particularly useful when testing an unsaturated soil with a high matric suction. The concepts behind the axis translation are explained in Chapter 3. Basically, both the pore-air and pore-water pressures are translated into the positive pressure range. The translation of pore-air pressure can be considered as an artificial increase in the atmospheric pressure under which the test is performed (Bishop *et al.* 1960). As a result, the negative gauge pore-water pressure is also raised by an equal amount to a positive gauge pressure. In this way, the matric suction of the soil specimen remains constant regardless of the magnitude of the pore-air pressure. A fine, porous disk with an air entry value greater than that of the matric suction of the soil should be used in order to prevent air from entering the water compartment.

The application of the axis-translation technique to a soil specimen with a matric suction of 101 kPa is illustrated in Fig. 10.2. The pore-water pressure is measured below a saturated porous disk which has an air entry value of 202 kPa. An air pressure of 202 kPa is applied directly to the specimen in order to raise the pore-air (and pore-water) pressures. In this example, the pore-air pressure, $u_a$, is raised to 202 kPa, which in turn increases the pore-water pressure by an equal amount. As a result, the pore-water pressure, $u_w$, is now at a positive value of 101 kPa, and there is no problem associated with the cavitation of the water in the measuring system. The pore-water pressure can now be measured since there is continuity between the pore-water and the water in the measuring system. The matric suction of the soil [Fig. 10.2(a)] remains constant at 101 kPa throughout the axis-translation procedure. However, the pore-air pressure which is used as a reference pressure has been translated from 0 to 202 kPa.

The use of the axis-translation technique over a long period of testing does not guarantee that the water pressure measuring system (i.e., compartment) will remain free of air bubbles. Unsaturated soil testing often requires an extended period of time due to the low coefficient of permeability of the soil. As the test progresses, pore-air diffuses through the water in the high air entry disk and appears as air bubbles beneath the disk (see Chapters 2 and 6). A diffused air volume indicator (DAVI) can be used in conjunction with the measuring system in order to flush and measure the volume of diffused air (Chapter 6). The diffused air volume must be used to correct the water volume change

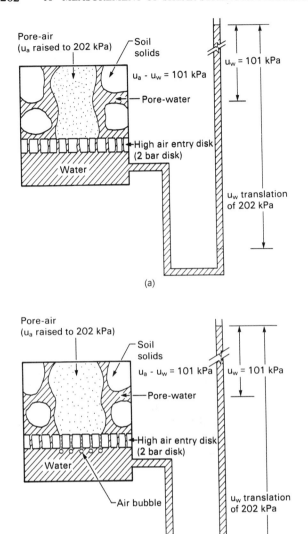

**Figure 10.2** Measurements of pore-water pressure in an unsaturated soil specimen using the axis-translation technique (not to scale). (a) Axis-translation of 101 kPa; (b) air diffusion through the high air entry disk.

measurements under drained testing conditions. Under undrained testing conditions, it is not possible to correct for pressure changes resulting from air diffusion.

The use of the axis-translation technique in the shear strength measurement of unsaturated soils was advocated by Bishop and Blight (1963). The results of an "unconfined triaxial" test on a compacted Selset clay specimen, using this technique, are presented in Fig. 10.3. The unconfined triaxial test was a special case of a constant water content test where the net confining pressure, $(\sigma_3 - u_a)$, was maintained at zero pressure. The test was performed by applying an all-around air pressure, $\sigma_3$, equal to the pore-air pressure, $u_a$. However, the all-around air pressure

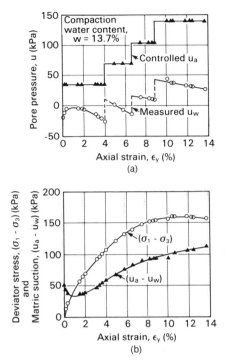

**Figure 10.3** "Unconfined triaxial" test on a compacted Selset clay specimen at various pore-air pressures. (a) Pore pressures versus axial strain curve; (b) deviator stress and matric suction versus strain curves (from Bishop and Blight, 1963.)

was varied using a series of steps, as shown in Fig. 10.3(a). The air pressure increase in each step was applied gradually in order for the pore-air pressure to equalize throughout the specimen. The pore-water pressure was measured throughout the test, and the measurements indicate that the pore-water pressure also increased in a stepwise manner [Fig. 10.3(a)]. The results indicate monotonic curves of deviator stress and matric suction versus axial strain regardless of the stepwise axis-translation during the test [Fig. 10.3(b)]. The shear strength testing does not appear to be affected by the translation of the reference axis.

The effect of axis translation on the measured shear strength of a soil is also illustrated in Fig. 10.4. Two unconfined triaxial tests were conducted on a compacted Talybont clay by Bishop and Blight (1963). The first test was performed using an axis translation such that the gauge pore-air pressure was elevated to 483 kPa and the pore-water pressure was measured. The pore-water pressure was approximately 69 kPa, which indicated a matric suction of approximately 414 kPa during the test [Fig. 10.4(b)]. The second test was performed under atmospheric air pressure conditions (i.e., $u_a = 0$). The stress versus strain curves obtained from both tests are essentially identical up to the maximum deviator stress and differ slightly subsequent to the peak strength. In other words, the measured shear strength was not affected by the axis-translation procedure. The results were also intended to illustrate that the pore-

10.1 SPECIAL DESIGN CONSIDERATIONS    263

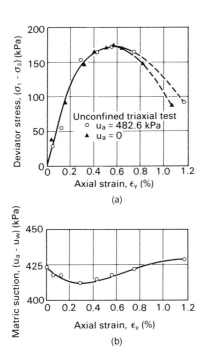

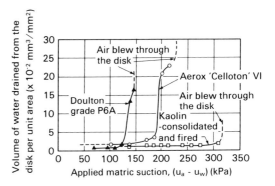

**Figure 10.5** Air passage characteristics of three high air entry disks (from Bishop and Henkel, 1962).

**Figure 10.4** "Unconfined triaxial" tests on a compacted Talybont clay with and without axis translation. (a) Deviator stress versus axial strain curve; (b) matric suction versus axial strain during the test with axis translation (from Bishop and Blight, 1963.)

water in the soil was able to withstand high tensions without rupturing even though the limit is $-100$ kPa for the pore-water pressure measuring system. Since the soil matric suction in both tests was about 414 kPa, the pore-water pressure in the second test would be $-414$ kPa gauge pressure or $-313$ kPa absolute pressure.

### 10.1.2 Pore-Water Pressure Control or Measurement

The pore-water pressure is controlled at a prescribed value when the water phase is maintained under drained conditions during a shear strength test (e.g., consolidated drained test; see Chapter 9). The measurement of pore-water pressure is commonly performed when the water phase is undrained throughout the test (e.g., constant water content or in a consolidated undrained test; see Chapter 9). The key element for both the control and measurement of pore-water pressure is the high air entry ceramic disk. The high air entry disk acts as a semi-permeable membrane that separates the air and water phases. The separation of the air and water phases can be properly achieved only when the air entry value of the disk is greater than the matric suction of the soil. The air entry value of the disk depends on the maximum pore size in the disk. The maximum pore size can be computed using Kelvin's equation (Fig. 10.5):

$$(u_a - u_w)_d = \frac{2T_s}{R_s} \qquad (10.1)$$

where

$(u_a - u_w)_d =$ air entry value of the high air entry disk
$T_s =$ surface tension of the contractile skin or the air-water interface (e.g., $T_s = 72.75$ mN/m at 20°C)
$R_s =$ radius of curvature of the contractile skin or the pore radius.

The air entry value in Eq. (10.1) refers to the maximum matric suction to which the high air entry disk can be subjected before free air passes through the disk. The maximum matric suction is associated with the minimum radius of curvature, $R_s$, which in this case is equal to the radius of the largest pore in the disk. Therefore, the air entry value of a porous ceramic disk is controlled by the pore sizes in the disk. This, in turn, is controlled by the preparation and sintering process used to manufacture the ceramic disk. The smaller the pore size in a disk, the greater will be its air entry value (see Fig. 10.1). Several types of high air entry disks used for unsaturated soils research at Imperial College, London, are listed in Table 10.1, together with their corresponding air entry values. Figure 10.5 shows the air passage characteristics of three of the disks given in Table 10.1. The plots indicate the matric suction value at which air starts to pass through the disk. This is the air entry value of the disk.

The properties of high air entry disks manufactured by Soilmoisture Equipment Corporation in Santa Barbara, CA, are tabulated in Table 10.2. The disks are named according to their air entry values, expressed in the unit of bars. One bar is equal to 100 kPa. The water coefficient of permeability of a disk can be measured by mounting the disk in a triaxial apparatus and placing water above the disk. The disk must be sealed on the sides in order to prevent the passage of water around its circumference. An air pressure can then be applied to the water, producing a gradient across the high air entry disk. The volume of water flowing through the disk can be measured using a water volume change indicator. Details on the equipment are presented later in this chapter. The flow of water is plotted against

Table 10.1 High Air Entry Disks at Imperial College (from Blight, 1961)

| Type | Porosity, $n$ (%) | Coefficient of Permeability, $k_d$ (m/s) | Air Entry Value, $(u_a - u_w)_d$ (kPa) |
|---|---|---|---|
| Doulton Grade P6A | 23 | $2.1 \times 10^{-9}$ | 152 |
| Aerox "Celloton" Grade VI | 46 | $2.9 \times 10^{-8}$ | 214 |
| Kaolin-consolidated from a slurry and fired | 45 | $6.2 \times 10^{-10}$ | 317 |
| Kaolin-dust pressed and fired | 39 | $4.5 \times 10^{-10}$ | 524 |

elapsed time, as shown in Fig. 10.6. The plot shows a straight line, indicating steady-state water seepage through the ceramic disk. The volume of water divided by the cross-sectional area of the disk and the elapsed time gives the coefficient of permeability of the disk. In general, the coefficient of permeability of the disk decreases with an increasing air entry value.

The selection of a high air entry disk for testing an unsaturated soil should be primarily based upon the maximum possible matric suction that can occur during the test. First, the initial matric suction of the specimen should be taken into consideration (e.g., for a constant water content or an undrained test). The initial value of matric suction could be measured or estimated (Chapter 4). Second, the desired range of matric suction that will be imposed on the soil specimen during a consolidated drained test should not exceed the selected air entry value. Third, the likely matric suction increase in the specimen during an undrained test should be considered when selecting the high air entry disk.

The water coefficient of permeability and the thickness of a high air entry disk affect the time required for the pore-water pressure to equalize across high air entry disk. This information is of importance when assessing the time for consolidation prior to shear and the time required for shearing the specimen (Chapter 9). It is also desirable to have the highest possible coefficient of permeability (i.e., the largest possible pore size) for the high air entry disk. This will ensure the most rapid equilibration of the pore-water pressure in an undrained test. It will also minimize impeded drainage of the pore-water in a drained test.

The use of a thin air entry disk also reduces the time required for equilibration throughout the specimen. However, a thin disk means that there is a short path for the diffused air to reach the water compartment below the disk. In other words, the use of a thin disk results in a greater accumulation of diffused air in the water compartment. A thin disk may also crack easily, particularly if care is not taken to ensure that the total stress and the pore-air pressure are applied to the specimen before pressurizing the water compartment beneath the disk. The removal of all pressures from the top of the disk while a water pressure is applied below the disk can also produce an upward bending moment that will crack the ceramic disk. Experience has shown that an excess uplift pressure greater than about

Table 10.2 High Air Entry Disks Manufactured by Soilmoisture Equipment Corporation

| Type | Approximate Pore Diameter ($\times 10^{-3}$ mm) | Coefficient of Permeability, $k_d$ (m/s) | Air Entry Value Range, $(u_a - u_w)_d$ (kPa) |
|---|---|---|---|
| 1/2 bar (high flow) | 6.0 | $3.11 \times 10^{-7}$ | 48–62 |
| 1 bar | 2.1 | $3.46 \times 10^{-9}$ | 138–207 |
| 1 bar (high flow) | 2.5 | $8.6 \times 10^{-8}$ | 131–193 |
| 2 bar | 1.2 | $1.73 \times 10^{-9}$ | 241–310 |
| 3 bar | 0.8 | $1.73 \times 10^{-9}$ | 317–483 |
| 5 bar | 0.5 | $1.21 \times 10^{-9}$ | >550 |
| 15 bar | 0.16 | $2.59 \times 10^{-11}$ | >1520 |

10.1 SPECIAL DESIGN CONSIDERATIONS 265

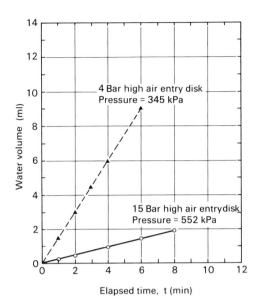

**Figure 10.6** Steady-state seepage of water through a high air entry disk.

70 kPa on a 3.2 mm thick ceramic disk (63.5 mm in diameter) can produce cracks in the disk (Fredlund, 1973). The cracks may not be visible to the naked eye, but are evident during an air entry or permeability test on the disk. A 6.4 mm thick disk can withstand higher uplift pressures, but should be checked for cracks whenever subjected to an excess pressure greater than about 70 kPa.

The installation of a high air entry disk into the base plate of a triaxial cell is illustrated in Fig. 10.7. A conventional triaxial base plate requires modification in order to accommodate the pore-air pressure channel (i.e., valve $C$), the high air entry disk, and the grooved water compartment below the disk. The grooves inside the water compartment serve as water channels for flushing air bubbles that may be trapped or have accumulated as a result of diffusion. The high air entry disk should be sealed properly into the base pedestal using an epoxy resin along its periphery. A tight seal between the disk and the base pedestal ensures that air will not leak into the water compartment.

Valve $A$ in the base plate (Fig. 10.7) is used to control

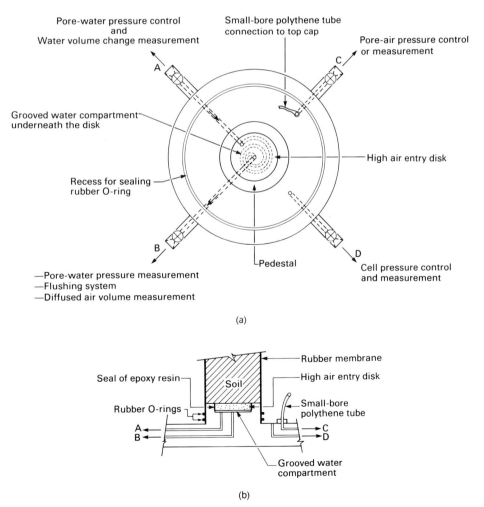

**Figure 10.7** Triaxial base plate for unsaturated soil testing. (a) Plan view of the base plate with its outlet ports; (b) cross-section of a base plate with a high air entry disk.

the pore–water pressure and to measure the water volume change during a *drained* shear test. Valve A can be connected to a twin-burette volume change indicator (Bishop and Donald, 1961; Bishop and Henkel, 1962; Head, 1986). Valve B on the base plate is used to measure the pore–water pressure during an *undrained* shear test. It can also be used to measure the prescribed pore–water pressure during a drained test. The pore–water pressure can either be measured using a null indicator (Bishop and Henkel, 1962) or a pressure transducer. Pressure transducers are now most common for measuring pore–water pressure (Head, 1986).

A diffused air volume indicator can be connected to valve B. It is used to measure the volume of diffused air accumulating below the high air entry disk. Valve C is connected to the pore–air pressure control or measuring system. The cell pressure is controlled through valve D.

The installation of a high air entry disk in the base of a direct shear box is shown in Fig. 10.8. The direct shear test is performed under drained conditions with controlled pore–air and pore–water pressures. The water compartment below the high air entry disk is connected to the measuring systems through steel tubing. The high air entry disk must be saturated prior to being placed in contact with a soil specimen.

It is preferable *not* to boil the disk and base plate in order to produce saturation. Boiling may produce fine cracks between the epoxy resin and the fine porous disk due to different thermal properties. The high air disk can best be saturated by first passing water through the disk and then driving the air into solution. Once the disk has been saturated, it appears to remain saturated as long as it is kept in contact with water. However, care should be taken to ensure the initial saturation of a newly installed porous disk.

### *Saturation Procedure for a High Air Entry Disk*

The following procedure is suggested to ensure saturation of a high air entry disk (Fredlund, 1973). A metal ring can be placed around the base pedestal and filled with distilled, deaired water to a height of about 25 mm above the ceramic disk. The triaxial cell or the direct shear box chamber is then put in place, and the water is subjected to an air pressure of approximately 600 kPa. Water is allowed to flow through the porous disk for approximately 1 h.

The air bubbles collected below the porous disk are periodically flushed, and the valves connecting the water compartment to the measuring system are then closed. The water above, inside, and below the porous disk takes on a pressure equal to the applied air pressure. The pressure is applied for approximately 1 h, during which time the air in the porous disk dissolves in the water. The valves connected to the water compartment are again opened for approximately 10 min to allow the water in the disk to flow into the compartment. The air bubbles are then flushed from below the porous disk. The above procedure is repeated about six times, after which the porous disk should be saturated. The disk should remain covered with water until a soil specimen is ready to be placed on the disk. This procedure for saturation of the ceramic disk is similar to that used by Bishop and Henkel (1962).

The permeability of the porous disk can be measured during the saturation process by recording the quantity of water passing through the disk. The coefficient of permeability is of interest in evaluating the compliance of the pore pressure measuring system and ensuring that the ceramic disk is not cracked.

### 10.1.3 Pressure Response Below the Ceramic Disk

It is important to understand the mechanisms associated with the pore–water pressure measuring system. There are numerous factors which affect the response of the measurement of pore–water pressure (e.g., time delays). Attempts have been made by numerous researchers (Whitman *et al.*, 1961; Gibson, 1963) to incorporate the flexibility of the measuring system (i.e., compliance) into the analysis of pore pressure measurements. Research has been directed mainly towards analyzing pore pressure measurements at the base of oedometer specimens.

The flexibility of the pressure measuring system gives an indication of how rapidly the system can respond to an applied pressure. A highly flexible pressure measuring system has a slow response. The flexibility of a measuring system depends upon the compressibility of the compartment and the fluids in the measuring system. It appears that the compressibility of the fluid in the measuring system (e.g., air–water mixture) to a large extent controls the response of the measuring system (Fredlund and Morgenstern, 1973). The measuring system becomes significantly more flexible with increasing quantities of air in the system.

Whitman *et al.* (1961) used an electrical analog to simulate the flexibility of the pore pressure measuring systems. Gibson (1963) theoretically studied the time lag in an open and closed pore pressure measuring system at the base of a clay layer. The time lag was attributed to the permeability and compressibility of the soil and the flexibility of the measuring system. These factors were combined to form a stiffness ratio. The stiffness ratio was shown to influence the distribution of pore–water pressures throughout a specimen during one-dimensional consolidation (Perloff *et al.* 1965; Northey and Thomas, 1965; Christie, 1965). Attempts to correlate the compliance theory with laboratory test results have met with limited success. In the above studies, the flexibility of the measuring system was assumed to involve a linear compressibility of fluid in the measuring system and a linear expansibility of the components comprising the measuring system.

A theoretical model describing the pore–water pressure response below a high air entry disk was presented by Fredlund and Morgenstern (1973). The model was used to characterize the transducer response below a high air entry disk mounted in a triaxial apparatus (Fig. 10.9). Consider a chamber pressure, $u_c$, applied to water in the triaxial cell.

10.1 SPECIAL DESIGN CONSIDERATIONS  267

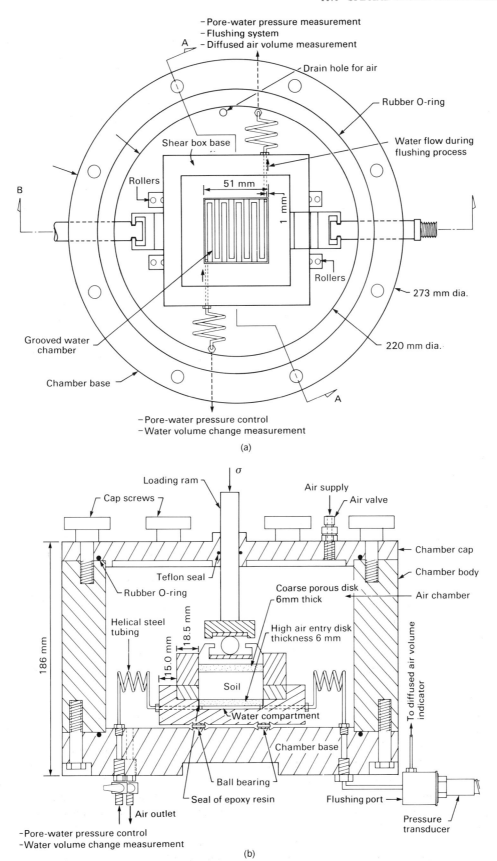

**Figure 10.8** Modified direct shear equipment for testing an unsaturated soil specimen. (a) Plan view of the pressure chamber for a direct shear box; (b) cross-sectional view A-A of a direct shear box and pressure chamber (from Gan, 1986.)

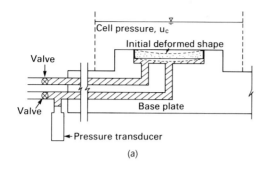

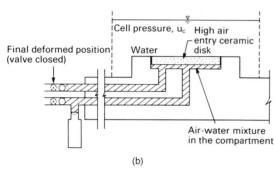

**Figure 10.9** Characterization of the pore-water pressure measuring system in the base plate of a triaxial cell. (a) Instant after closing valves on the base plate; (b) pressure measuring system at equilibrium.

One valve in the base plate is initially left open to maintain atmospheric pressure in the compartment below the disk. A hydraulic head gradient is established across the disk, and water flows through the disk into the compartment. After steady-state flow is established, the valves are closed, and further flow of water causes a compression of the air-water mixture in the compartment [Fig. 10.9(a)]. The compression of the air-water mixture results in a build-up of the measured pressure. As the pressure in the compartment increases, the hydraulic head gradient across the disk decreases, causing a reduction in water flow to the compartment. Equilibrium is established when the pressure in the compartment is equal to the pressure applied to the water in the cell (i.e., $u_c$), as illustrated in Fig. 10.9(b).

During the equalization process, water flows from above to below the high air entry disk. In other words, water flows from the triaxial cell through the disk into the compartment. Volume change in the compartment is caused by expansion of the compartment and compression of the air-water mixture in the compartment. The first factor depends on the rigidity of the components of the measuring system, the thickness and method of mounting the ceramic disk, and the deflection of the transducer membrane. All of these factors tend to produce an increase in the volume of the compartment under pressure. This volume change, along with the compression of the air-water mixture in the compartment, is equal to the volume of water flowing into the compartment during the equalization process.

The volume of water flowing into the compartment can be computed using Darcy's law. The hydraulic head gradient is continuously changing as the pressure in the compartment increases. Therefore, the rate of water inflow varies with time. At a particular elapsed time, the volume of water flowing into the compartment over a finite period of time can be written in the following form:

$$\Delta V_w = k_d i_{ave} A_d (t_j - t_i) \quad (10.2)$$

where

$\Delta V_w$ = volume of water flowing into the compartment during a finite time period at a particular elapsed time
$k_d$ = coefficient of permeability of the high air entry disk
$i_{ave}$ = average hydraulic head gradient during the finite time period
$A_d$ = cross-sectional area of the high air entry disk
$t_j$ = time at the end of time period
$t_i$ = time at the start of time period.

The average hydraulic head gradient, $i_{ave}$, can be obtained by comparing the applied pressure, $u_c$, and the pressures measured on the transducer at the start and the end of a selected time period, namely, $u_i$ and $u_j$, respectively:

$$i_{ave} = \frac{1}{\rho_w g} \left\{ \frac{\dfrac{(u_c - u_i)}{L_d} + \dfrac{(u_c - u_j)}{L_d}}{2} \right\} \quad (10.3)$$

where

$u_c$ = pressure applied to the water in the triaxial cell
$\rho_w$ = density of water
$g$ = gravitational acceleration
$u_i$ = compartment pressure at time, $t_i$
$u_j$ = compartment pressure at time, $t_j$
$L_d$ = thickness of the high air entry disk.

Substituting Eq. (10.3) into Eq. (10.2) gives

$$\Delta V_w = \frac{k_d A_d}{\rho_w g L_d} \left\{ u_c - \frac{(u_i + u_j)}{2} \right\} (t_j - t_i). \quad (10.4)$$

The volume change in the compartment can also be written as the sum of the compression of the air-water mixture and the expansion of the compartment. The expansion of the compartment is assumed to be negligible in comparison to the compression of the air-water mixture. Initially, the compartment is assumed to contain a mixture of air and water which undergoes compression as water flows into the compartment through the high air entry disk. Volume change in the compartment can be expressed using the compressibility of the air-water mixture and the pressures measured on the transducer at the start and end of the time period:

$$\Delta V_w = C_{aw} V_c (u_j - u_i) \quad (10.5)$$

where

$C_{aw}$ = compressibility of the air–water mixtures
$V_c$ = volume of the compartment.

The volume of the compartment, $V_c$, in Eq. (10.5) is essentially a constant. The compressibility of the air–water mixture is continuously changing. Let us also assume that no more air comes out of solution in the compartment, and that the equalization time involved is insufficient for free air to dissolve in the water. In other words, volume changes are due to the compression of free air. Computations based on a coefficient of diffusion of $2.0 \times 10^{-9}$ m$^2$/s show that the above assumption is reasonable for elapsed times less than 10 min. In addition, air and water pressures in the compartment are assumed to be equal. The compressibility of the air–water mixture in the compartment can be written as follows (Chapter 8):

$$C_{aw} = SC_w + (1 - S)/\bar{u}_a \quad (10.6)$$

where

$S$ = degree of saturation of the compartment
$C_w$ = compressibility of water
$\bar{u}_a$ = absolute compartment air pressure.

The air pressure in Eq. (10.6) can be taken as the average absolute pressure measured at the start and end of the time period (i.e., $(\bar{u}_i + \bar{u}_j)/2$). Theoretical curves for air–water mixture compressibility can be generated from Eq. (10.6) assuming various percentages for the initial air volume in the compartment (Fig. 10.10). The curves are generated using an initial atmospheric air pressure under isothermal conditions of 20°C.

The compressibility of the air–water mixture can be computed from Eq. (10.6) for various percentages of initial air volume. The air pressure is then increased using a pressure increment, and the air volume is subsequently reduced. A new compressibility is computed using the present air pressure and degree of saturation [and Eq. (10.6)]. The computation is repeated until the desired maximum air pressure is reached.

The compressibility equation for the air–water mixture in the compartment [Eq. (10.6)] has been experimentally studied (Fredlund and Morgenstern, 1973). Two tests were performed by altering the pressure applied to the compartment and monitoring the resulting volume change. The measured volume change for each pressure increment was used to calculate the compressibility of the air–water mixture [Eq. (10.5)]. The experimental results agree closely with the theoretical curves obtained from Eq. (10.6). The position of the experimental curves indicates the initial percentage of air in the compartment.

The theoretical response of the pressure transducer below the high air entry disk can be simulated by equating the volume change in the compartment [Eq. (10.5)] to the volume of water flowing into the compartment [Eq. (10.4)]:

$$(t_j - t_i) = \frac{C_{aw} C_f (u_j - u_i)}{\left\{ u_c - \frac{(u_i + u_j)}{2} \right\}} \quad (10.7)$$

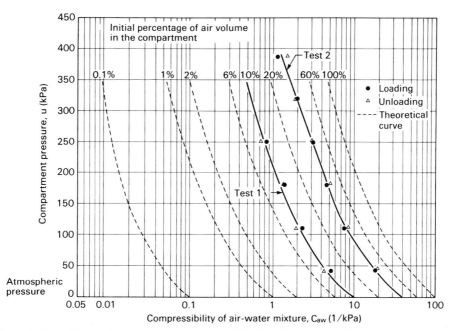

**Figure 10.10** Measured and computed compressibilities of air–water mixtures in the compartment.

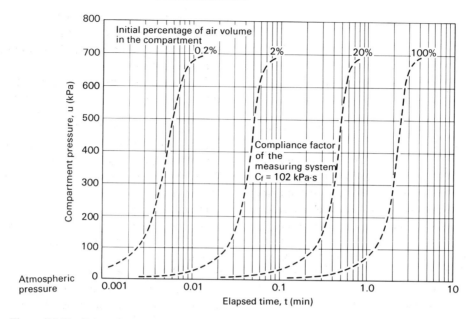

**Figure 10.11** Pressure response curves for various initial percentages of air in the compartment.

where

$C_f$ = compliance factor for the measuring system [i.e., $\rho_w g V_c L_d / (k_d A_d)$].

The compliance factor, $C_f$, contains the volume of the compartment, $V_c$, the dimensions of the high air entry disk, $L_d$ and $A_d$, and the permeability of the disk, $k_d$. Figure 10.11 presents pressure response curves for a measuring system with a compliance factor, $C_f$, of 102 kPa · s. The curves correspond to four different percentages of initial air volume in the compartment. The response times are computed using Eq. (10.7) by increasing the compartment pressure incrementally from an initial atmospheric pressure condition to an applied pressure, $u_c$, of 690 kPa. The results indicate a congruent shift in the response curves as the initial air volume increases. In other words, the time for a measuring system to respond increases with an increasing initial air volume. The effect of varying the applied cell pressure, $u_c$, is examined for a measuring system with a compliance factor of 1020 kPa · s and an initial air volume of 2% (Fig. 10.12). The slope of the steepest portion of the response curve decreases, and the time required for

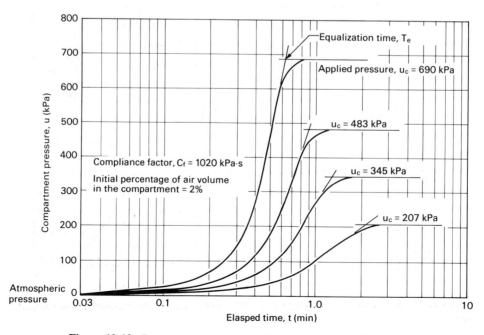

**Figure 10.12** Pressure response curves for various applied pressures.

10.1 SPECIAL DESIGN CONSIDERATIONS    271

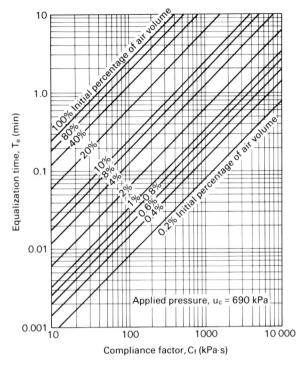

**Figure 10.13** Time for pressure equalization for an applied pressure of 690 kPa.

point represents the lateral shift of the response curve as the applied pressure is changed.

The equalization time depends on the initial air volume in the compartment, the compliance factor for the system, and the applied cell pressure. Figure 10.13 shows a plot of the logarithm of equalization time versus the logarithm of the compliance factor for an applied cell pressure of 690 kPa. Lines of equal initial air content are linear on this plot. The plot shows an increase in the equalization time as the compliance factor of the system increases or the initial air content in the system increases.

Typical experimental data are presented in Fig. 10.14. The experiments were performed by applying a pressure to water above a high air entry disk and measuring the build-up of pressure below the disk, as illustrated in Fig. 10.9. The experimental data agree closely with the theoretical characteristic curves. The compressibility of the air-water mixture, $C_{aw}$, in the measuring system can be computed using the experimental data [Eq. (10.7)]. In this case, the compliance factor, $C_f$, needs to be computed by measuring the volume of the compartment and the dimensions of the ceramic disk. The compressibilities obtained from the experimental data are plotted in Fig. 10.15, together with theoretical compressibility curves generated using Eq. (10.6). The experimental plots show the same shape as the theoretical curves up to a response of approximately 80%. At this point, the curves tend to the right, indicating an increase in the compressibility. This phenomenon may be attributable to air dissolving in water with increased elapsed time. In addition, minute volume changes associated with the seating of the components of the compartment (e.g., O-rings in the valves) may be significant during the final stages of equalization.

The study of the pore pressure response below high air

pressure equalization increases as the applied cell pressure decreases.

Two variables are required to characterize the theoretical response curves. These are the slope of the straight line portion of the response curve on a semi-logarithm plot, and the point of intersection of the extended straight line portion of the response curve and the horizontal line of 100% response (i.e., equalization time, $T_e$ in Fig. 10.12). This

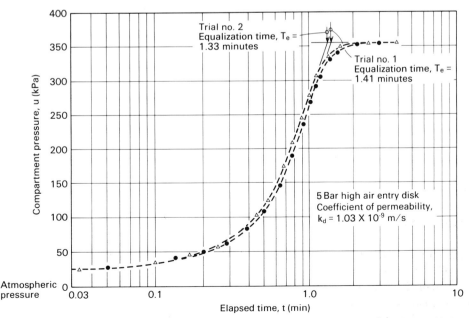

**Figure 10.14** Pore-water pressure response in a measuring system below a high air entry disk.

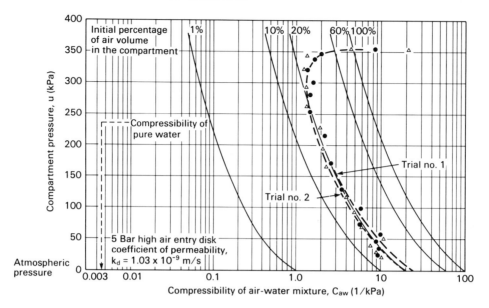

**Figure 10.15** Compressibility of the fluid in the pore pressure measuring system.

entry disks indicates that two primary factors are involved in measuring pore pressure response. First, the high air entry disks result in an impeded response. Second, even though attempts are made to saturate the compartment below the high air entry disk, there is still an air-water mixture which responds with a nonlinear compressibility.

### 10.1.4 Pore-Air Pressure Control or Measurement

Pore-air pressure is controlled at a specified pressure when performing a drained shear test (e.g., consolidated drained or constant water content test; see Chapter 9). Pore-air pressure is measured when performing a shear test in an undrained condition (e.g., consolidated undrained test; see Chapter 9). The control or measurement of pore-air pressure is conducted through a porous element which provides continuity between the air voids in the soil and the air pressure control or measuring system. The porous element must have a low attraction for water or a low air entry value in order to prevent water from entering the pore-air pressure system. The porous element can be fiberglass cloth disk (Bishop, 1961a) or a coarse porous disk (Ho and Fredlund, 1982b).

An arrangement for pore-air pressure control using triaxial equipment is shown in Fig. 10.16. A 3.2 mm thick, coarse corundum disk is placed between the soil specimen and the loading cap. The disk is connected to the pore-air pressure control through a hole drilled in the loading cap, connected to a small-bore polythene tube. The pore-air pressure can be controlled at a desired pressure using a pressure regulator from an air supply. The plumbing arrangement for controlling the pore-air pressure for a modified direct shear apparatus is illustrated in Fig. 10.8. The pore-air pressure is controlled by pressurizing the air chamber through the air valve located on the chamber cap.

The measurement of pore-air pressure can be achieved using a small pressure transducer, preferably mounted on the loading cap. When measuring pore-air pressure, the volume of the measuring system should be kept to a minimum in order to obtain accurate measurements. Pore-air pressure is also difficult to measure because of the ability of air to diffuse through rubber membranes, water, polythene tubing, and other materials. This is particularly true when considering the long time required in testing unsaturated soils.

An alternative pore pressure measuring system, used at the U.S. Bureau of Reclamation, is illustrated in Fig. 10.17 (Gibbs and Coffey, 1969). The pore-water pressure is measured at the top of the specimen through a high air entry disk. The pore-air pressure is measured at the bottom of the specimen through a saturated coarse porous disk and

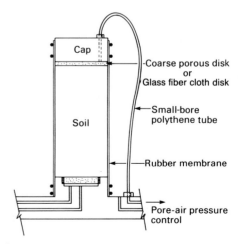

**Figure 10.16** Pore-air pressure control system.

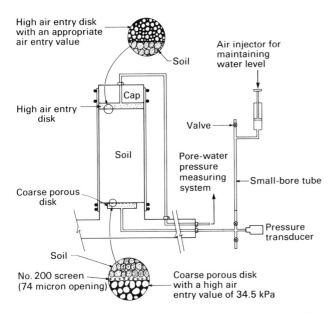

**Figure 10.17** Pore-air pressure measurements system used by the U.S. Bureau of Reclamation (from Gibbs and Coffey, 1969).

a fine screen (i.e., no. 200 mesh), as shown in Fig. 10.17. A slightly negative water pressure (i.e., −3.5 kPa) is applied to the saturated coarse porous disk. The pore-air pressure measurement is based on the separation of the menisci between the pore-water in the soil specimen and the water in the coarse porous disk. The fine screen layer is placed between the specimen and the coarse disk to ensure the separation between the pore-water and the water in the disk. The screen is sprayed with "silicone grease" to reduce its surface tension. As a result, the voids provided by the screen are filled with air which is in equilibrium with the pore-air pressure.

The negative water pressure in the coarse disk is kept constant by maintaining the water level in the small-bore tube (Fig. 10.17). This negative pressure is the zero reading from which the pore-air pressure is measured. Any change in the pore-air pressure due to the application of total stress (i.e., loading) will be reflected in the air pressure above the coarse disk, and subsequently in the water pressure inside the disk. However, the water pressure in the disk is maintained constant, even when pressure is supplied through an air pressure regulator. In other words, the water in the coarse disk acts as a flexible membrane when measuring the pore-air pressure through a null type system. The pressure required to maintain the water level in the tube is a measure of the pore-air pressure in the specimen. The principle of this system is similar to that of a null indicator.

Pore-air pressure measurements using the above technique appear to have produced reasonable results when compared to theoretical predictions (Knodel and Coffey, 1966; Gibbs and Coffey, 1969a, 1969b). Some of the experimental results obtained using this method are presented

in Chapter 8. However, water migration between the soil specimen and the coarse disk may occur, and this would lead to erroneous measurements. In addition, the compression of the coarse disk during the application of a total stress may affect the water pressure in the disk, which in turn affects the pore-air pressure measurements. Further study is required to clarify all concerns associated with this technique of pore-air pressure measurement.

### 10.1.5 Water Volume Change Measurement

The conventional twin-burette volume change indicator requires modifications prior to its use in testing unsaturated soil specimens. Greater accuracy is necessary because of the small water volume changes associated with unsaturated soil testing. A small-bore burette (e.g., 10 ml volume) can be used as the central tube in order to achieve a volume measurement accuracy of 0.01 ml. Leakage has to be essentially eliminated due to the long time periods involved in testing. The diffusion of pore-air through the pore-water and the rubber membrane can be greatly reduced by using two sheets of slotted aluminum foil (Dunn, 1965). Silicone grease can be placed between the two aluminum sheets. Rubber membranes can be placed next to the soil specimen and on the outside of the aluminum sheets. Further details on the control of leakage in the triaxial test are given by Poulos (1964).

Figure 10.18 shows a layout of the plumbing associated with a twin-burette volume change indicator manufactured by Wykeham Farrance. The water volume change can be measured under a controlled backpressure by connecting an air pressure regulator to the twin-burette. Swagelock[3] fit-

[3]Swagelock is manufactured by Crawford Fitting Company, Niagara Falls, Canada.

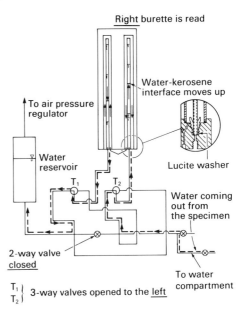

**Figure 10.18** Direction of water movement for flow out from the specimen with the three-way valves opened to the left.

tings on copper tubing are used for the plumbing. A small Lucite washer can be placed at the base of the burette to prevent the tubes from popping out when tightening down the rubber sleeve around the base of the burette. The above design has been found to satisfactorily eliminate leakage in the twin-burette volume change indicator (Fredlund, 1973). The twin-burette volume change indicator can be connected to either a triaxial or direct shear apparatus.

Two three-way Whitey[4] valves and one two-way Whitey[4] valve are used to enable the direction of water flow to be reversed, and therefore continuously monitored. The indicator can also be bypassed in order to flush diffused air from the water compartment below the high air entry disk. Volume changes due to the compressibility of the indicator and the fluid in the indicator, as well as water loss through tubes and valves, can be essentially eliminated by always reading the burette opposite the direction that the three-way valves are opened. When this procedure is not adhered to, errors on the order of 0.1 ml may occur over the duration of one day (Fredlund, 1973).

If the three-way valves in Fig. 10.18 (i.e., $T_1$ and $T_2$) are simultaneously opened to the *left*, the *right* burette should be read. The two-way valve is closed during the volume change measurement. The direction of water flow indicated by the arrow heads in Fig. 10.18 corresponds to the condition where water is coming *out* of the soil specimen. The water-kerosene (with red dye) interface in the right burette moves *upward*. The opposite condition occurs when water flows into the specimen.

When the water-kerosene interface in the right burette comes near the bottom of the burette, the three-way valves can be simultaneously reversed to the *right*. Readings should then be taken on the *left* burette, as illustrated in Fig. 10.19. The direction of water flow shown in Fig. 10.19 still corresponds to the condition where water is coming out of the specimen. The water-kerosene interface in the left burette moves *upward*. The opposite direction of water flow occurs when water flows into the specimen.

The twin-burette volume change indicator measures the water volume change plus the volume of diffused air moving into the compartment below the high air entry disk. The volume of diffused air in the compartment and in the water lines can be measured periodically and subtracted from the total water volume change. The diffused air volume indicator (i.e., DAVI) and techniques for measuring the diffused air volume are explained in Chapter 6. During the measurement of the diffused air volume, the three-way valves on the twin-burette volume change indicator are closed, as shown in Fig. 10.20. In other words, the water volume change indicator is temporarily bypassed when the diffused air volume is measured. At the same time, the two-way valve is opened in order to maintain a constant water pressure in the compartment. A pressure gradient across the base plate is then applied momentarily by opening the valve to the diffused air volume indicator. As a result, air is flushed from the compartment (Fig. 10.20) and forced into the diffused air volume indicator.

Other types of volume change indicators could also be used to measure water volume change. In each case, long-term tests should be performed to establish the reliability and accuracy of the equipment.

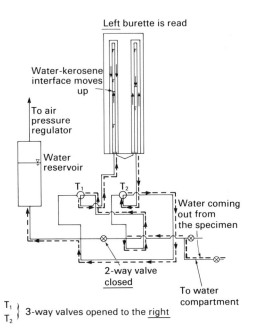

**Figure 10.19** Direction of water movement for flow out from a specimen with the three-way valves opened to the right.

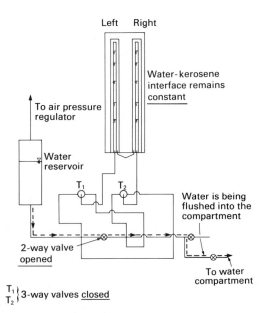

**Figure 10.20** Valve configuration when flushing diffused air from the water compartment.

---

[4]Whitey valves are manufactured by Whitey Company, Niagara Falls, Canada.

### 10.1.6 Air Volume Change Measurement

The overall volume change of an unsaturated soil specimen is equal to the sum of the water and air volume changes. The soil particles are assumed to be incompressible. The measurements of any two of the above volume changes (i.e., overall, water, and air volume changes) are sufficient to describe the volume change behavior of an unsaturated soil. The overall and the water volume changes are generally measured while the air volume change is computed as the difference between the measured volume changes. In addition, air volume change is difficult to measure due to its high compressibility and sensitivity to temperature change.

Figure 10.21 illustrates the use of two burettes to measure air volume change under atmospheric air pressure conditions (Bishop and Henkel, 1962; Matyas, 1967). Air moves out of a soil specimen, through a coarse porous disk, and is collected in a graduated burette. Both burettes can be adjusted to maintain the air–water interface at the specimen midheight. As a result, the pore-air pressure is maintained at atmospheric pressure. The changing elevation of the air–water interface in the graduated burette indicates the air volume change in the specimen. Water is prevented from entering the burette because the gauge pore-water pressure is negative. In addition, the coarse porous disk has a low air entry value. Water losses due to evaporation from the open burette can be prevented by covering the water surface with a layer of light oil (Head, 1986) or replacing the water entirely with a light oil (Matyas, 1967).

The above method for measuring air volume change is somewhat cumbersome. The apparatus shown is limited to measuring at atmospheric conditions. However, the apparatus could be extended to operate under an applied backpressure. The apparatus has been primarily used to measure the air coefficient of permeability.

### 10.1.7 Overall Volume Change Measurement

In a saturated soil, the overall volume change of the soil specimen is equal to the water volume change. For an unsaturated soil, the water volume change constitutes only a part of the overall volume change of a specimen. The overall volume change measurement must therefore be made independently of the water volume change measurement.

It would appear that the overall volume change could be measured by surrounding the specimen with an impermeable membrane and filling the cell with a pressurized fluid. The cell fluid could be connected to a twin-burette volume change indicator to measure volume change due to the compression or expansion of the soil specimen. However, it is difficult for this type of measurement to be accurate. There will generally be significant errors caused by leakage, diffusion, or volume changes of the cell fluid due to pressure and temperature variations. The fluid that has been most successfully used in this manner is mercury. It is also prudent to use a double-walled triaxial cell. However, mercury is hazardous to health, and its use should be avoided if possible. Fluids other than mercury have also been used with success in conjunction with double-walled triaxial cells (Wheeler and Sivakumar, 1992).

In a triaxial test, the overall volume change is commonly obtained by measuring the vertical deflection and radial deformation during the test. In a direct shear test, only the vertical deflection measurement is required since the soil specimen is confined laterally. The vertical deflection of the soil specimen can be measured using a conventional dial gauge or an LVDT (i.e., linear variable differential transformer). The LVDT's have an accuracy comparable to that of dial gauges. Some LVDT's can be submerged in oil or water. Various applications of LVDT's in triaxial testing are described by Head (1986).

Noncontacting transducers have been used increasingly during triaxial testing (Cole, 1978; Khan and Hoag, 1979; Drumright, 1987). The device consists of a sensor and an aluminum target (Fig. 10.22). The sensor is a displacement transducer[5] clamped to a post and connected to the electronic measuring system through a port in the base plate. The aluminum target can be attached to the rubber membrane (i.e., using silicon grease) near the midheight of the specimen. Three sensors and targets can be placed around the circumference of the specimen at 120° intervals (Fig. 10.22).

The noncontacting transducers operate on an eddy current loss principle. An eddy current is induced in the aluminum target by a coil in the sensor. The magnitude of the induced eddy current is a function of the distance between the sensor and the aluminum target. As the specimen deforms radially, the distance between the aluminum target and the sensor changes, causing a change in the magnitude of the eddy current generated. The impedance of the coil then changes, resulting in a change in the DC voltage output.

Figure 10.23 shows the calibration and the installation requirements for a noncontacting, button-type radial defor-

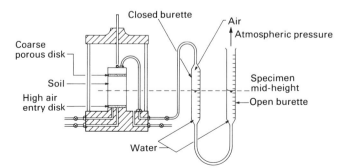

**Figure 10.21** Air volume change measurement under atmospheric conditions in a triaxial apparatus.

---
[5]Manufactured by Kaman Science Corporation, Colorado Springs, CO.

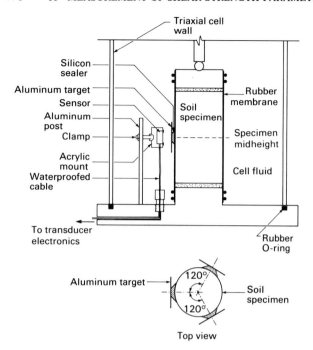

Figure 10.22 Installation of a noncontacting radial deformation transducer (from Drumright, 1987).

mation transducer. The transducer shown in Fig. 10.23 has a measuring range of 4 mm plus an additional 20% nominal offset. The offset or zero position gives the minimum distance between the sensor face and the aluminum target when the reading is zero [Fig. 10.23(a)]. The output is lin-

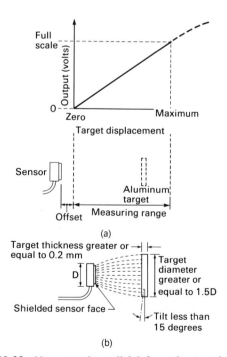

Figure 10.23 Noncontacting radial deformation transducer, KD-2310 series, model 4SB, shielded, button-type. (a) Calibration technique for noncontacting, radial deformation transducer; (b) installation requirements (from Kaman Science Corporation).

early proportional to the distance between the sensor and the aluminum target. The transducer has a high resolution equal to 0.01% of its range or 0.0004 mm. Requirements associated with the size, thickness, and orientation of the aluminum target are shown in Fig. 10.23(b).

The noncontacting transducer can operate using various cell fluids, such as air, water, and oil, with essentially the same sensitivity (Khan and Hoag, 1979). The transducers are not affected by cell pressure or temperature. Other devices and systems have also been used for measuring radial deformation, but details are not presented herein.

### 10.1.8 Specimen Preparation

Unsaturated soil specimens obtained from either undisturbed or compacted samples can be used for shear strength testing. Generally, the soil specimen has a high initial matric suction, while the test may be performed under a lower matric suction. For example, a multistage test on an unsaturated soil is commonly commenced at a low matric suction, with further stages conducted at higher matric suctions (Ho and Fredlund, 1982b; Gan et al. 1987). For this reason, it is sometimes necessary to relax the high initial matric suction in the specimen prior to performing the test. One way to reduce the soil matric suction is to impose pore-air and pore-water pressures which will result in a low matric suction. In order to reach equilibrium, water from the compartment below the high air entry disk must flow upward into the specimen. The equilibration process, however, may require a long time due to the low permeability of the high air entry disk. This is particularly true when the initial matric suction is much higher than the desired value for commencing the test. Therefore, the relaxation of the initial matric suction is usually accomplished by wetting the specimen from the top through the coarse porous disk. The relaxation of the initial matric suction is not required for some tests, such as undrained and unconfined compression tests.

A procedure used to relax the initial matric suction for multistage *triaxial testing* was outlined by Ho and Fredlund (1982b). The specimen is first trimmed to the desired diameter and height, and then mounted on the presaturated high air entry disk. During setup, appropriate measurements of the volume-mass properties of the specimen are made. A coarse porous disk and the loading cap are placed on top of the specimen. The specimen is then enclosed using two rubber membranes. The specimen has a composite membrane consisting of two slotted aluminum foil sheets between rubber membranes. The purpose of the aluminum foil is to greatly minimize air diffusion from the specimen. O-rings are placed over the membranes on the bottom pedestal. Spacers (i.e., pieces of 3.2 mm plastic tubing) are inserted between the membranes and the loading cap to allow air within the specimen to escape while water is added to the surface of the specimen. The Lucite cylinder of the triaxial cell is installed, and the cell is filled with water to a level partway up the specimen. Water for the specimen can either be added manually or through the air pressure line connected to the loading cap. In this case the air pressure line is temporarily connected to a water reservoir.

The specimen is left for several hours to allow the distribution of water throughout the specimen. The relaxation process is continued until air can no longer be seen escaping from around the top of the specimen. At the end of the process, the soil matric suction will be essentially zero.

The above procedure is conducted with the Lucite cylinder installed around the specimen while the top of the triaxial cell is detached. It is possible to now remove the plastic spacers between the membranes and the loading cap, and to place the top O-rings around the loading cap. The line connected to the loading cap can now be disconnected from the water reservoir and connected to the air pressure control system.

A low matric suction value can now be imposed on the specimen and time allowed for equalization. After pressure equalization between the applied pressures and the soil pressures, the soil specimen is ready to be tested.

The procedure used to prepare a specimen for a *direct shear* test has been reported by Gan (1986); Escario and Sáez (1986). The two halves of the direct shear box are sealed together using vacuum grease. The outside of the bottom half should be greased with vacuum grease before being mounted on the shear box base. The vacuum grease ensures that water will flow only towards the high air entry disk. It is important to *not* smear vacuum grease onto the surface of the high air entry disk. Vacuum grease blocks the fine pores of the high air entry disk, and disrupts the flow of water through the disk.

The soil specimen is mounted into the shear box, and the coarse porous stone and loading cap are installed. The initial matric suction in the soil specimen can be relaxed by adding water to the top of the specimen.

### 10.1.9 Backpressuring to Produce Saturation

The shear strength parameters, $c'$ and $\phi'$, can be obtained from tests on saturated specimens. Initially, unsaturated specimens, either undisturbed or compacted, must be saturated prior to testing. Saturation is commonly achieved by incrementally increasing the pore-water pressure, $u_w$. At the same time, the confining pressure, $\sigma_3$, is increased incrementally in order to maintain a constant effective stress, $(\sigma_3 - u_w)$, in the specimen. As a result, the pore-air pressure increases and the pore-air volume decreases by compression and dissolution into the pore-water. The simultaneous pore-water and confining pressure increases are referred to as a "backpressuring the soil specimen." The backpressure is essentially an axis-translation technique. In other words, the axis-translation technique used for unsaturated soils is similar to the backpressure concept used for saturated soil. The concepts and the techniques associated with backpressuring are briefly outlined in this section, while reference is made to Bishop and Henkel (1962) and Head (1986) for detailed explanations.

An equation was derived in Chapter 2 for the pore-air pressure increase required to dissolve free air in water (i.e., saturation) using undrained compression. Saturation was achieved by increasing the confining pressure and maintaining undrained conditions for the pore-air and pore-water. The water content of the specimen remained constant, while the total volume of the soil decreased due to the compression of the pore-air. The disadvantage with this procedure is related to the volume change which the specimen undergoes. The equation for the change in pore-air pressure has the following form:

$$\Delta u_a = \frac{(1 - S_0)}{S_0 h} \bar{u}_{a0} \qquad (10.8)$$

where

$\Delta u_a$ = pore-air pressure increase required to saturate the soil specimen
$S_0$ = initial degree of saturation
$h$ = volumetric coefficient of solubility
$\bar{u}_{a0}$ = absolute initial pore-air pressure.

Equation (10.8) gives the theoretical additional pore-air pressure required to saturate a soil specimen which has an initial degree of saturation, $S_0$.

The more common method for saturating a soil specimen is to backpressure deaired water into the specimen. Consequently, the pore-air is compressed and dissolved, as illustrated in Fig. 10.24. The confining pressure is also increased to maintain a constant effective stress. The saturation process is performed such that the water content increases as the degree of saturation is increased.

The pore-air pressure increase can be assumed to be the backpressure required to increase the degree of saturation in the specimen, while the pore-air pressure is assumed to be equal to the pore-water pressure. Consider a soil specimen with an initial degree of saturation of $S_0$ and an initial absolute pore-air pressure of $\bar{u}_{a0}$ (Fig. 10.24). Deaired water under a backpressure is forced into the specimen in order to increase the degree of saturation to some arbitrary value, $S$. The total volume of the soil and the pore voids' volume are assumed to remain constant during the saturation process. The absolute pore-air pressure increases to $(\bar{u}_{a0} + \Delta u_a)$. The pore-air pressure increase can be computed by applying Boyle's law to the volume of free and dissolved air. The volume of air versus pore-air pressure can be computed as follows:

$$(1 - S_0 + hS_0)\bar{u}_{a0} = (1 - S + hS)(\bar{u}_{a0} + \Delta u_a)$$
$$(10.9)$$

where

$S$ = final degree of saturation.

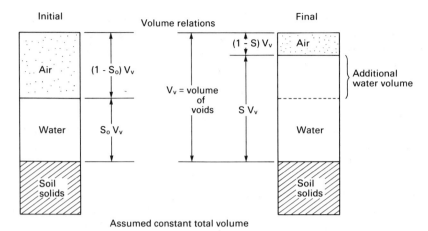

**Figure 10.24** Saturation process by forcing deaired water under backpressure into the soil specimen.

Rearranging Eq. (10.9) results in an expression for the pore–air pressure increase required to backpressure the soil (Lowe and Johnson, 1960):

$$\Delta u_a = \frac{(S - S_0)(1 - h)}{1 - S(1 - h)} \bar{u}_{a0}. \qquad (10.10)$$

A comparison between Eqs. (10.8) and (10.10) is shown in Fig. 10.25. All curves are drawn using an initial absolute pore–air pressure of 101.3 kPa and a volumetric coefficient of solubility of 0.02 at 20°C. There are three curves plotted using Eq. (10.10) for increasing the degree of saturation using an applied backpressure. The three curves correspond to three different final degrees of saturation (i.e., 99, 99.5, and 100%). Saturation by undrained compression [i.e., Eq. (10.8)] appears to require a significantly higher pressure increase than saturation using a backpressure directly applied to the water. This is particularly true when the initial degree of saturation is less than 95%. This difference may be attributed to the different water content conditions associated with each saturation process. The water volume, $V_w$, remains essentially constant during the saturation process by undrained compression, while the volume of water increases during the saturation process using an applied backpressure. As a result, the application of a backpressure provides a greater volume of water for the dissolution of free air.

Saturation by compression of the specimen is not as efficient as applying a backpressure to the water phase. This occurs because part of the applied total stress is taken by the soil structure, and part is taken by the fluid phase. It should also be noted that saturation by undrained compression may alter the soil structure due to volume change. In reality, however, the backpressures required for saturation may be lower than the values shown in Fig. 10.25, particularly for compacted specimens at low degrees of saturation (Bishop and Henkel, 1962).

The incremental application of backpressure is discussed by Head (1986). The backpressure increment is applied after the cell pressure increment has been applied to the specimen. Typically, the first two cell pressure increments can be 50 kPa, and the subsequent increments can be 100 kPa. Figure 10.26 illustrates an incremental procedure for the backpressure application, and the pore–water pressure response from the soil specimen. In the case shown, an effective stress of 10 kPa is maintained on the specimen. The tangent $B_w$ pore–water pressure parameter (Chapter 8) is checked by measuring the pore–water pressure response to a cell pressure increment after each stage of loading. Saturation is usually assumed to be complete when the $B_w$ parameter approaches unity. The saturation of compacted specimens is generally achieved at backpressures in the range of 400–750 kPa (Bishop and Henkel, 1962). Some soils however, may become saturated at $B_w$ parameters less than 1.0 when the compressibility of the soil structure is extremely low.

Theoretical values of the $B_w$ pore pressure parameter for

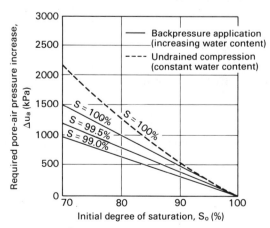

**Figure 10.25** Pore–air pressure increase required to saturate a soil specimen by two different methods (from Head, 1986).

10.2 TEST PROCEDURES FOR TRIAXIAL TESTS 279

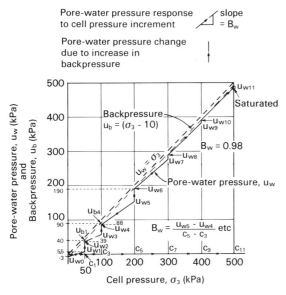

**Figure 10.26** Incremental application of cell pressure and backpressure, and the results of a pore-water pressure response test while saturating a specimen (from Head, 1986).

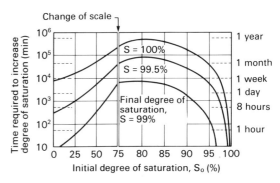

**Figure 10.28** Theoretical times required to increase the degree of saturation using appropriate backpressures corresponding to the initial degrees of saturation (from Black and Lee, 1973).

soils at various degrees of saturation, with various compressibilities, are presented by Black and Lee (1973) (Fig. 10.27). The application of a backpressure causes the compression of pore-air in accordance with Boyle's law, and the dissolution of pore-air into pore-water in accordance with Henry's law (Chapter 2). The pore-air compression is essentially instantaneous, causing an increase in the degree of saturation. On the other hand, the dissolution of pore-air into pore-water requires a longer time due to a relatively low coefficient of diffusion. Therefore, time must be allowed for equilibration after each backpressure increment. Theoretical times required to increase the degree of saturation are given in Fig. 10.28. Results are plotted for final degrees of saturation of 99, 99.5, and 100%. The plot shows that the time required for increasing the degree of saturation reaches a maximum value for soils at an initial degree of saturation around 80%. The required time decreases significantly at initial degrees of saturation higher than 95%.

## 10.2 TEST PROCEDURES FOR TRIAXIAL TESTS

This section provides a general description of the test procedures associated with various triaxial shear tests on unsaturated specimens. Tests can be performed in a triaxial cell which has been modified in accordance with the special design considerations explained in previous sections of this chapter. Figure 10.29 shows an assemblage of a modified triaxial cell. The measurements of the vertical deflection and the radial deformation are not shown. The layout of the plumbing for the control board is illustrated in Fig. 10.30. The pore-air pressure line shown in both figures controls the pore-air pressure. In the case where the pore-air pressure is measured, a pressure transducer can be installed in the loading cap, and the attached wires can be connected to the data acquisition system through the base plate.

The soil specimen should be prepared, and then several procedural checks should be conducted. The high air entry disk should be saturated. Attempts should be made to thoroughly flush water through the compartment in the base plate and all the connecting lines to ensure the expulsion of air bubbles. The volume change measuring devices, including the diffused air volume indicator, should be initialized (Fredlund, 1972).

The initial confining air and water pressures to be applied to the soil specimen can be set on the pressure regulators prior to preparing the specimen. This minimizes the time between the placement of the specimen on the high air entry disk and the application of the pressures. The confining air and water pressures are applied to the specimen through valves *D*, *C*, and *A*, respectively (Fig. 10.30). An initial water pressure of 30 kPa or greater is desirable in order to provide sufficient pressure to flush air from the base plate.

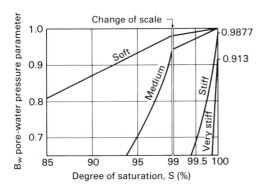

**Figure 10.27** Theoretical values of $B_w$ pore-water pressure parameters at various degrees of saturation and compressibility (from Black and Lee, 1973).

**280** 10 MEASUREMENT OF SHEAR STRENGTH PARAMETERS

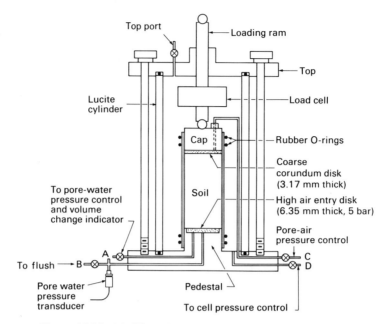

**Figure 10.29** Modified triaxial cell for testing unsaturated soils.

The diffused air volume can be measured using the diffused air volume indicator.

### 10.2.1 Consolidated Drained Test

The *consolidation* (or stress equalization) of the soil specimen is performed by applying a prescribed confining pressure, $\sigma_3$, pore-air pressure, $u_a$, and pore-water pressure, $u_w$. The confining pressure, pore-air, and pore-water pressures are applied by opening valves $D$, $C$, and $A$, (Fig. 10.30) in this order. Valves $B$ and $E$ are always closed during the test, except during the flushing of diffused air from the base plate. The water pressure, applied to the base plate, is registered on a transducer.

The vertical deflection and the radial deformation are periodically monitored to measure the overall volume change of the specimen. The volume of water flowing into (and out from) the specimen is recorded on the twin-burette volume change indicator. Therefore, the three-way valves, $T_1$

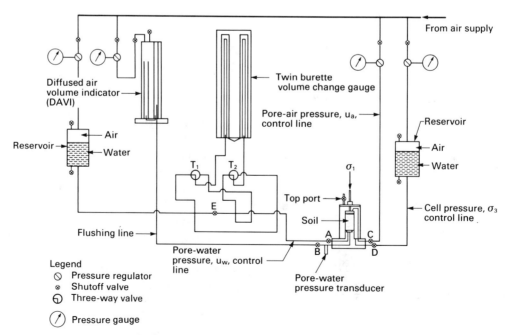

**Figure 10.30** Schematic diagram of the control board and plumbing layout for the modified triaxial apparatus.

and $T_2$, are always open during the test, except during the process of flushing diffused air from the base plate. The air volume change is generally not measured. Consolidation is assumed to have reached an equilibrium condition when there is no longer a tendency for the overall volume change or the flow of water from the specimen.

Upon attaining an equilibrium condition under the applied pressures (i.e., $\sigma_3$, $u_a$, and $u_w$), the specimen is *sheared* by compression at an appropriate strain rate (see Chapter 9). The magnitude of the axial load applied to the specimen can be recorded using a load cell. The axial load is converted to a deviator stress, $(\sigma_1 - \sigma_3)$. The shearing process is conducted under drained conditions for the applied pore-air and pore-water by leaving valves $C$ and $A$ (Fig. 10.30) open. The overall and water volume changes are monitored throughout the shear process. The shearing process is terminated when the selected failure criterion (e.g., maximum deviator stress) has been achieved (see Chapter 9).

Diffused air is generally flushed from the base plate once a day during both consolidation and shearing. The frequency of the diffused air measurement depends primarily on the applied air pressure. For a low applied air pressure, the diffused air volume can be measured less frequently. In any case, the diffused air volume should be measured prior to changing applied pressures. The water volume change correction, due to the diffused air volume, becomes necessary whenever tests extend over a period of several days.

The diffused air in the base plate can be flushed into the diffused air volume indicator, DAVI, by applying a pressure differential of 7-70 kPa between the base plate and the diffused air volume indicator. Each apparatus needs to be tested to assess the differential pressure at which diffused air can readily be removed from the base compartment. It is desirable not to significantly alter the water pressure in the base plate. Therefore, it may be necessary to elevate the air backpressure in the diffused air volume indicator. Having backpressurized the diffused air volume indicator, the three-way valves, $T_1$ and $T_2$, are closed, and valve $E$ is opened in order to bypass the twin-burette volume change indicator. The water pressure in the base plate is maintained through valve $E$. Subsequently, valve $B$ is opened and closed, causing surges of water to flow through the base. Diffused air moves into the diffused air volume indicator, and displaces the water in the burette. A few seconds may be required between each surge to allow the air to rise in the burette. The water pressure in the base plate only deviates momentarily from its set value when using this procedure. The computation of the diffused air volume from the readings on the indicator is described in Chapter 6. After measuring the diffused air volume, valves $B$ and $E$ are closed and valves $T_1$ and $T_2$ are turned to their previous direction. In other words, the twin-burette water volume change indicator is reconnected to the base plate.

In the case of a multistage test, the above procedure (i.e., consolidation and shearing) is repeated during each stage of the test. The consolidation for each stage can be commenced either at zero deviator stress or while maintaining the maximum deviator stress obtained from the previous stage (see Chapter 9). The deviator stress can be brought to zero by releasing the axial load to zero. The shearing process at each stage should be stopped when the maximum deviator stress is imminent, except for the last stage where the specimen can be sheared to a large strain.

### 10.2.2 Constant Water Content Test

The initial consolidation process is carried out in the same manner for both the constant water content test and the consolidated drain test. When equilibrium conditions have been achieved under the applied pressures (i.e., $\sigma_3$, $u_a$, and $u_w$), the soil specimen is sheared under drained conditions for the pore-air phase and undrained conditions for the pore-water phase. The pore-air pressure is maintained at the value to which the specimen was subjected during consolidation. That is, valve $C$ (Fig. 10.30) remains open during consolidation and shear. On the other hand, valves $A$ and $B$ are closed during shear in order to produce undrained pore-water conditions. The pore-water pressure is measured by the pressure transducer mounted on the base plate.

During shear, under undrained water phase conditions, the diffused air volume should also be measured. In this case, the water pressure in the base plate should be recorded prior to the flushing process and reset after flushing. The water in the pore-water pressure control line should first be subjected to the same pressure as recorded in the base plate. Valves $A$, $T_1$, and $T_2$ should remain closed while valve $E$ is opened when adjusting the water line pressure. The air backpressure in the diffused air volume indicator should be adjusted to a pressure slightly lower than the recorded water pressure in the base plate while valve $B$ remains closed. When valve $A$ is opened, the water in the base plate will quickly equalize to the pore-water pressure control line. The diffused air is then removed from the base plate by momentarily opening valve $B$, which produces a pressure difference across the base plate. Valves $A$ and $B$ are closed at the end of the diffused air volume measurement. The undrained pore-water pressure is then returned to the value existing prior to the flushing process. If the diffused air removal is performed in a short period of time, disturbance to the undrained condition of the soil specimen should be minimal.

### 10.2.3 Consolidated Undrained Test with Pore Pressure Measurements

The soil specimen is first consolidated following the procedure described for the consolidated drained test. After equilibrium conditions have been established under the applied pressures (i.e., $\sigma_3$, $u_a$, and $u_w$), the soil specimen is sheared under undrained conditions with respect to the air and water phases. Undrained conditions during shear are achieved by closing valves A, B, and C (Fig. 10.30).

The pore-water pressure developed during shear can be measured on the pressure transducer mounted on the base plate. A pore-air pressure transducer should be mounted on the loading cap, if possible, for measuring pore-air pressure changes. However, it is difficult to maintain an undrained condition for the pore-air due to its ability to diffuse through the pore-water, the rubber membrane, and the water in the high air entry disk.

The diffused air volume can be measured in a manner similar to that used during the constant water content test. Problems associated with air diffusion are the main reason why few consolidated undrained tests with pore-air and pore-water pressure measurements have been performed.

### 10.2.4 Undrained Test

The procedure for performing an undrained test on an unsaturated soil specimen is similar to the procedure used for performing an undrained test on a saturated soil specimen. The unsaturated soil specimen is tested at its initial water content or matric suction. In other words, the initial matric suction in the specimen is *not* relaxed or changed prior to commencing the test.

There is no consolidation process allowed since the confining pressure, $\sigma_3$, is applied under undrained conditions for both the pore-air and pore-water phases. The specimen is axially compressed under undrained conditions with respect to both the air and water phases. The test is usually run at a strain rate of 0.017–0.03%/s, and no attempt is made to measure the pore-air and pore-water pressures. Conventional triaxial equipment can be used to perform the undrained test on unsaturated soils. The porous disks are usually replaced by metal or plastic disks on the top and bottom of the specimen. The specimen is enclosed in a rubber membrane during the test. The undrained test results on unsaturated soils can be interpreted in accordance with the theory explained in Chapter 9.

### 10.2.5 Unconfined Compression Test

The unconfined compression test procedure is similar to the undrained test procedure, except that no confining pressure is applied to the specimen (i.e., $\sigma_3$ is equal to zero). The test is commonly performed in a simple loading frame by applying an axial load to the soil specimen. The interpretation of the unconfined compression test results on unsaturated soils is discussed in Chapter 9.

## 10.3 TEST PROCEDURES FOR DIRECT SHEAR TESTS

The consolidated drained direct shear test on an unsaturated soil specimen can be conducted using the modified direct shear apparatus shown in Fig. 10.8. A cross-sectional view of the direct shear equipment is shown in Fig. 10.31. The soil specimen is sheared by moving the lower portion of the shear box relative to the upper portion of the box. This is the same procedure as is used in the operation of a conventional direct shear apparatus. A motor that provides a constant horizontal shear displacement rate is con-

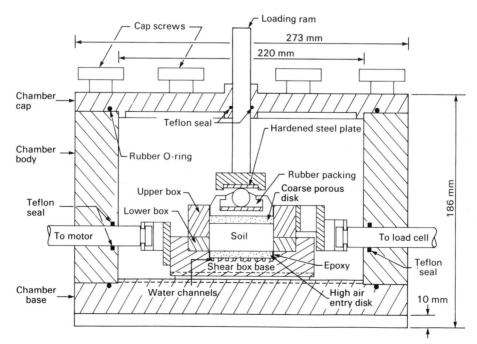

**Figure 10.31** Modified direct shear apparatus for testing unsaturated soils (from Gan and Fredlund, 1988).

nected to the shear box base. The shear box base is seated on a pair of rollers that can move along a pair of grooved tracks on the chamber base. The top box is connected to a load cell which measures the shear load resistance. The gap between the two halves of the shear box is filled with vacuum grease prior to mounting the specimen in the shear box.

The plumbing layout for the control board of the modified direct shear apparatus is illustrated in Fig. 10.32. The saturation of the high air entry disk, the relaxation of the initial matric suction in the specimen, and the flushing of entrapped air from the base plate and its connecting lines should be performed prior to commencing the test. At the same time, the initial air and water pressures to be applied to the soil specimen can be set on the pressure regulators while valves $A$, $B$, and $C$ are closed.

The procedure for conducting the consolidated drained, direct shear test is similar to the consolidated drained, triaxial text procedure explained in the previous section. After installing the chamber cap, the predetermined vertical normal load, air pressure, and water pressure are applied to the specimen, in this sequence. The vertical normal load is applied through the loading ram, while the air and water pressures are applied by opening valves $C$ and $A$ (Fig. 10.32), respectively. Valve $B$ remains closed during the test, except when measuring the diffused air volume. It is important to ensure that there are no leaks in the system. For example, the leakage of air from the chamber surrounding the specimen will cause a continuous water vapor loss from the specimen. The applied water pressure to the base plate can be measured on the pore-water pressure transducer mounted on the base plate. Measurements of vertical deflection and water movement from the specimen can be taken at various time increments. Water movement is observed on the twin-burette volume change indicator. In this case, valves $T_1$ and $T_2$ are opened while valve $D$ is closed throughout the test, except during the flushing process.

Consolidation under the applied vertical normal stress, the air pressure, and the water pressure is assumed to have reached equilibrium when there is no further tendency for overall volume change and water volume change.

After equilibration has been reached, the soil specimen is sheared at an appropriate horizontal shear displacement rate (Chapter 9). The horizontal shear load resistance is measured using a load cell. Similarly, readings are taken on the vertical deflection, the horizontal shear displacement, and the water volume change during shear. Shearing can be terminated either when the horizontal shear stress resistance has reached its peak value or when the horizontal shear displacement has reached a designated limiting value (Chapter 9). In the case of a multistage test, the shearing process for each stage should be stopped when the peak horizontal shear stress is imminent.

The monitoring of the diffused air volume follows the procedure explained for the consolidated drained triaxial test. During the flushing process, valves $T_1$ and $T_2$ are closed while valve $D$ is opened. Valve $B$ is opened momentarily to establish a pressure difference across the base plate. A surging of water through the base plate forces air in the base plate into the diffused air volume indicator for measurement. The diffused air volume measurement should

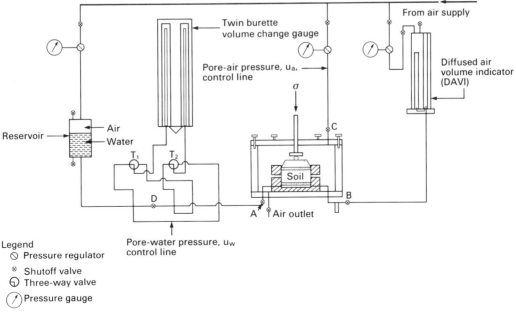

**Figure 10.32** Schematic diagram showing the plumbing layout for the control board of the modified direct shear apparatus.

be performed once or twice a day, or more frequently when high air pressures are used. The measured water volume changes should be adjusted in accordance with the diffused air volume measurements.

## 10.4 TYPICAL TEST RESULTS

The theory associated with various test methods was given in Chapter 9, while the equipment and procedures for testing were described earlier in this chapter. The test result presentation consists mainly of the data on the shear stress versus matric suction relationship (i.e., $\tau$ versus $(u_a - u_w)$ plane). The failure envelope on the shear strength versus matric suction plane is used to obtain the $\phi^b$ shear strength parameter. The nature of the shear strength versus net normal stress failure envelope at saturation (i.e., the $c'$ and $\phi'$ parameters) has been well explained in many soil mechanics publications.

Laboratory test results obtained from undisturbed and compacted, unsaturated soil specimens are presented in this section. However, only results from "identical" undisturbed or compacted soil specimens having the same initial dry density and water content can be analyzed to obtain the $\phi^b$ shear strength parameter.

Triaxial test data are presented in the following sections, followed by direct shear test data. The triaxial test data are categorized as: 1) consolidated drained test results, and 2) constant water content test results. Both cases are used to illustrate a linear $\phi^b$ shear strength parameter. Similar data are then presented illustrating a nonlinear relationship between shear strength and matric suction. These are followed by undrained and unconfined compression test data.

### 10.4.1 Triaxial Test Results

#### Consolidated Drained Triaxial Tests

A series of multistage, consolidated drained, triaxial tests on undisturbed specimens was performed by Ho and Fredlund (1982a). The specimens were from two residual soil deposits in Hong Kong, namely, decomposed granite and decomposed rhyolite. The decomposed granite specimens are mainly a silty sand, with an average specific gravity, $G_s$, of 2.65. The decomposed rhyolite specimens are essentially a sandy silt, having an average specific gravity of 2.66. The mineral compositions of these two soils are similar. Both soils are brittle and highly variable. Undisturbed specimens were sampled from boreholes and open cuts (i.e., block specimens).

Seventeen undisturbed specimens, 63.5 mm in diameter and approximately 140 mm in height, were tested. The tests were conducted in accordance with the *consolidated drained* triaxial testing procedure. For most tests, the deviator stress was removed once a maximum value was obtained for a particular stage (i.e., cyclic loading), while a new set of stresses were applied for the next stage. Some tests were performed with the stress changes between stages being applied, while leaving a constant strain rate being applied to the specimen (i.e., sustained loading) (see Chapter 9). The strain rate used in the testing program ranged from $1.7 \times 10^{-5}$ to $6.7 \times 10^{-5}\%/s$. A 5 bar high air entry disk (i.e., 505 kPa) was used for all tests. Angles of friction, $\phi'$, of 33.4° and 35.3° were obtained for the decomposed granite and rhyolite, respectively, from triaxial tests on saturated specimens.

Figure 10.33 presents typical test results from a decomposed granite specimen using the cyclic loading procedure. The test was performed by maintaining a constant net confining pressure, $(\sigma_3 - u_a)$, and varying the matric suction, $(u_a - u_w)$. The failure envelope was assumed to be a planar surface. Similar typical results from two rhyolite specimens are shown in Figs. 10.34 and 10.35. The results in Fig. 10.34 illustrate the cyclic loading procedure. The results in Fig. 10.35 illustrate the sustained loading proce-

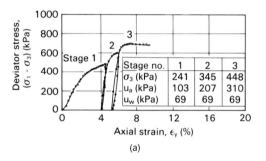

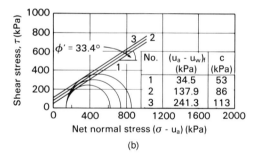

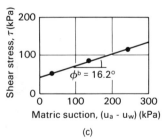

**Figure 10.33** Stress versus strain curves and two-dimensional presentations of the failure envelope for decomposed granite specimen no. 10. (a) Deviator stress versus strain curve; (b) failure envelope projected onto the $\tau$ versus $(\sigma - u_a)$ plane; (c) intersection line between the failure envelope and the $\tau$ versus $(u_a - u_w)$ plane at a zero net normal stress (i.e., $(\sigma_f - u_a)_f = 0$) (from Ho and Fredlund, 1982a).

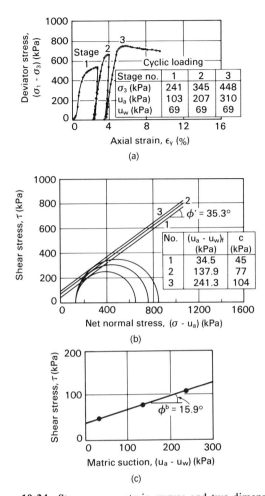

Figure 10.34 Stress versus strain curves and two-dimensional presentations of the failure envelope for decomposed rhyolite specimen no. 11C. (a) Deviator stress versus strain curve; (b) failure envelope projected onto the $\tau$ versus $(\sigma - u_a)$ plane; (c) intersection line between the failure envelope and the $\tau$ versus $(u_a - u_w)$ plane at zero net normal stress (i.e., $(\sigma_f - u_a)_f = 0$) (from Ho and Fredlund, 1982a).

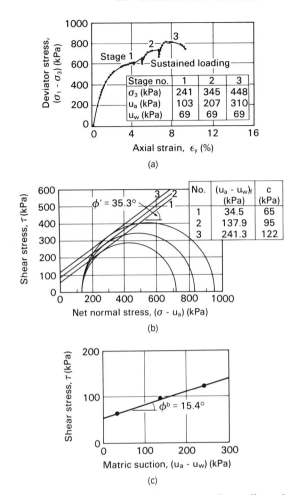

Figure 10.35 Stress versus strain curves and two-dimensional presentations of the failure envelope for decomposed rhyolite specimen no. 11D. (a) Deviator stress versus strain curve; (b) failure envelope projected onto the $\tau$ versus $(\sigma - u_a)$ plane; (c) intersection line between the failure envelope and the $\tau$ versus $(u_a - u_w)$ plane at zero net normal stress (i.e., $(\sigma_f - u_a)_f = 0$) (from Ho and Fredlund, 1982a).

dure. The average $\phi^b$ angles from all of the test results were found to be 15.3° for the decomposed granite and 13.8° for the rhyolite. It was observed that the soil structure of a specimen could be disturbed to a certain degree as the multistage test progressed. As a result, the measured peak, deviator stress for the last stage (i.e., stage no. 3) may actually be smaller than that obtained from a specimen under the same stress conditions, using a single-stage test. In this regard, the cyclic loading procedure appeared to be preferable to the sustained loading procedure in reducing soil structure disturbance. Part of the reduction in strength may also be due to nonlinearity in the shear strength versus matric suction relationship.

Two multistage triaxial tests on compacted specimens were reported by Krahn *et al.* (1987). The soil was sampled from a railway embankment at Notch Hill, British Columbia, and consisted of 10% clay, 85% silt, and 5% fine sand. The optimum water content was 21.5%, and the maximum dry density was 1590 kg/m³ when the soil was compacted in accordance with the standard AASHTO procedure. Specimens with a diameter of 38 mm and a height of 75 mm were trimmed for triaxial testing from the compacted soil. Consolidated undrained triaxial tests were performed on four compacted, saturated specimens with pore-water pressure measurements. The test results on the saturated specimens showed an angle of internal friction, $\phi'$, of 35° and an effective cohesion, $c'$, equal to 0.0.

The multistage triaxial tests on the unsaturated, compacted specimens were conducted using the *consolidated drained* test procedure (Chapter 9). The tests were conducted at a constant net confining pressure, while varying the matric suction. The test results obtained from two spec-

**286** 10 MEASUREMENT OF SHEAR STRENGTH PARAMETERS

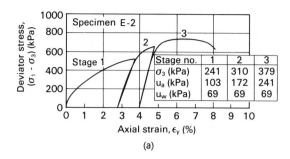

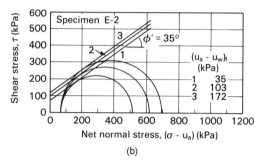

**Figure 10.36** Stress versus strain curves and two-dimensional presentations of the failure envelope for Tappen-Notch Hill Silt specimen no. E-2. (a) Deviator stress versus strain curve; (b) failure envelope projected onto the $\tau$ versus $(\sigma - u_a)$ plane.

imens are shown in Figs. 10.36 and 10.37. The combined results [Fig. 10.37(c)] indicate that the soil has a $\phi^b$ angle of 16° when a planar failure envelope is assumed.

### Constant Water Content Triaxial Tests

Constant water content or CW triaxial tests on a compacted shale and a compacted boulder clay were performed by Bishop *et al.* (1960) (Figs. 10.38 and 10.39, respectively). The shale had a clay fraction of 22%, and was compacted at a water content of 18.6%. A series of triaxial tests on the saturated specimens of the compacted shale gave an angle of internal friction, $\phi'$, of 24.8° and an effective cohesion, $c'$, of 15.8 kPa. The boulder clay had a clay fraction of 18%, and was compacted at a water content of 11.6%. The saturated boulder clay showed an effective angle of internal friction, $\phi'$, of 27.3° and an effective cohesion, $c'$, of 9.6 kPa. The tests on the compacted boulder clay were performed at a strain rate of $3.5 \times 10^{-5}\%/\text{s}$, and 15% strain was considered to represent failure. Assuming a planar failure envelope, the $\phi^b$ angle was 18.1° for the compacted shale and 22.0° for the boulder clay (Figs. 10.38 and 10.39).

### Nonlinear Shear Strength Versus Matric Suction

The significance of assuming a nonlinear failure envelope with respect to matric suction has been illustrated by Fredlund *et al.* (1987) (see Chapter 9). For example, the analyses of the triaxial test results on the compacted Dhanauri clay using a planar and a curved failure envelope were

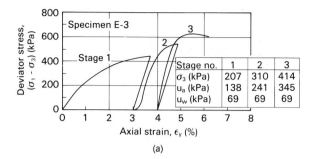

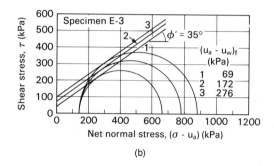

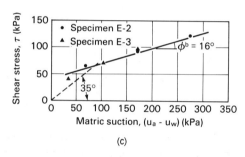

**Figure 10.37** Stress versus strain curves and two-dimensional presentations of the failure envelope for Tappen-Notch Hill Silt specimen no. E-3. (a) Deviator stress versus strain curve; (b) failure envelope projected onto the $\tau$ versus $(\sigma - u_a)$ plane; (c) intersection line between the failure envelope and the $\tau$ versus $(u_a - u_w)$ plane at zero net normal stress (i.e., $(\sigma_f - u_a)_f = 0$) (from Krahn, Fredlund and Klassen, 1987).

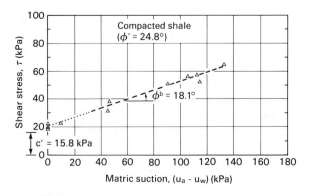

**Figure 10.38** Intersection line between the failure envelope and the $\tau$ versus $(u_a - u_w)$ plane for a compacted shale (data from Bishop, Alpan, Blight and Donald, 1960).

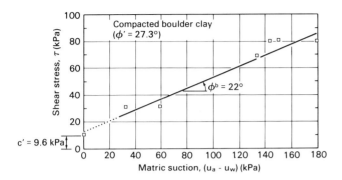

**Figure 10.39** Intersection line between the failure envelope and the $\tau$ versus $(u_a - u_w)$ plane for a compacted boulder clay (data from Bishop, Alpan, Blight and Donald, 1960).

compared. *Consolidated drained*, CD, and *constant water content*, CW, triaxial tests on compacted Dhanauri clay at two densities were conducted by Gulhati and Satija (1981). The Dhanauri clay consisted of 5% sand, 70% silt, and 25% clay. The soil had a liquid limit of 48.5% and a plastic limit of 25%. The saturated effective shear strength parameters (i.e., $c'$ and $\phi'$) for specimens compacted at two different densities were obtained from consolidated, undrained triaxial tests (Table 10.3). The consolidated drained and constant water content tests on the unsaturated, compacted specimens were performed at a strain rate of $1.3 \times 10^{-4}$ and $6.7 \times 10^{-4}$ %/s, respectively.

The test results on the unsaturated specimens were analyzed by Ho and Fredlund (1982a) using a planar failure envelope, and their results are summarized in Table 10.3. It appears that the linear interpretation of the failure envelope results in different $c'$ and $\phi^b$ parameters for the same soil tested using different procedures (i.e., CD and CW tests). In other words, the results give the impression that different test procedures may produce different shear strength parameters. For this case, the assumption of a planar failure envelope when analyzing the data causes a problem of nonuniqueness in the shear strength parameters. In addition, the $c'$ values obtained from the analysis do not agree with values obtained from triaxial tests on saturated specimens (Table 10.3).

The problem of nonuniqueness in the failure envelope necessitates a reevaluation of the shear strength data presented by Satija (1978). A reanalysis was performed assuming a curved failure envelop with respect to the matric suction axis (Fredlund *et al.* 1987). Figures 10.40 and 10.41 present the results for compacted Dhanauri clay at low and high densities, respectively. The results are plotted on the shear strength versus matric suction plane corresponding to a zero net normal stress at failure (i.e., $(\sigma_f - u_a)_f = 0$). The shear strength parameters, $c'$ and $\phi'$, obtained from the consolidated undrained tests on the saturated specimens (see Table 10.3) were used in the reanalysis. The curved failure envelopes have a cohesion intercept of $c'$ and a slope angle, $\phi^b$, equal to $\phi'$, starting at zero matric suction. The $\phi^b$ angle begins to decrease significantly at matric suction values greater than 50 kPa for the low-density specimens. The decrease in $\phi^b$ begins at matric suction values of 75–100 kPa for the high-density specimens. For the low-density specimens, the $\phi^b$ angle reaches a relatively constant value of 11° when the matric suction exceeds 150 kPa [Fig. 10.40(b)]. The $\phi^b$ angles for the high-density specimens reach a relatively constant value of 9° when the matric suction exceeds 300 kPa [Fig. 10.41(b)].

There is good agreement between the failure envelopes for the consolidated drained and constant water content test

**Table 10.3 Triaxial Tests on Compacted Dhanauri Clay (Data from Satija, 1978)**

| Initial Volume-Mass Properties | CU Tests on *Saturated* Specimens | | *Analysis* of Test Results on *Unsaturated* Specimens (Ho and Fredlund, 1982a) | | |
|---|---|---|---|---|---|
| | $c'$ (kPa) | $\phi'$ (degrees) | Type of Test | $c'$ (kPa) | $\phi^b$ (degrees) |
| Low density $\rho_d = 1478$ kg/m³ $w = 22.2\%$ | 7.8 | 29 | CD | 20.3 | 12.6 |
| | | | CW | 11.3 | 16.5 |
| High density $\rho_d = 1580$ kg/m³ $w = 22.2\%$ | 7.8 | 28.5 | CD | 37.3 | 16.2 |
| | | | CW | 15.5 | 22.6 |

Note: $\rho_d$ = dry density, $w$ = water content, CU = consolidated undrained, CD = consolidated drained, CW = constant water content.

288   10 MEASUREMENT OF SHEAR STRENGTH PARAMETERS

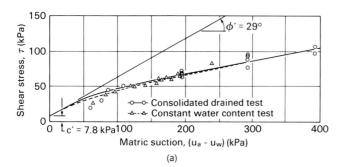

**Figure 10.40** Nonlinearity in the failure envelope with respect to matric suction for Dhanauri clay compacted to a low density. (a) Curved failure envelopes for Dhanauri clay compacted to a low density; (b) nonlinear relationship between $\phi^b$ and matric suction.

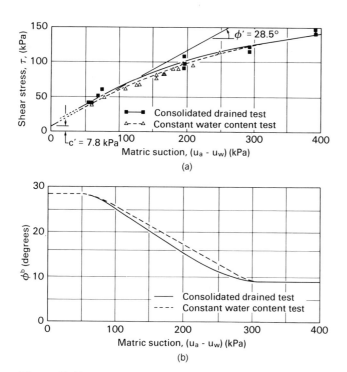

**Figure 10.41** Nonlinearity in the failure envelope with respect to matric suction for Dhanauri clay compacted to a high density. (a) Curved failure envelopes for Dhanauri clay compacted to a high density (data from Satija, 1978); (b) nonlinear relationship between $\phi^b$ and matric suction.

results when assuming a curved failure envelope with respect to matric suction. In other words, the assumption of a curved failure envelope leads to a unique failure envelope for the same soil tested using different stress paths or procedures. The uniqueness of the curved failure envelope is demonstrated at both densities.

It should be noted, however, that specimens prepared at different densities should be considered as different soils. Several procedures for accommodating the nonlinear failure envelope in engineering applications are described in Chapter 9.

### Undrained and Unconfined Compression Tests

Six series of undrained and unconfined compression (i.e., UC) tests on unsaturated, compacted specimens were performed by Chantawarangul (1983). The soil was a clayey sand consisting of 52% sand, 18% silt, and 30% clay. The soil had a liquid limit of 30%, a plastic limit of 19%, and a shrinkage limit of 16%. The soil was compacted using a miniature Harvard apparatus to give high- and low-density specimens at various water contents. In general, the water contents were on the dry side of optimum. The high- and low-density specimens correspond to a dry density, $\rho_d$, of approximately 1800 and 1700 kg/m$^3$, respectively.

The specimens were sheared under undrained conditions at a constant strain rate of 0.0017%/s. The test results are

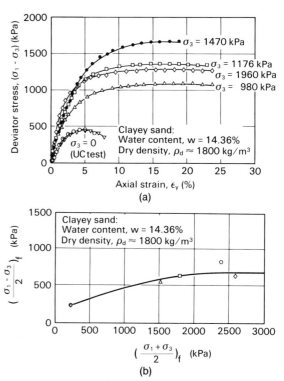

**Figure 10.42** Undrained triaxial and unconfined compression tests on a clayey sand compacted to a high density. (a) Deviator stress versus strain for various confining pressures; (b) total stress point envelope (from Chantawarangul, 1983).

10.4 TYPICAL TEST RESULTS    289

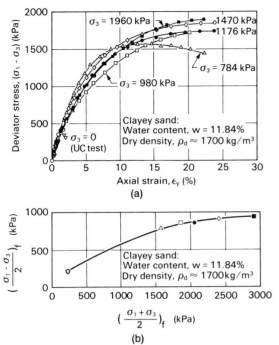

Figure 10.43 Undrained triaxial and unconfined compression tests on a clayey sand compacted to a low density. (a) Deviator stress versus strain for various confining pressures; (b) total stress point envelope (from Chantawarangul, 1983).

presented in Figs. 10.42 and 10.43 for the high- and low-density specimens, respectively. The results show a curved total stress point envelope which becomes a horizontal envelope at high confining pressures (see explanation in Chapter 9). The total stress point envelopes for specimens at various water contents are plotted in Fig. 10.44(a) for the high-density specimens, and in Fig. 10.44(b) for the low-density specimens. The envelopes also show a decrease in shear strength as the water content in the specimen increases.

### 10.4.2 Direct Shear Test Results

Multistage direct shear tests have been performed on saturated and unsaturated specimens of a compacted glacial till by Gan et al. (1987). The glacial till was sampled from the Indian Head area in Saskatchewan, and only material passing the no. 10 sieve was used to form specimens for testing. The soil consisted of 28% sand, 42% silt, and 30% clay. The liquid and plastic limits of the soil are 35.5% and 16.8%, respectively. Prior to testing, the soil was compacted in accordance with the AASHTO standard. The maximum dry density and the optimum water content are 1815 kg/m$^3$ and 16%, respectively.

The shear strength parameters, $c'$ and $\phi'$, were obtained from several single-stage and multistage direct shear tests on compacted specimens that had been saturated. The Indian Head glacial till was found to have an effective cohesion intercept of 10 kPa and an effective angle of internal friction of 25.5° (Fig. 10.45). Multistage direct shear test results on saturated specimens showed that a shear displacement rate of 1.2 mm was sufficient to mobilize the peak shear strength (Gan, 1986). Therefore, a shear displacement of 1.2 mm on specimens of 50 × 50 mm was selected as the failure criterion for subsequent multistage direct shear tests.

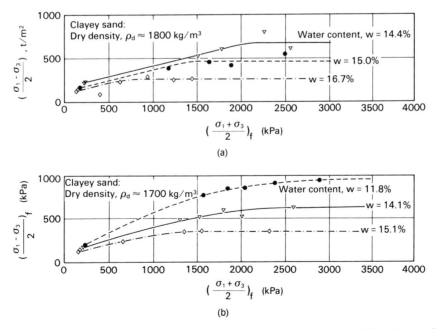

Figure 10.44 Total stress point envelopes obtained from undrained triaxial and unconfined compression tests. (a) Total stress point envelopes for high density specimens; (b) total stress point envelopes for low density specimens (from Chantawarangul, 1983).

**290** 10 MEASUREMENT OF SHEAR STRENGTH PARAMETERS

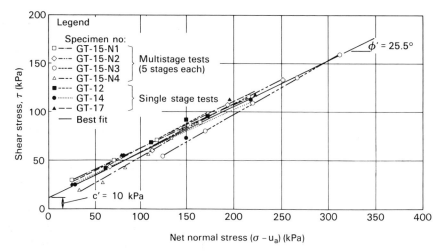

**Figure 10.45** Mohr-Coulomb failure envelopes for direct shear tests on saturated glacial till (from Gan, Fredlund and Rahardjo, 1988).

Five multistage, *consolidated drained* direct shear tests were performed on five compacted specimens. The initial volume–mass properties of the five specimens are tabulated in Table 10.4. The tests were run using the axis-translation technique on a modified direct shear apparatus (Gan and Fredlund, 1988). A displacement rate of $1.7 \times 10^{-4}$ mm/s was selected. Each specimen had three–seven stages of shearing. The tests were performed by maintaining a constant net normal stress, $(\sigma_n - u_a)$, of 72 kPa while varying the matric suction, $(u_a - u_w)$, between stages (Table 10.4).

The matric suction ranged from 0 to 500 kPa. As a result, the shear strength versus matric suction failure envelope was obtained, and the $\phi^b$ parameter could be computed.

Figures 10.46 and 10.47 show typical plots of water volume change and vertical deflection during consolidation prior to shearing. Matric suction equalization was generally attained in about one day. Typical results from the multistage direct shear tests on unsaturated specimens are illustrated in Figs. 10.48 and 10.49 for two specimens. The vertical deflection versus horizontal displacement curves

**Table 10.4 Multistage Direct Shear Tests on Unsaturated Glacial Till Specimens (from Gan et al. 1987)**

| Specimen No. | GT-16-N1 | | GT-16-N2 | | GT-16-N3 | | GT-16-N4 | | GT-16-N5 | |
|---|---|---|---|---|---|---|---|---|---|---|
| Initial Properties: | | | | | | | | | | |
| Void ratio, $e_0$ | 0.77 | | 0.53 | | 0.69 | | 0.51 | | 0.54 | |
| Degree of saturation, $S_0$ (%) | 42 | | 59 | | 48 | | 65 | | 61 | |
| Water content, $w_0$ (%) | 11.8 | | 11.5 | | 12.3 | | 12.2 | | 12.1 | |
| Stress State at Each Stage (kPa) | GT-16-N1 | | GT-16-N2 | | GT-16-N3 | | GT-16-N4 | | GT-16-N5 | |
| | $\sigma - u_a$ | $u_a - u_w$ | $\sigma - u_a$ | $u_a - u_w$ | $\sigma - u_a$ | $u_a - u_w$ | $\sigma - u_a$ | $u_a - u_w$ | $\sigma - u_a$ | $u_a - u_w$ |
| Stage No. 1 | 70.94 | 37.86 | 71.28 | 176.95 | 72.83 | 23.45 | 72.55 | 16.62 | 73.73 | 0.85 |
| 2 | 71.29 | 176.89 | 71.58 | 314.7 | 72.84 | 79.04 | 72.61 | 60.69 | 72.58 | 33.89 |
| 3 | 71.58 | 315.25 | 71.99 | 453.53 | 72.68 | 448.05 | 72.59 | 120.3 | 72.59 | 78.07 |
| 4 | | | | | 72.68 | 448.05 | 72.56 | 239.83 | 72.57 | 126.2 |
| 5 | | | | | | | 72.58 | 347.7 | 72.55 | 204.53 |
| 6 | | | | | | | 72.53 | 494.5 | 72.57 | 321.89 |
| 7 | | | | | | | | | 72.20 | 504.34 |

10.4 TYPICAL TEST RESULTS    291

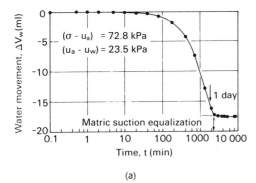

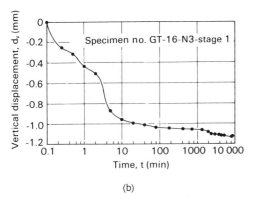

**Figure 10.46** Water volume change and consolidation of specimen no. GT-16-N3 during matric suction equalization. (a) Water volume change versus time curve; (b) vertical displacement of the specimen versus time (from Gan and Fredlund, 1988).

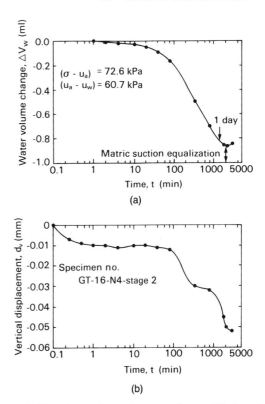

**Figure 10.47** Water volume change and consolidation of specimen no. GT-16-N4 during matric suction equalization. (a) Water volume change versus time curve; (b) vertical displacement of the specimen versus time (from Gan, Fredlund and Rahardjo, 1988).

[Figs. 10.48(b) and 10.49(b)] generally show that the soil dilated during shear, except during the initial stages at low matric suctions. As the matric suction was increased, the curves showed an increase in the specimen height with increasing horizontal displacement [Figs. 10.48(b) and 10.49(b)].

The shear stress normalized with respect to matric suction is plotted versus horizontal displacement in Figs. 10.48(c) and 10.49(c). The curves show a decrease in the peak normalized stress with increasing matric suction. These peak values appear to approach a relatively low but constant value at high matric suctions.

A typical plot of shear stress versus matric suction is shown in Fig. 10.50(a). The shear stress plotted corresponds to a shear displacement of 1.2 mm. The line joining the data points forms the shear stress versus matric suction failure envelope. The envelope corresponds to an average net normal stress of 72 kPa at failure [Fig. 10.50(a)]. The test results on the Indian Head glacial till exhibit significant nonlinearity in the failure envelope with respect to the matric suction. The varying $\phi^b$ angles along the curved failure envelope are plotted with respect to matric suction in Fig. 10.50(b).

Figure 10.51(a) presents a summary of the results obtained from five unsaturated specimens tested using the multistage direct shear test (Table 10.4). The results fall within a band, forming curved failure envelopes. The $\phi^b$ angles corresponding to the failure envelopes are plotted in Fig. 10.51(b) with respect to matric suction. The $\phi^b$ angles commence at a value equal to $\phi'$ (i.e., 25.5°) at matric suctions close to zero, and decrease significantly at matric suctions in the range of 50–100 kPa. The $\phi^b$ angles reach a fairly constant value ranging from 5° and 10° when the matric suction exceeds 250 kPa [Fig. 10.51(b)]. The scatter in the failure envelopes (Figs. 10.45 and 10.51) appears to be primarily due to slight variations in the initial void ratios of the soil specimens.

The nonlinearity of the failure envelope was also observed by Escario and Sáez (1986) from direct shear tests on three compacted soils. The properties, initial conditions, consolidation time, and displacement rate associated with the three soils are tabulated in Table 10.5. The tests were performed in a modified direct shear apparatus similar to that explained in the previous section. The *consolidated drained* testing procedure was used, along with the axis-translation technique. Figures

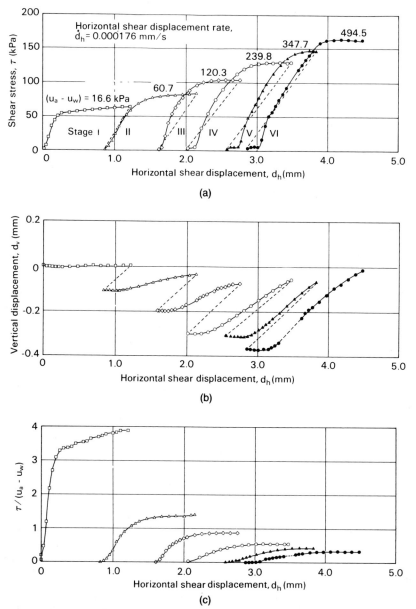

**Figure 10.48** Multistage direct shear tests results on unsaturated glacial till specimen no. GT-16-N4. (a) Shear stress versus horizontal displacement curves; (b) vertical displacement versus horizontal displacement curves; (c) $\tau/(u_a - u_w)$ versus horizontal displacement curves (from Gan, Fredlund and Rahardjo, 1988).

10.4 TYPICAL TEST RESULTS    293

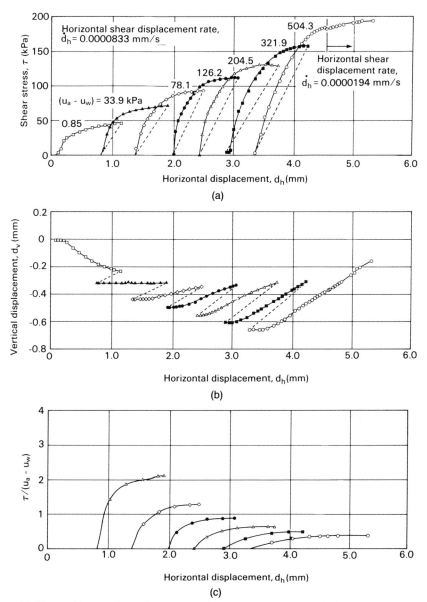

**Figure 10.49** Multistage direct shear test results on unsaturated glacial till specimen no. GT-16-N5. (a) Shear stress versus horizontal displacement curves; (b) vertical displacement versus horizontal displacement curves; (c) $\tau/(u_a - u_w)$ versus horizontal displacement curves (from Gan, Fredlund and Rahardjo, 1988).

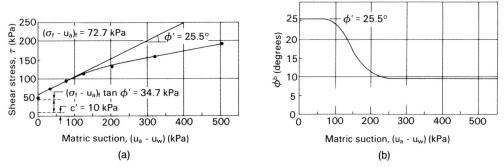

**Figure 10.50** Failure envelope obtained from unsaturated glacial till specimen no. GT-16-N5. (a) Failure envelope on the $\tau$ versus $(u_a - u_w)$ plane; (b) relationship between the $\phi^b$ values and matric suction (from Gan and Fredlund, 1988).

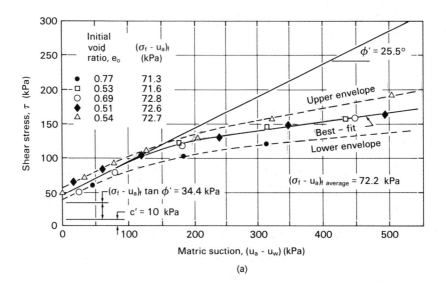

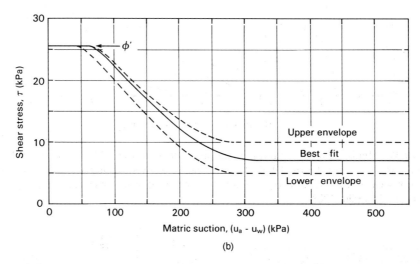

**Figure 10.51.** Failure envelopes obtained from unsaturated glacial till specimens. (a) Failure envelopes on the $\tau$ versus $(u_a - u_w)$ plane; (b) the $\phi^b$ values corresponding to the upper, lower and best-fit failure envelopes (from Gan, Fredlund and Rahardjo, 1988).

**Table 10.5 Direct Shear Tests on Unsaturated, Compacted Soils (from Escario and Sáez, 1986)**

| Properties | Madrid Grey Clay | Red Clay of Guadalix de la Sierra | Madrid Clayey Sand |
|---|---|---|---|
| Liquid limit | 71 | 33 | 32 |
| Plasticity index | 35 | 13.6 | 15 |
| Sieve analysis: % passing | | | |
| 10 | — | — | 100 |
| 16 | — | 100 | 94 |
| 40 | 100 | 97 | 48 |
| 200 | 99 | 86.5 | 17 |
| AASHTO standard compaction | | | |
| $\rho_d$ max (kg/m$^3$) | 1330 | 1800 | 1910 |
| $w$ optimum (%) | 33.7 | 17 | 11.5 |
| Initial conditions | | | |
| $\rho_{d0}$ (kg/m$^3$) | 1330 | 1800 | 1910 |
| $w_0$ (%) | 29 | 13.6 | 9.2 |
| $(u_a - u_w)_0$ (kPa) | 8.5 | 2.8 | 0.7 |
| Consolidation time under applied total stress and matric suction (days) | 4 | 4 | 4 |
| Displacement rate, $d_h$ (mm/s) | $2.8 \times 10^{-5}$ | $2.8 \times 10^{-5}$ | $2.8 \times 10^{-5}$ |
| Time to failure, $t_f$ (days) | 2.5–3 | 2–3 | 1–2 |

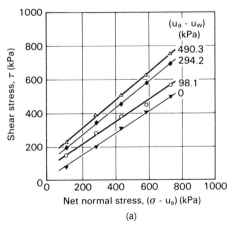
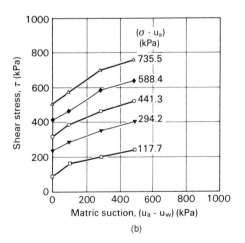

**Figure 10.52** Direct shear tests on compacted red clay of Guadalix de la Sierra. (a) Horizontal projections of the failure envelope onto the $\tau$ versus $(\sigma - u_a)$ plane; (b) horizontal projections of the failure envelope onto the $\tau$ versus $(u_a - u_w)$ plane (from Escario and Sáez, 1986).

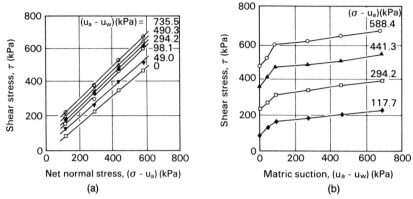

**Figure 10.53** Direct shear tests on compacted Madrid clayey sand. (a) Horizontal projections of the failure envelope onto the shear stress versus $(\sigma - u_a)$ plane; (b) horizontal projections of the failure envelope onto the shear stress versus $(u_a - u_w)$ plane (from Escario and Sáez, 1986).

10.52 and 10.53 present the results from compacted red clay of Guadalix de la Sierra and Madrid clayey sand specimens, respectively. The test results on Madrid grey clay were presented previously in Chapter 9. All of the above results exhibit nonlinearity in the failure envelope with respect to the matric suction axis. At the same time, the failure envelopes, with respect to the net normal stress, are essentially linear [i.e., Figs. 10.52(a) and 10.53(a)]. From a practical engineering standpoint, it should be noted that the applied matric suctions shown in the above test programs extended over an extremely wide range (e.g., 0 to 700 kPa).

# CHAPTER 11

# *Plastic and Limit Equilibrium*

Deformation problems encountered in soil mechanics can be divided into two categories in accordance with the stress level involved. When the deviator stress levels are relatively low, the problems are considered to be in the elastic range and are analyzed using elasticity theories. When the stress levels are relatively high, the problems are considered to be in the plastic range and are analyzed using plasticity theories. The two categories can be visualized on an idealized representation of a stress versus strain curve (Fig. 11.1).

This chapter considers problems based on the assumption that the soil behaves perfectly plastic. When the state of plastic equilibrium is limited to a specific, thin zone, the problems are referred to as limit equilibrium analyses. The thin plastic zone is called a slip surface or a slip plane. These assumptions are, of course, gross simplifications of the real soil behavior, but have formed a useful categorization of the types of analyses common to soil mechanics practice. These categories provide theoretical limits within which the behavior of a soil mass is studied.

There are three main types of soil mechanics analyses common to plastic equilibrium. These are: 1) lateral earth pressure analyses, 2) bearing capacity analyses, and 3) slope stability analyses. Each satisfies the equations of equilibrium and a condition of failure (i.e., failure criterion). Typical soil mechanics analyses associated with each of these categories are derived in this chapter. In each case, the soil is assumed to have negative pore-water pressures (or matric suctions).

The plasticity analyses can be subdivided into those situations where the pore–water pressure effects are simulated in some manner during the testing of the soil (i.e., total stress approach), and those situations where pore pressure designations become a part of the analysis (i.e., stress state variable approach). Most of the consideration in this chapter is given to the latter situation where the actual or predicted pore pressures are designated.

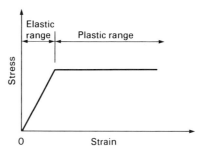

**Figure 11.1** Idealized elastic-plastic behavior giving rise to two categories of deformation analysis.

## 11.1 EARTH PRESSURES

Some of the earliest work in soil mechanics dealt with earth pressures on retaining walls. However, there is little information on the earth pressures exerted on engineering structures by unsaturated soils. Pressures exerted by expansive soils have been of concern, but a general earth pressure theory for these soils has only recently been proposed. It is not sufficient to say that cohesive backfill should not be used behind earth retaining type structures. Experience indicates that problems associated with the performance of earth structures can often involve compacted, clayey soils. Ireland (1964) showed that 68% of unsatisfactory retaining wall performance considered in his study used either clay as a backfill or were founded upon clay.

Some of the problems encountered with earth retaining structures result from the tendency of clayey expansive soils to undergo substantial changes in volume as a consequence of changing environmental conditions. There also are numerous situations where unsaturated soils are used as backfill or where structural members are cast in place against the soil. It is apparent that only limited consideration has been given to the behavior of the retaining structure under these circumstances.

This section presents a simplistic but theoretical analysis of earth pressures for soils with negative pore–water pressures. Appropriate expressions for the critical height of vertical or nearly vertical cuts are presented. The groundwater conditions are assumed to represent steady-state conditions.

Prior to discussing active and passive earth pressures in a soil mass, it is of value to discuss at rest earth pressure conditions. The meaning of the at rest condition is of particular interest since negative pore–water pressures cause a soil mass to shrink and crack near ground surface. Even the meaning of the term "at rest" must be reassessed. The mechanism behind the formation of desiccation cracks is of particular interest. This topic is given consideration prior to discussion of the active and passive earth pressure states.

### 11.1.1 At Rest Earth Pressure Conditions

The total vertical stress in a level soil mass is computed in the same manner for both saturated and unsaturated soils (Fig. 11.2). The total vertical stress, $\sigma_v$, is called the overburden pressure:

$$\sigma_v = \int_0^H \rho g \, dy \tag{11.1}$$

where

- $\rho$ = total density
- $g$ = gravitational acceleration
- $y$ = vertical distance from ground surface
- $H$ = depth of soil under consideration.

For a homogeneous soil mass, the total vertical stress can be written

$$\sigma_v = \rho g \, H. \tag{11.2}$$

The pore–air pressure is generally at equilibrium with atmospheric pressure. The pore–water pressure above the groundwater table can either be estimated or measured. In some cases, the estimate can be based on hydrostatic conditions.

The horizontal pressure at any depth below ground surface can be written as a ratio of the vertical pressure. Each of the pressures can be referenced to the pore–air pressure (or atmospheric pressure). Let us define the coefficient of earth pressure at rest, $K_0$, as follows:

$$K_0 = \frac{(\sigma_h - u_a)}{(\sigma_v - u_a)}. \tag{11.3}$$

It is difficult to theoretically quantify the coefficient of earth pressure at rest due to complexities arising from the stress history to which the soil mass has been subjected. However, consideration of elastic equilibrium within a soil mass can provide some insight into the coefficient of earth pressure at rest. Elastic equilibrium can also provide some indication of the depth of potential cracking in a soil mass.

The constitutive relations for the soil structure, in elasticity form, are presented in Chapter 12, and are used herein for the elastic equilibrium analysis. The stress versus strain equation in the vertical direction for a homogeneous, isotropic, unsaturated soil is written

$$\epsilon_v = \frac{(\sigma_v - u_a)}{E} - \frac{2\mu}{E}(\sigma_h - u_a) + \frac{(u_a - u_w)}{H} \tag{11.4}$$

where

- $\epsilon_v$ = normal strain in the vertical direction
- $\sigma_v$ = total normal stress in the vertical direction
- $\sigma_h$ = total normal stress in the horizontal direction
- $\mu$ = Poisson's ratio
- $E$ = elastic modulus with respect to a change in $(\sigma - u_a)$
- $H$ = elastic modulus with respect to a change in $(u_a - u_w)$
- $u_a$ = pore-air pressure
- $u_w$ = pore-water pressure.

The stress versus strain equation in the horizontal direction is written

$$\epsilon_h = \frac{(\sigma_h - u_a)}{E} - \frac{\mu}{E}(\sigma_v + \sigma_h - 2u_a) + \frac{(u_a - u_w)}{H}. \tag{11.5}$$

Equation (11.5) applies to both horizontal directions. For the at rest or $K_0$ condition in an intact, homogeneous, unsaturated soil mass, the strain in the horizontal directions can be set to zero (i.e., $\epsilon_h = 0$). The net horizontal stress can be written in terms of the vertical stress from Eq. (11.5):

$$(\sigma_h - u_a) = \frac{\mu}{1-\mu}(\sigma_v - u_a)$$

$$- (1 - \mu)\frac{E}{H}(u_a - u_w). \tag{11.6}$$

Equation (11.6) can be normalized to the net vertical stress, and the equation takes the form for the coefficient

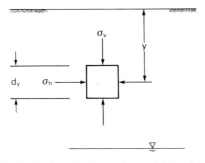

**Figure 11.2** Vertical and horizontal stress designation in a soil mass.

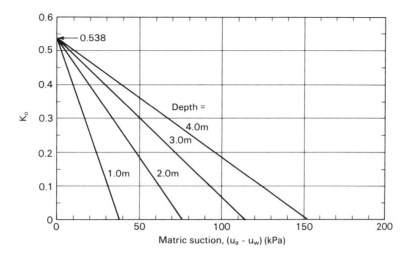

**Figure 11.3** The relationship between the coefficient of earth pressure at rest, $K_0$, and matric suction.

of earth pressure at rest:

$$K_0 = \frac{\mu}{1 - \mu} - \frac{E}{(1 - \mu)H} \frac{(u_a - u_w)}{(\sigma_v - u_a)}. \quad (11.7)$$

Equation (11.7) reverts to the form for a saturated soil when the matric suction goes to zero. When matric suction is present in the soil, the horizontal stress is reduced. The reduction is also a function of the depth under consideration. At shallow depths, a relatively small matric suction will cause the net horizontal stress to go to zero and tend to go negative. If the soil cannot sustain any tensile strain, cracking of the soil will occur, commencing at ground surface.

Typical $K_0$ values for the first loading of a clay would range between approximately 0.3 and 0.7, depending upon Poisson's ratio. Let us consider the slow drying of a lacustrine deposit. For illustrative purposes, the following properties are assumed: $\mu = 0.35$, $E/H = 0.17$, and $\rho = 1886$ kg/m$^3$. Figure 11.3 illustrates the relationship between the coefficient of earth pressure at rest and matric suction for various overburden pressures. When the soil is saturated with zero pore-water pressure, the at rest coefficient of earth pressure is 0.538. The at rest coefficient then decreases as the matric suction of the soil increases. This is true for all depths, but the rate of reduction in the at rest coefficient is greater at shallow depths. An at rest coefficient, $K_0$, of zero indicates a tendency for cracking.

The above example is a simplification which does not take into consideration the effects of previous wetting and drying, loading and unloading, to which a soil deposit may have been subjected. Figure 11.4 illustrates typical horizontal and vertical effective stress paths where a saturated soil is subjected to a history of sedimentation, followed by erosion and subsequent reloading. The stress paths can become even more complex for unsaturated soils subjected to cycles of drying and wetting. The coefficients of earth pressure can go from as low as zero to as high as the coefficient of passive earth pressure.

If all of the elastic parameters were known, and the above analyses were applied to a soil that had undergone a complex stress history, the coefficient of earth pressure should be a tangent value as opposed to a secant value (Fig. 11.4).

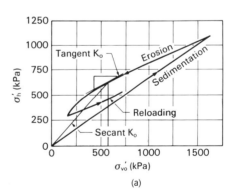

(a)

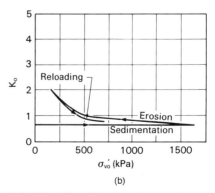

(b)

**Figure 11.4** The effect of a changing overburden stress during sedimentation, erosion, and reloading of a saturated soil. (a) The effective horizontal stress for various effective vertical stresses; (b) the coefficient of earth pressure at rest, $K_0$, for various stress histories (from Morgenstern and Eisenstein, 1970).

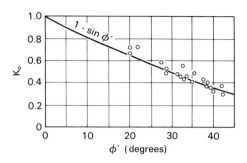

**Figure 11.5** Relationship between the effective angle of internal friction and the coefficient of earth pressure at rest, $K_0$ (from Bishop, 1958).

The engineer is generally interested in the secant value. The above analysis is most relevant immediately following sedimentation. Even in this case, it must be recognized that elastic parameters are difficult to assess accurately. It is primarily for this reason that empirical expressions have been proposed for the at rest coefficient of earth pressure.

Bishop (1959) presented the results of a comprehensive laboratory study into the coefficient of earth pressure at rest, which for all practical purposes supported Jaky's (1944) earth pressure expression (Fig. 11.5):

$$K_0 = 1 - \sin\phi' \quad (11.8)$$

where

$\phi'$ = effective angle of internal friction.

The equation applies for initial or first loading of the soil. Others have also lent support for this equation (Simons, 1958; Brooker and Ireland, 1965). Test results obtained by Bishop (1957, 1958) and Simons (1958) are shown in Table 11.1.

The compaction of granular soils against an unyielding wall can produce horizontal pressures greater than the vertical pressure (i.e., $K_0$ greater than 1.0). Expansive soils can also exert lateral pressures greater than the vertical pressure. It is even possible to reach the passive pressure state, at which time the soil fails in shear. This condition is illustrated in the oedometer specimen shown in Fig. 11.6. The $K_0$ value for soils has also been shown to be a function of the overconsolidation ratio of the soil. An increase in the overconsolidation ratio produces an increase in $K_0$ (Ladd, 1971).

### 11.1.2 Estimation of Depth of Cracking

The coefficient of earth pressure at rest can be used to give an indication of the depth of cracking in a soil. The assumption is made that the at rest coefficient of earth pressure, $K_0$, is zero at the bottom of a crack. The above analysis assumes that the soil cannot sustain any tensile strain prior to failing.

Let us consider a desiccated soil with vertical cracks extending to a depth, $y_c$, as shown in Fig. 11.7. At the bottom of the crack, the net horizontal stress is zero (i.e., $(\sigma_h - u_a) = 0$). At this point, Eq. (11.6) becomes

$$(\sigma_v - u_a)_c = \frac{E}{\mu H}(u_a - u_w)_c \quad (11.9)$$

where the subscript, $c$, refers to the bottom of the crack. Equation (11.9) indicates that the crack depth for a homogeneous soil depends on the matric suction and the elastic parameters of the soil.

Numerous assumptions could be made concerning the matric suction variation with respect to depth. One typical matric suction profile is illustrated in Fig. 11.8, which shows the negative pore–water pressure as a linear function of the distance above the groundwater table (profile $A$). The variable, $f_w$, is used to permit the pore–water pressure to be represented as a percentage of the hydrostatic profile where a value greater than 1.0 signifies pore–water pres-

**Table 11.1 Coefficient of Earth Pressure At Rest According to the Test Results of Bishop (1957, 1958) and Simons (1958)**

| Type of Soil | Liquid Limit, $w_L$ | Plastic Limit, $w_p$ | Plasticity Index, $I_p$ | Activity | $K_0$ |
|---|---|---|---|---|---|
| Loose, saturated sand | — | — | — | — | 0.46 |
| Dense, saturated sand | — | — | — | — | 0.36 |
| Compacted residual clay | — | — | 9.3 | 0.44 | 0.42 |
| Compacted residual clay | — | — | 31 | 1.55 | 0.66 |
| Undisturbed, organic, silty clay | 74.0 | 28.6 | 45.4 | 1.2 | 0.57 |
| Remoulded kaolin | 61 | 38 | 23 | 0.32 | 0.66 |
| Undisturbed marine clay | 37 | 21 | 16 | 0.21 | 0.48 |
| Quick clay | 34 | 24 | 10 | 0.18 | 0.52 |

**Figure 11.6** Passive-type failure of compacted clay till resulting from swelling in an oedometer ring.

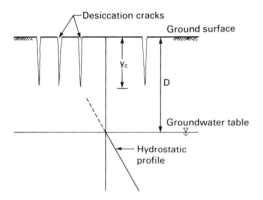

**Figure 11.7** A typical desiccated soil with cracks extending down from ground surface.

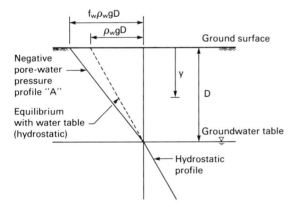

**Figure 11.8** Idealized matric suction profile "$A$" with the matric suction varying linearly to the water table.

sures that are more negative than the hydrostatic profile. This type of profile would represent a net upward flux. The matric suction at any depth, $d$, in the soil can be written

$$(u_a - u_w)_y = f_w \rho_w g (D - y) \quad (11.10)$$

where

$D$ = distance from ground surface to the water table.

The net vertical stress at any depth, $y$, in a homogeneous deposit can be written

$$(\sigma_v - u_a)_y = \rho g\, y. \quad (11.11)$$

Substituting Eqs. (11.10) and (11.11) into Eq. (11.9) gives an equation for the depth of cracking, $y_c$:

$$y_c = \frac{D}{1 + \dfrac{\mu \rho H}{f_w \rho_w E}}. \quad (11.12)$$

The above analysis assumes that the soil has little tensile strength. Lau (1987) demonstrated that including the tensile strain at failure in the above analysis will decrease the predicted crack depth by about 0.1 m.

To illustrate the form of Eq. (11.12), let us assume that the soil has a total density of 1886 kg/m$^3$ and a matric suction profile equivalent to hydrostatic conditions. Figure 11.9 shows the depth of cracking as a ratio of the distance to the water table for various elasticity parameter, $E/H$, ratios. Lau (1987) showed that for an initially saturated clay, the $E/H$ ratios are typically in the range of 0.15–0.20. For a Poisson's ratio of 0.35, the anticipated depth of cracking would be approximately 20% of the distance to the water table. The depth of cracking increases for the case of larger suction values. For example, when $f_w$ is equal to 2.0, the depth of cracking for the above conditions increases to 34% of the distance to the water table.

Little research has been done to verify the above analysis for the depth of cracking. At this point, the analysis primarily provides an insight into the physics related to the problem. The analysis has assumed a simplistic stress history for the soil.

### 11.1.3 Extended Rankine Theory of Earth Pressures

The active and passive earth pressures for an unsaturated soil can be determined by assuming that the soil is in a state of plastic equilibrium. Let us first review the stresses in a soil mass when the surfaces of failure are planar. The major and minor principal planes at all points are assumed to have similar directions. The solution is known as Rankine's earth pressure theory. For an unsaturated soil, it is necessary to extend some of the conventional concepts. For this reason, the theory is called the extended Rankine theory of earth pressures.

Figure 11.10 shows a vertical, frictionless plane passed through a soil mass of infinite depth. An element of unsaturated soil at any depth is subjected to a vertical stress, $\sigma_v$, and a horizontal stress, $\sigma_h$. These planes are assumed to be principal planes, and the vertical and horizontal stresses are the principal stresses. The ground surface is horizontal, and the vertical stress is written in terms of the soil density.

The state of stress for the unsaturated soil element is shown on an extended Mohr diagram in Fig. 11.11. The equation corresponding to the limiting or failure condition (i.e., shear strength equation) can be written as follows (see

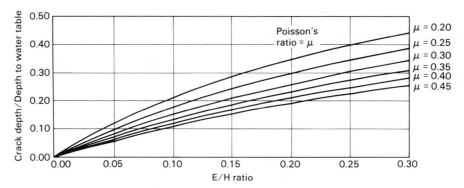

**Figure 11.9** $E/H$ ratio versus the ratio of the crack depth to the depth to water table for matric suctions corresponding to hydrostatic conditions (i.e., $f_w = 1.0$).

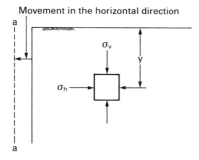

**Figure 11.10** Stresses on an element in a soil mass behind a frictionless wall when there is horizontal movement.

Chapter 9):

$$\tau = c' + (\sigma_n - u_a)\tan\phi' + (u_a - u_w)\tan\phi^b \quad (11.13)$$

where

- $\tau$ = shear strength
- $c'$ = effective cohesion
- $\sigma_n$ = total normal stress on the failure plane
- $\phi^b$ = angle of friction with respect to changes in matric suction.

Equation (11.13) can be written in a form similar to that used for saturated soils:

$$\tau = c + (\sigma_n - u_a)\tan\phi'. \quad (11.14)$$

The total cohesion, $c$, is written as

$$c = c' + (u_a - u_w)\tan\phi^b. \quad (11.15)$$

Using this form for the shear strength of an unsaturated soil has the advantage that derivations relevant to saturated soils can readily be modified to accommodate situations where the soil is unsaturated. It is simply necessary to remember that the cohesion of a soil consists of two components (i.e., effective cohesion and the matric suction component).

The initial, total vertical stress in a soil mass is equal to the overburden pressure, $\rho g y$. The total horizontal pressure is equal to the coefficient of earth pressure at rest multiplied by the overburden pressure. The assumption is made that the soil has an initial matric suction equal to $(u_a - u_w)_0$. It is also assumed that the pore-water pressure is controlled by environmental conditions which remain constant during

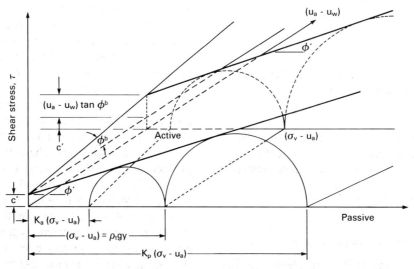

**Figure 11.11** Active and passive earth pressures for a soil with matric suction.

the analysis. Changes in total stress are assumed not to influence the pore pressures.

## Active Earth Pressure

Let us suppose that the wall, $a$–$a$, in Fig. 11.10 is allowed to move away from the soil mass. The horizontal stress is reduced until a limiting value corresponding to the plastic equilibrium state is attained. Thus, failure is obtained by reducing the horizontal stress. As a result, the horizontal stress must be the minor principal stress, while the vertical stress is the major principal stress. The horizontal stress at any point corresponding to the active state can be computed from the vertical stress and the failure criterion for the soil.

Figure 11.11 illustrates how the active and passive pressures in a soil change as the matric suction changes. As the matric suction increases, the active pressure is shown to decrease. In other words, as the pore-water pressure in the soil goes more negative, the soil becomes stronger. This means that less force would be carried by the retaining wall.

Let us consider a vertical plane corresponding to a specific matric suction, as shown in Fig. 11.10. An element from a depth, $y$, has an overburden stress, $\sigma_v$. This element is shown in Fig. 11.12, along with the definition of pertinent variables. If the wall moves away from the soil, the active earth pressure will be developed, and it is designated as $(\sigma_h - u_a)$. The horizontal pressure can be written in terms of the vertical pressure, $(\sigma_v - u_a)$, by considering the geometrics of the Mohr circle:

$$\sin \phi' = \frac{((\sigma_v - u_a) - (\sigma_h - u_a))/2}{\frac{(\sigma_h - u_a) + (\sigma_v - u_a)}{2} + c \cot \phi'} \quad (11.16)$$

where

$c$ = total cohesion (i.e., $c = c' + (u_a - u_w) \tan \phi^b$).

Rearranging Eq. (11.16) and solving for $(\sigma_h - u_a)$ gives

$$(\sigma_h - u_a) = (\sigma_v - u_a) \frac{1 - \sin\phi'}{1 + \sin\phi'} - 2c \frac{\cos\phi'}{1 + \sin\phi'}. \quad (11.17)$$

The following trigonometric relation can be used to simplify Eq. (11.17):

$$\frac{\cos\phi'}{1 + \sin\phi'} = \sqrt{\frac{1 - \sin\phi'}{1 + \sin\phi'}}. \quad (11.18)$$

Equation (11.17) can now be written

$$(\sigma_h - u_a) = (\sigma_v - u_a) \frac{1 - \sin\phi'}{1 + \sin\phi'} - 2c \sqrt{\frac{1 - \sin\phi'}{1 + \sin\phi'}}. \quad (11.19)$$

The trigonometric function appearing in Eq. (11.19) can be written in terms of the angle of the slip planes from a vertical plane:

$$\frac{1 - \sin\phi'}{1 + \sin\phi'} = \tan^2\left[45 - \frac{\phi'}{2}\right] \quad (11.20)$$

Terzaghi and Peck (1967) used the variable, $N_\phi$, to designate the above trigonometric relation:

$$\frac{1}{N_\phi} = \tan^2\left[45 - \frac{\phi'}{2}\right]. \quad (11.21)$$

The active pressure, $(\sigma_h - u_a)$, for an element of soil at any depth can be written

$$(\sigma_h - u_a) = (\sigma_v - u_a) \frac{1}{N_\phi} - 2c \frac{1}{\sqrt{N_\phi}}. \quad (11.22)$$

Equation (11.22) can also be rewritten by taking into ac-

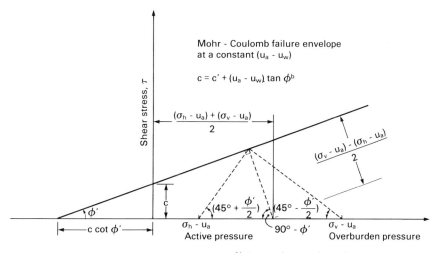

**Figure 11.12** Mohr circle construction for the active case.

count the components of cohesion:

$$(\sigma_h - u_a) = (\sigma_v - u_a)\frac{1}{N_\phi} - 2c'\frac{1}{\sqrt{N_\phi}}$$
$$- 2(u_a - u_w)\tan\phi^b \frac{1}{\sqrt{N_\phi}}. \quad (11.23)$$

### Coefficient of Active Earth Pressure

Let us define the coefficients of active earth pressure as the ratio of the net horizontal pressure to the net vertical pressure:

$$K_a = \frac{(\sigma_h - u_a)}{(\sigma_v - u_a)}. \quad (11.24)$$

Referring to Eq. (11.23), the coefficient of active earth pressure can also be written as

$$K_a = \frac{1}{N_\phi} - \frac{2c'}{(\sigma_v - u_a)}\frac{1}{\sqrt{N_\phi}}$$
$$- 2\frac{(u_a - u_w)}{(\sigma_v - u_a)}\tan\phi^b \frac{1}{\sqrt{N_\phi}}. \quad (11.25)$$

### Active Earth Pressure Distribution (Constant Matric Suction with Depth)

The horizontal pressure corresponding to the active state can be computed for various depths and plotted as shown in Fig. 11.13. For the active case, conjugate planes are formed in the soil mass at angles of $45 + \phi'/2$ to the horizontal, as shown in Fig. 11.13(b). The saturated soil case is designated by use of the effective cohesion. Let us suppose that the matric suction were a constant value with depth. Then the total cohesion is also a constant with respect to depth, and the active pressure distribution is translated to the left, parallel to the saturated soil case. Figure 11.14 shows the breakdown of the active pressure into its three components.

### Tension Zone Depth

The tension zone depth, $y_t$, against the wall can be computed by setting the total horizontal pressure to zero and assuming an atmospheric air pressure (i.e., $u_a = 0$) in Eq. (11.22) or (11.23):

$$y_t = \frac{2c'}{\rho g}\sqrt{N_\phi} + 2\frac{(u_a - u_w)\tan\phi^b}{\rho g}\sqrt{N_\phi}. \quad (11.26)$$

The tension zone depth, $y_t$, is equal to the depth of the vertical cracking, $y_c$, when the tensile strength of the soil is assumed to be negligible. The tension zone depth increases as the matric suction of the soil increases. This depth corresponds to the zone which would pull away from the wall as the wall is moved horizontally away from the soil. Figure 11.15 illustrates how matric suction causes a soil to pull away from the wall.

### Active Earth Pressure Distribution (Linear Decrease in Matric Suction to the Water Table)

Let us assume that the matric suction in the soil decreases with depth to a value of zero at the water table (Fig. 11.16). The matric suction at ground surface is designated as a ratio of the hydrostatic pressure condition using the factor, $f_w$. The negative pore–water pressure at ground surface for hydrostatic conditions can be written as a function of the distance from the groundwater table:

$$(u_a - u_w)_h = \rho_w g D \quad (11.27)$$

where

$(u_a - u_w)_h$ = matric suction at ground surface
$D$ = depth from ground surface to the water table.

A simple relationship can be used to define the variation in matric suction with depth for this profile. For a depth,

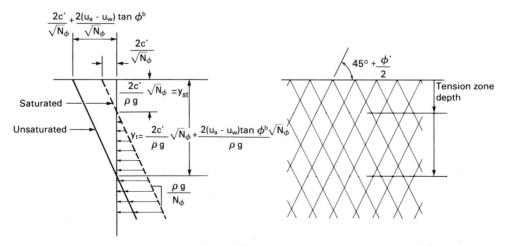

**Figure 11.13** Rankine's active earth pressure distribution for a saturated soil and a soil with a constant matric suction.

11.1 EARTH PRESSURES   305

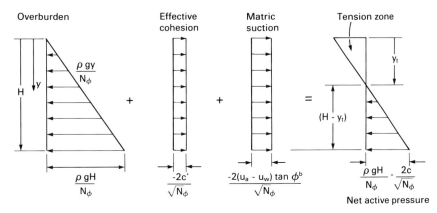

**Figure 11.14** Components of the active pressure distribution when the matric suction is constant with respect to depth.

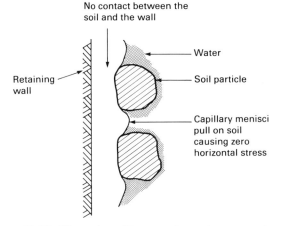

**Figure 11.15** Illustration of how matric suction causes the soil to pull away from a retaining wall.

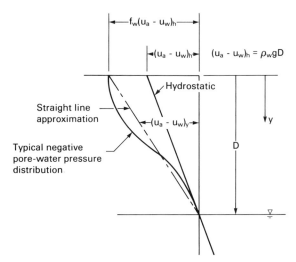

**Figure 11.16** Procedure used to designate the matric suction profile.

$y$, less than or equal to $D$, the matric suction is equal to

$$(u_a - u_w)_y = f_w(u_a - u_w)_h \left(1 - \frac{y}{D}\right). \quad (11.28)$$

The active pressure, $p_a$, at any depth above the water table is equal to

$$p_a = (\sigma_v - u_a)\frac{1}{N_\phi} - \frac{2c'}{\sqrt{N_\phi}}$$
$$- \frac{2f_w(u_a - u_w)_h \tan\phi^b}{\sqrt{N_\phi}}\left(1 - \frac{y}{D}\right). \quad (11.29)$$

The active pressure distribution diagram, along with the plot of each of the components is shown in Fig. 11.17.

The tension zone depth, $y_t$, can be computed by setting the total horizontal stress to zero and assuming an atmospheric air pressure (i.e., $u_a = 0$) in Eq. (11.27):

$$y_t = \frac{2c'\sqrt{N_\phi} + 2f_w(u_a - u_w)_h \tan\phi^b \sqrt{N_\phi}}{\rho g + \frac{2\sqrt{N_\phi}}{D}f_w(u_a - u_w)_h \tan\phi^b} \quad (11.30)$$

where $y_t$ must be less than $D$.

### Active Earth Pressure Distribution When the Soil has Tension Cracks

The soil behind a retaining wall often has tension cracks. It should be noted that the depth of tension cracks in the soil must be considered as being analytically independent from the tension zone depth against the retaining wall. The depth of cracking is given the variable, $y_c$, and the soil above this depth can be considered as a surcharge load applied to the underlying soil (Fig. 11.18). The matric suction at the bottom of the tension cracks is expressed as a ratio of the hydrostatic pressure condition by using the factor, $f_w$. Equation (11.28) applies to the case under consideration as long as the depth is below the bottom of the tension cracks.

The surcharge load, $q_s$, must be applied *below a depth of $y_c$*, and is equal to the overburden pressure (i.e., $q_s = \rho g y_c$). An appropriate total density, $\rho$, must be used for the upper soil with the tension cracks. The active pressure above the water table can be derived in a manner similar to the previous cases. The active pressure at any depth can

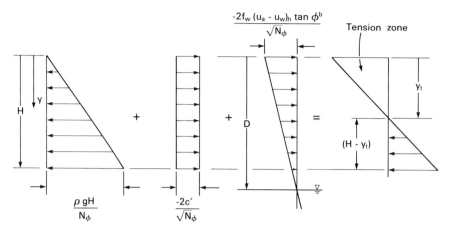

**Figure 11.17** Components of the active pressure distribution when the matric suction decreases linearly with depth.

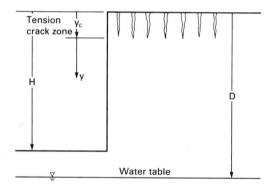

**Figure 11.18** Designation of variables when the backfill soil has tension cracks.

be written as

$$p_a = \frac{(\sigma_v - u_a)}{N_\phi} - \frac{2c'}{\sqrt{N_\phi}} - 2f_w \frac{(u_a - u_w)_h \tan\phi^b}{\sqrt{N_\phi}} \cdot \left[1 - \frac{y}{D - y_c}\right] + \frac{q_s}{N_\phi}. \quad (11.31)$$

Equation (11.31) is an extension of Eq. (11.29), and shows the role of a surcharge load in affecting the active earth pressure. The components which make up the active earth pressure diagram for all depths are shown in Fig. 11.19. Other types of surcharge load can be applied in a similar manner.

The tension zone depth, $y_t$, for this case is computed as the depth where the horizontal stress is zero:

$$y_t = \frac{2c'\sqrt{N_\phi} + 2f_w(u_a - u_w)_h \tan\phi^b \sqrt{N_\phi} - q_s}{\rho g + 2\dfrac{\sqrt{N_\phi}}{(D - y_c)} f_w(u_a - u_w)_h \tan\phi^b}. \quad (11.32)$$

Equation (11.32) applies as long as the tension zone depth, $y_t$, is less than the distance $(D - y_c)$.

Other assumptions could be made regarding the distribution of the pore–water pressures with respect to depth. The same type of formulation can be used to compute the active earth pressure state at any depth.

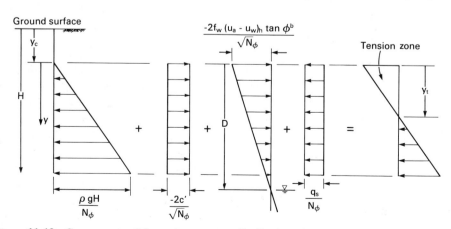

**Figure 11.19** Components of the active pressure distribution when the soil has tension cracks and the matric suction decreases linearly with depth.

## Passive Earth Pressure

If the wall in Fig. 11.10 is moved towards the soil (i.e., the soil is compressed), the horizontal pressure is increased to be greater than the vertical pressure. When failure of the soil mass is attained, the total horizontal stress is the major principal stress, and the total vertical stress is the minor principal stress. The horizontal stress corresponding to the passive state can be computed from the vertical stress and the failure criterion for the soil.

Figure 11.11 illustrates the passive pressure in a soil mass as a function of matric suction. As the matric suction increases, the passive pressure is shown to increase.

Let us consider a vertical plane corresponding to a specific matric suction plane, as shown in Fig. 11.11. An element from a depth, $y$, has an overburden pressure, $\sigma_v$. This element is shown in Fig. 11.20, along with the definition of pertinent variables. If the wall moves into the soil mass, the passive pressure can be defined as $(\sigma_h - u_a)$. The horizontal pressure can be derived in terms of the vertical pressure, $(\sigma_v - u_a)$, in a manner similar to the active pressure derivation:

$$\sin\phi' = \frac{\dfrac{(\sigma_h - u_a) - (\sigma_v - u_a)}{2}}{\dfrac{(\sigma_h - u_a) + (\sigma_v - u_a)}{2} + c \cot\phi'} \quad (11.33)$$

where

$c$ = total cohesion (i.e., $c = c' + (u_a - u_w) \tan\phi^b$).

Rearranging Eq. (11.33), the horizontal stress can be written

$$(\sigma_h - u_a) = (\sigma_v - u_a)\frac{1 + \sin\phi'}{1 - \sin\phi'} - 2c\frac{\cos\phi'}{1 - \sin\phi'}.$$

$$(11.34)$$

Similar trigonometric functions to those used in the active pressure analysis can be used to rewrite Eq. (11.34):

$$(\sigma_h - u_a) = (\sigma_v - u_a)N_\phi + 2c' \sqrt{N_\phi}$$
$$+ 2(u_a - u_w)\tan\phi^b \sqrt{N_\phi} \quad (11.35)$$

where

$N_\phi = (1 + \sin\phi')/(1 - \sin\phi')$ or $\tan^2(45 + \phi'/2)$.

## Coefficient of Passive Earth Pressure

The coefficient of passive earth pressure can be written as the ratio of the net horizontal pressure to the net vertical pressure. Dividing Eq. (11.35) by the net vertical pressure gives

$$K_p = N_\phi + \frac{2c' \sqrt{N_\phi}}{(\sigma_v - u_a)} + \frac{2(u_a - u_w)\tan\phi^b \sqrt{N_\phi}}{(\sigma_v - u_a)}.$$

$$(11.36)$$

Equations (11.25) and (11.36) show that both the active and passive earth pressure coefficients vary with the overburden pressure.

## Passive Earth Pressure Distribution (Constant Matric Suction with Depth)

The horizontal pressure corresponding to the passive state can be computed for various depths and plotted as shown in Fig. 11.21. For the passive case, conjugate planes are formed at $45 - \phi'/2$ to the horizontal, as shown in Fig. 11.21. The saturated soil case is designated by the use of the effective cohesion term. Since the total cohesion is assumed to be a constant with respect to depth, the passive pressure is translated to the right as the suction is increased. Figure 11.22 shows a breakdown of the passive pressure distribution into its three components.

The entire soil mass is in a state of compression for passive pressure conditions. At *ground surface*, the total horizontal pressure is a function of the total cohesion:

$$p_p = 2c' \sqrt{N_\phi} + 2(u_a - u_w)\tan\phi^b \sqrt{N_\phi}. \quad (11.37)$$

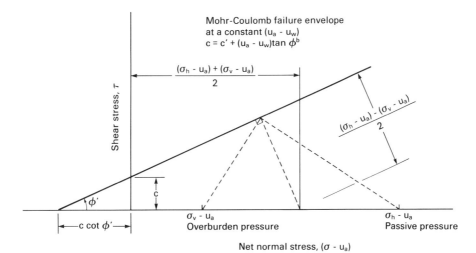

**Figure 11.20** Mohr-circle construction for the passive earth pressure case.

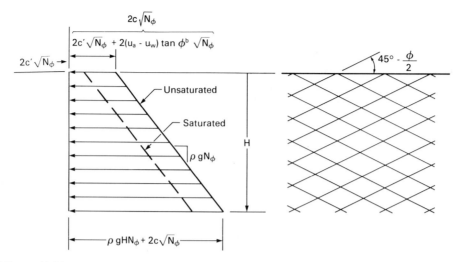

**Figure 11.21** Rankine's passive earth pressure distribution for a saturated soil and a soil with a constant matric suction.

### Passive Earth Pressure Distribution (Linear Decrease in Matric Suction to the Water Table)

It is possible to select various distributions for matric suction with respect to depth. Let us assume that the matric suction linearly decreases with depth to a value of zero at the water table (Fig. 11.16). The matric suction at any depth can be written as a linear equation [Eq. (11.28)]. For this matric suction distribution, the passive pressure, $p_p$, can be written as follows:

$$p_p = (\sigma_v - u_a)N_\phi + 2c'\sqrt{N_\phi}$$
$$+ 2f_w(u_a - u_w)_h \tan\phi^b \left(1 - \frac{y}{D}\right)\sqrt{N_\phi}. \quad (11.38)$$

The passive pressure distribution, along with a plot of the components is shown in Fig. 11.23. The existence of tension cracks in the soil is not of relevance for the passive pressure case. The assumption is made that the cracks would close and the soil mass would become intact as the passive pressure is applied.

### Deformations Associated with Active and Passive States

Studies by Terzaghi (1954) showed the relationship between the movement of a wall and the earth pressure coefficients for dense and loose sands (Fig. 11.24). Most unsaturated soils behave similar to a dense soil in that relatively low displacements are required to develop the active and passive states. The results indicate that displacements as low as $0.001H$ are required to develop the active state in a dense soil. More displacement is required to develop the passive state.

Lambe and Whitman (1968) presented triaxial test results on a dense sand, and summarized the percent strains required for the active and passive states. It was concluded

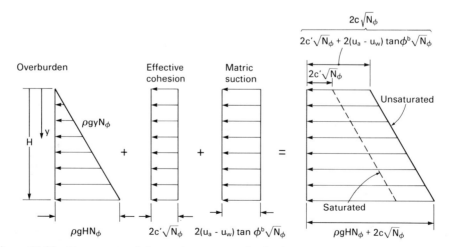

**Figure 11.22** Components of the passive pressure distribution when the matric suction is constant with respect to depth.

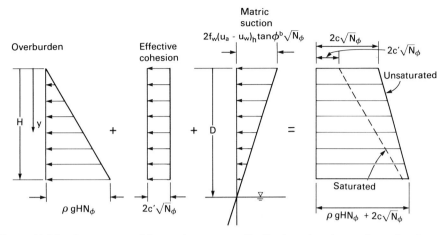

**Figure 11.23** Components of the passive pressure distribution when the matric suction decreases linearly with depth.

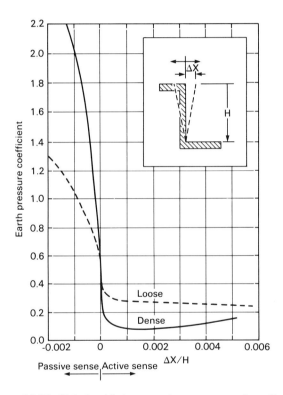

**Figure 11.24** Relationship between the movement of a wall and the earth pressures developed for different densities (from Terzaghi, 1954).

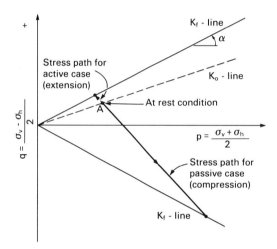

**Figure 11.25** Stress paths for Rankine's active and passive conditions.

that low strains on the order of 0.5% can produce the active state. In compressive loading, strains of 0.5% develop about one-half of the passive resistance. As much as 2% was required to reach the full passive resistance for a dense sand. Figure 11.25 illustrates the difference between the active and passive stress paths. The larger stress changes associated with the development of the passive state is one of the reasons for a larger strain at failure.

The strains at failure in unsaturated soils could be quite different from those mentioned above because of unusual $K_0$ conditions. Little research has been done on this subject for unsaturated soils.

### 11.1.4 Total Lateral Earth Force

Retaining walls serve to retain the backfill placed behind the wall (Fig. 11.26). If the walls were fixed, the earth pressure would tend to be close to the at rest pressure state. However, movement of the wall away from the soil gives rise to the development of the active earth force against the wall. Movement of the wall into the soil results in the development of the passive earth force against the wall. If the wall is vertical and smooth, the lateral force is equal to the sum of the active or passive pressures at all depths. There may be an active earth force developed against one side of the wall, and a passive earth force developed against the other side, as shown in Fig. 11.26.

When there is friction between the soil and the wall,

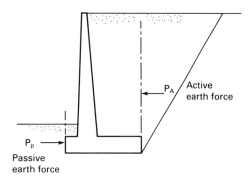

**Figure 11.26** Earth pressures acting on a retaining wall.

analyses such as those proposed by Coulomb can be used. The shape of the slip surface is changed, and is dependent upon the vertical movement of the wall as well as the horizontal movement. Only the case of an unsaturated soil placed against a smooth, vertical wall is considered in this chapter. However, unsaturated soil analyses involving a wall with friction are essentially the same as those for a saturated soil. The difference lies in the fact that the cohesion has a component related to the matric suction of the soil.

*Active Earth Force*

The active earth force, $P_A$, against a smooth wall is equal to the active earth pressure integrated from the bottom of the tension zone depth, $y_t$, to the bottom of the wall, $H$. In other words, it is possible to integrate the active pressure equation over the entire depth. The upper portion of the net pressure diagram is in a state of tension. It can be assumed that the soil in this zone cannot adhere to the retaining wall. Figure 11.27 illustrates the components of the pressure diagram and the limits of integration on the active pressure diagram:

$$P_A = \int_{y_t}^{H} p_a \, dy. \quad (11.39)$$

In order for Eq. (11.39) to yield a positive total force, the depth of the tension cracks, $y_t$, must be less than $H$. Let us consider the case where the matric suction is a constant value with respect to depth:

$$P_A = \int_{y_t}^{H} \frac{\rho g y \, dy}{N_\phi} - \int_{y_t}^{H} \frac{2c' \, dy}{\sqrt{N_\phi}}$$
$$- \int_{y_t}^{H} \frac{2(u_a - u_w) \tan\phi^b \, dy}{\sqrt{N_\phi}}. \quad (11.40)$$

Integrating Eq. (11.40) gives an equation which is an extension of the active force equation for saturated soils:

$$P_A = \frac{\rho g}{N_\phi} \frac{(H^2 - y_t^2)}{2} - \frac{2c'}{\sqrt{N_\phi}} (H - y_t)$$
$$- \frac{2(u_a - u_w) \tan\phi^b}{\sqrt{N_\phi}} (H - y_t) \quad (11.41)$$

As the matric suction increases, the active force decreases. It is possible to reach a condition where the active force is zero. This situation is referred to as an unsupported excavation, and is dealt with later in this chapter.

Equation (11.41) assumes that the soil density is a constant with respect to depth. Figure 11.28 illustrates a typical active earth force against a retaining wall as a function of the matric suction of the soil. Two heights of wall are considered (i.e., 8 and 12 m). The density of the soil is assumed to be 1800 kg/m³. The shear strength parameters of the soil are an effective cohesion of 5 kPa, an effective angle of internal friction of 22°, and a $\phi^b$ angle of 14°. As the matric suction is increased from 0 to 100 kPa on the 8 m high wall, the total active pressure is reduced from 206 to 35 kN/m, respectively. In other words, the matric suction has a large influence on the active force against a retaining wall.

The matric suction in a soil may not remain at a constant value with time, and therefore, the active force against the wall may vary. The highest force will occur when the matric suction goes to zero. The results illustrate the importance of matric suction being maintained in the soil. At the

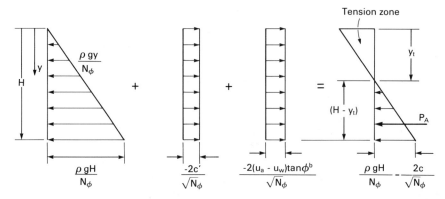

**Figure 11.27** Components of the active earth pressure diagram showing the limits of integration.

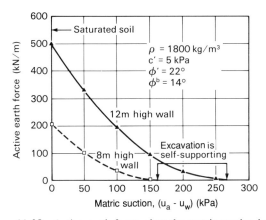

**Figure 11.28** Active earth force when the matric suction is constant with respect to depth.

same time, it is readily recognized that it is difficult to analytically predict the long-term matric suction in the soil.

Equations for the active earth force can also be written for other situations. Let us consider the case where matric suction decreases linearly with depth. The water table is assumed to be 1 m below the base of the wall. The negative pore–water pressures are assumed to decrease in a hydrostatic manner (i.e., $f_w = 1.0$):

$$P_A = \frac{\rho g (H^2 - y_t^2)}{2N_\phi} - \frac{2c'(H - y_t)}{\sqrt{N_\phi}}$$
$$- \left\{\frac{2f_w(u_a - u_w)_h \tan\phi^b}{\sqrt{N_\phi}}\right\}\left(H - \frac{H^2}{2D} - y_t - \frac{y_t^2}{2D}\right). \quad (11.42)$$

Typical results using Eq. (11.42) are shown in Fig. 11.29. The soil properties and the heights of the wall are the same as for the previous example. The results show that when matric suction decreases linearly with depth (which is often the case), the active earth force does not decrease as rapidly with a changing matric suction as when the suctions are constant with depth. Equation (11.42) represents the compressive components of the active force diagram.

A surcharge term must be included in Eq. (11.42) when the soil has tension cracks to a depth, $y_c$. Figure 11.30 shows the variation in the active earth force for the above example with tension cracks 2 m deep. The results show that when other variables remain constant, tension cracks increase the active earth force. In all of the above cases, the active earth force corresponding to zero matric suction refers to a saturated soil with zero pore–water pressures.

The point of application of the resultant force can be computed by considering the line of action of the force associated with each component of the active earth pressure diagram. The point of application becomes lower on the wall as the matric suction of the soil increases. When the tension zone depth is equal to the height of the wall, the active earth force is zero. That is, the soil should stand without any support if the height of the vertical excavation is less than the tension zone depth.

*Passive Earth Force*

The passive earth force, $P_p$, against a smooth wall is equal to the passive pressure integrated over the depth under consideration. When the matric suction is constant with respect to depth, Eq. (11.35) can be integrated over its entire depth:

$$P_p = \frac{\rho g N_\phi H^2}{2} + 2c'\sqrt{N_\phi}\, H$$
$$+ 2(u_a - u_w)\tan\phi^b \sqrt{N_\phi}\, H. \quad (11.43)$$

If the matric suction decreases linearly with respect to depth, Eq. (11.38) can be integrated over the entire depth:

$$P_p = \frac{\rho g N_\phi H^2}{2} + 2c'\sqrt{N_\phi}\, H$$
$$+ 2f_w(u_a - u_w)_h \tan\phi^b \sqrt{N_\phi}\left(H - \frac{H^2}{2D}\right). \quad (11.44)$$

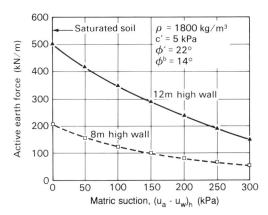

**Figure 11.29** Active earth force when the matric suction decreases linearly to a water table one meter below the base of the wall.

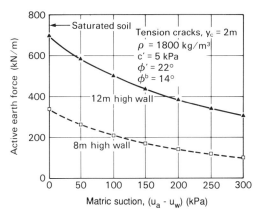

**Figure 11.30** Active earth force when the soil has tension cracks and a linear decrease in the matric suction to a water table one meter below the base of the wall.

Tension cracks are assumed to close for the passive earth force case. Equations (11.43) and (11.44) can be solved using typical soil properties in order to illustrate the effect of matric suction on the passive earth force.

The soil properties and wall dimensions used to illustrate the variation in active earth force with matric suction are again used to compute the passive earth force. Figure 11.31 shows the passive earth force for the case of a constant matric suction with respect to depth. The passive earth force is shown to increase significantly as the matric suction of the soil increases. The point of application of the resultant force can be computed using the line of action of the force associated with each component of the passive earth pressure diagram.

Figure 11.32 shows the passive earth force developed against a smooth wall when the suction decreases linearly with depth. The water table is 1 m below the bottom of the wall, and the pore-water pressure distribution is hydrostatic. It can be seen that the passive earth force in this case is less sensitive to suction changes than when the suction is a constant value with respect to depth.

### 11.1.5 Effect of Changes in Matric Suction on the Active and Passive Earth Pressure

Changes in the environment may result in an ingress of water around the retaining wall backfill. The water results in a change in the active or passive pressure. For the active earth pressure case, a decrease in matric suction results in an increase in the pressure against the wall. The magnitude of the pressure developed depends on whether the wall moves in response to the pressure change. If the wall cannot move or does not move a sufficient amount, the pressure developed against the wall can become even greater than the computed active pressure corresponding to the saturated soil case.

Let us consider a clay compacted behind a retaining wall, as shown in Fig. 11.33. The changes in stresses at a specific depth can be visualized using an extended Mohr-Cou-

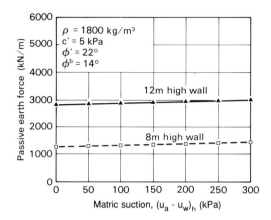

**Figure 11.32** Passive earth force when the matric suction decreases linearly to a water table one meter below the base of the wall.

lomb diagram (Fig. 11.34). Circle $A$ shows the net vertical and the net horizontal pressure at a depth, $y$, for the active pressure state. As the water content in the soil is increased, the matric suction decreases. Under unrestricted movement of the wall, the horizontal pressure will follow a stress path over to the stress circle $A_1$. The magnitude of the pressure against the wall is increased to the active pressure state for a saturated soil.

Let us suppose, on the other hand, that the retaining structure is fixed, and the initial stress state is represented by the at rest condition shown by circle $C$. As the soil becomes wet and the matric suction decreases, the horizontal pressure change will follow the stress path to circle $C_1$. At this point, the horizontal pressure is greater than the vertical pressure.

Let us also consider the possibility where the initial at rest coefficient of earth pressure is 1.0. The stress changes in the soil could now follow a path to circle $D_1$, where the final horizontal pressure is even higher than that shown for the previous stress case (i.e., $C_1$).

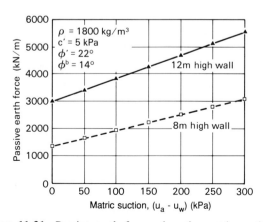

**Figure 11.31** Passive earth force when the matric suction is constant with respect to depth.

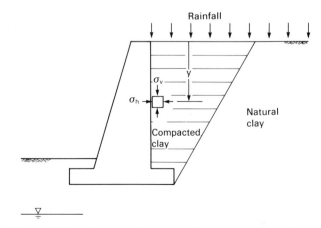

**Figure 11.33** A compacted clay subjected to a change in the surrounding environment.

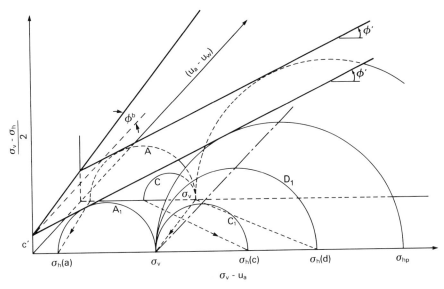

**Figure 11.34** Lateral pressures on a wall restricted to limited movement.

The exact stress path followed by the soil during wetting is dependent upon: 1) the initial at rest pressure state, 2) deformation moduli for the soil, and 3) the rigidity of the retaining structure. In other words, the problem is one involving soil–structure interaction. It would be necessary to use a numerical technique such as the finite element method in order to more closely model the stress changes.

There are undoubtedly conditions in practice where backfill with a high matric suction is placed against a rigid wall or where a heavy structural member is cast against a clay with an at rest coefficient greater than 1.0. In these circumstances, the horizontal stresses can become as large as the passive resistance of the soil. Generally, this would correspond to the passive resistance of the saturated soil. When this happens, either the structural member may fail or the soil may fail in shear. If a structural member is designed so that it is not free to move, it must be designed to resist the passive earth pressure.

### Relationship Between Swelling Pressures and the Earth Pressures

It can be asked, "What is the relationship between the above earth pressures and the swelling pressure of a soil?" Suppose the swelling pressure of the soil was measured in an oedometer using the "constant volume" procedure (Chapter 13). The measured swelling pressure corresponds to conditions of no volume change. The swelling pressure will be a function of the initial stress state and the change in matric suction. However, it is difficult to relate the swelling pressure to the active or passive earth pressure states. The stress paths followed in the two situations are different.

In the earth retaining structure, vertical stress will be the overburden pressure. Its magnitude is constant, while in the laboratory oedometer test, all vertical and horizontal movement is restricted (i.e., using the "constant volume" testing technique). To more closely simulate the *in situ* retaining structure, it would be necessary to subject the laboratory specimen to a vertical pressure corresponding to a particular depth and immerse the specimen in water. It would then be necessary to measure the lateral swelling pressure of the soil in order to obtain an indication of the horizontal pressure against a retaining structure. In other words, it is difficult to apply conventional swelling pressure measurements to retaining wall design.

The stress paths shown in Fig. 11.34 assist in understanding why walls with cohesive backfills may undergo large movements during the life of the structure. As the soil dries, a crack may open between the soil and the wall. Dust and debris collect in the opening, partially filling the crack. As the matric suction decreases during wetter seasons, the soil swells and pushes the wall until the resistance of the wall is in equilibrium with the soil mass. This cycle may be repeated many times over the years, gradually moving the wall.

### 11.1.6 Unsupported Excavations

Frequently, the slopes of temporary cuts or excavations in cohesive soils are allowed to stand unsupported in a vertical or near vertical position. If the slope is assumed to fail along a planar surface, the critical height of a vertical slope, $H_c$, can be computed by summing the forces acting on a sliding wedge (Fig. 11.35) (Pufahl *et al.* 1983). The plane of failure is assumed to occur at an angle, $45 + \phi'/2$, to the horizontal. The shear force mobilized along the failure surface must take into account the strength due to matric suction:

$$S_m = c'\beta + (N - u_a\beta)\tan\phi' + (u_a - u_w)\beta\tan\phi^b$$

(11.45)

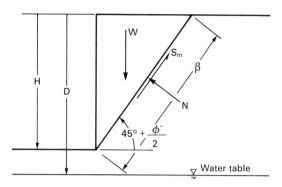

**Figure 11.35** Unsupported vertical slope in an intact soil mass with matric suction.

where

$\beta$ = length of the failure surface.

It is possible to obtain the critical height from the calculation of the tension zone depth (Fig. 11.36). At a depth, $y_t$, the net horizontal stress is equal to zero. At a depth less than $y_t$, the pressure against the wall is negative, provided a crack does not open up between the wall and the soil. Integrating over the entire depth, it can be observed that the total wall force will be zero when the height of the slope is equal to twice the tension zone depth (Terzaghi and Peck, 1967).

$$H_c = 2y_t. \quad (11.46)$$

If the height of the vertical bank is smaller than $H_c$, the bank should be able to stand without lateral support. This analysis assumes that the soil can sustain tensile stress without cracking. The unsupported height, $H_c$, will be reduced when tension cracks are present.

### Effect of Tension Cracks on the Unsupported Height

If the upper zone of the soil profile is weakened by tension cracks of a depth, $y_c$, the maximum unsupported height of a vertical bank, $H_c$, can be computed (Fig. 11.37) by summing forces parallel and perpendicular to the sliding surface. It is also possible to recognize that the critical height is equal to twice the tension zone depth [i.e., Eq. (11.46)].

Figure 11.38 shows the critical height of a vertical bank plotted versus the matric suction. The assumption is made that the matric suction is a constant value with depth. Figure 11.39 shows the critical height for the situation where matric suction decreases linearly with depth. The matric suction distributions correspond to the case where the water table is at 9 and 13 m below ground surface. The matric suctions shown on the abscissa correspond to the matric suction at ground surface. Hydrostatic conditions corresponding to a surface matric suction of 88 kPa relate to a water table depth of 9 m, and those of a surface matric suction of 128 kPa relate to a water table depth of 13 m. All other soil properties are the same as those previously used for the active and passive earth pressure examples.

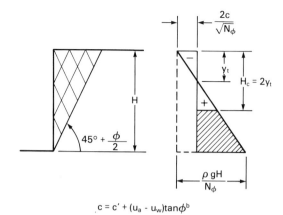

$c = c' + (u_a - u_w)\tan\phi^b$

**Figure 11.36** Illustration of the critical height of a slope.

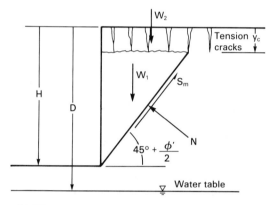

**Figure 11.37** Unsupported vertical slope in an unsaturated soil mass with tension cracks.

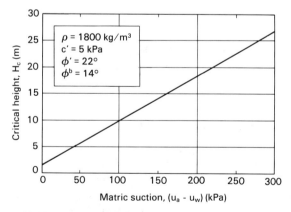

**Figure 11.38** Critical height versus matric suction for the case of constant matric suction versus depth.

The critical height of a vertical slope in an unsaturated soil increases substantially as the matric suction is increased. Since matric suction is quite effective in increasing the critical height of a slope (or in increasing the factor of safety of a slope), an attempt should be made to maintain the matric suction in the soil behind an excavated slope.

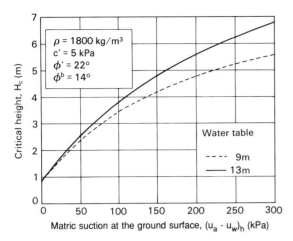

**Figure 11.39** Critical height versus surface matric suction for the case where the matric suction decreases with depth.

Such obvious measures as covering the bank adjacent to an excavation with an impermeable membrane and diverting surface runoff during heavy rains would be beneficial. Tensiometers or other gauges that can measure negative pore-water pressure could be used to monitor changes in soil suction near the base of tension cracks. Such a technique would provide a rational approach to the stability of unsupported excavations.

## 11.2 BEARING CAPACITY

The bearing capacity of unsaturated soils, like other plasticity problems, can be viewed as an extension of saturated soil mechanics. The unsaturated soil can be visualized as having a cohesion consisting of two components. One component is the effective cohesion, and the other component is due to matric suction [Eq. (11.15)]. With this concept in mind, the conventional bearing capacity theory is applicable to unsaturated soils.

Prandtl (1921) presented the case of a strip footing with a smooth base located at ground surface (Fig. 11.40). The load, $Q$, was increased until the strip footing penetrated the soil indefinitely. The unit pressure, $q_f$, at this point is called the ultimate bearing capacity. This gives rise to an active pressure zone immediately below the footing and a passive zone where the soil pushes laterally and upward. The intermediate portion of the slip surface is defined by a logarithmic spiral. The soil is assumed to be weightless. The rigorous and demanding nature of the solution to this idealized problem has led to the consideration of other approximate solutions.

### 11.2.1 Terzaghi Bearing Capacity Theory

The Terzaghi bearing capacity formulation assumes that the base of the footing is rough, and that the slip surface is bounded by a straight line and a logarithmic spiral [Fig. 11.41(a)]. The load, $Q$, at failure is calculated by considering the forces on the sliding mass of soil (Terzaghi, 1943).

For foundations embedded at some depth, the soil above the footing base was treated as a surcharge pressure [Fig. 11.41(b)]. The bearing capacity solution consists of three components. One component, $q_\gamma$, takes into account the weight of the soil and the passive earth pressure block. This portion of bearing capacity can be written as follows:

$$q_\gamma = \tfrac{1}{2}\rho g B N_\gamma \quad (11.47)$$

where

$N_\gamma$ = proportionally bearing capacity factor.

The second component of the bearing capacity equation is related to the cohesion of the soil:

$$q_c = c N_c \quad (11.48)$$

where

$N_c$ = cohesion bearing capacity factor
$c$ = total cohesion of the soil.

The cohesion value used in the analysis is dependent on how the shear strength of the soil is defined. Section 11.2.2 discusses how the cohesion can be measured, interpreted, and applied in the bearing capacity equation.

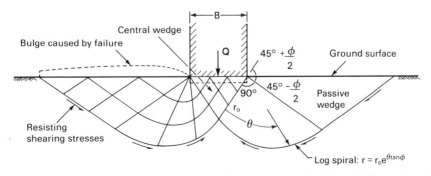

**Figure 11.40** Idealized bearing capacity for failure geometry using Prandtl's log spiral (from Prandtl, 1921).

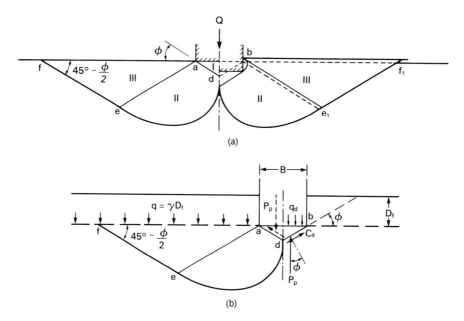

**Figure 11.41** Boundaries of zone of plastic equilibrium after failure of soil beneath a continuous footing. (a) Rough base; (b) rough base and surcharge (from Terzaghi, 1943).

The third component of the bearing capacity equation takes into account the surcharge effect of the soil above the base of the footing:

$$q_q = \rho g D_f N_q \quad (11.49)$$

where

$N_q$ = surcharge bearing capacity factor.

The total, ultimate bearing capacity of a soil can be expressed as the sum of the above-mentioned components:

$$q_f = \tfrac{1}{2}\rho g B N_\gamma + c N_c + \rho g D_f N_q. \quad (11.50)$$

$N_\gamma$, $N_c$, and $N_q$ are dimensionless coefficients that are a function of the angle of internal friction. The bearing capacity factors computed by Terzaghi (1943) are shown in Fig. 11.42. The bearing capacity equation (11.50) was derived for strip footings, and it can be further refined to accommodate various shapes of the footing:

$$q_f = \lambda_\gamma \rho g \frac{B}{2} N_\gamma + \lambda_c c N_c + \rho g D_f N_q \quad (11.51)$$

where

$\lambda_\gamma$, $\lambda_c$ = shape factors.

For circular footings, the shape factors are as follows (Terzaghi, 1943):

$$\lambda_\gamma = 0.6 \quad \text{and} \quad \lambda_c = 1.3.$$

For rectangular footings, the shape factors are as follows

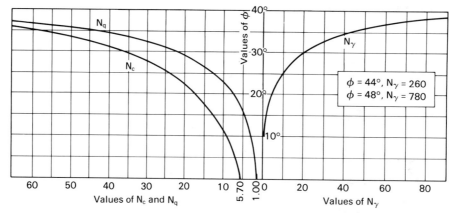

**Figure 11.42** Chart showing the relationship between $\phi$ and Terzaghi's bearing capacity factors (from Terzaghi, 1943).

(Skempton, 1951):

$$\lambda_\gamma = 1 - 0.2\frac{B}{L} \quad \text{and} \quad \lambda_c = 1 + 0.2\frac{B}{L}$$

where

$B$ = width of the footing
$L$ = length of the footing.

The ultimate bearing capacity can be reduced to an allowable bearing capacity by the use of an appropriate factor of safety. Shallow foundations are of most interest with respect to unsaturated soils. The above discussion illustrates that the same theory applies for both saturated and unsaturated soils. The most important aspect in using the theory for unsaturated soils is the assessment of appropriate shear strength parameters and a design matric suction value.

### 11.2.2 Assessment of Shear Strength Parameters and a Design Matric Suction

The shear strength parameters used in practice in conjunction with bearing capacity design for saturated soils have ranged from effective shear strength parameters to the undrained shear strength, $c_u$. In other words, the design is either an effective stress or a total stress type of analysis. Likewise, for unsaturated soils, it is possible to consider a stress state variable approach or a total stress approach.

*Stress State Variable Approach*

The shear strength parameters for a soil with matric suction are the: 1) effective angle of internal friction, $\phi'$, 2) effective cohesion, $c'$, and 3) angle of shear strength change with respect to matric suction, $\phi^b$. These parameters are based on the assumption that the failure surface is planar (see Chapter 9). However, two of the shear strength parameters can be combined, with the result that the shear strength equation has a form similar to that used for saturated soils.

The required shear strength parameters for a conventional bearing capacity design are now the total cohesion, $c$, and the effective angle of internal friction, $\phi'$. In order to obtain the cohesion value, it is necessary to know the $\phi^b$ angle and have an estimate for the design matric suction value. Procedures for measuring the $\phi^b$ angle, along with typical values, were presented in Chapter 10.

It is difficult to assess an appropriate design matric suction value. If the matric suction of the soil were measured, it would be appropriate to use this value only if there were some assurance that this value would be maintained. In some cases, it may be reasonable to apply a factor of safety to the measured matric suction in order to obtain a design matric suction value. Another option is to consider the hydrostatic condition with respect to the groundwater table (Fig. 11.43).

Footings are commonly placed well above the groundwater table. If adequate surface and subsurface drainage is provided around the structure, it may be reasonable to as-

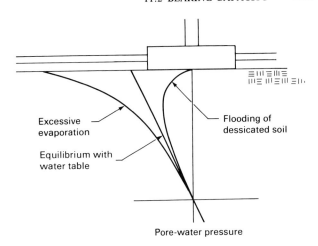

**Figure 11.43** Typical schematic pore–water pressure profiles below a spread footing.

sume that negative pore-water pressures will be maintained immediately below a footing. It should be realized that there may be a fluctuation in the groundwater table as a result of building the structure. In some cases, the groundwater table may be lowered, but more commonly, the water table will rise due to excessive watering of the vegetation surrounding the building.

There are many situations where the groundwater table is far below the ground surface and the hydrostatic profile is not reasonable for design purposes. In this case, measurements of the *in situ* suction below the footings of existing structures in the vicinity can prove to be of value. These values of matric suction are generally found to be a function of the microclimate in the vicinity of the structure (Richards, 1967). In many situations, the long-term suctions are 1 atm or greater. These suctions can be relied upon to contribute to the shear strength of the soil. The decision regarding what value to use becomes dependent upon local experience and the microclimate in a particular region.

Figure 11.44 illustrates the effect of various matric suction values on bearing capacity. The following soil parameters are selected for the analysis: 1) the effective angle of internal friction, $\phi'$, is 20°, 2) the effective cohesion, $c'$, is 5 kPa, and 3) the friction angle with respect to matric suction, $\phi^b$, is 15°. The density of the soil is 1830 kg/m³. The design is for strip footings with a width of 0.5 and 1.0 m. The footings are assumed to be at a depth of 0.5 m. The bearing capacity factors from Fig. 11.42 are as follows: $N_\gamma = 5.0$, $N_c = 17.5$, and $N_q = 8.0$.

The computations show that for a footing width of 0.5 m and no matric suction in the soil, the bearing capacity is 182 kPa. Of this capacity, about 48% arises from the effective cohesion of the soil. When the matric suction is increased to 100 kPa (i.e., total cohesion equal to 32 kPa), the bearing capacity increases to 655 kPa. Now, 85% of the bearing capacity is due to the total cohesion component. The results show a similar trend for a wider footing.

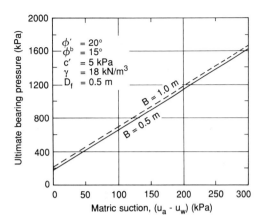

**Figure 11.44** Bearing capacity of a strip footing for various matric suction values.

The main observation is that the matric suction dramatically increases the bearing capacity of the soil. When attempting to arrive at a suitable design matric suction, it is of value to construct a plot similar to Fig. 11.44.

### Total Stress Approach

The total stress approach is, in essence, common to geotechnical engineering practice. However, the engineer does not normally view his design as one involving the behavior of soils with matric suction. Let us assume that the site under consideration involves a clayey soil with a groundwater table well below the proposed depth for the footings. Typical engineering practice can be described as follows. A field investigation is conducted in which samples are obtained at predetermined depths. The samples are brought to the laboratory where they are extruded and tested for their unconfined compressive strength. The data are interpreted, and a design compressive strength is selected.

The compressive strength is divided by 2 to give an undrained shear strength for the soil, $c_u$. The angle of friction is taken as zero, and the bearing capacity equation is solved. It is easy for the engineer to lose sight of the fact that the soil has a matric suction (or negative pore-water pressure) which is holding the specimen together. The matric suction in the specimen tested in the laboratory is a function of the *in situ* negative pore-water pressure and the change in pore-water pressure resulting from unloading the soil during sampling. The change in pore-water pressure during sampling is generally small relative to the *in situ* negative pore-water pressure.

The shear strength measured in the laboratory reflects the *in situ* matric suction of the soil. The extended Mohr-Coulomb failure envelope can be used to visualize the relationship between the unconfined compression test results and the shear strength defined in terms of matric suction. Figure 11.45 illustrates a possible stress path followed during an unconfined compression test (i.e., stress path $\overline{AB}$) where the matric suction is assumed to remain constant during the

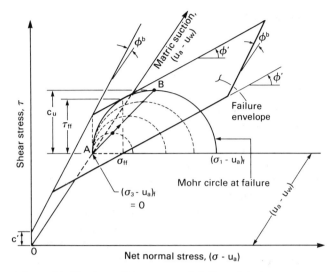

**Figure 11.45** A possible stress path followed during an unconfined compression test of an unsaturated soil.

test. The unconfined compression test results, translated onto the strength envelope, represent the stress state at failure. Mathematically, the undrained shear strength, $c_u$, can be written in an approximate sense, in terms of the extended Mohr-Coulomb failure envelope:

$$c_u \approx c' + (\sigma_f - u_a)_f \tan\phi' + (u_a - u_w)_f \tan\phi^b \quad (11.52)$$

where

$c_u$ = undrained shear strength.

The *in situ* matric suction can increase or decrease in response to changes in the climatic conditions such as evaporation and precipitation. As a result, the undrained shear strength will also change, and its change can be expressed as follows:

$$\Delta c_u = \Delta(u_a - u_w) \tan\phi^b \quad (11.53)$$

where

$\Delta c_u$ = change in undrained shear strength due to matric suction change

$\Delta(u_a - u_w)$ = change in matric suction due to drying and wetting.

The bearing capacity of clayey soils is often computed using the undrained shear strength, $c_u$, in accordance with the total stress approach (i.e., the $\phi = 0$ approach). Applying the ($\phi = 0$) condition to Eq. (11.50) gives the ultimate bearing capacity of clay in terms of its undrained shear strength (i.e., $q_f = c_u N_c$).

Let us consider a clay with an initially measured undrained shear strength of $c_{u0}$ and an initial ultimate bearing capacity of $q_{f0}$ (i.e., $N_c c_{u0}$).

In the field, a change in matric suction, $\Delta(u_a - u_w)$ (increase or decrease), will result in a change in the undrained shear strength, $\Delta c_u$, as expressed in Eq. (11.53).

## 11.2 BEARING CAPACITY 319

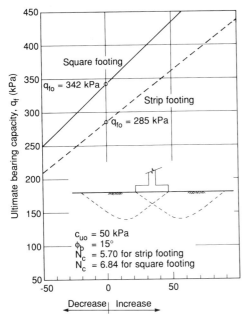

**Figure 11.46** Variation in the ultimate bearing capacity of a clay due to matric suction changes (from Rahardjo and Fredlund, 1992).

increase the bearing capacity, while a decrease in the matric suction reduces the bearing capacity.

The percent change in the ultimate bearing capacity can be related to the change in matric suction as follows:

$$\frac{\Delta q_f}{q_{f0}} = \frac{\Delta(u_a - u_w)}{c_{u0}} \tan\phi^b \quad (11.54)$$

where

$\Delta q_f / q_{f0}$ = percent change in the ultimate bearing capacity

$\Delta(u_a - u_w)/c_{u0}$ = percent change in the matric suction with respect to the initial undrained shear strength.

Equation (11.54) is plotted in Fig. 11.47 for various $\phi^b$ values. The relationship is applicable to all shapes of footing since it is not a function of $N_c$. For a $\phi^b$ value of 15°, the ultimate bearing capacity will increase or decrease by 27% when the matric suction changes as much as the initial undrained shear strength (i.e., $\Delta(u_a - u_w) = 100\% \, c_{u0}$). The higher the $\phi^b$ values, the higher will be the percent change in the ultimate bearing capacity.

### 11.2.3 Bearing Capacity of Layered Systems

There are man-made earth structures where the concepts of bearing capacity are of interest. These involve the design of highway, railway, and airport systems. Highways (and airport runways) commonly consist of an asphalt layer, a base, and a subbase layer overlying the subgrade soil. The railway system consists of the track structure (i.e., rails and ties) resting on a ballast and subballast layer. In both cases, it is the bearing capacity of the subgrade which is of primary importance.

Throughout each year, the environmental conditions change surrounding the highway, railway, or airport, resulting in a change in the matric suction in the subgrade. "Road bans" are placed on highways in the spring of the year because of the low matric suction in the subgrade, and thus a low bearing capacity of the road.

As a result, a change in the ultimate bearing capacity, $\Delta q_f$ (i.e., $N_c \Delta c_u$), will occur, and the final bearing capacity can be written as the sum of the initial capacity and its change (i.e., $q_f = q_{f0} + \Delta q_f$).

Figure 11.46 illustrates the possible variation in the ultimate bearing capacity of a clay due to matric suction changes. The clay has an initial measured undrained shear strength, $c_{u0}$, of 50 kPa and a $\phi^b$ angle of 15°. The initial computed bearing capacity of the clay, $q_{f0}$, is equal to 285 kPa for a strip footing or 342 kPa for a square footing. An average change in matric suction is assumed in order to compute the changes in the ultimate bearing capacity of the clay. The $\phi^b$ angle is assumed to remain constant. Figure 11.46 shows that an increase in the soil matric suction will

Highway, railway, and airport designs have not com-

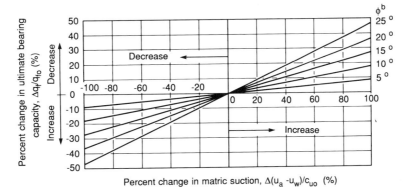

**Figure 11.47** Variation in the ultimate bearing capacity with respect to a variation in matric suction in a clay (from Rahardjo and Fredlund, 1992).

monly been viewed as bearing capacity-type designs. However, this appears to be largely related to the lack of understanding on how to assess the shear strength of a compacted, subgrade soil with matric suction. There has more recently been a renewed interest in a bearing capacity approach to these problems, particularly where the granular base coarse is relatively thin.

The wheel loads applied to a highway surface are transmitted through the base and subbase to the underlying subgrade. The theory of elasticity can be used to compute the stresses transmitted to the top of the subgrade. If these stresses are high, the subgrade may undergo a bearing capacity type of failure. Broms (1965) suggested the use of a modified bearing capacity approach to the design of highway structures.

Train loads are transmitted through the rail-tie system and the ballasts to the subgrade. This becomes a complex loading system for which it is necessary to perform a stress analysis. GEOTRACK (Chang *et al.* 1980) is a computer program developed for computing stresses at various locations below a railway track structure. The stresses applied to the subgrade are of primary interest. Hanna and Meyerhof (1980) analyzed a layered system similar to a railway structure, and suggested modifications and simplifications that could be applied to a conventional bearing capacity analysis in order to accommodate the rail system.

The shear strength of the subgrade of a highway, railway, or airport structure can be expressed in terms of the previously proposed equation for an unsaturated soil [i.e., Eq. (11.13)]. An important aspect for the description of the shear strength is the assessment of the design matric suction. For design purposes, it is necessary to have an indication of how low the matric suction may go during any year. Even a matric suction below 100 kPa can produce a substantial increase in the bearing capacity of a soil. Studies by van der Raadt (1988) and Sattler (1989) have been directed towards a better understanding of seasonal variations of matric suction values for subgrades.

## 11.3 SLOPE STABILITY

Slope stability analyses have become a common analytical tool for assessing the factor of safety of natural and man-made slopes. Any one of numerous two-dimensional, limit equilibrium methods of slices is generally used in practice. These methods are based upon the principles of statics (i.e., static equilibriums of forces and/or moments), without giving any consideration to the displacement in the soil mass. Several basic assumptions and principles used in formulating these limit equilibrium analyses are outlined prior to deriving the general factor of safety equations.

Effective shear strength parameters (i.e., $c'$ and $\phi'$) are generally used when performing slope stability analyses on soils which are saturated. The shear strength contribution from the negative pore-water pressures above the ground-water table are usually ignored by setting their magnitudes to zero. The difficulties associated with the measurement of negative pore-water pressures and their incorporation into the slope stability analysis are the primary reasons for this practice. It may be a reasonable assumption to ignore negative pore-water pressures for many situations where the major portion of the slip surface is below the groundwater table. However, for situations where the groundwater table is deep or where the concern is over the possibility of a shallow failure surface (Fig. 11.48), negative pore-water pressures can no longer be ignored.

In recent years, there has developed a better understanding of the role of negative pore-water pressures (or matric suctions) in increasing the shear strength of the soil. Recent developments have led to several devices which can be used to better measure the negative pore-water pressures. Therefore, it is now appropriate to perform slope stability analyses which include the shear strength contribution from the negative pore-water pressures. These types of analyses are an extension of conventional limit equilibrium analyses.

Several aspects of a slope stability study remain the same for soils with positive pore-water pressures (e.g., saturated soils) and soils with negative pore-water pressures (e.g., unsaturated soils). For example, the nature of the site investigation, the identification of the strata, and the measurement of the total unit weight remain the same in both situations. On the other hand, extensions to conventional testing procedures are required with respect to the characterization of the shear strength properties of the soil. The analytical tools used to incorporate pore-water pressures and calculate the factor of safety also need to be extended.

### 11.3.1 Location of the Critical Slip Surface

A study of the stability of a slope with negative pore-water pressures involves the following steps: 1) a survey of the elevation of the ground surface on a selected section perpendicular to the slope, 2) the advancement of several boreholes to identify the stratigraphy and obtain undisturbed soil samples, 3) the laboratory testing of the undisturbed soil specimens to obtain suitable shear strength parameters for each stratigraphic unit (i.e., $c'$, $\phi'$, and $\phi^b$ parameters), and 4) the measurement of negative pore-water pressures above the groundwater table. These steps provide the input data for performing a stability analysis. However, the location and shape of the most critical slip surface is an unknown (Fig. 11.48). Some combination of actuating and resisting forces along a slip surface of unknown shape and location will produce the lowest factor of safety. Of course, in the case of an already failed slope, the location of the slip surface is known.

In design, the shape of the unknown slip surface is generally assumed, while its location is determined by a trial-and-error procedure. If the shape of the slip surface is assumed to be circular, a grid of centers can be selected, and

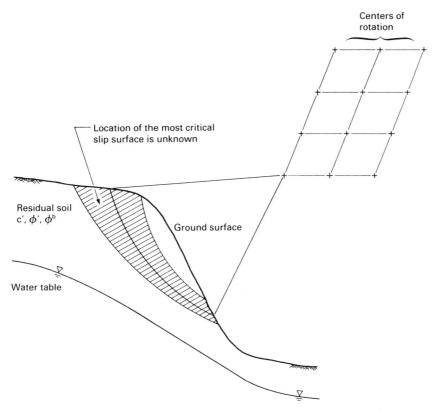

**Figure 11.48** A steep natural slope with a deep groundwater table.

the radius varied at each center, providing a coverage of all possible conditions (Fig. 11.48). When the slip surface takes on a composite shape (i.e., part circular and part linear), it is still possible to use a grid of centers and varying radii in order to search for a critical slip surface (Fig. 11.49; Fredlund, 1981). In addition, a general shape can also be assumed using a series of straight lines to define the slip surface.

Various automatic search routines have been programmed to reduce the number of computations. Some routines start with a center at an assumed point, and seek the critical center by moving in a zigzag manner (Wright, 1974). Others use an initially coarse grid of centers which rapidly converges to the critical center (Fredlund, 1981).

There is probably no analysis conducted by geotechnical engineers which has received more programming attention than the limit equilibrium methods of slices used to compute a factor of safety (Fredlund, 1980). The main reasons appear to be as follows. First, the limit equilibrium method has proved to be a useful and reasonably reliable tool in assessing the stability of slopes. Its "track record" is impressive for most cases where the shear strength properties of the soil and the pore-water pressure conditions have been properly assessed (Sevaldson, 1956; Kjaernsli and Simons, 1962; Skempton and Hutchison, 1969; Chowdhury, 1980). Second, the limit equilibrium methods of slices require a limited amount of input information, but can quickly perform extensive trial-and-error searches for the critical slip surface.

**11.3.2 General Limit Equilibrium (GLE) Method**

The General Limit Equilibrium method (i.e., GLE) provides a general theory wherein other methods can be viewed as special cases. The elements of statics used in the GLE method for deriving the factor of safety are the summation of forces in two directions and the summations of moments about a common point (Fredlund et al. 1981).

These elements of statics, along with the failure criteria, are insufficient to make the slope stability problem determinate (Morgenstern and Price, 1965; Spencer, 1967). Either additional elements of physics or an assumption regarding the direction or magnitude of some of the forces is required to render the problem determinate. The GLE method utilizes an assumption regarding the direction of the interslice forces. This approach has been widely adopted in limit equilibrium methods (Fredlund and Krahn, 1977). The various limit equilibrium slope stability methods that follow this approach have been demonstrated to be special cases of the GLE method (Fredlund et al. 1981).

Calculations for the stability of a slope are performed by dividing the soil mass above the slip surface into vertical slices. The forces acting on a slice within the sliding soil mass are shown in Figs. 11.50 and 11.51 for a circular and a composite slip surface, respectively. The forces are des-

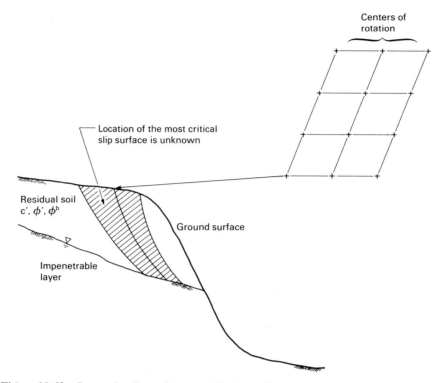

**Figure 11.49** Composite slip surfaces resulting from the presence of an impenetrable layer.

ignated for a unit width (i.e., perpendicular direction to motion) of the slope. The variables are defined as follows:

- $W$ = the total weight of the slice of width "$b$" and height "$h$"
- $N$ = the total normal force on the base of the slice
- $S_m$ = the shear force mobilized on the base of each slice
- $E$ = the horizontal interslice normal forces (the "$L$" and "$R$" subscripts designate the left and right sides of the slice, respectively)
- $X$ = the vertical interslice shear forces (the "$L$" and "$R$" subscripts designate the left and right sides of the slice, respectively)
- $R$ = the radius for a circular slip surface or the moment arm associated with the mobilized shear force, $S_m$, for any shape of slip surface
- $f$ = the perpendicular offset of the normal force from the center of rotation or from the center of moments
- $x$ = the horizontal distance from the centerline of each

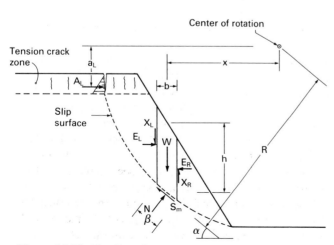

**Figure 11.50** Forces acting on a slice through a sliding mass with a circular slip surface.

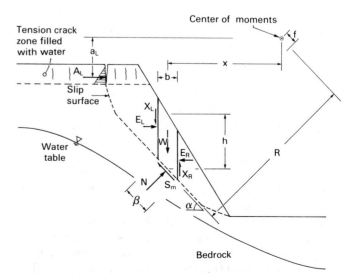

**Figure 11.51** Forces acting on a slice through a sliding mass with a composite slip surface.

slice to the center of rotation or to the center of moments

$h$ = the vertical distance from the center of the base of each slice to the uppermost line in the geometry (i.e., generally ground surface)

$a$ = the perpendicular distance from the resultant external water force to the center of rotation or to the center of moments; the "$L$" and "$R$" subscripts designate the left and right sides of the slope, respectively

$A$ = the resultant external water forces; the "$L$" and "$R$" subscripts designate the left and right sides of the slope, respectively

$\alpha$ = the angle between the tangent to the center of the base of each slice and the horizontal; the sign convention is as follows: when the angle slopes in the same direction as the overall slope of the geometry, $\alpha$ is positive, and vice versa

$\beta$ = sloping distance across the base of a slice.

The examples shown in Figs. 11.50 and 11.51 are typical of a steep slope with a deep groundwater table. The crest of the slope is highly desiccated, and there are tension cracks filled with water. The tension crack zone is assumed to have no shear strength, and the presence of water in this zone produces an external water force, $A_L$. As a result, the assumed slip surface in the tension crack zone is a vertical line. The depth of the tension crack is generally estimated or can be approximated analytically (Spencer, 1968; Spencer, 1973). The weight of the soil in the tension crack zone acts as a surcharge on the crest of the slope. The external water force, $A_L$, is computed as the hydrostatic force on a vertical plane. An external water force can also be present at the toe of the slope as a result of partial submergence. This water force is designated as $A_R$.

*Shear Force Mobilized Equation*

The mobilized shear force at the base of a slice can be written using the shear strength equation for an unsaturated soil presented in Chapter 9:

$$S_m = \frac{\beta}{F} \{c' + (\sigma_n - u_a)\tan\phi' + (u_a - u_w)\tan\phi^b\}$$

(11.55)

where

$\sigma_n$ = total stress normal to the base of a slice

$F$ = factor of safety which is defined as the factor by which the shear strength parameters must be reduced in order to bring the soil mass into a state of limiting equilibrium along the assumed slip surface.

The factor of safety for the cohesive parameter (i.e., $c'$) and the frictional parameters (i.e., $\tan\phi'$ and $\tan\phi^b$) are assumed to be equal for all soils involved and for all slices.

The components of the mobilized shear force at the base of a slice are illustrated in Fig. 11.52. The contributions from the total stress and the negative pore-water pressures are separated using the $\phi'$ and $\phi^b$ angles, respectively.

It is possible to consider the matric suction term as part of the cohesion of the soil. In other words, the matric suction can be visualized as increasing the cohesion of the soil. As a result, the conventional factor of safety equations does not need to be rederived. The mobilized shear force at the base of a slice, $S_m$, will have the following form:

$$S_m = \frac{\beta}{F} \{c + (\sigma_n - u_a)\tan\phi'\}$$

(11.56)

where

$c$ = total cohesion of the soil, which has two components (i.e., $c' + (u_a - u_w)\tan\phi^b$).

This approach has the advantage that the shear strength equation retains its conventional form. It is therefore possible to utilize a computer program written for saturated soils to solve unsaturated soil problems. When this is done, the soil in the negative pore-water pressure region must be subdivided into several discrete layers, with each layer having a constant cohesion. The pore-air and pore-water pressures must be set to zero within the computer program. This approach has the disadvantage that the cohesion is not a continuous function, and the appropriate cohesion values for each soil layer must be manually computed.

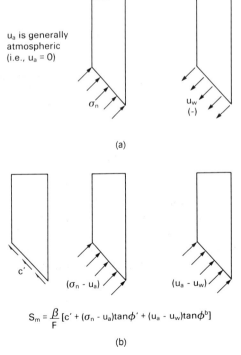

**Figure 11.52** Pressure and shearing resistance components at the base of a slice. (a) Pressure components on the base of a slice; (b) contributors to the shear resistance.

# 11 PLASTIC AND LIMIT EQUILIBRIUM

The formulations presented in the next sections are the revised derivations for the factor of safety equations that directly incorporate the shear strength contribution from the negative pore-water pressures. The mobilized shear force defined using Eq. (11.55) is used throughout the derivation. The effect of partial submergence at the toe of the slope, the effect of seismic loading, and external line loads are not incorporated in the derivations.

### Normal Force Equation

The normal force at the base of a slice, $N$, is derived by summing forces in the vertical direction:

$$W - (X_R - X_L) - S_m \sin\alpha - N \cos\alpha = 0. \quad (11.57)$$

Substituting Eq. (11.55) into Eq. (11.57) and replacing the $(\sigma_n \beta)$ term with $N$ gives

$$W - (X_R - X_L) - \left\{ \frac{c' \beta}{F} + \frac{N \tan\phi' \beta}{F} - \frac{u_a \tan\phi' \beta}{F} + \frac{(u_a - u_w) \tan\phi^b \beta}{F} \right\} \sin\alpha - N \cos\alpha = 0 \quad (11.58)$$

or

$$N \left( \cos\alpha + \frac{\sin\alpha \tan\phi'}{F} \right) = W - (X_R - X_L) - \frac{c' \beta \sin\alpha}{F} + u_a \frac{\beta \sin\alpha}{F} (\tan\phi' - \tan\phi^b) + u_w \frac{\beta \sin\alpha}{F} \tan\phi^b. \quad (11.59)$$

Rearranging Eq. (11.59) gives rise to the normal force equation:

$$N = \frac{W - (X_R - X_L) - \dfrac{c' \beta \sin\alpha}{F} + u_a \dfrac{\beta \sin\alpha}{F} (\tan\phi' - \tan\phi^b) + u_w \dfrac{\beta \sin\alpha}{F} \tan\phi^b}{m_\alpha} \quad (11.60)$$

where

$$m_\alpha = \cos\alpha + (\sin\alpha \tan\phi')/F.$$

The factor of safety, $F$, in Eq. (11.60) is equal to the moment equilibrium factor of safety, $F_m$, when solving for moment equilibrium, and is equal to the force equilibrium factor of safety, $F_f$, when solving for force equilibrium. In most cases, the pore-air pressure, $u_a$, is atmospheric, and as a result, Eq. (11.60) reduces to the following form:

$$N = \frac{W - (X_R - X_L) - \dfrac{c' \beta \sin\alpha}{F} + u_w \dfrac{\beta \sin\alpha}{F} \tan\phi^b}{m_\alpha}. \quad (11.61)$$

If the base of the slice is located in the saturated soil, the $(\tan\phi^b)$ term in Eq. (11.61) becomes equal to $(\tan\phi')$. Equation (11.61) then reverts to the conventional normal force equation used in saturated slope stability analysis. Computer coding for solving Eq. (11.61) can be written such that the angle $\phi^b$ is used whenever the pore-water pressure is negative, while the angle $\phi'$ is used whenever the pore-water pressure is positive. The $\phi^b$ angle can also be considered to be equal to $\phi'$ at low matric suction values, while a lower $\phi^b$ angle is used at high matric suctions (Fredlund et al., 1987).

The vertical interslice shear forces, $X_L$ and $X_R$, in the normal force equation can be computed using an interslice force function, as described later.

### Factor of Safety with Respect to Moment Equilibrium

Two independent factor of safety equations can be derived: one with respect to moment equilibrium, and the other with respect to horizontal force equilibrium. Moment equilibrium can be satisfied with respect to an arbitrary point above the central portion of the slip surface. For a circular slip surface, the center of rotation is an obvious center for moment equilibrium. The center of moments would appear to be immaterial when both force and moment equilibria are satisfied, as in the case for the complete General Limit Equilibrium method. When only moment equilibrium is satisfied, the computed factor of safety varies slightly with the point selected for the summation of moments.

Let us consider moment equilibrium for a composite slip surface (Fig. 11.51) with respect to the center of rotation of the circular portion:

$$A_L a_L + \sum Wx - \sum Nf - \sum S_m R = 0. \quad (11.62)$$

Substituting Eq. (11.55) for the $S_m$ variable into Eq. (11.62) and replacing the $(\sigma_n \beta)$ term with $N$ yields

$$A_L a_L + \sum Wx - \sum Nf$$
$$= \frac{1}{F_m} \sum [c' \beta R + \{N \tan\phi' - u_a \tan\phi' \beta + (u_a - u_w) \tan\phi^b \beta\} R] \quad (11.63)$$

where

$F_m$ = factor of safety with respect to moment equilibrium.

Rearranging Eq. (11.63) yields

$$F_m = \frac{\sum \left[ c'\beta R + \left\{ N - u_w \beta \frac{\tan\phi^b}{\tan\phi'} - u_a \beta \left(1 - \frac{\tan\phi^b}{\tan\phi'}\right) \right\} R \tan\phi' \right]}{A_L a_L + \sum Wx - \sum Nf}. \quad (11.64)$$

In the case where the pore-air pressure is atmospheric (i.e., $u_a = 0$), Eq. (11.64) has the following form:

$$F_m = \frac{\sum \left[ c'\beta R + \left\{ N - u_w \beta \frac{\tan\phi^b}{\tan\phi'} \right\} R \tan\phi' \right]}{A_L a_L + \sum Wx - \sum Nf}. \quad (11.65)$$

When the pore-water pressure is positive, the $\phi^b$ value can be set equal to the $\phi'$ value. Equation (11.64) can also be simplified for a circular slip surface (Fig. 11.50) as follows:

$$F_m = \frac{\sum [c'\beta + \{N - u_w \beta - u_a \beta\} \tan\phi']R}{A_L a_L + \sum Wx}. \quad (11.66)$$

For a circular slip surface, the radius, $R$, is constant for all slices, and the normal force, $N$, acts through the center of rotation (i.e., $f = 0$).

### Factor of Safety with Respect to Force Equilibrium

The factor of safety with respect to force equilibrium is derived from the summation of forces in the horizontal direction for all slices:

$$-A_L + \sum S_m \cos\alpha - \sum N \sin\alpha = 0. \quad (11.67)$$

The horizontal interslice normal forces, $E_L$ and $E_R$, cancel when summed over the entire sliding mass. Substituting Eq. (11.55) for the mobilized shear force, $S_m$, into Eq. (11.67) and replacing the $(\sigma_n \beta)$ term with $N$ give

$$\frac{1}{F_f} \sum [c'\beta \cos\alpha + \{N \tan\phi' - u_a \tan\phi' \beta + (u_a - u_w) \tan\phi^b \beta\}] = A_L + \sum N \sin\alpha \quad (11.68)$$

where

$F_f$ = factor of safety with respect to force equilibrium.

Rearranging Eq. (11.68) yields

$$F_f = \frac{\sum \left[ c'\beta \cos\alpha + \left\{ N - u_w \beta \frac{\tan\phi^b}{\tan\phi'} - u_a \beta \left(1 - \frac{\tan\phi^b}{\tan\phi'}\right) \right\} \tan\phi' \cos\alpha \right]}{A_L + \sum N \sin\alpha}. \quad (11.69)$$

In the case where the pore-air pressure is atmospheric (i.e., $u_a = 0$), Eq. (11.69) reverts to the following form:

$$F_f = \frac{\sum \left[ c'\beta \cos\alpha + \left\{ N - u_w \beta \frac{\tan\phi^b}{\tan\phi'} \right\} \tan\phi' \cos\alpha \right]}{A_L + \sum N \sin\alpha}. \quad (11.70)$$

When the pore-water pressure is positive, the $\phi^b$ value is equal to the $\phi'$ value. Equation (11.70) remains the same for both circular and composite slip surfaces.

### Interslice Force Function

The interslice normal forces, $E_L$ and $E_R$, are computed from the summation of horizontal forces on each slice (Fig. 11.53):

$$E_R - E_L = N \cos\alpha \tan\alpha - S_m \cos\alpha. \quad (11.71)$$

Substituting Eq. (11.57) for the ($N \cos\alpha$) term in Eq. (11.71) gives the following equation:

$$E_R - E_L = \{W - (X_R - X_L) - S_m \sin\alpha\} \tan\alpha - S_m \cos\alpha. \quad (11.72)$$

Rearranging Eq. (11.72) gives

$$E_R = E_L + \{W - (X_R - X_L)\} \tan\alpha - \frac{S_m}{\cos\alpha}. \quad (11.73)$$

The interslice normal forces are calculated from Eq. (11.73) by integrating from left to right across the slope (Fig. 11.53). The procedure is further explained in the next section. The left interslice normal force on the first slice is equal to any external water force which may exist, $A_L$, or it is set to zero when there is no water present in the tension crack zone.

The assumption is made that the interslice shear force, $X$, can be related to the interslice normal force, $E$, by a mathematical function (Morgenstern and Price, 1965):

$$X = \lambda f(x) E \quad (11.74)$$

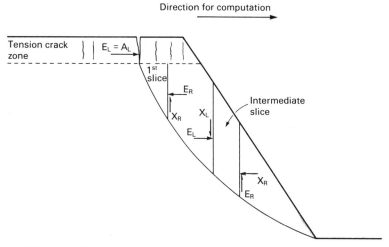

**Figure 11.53** Convention for the designation of the interslice forces.

where

- $f(x)$ = a functional relationship which describes the manner in which the magnitude of $X/E$ varies across the slip surface
- $\lambda$ = a scaling constant which represents the percentage of the function, $f(x)$, used for solving the factor of safety equations.

Some functional relationships, $f(x)$, that can be used for slope stability analyses are illustrated in Fig. 11.54. Basically, any shape of function can be assumed in the analysis. However, an unrealistic assumption of the interslice force function can result in convergence problems associated with solving the nonlinear factor of safety equations (Ching and Fredlund, 1983). Morgenstern and Price (1967) suggested that the interslice force function should be related to the shear and normal stresses on vertical slices through the soil mass. In 1979, Maksimovic (1979) used the finite element method and a nonlinear characterization of the soil to compute stresses in a soil mass. These stresses were then used in the limit equilibrium slope stability analysis.

A generalized interslice force function, $f(x)$, has been proposed by Fan *et al.* 1986. The function is based on two-dimensional finite element analyses of a linear elastic continuum using constant strain triangular elements. The normal stresses in the x-direction and the shear stresses in the y-direction were integrated along vertical planes within a sliding mass in order to obtain normal and shear forces, respectively. The ratio of the shear force to the normal force was plotted for each vertical section to provide a distribution for the direction of the resultant interslice forces. Figure 11.55 illustrates a typical interslice force function for one slip surface through a relatively steep slope.

The analysis of many slopes showed that the interslice force function could be approximated by an extended form of an error function equation. Inflection points were close to the crest and toe of the slope. The slope of the resultant interslice forces was steepest at the midpoint of the slope, and tended towards zero at some distances behind the crest and beyond the toe. The mathematical form for the empirical interslice force function can be written as follows:

$$f(x) = Ke^{-(C^n \omega^n)/2} \quad (11.75)$$

where

- $e$ = base of the natural logarithm
- $K$ = magnitude of the interslice force function at midslope (i.e., maximum value)
- $C$ = variable to define the inflection points
- $n$ = variable to specify the flatness or sharpness of curvature of the function
- $\omega$ = dimensionless x-position relative to the midpoint of the slope.

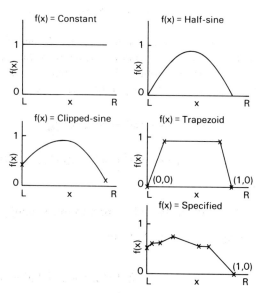

L = Left dimensional x-coordinate of the slip surface
R = Right dimensional x-coordinate of the slip surface

**Figure 11.54** Various possible interslice force functions.

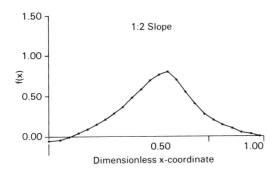
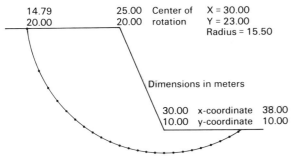

**Figure 11.55** The interslice force function for a deep-seated slip surface through a one horizontal to two vertical slope.

Figure 11.56 shows the definition of the dimensionless distance, $\omega$. The factor, $K$, in the interslice force function equation [i.e., Eq. (11.75)], is a variable related to the average inclination of the slope and the depth factor, $D_f$, for the slope surface under consideration:

$$K = \exp\{D_i + D_s(D_f - 1.0)\} \quad (11.76)$$

where

$D_f$ = depth factor (defined in Fig. 11.57)
$D_i$ = the natural logarithm of the intercept on the ordinate when $D_f = 1.0$
$D_s$ = slope of the depth factor versus $K$ relationship for a specific slope

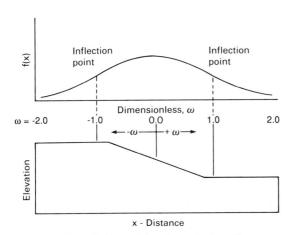

**Figure 11.56** Definition of the dimensionless distance, $\omega$.

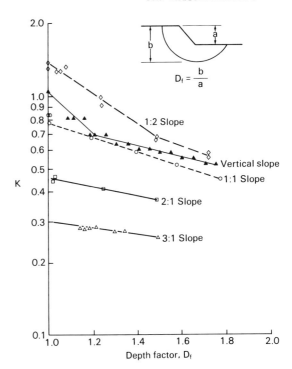

**Figure 11.57** The interslice side force ratio, $K$, at midslope versus the depth factor, $D_f$.

Eq. (11.76) is shown graphically in Fig. 11.57. Slip surfaces passing through or below a vertical slope are considered as a special case. The relationship between the factor, $K$, and depth factor, $D_f$, for vertical slopes is also shown in Fig. 11.57.

The "finite element" based functions have been computed for slip surfaces which are circular. However, the shape of the function should, in general, be satisfactory for composite slip surfaces.

The magnitude of "$C$" varies with the slope angle, as does the variable "$n$" (Fig. 11.58).

### Procedures for Solving the Factors of Safety Equation

The factor of safety equations with respect to moment and force equilibriums (i.e., Eqs. (11.64) and (11.69), respectively) are nonlinear. The factors of safety, $F_m$ or $F_f$, appear on both sides of the equations, with the factor of safety being included through the normal force equation [i.e., Eq. (11.60)]. The nonlinear factor of safety equations can be solved using an iterative technique. The factors of safety with respect to moment and force equilibriums can be calculated when the normal force, $N$, on each slice is known. The computation of the normal force [i.e., Eq. (11.60)] requires a magnitude for the interslice shear forces, $X_L$ and $X_R$, and an estimate of the factor of safety, $F$.

For the first iteration, the factor of safety, $F$, in the normal force equation can be set to 1.0 or estimated from the Ordinary method (Fredlund, 1985c). The interslice shear forces can be set to 0.0 for the first iteration when com-

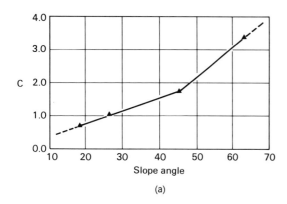

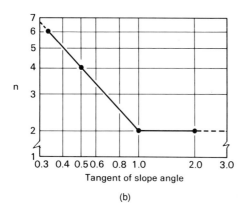

**Figure 11.58** Values of the "C" and "n" coefficients versus the slope angle. (a) C coefficient versus the slope angle; (b) n coefficient versus the tangent of the slope angle.

puting the normal force. The computed normal force is then used to calculate the factors of safety with respect to moment and force equilibriums (i.e., $F_m$ and $F_f$ in Eqs. (11.64) and (11.69), respectively). This results in initial values for the factors of safety.

The next step is to compute the interslice normal forces, $E_L$ and $E_R$, in accordance with Eq. (11.73). There are two sets of interslice force calculations: one associated with moment equilibrium, and the other associated with force equilibrium. The interslice force calculation with respect to moment equilibrium uses the moment equilibrium factor of safety, $F_m$, in computing the mobilized shear force, $S_m$, in Eq. (11.55). On the other hand, the interslice force calculation associated with force equilibrium uses the force equilibrium factor of safety, $F_f$, in computing the mobilized shear force, $S_m$, in Eq. (11.55). The interslice shear forces in Eq. (11.60) are also set to zero for the first iteration. The computation commences from the first slice on the left-hand side of the slope (i.e., at the crest), and proceeds across the slope to the last slice at the toe (Fig. 11.53). The right interslice normal force, $E_R$, on the last slice will become zero when overall force equilibrium in the horizontal direction is fully satisfied.

The computed interslice normal forces, $E_L$ and $E_R$, can then be used in the calculation of the interslice shear forces, $X_L$ and $X_R$, for all slices, in accordance with Eq. (11.74). An interslice force function, $f(x)$, can be assumed from one of the functions shown in Fig. 11.54 or calculated using Eq. (11.75). The selected interslice force function, along with a specified λ value, is used for the entire iterative procedure until convergence is achieved.

For the next iteration, the computed moment equilibrium factor of safety, $F_m$, and the corresponding interslice forces are used to recalculate new values for the normal force, $N$, and the moment of equilibrium factor of safety, $F_m$. The updated values for the normal force, $N$, and the moment equilibrium factor of safety, $F_m$, are then used to revise the interslice normal forces and interslice shear forces (i.e., $E_L$, $E_R$, $X_L$, and $X_R$) associated with moment equilibrium.

The computed force equilibrium factor of safety, $F_f$, and the corresponding interslice forces from the first iteration are used to revise the magnitudes of the following variables: $N$, $F_f$, $E_L$, $E_R$, $X_L$, and $X_R$, associated with force equilibrium.

The revised factor of safety values, $F_m$ and $F_f$, are then compared with the corresponding values from the previous iteration. Calculations are stopped when the difference in the factor of safety between two successive iterations is less than the desired tolerance. If the difference in either factor of safety, $F_f$ or $F_m$, is greater than the tolerance, the above procedure is repeated until convergence is attained for both factors of safety.

When the solution has converged, moment and force equilibrium factors of safety corresponding to the selected interslice force function, $f(x)$, and the selected λ value are obtained. The analysis can proceed using the same interslice force function, $f(x)$, but varying the λ value. Several factor of safety values, $F_m$ and $F_f$, associated with different λ values can be obtained and plotted as shown in Fig. 11.59. The moment equilibrium factor of safety, $F_m$, does not vary significantly with respect to the λ values as compared to the force equilibrium factor of safety, $F_f$. Curves joining the $F_m$ and $F_f$ data intersect at a point where total equilibrium (i.e., moment and force equilibrium) is satisfied.

### Pore-Water Pressure Designation

Pore-water pressures are often designated in terms of a pore pressure coefficient, $r_u$, for analysis purposes (Bishop and Morgenstern, 1960):

$$r_u = \frac{u_w}{g \sum \rho_i h_i} \quad (11.77)$$

where

$r_{uw}$ = water pore pressure coefficient
$h_i$ = thickness of each soil layer
$\rho_i$ = density of each soil layer.

The pore pressure coefficient is generally considered as a positive value. However, it can also be used to represent

## 11.3 SLOPE STABILITY 329

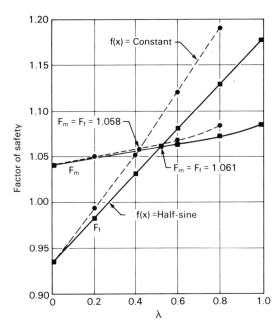

**Figure 11.59** Variation of moment and force equilibrium factors of safety with respect to lambda, λ.

negative pore-water pressures, as well as pore-air pressures:

$$r_{ua} = \frac{u_a}{g \sum \rho_i h_i} \qquad (11.78)$$

where

$r_{ua}$ = air pore pressure coefficient.

Figure 11.60 illustrates how the pore pressure coefficient can be used when the pore-water pressure is negative. In this case, the pore pressure coefficient is also negative.

The water pore pressure coefficient at the phreatic line is equal to zero. A water pore pressure coefficient of +0.5 indicates the verge of artesian pressure conditions since the density of water is approximately one half the density of soil. At points above the phreatic line, the pore-water pressure becomes increasingly negative. At the same time, the overburden pressure is decreasing. As a result, it is possible for the pore pressure coefficient to become highly negative. Let us assume that the pore-water pressure is −200

kPa at a depth of 1 m (i.e., $\rho g h = 20$ kPa for $\rho = 2000$ kg/m$^3$). This gives rise to a water pore pressure coefficient of −10. The water pore pressure coefficient can tend to a negative, infinite number as ground surface is approached. In other words, the water pore pressure coefficient becomes a highly variable term as ground surface is approached.

Figure 11.60 shows the cross section of a dam under steady-state seepage conditions. One equipotential line is selected, and the water pore pressure coefficients are computed at various depths and plotted in Fig. 11.61. There is essentially a linear change in the pore pressure coefficient until the ground surface is approached. At this point, the coefficient becomes highly negative. The nonlinearity of the water pore pressure coefficient near ground surface somewhat limits its use as a means of designating negative pore-water pressures.

The air pore pressure coefficient in natural soil deposits is always close to zero due to its contact with the atmosphere. In compacted earth fills, the pore-air pressures may become positive due to the weight of the overlying soil layers. The air pore pressure coefficient will be positive, but generally quite small.

The water and air pore pressure coefficients are similar in form to the $B$ pore pressure parameters developed in Chapter 8. However, there are some differences. First, the pore pressure coefficients are generally used in conjunction with relating field experience. Second, they are used to compute changes in the pore pressures referenced to the total overburden pressure, as opposed to being referenced to changes in the principal stresses. Third, the pore pressure coefficient has been used in two ways. The above description has defined the pore pressure coefficient with respect to a point in an earth mass. However, the pore pressure coefficient is also used as an average value for an entire soil region. Bishop and Morgenstern (1960) suggested procedures for obtaining an average pore pressure coefficient over a region. This value was then used to compute the pore-water pressure in a slope stability analysis.

There are other procedures which can be used to designate pore-water pressures for a slope stability analysis. Pore pressure changes can also be written in terms of $A$ and $B$ pore pressure parameters. The $B$ pore pressure parameter represents the change in pore pressure due to isotropic or all-around loading. The $A$ pore pressure parameter repre-

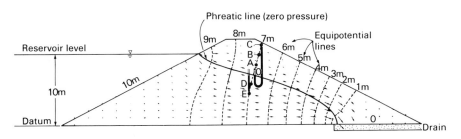

**Figure 11.60** Steady-state seepage conditions in an earth-fill dam.

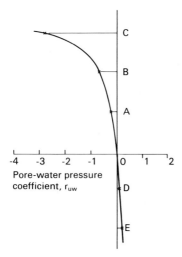

**Figure 11.61** Water-pore pressure coefficient along the 7 m equipotential line.

sents the change in pore pressure due to deviatoric loading. The pore pressure parameters for the water and air phases are presented in Chapter 8 (i.e., $B_w$, $A_w$, $B_a$, and $A_a$).

Hilf's analysis (1948) and Bishop's analysis (1956) also provide relationships between pore-water pressure and overburden pressure. These analyses make use of the compressibility of air and its solubility in water, along with the compressibility of the soil structure, to obtain what is essentially a nonlinear pore pressure coefficient (see Chapter 8).

A grid of pore-water pressure values can be superimposed over a cross section under consideration (Fig. 11.62). The pore-water pressures can be either positive or negative, and an interpolation technique can be used in order to obtain the pore-water pressure at any designated point.

Negative pore-water pressures can also be contoured as a series of lines (Fig. 11.63). An interpolation procedure can be used to obtain the pore-water pressure for points between the contours. Some of the above procedures are later described in further detail.

Piezometric lines can also be used to designate the pore-water pressures in a slope (Fig. 11.64). The vertical distance from the piezometric line down to a point below the line is equal to the positive pore-water pressure head (i.e., $u_w = h_w \rho_w g$). On the other hand, the vertical distance from the piezometric line up to a point above the line can be considered as the negative pore-water pressure head (i.e., $u_w (-) = h_w \rho_w g$). When slopes are steep and the gradient along the water table is high, this procedure can lead to pore-water pressures which are in considerable error.

### 11.3.3 Other Limit Equilibrium Methods

The General Limit Equilibrium (GLE) method can be specialized to correspond to various limit equilibrium methods. The various methods of slices can be categorized in terms of the conditions of statical equilibrium satisfied and the assumption used with respect to the interslice forces. Table 11.2 summarizes the conditions of statical equilibrium satisfied by the various methods of slices. The statics used in each of the methods of slices for computing the factor of safety are summarized in Table 11.3. Most methods use either moment equilibrium or force equilibrium in the calculation for the factor of safety. The Ordinary and Simplified Bishop methods use moment equilibrium, while the Janbu Simplified, Janbu Generalized, Lowe and Karafiath, and the Corps of Engineers methods use force equilibrium in computing the factor of safety. On the other hand, the Spencer and Morgenstern-Price methods satisfy both moment and force equilibriums in computing the factor of safety. In this respect, these two methods are similar in principle to the GLE method which satisfies force and moment equilibriums in calculating the factor of safety.

The GLE method can be used to simulate the various methods of slices by using the appropriate interslice force assumption. The interslice force assumptions used for simulating the various methods are given in Table 11.3.

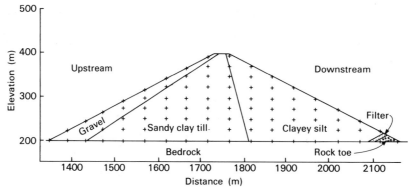

**Figure 11.62** Grid of pore pressures heads superimposed over the geometry.

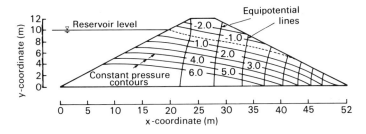

**Figure 11.63** Contours used to designate pore–water pressure heads.

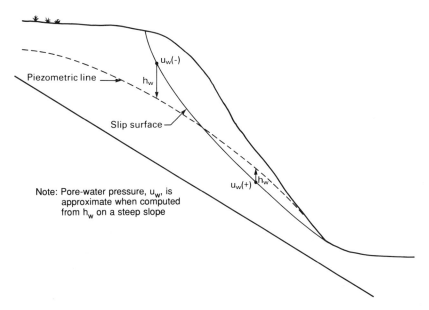

**Figure 11.64** Piezometric line for designating pore–water pressures.

**Table 11.2 Elements of Statical Equilibrium Satisfied by Various Limit Equilibrium Methods**

| Method | Force Equilibrium | | Moment Equilibrium |
|---|---|---|---|
| | 1st Direction[a] (e.g., Vertical) | 2nd Direction[a] (e.g., Horizontal) | |
| Ordinary or Fellenius | Yes | No | Yes |
| Bishop's Simplified | Yes | No | Yes |
| Janbu's Simplified | Yes | Yes | No |
| Janbu's Generalized | Yes | Yes | [b] |
| Spencer | Yes | Yes | Yes |
| Morgenstern–Price | Yes | Yes | Yes |
| Corps of Engineers | Yes | Yes | No |
| Lowe–Karafiath | Yes | Yes | No |

[a] Any of two orthogonal directions can be selected for the summation of forces.
[b] Moment equilibrium is used to calculate interslice shear forces.

**Table 11.3 Comparison of Commonly Used Methods of Slices**

| Method | Factors of Safety | | Interslice Force Assumption |
|---|---|---|---|
| | Moment Equilibrium | Force Equilibrium | |
| Ordinary | x | | $X = 0, E = 0$ |
| Bishop's Simplified | x | | $X = 0, E \geq 0$ |
| Janbu's Simplified | | x | $X = 0, E \geq 0$ |
| Janbu's Generalized | | x | $X_R = E_R \tan \alpha_t - (E_R - E_L)t_R/b$ [a] |
| Spencer | x | x | $X/E = \tan \theta$ [b] |
| Morgenstern–Price | x | x | $X/E = \lambda f(x)$ |
| Lowe–Karafiath | | x | $X/E$ = Average slope of ground and slip surface |
| Corps of Engineers | | x | $X/E$ = Average ground surface slope |

[a] $\alpha_t$ = angle between the line of thrust across a slice and the horizontal.
$t_R$ = vertical distance from the base of the slice to the line of thrust on the right side of the slice.
[b] $\theta$ = angle of the resultant interslice force from the horizontal.

### 11.3.4 Numerical Difficulties Associated with the Limit Equilibrium Method of Slices

Most problems associated with nonconverging solutions can be traced to one of three possible conditions (Ching and Fredlund, 1983). First, an unrealistic assumption regarding the shape of the slip surface can produce mathematical instability. Second, high cohesion values may result in a negative normal force and produce mathematical instability. Third, the assumption used to render the analysis determinate may impose unrealistic conditions and prevent convergence.

The normal force at the base of a slice [Eq. (11.60)] may become unreasonable due to the unrealistic value of $m_\alpha$ (Whitman and Bailey, 1967). Unrealistic $m_\alpha$ values commonly occur as a result of an assumed slip surface, which is inconsistent with the earth pressure theory. When the $m_\alpha$ term approaches zero, the normal force at the base of a slice will tend to infinity [Eq. (11.60)]. An unreasonably large normal force will affect the calculation of the factor of safety. The $m_\alpha$ problem can be resolved by limiting the inclination of the slip surface at the crest of the slope (i.e., the active zone) to the maximum obliquity for the active state (Fig. 11.65):

$$\alpha_{max} = 45° + \frac{\phi'}{2}. \quad (11.79)$$

Similarly, the inclination of the slip surface at the toe of the slope (i.e., the passive zone) should be limited to a maximum angle in accordance with the passive state (Fig. 11.65):

$$\alpha_{max} = 45° - \frac{\phi'}{2}. \quad (11.80)$$

A vertical tension crack zone at the crest of the slope can also be used to limit the inclination of the slip surface in order to alleviate the $m_\alpha$ problem. The slip surface will terminate at the base of the tension crack zone.

The problem of negative normal forces at the base of a slice is caused by a high cohesion value, and is relevant to slopes with highly negative pore-water pressures. This problem is particularly significant for relatively shallow slip surfaces where the cohesive component dominates the shear strength of the soil. Figure 11.66 illustrates the effect on the normal stress along the slip surface of increasing cohesion on a steep slope with a shallow slip surface. The cohesion can be considered to increase with increasing matric suction (i.e., $c = c' + (u_a - u_w)\tan\phi^b$).

The increase in cohesion has been shown in Fig. 11.66 to result in negative normal stresses. The negative normal

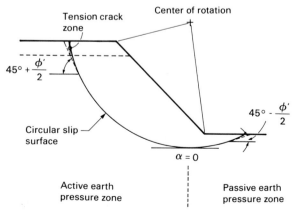

**Figure 11.65** Limiting inclination angles for the slip surface at the crest and toe of the slope.

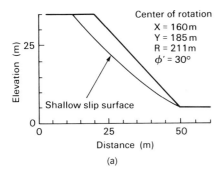

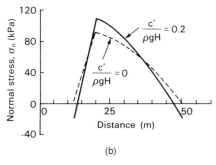

**Figure 11.66** Effect of increasing cohesion values on the normal stress distribution. (a) Steep slope with a shallow slip surface; (b) normal stress distribution along the slip surface.

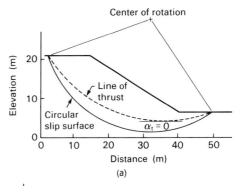

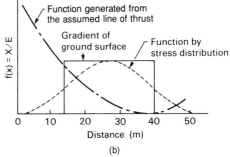

**Figure 11.67** Illustration of an interslice force function generated from an unrealistic assumption for the line of thrust. (a) Geometry showing the line of thrust; (b) possible interslice force functions.

forces are the result of having mobilized a large shearing force, $S_m$, due to the high cohesion values. The shearing force has a positive sign, indicating an opposite direction to the sliding direction. In order for the force polygon to close (or force equilibrium to be satisfied), the normal force has to become negative. Spencer (1968, 1973) suggested that a tension crack zone should be located at the crest of the slope in order to reduce the large mobilized shearing force. The depth of the tension crack zone may extend through the region with negative normal forces.

Nonconvergence can be encountered with any limit equilibrium method which uses an unrealistic assumption regarding the interslice conditions. This problem appears to be attributable to unreasonable assumption regarding the line of thrust (Ching, 1981). The moment equilibrium equation can be used to generate the equivalent of an interslice force function based on an assumed "line of thrust." The shape of the resulting function can be unrealistic when compared to an elastic analysis (Fan, 1983) as illustrated in Fig. 11.67. The steepness of the function at the ends produces high interslice shear forces which may exceed the weight of the slice. It is suggested that the assumption used in a slope stability analysis should be somewhat consistent with the stresses resulting from gravity.

### 11.3.5 Effects of Negative Pore-Water Pressure on Slope Stability

All of the components associated with performing slope stability analyses in situations where the pore-water pressures are negative, have been discussed. One procedure which can be used for performing slope stability analysis involves the incorporation of the matric suction into the cohesion of the soil (Ching et al. 1984). This will be referred to as the "total cohesion" method. The second procedure involves rederiving the factor of safety equations in order to accommodate both positive and negative pore-water pressures (Fredlund, 1987a, 1989; Fredlund and Barbour, 1990; Rahardjo and Fredlund, 1991). A nonlinear shear strength versus matric suction relationship can also be incorporated in the slope stability analysis (Rahardjo, Fredlund and Vanapalli, 1992).

### The "Total Cohesion" Method

In the "total cohesion" method, the soil cohesion, $c$, is considered to increase as the matric suction of the soil increases. The increase in the cohesion due to matric suction (i.e., $(u_a - u_w) \tan \phi^b$) is illustrated in Fig. 11.68 for various $\phi^b$ angles.

Another form for the relationship between negative pore-water pressures and cohesion is illustrated in Fig. 11.69. Here, matric suctions are presented as a percentage of the hydrostatic negative pore-air pressures above the water table. Matric suction is multiplied by ($\tan \phi^b$) to give an equivalent increase in the cohesion for a $\phi^b$ angle of 15° (Fig. 11.69).

The increase in the factor of safety due to negative pore-water pressures (or matric suction) is illustrated in Figs. 11.70 and 11.71. The shear strength contribution from matric suction is incorporated into the designation of the cohe-

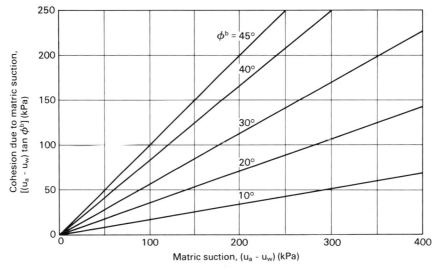

**Figure 11.68** The component of cohesion due to matric suction for various $\phi^b$ angles.

sion of the soil (i.e., $c = c' + (u_a - u_w) \tan \phi^b$). It can readily be appreciated that the factor of safety of a slope can decrease significantly when the cohesion due to matric suction is decreased during a prolonged wet period.

### Two Examples Using the "Total Cohesion" Method

The following two example problems illustrate the application of the "total cohesion" method in analyzing slopes with negative pore-water pressures. The example problems involve studies of steep slopes in Hong Kong. The soil stratigraphy was determined from numerous borings. The shear strength parameters (i.e., $c'$, $\phi'$, and $\phi^b$) were obtained through the testing of undisturbed soil samples in the laboratory. Negative pore-water pressures were measured *in situ* using tensiometers. Slope stability analyses were performed to assess the effect of matric suction changes on the factor of safety. Also, parametric-type analyses were conducted using various percentages of the hydrostatic negative pore-water pressures.

**Example no. 1.** The site plan of example no. 1 is shown in Fig. 11.72. The site consists of a row of residential buildings with a steep cut slope at the back. The slope has an average inclination angle of 60° to the horizontal and a maximum height of 35 m. The slope has been protected from infiltration of surface water by a layer of soil

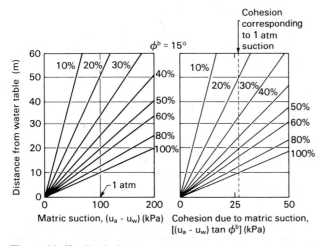

**Figure 11.69** Equivalent increase in cohesion for various matric suction profiles.

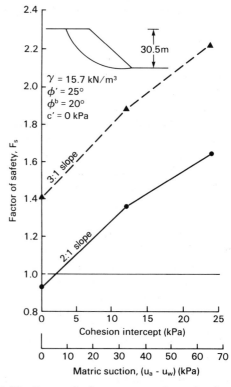

**Figure 11.70** Factor of safety versus matric suction for a simple slope.

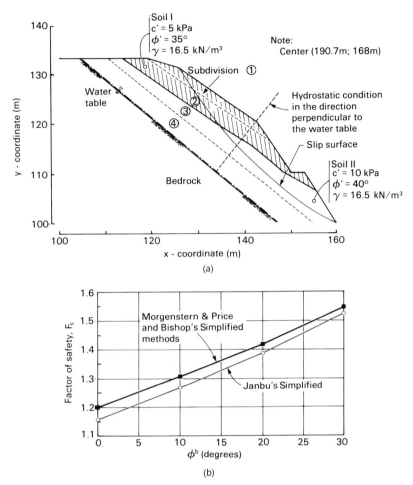

**Figure 11.71** Factor of safety of a steep slope versus cohesion increase due to matric suction. (a) Example of a typical steep slope in Hong Kong (from Sweeney and Robertson, 1979); (b) increase in factor of safety due to an increase in $\phi^b$ angle.

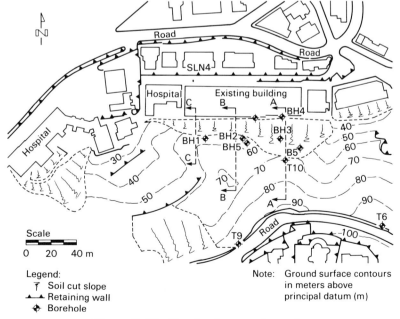

**Figure 11.72** Site plan for example no. 1.

cement and lime plaster which is locally referred to as "chunam" plaster. Small but dangerous failures have occurred periodically at the crest of the cut slope. This condition prompted a detailed investigation.

Three cross sections of the cut slope (i.e., *A-A*, *B-B*, and *C-C*) were analyzed, as illustrated in Fig. 11.73. The stratigraphy consists primarily of weathered granite. A 4–5 m layer of granitic colluvium is underlain by a 10 m layer of completely to highly weathered granite. The bedrock is situated 20–30 m below the surface. The water table is located well into the bedrock. It is estimated that the water table may rise by 5–8 m under the influence of heavy rainfalls, with return periods of 10 and 1000 years, respectively. However, the fluctuation of this deep groundwater table does not directly affect the slope stability analysis. Potential failures in a steep slope, such as the slope being analyzed, are primarily associated with relatively shallow slip surfaces.

Triaxial tests on undisturbed core samples were conducted, and the results are presented in Table 11.4. The average measured $\phi^b$ angle for the soil was 15° (Ho and Fredlund, 1982a).

*In situ* measurements of matric suction were conducted from the face of the slope using tensiometers. Two typical matric suction profiles measured near section *A-A* are shown in Fig. 11.74. The suction profiles showed considerable variation, corresponding to different microclimatic conditions near the proximity of the slope face. No matric suction measurements were made near the upper part of the slope.

Slope stability analyses were performed on the three cross sections (i.e., sections *A-A*, *B-B*, and *C-C*), as shown in Fig. 11.73. The GLE method was used for all analyses. The vertical interslice shear forces were assumed to be zero, and only moment equilibrium was used to solve for the factor of safety. The analyses were performed using circular slip surfaces, and all critical slip surfaces were found to pass through the toe of the slope.

For the first analysis, the effect of matric suction was ignored (i.e., $c = c'$), and the lowest factors of safety for the three cross sections were computed (Table 11.5). All three sections show critical factors of safety less than 1.0, indicating unstable slope conditions. However, the slope has remained stable. It is assumed that the additional shear strength contribution from matric suction has played a significant role in the overall stability of the slope.

For subsequent analyses, matric suction was taken into account as part of the cohesion component of shear strength. Each of the cross sections was divided into substrata drawn parallel to the water table in order to account for the changing matric suction with depths. The subdivisions for cross section *A-A* are shown in Fig. 11.75. Each of the substrata was 5 m thick and assumed to have an independent total cohesion, $c$. The equivalent increase in cohesion (i.e., $(u_a - u_w) \tan\phi^b$) for each substratum was computed from the matric suction profile and a $\phi^b$ angle of 15°.

A parametric study was conducted using various percentages of the hydrostatic negative pore-water pressure. The corresponding matric suctions and the equivalent increases in cohesion are shown in Fig. 11.69. The matric suction used in each analysis was limited to 101.3 kPa (i.e., 1 atm), corresponding to a cohesion increase of 27 kPa. The results of the parametric study on the three cross sec-

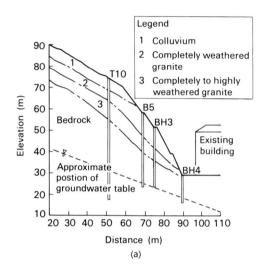

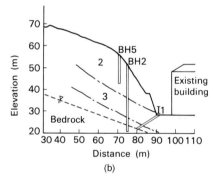

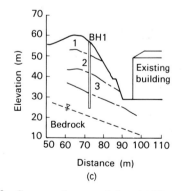

**Figure 11.73** Cross sections used for stability analyses in example no. 1. (a) Section *A-A*; (b) section *B-B*, (c) section *C-C*.

**Table 11.4 Strength Properties for Soils of Example Problem 1**

| Soil Type | Unit Weight (kN/m³) | $c'$ (kPa) | $\phi'$ (degree) |
|---|---|---|---|
| Colluvium | 19.6 | 10.0 | 35.0 |
| Completely weathered granite | 19.6 | 15.1 | 35.2 |
| Completely to highly weathered granite | 19.6 | 23.5 | 41.5 |

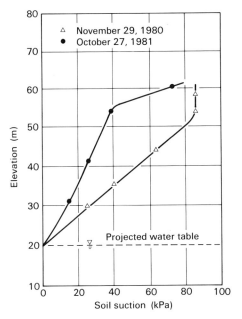

**Figure 11.74** *In situ* measurements of matric suction near section A-A for example no. 1 (from Sweeney [468]).

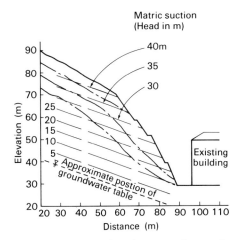

**Figure 11.75** Subdivision of section A-A of example no. 1 for the slope stability study.

tions are shown in Fig. 11.76 for various percentages of matric suction. A factor of safety of 1.0 corresponds to a matric suction profile of approximately 10–20% of hydrostatic conditions. A significant increase in the factor of safety (i.e., approximately 25%) is obtained when the matric suction profile is increased up to 40%. The smaller increase in the factor of safety beyond the 40% profile is attributable to the maximum limit of 101.3 kPa placed on the matric suction value. The position of the critical centers of rotation corresponding to various matric suction profiles is depicted in Fig. 11.77 for section A-A. The critical slip surface tends to penetrate deeper into the slope as the cohesion of the soil increases. The increase in cohesion is due to the increase in matric suction.

Slope stability analyses were also performed on example

**Table 11.5 Results of Slope Stability Analyses on Example Problem 1 Without the Effect of Matric Suction**

| | Center or Rotation[a] (meters) | | | Factor of Safety |
|---|---|---|---|---|
| Section | x-coordinate | y-coordinate | Radius | |
| A–A | 232.5 | 190.0 | 216.0 | 0.864 |
| B–B | 143.8 | 120.0 | 89.5 | 0.910 |
| C–C | 171.6 | 118.1 | 120.8 | 0.881 |

[a]Critical center of rotation.

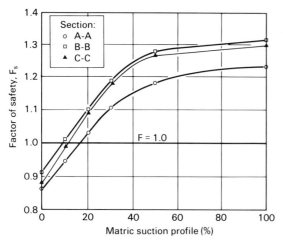

**Figure 11.76** Results of the parametric slope stability study using example no. 1.

**Table 11.6 Results of Slope Stability Analyses on Example Problem 1 with the Effect of Matric Suction as Measured *In Situ***

A. Suction Profile (November 29, 1980)

| Section | Center of Rotation (meters) x-coordinate | y-coordinate | Radius | Factor of Safety |
|---|---|---|---|---|
| A–A | 176.3 | 141.9 | 143.0 | 1.072 |
| B–B | 133.1 | 117.5 | 81.4 | 1.143 |
| C–C | 138.8 | 96.3 | 83.1 | 1.132 |

B. Suction Profile (October 27, 1981)

| Section | Center of Rotation (meters) x-coordinate | y-coordinate | Radius | Factor of Safety |
|---|---|---|---|---|
| A–A | 201.3 | 167.5 | 178.6 | 0.984 |
| B–B | 165.0 | 125.0 | 122.2 | 1.046 |
| C–C | 156.9 | 108.8 | 104.1 | 1.014 |

problem no. 1 using the measured matric suctions shown in Fig. 11.74. The equivalent increase in cohesion for each substratum was computed from the matric suction profiles, with a maximum measured value of 85 kPa. The results are presented in Table 11.6. The overall factors of safety are 1.10 and 1.01, based on the matric suction profiles measured on November 29, 1980 and October 27, 1981, respectively.

**Example no. 2.** The site plan for example no. 2 is shown in Fig. 11.78. A steep and high cut slope exists behind a residential building. A proposed high-rise residential building above the slope prompted a detailed investigation for the stability of the slope under the new conditions. The cut slope under consideration is below a major access road, and the cross section of concern, A–A, is shown in Fig. 11.79. The slope is inclined at 60° to the horizontal, and has an average height of 30 m. The stratigraphy consists entirely of weathered rhyolite. The water table lies well below the ground surface. It is estimated that the water table will rise less than 5–8 m under the influence of a heavy rainfall, with return periods of 10 and 1000 years, respectively. As a result, the deep water table does not directly affect the stability analysis associated with relatively shallow slip surfaces.

The shear strength parameters of the soils comprising example no. 2 were measured in the laboratory on undisturbed samples. The properties of the soils are summarized in Table 11.7. *In situ* matric suction measurements were obtained using tensiometers installed vertically along an exploratory caisson shaft near the cut slope. The matric suction profiles measured through the rainy season of 1980 are plotted in Fig. 11.80. The profiles remain essentially constant, with some variations near the surface due to infiltration as well as fluctuations of the groundwater table.

Slope stability analyses were performed on section A–A (Fig. 11.79) using circular slip surfaces passing through the toe. The analytical procedure was the same as that followed in example no. 1. The critical factor of safety for the cut slope, without taking into consideration the matric suction of the soil (i.e., $c = c'$), is approximately 1.05. This low factor of safety indicates the imminence of an unstable condition, although no sign of distress was ob-

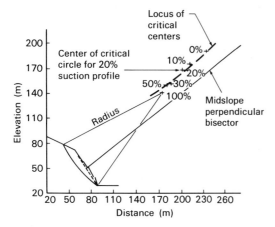

**Figure 11.77** Location of the critical centers of rotation for various matric suction profiles on section A–A.

11.3 SLOPE STABILITY   339

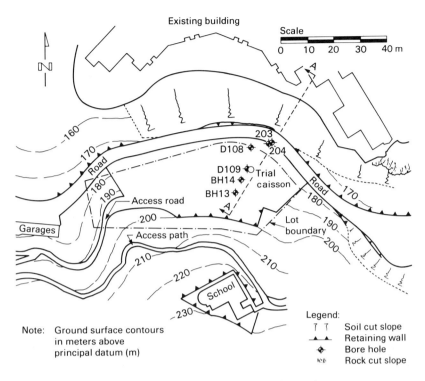

**Figure 11.78** Site plan for example no. 2.

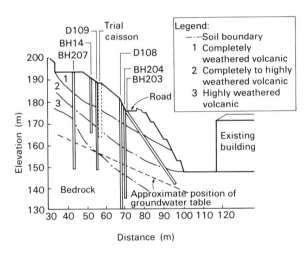

**Figure 11.79** Section *A-A* for example no. 2.

served. The next analysis was performed by taking the matric suction into consideration (i.e., $c = c' + (u_a - u_w)$ tan $\phi^b$). The matric suction profile of September 2, 1980 was used in the analysis. In this case, the cross section was divided into substrata drawn parallel to the water table in order to account for the varying matric suctions with depths. The critical factor of safety corresponding to this matric suction profile is 1.25. In other words, matric suctions have resulted in approximately a 20% increase in the factor of safety.

A parametric study was also conducted on section *A-A* in order to investigate the effect of changes in the matric suction profile and the groundwater table on the factor of safety. The various matric suction profiles shown in Fig. 11.69 were used in the study. The groundwater table position was varied in accordance with the heavy rainfalls of

**Table 11.7 Strength Properties for Soils of Example Problem 2**

| Soil Type | Unit Weight ($kN/m^3$) | $c'$ (kPa) | $\phi'$ (degree) | $\phi^b$ (degree) |
|---|---|---|---|---|
| Completely weathered rhyolite | 18.4 | 10.1 | 42.6 | 12.0 |
| Completely to highly weathered rhyolite | 21.4 | 12.0 | 43.9 | 12.0 |

## 340   11   PLASTIC AND LIMIT EQUILIBRIUM

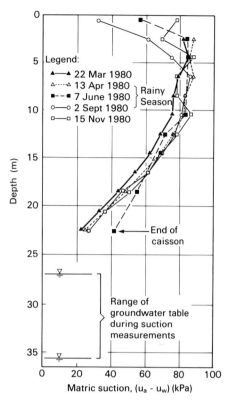

**Figure 11.80** *In situ* measurements of matric suction throughout 1980 for example no. 2 (from Sweeney, 1982).

10 and 1000 year return periods. The results of the parametric study are summarized in Fig. 11.81. The results show that the factor of safety is more sensitive to changes in the matric suction profile than to changes in the position of the groundwater table.

### The "Extended Shear Strength" Method

The application of the slope stability formulation using the "extended shear strength" equations is illustrated in the following example (Ng, 1988). A typical cross section for a steep slope in Hong Kong is selected for purposes of il-

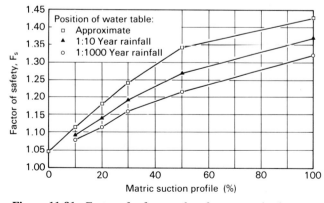

**Figure 11.81** Factor of safety results of a parametric slope stability study on example no. 2.

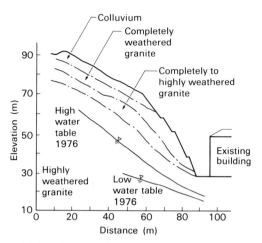

**Figure 11.82** Cross section of a steep slope in residual soil.

lustration. The role of the negative pore-water pressures is shown by computing the factor of safety of the slope for a variety of pore-water pressure conditions. The slope is subjected to a flux at ground surface in order to simulate a severe rainfall condition. As a result, the pore-water pressure increases (i.e., matric suction decreases). The factor of safety of the slope is shown to decrease with decreasing matric suctions. The influence of various $\phi^b$ values on the factor of safety is also illustrated.

The analysis pertaining to transient water flow is described in Chapters 15 and 16. In this section, the transient water flow equations are used to calculate the negative pore-water pressure changes that occur as a result of heavy rainfalls. The pore-air pressures are assumed to remain atmospheric (i.e., $u_a = 0$) in both the seepage and slope stability analyses.

**General layout of problems and soil properties.** The example slope has an inclination angle of approximately 60° to the horizontal. Its height is about 38 m. The cross section of the slope along with its stratigraphy is shown in Fig. 11.82. The slope consists primarily of the residual soil, decomposed granite. The upper 5 m is a layer of colluvium overlaying a layer of completely decomposed granite. Be-

**Table 11.8   Summary of Saturated Coefficients of Permeability for the Soils in the Example**

| Soil Type | Selected Permeability, $k_s$ (m/s) |
| --- | --- |
| Colluvium | $3 \times 10^{-5}$ |
| Completely decomposed granite | $7 \times 10^{-6}$ |
| Completely to highly decomposed granite | $6 \times 10^{-6}$ |
| Highly decomposed granite | $5 \times 10^{-6}$ |

## 11.3 SLOW STABILITY

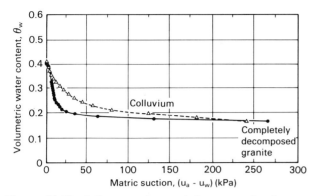

**Figure 11.83** Soil-water characteristic curves for the completely decomposed granite and the colluvium.

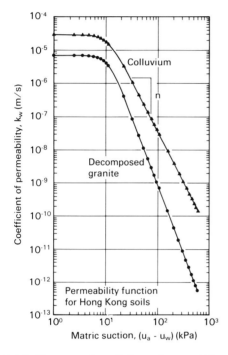

**Figure 11.84** Unsaturated coefficient of permeability functions for decomposed granite and colluvium.

low this is a layer of completely to highly decomposed granite. A thick layer of highly decomposed granite lies at the bottom of these strata. The face of the slope is covered by a thin layer of low permeability, cement–lime stabilized, decomposed granite called "chunam."

Several soil properties related to permeability and shear strength are required for the analyses. The saturated coefficients of permeability are presented in Table 11.8.

The unsaturated coefficient of permeability functions can be predicted using the saturated coefficients of permeability, along with the soil-water characteristic curves of the soil (Green and Corey, 1971b). The soil-water characteristic curves for the colluvium and the completely decomposed granite are shown in Fig. 11.83. Data from the soil-water characteristic curves, along with the saturated coefficients of permeability are used to establish the relationship between the unsaturated coefficients of permeability, $k_w$, and matric suction (Lam, 1984). The computed values can then be fitted to the unsaturated permeability function proposed by Gardner (1958a,b).

$$k_w = \frac{k_s}{1 + a\left\{\dfrac{(u_a - u_w)}{\rho_w g}\right\}^n}. \qquad (11.81)$$

The "$a$" and "$n$" values in Eq. (11.81) were found to be approximately 0.1 and 3.0, for both colluvium and decomposed granite, respectively (Ng, 1988). Using these values along with the saturated coefficients of permeability, the unsaturated permeability functions can be drawn as shown in Fig. 11.84.

The coefficient of water volume change, $m_2^w$, for the colluvium and the decomposed granite was computed to be approximately $3 \times 10^{-3}$ kPa$^{-1}$. The shear strength parameters and the unit weights for the soils involved are summarized in Table 11.9. These properties are required when performing the slope stability analysis. The $\phi^b$ angle for each material was assumed to be a percentage of the effective angle of internal friction, $\phi'$. The percentage of the $\phi^b$

**Table 11.9 Summary of Shear Strength Parameters and Total Unit Weights for the Soils in the Example**

| Soil Type | Cohesion, $c'$ (kPa) | Effective Angle of Internal Friction, $\phi'$ (degrees) | Total Unit Weight, $\gamma$ (kN/m$^3$) |
|---|---|---|---|
| Colluvium | 10 | 35 | 19.6 |
| Completely decomposed granite | 10 | 38 | 19.6 |
| Completely to highly decomposed granite | 29 | 33 | 19.6 |
| Highly decomposed granite | 24 | 41.5 | 19.6 |

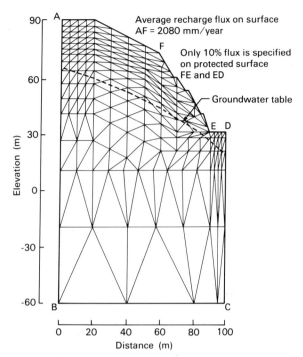

**Figure 11.85** Finite element mesh of the steep cut slope and the initial boundary conditions.

angle varied during the parametric study of the slope (i.e., $\phi^b$ is 0, 25, 50, 75, and 100% of $\phi'$).

**Initial conditions for the seepage analysis.** The steady-state and transient seepage analyses are conducted using the finite element method (see Chapters 7 and 16). The slope cross section is first discretized into elements, as shown in Fig. 11.85. The initial conditions are assigned around the boundary of the finite element mesh of the slope. A zero flux condition is imposed along the bottom boundary, *BC*. The left-hand and right-hand boundaries (i.e., *AB* and *DC*, respectively) consist of a constant head boundary below the water table and a zero flux boundary above the water table. The constant head boundary below the water table is equal to the initial elevation of the water table.

The surface boundary, *AFED*, is specified as a flux boundary. The applied flux is equal to the average annual rainfall in Hong Kong, which is about 2080 mm/year or $6.6 \times 10^{-8}$ m/s (Anderson, 1983). However, only 10% intake of the applied flux is allowed to enter on the steep cut boundary, *FED*, which is protected by chunam. It is assumed that when the applied flux is in excess of the amount of water that can be taken in by the soil, water will not be allowed to pond at the surface boundaries. It is anticipated that a seepage face may develop with time from the base of the slope along the boundary, *EF*.

The pore–water pressure contours and the water table under steady-state flux conditions are shown in Fig. 11.86. The pore–water pressure profiles along the cross section *X-X* are illustrated in Fig. 11.87. The matric suction de-

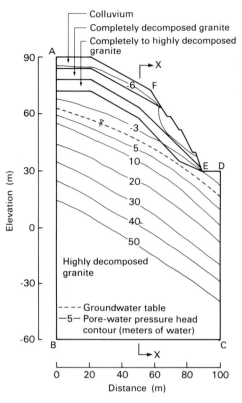

**Figure 11.86** Initial groundwater condition and pore–water pressure head contours.

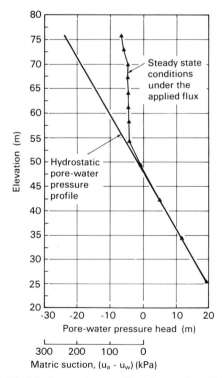

**Figure 11.87** Matric suction profiles for section *X-X* under steady-state flux conditions.

11.3 SLOPE STABILITY 343

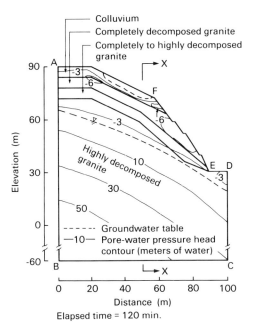

**Figure 11.88** Positions of groundwater table and pore-water pressure head contours at an elapsed time of 120 min.

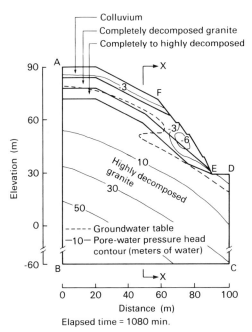

**Figure 11.90** Position of groundwater table and pore-water pressure head contours at an elapsed time of 1080 min.

viates from the hydrostatic condition as a result of the applied flux, which produces a recharge condition. This recharge condition is maintained even in the dry season, as indicated by field studies (Leach and Herbert, 1982; Sweeney, 1982).

The computed pore-water pressures under the steady-state flux conditions (Figs. 11.86 and 11.87) can be used to analyze the stability of the slope. The GLE method was used for all analyses. The analyses were performed on circular slip surfaces. The results are presented in Fig. 11.92 for various $(\phi^b/\phi')$ ratios. When the negative pore-water

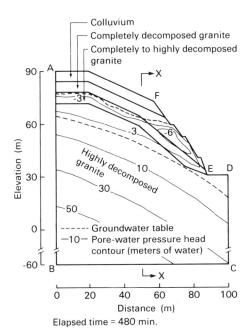

**Figure 11.89** Positions of groundwater table and pore-water pressure head contours at an elapsed time of 480 min.

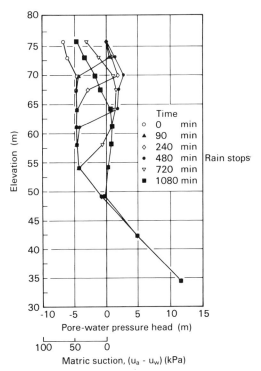

**Figure 11.91** Matric suction profiles for section $X$-$X$ at various elapsed times.

# 344   11 PLASTIC AND LIMIT EQUILIBRIUM

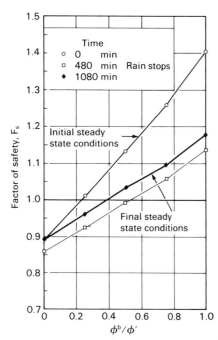

**Figure 11.92** Factors of safety with respect to the $\phi^b/\phi'$ ratio for various seepage conditions.

pressures are ignored (i.e., $\phi^b/\phi' = 0$), the factor of safety is equal to 0.9. The absence of signs of distress on the slope would indicate that the factor of safety is greater than 1.0. The negative pore-water pressures must be contributing to the shear strength of the soil, which in turn increases the factor of safety. The factor of safety ranges from 1.0 to 1.4 as the ($\phi^b/\phi'$) ratio varies from 0.25 to 1.0, respectively. This range in factor of safety is in agreement with the range obtained using the "total cohesion" method corresponding to various matric suction profiles.

**Seepage and slope stability results under high-intensity rainfall conditions.** The following transient analyses illustrate the effect of a high-intensity rainfall for a long duration on the stability of the steep slope shown in Fig. 11.82. The rainfall rate is selected as $1.30 \times 10^{-5}$ m/s for a duration of 480 min. The pore-water pressure distributions corresponding to the simulated steady-state flux conditions are used as the initial conditions for the transient seepage analyses during this high-intensity rainfall. The subsequent pore-water pressure distributions at various elapsed times after the commencement of the high-intensity rainfall are shown in Figs. 11.88–11.90. The upper portion of the slope becomes saturated starting at the ground surface, $AF$, which is not protected by chunam (see Fig. 11.88). The saturation front has penetrated to a depth of approximately 12 m and extends to the region below the protected surface, $FE$, when the rainfall stops after 480 min (see Fig. 11.89). The final steady-state conditions are achieved at 1080 min of elapsed time. At this time, the saturation front has moved deeper as the water is redistributed deeper within the slope. As a result, the groundwater table corresponding to the final steady-state condition (Fig. 11.90) is higher than its position corresponding to the initial conditions (Fig. 11.86).

Fig. 11.91 illustrates the pore-water pressure profiles for section $X$–$X$ at various elapsed times. The negative pore-water pressures at depth increase to zero (i.e., the matric suction is dissipated) as the saturation front moves downward. Under final steady-state conditions, the upper 10 m of the slope has gained back only a fraction of its initial matric suction, while the lower part of the slope has lost its initial matric suctions.

The slope stability analyses can be performed for various elapsed times using the corresponding pore-water pressure

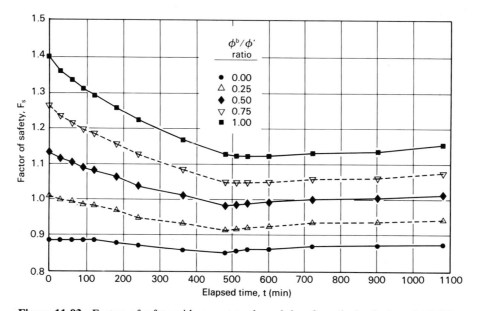

**Figure 11.93** Factors of safety with respect to elapsed time from the beginning of rainfall.

distributions. The changing factors of safety during the rainfall period are plotted in Figs. 11.92 and 11.93 for various ratios of ($\phi^b/\phi'$). The factors of safety continue to decrease until the rainfall stops after 480 min. The decrease in the factor of safety becomes more substantial as the ($\phi^b/\phi'$) ratio increases. This can be attributed to the fact that the critical slip surface is shallow, and the mobilized shear force is significantly affected by contributions from the negative pore–water pressures. The above scenario illustrates a possible catastrophic failure of a steep slope as it relates to a loss of matric suction during a heavy, prolonged rainfall.

The factor of safety on a fixed slip surface increases again after 480 min of elapsed time when the rainfall has stopped and the water has moved deeper into the slope. However, the critical slip surface during and after the rainfall may be different. The critical slip surface may go deeper as the wetting front moves into the slope. The increase in the factor of safety appears to occur at a slower rate than the decrease in the factor of safety during the rainfall period.

# CHAPTER 12

# Volume Change Theory

Several types of constitutive relations for an unsaturated soil are mentioned in Table 12.1. The volume change constitutive relations are simply one of several constitutive relations used in soil mechanics. The constitutive equations for volume change relate the deformation state variables to the stress state variables. The shear strength equation presented in Chapter 9 is a constitutive equation that relates shear stress to the normal stress state variables. A constitutive relation requires soil properties which must generally be evaluated experimentally. The soil properties used in the shear strength equation are the effective cohesion, $c'$, the effective angle of internal friction, $\phi'$, and the $\phi^b$ angle.

A brief literature review on various volume change theories and experiments for unsaturated soils is presented in this chapter. The volume change concepts for an unsaturated soil are outlined. The deformation state variables are selected to maintain a consistency with continuity (i.e., conservation of mass) for an unsaturated soil. The stress state variables for an unsaturated soil were described in Chapter 3.

Several forms of the volume change constitutive equations are presented in this chapter. Experiments that have been used to verify the constitutive equations are also described. Soil properties used in the volume change constitutive equations come under the general term "volumetric deformation coefficients." The relationships among the various volumetric coefficients are presented near the end of this chapter.

## 12.1 LITERATURE REVIEW

In 1941, Biot (1941) presented a three-dimensional consolidation theory based on the assumption that the soil was isotropic and behaved in a linear elastic manner. The soil was assumed to be unsaturated in that the pore–water contained occluded air bubbles. Two constitutive relationships were proposed in order to completely describe the deformation state of the unsaturated soil. One constitutive relationship was formulated for the soil structure, and the other constitutive relationship was for the water phase. Two independent stress variables were used in the formulations. In total, four volumetric deformation coefficients were required to link the stress and deformation states.

Attempts to link the deformation behavior of an unsaturated soil with a single-valued effective stress equation (Bishop, 1959) have resulted in limited success (Jennings and Burland, 1962). Oedometer and all-around compression tests have been performed on unsaturated and saturated soils ranging from silty sands to silty clays. The results have indicated that there was not a unique relationship between volume change and effective stress for most soils, particularly below a critical degree of saturation. The critical degree of saturation appeared to be approximately 20% for silts and sands, and as high as 85–90% for clays.

Coleman (1962) separated the components of Bishop's effective stress equation, and proposed one set of constitutive relationships for the soil structure and another for the water phase. The volumetric deformations of an unsaturated soil specimen under triaxial test loading were considered. The proposed volume change constitutive relation associated with the soil structure was as follows:

$$-\frac{dV}{V} = -C_{21}(du_w - du_a) + C_{22}(d\sigma_m - du_a) + C_{23}(d\sigma_1 - d\sigma_3) \tag{12.1}$$

where

$dV$ = overall volume change of a soil element
$V$ = current overall volume of the soil element
$u_a$ = pore–air pressure
$u_w$ = pore–water pressure
$\sigma_1$ = total axial normal stress (i.e., major principal stress)
$\sigma_3$ = total confining pressure (i.e., minor principal stress)

## 12.1 LITERATURE REVIEW

**Table 12.1 Several Types of Constitutive Relations for Unsaturated Soil**

| Constitutive Relations | Description | Reference |
| --- | --- | --- |
| Stress versus volume–mass | 1) Relates the stress state variables to strains, deformations, and volume–mass properties such as void ratio, water content, and degree of saturation | Chapter 12 |
| | 2) Density equations for air–water mixtures | Chapter 8 |
| | 3) Compressibility equations for air–water mixtures | Chapter 8 |
| Stress versus stress | 1) Pore pressure parameters relating pore pressures to normal stress under undrained loading conditions | Chapter 8 |
| | 2) Strength equations relating shear stress to stress state variables | Chapter 9 |
| Stress gradient versus flow rate | 1) Flow laws for the pore-air and pore-water | Chapter 5 |

$\sigma_m$ = mean total normal stress [i.e., $1/3\,(\sigma_1 + 2\sigma_3)$]

$C_{21}, C_{22}, C_{23}$ = soil parameters associated with the soil structure volume change.

The compressibility parameters, $C_{21}, C_{22}, C_{23}$, depend solely upon the current values of $(u_w - u_a)$, $(\sigma_m - u_a)$, and $(\sigma_1 - \sigma_3)$ and the stress history of the soil. The constitutive relation for the volume change associated with the water phase was written as

$$-\frac{dV_w}{V} = -C_{11}\,(du_w - du_a) + C_{12}\,(d\sigma_m - du_a) + C_{13}\,(d\sigma_1 - d\sigma_3) \qquad (12.2)$$

where

$dV_w$ = change in the volume of water in the soil element

$C_{11}, C_{12}, C_{13}$ = soil parameters associated with the change in the volume of water in the soil element.

The formulation by Coleman (1962) assumes that a change in the deviatoric stress also produces volume changes.

Some of the difficulties in using a single-valued effective stress variable to describe the deformation behavior of an unsaturated soil were explained by Bishop and Blight (1963). It was concluded that the stress versus deformation paths of both stress components [i.e., $(\sigma - u_a)$ and $(u_a - u_w)$] must be taken into consideration in an independent manner. Bishop and Blight (1963) proposed that volume change data be plotted against the $(\sigma - u_a)$ and $(u_a - u_w)$ stress variables in a three-dimensional form. In 1965, Burland (1965) restated that volume changes in unsaturated soils should be independently related to the $(\sigma - u_a)$ and $(u_a - u_w)$ stress state variables.

Aitchison (1967) again pointed out the importance of mapping volume changes with respect to the independent stress variables. Later in 1969, Aitchison presented typical volume change curves obtained by independently following paths of $(\sigma - u_a)$ and $(u_a - u_w)$ versus deformation.

Matyas and Radhakrishna (1968) introduced the concept of state parameters for an unsaturated soil. The state parameters consisted of stress variables (e.g., $\sigma_m = (\sigma_1 + 2\sigma_3)/3 - u_a$, $(\sigma_1 - \sigma_3)$, and $(u_a - u_w)$ for triaxial compression), along with the initial void ratio and degree of saturation (i.e., $e_0$ and $S_0$). Tests were performed on "identical" soil specimens compacted at the same water content and dry density. For isotropic compression, the stress parameters reduced to $(\sigma_3 - u_a)$ and $(u_a - u_w)$. The void ratio and degree of saturation were used to represent the deformation state of the soil.

Three-dimensional state surfaces were formed with void ratio and degree of saturation plotted against the independent state parameters, $(\sigma - u_a)$ and $(u_a - u_w)$. These state surfaces are, in essence, constitutive surfaces. Matyas and Radhakrishna (1968) experimentally tested the uniqueness of the constitutive surfaces. Isotropic and $K_0$ compression tests were performed on mixtures of 80% flint powder and 20% kaolin. The total, pore-air, and pore-water pressures were controlled during the tests. The constitutive surfaces of void ratio and degree of saturation versus $(\sigma - u_a)$ and $(u_a - u_w)$ stress variables were defined using different stress paths to test their uniqueness.

The void ratio results (Matyas and Radhakrishna, 1968) produced a single warped surface, with the soil structure always decreasing in volume as the $(u_a - u_w)$ stress was decreased or as the $(\sigma - u_a)$ stress was increased [Fig.

12.1(a)]. The results indicated that the soil had a metastable soil structure which collapsed as a result of a gradual reduction in matric suction, $(u_a - u_w)$. A soil with a stable structure would have swelled when the matric suction was decreased. In spite of the collapse phenomenon, the results show a unique constitutive surface for the soil structure [Fig. 12.1(a)] provided the deformation paths resulted in an increasing degree of saturation. When other paths were followed which involved wetting and drying, the void ratio versus stress constitutive surface was not found to be completely unique. This restriction on the path appeared to be related to hysteresis associated with wetting and drying. These paths introduce certain nonunique characteristics in the soil structure constitutive surface (Matyas and Radhakrishna, 1968).

The constitutive surface for the water phase, represented by the degree of saturation, was not found to be unique [Fig. 12.1(b)]. However, once again there was wetting and drying prior to moving towards saturation.

In 1969, Barden et al. studied the volume change characteristics of unsaturated soils under $K_0$-loading conditions. The tests were performed on low to high plasticity illite clay specimens (i.e., Westwater and Derwent clays). The total, pore-air, and pore-water pressures were controlled while investigating the effect of various stress paths during $K_0$-loading. In all cases, the net normal stress $(\sigma - u_a)$ was increased subsequent to the initial conditions. In most cases, the matric suction, $(u_a - u_w)$, was increased subsequent to the initial state; however, in a few cases, the suction was decreased. The results indicated that the overall volume change of the specimen was stress path dependent, being a function of whether the soil was going towards saturation or away from saturation. Hysteresis between the saturation and desaturation processes was considered as the major cause of stress path dependence. It was concluded that the volume change behavior of an unsaturated soil was best analyzed in terms of separate components of stress, $(\sigma - u_a)$ and $(u_a - u_w)$.

Subsequently, several other researchers have suggested the use of net normal stress and matric suction as stress variables for describing volume change behavior (Aitchison and Woodburn, 1969; Brackley, 1971; and Aitchison and Martin, 1973). The role of $(\sigma - u_a)$ and $(u_a - u_w)$ as stress state variables for an unsaturated soil was later demonstrated by Fredlund (1974) and Fredlund and Morgenstern (1977). A stress analysis based on multiphase continuum mechanics showed that any two of three independent stress variables [i.e., $(\sigma - u_a)$, $(u_a - u_w)$, and $(\sigma - u_w)$] could be used to describe the stress state. Therefore, it became understandable why $(\sigma - u_a)$ and $(u_a - u_w)$ had been successfully used to describe the volume change characteristics of an unsaturated soil.

In 1977, Fredlund and Morgenstern proposed semi-empirical constitutive relations for an unsaturated soil using any two of the three independent stress state variables. The proposed equations are similar in form to those proposed by Biot (1941) and Coleman (1962). The deformation state variables required to describe volume changes satisfied continuity requirements for a multiphase continuum (Fredlund, 1973, 1974). The stress and deformation state variables were combined using suitable constitutive relations for the soil structure, air phase, and water phase. However, only two of the three constitutive relations are required for the complete description of volume changes. Generally, the constitutive relations for the soil structure and the water phase are used in volume change analyses. In engineering practice, volume changes associated with the soil structure and the water phase are often written in terms of void ratio change and water content change. Volume changes associated with the air phase are computed as the difference between the soil structure and water volume changes.

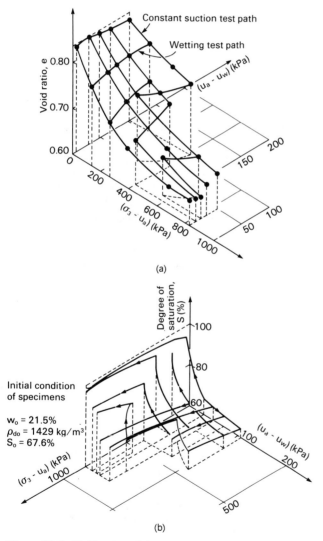

**Figure 12.1** Void ratio and degree of saturation constitutive surfaces for a mixture of flint and kaolin under isotropic loading conditions. (a) Void ratio constitutive surface, (b) degree of saturation constitutive surface (from Matyas and Radhakrishna, 1968).

The proposed constitutive relations were presented graphically to form constitutive surfaces by plotting the deformation state variable with respect to two independent stress state variables. The proposed constitutive surfaces were experimentally tested for uniqueness near a point (Fredlund and Morgenstern, 1976). Four series of experiments were performed involving undisturbed Regina clay and compacted kaolin. The specimens were tested under $K_0$- and isotropic loading conditions, using a modified Anteus oedometer and a triaxial apparatus, respectively. The total, pore–air, and pore–water pressures were controlled independently during the tests. The results indicated uniqueness as long as the deformation conditions were monotonic. Uniqueness of the constitutive surfaces under larger stress increments was previously demonstrated experimentally by Matyas and Radhakrishna (1968) and Barden et al. (1969).

The proposed constitutive relations can be expressed in an elasticity form, a compressibility form, or a soil mechanics form (Fredlund, 1979, 1982, 1985a). Each form of the constitutive equations has a role to play in geotechnical practice.

Alonso and Lloret (1982) conducted an analytical study on the behavior of an unsaturated soil under undrained loading conditions. Two equations for predicting the changes associated with the overall volume and degree of saturation of the soil were proposed. In 1985, Lloret and Alonso (1985) presented a number of linear and nonlinear functions for describing the constitutive surfaces of an unsaturated soil under $K_0$- and isotropic loading conditions. The constitutive surfaces for the soil structure and the water phase were expressed in terms of void ratio and degree of saturation. Data from published test results were used to determine the best-fit functions through the use of optimization techniques.

## 12.2 CONCEPTS OF VOLUME CHANGE AND DEFORMATION

Volume changes in an unsaturated soil can be expressed in terms of deformations or relative movement of the phases of the soil. It is necessary to establish deformation state variables that are consistent with multiphase continuum mechanics principles. A change in the relative position of points or particles in a body forms the basis for establishing deformation state variables. These variables should produce the displacements of the body under consideration when integrated over the body. This concept applies to a single or multiphase system, and is independent of the physical properties.

Two sets of deformation state variables are required to adequately describe the volume changes associated with an unsaturated soil. The deformation state variables associated with the soil structure and the water phase are commonly used in a volume change analysis (Biot, 1941; Coleman, 1962; Matyas and Radhakrishna, 1968). Void ratio changes or porosity changes can also be used as deformation state variables representing the soil structure deformation. On the other hand, changes in water content can be considered as the deformation state variable for the water phase.

Referring to continuum mechanics kinematics, there are several ways to describe the relative movement or deformation in a phase. Only two of these descriptions are relevant to unsaturated soils. For a *referential* description, the position of each particle is described as a function of its *initial* position and time. In other words, the original position and time are the independent variables. A fixed element of mass is chosen, and its motion is traced. When the reference configuration is the initial configuration, the description is generally called the Lagrangian description. The time variable disappears when equilibrium conditions are achieved. The referential description is commonly used in problems involving the elasticity of solids where the initial geometry, boundary, and loading conditions are specified.

For a *spatial* description, the position of each particle is described as a function of its *current* position and time. A fixed region in space is chosen instead of an element of mass. The spatial description is generally used in fluid mechanics, and is commonly referred to as the Eulerian description. The time variable vanishes under steady-state flow conditions. The Lagrangian and the Eulerian descriptions give the same results for cases with infinitesimal deformations.

### 12.2.1 Continuity Requirements

A *saturated* soil is visualized as a fluid–solid multiphase. The soil particles form a structure with voids filled with water. Under an applied stress gradient, the soil structure deforms and the volume changes. The soil structure volume change represents the overall volume change of the soil. It must be equal to the sum of volume changes associated with the solid phase (i.e., soil particles) and the fluid phase (i.e., water). This equality concept is referred to as the "continuity requirement." The continuity requirement is a volumetric restriction that prevents "gaps" between the phases of a deformed multiphase system in order to ensure the conservation of mass. Volume changes in a saturated soil are primarily the result of water flowing in or out of the soil since the particles are essentially incompressible.

An *unsaturated* soil can be visualized as a mixture with two phases that come to equilibrium under applied stress gradients (i.e., soil particles and contractile skin) and two phases that flow under applied stress gradients (i.e., air and water). Consider an element of soil that deforms under an applied stress gradient. The total volume change of the soil element must be equal to the sum of volume changes associated with each phase. If the soil particles are assumed to be incompressible, the continuity requirement for the unsaturated soil can be stated as follows:

$$\frac{\Delta V_v}{V_0} = \frac{\Delta V_w}{V_0} + \frac{\Delta V_a}{V_0} + \frac{\Delta V_c}{V_0} \quad (12.3)$$

where

- $V_0$ = initial overall volume of an unsaturated soil element
- $V_v$ = volume of soil voids
- $V_w$ = volume of water
- $V_a$ = volume of air
- $V_c$ = volume of contractile skin.

Assuming that the contractile skin volume change is internal to the element, the continuity requirement reduces to

$$\frac{\Delta V_v}{V_0} = \frac{\Delta V_w}{V_0} + \frac{\Delta V_a}{V_0}. \quad (12.4)$$

The above continuity requirement shows that the volume changes associated with any two of the three variables must be measured, while the third volume change can be computed. In practice, the overall and water volume changes are usually measured, while the air volume change is calculated. Suitable deformation state variables can now be defined to be consistent with the continuity requirement.

### 12.2.2 Overall Volume Change

The overall or total volume change of a soil refers to the volume change of the soil structure. Consider a two-dimensional representation of a *referential* element of unsaturated soil, as shown in Fig. 12.2. The element is referential with respect to a fixed mass of soil particles. The element has infinitesimal dimensions of $dx$, $dy$, and $dz$ in the $x$-, $y$-, and $z$-directions, respectively. Only the $x$- and $y$-directions are shown in Fig. 12.2.

The soil element is assumed to undergo translations of $u$, $v$, and $w$ from their original $x$-, $y$-, and $z$-coordinate positions, respectively. The final position of the element becomes $(x + u, y + v,$ and $z + w)$. The element is assumed to deform in response to an applied stress gradient. The deformation consists of a change in length and a rotation of the element sides with respect to each other, as illustrated in Fig. 12.2. The changes in length in the $x$-, $y$-, and $z$-directions can be written as $(\partial u/\partial x)dx$, $(\partial v/\partial y)dy$, and $(\partial w/\partial z)dz$. Defining *normal strain*, $\epsilon$, as a change in length per unit length, the normal strains of the soil structure in the $x$-, $y$-, and $z$-directions can be expressed as

$$\epsilon_x = \frac{\partial u}{\partial x} \quad (12.5)$$

$$\epsilon_y = \frac{\partial v}{\partial y} \quad (12.6)$$

$$\epsilon_z = \frac{\partial w}{\partial z} \quad (12.7)$$

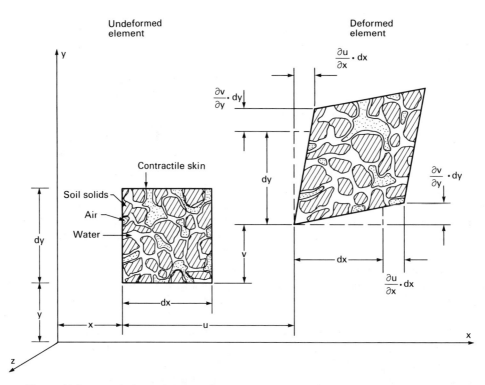

**Figure 12.2** Translation and deformation of a two-dimensional element of unsaturated soil.

where

$\epsilon_x$ = normal strain in the *x*-direction
$\epsilon_y$ = normal strain in the *y*-direction
$\epsilon_z$ = normal strain in the *z*-direction.

The above normal strains are positive for an increase in length, and negative for a decrease in length. The distortion of the element is expressed in terms of shear strain, which corresponds to two orthogonal directions. *Shear strain*, $\gamma$, is defined as the change in the original right angle between two axes (Chou and Pagano, 1967). The angle is measured in radians. A positive shear strain indicates that the right angle between the positive directions of the two axes decreases. The shear strain components of a three-dimensional element are formulated as

$$\gamma_{xy} = \frac{\partial u}{\partial y} + \frac{\partial v}{\partial x} \qquad (12.8)$$

$$\gamma_{yz} = \frac{\partial v}{\partial z} + \frac{\partial w}{\partial y} \qquad (12.9)$$

$$\gamma_{zx} = \frac{\partial w}{\partial x} + \frac{\partial u}{\partial z} \qquad (12.10)$$

where

$\gamma_{xy}$ = shear strain on the *z*-plane (i.e., $\gamma_{xy} = \gamma_{yx}$)
$\gamma_{yz}$ = shear strain on the *x*-plane (i.e., $\gamma_{yz} = \gamma_{zy}$)
$\gamma_{zx}$ = shear strain on the *y*-plane (i.e., $\gamma_{zx} = \gamma_{xz}$).

The normal and shear strains of the soil structure can be written as a deformation tensor:

$$\begin{bmatrix} \epsilon_x & \frac{1}{2}\gamma_{xy} & \frac{1}{2}\gamma_{xz} \\ \frac{1}{2}\gamma_{yx} & \epsilon_y & \frac{1}{2}\gamma_{yz} \\ \frac{1}{2}\gamma_{zx} & \frac{1}{2}\gamma_{zy} & \epsilon_z \end{bmatrix}. \qquad (12.11)$$

The sum of the normal strain components is called volumetric strain:

$$\epsilon_v = \epsilon_x + \epsilon_y + \epsilon_z \qquad (12.12)$$

where

$\epsilon_v$ = volumetric strain.

The volumetric strain is equal to the difference between the volumes of the voids in the element before and after deformation, $\Delta V_v$, referenced to the initial volume of the element, $V_0$:

$$\epsilon_v = \frac{\Delta V_v}{V_0}. \qquad (12.13)$$

The volumetric strain, $\epsilon_v$, can be used as a deformation state variable for the soil structure. It defines the soil structure volume change resulting from deformation.

### 12.2.3 Water and Air Volume Changes

The unsaturated soil element shown in Fig. 12.2 can be used to describe net changes in the fluid volumes (i.e., air and water phases). For this purpose, the element is considered as a *spatial* element for the water and air phases. The change in the volume of fluid is defined as the difference between the fluid volumes in the deformed and the undeformed elements (Fig. 12.2). The fluid change per unit initial volume of the soil element can be used as the deformation state variables for the fluid phases. The deformation variable can be written as $(\Delta V_w/V_0)$ for the water phase and $(\Delta V_a/V_0)$ for the air phase.

## 12.3 CONSTITUTIVE RELATIONS

Constitutive relations for an unsaturated soil can be formulated by linking selected deformation state variables to appropriate stress state variables. The stress state variables were previously established (Chapter 3). The deformation state variables must satisfy the continuity requirement. The linking of deformation and stress state variables results in the incorporation of volumetric deformation coefficients. Several forms of constitutive relations are discussed in the following sections. The established constitutive relations can be used to predict volume changes due to changes in the stress state.

### 12.3.1 Elasticity Form

There are two approaches that can be used in establishing the stress versus deformation relationships. These are the "mathematical" approach and the "semi-empirical" approach. In the mathematical approach, each component of the deformation state variable tensor is expressed as a linear combination of the stress state variables or vice versa. In other words, the relationship between the stress and deformation state variables is expressed by a series of linear equations. The problem with this approach is that it involves the assessment of a large number of soil properties.

The semi-empirical approach involves several assumptions which are based on experimental evidence from observing the behavior of many materials (Chou and Pagano, 1967). These assumptions are that: 1) normal stress does not produce shear strain; 2) shear stress does not cause normal strain; and 3) a shear stress component, $\tau$, causes only one shear strain component, $\gamma$. In addition, the principle of superposition is assumed to be applicable to cases involving small deformations.

The semi-empirical approach is more commonly used in conventional soil mechanics, and is used herein to formulate the constitutive relations for unsaturated soils. The constitutive equations must be tested experimentally to ensure uniqueness. For an elastic solid with a positive definite strain energy function, the uniqueness theorem states that there exists a one-to-one correspondence between elastic deformations and stresses (Fung, 1965).

In *saturated* soil mechanics, the constitutive relations for the soil structure can be formulated in accordance with the generalized Hooke's law using the effective stress variable, $(\sigma - u_w)$. For an isotropic and linearly elastic soil structure, the constitutive relations in the $x$-, $y$-, and $z$-directions have the following form:

$$\epsilon_x = \frac{(\sigma_x - u_w)}{E} - \frac{\mu}{E}(\sigma_y + \sigma_z - 2u_w) \quad (12.14)$$

$$\epsilon_y = \frac{(\sigma_y - u_w)}{E} - \frac{\mu}{E}(\sigma_x + \sigma_z - 2u_w) \quad (12.15)$$

$$\epsilon_z = \frac{(\sigma_z - u_w)}{E} - \frac{\mu}{E}(\sigma_x + \sigma_y - 2u_w) \quad (12.16)$$

where

- $\sigma_x$ = total normal stress in the $x$-direction
- $\sigma_y$ = total normal stress in the $y$-direction
- $\sigma_z$ = total normal stress in the $z$-direction
- $E$ = modulus of elasticity or Young's modulus for the soil structure
- $\mu$ = Poisson's ratio.

The sum of the normal strains, $\epsilon_x$, $\epsilon_y$, and $\epsilon_z$, constitutes the volumetric strain, $\epsilon_v$ [Eq. (12.12)]. For a saturated soil, the overall volume change of the soil is equal to the water volume change since soil particles are essentially incompressible.

The constitutive relations for an *unsaturated* soil can be formulated as an extension of the equations used for a saturated soil, using the appropriate stress state variables (Fredlund and Morgenstern, 1976; Fredlund, 1979). Let us assume that the soil behaves as an isotropic, linear elastic material. The following constitutive relations are expressed in terms of the stress state variables, $(\sigma - u_a)$ and $(u_a - u_w)$. The formulation is similar in form to that proposed by Biot in 1941. The soil structure constitutive relations associated with the normal strains in the $x$-, $y$-, and $z$-directions are as follows:

$$\epsilon_x = \frac{(\sigma_x - u_a)}{E} - \frac{\mu}{E}(\sigma_y + \sigma_z - 2u_a)$$
$$+ \frac{(u_a - u_w)}{H} \quad (12.17)$$

$$\epsilon_y = \frac{(\sigma_y - u_a)}{E} - \frac{\mu}{E}(\sigma_x + \sigma_z - 2u_a)$$
$$+ \frac{(u_a - u_w)}{H} \quad (12.18)$$

$$\epsilon_z = \frac{(\sigma_z - u_a)}{E} - \frac{\mu}{E}(\sigma_x + \sigma_y - 2u_a)$$
$$+ \frac{(u_a - u_w)}{H} \quad (12.19)$$

where

$H$ = modulus of elasticity for the soil structure with respect to a change in matric suction, $(u_a - u_w)$.

The constitutive equations associated with the shear deformations are

$$\gamma_{xy} = \frac{\tau_{xy}}{G} \quad (12.20)$$

$$\gamma_{yz} = \frac{\tau_{yz}}{G} \quad (12.21)$$

$$\gamma_{zx} = \frac{\tau_{zx}}{G} \quad (12.22)$$

where

- $\tau_{xy}$ = shear stress on the $x$-plane in the $y$-direction (i.e., $\tau_{xy} = \tau_{yx}$)
- $\tau_{yz}$ = shear stress on the $y$-plane in the $z$-direction (i.e., $\tau_{yz} = \tau_{zy}$)
- $\tau_{zx}$ = shear stress on the $z$-plane in the $x$-direction (i.e., $\tau_{zx} = \tau_{xz}$)
- $G$ = shear modulus.

The modulus of elasticity, $E$, in the above equations is defined with respect to a change in the net normal stress, $(\sigma - u_a)$. The above constitutive equations can also be applied to situations where the stress versus strain curves are nonlinear. Figure 12.3 shows a typical stress versus strain curve. An incremental procedure using small increments of stress and strain can be used to apply the linear elastic formulation to a nonlinear stress versus strain curve. The nonlinear stress versus strain curve is assumed to be linear within each stress and strain increment. The elastic moduli,

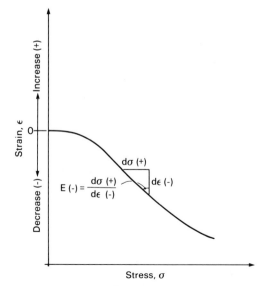

**Figure 12.3** Nonlinear stress versus strain curve.

$E$ and $H$, have negative signs, as indicated in Fig. 12.3, and may vary in magnitude from one increment to another. The soil structure constitutive relations associated with the normal strains can be written in an *incremental* form:

$$d\epsilon_x = \frac{d(\sigma_x - u_a)}{E} - \frac{\mu}{E} d(\sigma_y + \sigma_z - 2u_a)$$
$$+ \frac{d(u_a - u_w)}{H} \quad (12.23)$$

$$d\epsilon_y = \frac{d(\sigma_y - u_a)}{E} - \frac{\mu}{E} d(\sigma_x + \sigma_z - 2u_a)$$
$$+ \frac{d(u_a - u_w)}{H} \quad (12.24)$$

$$d\epsilon_z = \frac{d(\sigma_z - u_a)}{E} - \frac{\mu}{E} d(\sigma_x + \sigma_y - 2u_a)$$
$$+ \frac{d(u_a - u_w)}{H}. \quad (12.25)$$

Equations (12.23), (12.24), and (12.25) represent the general elasticity constitutive relations for the soil structure. The left-hand side of the equations refers to a change in the deformation state variable, while its right-hand side contains changes in the stress state variables. A change in the volumetric strain of the soil for each increment, $d\epsilon_v$, can be obtained by summing the changes in normal strains in the $x$-, $y$-, and $z$-directions:

$$d\epsilon_v = d\epsilon_x + d\epsilon_y + d\epsilon_z \quad (12.26)$$

where

$d\epsilon_v$ = volumetric strain change for each increment.

Substituting Eqs. (12.23), (12.24), and (12.25) into Eq. (12.26) gives

$$d\epsilon_v = 3\left(\frac{1-2\mu}{E}\right) d\left(\frac{\sigma_x + \sigma_y + \sigma_z}{3} - u_a\right)$$
$$+ \frac{3}{H} d(u_a - u_w). \quad (12.27)$$

Equation (12.27) can be simplified to the following form:

$$d\epsilon_v = 3\left(\frac{1-2\mu}{E}\right) d(\sigma_{\text{mean}} - u_a) + \frac{3}{H} d(u_a - u_w)$$
$$(12.28)$$

where

$\sigma_{\text{mean}}$ = average total normal stress (i.e., $(\sigma_x + \sigma_y + \sigma_z)/3$).

The volumetric strain change, $d\epsilon_v$, is equal to the volume change of the soil element divided by the initial volume of the element:

$$d\epsilon_v = \frac{dV_v}{V_0}. \quad (12.29)$$

The initial volume, $V_0$, refers to the volume of the soil element at the start of the volume change process. Therefore, $V_0$ remains constant for all increments. At the end of each increment, the volumetric strain change, $d\epsilon_v$, can be computed from Eq. (12.28), and the volume change of the soil element, $dV_v$, is obtained from Eq. (12.29). The summation of the volumetric strain changes for each increment gives the final volumetric strain of the soil:

$$\epsilon_v = \Sigma d\epsilon_v. \quad (12.30)$$

*Water Phase Constitutive Relation*

The soil structure constitutive relationship is not sufficient to completely describe the volume changes in an unsaturated soil. Either an air or water phase constitutive relation must be formulated. It is suggested that the water phase is most suitable for formulating the second constitutive relationship. The water phase constitutive relation describes the change in the volume of water present in the referential soil structure element under various stress conditions. The water itself is assumed to be incompressible, and the equation accounts for the net inflow or outflow from the element. The water phase constitutive relation can be formulated in a semi-empirical manner on the basis of a linear combination of the stress state variables. In an incremental form, the constitutive equation can be written

$$\frac{dV_w}{V_0} = \frac{3}{E_w} d(\sigma_{\text{mean}} - u_a) + \frac{d(u_a - u_w)}{H_w} \quad (12.31)$$

where

$E_w$ = water volumetric modulus associated with a change in $(\sigma - u_a)$

$H_w$ = water volumetric modulus associated with a change in $(u_a - u_w)$.

The summation of water volume changes at each increment gives the final change in the volume of water:

$$\frac{\Delta V_w}{V_0} = \frac{\Sigma dV_w}{V_0}. \quad (12.32)$$

*Change in the Volume of Air*

The change in the volume of air in an element can be computed as the difference between the soil structure and water volume changes. The continuity requirement [i.e., Eq. (12.4)] can also be written in an incremental form using the volumetric strain change, $d\epsilon_v$:

$$d\epsilon_v = \frac{dV_w}{V_0} + \frac{dV_a}{V_0}. \quad (12.33)$$

The constitutive relationships for an unsaturated soil can be presented graphically in the form of constitutive surfaces (Fig. 12.4). The deformation state variable is plotted with respect to the $(\sigma_{\text{mean}} - u_a)$ and $(u_a - u_w)$ stress state variables. The coefficients used in the constitutive equations are the slopes of the constitutive surface at a point. The slopes are with respect to both axes. For example, the slopes of the soil structure constitutive surface at a point are equal to $(3(1 - 2\mu)/E)$ and $(3/H)$ with respect to the $(\sigma_{\text{mean}} - u_a)$ and $(u_a - u_w)$ axes, respectively [Eq. (12.28) and Fig. 12.4(a)]. The coefficients on the constitutive surface are referred to as "volumetric deformation coefficients." These coefficients vary from one stress state to another for a curved constitutive surface. Similarly, the slopes on the water phase constitutive surface at a point are $(3/E_w)$ and $(1/H_w)$ with respect to the $(\sigma_{\text{mean}} - u_a)$ and $(u_a - u_w)$ axes, respectively [Eq. (12.31) and Fig. 12.4(b)].

The above constitutive relations can be formulated for a general, three-dimensional loading in the x-, y-, and z-directions. The formulations can also be applied to special loading conditions, as depicted in Fig. 12.5. The constitutive equations for each loading condition can be derived from the general formulations, and they are presented in an incremental form in the following sections.

### Isotropic Loading

For isotropic loading, the total stress increments in the three directions are equal (i.e., $d\sigma_x = d\sigma_y = d\sigma_z = d\sigma_3$). No shear stress is developed in the soil. Substituting the above condition into Eq. (12.28) gives the soil structure constitutive equation for isotropic loading:

$$d\epsilon_v = 3\left(\frac{1 - 2\mu}{E}\right) d(\sigma_3 - u_a) + \frac{3}{H} d(u_a - u_w) \quad (12.34)$$

where

$\sigma_3$ = total isotropic stress.

The inverse of the coefficients in the first term [i.e., $E/(3(1 - 2\mu))$] is commonly called the "bulk modulus" (Lambe and Whitman, 1979).

The constitutive equation for the water phase can be derived from Eq. (12.31):

$$\frac{dV_w}{V_0} = \frac{3}{E_w} d(\sigma_3 - u_a) + \frac{d(u_a - u_w)}{H_w}. \quad (12.35)$$

The soil will undergo equal deformation in all directions (or isotropic compression) when subjected to an equal all-around pressure, provided the soil properties are isotropic (Fig. 12.5).

### Uniaxial Loading

For uniaxial loading, a total stress increment is applied to the soil in one direction (e.g., the vertical direction), as illustrated in Fig. 12.5. It is assumed that no shear stress is developed on the x-, y-, and z-planes. The stress increase is applied in the y-direction, $d\sigma_y$, while the total stress change in the other directions is zero (i.e., $d\sigma_x = d\sigma_z = 0$). The soil compresses in the y-direction, and expands in the x- and z-directions. Applying these stress conditions to Eq. (12.28) gives the soil structure constitutive equation for uniaxial loading:

$$d\epsilon_v = 3\left(\frac{1 - 2\mu}{E}\right) d\left(\frac{1}{3}\sigma_y - u_a\right) + \frac{3}{H} d(u_a - u_w). \quad (12.36)$$

The water phase constitutive equation for uniaxial loading is obtained from Eq. (12.31):

$$\frac{dV_w}{V_0} = \frac{3}{E_w} d\left(\frac{1}{3}\sigma_y - u_a\right) + \frac{d(u_a - u_w)}{H_w}. \quad (12.37)$$

### Triaxial Loading

The triaxial loading conditions shown in Fig. 12.5 can be considered as a superposition of isotropic and uniaxial loading. Isotropic loading applies to an all-around pressure

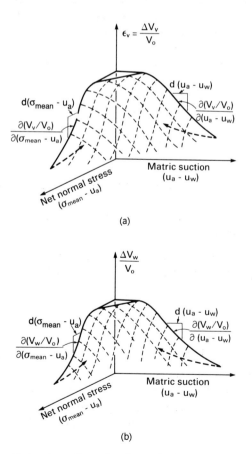

**Figure 12.4** Three-dimensional constitutive surfaces for an unsaturated soil. (a) Soil structure constitutive surface; (b) water phase constitutive surface.

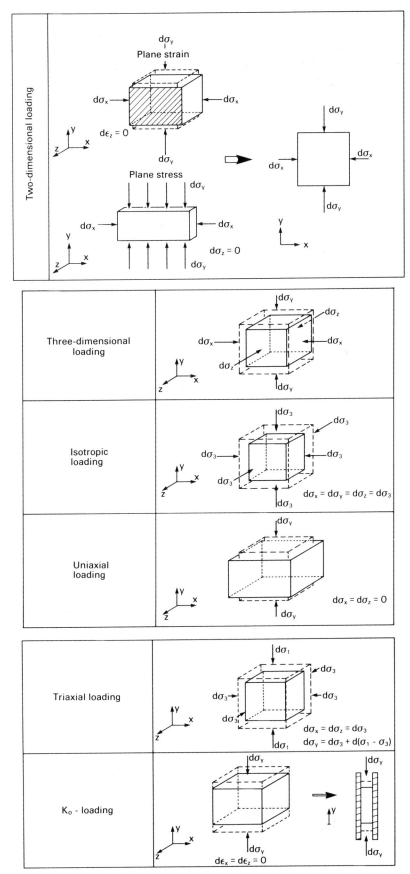

**Figure 12.5** Various loading conditions and their associated deformations.

change of $d\sigma_3$, whereas the uniaxial pressure change applies to a deviator stress, $d(\sigma_1 - \sigma_3)$, in the vertical or y-direction. The x-, y-, and z-planes are assumed to be principal planes. The above stress conditions [i.e., $d\sigma_x = d\sigma_z = d\sigma_3$ and $d\sigma_y = d\sigma_3 + d(\sigma_1 - \sigma_3)$] can be substituted into Eq. (12.28) go give the soil structure constitutive equation for triaxial loading:

$$d\epsilon_v = 3\left(\frac{1 - 2\mu}{E}\right) d\left\{\frac{\sigma_3 + (\sigma_3 + \sigma_1 - \sigma_3) + \sigma_3}{3} - u_a\right\}$$
$$+ \frac{3}{H} d(u_a - u_w). \qquad (12.38)$$

Rearranging Eq. (12.38) gives

$$d\epsilon_v = 3\left(\frac{1 - 2\mu}{E}\right)\left[d(\sigma_3 - u_a) + d\left\{\frac{1}{3}(\sigma_1 - \sigma_3)\right\}\right]$$
$$+ \frac{3}{H} d(u_a - u_w). \qquad (12.39)$$

The water phase constitutive relation can be written by substituting the triaxial stress conditions into Eq. (12.31):

$$\frac{dV_w}{V_0} = \frac{3}{E_w}\left[d(\sigma_3 - u_a) + d\left\{\frac{1}{3}(\sigma_1 - \sigma_3)\right\}\right]$$
$$+ \frac{d(u_a - u_w)}{H_w}. \qquad (12.40)$$

The $du_a$ and $d(u_a - u_w)$ terms in Eqs. (12.39) and (12.40) refer to the pore–air pressure and matric suction increments, respectively, during triaxial loading. For triaxial testing in the laboratory, isotropic loading is generally first applied, followed by uniaxial compression. Therefore, it may be useful to separate the pore–air pressure and matric suction increments during the two loading conditions. This can be done by superimposing Eq. (12.34) (i.e., isotropic loading) and Eq. (12.36) (i.e., uniaxial loading) to form the soil structure constitutive relation for triaxial loading:

$$d\epsilon_v = 3\left(\frac{1 - 2u}{E}\right)\left[d(\sigma_3 - u_{ai})\right.$$
$$\left. + d\left\{\frac{1}{3}(\sigma_1 - \sigma_3) - u_{au}\right\}\right] + \frac{3}{H}\{d(u_a - u_w)_i$$
$$+ d(u_a - u_w)_u\} \qquad (12.41)$$

where

$du_{ai}$ = pore–air pressure increment during isotropic loading
$du_{au}$ = pore–air pressure increment during uniaxial loading
$du_{wi}$ = pore–water pressure increment during isotropic loading
$du_{wu}$ = pore–water pressure increment during uniaxial loading.

Equation (12.41) can be used to compute the volumetric strain increment during triaxial loading. During isotropic compression, only the first and third terms are used since the other terms drop out. On the other hand, the second and fourth terms are used to compute the volumetric strain increment during uniaxial compression, while the other terms become zero. In the end, the volumetric strain increments during isotropic and uniaxial compression can be summed to give the total volumetric strain associated with triaxial loading. It should be noted that Eq. (12.41) reverts to Eq. (12.39) by substituting $(du_a = du_{ai} + du_{au})$ and $(du_w = du_{wi} + du_{wu})$ into Eq. (12.41).

Similarly, the water phase constitutive relation can be obtained by superimposing Eq. (12.35) (i.e., isotropic loading) and Eq. (12.37) (i.e., uniaxial loading):

$$\frac{dV_w}{V_0} = \frac{3}{E_w}\left[d(\sigma_3 - u_{ai}) + d\left\{\frac{1}{3}(\sigma_1 - \sigma_3) - u_{au}\right\}\right]$$
$$+ \frac{d(u_a - u_w)_i}{H_w} + \frac{d(u_a - u_w)_u}{H_w}. \qquad (12.42)$$

The first and third terms in Eq. (12.42) are used for isotropic loading, whereas the second and fourth terms are for uniaxial compression. The sum of the water volume changes during isotropic and uniaxial compression gives the total change in volume of water during triaxial loading. Substituting $(du_a = du_{ai} + du_{au})$ and $(du_w = du_{wi} + du_{wu})$ into Eq. (12.42) gives Eq. (12.40).

### $K_0$-Loading

For $K_0$-loading, a total stress increment of $d\sigma_y$ is applied in the vertical direction, while the soil is not permitted to deform laterally (i.e., $d\epsilon_x = d\epsilon_z = 0$). This loading condition occurs during *one-dimensional* consolidation where deformation is allowed only in the vertical direction (i.e., $d\epsilon_y$). The application of a total stress increment, $d\sigma_y$, is in the vertical direction. The strain conditions during the $K_0$-loading can be applied to Eqs. (12.23) and (12.25). Multiplying Eq. (12.23) by Poisson's ratio, $\mu$, and adding the result to Eq. (12.25) gives

$$\left(\frac{1 - \mu^2}{E}\right) d(\sigma_z - u_a) = \frac{\mu}{E}(1 + \mu) d(\sigma_y - u_a)$$
$$- \left(\frac{1 + \mu}{H}\right) d(u_a - u_w).$$
$$(12.43)$$

Equation (12.43) can be arranged as follows:

$$d(\sigma_z - u_a) = \left(\frac{\mu}{1 - \mu}\right) d(\sigma_y - u_a)$$
$$- \frac{E}{(1 - \mu)H} d(u_a - u_w). \qquad (12.44)$$

Similarly, Eq. (12.25) can be multiplied by Poisson's

ratio, $\mu$, and added to Eq. (12.23) to give

$$d(\sigma_x - u_a) = \left(\frac{\mu}{1 - \mu}\right) d(\sigma_y - u_a)$$
$$- \frac{E}{(1 - \mu)H} d(u_a - u_w). \quad (12.45)$$

Substituting Eqs. (12.44) and (12.45) into Eq. (12.24) gives

$$d\epsilon_y = \frac{(1 - \mu - 2\mu^2)}{E(1 - \mu)} d(\sigma_y - u_a)$$
$$+ \frac{(1 - \mu + 2\mu)}{H(1 - \mu)} d(u_a - u_w). \quad (12.46)$$

Volumetric strain, $d\epsilon_v$, is equal to the strain in the vertical direction, $d\epsilon_y$, for $K_0$ loading conditions. Eq. (12.46) for the soil structure can therefore be rewritten as follows:

$$d\epsilon_v = \frac{(1 + \mu)(1 - 2\mu)}{E(1 - \mu)} d(\sigma_y - u_a)$$
$$+ \frac{(1 + \mu)}{H(1 - \mu)} d(u_a - u_w). \quad (12.47)$$

The inverse of the coefficients in the first term (i.e., $E(1 - \mu)/\{(1 + \mu)(1 - 2\mu)\}$) is referred to as the "constrained modulus" (Lambe and Whitman, 1979).

The water phase constitutive relation is obtained in a similar manner [i.e., substituting Eqs. (12.44) and (12.45) into Eq. (12.31)]:

$$\frac{dV_w}{V_0} = \frac{(1 + \mu)}{E_w(1 - \mu)} d(\sigma_y - u_a)$$
$$+ \left\{\frac{1}{H_w} - \frac{2(E/H)}{E_w(1 - \mu)}\right\} d(u_a - u_w). \quad (12.48)$$

*Plane Strain Loading*

Many geotechnical problems can be simplified into a two-dimensional form using the concept of plane strain or plane stress loading (Fig. 12.5). If an earth structure is significantly long in one direction (e.g., the $z$-direction) in comparison to the other two directions (e.g., the $x$- and $y$-directions) and the loadings are applied only on the $x$- and $y$-planes, the structure can be modeled as a plane strain problem. The slope stability, retaining wall, and strip footing are problems commonly analyzed by assuming plane strain loading conditions. For plane strain conditions, the soil deformation in the $z$-direction is assumed to be negligible (i.e., $d\epsilon_z = 0$). Imposing a condition of zero strain in the $z$-direction in Eq. (12.25) gives

$$d(\sigma_z - u_a) = \mu \, d(\sigma_x + \sigma_y - 2u_a)$$
$$- (E/H) \, d(u_a - u_w). \quad (12.49)$$

Volumetric strain during plane strain loading is obtained by substituting Eq. (12.49) into Eq. (12.27):

$$d\epsilon_v = \frac{2(1 + \mu)(1 - 2\mu)}{E} d(\sigma_{\text{ave}} - u_a)$$
$$+ 2\left(\frac{1 + \mu}{H}\right) d(u_a - u_w) \quad (12.50)$$

where

$\sigma_{\text{ave}}$ = average total normal stress for two-dimensional loading (i.e., $(\sigma_x + \sigma_y)/2$).

Equation (12.50) can be used as the soil structure constitutive relation for plane strain loading.

The water phase constitutive equation is obtained from Eq. (12.31) by replacing the $d(\sigma_z - u_a)$ term with Eq. (12.49):

$$\frac{dV_w}{V_0} = 2\left(\frac{1 + \mu}{E_w}\right) d(\sigma_{\text{ave}} - u_a)$$
$$+ \left\{\frac{1}{H_w} - \frac{(E/H)}{E_w}\right\} d(u_a - u_w). \quad (12.51)$$

*Plane Stress Loading*

For plane stress conditions, the change in the total stress in the $z$-direction is assumed to be negligible (i.e., $d\sigma_z = 0$; Fig. 12.5). The soil structure constitutive relation for the plane stress loading is computed by setting $d\sigma_z$ equal to zero in Eq. (12.28):

$$d\epsilon_v = 2\left(\frac{1 - 2\mu}{E}\right) d\left(\sigma_{\text{ave}} - \frac{3}{2} u_a\right)$$
$$+ \frac{3}{H} d(u_a - u_w). \quad (12.52)$$

The constitutive equation for the water phase under plane stress loading is obtained from Eq. (12.31) by setting $d\sigma_z$ equal to zero:

$$\frac{dV_w}{V_0} = \frac{2}{E_w} d\left(\sigma_{\text{ave}} - \frac{3}{2} u_a\right) + \frac{d(u_a - u_w)}{H_w}. \quad (12.53)$$

### 12.3.2 Compressibility Form

In the preceding sections, the constitutive relations for an unsaturated soil were formulated using a linear elasticity form. These constitutive equations can be rewritten in a compressibility form more common to soil mechanics. The compressibility form of the constitutive equation for the soil structure of a *saturated* soil is written as

$$d\epsilon_v = m_v d(\sigma - u_w) \quad (12.54)$$

where

$m_v$ = coefficient of volume change.

The compressibility form for the soil structure constitutive equation for an *unsaturated* soil under general, three-dimensional loading is as follows:

$$d\epsilon_v = m_1^s d(\sigma_{\text{mean}} - u_a) + m_2^s d(u_a - u_w) \quad (12.55)$$

where

$m_1^s$ = coefficient of volume change with respect to net normal stress [i.e., $3(1 - 2\mu)/E$; Eq. (12.28)]
$m_2^s$ = coefficient of volume change with respect to matric suction [i.e., $3/H$; Eq. (12.28)].

The $m_1^s$ and $m_2^s$ coefficients of volume change in Eq. (12.55) can be called "compressibilities" since they are essentially a ratio between changes in volumetric strain and stress variables.

The negative signs for the coefficients of volume change, $m_1^s$ and $m_2^s$, are the result of the $E$ and $H$ moduli being negative for typical soils (Fig. 12.3). The $m_1^s$ and $m_2^s$ coefficients vary in a nonlinear manner, but can be considered as being constant for a small increment of stress or strain.

The compressibility form of the water phase constitutive equation for an unsaturated soil under three-dimensional loading can be written as

$$\frac{dV_w}{V_0} = m_1^w d(\sigma_{\text{mean}} - u_a) + m_2^w d(u_a - u_w) \quad (12.56)$$

where

$m_1^w$ = coefficient of water volume change with respect to net normal stress [i.e., $3/E_w$; Eq. (12.31)]
$m_2^w$ = coefficient of water volume change with respect to matric suction [i.e., $1/H_w$; Eq. (12.31)].

Similar constitutive equations in a compressibility form can be written for specific loading conditions. Table 12.2 presents the $m_1^s$, $m_2^s$, $m_1^w$, and $m_2^w$ coefficients for the loading conditions described in the previous section. These coefficients are regarded as another form of the soil volumetric deformation coefficients.

### 12.3.3 Volume–Mass Form (Soil Mechanics Terminology)

The volume–mass properties of a soil can also be used in the formulation of constitutive equations for an unsaturated soil. A change in void ratio is commonly used as the deformation state variable for a *saturated* soil, giving rise to the following constitutive equation:

$$de = a_v d(\sigma - u_w) \quad (12.57)$$

where

$a_v$ = coefficient of compressibility.

For an unsaturated soil, void ratio and gravimetric water content can be used as the deformation state variables for the soil structure and water phase, respectively. Using soil mechanics terminology, the change in void ratio, $de$, of an *unsaturated* soil under general, three-dimensional loading can be written as

$$de = a_t d(\sigma_{\text{mean}} - u_a) + a_m d(u_a - u_w) \quad (12.58)$$

where

$a_t$ = coefficient of compressibility with respect to a change in net normal stress, $d(\sigma_{\text{mean}} - u_a)$
$a_m$ = coefficient of compressibility with respect to a change in matric suction, $d(u_a - u_w)$.

Equation (12.58) is equivalent to a soil structure constitutive relation written using soil mechanics terminology.

The water phase constitutive equation can be expressed as a change in water content. For three-dimensional loading, the water phase constitutive relation has the following form:

$$dw = b_t d(\sigma_{\text{mean}} - u_a) + b_m d(u_a - u_w) \quad (12.59)$$

where

$b_t$ = coefficient of water content change with respect to a change in net normal stress, $d(\sigma_{\text{mean}} - u_a)$
$b_m$ = coefficient of water content change with respect to a change in matric suction, $d(u_a - u_w)$.

The above constitutive equations can be visualized as constitutive surfaces on a three-dimensional plot, with each abscissa representing one of the stress state variables, and the ordinate representing the soil volume–mass property [Fig. 12.6(a)]. The $a_t$, $a_m$, $b_t$, and $b_m$ coefficients become another form of the volumetric deformation coefficients. These plots can be reduced to two-dimensional plots which graphically show the relationship between the various volumetric deformation coefficients [Fig. 12.6(b)]. The relationships between various coefficients are explained later in this chapter. Similar constitutive relationships can be formulated for unloading conditions.

### 12.3.4 Use of $(\sigma - u_w)$ and $(u_a - u_w)$ to Formulate Constitutive Relations

The constitutive equations for an unsaturated soil have been formulated using the stress state variables, $(\sigma - u_a)$ and $(u_a - u_w)$. Other combinations of stress state variables, such as $(\sigma - u_w)$ and $(u_a - u_w)$, can also be used. As an example, Eq. (12.58) can be rewritten using the $(\sigma - u_w)$ and $(u_a - u_w)$ stress state variables:

$$de = a_{t2} d(\sigma_{\text{mean}} - u_w) + a_{m2} d(u_a - u_w) \quad (12.60)$$

where

$a_{t2}$ = coefficient of compressibility with respect to $(\sigma_{\text{mean}} - u_w)$ when using the $(\sigma - u_w)$ and $(u_a - u_w)$ stress state variables
$a_{m2}$ = coefficient of compressibility with respect to matric suction when using the $(\sigma - u_w)$ and $(u_a - u_w)$ stress state variables.

The water phase constitutive equation can also be written

**Table 12.2 Coefficients of Volume Change for Various Loading Conditions**

| Loading | Deformation State Variable | $m_1^s$ $m_1^w$ | First Stress State Variable, $d(\sigma - u_a)$ | $m_2^s$ $m_2^w$ | Second Stress State Variable, $d(u_a - u_w)$ |
|---|---|---|---|---|---|
| Three-dimensional (General) | $d\epsilon_v$ | $3\left(\dfrac{1-2\mu}{E}\right)$ | $d(\sigma_{\text{mean}} - u_a)$ | $\dfrac{3}{H}$ | $d(u_a - u_w)$ |
|  | $\dfrac{dV_w}{V_0}$ | $\dfrac{3}{E_w}$ | $d(\sigma_{\text{mean}} - u_a)$ | $\dfrac{1}{H_w}$ | $d(u_a - u_w)$ |
| Isotropic | $d\epsilon_v$ | $3\left(\dfrac{1-2\mu}{E}\right)$ | $d(\sigma_3 - u_a)$ | $\dfrac{3}{H}$ | $d(u_a - u_w)$ |
|  | $\dfrac{dV_w}{V_0}$ | $\dfrac{3}{E_w}$ | $d(\sigma_3 - u_a)$ | $\dfrac{1}{H_w}$ | $d(u_a - u_w)$ |
| Uniaxial | $d\epsilon_v$ | $3\left(\dfrac{1-2\mu}{E}\right)$ | $d\left(\tfrac{1}{3}\sigma_y - u_a\right)$ | $\dfrac{3}{H}$ | $d(u_a - u_w)$ |
|  | $\dfrac{dV_w}{V_0}$ | $\dfrac{3}{E_w}$ | $d\left(\tfrac{1}{3}\sigma_y - u_a\right)$ | $\dfrac{1}{H_w}$ | $d(u_a - u_w)$ |
| Triaxial (General case) | $d\epsilon_v$ | $3\left(\dfrac{1-2\mu}{E}\right)$ | $d(\sigma_3 - u_a) + d\left\{\tfrac{1}{3}(\sigma_1 - \sigma_3)\right\}$ | $\dfrac{3}{H}$ | $d(u_a - u_w)$ |
|  | $\dfrac{dV_w}{V_0}$ | $\dfrac{3}{E_w}$ | $d(\sigma_3 - u_a) + d\left\{\tfrac{1}{3}(\sigma_1 - \sigma_3)\right\}$ | $\dfrac{1}{H_w}$ | $d(u_a - u_w)$ |
| Triaxial (Separation of isotropic and uniaxial components) | $d\epsilon_v$ | $3\left(\dfrac{1-2\mu}{E}\right)$ | $d(\sigma_3 - u_{ai}) + d\left\{\tfrac{1}{3}(\sigma_1 - \sigma_3) - u_{au}\right\}$ | $\dfrac{3}{H}$ | $d(u_a - u_w)_i + d(u_a - u_w)_u$ |
|  | $\dfrac{dV_w}{V_0}$ | $\dfrac{3}{E_w}$ | $d(\sigma - u_{ai}) + d\left\{\tfrac{1}{3}(\sigma_1 - \sigma_3) - u_{au}\right\}$ | $\dfrac{1}{H_w}$ | $d(u_a - u_w)_i + d(u_a - u_w)_u$ |
| $K_0$-loading (One-dimensional) | $d\epsilon_v$ | $\dfrac{(1+\mu)(1-2\mu)}{E(1-\mu)}$ | $d(\sigma_y - u_a)$ | $\dfrac{(1+\mu)}{H(1-\mu)}$ | $d(u_a - u_w)$ |
|  | $\dfrac{dV_w}{V_0}$ | $\dfrac{(1+\mu)}{E_w(1-\mu)}$ | $d(\sigma_y - u_a)$ | $\dfrac{1}{H_w} - \dfrac{2(E/H)}{E_w(1-\mu)}$ | $d(u_a - u_w)$ |
| Plane strain (Two-dimensional) | $d\epsilon_v$ | $\dfrac{2(1+\mu)(1-2\mu)}{E}$ | $d(\sigma_{\text{ave}} - u_a)$ | $2\dfrac{(1+\mu)}{H}$ | $d(u_a - u_w)$ |
|  | $\dfrac{dV_w}{V_0}$ | $2\dfrac{(1+\mu)}{E_w}$ | $d(\sigma_{\text{ave}} - u_a)$ | $\dfrac{1}{H_w} - \dfrac{(E/H)}{E_w}$ | $d(u_a - u_w)$ |
| Plane stress (Two-dimensional) | $d\epsilon_v$ | $2\dfrac{(1-2\mu)}{E}$ | $d\left(\sigma_{\text{ave}} - \tfrac{3}{2}u_a\right)$ | $\dfrac{3}{H}$ | $d(u_a - u_w)$ |
|  | $\dfrac{dV_w}{V_0}$ | $\dfrac{2}{E_w}$ | $d\left(\sigma_{\text{ave}} - \tfrac{3}{2}u_a\right)$ | $\dfrac{1}{H_w}$ | $d(u_a - u_w)$ |

*Note:* $\sigma_{\text{mean}} = (\sigma_x + \sigma_y + \sigma_z)/3$; $\sigma_{\text{ave}} = (\sigma_x + \sigma_y)/2$.

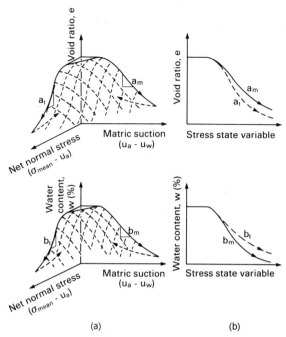

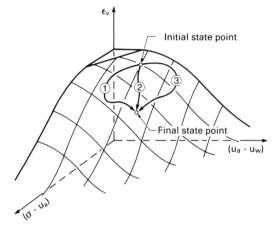

**Figure 12.6** Constitutive surfaces for an unsaturated soil expressed using soil mechanics terminology. (a) Three-dimensional void ratio and water content constitutive surfaces, (b) two-dimensional comparison showing volumetric deformation moduli.

in terms of the $(\sigma - u_w)$ and $(u_a - u_w)$ stress state variables:

$$dw = b_{t2} d(\sigma_{mean} - u_w) + b_{m2} d(u_a - u_w) \quad (12.61)$$

where

$b_{t2}$ = coefficient of water content change with respect to $(\sigma_{mean} - u_w)$ when using the $(\sigma - u_w)$ and $(u_a - u_w)$ stress state variables

$b_{m2}$ = coefficient of water content change with respect to matric suction when using the $(\sigma - u_w)$ and $(u_a - u_w)$ stress state variables.

The $(\sigma - u_a)$ and $(\sigma - u_w)$ stress state variables can similarly be used to formulate the constitutive relations.

## 12.4 EXPERIMENTAL VERIFICATIONS FOR UNIQUENESS OF CONSTITUTIVE SURFACES

The two constitutive surfaces for an unsaturated soil (i.e., soil structure and water phase) require experimental data in order to examine their uniqueness. The term "uniqueness" is used to indicate that there is one and only one relationship between the deformation and stress state variables. Figure 12.7 illustrates the meaning of the term "uniqueness" of a constitutive surface. Consider three identical soil specimens at the same initial state point. The stress state variables for the three specimens are then varied along three different paths. However, the final stress state variables for the three specimens are identical. If the con-

**Figure 12.7** Tests for the uniqueness of the constitutive surface for an unsaturated soil.

stitutive surface is unique, the final combination of stress state variables should produce the same deformation state. In other words, the final volumes for the three soil specimens should be equal and independent of the path.

Complete uniqueness of a constitutive relationship under loading and unloading conditions is virtually impossible. Hysteresis is the main cause of nonuniqueness. Hysteresis is associated with phases that behave as a solid such as the soil structure and the contractile skin in an unsaturated soil. The soil structure hysteresis during loading and unloading of a saturated soil is reflected by the different compression and rebound curves for the soil [Fig. 12.8(a)]. The two curves have a marked difference in their slopes.

Hysteresis associated with the contractile skin can be readily visualized from the drying and wetting curves of an incompressible material such as chalk [Fig. 12.8(b)]. In an unsaturated soil, intuitively it would appear that there should be hysteresis associated with the soil structure and the contractile skin. It appears that a reversal in the direction of deformation (i.e., an increase and then a decrease in volume) results in different constitutive surfaces. As a result, the volumetric deformation coefficients associated with a decreasing volume would be different from those associated with an increasing volume. However, the uniqueness of the volume increase and decrease constitutive surfaces can be verified independently. This involves tracing the surface through various loading paths, provided the deformation state variable always changes in the same direction (i.e., monotonic or unidirectional deformation). Therefore, the term "uniqueness" in soil mechanics is generally restricted to constitutive surfaces representing monotonic deformation.

The uniqueness of the soil structure constitutive surface can be verified for either a decrease or an increase in the soil volume. In addition, the uniqueness of the water phase constitutive surface can also be verified for either a decrease or an increase in the volume of water (i.e., drying or wetting surface).

## 12.4 EXPERIMENTAL VERIFICATIONS FOR UNIQUENESS OF CONSTITUTIVE SURFACES

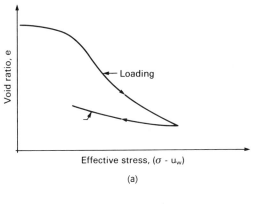

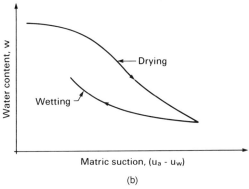

**Figure 12.8** Hysteresis associated with the soil structure and contractile skin in a soil. (a) Loading and unloading curves for a saturated soil; (b) drying and wetting curves for a chalk.

The uniqueness of a constitutive surface can first be explored locally using small stress state variable changes around a state point on the surface (Fredlund and Morgenstern, 1976). Subsequently, larger stress state variable changes can be used to test for uniqueness of the entire constitutive surface (Matyas and Radhakrishna, 1968; Barden et al. (1969). Both methods for verifying the uniqueness of a constitutive surface are briefly explained. The sign conventions associated with volumetric deformation properties are outlined prior to describing the results of tests for uniqueness.

### 12.4.1 Sign Convention for Volumetric Deformation Properties

The relationship between the deformation and stress state variable changes for a typical soil is illustrated in Fig. 12.3. A positive or negative sign is associated with the deformation or stress state variable change in order to indicate an increase or decrease. The following sign convention is suggested. A positive change in the state variable (i.e., deformation or stress) refers to an increase in the state variable, whereas a negative change indicates a decrease in the state variable. The signs of the deformation state variables and the stress state variables determine the sign for the volumetric deformation coefficients. As an example, the stable-structured soil shown in Fig. 12.3 would have a nega-

tive sign for the modulus of elasticity, $E$. This means that an increase in the net normal stress, $(+) d(\sigma - u_a)$, causes a decrease in volumetric strain, $(-) d\epsilon_v$ [i.e., $(-) E = (-) d\epsilon_v / (+) d(\sigma - u_a)$]. The elastic modulus, $H$, also has a negative sign [e.g., $(-) H = (-) d\epsilon_v / (+) d(u_a - u_w)$].

The relationship between the water volume change and stress state variable change for a stable-structured soil is similar to that shown in Fig. 12.3. Consequently, the water volumetric moduli, $E_w$ and $H_w$, would also have a negative sign. Therefore, it can be concluded that a *stable-structured* soil has *negative* volumetric deformation *moduli* associated with the soil structure and water phase. The negative sign applies similarly to the coefficients of volume change, $m_1^s$, $m_2^s$, $m_1^w$, and $m_2^w$, which are used in the compressibility form of the constitutive equations. The $a_t$, $a_m$, $b_t$, and $b_m$ coefficients in the volume-mass form of the constitutive equations are also negative for a stable-structured soil.

A collapsing soil is commonly referred to as a metastable-structured soil. A decrease in matric suction results in swelling for a stable-structured soil, whereas it may cause a volume decrease in a metastable-structured soil (Barden et al. (1969). In the case of a collapsing soil, the $m_2^s$ and $m_2^w$ coefficients have a positive sign [Eqs. (12.55) and (12.56)]. In other words, a *metastable*-structured soil refers to a soil that has one or more *positive* volumetric deformation *moduli*.

### 12.4.2 Verification of Uniqueness of the Constitutive Surfaces Using Small Stress Changes

Let us suppose that several so-called "identical" unsaturated soil specimens are subjected to the same total normal stress and the same pore-air and pore-water pressures. All specimens will be at the same state point in space and have the same initial volume-mass properties. Let each specimens then be subjected to different stress changes, while the volume changes are monitored. If the constitutive surface is essentially planar near a state point, the volumetric deformation moduli associated with any two orthogonal directions can be used to describe the deformation produced by other stress state variable changes.

The above test is somewhat fictitious since it would be extremely difficult to conduct. However, a simpler test could be considered. Suppose that three so-called "identical" specimens are prepared at the same initial volume-mass properties and subjected to the same initial stress state variables [Fig. 12.9(a)]. Let one specimen be subjected to a small increase in net normal stress, $d(\sigma - u_a)$, by increasing the total normal stress, while the matric suction, $(u_a - u_w)$, is maintained constant [i.e., 0-1 loading path in Fig. 12.9(a)]. The measured volumetric strain change, $d\epsilon_v$, can be used to compute the coefficient of volume change, $m_1^s$:

$$m_1^s = \frac{d\epsilon_v}{d(\sigma - u_a)}. \quad (12.62)$$

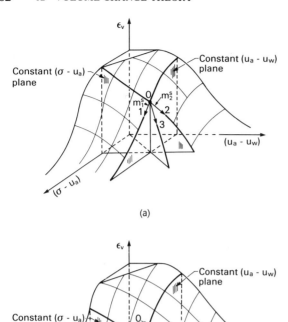

**Figure 12.9** Verification for uniqueness of the soil structure constitutive surface. (a) Constitutive surface verification using three identical soil specimens; (b) constitutive surface verification using a single soil specimen.

A second specimen could be subjected to a small increase in matric suction, $d(u_a - u_w)$, by decreasing the pore-water pressure, $u_w$, while maintaining a constant net normal stress [i.e., 0–2 loading path in Fig. 12.9(a)]. The measured volumetric strain change, $d\epsilon_v$, can be used to calculate the $m_2^s$ coefficient of volume change:

$$m_2^s = \frac{d\epsilon_v}{d(u_a - u_w)}. \qquad (12.63)$$

A third specimen could then be subjected to small increases in both the net normal stress, $d(\sigma - u_a)$, and the matric suction, $d(u_a - u_w)$. This can be achieved by increasing the total normal stress, $\sigma$, and decreasing the pore-water pressure, $u_w$. The anticipated volume change can be predicted from Eq. (12.55) using the above-measured values of $m_1^s$ and $m_2^s$ [Eqs. (12.62) and (12.63)]. The actual volume change can then be measured and compared with the computed value. A close agreement between the measured and predicted volume changes indicates "uniqueness" of the constitutive surface. The above example illustrates the uniqueness test for the soil structure constitutive surface. At the same time, the water phase constitutive surface can be verified if the water volume change measurements are used to compute the $m_1^w$ and $m_2^w$ coefficients.

The above test for uniqueness can be approximated by an even simpler procedure using a single soil specimen subjected to a series of small stress changes. The procedure is illustrated in Fig. 12.9(b). Consider a specimen subjected first to a small increase in net normal stress along a constant matric suction plane [i.e., 0–1 loading path in Fig. 12.9(b)]. This is achieved by increasing the total normal stress. The measured volume changes are used to compute the $m_1^s$ and $m_1^w$ coefficients, respectively.

The specimen can then be further loaded by increasing its matric suction while maintaining a constant net normal stress [i.e., 1–2 loading path in Fig. 12.9(b)]. The matric suction is increased by decreasing the pore-water pressure. The measured volume changes, can be used to compute the $m_2^s$ and $m_2^w$ coefficients, respectively. The $m_1^s$, $m_1^w$, and $m_2^s$, and $m_2^w$ coefficients can then be used to predict the overall volume change and the water volume change of the specimen during any subsequent loadings. Although the first and second loadings are performed at slightly differing initial volume–mass properties, the coefficients are assumed to approximate the state point under consideration on the constitutive surface. This assumption is reasonable for small changes in the stress state variables.

The next step is to apply a small increment in both the net normal stress, $d(\sigma - u_a)$, and matric suction, $d(u_a - u_w)$ [i.e., 2–3 loading path in Fig. 12.9(b)]. This path is adhered to by increasing the total normal stress and decreasing the pore-water pressure. The anticipated overall volume change and change in volume of water as a result of the third loading can be predicted from Eqs. (12.55) and (12.56), respectively, using the coefficients of volume change obtained from the previous two loadings. The measured overall and water volume changes can then be compared with the predicted values. The constitutive surface at a point is considered unique if the measured and predicted deformations are essentially equal. This procedure can also be repeated at various state points on the constitutive surface.

Fredlund and Morgenstern (1976) experimentally verified the uniqueness of constitutive surfaces for unsaturated soils by using small changes in stress state variables. Four series of experiments were conducted using three specimens of undisturbed Regina clay and one specimen of compacted kaolin. The specimens were tested using a modified Anteus oedometer and a triaxial cell. The modified oedometer was used for one-dimensional or $K_0$-loading. The modified triaxial cell was used for isotropic loading. The total, pore-air, and pore-water pressures were controlled during the experiments. The water pressure was isolated by means of a high air entry disk placed at the bottom of the specimen. After equalization at selected pressures, each specimen was loaded in several stages in order to test the uniqueness of the constitutive surface, as outlined previously. The predicted and measured volume changes were assessed by plotting the predicted volume changes versus the measured volume changes.

Good agreement was observed between the measured and predicted volume changes for undisturbed Regina clay, as shown in Fig. 12.10. The specimen was tested under $K_0$-loading. The volume changes at two elapsed times (e.g., 1000 and 5000 min) are considered for each specimen. Uniqueness implies a slope of unity and an intercept of zero for the measured versus predicted volume change plot. Figure 12.10(a) demonstrates that all points obtained at two elapsed times essentially fall along the 45° line, indicating virtually perfect correlation. This agreement supports the uniqueness of the soil structure constitutive surface.

The agreement between the predicted and measured water volume changes [Fig. 12.10(b)] is not as good as that for the soil structure. This was attributed to the difficulty in measuring small water volume changes over long periods of time (Fredlund and Morgenstern, 1976). It was later discovered that some moisture was lost from the specimen by loss from the oedometer chamber.

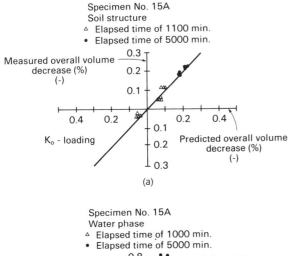

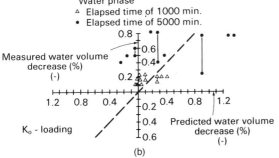

**Figure 12.10** Comparison of predicted and measured volume changes for undisturbed Regina clay under $K_0$-loading. (a) Soil structure constitutive surface; (b) water phase constitutive surface (from Fredlund and Morgenstern, 1976).

### 12.4.3 Verification of the Constitutive Surfaces Using Large Stress State Variable Changes

Matyas and Radhakrishna (1968) experimentally tested the void ratio and degree of saturation constitutive surfaces for uniqueness. Soil specimens consisting of 80% flint powder and 20% kaolin were prepared using static compaction with a similar initial water content, dry density, and compactive effort. The specimens were tested under isotropic and $K_0$-loading with controlled total, pore–air, and pore–water pressures. The tests were performed in a modified triaxial apparatus.

The constitutive surfaces were traced by following "constant suction" and "wetting" test paths. In the "constant suction" test, the specimens were either saturated or desaturated first, and then compressed at a constant matric suction. The saturation of specimens was conducted while holding the specimen at a constant overall volume in order to minimize soil structure disturbance. Some specimens were isotropically loaded, while others were subjected to $K_0$-loading. The "wetting" tests were conducted by decreasing the matric suction while allowing the specimens to imbibe water. Matric suction was decreased by increasing the pore–water pressure.

The constitutive surfaces obtained from isotropic loading are shown in Fig. 12.1. The "constant suction" test results showed a decrease in the soil structure compressibility at higher matric suctions [Fig. 12.1(a)]. This means that the soil structure is more rigid at high matric suctions. The "wetting" test results indicated a collapse-type behavior or a decrease in void ratio upon a reduction in matric suction [Fig. 12.1(a)]. This behavior is characteristic of a metastable soil structure. It was also found that the change in void ratio between any two state points was independent of the deformation versus stress path followed (Matyas and Radhakrishna, 1968). The soil structure constitutive surface [Fig. 12.1(a)] shows uniqueness for monotonic deformation. Monotonic deformation was obtained by following paths with increasing degrees of saturation and where the specimen was not allowed to swell. The degree of saturation constitutive surface, however, did not indicate complete uniqueness, as shown in Fig. 12.1(b). The non-uniqueness of the constitutive surface was attributed to incomplete saturation during the wetting process.

The constitutive surfaces obtained from $K_0$-loading are presented in Fig. 12.11. All test paths appeared to form a single warped void ratio surface [Fig. 12.11(a)]. A collapsing soil behavior was again observed. The degree of saturation was found to be more sensitive to stress state changes than was the void ratio constitutive surface [Fig. 12.11(b)]. The uniqueness of both constitutive surfaces was again restricted to monotonic deformation (Matyas and Radhakrishna, 1968).

Barden *et al.* (1969) investigated the uniqueness of the

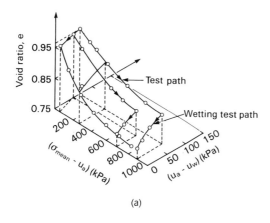

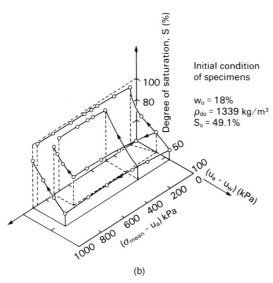

**Figure 12.11** Void ratio and degree of saturation constitutive surfaces for a mixture of flint and kaolin under $K_0$-loading. (a) Void ratio constitutive surface; (b) degree of saturation constitutive surface (from Matyas and Radhakrishna, 1968).

soil structure constitutive surface under $K_0$-loading conditions. In particular, the coefficients of volume change, $m_{1k}^s$ and $m_{2k}^s$, and their stress path dependency were studied. The soil structure constitutive equation for $K_0$-loading is as follows:

$$d\epsilon_v = m_{1k}^s \, d(\sigma_y - u_a) + m_{2k}^s \, d(u_a - u_w) \quad (12.64)$$

where

$m_{1k}^s$ = coefficient of volume change with respect to a change in net normal stress, $d(\sigma_y - u_a)$, for $K_0$-loading

$m_{2k}^s$ = coefficient of volume change with respect to a change in matric suction, $d(u_a - u_w)$, for $K_0$-loading.

The $m_{1k}^s$ and $m_{2k}^s$ coefficients expressed in terms of Young's modulus, $E$, and Poisson's ratio, $\mu$, are given in Table 12.2. Three illite clay soils of low to high plasticity were used in the study. Eleven groups of soil specimens were prepared using both dynamic or static compaction. The specimens were tested in a consolidation cell modified to control the total, pore-air, and pore-water pressures. The constitutive surface was traced following different test paths through various combinations of the stress state variables, $(\sigma_y - u_a)$ and $(u_a - u_w)$ (Fig. 12.12). In all cases, the pore-air pressure was held constant, while the total normal stress and the pore-water pressure were varied.

A preliminary study using smaller stress increments than those shown in Fig. 12.12 indicated that specimens with similar initial water contents would have similar volume changes independent of the test paths. In the remainder of the program, soil specimens in each group were prepared in an "identical" manner. Figure 12.13 presents the soil structure constitutive surface for the illite clay as plotted with respect to the $(\sigma_y - u_a)$ and $(u_a - u_w)$ axes. The results were obtained by following several loading paths from initial to final conditions (i.e., from point $A$ to point $H$; see Fig. 12.12). The constitutive surface exhibits uniqueness or loading path independence as long as the deformation is monotonic (Barden et al. (1969). Hysteresis between saturation and desaturation processes was considered to be the major cause of loading path dependence. An increase in the matric suction was found to increase the stiffness of a flocculent soil, compacted dry of optimum water content. However, matric suction had less effect on the stiffness for a soil with a dispersed structure, compacted wet of optimum water content. The soil structure stiffness increases with an increase in net normal stress, regardless of the type of soil structure.

The second independent constitutive surface was represented by the degree of saturation of the soil. Figure 12.14 shows the degree of saturation constitutive surface for a group of soil specimens. The surface was traced by following several loading paths from point $A$ to point $H$ (Fig. 12.12). Uniqueness was shown with respect to the degree of saturation, provided the deformation was monotonic.

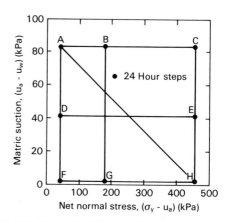

**Figure 12.12** Test paths followed during the study of the constitutive surface for unsaturated soil specimens (from Barden et al. 1969).

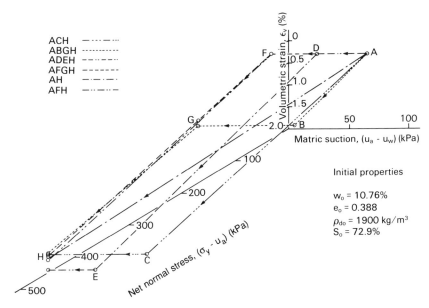

**Figure 12.13** Soil structure constitutive surface for an illite clay (from Barden *et al.* 1969a).

## 12.5 RELATIONSHIP AMONG VOLUMETRIC DEFORMATION COEFFICIENTS

The constitutive equations or surfaces for unsaturated soils have been formulated and presented in three forms. In the elasticity form, the $E$, $H$, $\mu$, $E_w$, and $H_w$ parameters are required to define the volumetric deformations associated with the soil structure and water phase constitutive equations [Eqs. (12.28) and (12.31)]. In the compressibility form, the soil structure and water phase constitutive surfaces makes use of the $m_1^s$, $m_2^s$, $m_1^w$, and $m_2^w$ coefficients [Eqs. (12.55) and (12.56)]. These coefficients of volume change can be expressed in terms of the above elastic parameters, as shown in Table 12.2. In soil mechanics terminology, the $a_t$, $a_m$, $b_t$, and $b_m$ coefficients are used to define slopes on the void ratio and water content constitutive surfaces [Eqs. (12.58) and (12.59)].

This section discusses the relationships among the various volumetric deformation coefficients. Theoretical and experimental methods that can be used to obtain these relationships are outlined. Emphasis is given to the coefficients used in the compressibility and soil mechanics forms

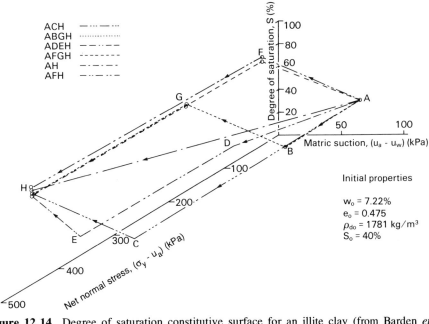

**Figure 12.14** Degree of saturation constitutive surface for an illite clay (from Barden *et al.* 1969a).

of the constitutive surfaces (i.e., $m_1^s$, $m_2^s$, $m_1^w$, $m_2^w$ and $a_t$, $a_m$, $b_t$, and $b_m$). These coefficient can, in general, be obtained from various laboratory soils tests.

### 12.5.1 Relationship of Volumetric Deformation Coefficients for the Void Ratio and Water Content Surfaces

The void ratio and water content surfaces for an unsaturated soil are illustrated in Fig. 12.6. The orientation of each point on the void ratio surface can be defined in terms of two slope angles. The first slope angle is referenced to the net normal stress, and is defined by the coefficient of compressibility, $a_t$:

$$a_t = \frac{\partial e}{\partial (\sigma - u_a)}. \quad (12.65)$$

The second slope angle is referenced to the matric suction and is defined by the coefficient of compressibility, $a_m$:

$$a_m = \frac{\partial e}{\partial (u_a - u_w)}. \quad (12.66)$$

Similarly, a point on the water content surface will have two slopes, which can be defined as $b_t$ and $b_m$ with respect to the net normal stress and matric suction, respectively:

$$b_t = \frac{\partial w}{\partial (\sigma - u_a)} \quad (12.67)$$

$$b_m = \frac{\partial w}{\partial (u_a - u_w)}. \quad (12.68)$$

Both the void ratio and water content surfaces are generally nonlinear. Therefore, the four coefficients, $a_t$, $a_m$, $b_t$, and $b_m$, vary over the constitutive surfaces. In other words, these coefficients are a function of the state point on the surface. However, the slopes on the constitutive surface have some limiting conditions, as observed from experimental results.

The $a_m$ coefficient is approximately equal to the $a_t$ coefficient when the degree of saturation is near 100% or when the matric suction goes to zero [Fig. 12.6(b)]. At saturation (i.e., on the plane of zero matric suction), the $a_t$ coefficient becomes the coefficient of compressibility, $a_v$, for a saturated soil.

There may be a slight difference between the $a_t$ and $a_m$ coefficients near saturation since $a_t$ can be measured for various types of loading, whereas $a_m$ is measured under isotropic loading. As the degree of saturation decreases, it has been found that the $a_t$ coefficient becomes greater than the $a_m$ coefficient. This shows that a change in total stress is more effective in changing void ratio than is a change in matric suction. It can be observed that the void ratio corresponding to the shrinkage limit of a soil is the minimum void ratio that can be attained under unconfined, maximum matric suction conditions.

The $b_m$ coefficient approaches the $b_t$ coefficient as the degree of saturation approaches 100%. At lower degrees of saturation, the $b_m$ coefficient will generally be greater than the $b_t$ coefficient [Fig. 12.6(b)]. In other words, a change in matric suction is more effective in changing the water content of the soil than is a change in total stress. This occurs because the matric suction is applied directly to the water phase. *On the saturation plane*, when matric suction is equal to zero, the $b_t$ coefficient can be related to the $a_t$ or $a_v$ coefficients using the basic volume–mass relationship (i.e., $Se = wG_s$ where $G_s$ is the specific gravity of the soil solids). Substituting a degree of saturation of 100% into the basic relationship illustrates the relationship between the moduli:

$$a_t = b_t G_s. \quad (12.69)$$

A two-dimensional plot of the constitutive surfaces allows a comparison of the above coefficients [i.e., $a_t$ versus $a_m$ and $b_t$ versus $b_m$; Fig. 12.6(b)]. The relationship among the four coefficients can also be illustrated in one plot (Fig. 12.15). Consider a saturated silt subjected to isotropic consolidation by increasing the total normal stress. The soil remains saturated during the consolidation process. Consolidation curve A in Fig. 12.15 represents the relationship between the void ratio and net normal stress, $(\sigma - u_a)$, as well as the water content (i.e., $wG_s$) and net normal stress. The pore-air pressure approaches the pore-water pressure for the saturated condition.

If the same soil is subjected to an increasing matric suction, the volume change will be the same as long as the soil remains saturated. Once the soil starts to desaturate, a matric suction change will not be as effective as a total stress change in producing a volume change. This is reflected by curve B [i.e., $e$ versus $(u_a - u_w)$] in Fig. 12.15, which gives less volume change than curve A [i.e., $e$ versus $(\sigma - u_a)$]. On the other hand, a matric suction increase is more effective than a net normal stress increase in removing water from the soil. This is illustrated by the water

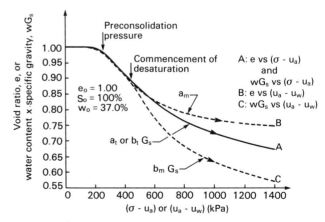

**Figure 12.15** Relationship among volumetric deformation moduli for an initially saturated soil.

content curve [i.e., $wG_s$ versus $(u_a - u_w)$], which shows a greater change in water content than curve $A$. Unlike curve $A$, the void ratio versus matric suction curve (i.e., curve $B$) starts to deviate from the water content versus matric suction curve (i.e., curve $C$) as the soil begins to desaturate. The separation of the two curves is due to the decreasing degree of saturation as the matric suction increases. On the other hand, there is essentially a constant degree of saturation during the total stress increment (i.e., $wG_s = e$ on curve $A$). The ratio of the ordinates for curves $C$ and $B$ indicates the degree of saturation (i.e., $S = wG_s/e$).

Curve $C$ eventually reaches a zero water content or zero degree of saturation. It has been observed from experimental data that the maximum suction a soil can reach is approximately 1 000 000 kPa at zero water content (Croney and Coleman, 1961). This value gives an upper limit for the suction that can be attained in a soil.

### 12.5.2 Relationship of Volumetric Deformation Coefficients for the Volume-Mass Form of the Constitutive Surfaces

The volumetric deformation coefficients used for the compressibility form of the constitutive surfaces can be defined in a manner similar to that described above. The soil structure and water phase constitutive surfaces are shown in Fig. 12.4. Lines through a point on the soil structure constitutive surface have slopes of $m_1^s$ and $m_2^s$ with respect to the net normal stress and matric suction, respectively:

$$m_1^s = \frac{\partial(V_v/V_0)}{\partial(\sigma - u_a)} \quad (12.70)$$

$$m_2^s = \frac{\partial(V_v/V_0)}{\partial(u_a - u_w)}. \quad (12.71)$$

Lines through a point on the water phase constitutive surface have slopes of $m_1^w$ and $m_2^w$ with respect to the net normal stress and matric suction, respectively:

$$m_1^w = \frac{\partial(V_w/V_0)}{\partial(\sigma - u_a)} \quad (12.72)$$

$$m_2^w = \frac{\partial(V_w/V_0)}{\partial(u_a - u_w)}. \quad (12.73)$$

The $m_1^s$, $m_2^s$, $m_1^w$, and $m_2^w$ coefficients are related to the $a_t$, $a_m$, $b_t$, and $b_m$ coefficients as follows:

$$m_1^s = \frac{a_t}{1 + e_0} \quad (12.74)$$

$$m_2^s = \frac{a_m}{1 + e_0} \quad (12.75)$$

$$m_1^w = \frac{b_t G_s}{1 + e_0} \quad (12.76)$$

$$m_2^w = \frac{b_m G_s}{1 + e_0} \quad (12.77)$$

where

$e_0$ = initial void ratio prior to deformation.

The $m_1^s$ is essentially equal to $m_1^w$ on the *saturation plane* (i.e., $S = 100\%$) since $a_t$ is essentially equal to $b_t G_s$ [Eq. (12.69)]. The $m_1^s$ coefficient for the saturated condition is commonly called the coefficient of volume change, $m_v$.

### 12.5.3 Laboratory Tests Used for Obtaining Volumetric Deformation Coefficients

The $a_t$, $a_m$, $b_t$, and $b_m$ coefficients are used to discuss the relationships that exist between the volumetric deformation coefficients. These coefficients vary from one state point to another along a nonlinear constitutive surface. A direct method to determine these coefficients at a specific state point is to measure their magnitude at the stress point under consideration. The experimental measurements required are similar to those conducted for the verification of the constitutive surfaces. It might require numerous specimens and a long period of testing if the entire constitutive surface were to be defined.

A simpler procedure is to assume that the constitutive surface is planar at a particular void ratio or water content, as shown in Figs. 12.16 and 12.17. Therefore, every point on the surface corresponding to an equal void ratio or water content plane has the same $a_t$ and $a_m$ coefficients or $b_t$ and $b_m$ coefficients, respectively. As a result, the $a_t$ and $b_t$ coefficients obtained from the saturation plane (i.e., $(u_a - u_w)$ equal to zero plane) can be used for other points on the surface as long as the void ratio (or water content) is constant. Similarly, the $a_m$ and $b_m$ coefficients obtained from the zero net normal stress plane (i.e., $(\sigma - u_a)$ equal to zero plane) can also be used for other points on the surface along a constant void ratio or water content plane. In other words, the values for $a_t$, $a_m$, $b_t$, and $b_m$ obtained on the saturation and zero net normal stress planes are assumed to apply for the entire constitutive surface.

The above method of determining the volumetric coefficients is much simpler than defining the entire constitutive surface. However, it may be inferior to the use of a direct determination at individual state points. The applicability of this approach will depend primarily on the character of the constitutive surfaces as established through laboratory experiments. The required level of accuracy for the problem at hand must also be borne in mind when using the above assumptions. In general, the suggested procedure may be sufficiently accurate for many geotechnical applications.

The following discussion pertains to the laboratory tests which can be used to obtain the $a_t$, $a_m$, $b_t$, and $b_m$ coefficients. The $a_t$ coefficients can be obtained from curve $A$ in Fig. 12.16, which is the result of a consolidation test on a

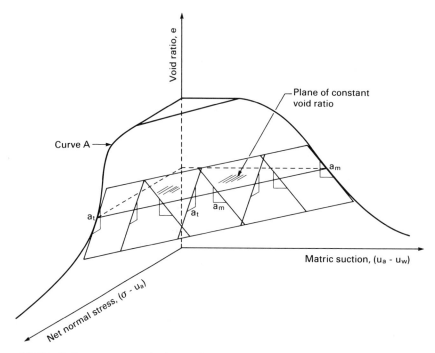

**Figure 12.16** Soil structure constitutive surface for monotonic loading, assuming a planar surface at a specific void ratio.

soil which has been saturated. Figure 12.18 shows a typical compression curve for a compacted soil which has been saturated while maintaining a constant volume prior to its decrease in volume due to loading. The $b_t$ coefficient in Fig. 12.17 can be computed as the $a_t$ coefficient divided by the specific gravity of the solids.

Curve $B$ in Fig. 12.17 is a soil–water characteristic curve that can be obtained using a pressure plate type of test (Chapter 13). Figure 12.19 illustrates both the drying and wetting portions of a typical soil–water characteristic curve. The $b_m$ coefficient can be computed from the drying portion of the soil–water characteristic curve.

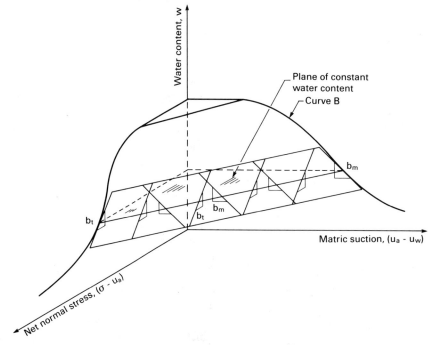

**Figure 12.17** Water phase constitutive surface for monotonic loading, assuming a planar surface at a specific water content.

12.5 RELATIONSHIP AMONG VOLUMETRIC DEFORMATION COEFFICIENTS 369

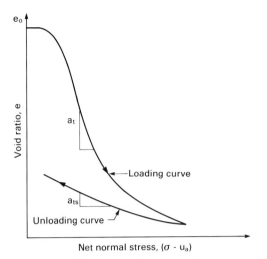

**Figure 12.18** Typical compression curves for a saturated soil plotted to an arithmetic scale.

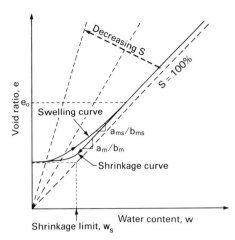

**Figure 12.20** Typical shrinking and swelling curves for a soil.

The $a_m$ coefficient can be related to the $b_m$ coefficient by use of a shrinkage-type test. A shrinkage test relates the void ratio to the water content at various matric suctions during a drying process with zero external loads. Figure 12.20 shows a typical shrinkage curve obtained from a shrinkage limit test. The slope of the shrinkage curve (i.e., $de/dw$ or $[\{\partial e/\partial(u_a - u_w)\}/\{\partial w/\partial(u_a - u_w)\}]$ defines the ratio between the $a_m$ and $b_m$ coefficients.

Details on the laboratory procedures associated with the determination of the volumetric deformation coefficients are discussed in Chapter 13. Techniques used to relate the volumetric coefficients are also demonstrated in Chapter 13.

### 12.5.4 Relationship of Volumetric Deformation Coefficients for Unloading Surfaces

Hysteresis causes the constitutive surfaces obtained from loading (and drying) to be different from the surfaces obtained from unloading (and wetting). Consequently, the volumetric deformation coefficients associated with each surface will be different. For example, the loading curve for a saturated soil has a compression index, $C_c$, whereas the unloading curve has a swelling index, $C_s$. The void ratio and water content surfaces for monotonic unloading are illustrated in Figs. 12.21 and 12.22, respectively. Coefficients of compressibility corresponding to the unloading surfaces can be further subscripted with an "$s$" (i.e., $a_{ts}$, $a_{ms}$, $b_{ts}$, and $b_{ms}$).

The same procedure used to obtain the coefficients for the loading surfaces can also be used to obtain the coefficients for the unloading surface. The only difference lies in the direction of deformation for unloading. The $a_{ts}$ coefficient can be computed from curve $A_s$ in Fig. 12.21. The curve represents the rebound curve on the saturation plane. The $b_{ts}$ coefficient in Fig. 12.22 is calculated as $a_{ts}$ divided by $G_s$.

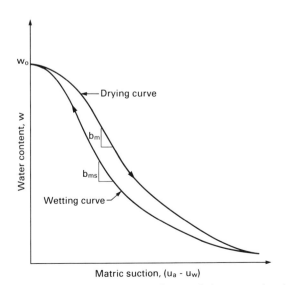

**Figure 12.19** Typical soil–water characteristic curves showing drying and wetting of the soil.

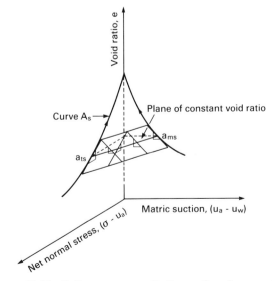

**Figure 12.21** Soil structure constitutive surface for monotonic unloading.

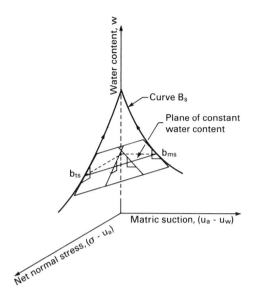

**Figure 12.22** Water phase constitutive surface for monotonic unloading.

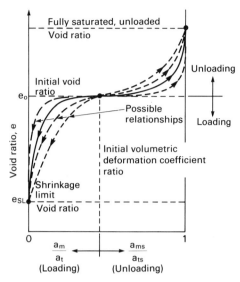

**Figure 12.23** Possible relationships between volumetric deformation coefficients during loading and unloading for an initially unsaturated soil.

Curve $B_s$ in Fig. 12.22 represents the wetting portion of the soil-water characteristic curve. The $b_{ms}$ coefficient can be obtained from the wetting curve. The $a_{ms}$ coefficient is then calculated from the $a_{ms}/b_{ms}$ ratio obtained from the slope of the unconfined swelling test presented in Fig. 12.20 (i.e., the opposite of a shrinkage curve). The slope of the swelling curve is defined as $de/dw$ or $[\{\partial e/\partial(u_a - u_w)\}/\{\partial w/\partial(u_a - u_w)\}]$, which is equal to $a_{ms}/b_{ms}$.

### 12.5.5 Relationship of Volumetric Deformation Coefficients for Loading and Unloading Surfaces

The relationship between the volumetric deformation coefficients associated with the loading and unloading void ratio surfaces is illustrated in Fig. 12.23. The moduli relationships are expressed in terms of the $a_m/a_t$ and $a_{ms}/a_{ts}$ ratios for loading and unloading, respectively. This relationship is derived from Fig. 12.15 for the loading surface. The $a_m$ and $a_t$ coefficients are equal (i.e., $(a_m/a_t)$ is approximately 1) when the soil is saturated.

As the soil desaturates, the $a_t$ coefficient becomes larger than the $a_m$ coefficient (i.e., $(a_m/a_t)$ becomes less than 1). However, the $a_m$ coefficient reaches a limiting value of zero as the water content in the soil reaches the shrinkage limit (Fig. 12.20). At this point, the $a_m/a_t$ ratio becomes zero. Therefore, it can be concluded that the $a_m/a_t$ ratio ranges from zero to one.

Consider an initially saturated soil at an initial void ratio, $e_0$, as shown in Fig. 12.23. At these initial conditions, the soil has an $a_m/a_t$ ratio which is greater than zero and less than one. As the soil is loaded, the void ratio decreases, and the $a_m/a_t$ ratio reduces. The loading may continue until the $a_m/a_t$ ratio reaches zero and the void ratio reaches a value corresponding to the shrinkage limit. Possible relationships between the void ratio and the $a_m/a_t$ ratio are shown in Fig. 12.23.

The soil may also be unloaded, causing the void ratio and the $a_{ms}/a_{ts}$ ratio to increase. A maximum $a_{ms}/a_{ts}$ ratio of one is obtained when the void ratio reaches a saturated condition. Similar relationships to Fig. 12.23 can also be generated for the ratio of the water content coefficients.

### 12.5.6 Constitutive Surfaces on a Semi-Logarithmic Plot

The constitutive surfaces can be plotted with respect to the logarithm of the stress state variables [Figs. 12.24(b) and 12.25(b)]. The logarithmic plots are linear over a relatively large stress range, on the extreme planes (i.e., the $\{\log (u_a - u_w) \approx 0\}$ plane and the $\{\log (\sigma - u_a) \approx 0\}$ plane). The slopes of the curves on these extreme planes are called indices. The volumetric deformation indices associated with the void ratio surface for loading conditions [Fig. 12.24(b)] are

$$C_t = \frac{\partial e}{\partial \{\log (\sigma - u_a)\}} \quad (12.78)$$

$$C_m = \frac{\partial e}{\partial \{\log (u_a - u_w)\}} \quad (12.79)$$

where

$C_t$ = compressive index with respect to net normal stress, $(\sigma - u_a)$
$C_m$ = compressive index with respect to matric suction, $(u_a - u_w)$.

For the unloading surface, the indices are subscripted with an "s" as $C_{ts}$ and $C_{ms}$.

12.5 RELATIONSHIP AMONG VOLUMETRIC DEFORMATION COEFFICIENTS 371

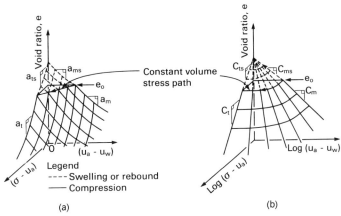

**Figure 12.24** Void ratio constitutive surface for an unsaturated soil. (a) Arithmetic plot of stress state variables versus void ratio; (b) semi-logarithmic plot of stress state variables versus void ratio.

The volumetric deformation indices associated with the water content surface for loading conditions [Fig. 12.25(b)] can be defined as

$$D_t = \frac{\partial w}{\partial \{\log (\sigma - u_a)\}} \quad (12.80)$$

$$D_m = \frac{\partial w}{\partial \{\log (u_a - u_w)\}} \quad (12.81)$$

where

$D_t$ = water content index with respect to net normal stress, $(\sigma - u_a)$

$D_m$ = water content index with respect to matric suction, $(u_a - u_w)$.

Similarly, the water content indices are subscripted with an "s" for the unloading surface (i.e., $D_{ts}$ and $D_{ms}$).

The $C_t$, $C_m$, $D_t$, and $D_m$ indices can be obtained from the same test data used to obtain the $a_t$, $a_m$, $b_t$, and $b_m$ coefficients. The difference between the soil properties lies in the manner in which the results are plotted. Figure 12.26 shows a typical compression curve for an unsaturated, compacted soil. The results show that the void ratio versus net normal stress curve can be linearized when a logarithmic scale is used for the stress state variable. The compressive index, $C_t$, can be computed from Fig. 12.26, and is commonly referred to as the compression index, $C_c$, for saturated soils. On the *saturation plane*, the $D_t$ index is related to $C_t$:

$$C_t = D_t G_s. \quad (12.82)$$

Figure 12.27 illustrates a typical soil–water characteristic curve plotted on a semi-logarithmic scale. A reasonably

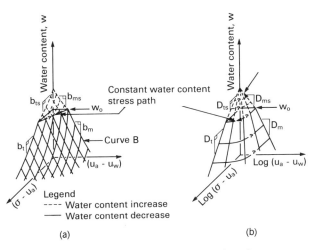

**Figure 12.25** Water content constitutive surface for an unsaturated soil. (a) Arithmetic plot of stress state variables versus water content; (b) semi-logarithmic plot of stress state variables versus water content.

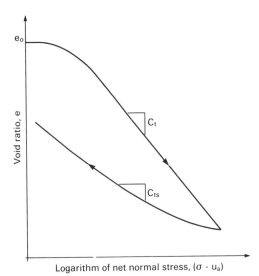

**Figure 12.26** Typical compression curve for a compacted soil plotted to a semi-logarithmic scale.

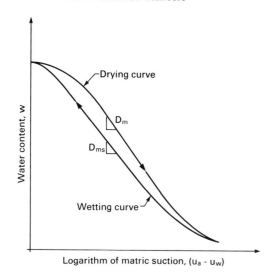

**Figure 12.27** Typical soil-water characteristics curves plotted to a semi-logarithmic scale.

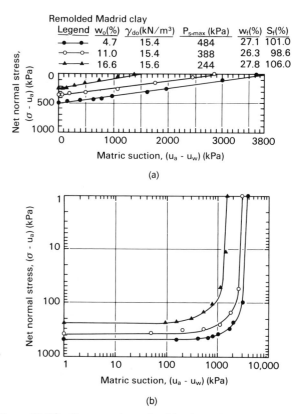

**Figure 12.28** Cross-sections of void ratio surfaces plotted using arithmetic and logarithmic scales. (a) Cross-sections of void ratio surfaces on an arithmetic scale. $\{(\sigma - u_a) \text{ and } (u_a - u_w) \text{ plane}\}$; (b) cross-section of the void ratio surfaces on a logarithmic scale $\{\log (\sigma - u_a) \text{ and } \log (u_a - u_w) \text{ plane}\}$ (modified from Escario, 1969).

linear curve is obtained over a wide stress change, as compared to the arithmetic plot in Fig. 12.19. The slope is equal to the $D_m$ index. The $C_m$ index can then be computed from the $C_m/D_m$ ratio obtained from the slope of the shrinkage curve (Fig. 12.20).

Using a conversion between a semi-logarithmic and arithmetic scale (Lambe and Whitman, 1979), the $C_t$, $C_m$, $D_t$, and $D_m$ indices can be written in terms of the $a_t$, $a_m$, $b_t$, and $b_m$ coefficients:

$$C_t = \frac{a_t(\sigma - u_a)_{\text{ave}}}{0.435} \quad (12.83)$$

$$C_m = \frac{a_m(u_a - u_w)_{\text{ave}}}{0.435} \quad (12.84)$$

$$D_t = \frac{b_t(\sigma - u_a)_{\text{ave}}}{0.435} \quad (12.85)$$

$$D_m = \frac{b_m(u_a - u_w)_{\text{ave}}}{0.435} \quad (12.86)$$

where

$(\sigma - u_a)_{\text{ave}}$ = average of the initial and final net normal stresses for an increment

$(u_a - u_w)_{\text{ave}}$ = average of the initial and final matric suctions for an increment.

The 0.435 constant arises from the logarithm of the natural log base taken to the base 10 (i.e., $\log_{10} 2.718$).

In spite of the linear curves on the extreme planes, the cross section of the constitutive surface on the logarithmic $\{\log (\sigma - u_a) \text{ and } \log (u_a - u_w)\}$ plane is no longer a series of straight lines as are found on the arithmetic $\{(\sigma - u_a) \text{ and } (u_a - u_w)\}$ plane. Figure 12.28 shows a comparison of the constitutive surface cross section at a constant void ratio when plotted using the arithmetic and logarithmic and logarithmic scales. An essentially linear cross section on the arithmetic plot becomes an asymptotic curve on the logarithmic plot.

The asymptotic cross-section curve illustrates the logarithmic form of the void ratio constitutive surface in Fig. 12.29(a). Ho (1988) proposed an approximate form of the void ratio constitutive surface, as shown in Fig. 12.29(b). The approximate surface consists of three planes, namely, planes I, II, and III. The three planes converge at a void ratio ordinate corresponding to nominal values of the stress state variables (i.e., $\log (\sigma - u_a) \approx 0$ and $\log (u_a - u_w) \approx 0$). Planes I and III are referred to as the orthogonal planes. Plane I is perpendicular to the $e$ versus $\log (\sigma - u_a)$ plane, and has a slope in plane I which is represented by the $C_t$ index. Plane III is perpendicular to the $e$ versus $\log (u_a - u_w)$ plane, and has a slope in plane III which is represented by the $C_m$ index. In other words, only one index is required to describe void ratio changes when the stress state variable changes occur within the regions defined by planes I and III. Therefore, the void ratio consti-

## 12.5 RELATIONSHIP AMONG VOLUMETRIC DEFORMATION COEFFICIENTS

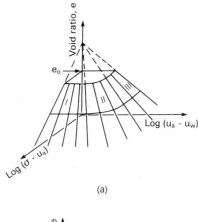

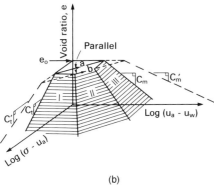

**Figure 12.29** Logarithmic forms for the void ratio constitutive surface for loading conditions. (a) Void ratio constitutive surface for loading when plotted on a logarithmic scale; (b) approximate form of the void ratio constitutive surface on a logarithmic plot (from Ho, 1988).

tutive equation describing plane I or plane III can be written in a general form as

$$de = C_t d \log (\sigma - u_a) + C_m d \log (u_a - u_w). \quad (12.87)$$

The water content constitutive equation describing plane I or plane III can be written as

$$dw = D_t d \log (\sigma - u_a) + D_m d \log (u_a - u_w). \quad (12.88)$$

Plane II represents a transition zone between planes I and III. This plane intersects both the void ratio versus log $(\sigma - u_a)$ plane and the void ratio versus log $(u_a - u_w)$ plane. These intersection lines define the slope of plane II, namely, the $C_t'$ and $C_m'$ indices. In this case, two indices are required to describe void ratio changes when the stress state variable changes occur within the region of plane II. Techniques to obtain the $C_t'$ and $C_m'$ indices based on the $C_t$ and $C_m$ indices are explained in Chapter 13. The constitutive equation describing plane II can be written as

$$de = C_t' d \log (\sigma - u_a) + C_m' d \log (u_a - u_w) \quad (12.89)$$

where

$C_t'$ = slope of the intersection line of plane II with the void ratio versus log $(\sigma - u_a)$ plane

$C_m'$ = slope of the intersection line of plane II with the void ratio versus log $(u_a - u_w)$ plane.

It should also be noted that line "b" associated with plane II is assumed to be parallel to line "a" which joins planes I and III [Fig. 12.29(b)]. This assumption is required in constructing the intersection lines of plane II on the extreme planes (see Chapter 13). The above approximation is also applicable to the unloading surface of the void ratio constitutive surface (Ho, 1988). The slope of the water content constitutive surface is not fully understood. It is a matter of speculation to assume that the water content constitutive surface is similar in shape to the void ratio constitutive surface. However, there does not presently appear to be any experimental data to substantiate this assumption.

# CHAPTER 13

# *Measurements of Volume Change Indices*

The application of the volume change theory presented in Chapter 12 requires the measurement of volume change coefficients and indices. These volume change coefficients and indices must be experimentally measured for each soil under investigation. Test procedures and equipments required for the measurements of the volume change coefficients and indices are outlined and discussed in this chapter. Test results on a compacted silt and a glacial till are used to illustrate the relationships between the various volume change coefficients and indices. Most of the results are presented in a semi-logarithmic form and can be used for determining the volume change indices. These indices can then be converted to other volume change coefficients using the relationships given in Chapter 12.

Techniques used to determine all volume change indices from the test results are also described in this chapter. The test procedures and equipment used in the measurement of the volume change properties are common to most soil mechanics laboratories. Some of the equipment (i.e., pressure plate apparatuses) are more common to the field of soil science and agronomy.

The test procedures and equipments for the loading constitutive surfaces are described prior to those for the unloading constitutive surfaces. The use of oedometer test results for assessing the *in situ* stress state in terms of the swelling pressure is briefly described. Procedural corrections pertinent to the determination of a corrected swelling pressure are outlined, and their importance to the prediction of heave is explained in Chapter 14.

## 13.1 LITERATURE REVIEW

Following is a review of typical test results that can be used to obtain the volume change coefficients associated with the constitutive surfaces. Figure 13.1 shows three-dimensional views of the void ratio and water content constitutive surfaces. The volume change coefficients corresponding to the loading conditions (i.e., $a_t$, $a_m$, $b_t$, $b_m$) are also shown.

Curve *A* in Fig. 13.1 is essentially a compression curve

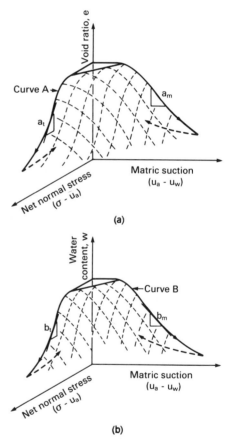

**Figure 13.1** Three-dimensional void ratio and water content constitutive surfaces for an unsaturated soil. (a) Void ratio constitutive surface; (b) water content constitutive surface.

obtained from an oedometer test on a soil in a saturated condition. There are two commonly used procedures for conducting oedometer tests on initially unsaturated specimens, namely, the "free-swell" and "constant volume" tests (Noble, 1966). These test procedures are generally used in the determination of the swelling pressure of the soil for the prediction of heave.

In the "free-swell" oedometer test, the specimen is al-

13.1 LITERATURE REVIEW    375

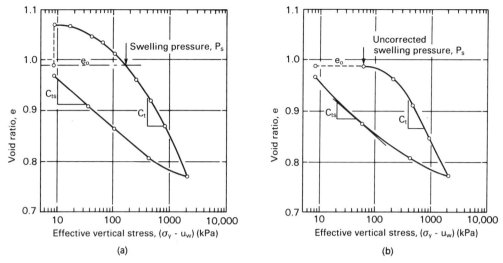

**Figure 13.2** Typical "free-swell" and "constant volume" one-dimensional oedometer test results. (a) "Free-swell" test procedure; (b) "constant volume" test procedure.

lowed to swell under a token pressure by submerging the specimen in distilled water [Fig. 13.2(a)]. After attaining an equilibrium condition, the soil specimen is then loaded and unloaded following the conventional oedometer test procedure. In the "constant volume" oedometer test, the applied load is increased in order to maintain the specimen at a constant volume after it has been submerged in distilled water [Fig. 13.2(b)]. When there is no further tendency for volume increase, the soil specimen is then loaded and unloaded following the conventional oedometer test procedure.

The slope of the loading curve plotted on a semi-logarithmic scale (Fig. 13.2) gives the compression index, $C_t$, which can be related to the coefficient of compressibility, $a_t$. Similarly, the slope of the unloading curve is equal to the swell index, $C_{ts}$, which can be related to the coefficient of swelling, $a_{ts}$. Figures 13.3 and 13.4 show typical results of "free-swell" and "constant volume" oedometer tests on a compacted Regina clay, respectively.

The coefficient of compressibility, $a_t$, is equal to the $a_v$ coefficient of compressibility commonly measured in saturated soil mechanics. Having determined the coefficient of compressibility, $a_t$, the coefficient of water content change, $b_t$ [Fig. 13.1(b)], can then be calculated. The coefficient of water content change, $b_t$, is equal to the coefficient of compressibility, $a_t$, multiplied by the specific gravity of the soil, $G_s$.

Curve $B$ in Fig. 13.1(b) is equivalent to a soil-water

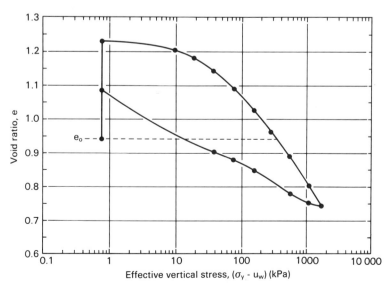

**Figure 13.3** "Free-swell" one-dimensional oedometer data for compacted Regina clay (from Gilchrist, 1963).

**376** 13 MEASUREMENTS OF VOLUME CHANGE INDICES

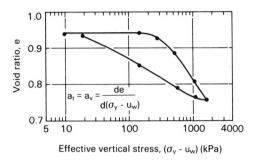

**Figure 13.4** "Constant volume" one-dimensional oedometer data for compacted Regina clay (from Gilchrist, 1963).

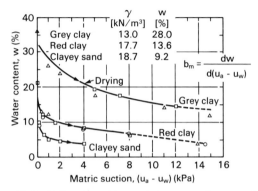

**Figure 13.5** Soil–water characteristic curves for several compacted soils (from Escario et al., 1989).

characteristic curve which can be obtained from a pressure plate test. Figures 13.5 and 13.6 present the soil–water characteristic curves for several compacted soils. The soil–water characteristic curve for a silty clay soil is shown in Fig. 13.7 on a semi-logarithmic plot. The characteristic curves exhibit hysteresis between the drying and wetting processes.

The slope of the soil–water characteristic curve plotted on an arithmetic scale is equal to the coefficient of water content change, $b_m$, when the drying curve (i.e., loading) is considered. The slope is equal to the coefficient, $b_{ms}$, when the wetting curve (i.e., unloading) is considered. Correspondingly, the semi-logarithmic plot (Fig. 13.7) results in the water content indices, $D_m$ and $D_{ms}$, for the drying and wetting curves.

The shrinkage and swelling relationships for a soil relate the void ratio to the water content at various matric suctions. Figure 13.8 shows a shrinkage curve relationships for a silty clay, corresponding to the soil–water characteristic curve shown in Fig. 13.7. The slope of the shrinkage curve defines the ratio between the $a_m$ and $b_m$ coefficients. Having determined the water content coefficient, $b_m$, from the drying curve in Fig. 13.7, the coefficient of volume change, $a_m$, can then be computed using the $a_m/b_m$ ratio obtained from the shrinkage curve in Fig. 13.8.

Several typical shrinkage relationships (i.e., void ratio versus water content) for compacted soils are presented in Fig. 13.9. The results show that the shrinkage curves are further from the saturation line as the water content or the degree of saturation of the soil decreases.

Figure 13.10 presents the shrinkage and swelling water content versus void ratio relationships for Durham clay. The slope of the swelling curve defines the ratio between the coefficient of swelling, $a_{ms}$, and the coefficient of water content, $b_{ms}$. Having determined the coefficient of water content, $b_{ms}$, from the wetting portion of the soil–water characteristic curve, the coefficient of swelling, $a_{ms}$, can be computed from the swelling curve.

## 13.2 TEST PROCEDURES AND EQUIPMENTS

Tests and equipments necessary for the measurements of the volume change coefficients are outlined and described in this section. The tests and equipments required for de-

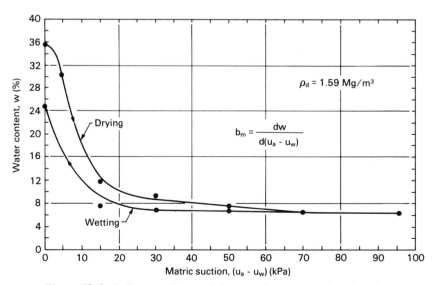

**Figure 13.6** Soil–water characteristic curve for compacted sand tailings.

13.2 TEST PROCEDURES AND EQUIPMENTS   377

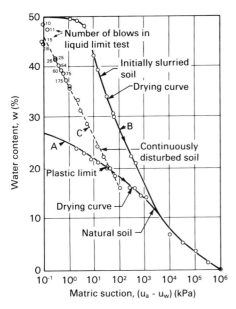

**Figure 13.7** Soil–water characteristic curve for a silty clay (from Croney and Coleman, 1954).

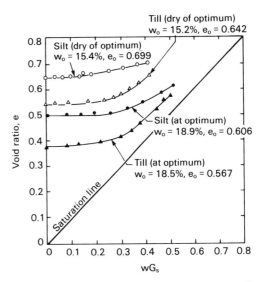

**Figure 13.9** Shrinkage relationships for a compacted silt and a glacial till (from Ho and Fredlund, 1989).

termining the volume change coefficients associated with the loading constitutive surface are discussed first. Conventional tests and equipments common to a soil mechanics laboratory can be used in the measurement of the volume change coefficients for an unsaturated soil (Ho and Fredlund, 1989; Rahardjo et al., 1990; and Ho et al., 1992).

There are, however, several procedural factors that need to be considered when testing unsaturated soils. These factors are identified when discussing the test procedures. Typical test results are given in order to illustrate the techniques used to determine the volume change coefficients. The last part of this section describes the tests and equipment required for the determination of the volume change coefficients associated with the unloading constitutive surface.

### 13.2.1 Loading Constitutive Surfaces

The relationship between the volume change indices associated with the loading conditions can best be visualized by presenting several intersection curves on one plot (Fig. 13.11). These data are referred to as intersection curves because they correspond to the planes where one of the stress state variables goes to zero. Intersection curves 1 and 2 for the void ratio surface [Fig. 13.11(a)] are combined in Fig. 13.11(b). The slopes of curves 1 and 2 are called the $C_t$ and $C_m$ volume change indices corresponding to the net normal stress and matric suction planes, respectively. Intersection curves 3 and 4 for the water content surface [Fig. 13.11(c)] are combined in Fig. 13.11(d). The water

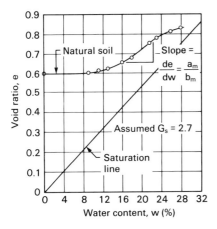

**Figure 13.8** Shrinkage relationship for a silty clay (from Croney and Coleman, 1954).

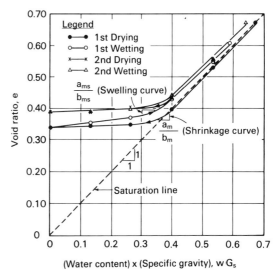

**Figure 13.10** Unconfined shrinking and swelling curves for Durham clay (from Haines, 1923).

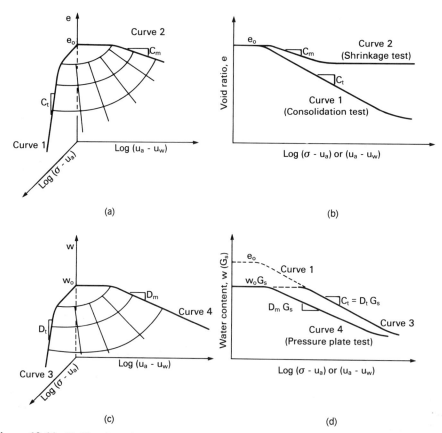

**Figure 13.11** Void ratio and water content constitutive relationships for an unsaturated soil. (a) Void ratio versus logarithm of stress state relationship; (b) intersection curves between void ratio surface and the $(\sigma - u_a)$ or $(u_a - u_w)$ plane; (c) water content versus logarithm of stress state relationship; (d) intersection curves between water content surface and the $(\sigma - u_a)$ or $(u_a - u_w)$ plane.

content, $w$, can be multiplied by the specific gravity, $G_s$, (i.e., $wG_s$) in order to obtain the same scale as used for plotting the void ratio data. This means that curves 3 and 4 in Fig. 13.11(c) are translated vertically by a magnitude of $G_s$ in Fig. 13.11(d). As a result, the slopes of curves 3 and 4 in Fig. 11(d) are the products of $G_s$ and the volume change indices (i.e., $(D_t G_s)$ and $(D_m G_s)$, respectively).

Curve 1 in Fig. 13.11 is essentially the consolidation curve for the soil in a saturated condition (i.e., $(u_a - u_w) = 0$ or $u_a = u_w$). The curve exhibits a linear relationship between the void ratio and the logarithm of net normal stress over a wide loading range. The slope of curve 1, $C_t$, is equal to the compression index, $C_c$, of the saturated soil. Curve 3 in Fig. 13.11(d) coincides with curve 1 from Fig. 13.11(b) since the water content multiplied by specific gravity is equal to the void ratio when the soil is saturated. Therefore, the water content index, $D_t$, can be computed as $(C_t/G_s)$.

Curve 4 in Fig. 13.11 is called the soil-water characteristic curve (i.e., drying curve), and can be obtained from a pressure plate test (see next sections). A shrinkage curve (explained in the next sections) combined with the drying portion of the soil-water characteristic curve can be used to construct curve 2 in Fig. 13.11. In other words, the four volume change curves and their corresponding indices (i.e., $C_t$, $D_t$, $D_m$, and $C_m$) can be obtained from routine soil tests.

The combined plot of curves 1, 2, 3, and 4 are presented in Fig. 13.12, which is essentially a combination of Fig. 13.11(b) and (d). The arrows in Fig. 13.12 indicate the direction of deformation for curves 2 and 4. These curves approach curve 1 when the initial degree of saturation of the soil increases. Curve 1 has been assumed to remain constant for conditions corresponding to various initial degrees of saturation. When the soil is saturated, the void ratio and the water content curves are the same, varying only with respect to net normal stress or the effective stress, $(\sigma - u_w)$ (i.e., curve 1 in Fig. 13.12).

*Oedometer tests*

Procedures for oedometer tests on unsaturated soil specimens (e.g., compacted specimens) are described in ASTM D4546. This ASTM standard describes three methods for inundating the soil specimen prior to performing the oedometer test. During inundation, the matric suction of

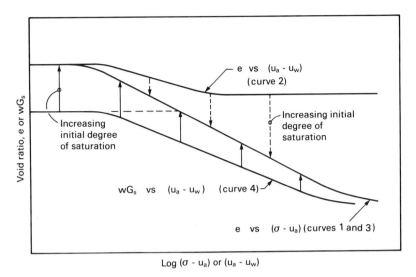

**Figure 13.12** Relationship between curves that define the mass and volume change behavior of an unsatured soil.

the soil is brought to zero, and the results can be used to calculate the swelling potential of the soil. The inundation can be conducted under "constant volume" [Fig. 13.2(b)] or "free-swell" [Fig. 13.2(a)] conditions. Curve 1 in Figs. 13.11 and 13.12 illustrate the oedometer test results using the "constant volume" inundation procedure at the beginning of the test. Having inundated the soil specimen and allowed time for equilibrium, the test can proceed using the conventional oedometer test procedure used for saturated specimens (ASTM D2435). The decreasing void ratios are plotted against the logarithms of effective stress, $(\sigma - u_w)$, to yield curves 1 and 3, as shown in Figs. 13.11 and 13.12.

### Pressure Plate Drying Tests

The soil-water characteristic curve (i.e., curve 4 and Figs. 13.11 and 13.12) of a soil relates the water content to the applied matric suction in the soil. In an unsaturated soil, the pore–air pressure, $u_a$, is usually atmospheric (i.e., $u_a = 0$) and the pore–water pressure, $u_w$, is negative. The difference between the pore–air and pore–water pressures is called the soil matric suction, $(u_a - u_w)$. In the laboratory, a matric suction can be applied to a soil specimen by maintaining a zero pore–water pressure and applying a positive pore–air pressure. Therefore, the matric suction in the soil specimen (i.e., $(u_a - u_w)$ where $u_w$ is maintained zero) can be varied by applying different air pressures to the specimen. This procedure is referred to as the axis-translation technique (Hilf, 1956), and is explained in Chapter 6.

Pressure plate extractors such as the type manufactured by Soilmoisture Equipment Corporation, Santa Barbara, CA, can be used to measure the soil–water characteristic curve. The extractors are commonly used to apply various matric suctions to the soil specimen, and the test is called a pressure plate test (ASTM D2325). The pressure plate extractor consists of a high air entry ceramic disk contained in an air pressure chamber. The high air entry disk is saturated, and is always in contact with water in a compartment below the disk. The compartment is maintained at zero water pressure.

Tempe pressure cells and volumetric pressure plate extractors are two other types of extractors which can be used for the low range of matric suction applications (i.e., maximum of 200 kPa). This equipment and its operational procedures are explained in Chapter 6. Similar pressure plate extractors for the high range of matric suction applications (i.e., up to 1500 kPa) are presented in Figs. 13.13, 13.14, and 13.15. The pressure membrane extractor utilizes cellulose membranes instead of a high air entry ceramic disk. In addition, a compression diaphragm is provided on the

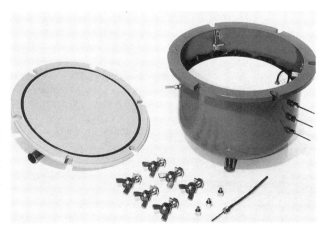

**Figure 13.13** A 5 bar ceramic plate extractor (i.e., maximum applied matric suction is 500 kPa). (Photograph courtesy of Soilmoisture Equipment Corporation, Santa Barbara, CA.)

**Figure 13.14** A 15 bar ceramic plate extractor (i.e., maximum applied matric suction is 1500 kPa). (Photograph courtesy of Soilmoisture Equipment Corporation, Santa Barbara, CA.)

lid of the pressure membrane extractor. A pressure regulator maintains a higher pressure behind the diaphragm than inside the extractor in order to keep the soil specimen in contact with the cellulose membrane during the test. This is particularly important in the high range of matric suction applications where the soil specimens may shrink considerably.

**Figure 13.15** Pressure membrane extractor. (Photograph courtesy of Soilmoisture Equipment Corporation, Santa Barbara, CA.)

A soil specimen is placed on top of the disk, and the airtight chamber is pressurized to the desired matric suction. The disk does not allow the passage of air as long as the applied matric suction does not exceed the air entry value of the disk. The air entry value of the disk is related to the diameter of the fine pores in the ceramic disk. Therefore, the air entry value of the disk and the strength of the chamber control the maximum air pressure (or matric suction) which can be applied to the specimen.

The application of matric suction to the soil causes the pore-water to drain to the water compartment through the disk. A burette can be connected to the compartment to measure the water volume changes if only one specimen is tested. At equilibrium, the soil will have a reduced water content corresponding to the increased matric suction. The water content at each equilibrium condition can be computed from the water volume change measurements. If more than one specimen are being tested, it is necessary to dismantle the chamber and measure the weight of the specimens after equilibrium at each applied pressure. This procedure is commonly used with 5 and 15 bar ceramic plate extractors when several specimens are tested simultaneously. Plots of the equilibrium water contents versus the logarithm of the corresponding matric suctions give rise to curve 4 in Figs. 13.11 and 13.12.

### Shrinkage Tests

A shrinkage curve for a soil shows the relationship between the void ratio and the water content at various matric suctions. A soil specimen can either be allowed to dry in the air or it can be subjected to various matric suctions using a pressure plate extractor. In either case, the void ratio and water content of the specimen can be measured at various equilibrium states. When the specimen is allowed to slowly dry in the air, the specimen is covered at various time intervals in order to allow the specimen to come to equilibrium.

Accurate measurements of the void ratio can be made following the techniques used in the shrinkage limit test (ASTM D427). The shrinkage test involves the measurement of the total volume of the specimen. The standard procedure for measuring the total volume involves the use of the mercury displacement technique. The total volume of a soil specimen is measured by immersing the specimen into a cup filled with mercury (Fig. 13.16). The volume of the displaced mercury during immersion can be converted to the total volume of the specimen.

Direct measurements of total volume can also be performed using calipers. The shrinkage curve can be constructed by plotting the decreasing void ratios against the decreasing water contents as the matric suction increases.

### Determination of Volume Change Indices

Two soils, a uniform silt and a glacial till, were tested by Ho (1988) and their index properties are presented in Table 13.1. An attempt was made to prepare soil specimens with

13.2 TEST PROCEDURES AND EQUIPMENTS    381

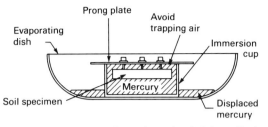

**Figure 13.16** Immersion of specimen in shrinkage limit test (from Head, 1984).

**Table 13.1 Index Properties of the Silt and Glacial Till Used in the Test Program**

|  | Silt | Glacial Till |
|---|---|---|
| Liquid limit | 26.7 | 33.2 |
| Plastic limit | 14.9 | 13.0 |
| Plasticity index | 11.8 | 20.2 |
| Percent sand sizes | 25.0% | 32.0% |
| Percent silt sizes | 52.0% | 39.0% |
| Percent clay sizes | 23.0% | 29.0% |
| Specific gravity, $G_s$ | 2.72 | 2.76 |
| Half standard effort compaction: | | |
| $w_{optimum}$ | 19.0% | 18.75% |
| $\gamma_{d\,maximum}$ | 16.65 kN/m$^3$ | 17.12 kN/m$^3$ |

nearly identical initial conditions. Each soil was oven-dried and hand-mixed with a predetermined quantity of distilled water. The wet soil was placed in a sealed plastic bag, and was left to cure in a constant humidity and temperature room. The difference in water content between batches of the same soil was controlled to within 0.5%. Specimens were formed by static compaction at one-half standard AASHTO compaction effort at either "dry of optimum" or "at optimum" initial water contents. The compaction characteristics of the silt and till are given in Table 13.1.

The volume change and water content change tests (i.e., oedometer, pressure plate and shrinkage tests) were conducted on compacted silt and glacial till specimens, with initial conditions corresponding to both the "dry of optimum" and "at optimum" conditions. The laboratory data are used to determine the volume change indices for the soils at both compaction conditions.

The data for silt specimens compacted "dry of optimum" are used to illustrate the technique for obtaining the volume change indices. The oedometer test results for the silt specimens are shown in Fig. 13.17. The oedometer tests were performed in accordance with the "constant volume" test method. The loading curves correspond to curves 1 and 3 in Figs. 13.11 and 13.12, and their slopes are equal to $C_t$ and $(D_t\,G_s)$, respectively. The soil-water characteristic curve for the silt (Fig. 13.18) was obtained from pressure plate tests. The water content is multiplied by the specific gravity, $G_s$, for the purpose of combining the void ratio and water content plots, as previously illustrated in Fig. 13.12 (i.e., curve 4). The slope of the curve is equal to $(D_m\,G_s)$.

The shrinkage curve for the compacted silt is shown in Fig. 13.19. The void ratios corresponding to the various water contents in Fig. 13.18 can be found using the shrinkage relationships in Fig. 13.19(a). As a result, the void ratio versus matric suction relationship (i.e., curve 2 in Figs. 13.11 and 13.12) can be constructed using Figs. 13.18 and 13.19(a). The slope of the shrinkage curve [i.e.,

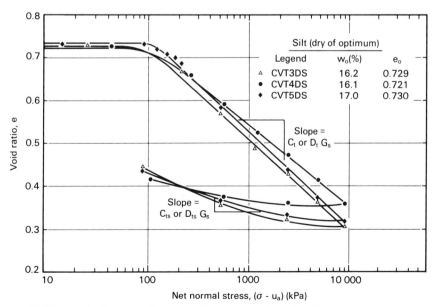

**Figure 13.17** Results from one-dimensional "constant volume" oedometer tests on a compacted silt.

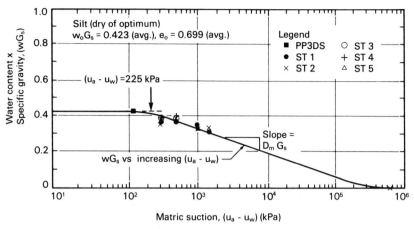

**Figure 13.18** Soil–water characteristic curve obtained from a pressure plate test on a silt compacted dry of optimum water content.

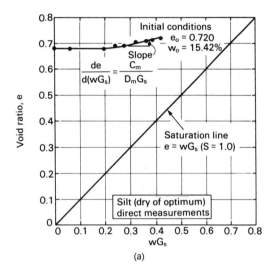

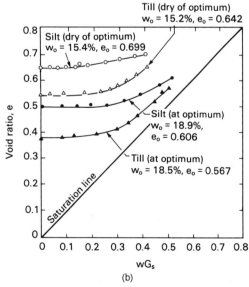

**Figure 13.19** Results from shrinkage tests on compacted silts and glacial tills. (a) Shrinkage test data for the compacted silt; (b) shrinkage test data for compacted silts and glacial tills.

$de/d(wG_s)$ or $(\partial e/\partial(u_a - u_w))/(\partial(wG_s)/\partial(u_a - u_w))]$ is equivalent to the ratio of volume change indices (i.e., $C_m/D_m G_s$).

The combined plot of Figs. 13.17, 13.18, and curve 2 [i.e., constructed from Figs. 13.18 and 13.19(a)] is depicted in Fig. 13.20, which illustrates the volume change characteristics of an unsaturated, compacted silt. The volume change indices (i.e., $C_t$, $C_m$, $D_t$, and $D_m$) can be computed from Fig. 13.20. Changes in void ratio and water content due to an increase in total stress or matric suction can now be predicted using the computed volume change indices.

The same test procedures were applied to other compacted silt and the glacial till specimens. Figure 13.19(b) summarizes the results of shrinkage tests on various compacted specimens. Typical volume change relationships for the compacted silt and glacial till are presented in Figs. 13.21, 13.22, and 13.23. The relationships are similar to that shown in Fig. 13.20. The computed volume change indices for the compacted silt and glacial till are tabulated in Table 13.2. These indices can be converted to other volume change coefficients such as "$m_1$ and $m_2$" or "$a$ and $b$," as explained in Chapter 12.

In summary, oedometer tests, pressure plate tests, and shrinkage tests are the experiments required to obtain the volume change indices corresponding to the loading of an unsaturated soil. These tests can be performed using conventional soil mechanics testing procedures. The test results give rise to the volume change relationships for an unsaturated soil.

### Determination of Volume Change Indices Associated with the Transition Plane

The entire void ratio constitutive surface in a semi-logarithmic form can be approximated by three planes, as illustrated in Fig. 13.24 and described in Chapter 12. The volume change indices, $C_t$ and $C_m$, are associated with or-

13.2 TEST PROCEDURES AND EQUIPMENTS 383

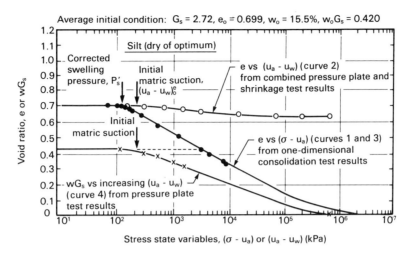

**Figure 13.20** Volume change relationships for the silt compacted dry of optimum water content.

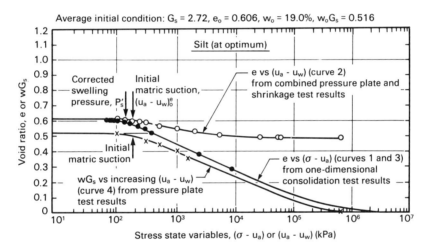

**Figure 13.21** Volume change relationships for the silt compacted at optimum water content.

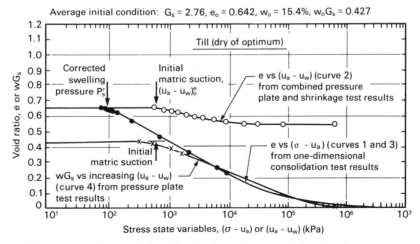

**Figure 13.22** Volume charge relationships for the till compacted dry of optimum water content.

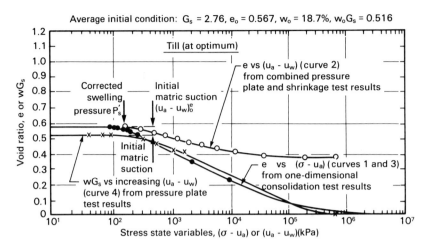

**Figure 13.23** Volume change relationships for the till compacted at optimum water content.

thogonal planes I and III, respectively, and can be determined from the test results presented in the previous sections. The volume change indices, $C'_t$ and $C'_m$, are associated with transition plane II, and can be determined graphically as outlined in this section. The procedure is applicable to stable-structured soils.

Figure 13.25 illustrates the graphical determination of the volume change indices, $C'_t$ and $C'_m$, based on the "constant volume" oedometer test results and the measured values of the $C_t$ and $C_m$ indices. The first step is to determine the corrected swelling pressure, $P'_s$, as detailed in the next section. Having determined the corrected swelling pressure, point $A$ in Fig. 13.25 can be plotted with a coordinate equal to $(\log P'_s)$ and $e_0$ where $e_0$ is the initial void ratio.

A line can be drawn through point $A$ with a slope of $C_t$ to intersect the convergence void ratio, $e^*$ (i.e., the point where the lines converge).

The second step is to determine the initial matric suction, $(u_a - u_w)^e_0$. A line can be drawn through the convergence void ratio, $e^*$, at a slope of $C_m$, as shown in Fig. 13.25. The line intersects the initial void ratio line (i.e., $e_0$) at the logarithm of the initial matric suction (i.e., $\log (u_a - u_w)^e_0$).

The magnitudes of $P'_s$ and $(u_a - u_w)^e_0$ are used to determine the location of points $B_1$ and $B_2$ along the constant void ratio plane (Fig. 13.26). The straight line of constant void ratio on the arithmetic plot [Fig. 13.26(a)] must be converted to a semi-logarithmic plot, as shown in Fig.

**Table 13.2 A Summary of the Experimentally Measured Volumetric Deformation Indices**[a]

| | One-Dimensional Loading | | | | | | One-Dimensional Unloading | | | | | |
|---|---|---|---|---|---|---|---|---|---|---|---|---|
| Soil Type[b] | $w_0 G_s$ | $e_0$ | $C_t$ or $D_t G_s$ | $C_m$ | $D_m G_s$ | Soil Type[b] | $w_0 G_s$ | $e_0$ | $C_{ts}$[c] | $D_{ts} G_s$[d] | $C_{ms}$ | $D_{ms} G_s$ |
| DS | 0.420 | 0.699 | 0.196 | 0.030 | 0.124 | DS | 0.419 | 0.702 | 0.040 | 0.126 | 0.033 | 0.263 |
| DS | 0.424 | 0.700 | | | | DS | 0.421 | 0.699 | | | | |
| OS | 0.516 | 0.606 | 0.177 | 0.082 | 0.158 | OS | 0.517 | 0.616 | 0.055 | 0.076 | 0.052 | 0.101 |
| OS | 0.511 | 0.609 | | | | OS | 0.514 | 0.609 | | | | |
| DT | 0.427 | 0.642 | 0.206 | 0.089 | 0.159 | DT | 0.436 | 0.693 | 0.066 | 0.084 | 0.056 | 0.122 |
| OT | 0.516 | 0.567 | 0.179 | 0.106 | 0.171 | OT | 0.523 | 0.571 | 0.037 | 0.057 | 0.024 | 0.060 |

*Notes:*
[a] All indices have a negative sign, as described by the sign convention in Chapter 12.
[b] "DS" stands for silt at dry of optimum initial water content. "OS" stands for silt at optimum initial water content. "DT" stands for glacial till at dry of optimum initial water content. "OT" stands for glacial till at optimum initial water content.
[c] Average slope of the unloading curve.
[d] Slope of the linear portion of the unloading curve.

13.2 TEST PROCEDURES AND EQUIPMENTS     385

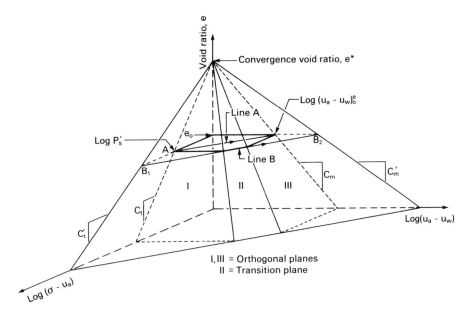

**Figure 13.24** Approximate form for the void ratio constitutive surface on a logarithmic plot.

13.26(b). Log $P'_s$ and log $(u_a - u_w)^e_0$ are joined by line "$A$," which constitutes the chord of the asymptotic curve in Fig. 13.26(b). Line "$B$," tangent to the asymptotic curve, is drawn parallel to line "$A$." Line "$B$" intersects the log $(\sigma - u_a)$ and log$(u_a - u_w)$ axes at points $B_1$ and $B_2$, respectively. As a result, the abscissas of points $B_1$ and $B_2$ along the initial void ratio line are known.

The third step is to draw lines extending from the convergence void ratio, $e^*$, to points $B_1$ and $B_2$ on the initial void ratio [Fig. 13.25]. The slopes of these lines are equal to the $C'_t$ and $C'_m$ indices associated with transition plane II (see Fig. 13.24).

The above procedure is used for obtaining the volume change indices associated with transition plane II on the *semi-logarithmic form* of the void ratio constitutive surface. In the *arithmetic form* of the constitutive surface, the volume change coefficients obtained from the extreme planes (i.e., $(\sigma - u_a) = 0$ plane and $(u_a - u_w) = 0$ plane)

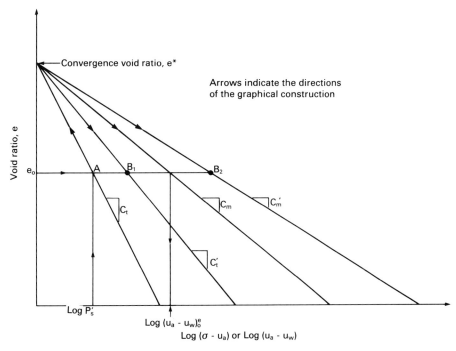

**Figure 13.25** Graphical determination of the volume change indices.

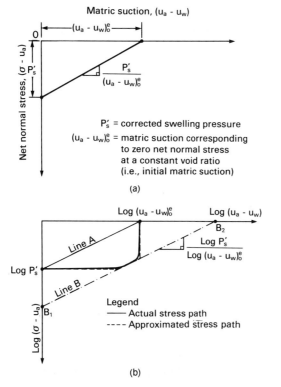

**Figure 13.26** Construction of lines *A* and *B* from "constant volume" oedometer test results. (a) "Constant volume" stress plane on an arithmetic scale; (b) "constant volume" stress plane on a logarithmic scale.

are assumed to be applicable to every state point along a constant void ratio plane or a constant water content plane. The significance of this assumption has been explained in Chapter 12.

## Typical Results from Pressure Plate Tests

Soil–water characteristic curves obtained from pressure plate tests are an important part of the water phase constitutive surface for an unsaturated soil. An unsaturated soil in the field is often subjected to more significant and frequent changes in matric suction, than in total stress. The soil undergoes processes of drying and wetting as a result of climatic changes. On the other hand, the applied total stress on the soil is seldom altered. Therefore, it is important to know the nature of the soil–water characteristic curve of an unsaturated soil in order to predict the water content changes when the soil is subjected to drying or wetting.

Croney and Coleman (1954) have summarized soil–water characteristic curves which illustrate the different behavior observed for incompressible and compressible soils. Figure 13.27 compares the soil–water characteristic curves of soft and hard chalks, which are considered relatively incompressible. The drying curves of these incompressible soils exhibit essentially constant water contents at low matric suctions and rapidly decreasing water contents at higher suctions. The point where the water content starts to decrease significantly indicates the air entry value of the soil. The data show that the hard chalk has a higher air entry value than the soft chalk. The high preconsolidation pressure during the formation of the hard chalk bed results in a smaller average pore size than for the soft chalk.

Another noticeable characteristic is that the drying curves for both hard and soft chalks become identical at high matric suctions (Fig. 13.27). This indicates that at high suctions, both soils have similar pore size distributions. There is a marked hysteresis between the drying and wetting curves for both soils.

The effect of initial water content on the drying curves

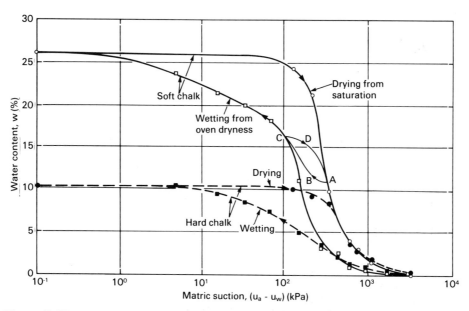

**Figure 13.27** Soil–water characteristic curves for soft and hard chalks with incompressible soil structures (from Croney and Coleman, 1954).

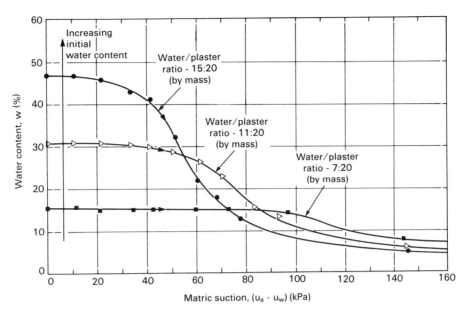

**Figure 13.28** Effects of initial water content on the drying curves of incompressible mixtures (from Croney and Coleman, 1954).

of incompressible mixtures is demonstrated in Fig. 13.28. An increase in the initial water content of the soil results in a decrease in the air entry value. This can be attributed to the larger pore sizes in the high initial water content mixtures. These soils drain quickly at relatively low matric suctions. As a result, the water content in the soil with the large pores is less than the water content in the soil with small pores at matric suctions beyond the air entry value. In other words, soils with a low initial water content (or small pore sizes) require a larger matric suction value in order to commence desaturation. There is then a slower rate of water drainage from the pores.

The initial dry density of incompressible soils has a similar effect on the soil-water characteristic curve, as was illustrated by the initial water contents. As the initial dry density of an incompressible soil increases, the pore sizes are small and the air entry value of the soil is higher, as illustrated in Fig. 13.29. The high-density specimens desaturate at a slower rate than the low-density specimens. As a result, the high-density specimens have higher water contents than the low-density specimens at matric suctions beyond their air entry values. In addition, the hysteresis associated with the high-density specimens is less than the hysteresis exhibited by the low-density specimens.

Croney and Coleman (1954) used the soil-water characteristic curve for London clay (Fig. 13.30) to illustrate the behavior of a compressible soil upon wetting and drying. The gradual decrease in water content upon drying results in the air entry value of the soil being indistinct. In this case, the shrinkage curve of the soil (Fig. 13.31) must be used together with the soil-water characteristic curve in order to determine the air entry value of the soil. The shrinkage curve clearly indicates the compressible nature of the soil. The total and water volume changes caused by an increase in matric suction are essentially equal until the water content reaches 22%. As a result, the shrinkage curve above a water content of 22% is parallel and close to the saturation line, indicating essentially a saturation condition. The soil starts to desaturate when the water content goes below 22%, causing the shrinkage curve to deviate from the saturation line. The void ratio of the soil reaches

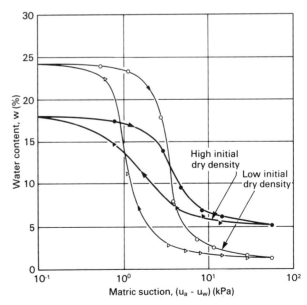

**Figure 13.29** Effect of initial dry density on the soil-water characteristic curves of a compacted silty sand (from Croney and Coleman, 1954).

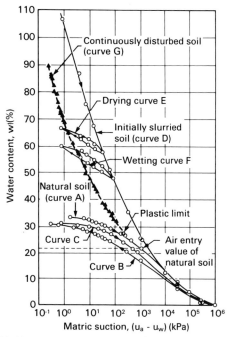

**Figure 13.30** Soil-water characteristic curves for London clay (from Croney and Coleman, 1954).

a limiting value (i.e., $e = 0.48$), corresponding to a water content of 0%. A water content of 22% corresponds to a matric suction in the natural soil of approximately 1000 kPa, as indicated by the soil-water characteristic curve (Fig. 13.30).

Some irreversible structural changes causing an irreversible volume change occur primarily during the first drying process, as indicated by curve $A$ in Fig. 13.30. Subsequent wetting and drying cycles follow curves $B$ and $C$ (Fig. 13.30), respectively. Curves $B$ and $C$ have lower water contents than curve $A$, with the difference indicating irreversible volume change.

Curve $D$ in Fig. 13.30 was obtained from initially slurried specimens where the soil structure was partially disturbed. Curve $A$ for the natural soil joined curve $D$ at a matric suction of 6300 kPa, indicating the maximum suction to which the clay has been subjected during its geological history. The maximum suction has a similar meaning to the preconsolidation pressure of a saturated soil, and this is explained in further detail in Chapter 14. The deviations of the natural soil curves $A$, $B$, and $C$ from the initially slurried soil curve $D$ represent the natural state of disturbance due to past drying and wetting cycles.

Another curve plotted in Fig. 13.30 is curve $G$ that relates the water content to the matric suction for disturbed soil specimens. Curve $G$ is not a soil-water characteristic curve since the points on the curve were obtained from soil specimens with different soil structures. A soil-water characteristic curve must be obtained from a single specimen or several specimens with "identical" initial soil structures. Curve $G$ appears to be unique for London clay, regardless of the matric suction of the soil. State points along the drying curve $D$ or curve $A$ will move to corresponding points on curve $G$ when disturbed at a constant water content. Similarly, state points along any wetting curve will move to curve $G$ when disturbed at a constant water content. Disturbance can take the form of remolding or thoroughly mixing the specimens. Similar relationships to curve $G$ have also been found for other soils (see Fig. 13.7). It can therefore be concluded that there is a unique relationship between water content and matric suction for a disturbed soil, regardless of its soil structure, initial matric suction, or its initial state path (Croney and Coleman, 1954).

A similar relationship is commonly observed in soils compacted at various water contents and dry densities (see Chapter 4). In other words, compacted soils have a unique relationship between water content and matric suction, regardless of the compacted dry density of the specimens.

### Determination of In Situ Stress State Using Oedometer Test Results

One-dimensional oedometer tests are most often used for the assessment of the *in situ* stress state and the swelling properties of expansive, unsaturated soils. The oedometer can only be used to perform testing in the net normal stress plane. Therefore, the assumption is made that it is possible to eliminate the matric suction from the soil and obtain the necessary soil properties and stress state values from the net normal stress plane. The "free-swell" and "constant volume" tests (Fig. 13.2) are two commonly used procedures which first eliminate the soil matric suction.

### "Constant Volume" Test

Let us first consider the "constant volume" oedometer test. In this procedure, the specimen is subjected to a token load and submerged in water. The release of the negative pore-water pressure to atmospheric conditions results in a ten-

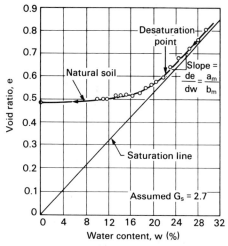

**Figure 13.31** Shrinkage curve for London clay (from Croney and Coleman, 1954).

dency for the specimen to swell. As the specimen tends to swell, the applied load is increased to maintain the specimen at a constant volume. This procedure is continued until the specimen exhibits no further tendency to swell. The applied load at this point is referred to as the "uncorrected swelling pressure," $P_s$. The specimen is then further loaded and unloaded in a conventional manner.

The test results are generally plotted as shown in Fig. 13.2(b). The actual stress paths followed during the test can be more fully understood by use of a three-dimensional plot, with each of the stress state variables forming an abscissa (Fig. 13.32). It is important to understand the stress paths in order to propose a proper interpretation of the test data. The void ratio and water content stress paths are shown for the situation where there is a minimum disturbance due to sampling. Even so, the loading path will display some curvature as the net normal stress plane is approached. In reality, the actual stress path will be even more affected by sampling (Fig. 13.33).

Geotechnical engineers have long recognized the effect of sample disturbance when determining the preconsolidation pressure for a saturated clay. In the oedometer test, it is impossible for the soil specimen to return to an *in situ* stress state after sampling without displaying some curvature in the void ratio versus effective stress plot (i.e., consolidation curve). However, only recently has the significance of sampling disturbance been recognized in the measurement of swelling pressure (Fredlund *et al.*, 1980).

Sampling disturbance causes the conventionally determined swelling pressure, $P_s$, to fall well below the "ideal" or "correct" swelling pressure, $P_s'$. The "corrected" swelling pressure represents the *in situ* stress state translated to the net normal stress plane. The "corrected" swelling pressure is equal to the overburden pressure plus the *in situ* matric suction translated onto the net normal stress plane. The translated *in situ* matric suction is called the "matric suction equivalent," $(u_a - u_w)_e$ (Yoshida *et al.*, 1982.) The magnitude of the matric suction equivalent will be equal to or lower than the *in situ* matric suction. The difference between the *in situ* matric suction and the matric suction equivalent is primarily a function of the degree of saturation of the soil. The engineer desires to obtain the "corrected" swelling pressure from an oedometer test in order to reconstruct the *in situ* stress conditions. The procedure to accounting for sampling disturbance is discussed later.

### "Free-Swell" Test

In the "free-swell" type of oedometer test, the specimen is initially allowed to swell freely, with only a token load applied (Fig. 13.2(a) and Fig. 13.34). The load required to bring the specimen back to its original void ratio is termed the swelling pressure. The stress paths being followed can best be understood using a three-dimensional plot of the stress variables versus void ratio and water content, as shown in Fig. 13.34. This test has the limitation that it allows volume change and incorporates hysteresis into the estimation of the *in situ* stress state. On the other hand, this testing procedure somewhat compensates for the effect of sampling disturbance.

### Correction for the Compressibility of the Apparatus

The following procedure is suggested for obtaining the "corrected" swelling pressure from "constant volume"

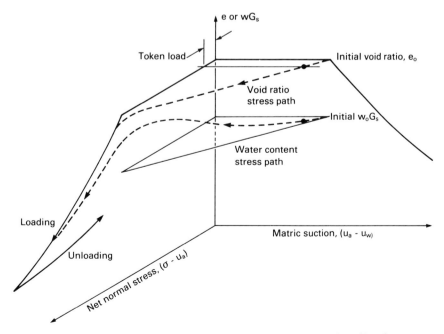

**Figure 13.32** "Ideal" stress path representation for a "constant volume" oedometer test.

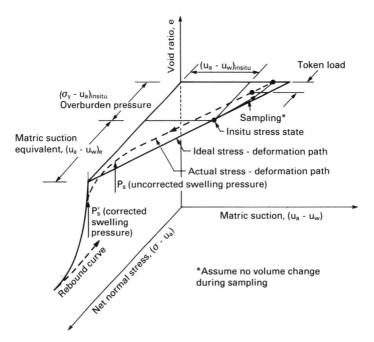

**Figure 13.33** Ideal and actual stress-deformation paths showing the effect of sampling disturbance.

oedometer test results. Detailed testing procedures are presented in ASTM D4546. When interpreting the laboratory data, an adjustment should be made to the data in order to account for the compressibility of the oedometer apparatus. Desiccated, swelling soils have a low compressibility, and the compressibility of the apparatus can significantly affect the evaluation of *in situ* stresses and the slope of the rebound curve (Fredlund, 1969).

Because of the low compressibility of the soil, the compressibility of the apparatus should be measured using a steel plug substituted for the soil specimen. The measured deflections should be subtracted from the deflections measured when testing the soil. Figure 13.35 shows the manner in which an adjustment should be applied to the laboratory data. The adjusted void ratio versus pressure curve can be sketched by drawing a horizontal line from the initial void ratio, which curves downward and joins the recompression curve adjusted for the compressibility of the apparatus.

### Correction for Sampling Disturbance

Second, a correction can now be applied for sampling disturbance. Sampling disturbance increases the compressibility of the soil, and does not permit the laboratory specimen to return to its *in situ* state of stress at its *in situ* void ratio. Casagrande (1936) proposed an empirical construction on the laboratory curve to account for the effect of

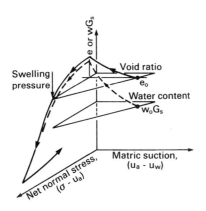

**Figure 13.34** Stress path representation for the "free-swell" type of oedometer test.

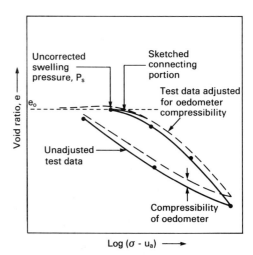

**Figure 13.35** Adjustment of laboratory test data for the compressibility of the oedometer apparatus.

sampling disturbance when assessing the preconsolidation pressure of a soil. Other construction procedures have also been proposed (Schmertmann, 1955). A modification of Casagrande's construction is suggested for determining the "corrected" swelling pressure.

The following procedure is suggested for the determination of the "corrected" swelling pressure. Locate the point of maximum curvature where the void ratio versus pressure curve bends downward onto the recompression branch (Fig. 13.36). At the point of maximum curvature, a horizontal line and a tangential line are drawn. The "corrected" swelling pressure is designated as the intersection of the bisector of the angle formed by these lines and a line parallel to the slope of the rebound curve which is placed in a position tangent to the loading curve.

The need for applying a correction to the swelling pressure measured in the laboratory, is revealed in numerous ways. First, it would be anticipated that such a correction is necessary as a result of early experience in determining the preconsolidation pressure for normally consolidated soils. Second, attempts to use the "uncorrected" swelling pressure in the prediction of total heave commonly result in predictions which are too low. Predictions using "corrected" swelling pressures may often be twice the magnitude of those computed when no correction is applied. Third, the analysis of oedometer results from desiccated deposits often produces results which are difficult to interpret if no correction is applied for sampling disturbance.

Figure 13.37 shows an average oedometer curve obtained from 34 tests performed on Regina clay. The deposit is of preglacial lacustrine origin, and the natural water contents are near the plastic limit (Fredlund *et al.*, 1980). The average liquid limit is 75%. The climate of the region is semi-arid, and there is no evidence of a regional groundwater table in the deposit. The soil is very stiff, and would be anticipated to have high swelling pressures. The oedometer results show, however, that if a correction for sam-

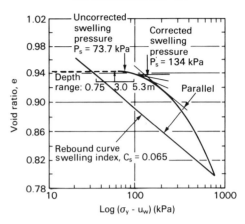

**Figure 13.37** Average data from oedometer tests on Regina clay illustrating the need for the swelling pressure correction.

pling disturbance is not applied, the swelling pressure is only slightly in excess of the average overburden pressure. This soil could easily be misinterpreted as a low swelling clay. However, swelling problems are common, with a total heave in the order of 5–15 cm. Samples from depths deeper than 5.5 m often show "uncorrected" swelling pressures less than the overburden pressure. In other words, the correction for sampling disturbance is imperative to the interpretation of the *in situ* stress state of the soil.

Figure 13.38 shows a comparison of "corrected" and "uncorrected" swelling pressure data from two soil deposits. The results indicate that it is possible for the "cor-

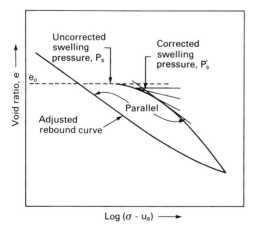

**Figure 13.36** Construction procedure to correct for the effect of sampling disturbance.

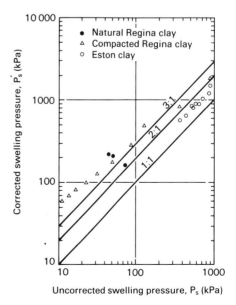

**Figure 13.38** Change in swelling pressure due to applying the correction for sampling disturbance.

rected'' swelling pressures to be more than 300% of the "uncorrected" swelling pressures.

### 13.2.2 Unloading Constitutive Surfaces

The unloading constitutive surfaces are illustrated in Fig. 13.39(a) and (c) for the void ratio and water content surfaces, respectively. The intersection curves 1 and 2 from the void ratio surface [Fig. 13.39(a)] are combined in Fig. 13.39(b). The slopes of curves 1 and 2 are called the $C_{ts}$ and $C_{ms}$ volume change indices, respectively. The intersection curves 3 and 4 from the water content surface [Fig. 13.39(c)] are combined in Fig. 13.39(d) using the variable, $wG_s$, as the ordinate. Therefore, the slopes of curves 3 and 4 in Fig. 13.39(d) are also the product of the specific gravity and the volume change indices (i.e., ($D_{ts} G_s$) and ($D_{ms} G_s$), respectively).

The following discussions outline the test procedures that can be used to obtain the four indices associated with the unloading constitutive surfaces (i.e., $C_{ts}$, $C_{ms}$, $D_{ts}$, and $D_{ms}$). Tests similar to those described in Section 13.2.1 are also applicable to the unloading surface when the tests are conducted in an unloading mode.

#### Unloading Tests after Compression

Curve 1 of the unloading surface can be obtained from the "free-swell" and "constant volume" oedometer tests, as illustrated in Fig. 13.40. Curve 1 connects the void ratio ordinates at the end of the "free-swell" tests. Curve 1 is essentially parallel to the rebound curve, corresponding to the unloading portion of the test at a lower void ratio (Fig. 13.41). The rebound curves are approximately parallel to one another and can be linearized on a semi-logarithmic scale (Schmertmann, 1955; Holtz and Gibbs, 1956; Gilchrist, 1963; Noble, 1966; Lambe and Whitman, 1979; Lidgren, 1970; Chen, 1975). The slope of the rebound curve is referred to as the swell index, $C_s$, which is significantly smaller than the compression index, $C_c$.

The slope of curve 1 (i.e., the $C_{ts}$ index) can be considered to be essentially equal to the $C_s$ index from the rebound curves. As a result, the $C_{ts}$ index from the unloading constitutive surface is obtained by performing an unloading test after completion of the compression portion, in accordance with the conventional test procedures for saturated specimens (ASTM D2435, 1985).

The swelling index, $C_s$, will generally range between 10-

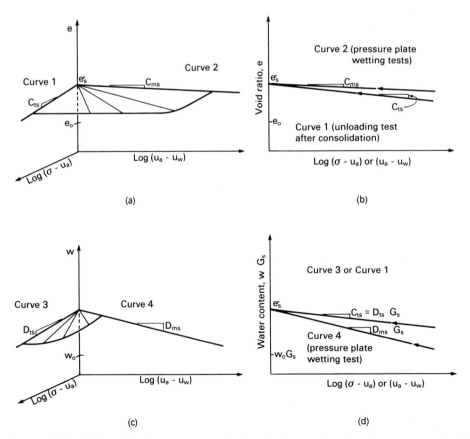

**Figure 13.39** Void ratio and water content relationship during unloading of an unsaturated soil. (a) Void ratio relationship for unloading; (b) Intersection curves between void ratio surface and ($\sigma - u_a$) or ($u_a - u_w$) plane; (c) water content relationship for unloading; (d) Intersection curves between water content surface and ($\sigma - u_a$) or ($u_a - u_w$) plane.

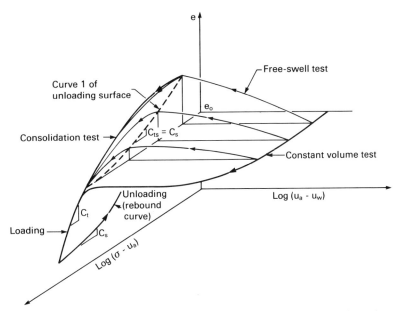

**Figure 13.40** Location of curve 1 of the unloading surface as obtained from the oedometer test results.

20% of the compressive index, $C_c$, for a particular soil. Figure 13.42 shows approximate swelling index values which have been correlated with the liquid limit and the rebound void ratio of a soil (NAVFAC, 1971). The plot is useful for obtaining an estimate of the swelling index.

Curve 3 in Fig. 13.39(d) coincides with curve 1 from Fig. 13.39(b) since $wG_s$ is equal to the void ratio, $e$, when the soil is saturated. Therefore, the unloading water content index, $D_{ts}$, can be computed as $(C_{ts}/G_s)$.

*Pressure Plate Wetting Tests*

Curve 4 in Fig. 13.39 is called the wetting portion of the soil–water characteristic curve. The wetting curve can be established by performing pressure plate tests on specimens after the drying portion has been completed, as explained in Section 13.2.1. The test procedures and equipments are similar to those used in the drying tests (ASTM D2325). The specimen is equilibrated to a lower matric suction by decreasing the air pressure in the pressure plate extractor. As a result, water from the compartment below the high air entry disk moves into the specimen, causing an increase in water content. The time required for water to be drawn into the specimen can be substantial and care must be taken to ensure that complete equilibrium has been attained. The equilibrium water contents are then plotted against the corresponding matric suctions to establish wet-

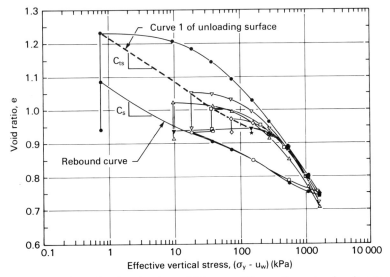

**Figure 13.41** Two-dimensional projections of "free-swell" one-dimensional oedometer data for compacted Regina clay (from Gilchrist, 1963).

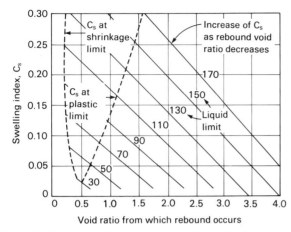

**Figure 13.42** Approximate correlation of swelling index versus rebound void ratio (from NAVFAC DM-7, 1971).

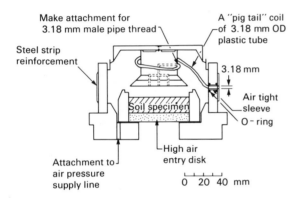

**Figure 13.43** Schematic layout of the modified loading cap of the Anteus oedometer.

ting curve 4, as shown in Fig. 13.39(d). The slope of the wetting curve 4 is equal to $(D_m G_s)$.

Curve 2 in Fig. 13.39 can also be constructed from the pressure plate test results when void ratio measurements are made at each point of matric suction equilibrium. The measurements can be made by reducing the air pressure in the extractor to zero, dismantling the extractor, and measuring the total volume of the specimen. The measurements must be made as quickly as possible in order to prevent changes in the water content of the soil. Having measured the total volume, the specimen is placed back into the extractor, and the test is continued at a lower matric suction value. The computed void ratios at each equilibrium point are plotted against the corresponding matric suctions to give curve 2 in Fig. 13.39(b). The slope of curve 2 is equal to the volume change index, $C_{ms}$.

### Free-Swell Tests

Curves 2 and 4 in Fig. 13.39 can also be obtained by conducting a "free-swell" oedometer test (see Fig. 13.40) with void ratio and water content measurements. In this case, more specialized equipment such as a modified Anteus oedometer for the $K_0$-loading condition (Fig. 13.43) is required. The modified Aneus oedometer allows the control of total, pore–air, and pore–water pressures and the measurement of total and water volume changes during the tests. The soil specimen can also be wetted by injecting water through hypodermic needles installed in the loading cap. This procedure was used by Ho (1988) in an attempt to expedite the entrance of water into the specimen.

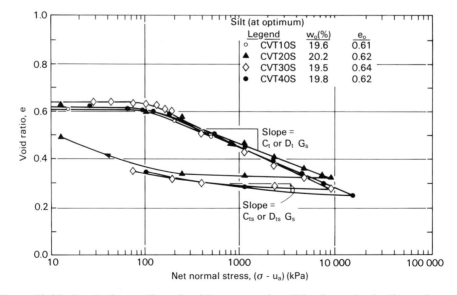

**Figure 13.44** Results for one-dimensional "constant volume" loading and unloading oedometer tests on silt compacted at optimum water content.

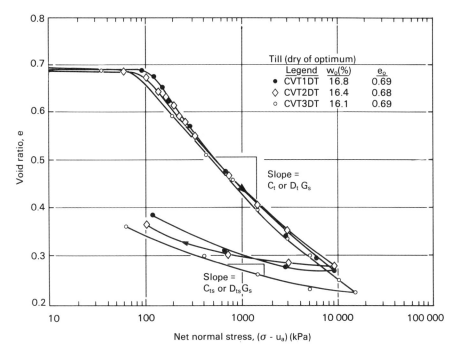

**Figure 13.45** Results for one-dimensional "constant volume" loading and unloading oedometer tests on till compacted dry of optimum water content.

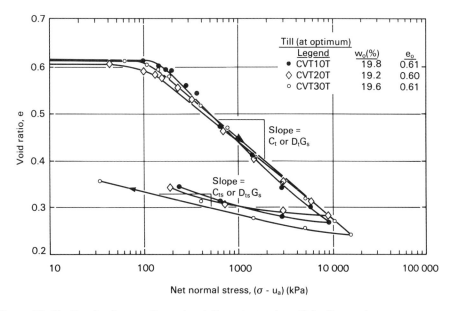

**Figure 13.46** Results for one-dimensional "constant volume" loading oedometer tests on till compacted at optimum water content.

### Determination of Volume Change Indices

The silt and glacial till described in Table 13.1 were also tested by Ho and Fredlund (1989) to determine the volume change indices associated with the unloading constitutive surfaces. The rebound curves from the unloading portion of the test are presented in Figs. 13.17 and 13.44 for the silt specimens, and in Figs. 13.45 and 13.46 for the glacial till specimens. The slopes of the rebound curves are equal to the volume change index $C_{ts}$ or $(D_{ts}G_s)$. The results of the "free-swell" tests with void ratio and water content measurements on the silt and glacial till specimens are presented in Figs. 13.47 and 13.48, respectively. The tests

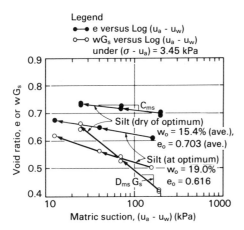

**Figure 13.47** Unloading portion results for one-dimensional "free-swell" oedometer tests on silt specimens.

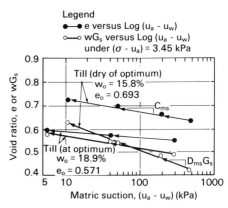

**Figure 13.48** Unloading portion results for one-dimensional "free-swell" oedometer tests on till specimens.

were conducted using the modified Anteus consolidometer. Void ratios and water contents are plotted against the logarithms of matric suction. The slope of the void ratio versus log $(u_a - u_w)$ curve is equal to the $C_{ms}$ index, while the slope of the $(wG_s)$ versus log $(u_a - u_w)$ curve is equal to the $(D_{ms}G_s)$ index. As a result, all four indices associated with the unloading surface (i.e., $C_{ts}$, $C_{ms}$, $D_{ts}$, and $D_{ms}$) are obtainable. These indices can be converted to other volume change coefficients, such as "$m_{1s}$ and $m_{2s}$" or the "$a_s$ and $b_s$" coefficients, as explained in Chapter 12.

# CHAPTER 14

## *Volume Change Predictions*

An unsaturated soil will undergo volume change when the net normal stress or the matric suction variable changes in magnitude. The volume change theory and the modulus measurements presented in Chapters 12 and 13, respectively, can be used to calculate volume changes in an unsaturated soil. Under a constant total stress, an unsaturated soil will experience swelling and shrinking as a result of matric suction variations associated with environmental changes. In collapsible soils, the collapse phenomenon occurs when the matric suction of the soil decreases.

In this chapter, the methodology for the prediction of heave in a swelling soil is described in detail. The stress history of a soil is an important factor to consider in understanding the swelling behavior. The formulations and example problems for heave prediction are presented and supplemented with two case histories. A detailed discussion on the factors influencing the amount of heave is also included. At the end, there is a brief note on collapsible soils and methods to predict the amount of collapse.

## 14.1 LITERATURE REVIEW

Expansive soils are found in many parts of the world, particularly in semi-arid areas. An expansive soil is generally unsaturated due to desiccation. Expansive soils also contain clay minerals that exhibit high volume change upon wetting. The large volume change upon wetting causes extensive damage to structures, in particular, light buildings and pavements. In the United States alone, the damage caused by the shrinking and swelling soils amounts to about $9 billion per year, which is greater than the combined damages from natural disasters such as floods, hurricanes, earthquakes, and tornadoes (Jones and Holtz, 1987). Therefore, the problems associated with swelling soils are of enormous financial proportions.

Table 14.1 summarizes examples of causes for foundation heave as a result of the changes in the water content of the soil. These changes can originate from the environment or from man-made causes. Nonuniform changes in water content will result in differential heaves which can cause severe damage to the structure. In fact, the differential heave experienced by a light structure is often of similar magnitude to the total heave.

The heave potential of a soil depends on the properties of the soil, such as clay content, plasticity index, and shrinkage limit. In addition, the heave potential depends upon the initial dryness or matric suction of the soil. Many empirical methods have been proposed to correlate the swelling potential of a soil to properties such as are shown in Table 14.2 and Fig. 14.1. These relationships are useful for identifying the swelling potential of a soil. In other words, these correlations reflect one component of the potential magnitude of heave.

The amount of total heave can also be written as a function of the difference between the present *in situ* stress state and some future stress state and the volume change indices for the soil. In general, the net normal stress state variable remains constant, while the matric suction stress state variable changes during the heave process. Matric suction changes result in changes in water content (Table 14.1). Therefore, total heave can be predicted by measuring the *in situ* matric suction and estimating or predicting the future matric suction in the field under specific environmental conditions. The volume change indices with respect to matric suction changes must be measured in accordance with the procedures outlined in Chapter 13.

There are several heave formulations related to the volume change indices which have been proposed by various researchers (Table 14.3). These formulations differ primarily in the manner in which strain and soil suction are defined.

The prediction of heave on the basis of matric suction measurements has not been extensively used due to difficulties associated with accurate measurements of matric suction and appropriate soil properties. More common are the methods for heave prediction based on one-dimensional

## Table 14.1 Examples of Causes for Foundation Heave Resulting from Soil Water Content Changes (from Headquarters, U.S. Department of the Army, 1983)

| | |
|---|---|
| Changes in field environment from natural conditions | 1) Significant variations in climate, such as long droughts and heavy rains, cause cyclic water content changes resulting in edge movement of structures.<br>2) Changes in depth to the water table lead to changes in soil water content. |
| Changes related to construction | 1) Covered areas reduce natural evaporation of moisture from the ground, thereby increasing the soil water content.<br>2) Covered areas reduce transpiration of moisture from vegetation, thereby increasing the soil water content.<br>3) Construction on a site where large trees were removed may lead to an increase of moisture because of prior depletion of soil water by extensive root systems.<br>4) Inadequate drainage of surface water from the structure leads to ponding and localized increases in soil water content. Defective rain gutters and downspouts contribute to localized increases in soil water content.<br>5) Seepage into foundation subsoils at soil/foundation interfaces and through excavations made for basements or shaft foundations leads to increased soil water content beneath the foundation.<br>6) Drying of exposed foundation soils in excavations and reduction in soil surcharge weight increases the potential for heave.<br>7) Aquifers tapped provide water to an expansive layer of soil. |
| Usage effects | 1) Watering of lawns leads to increased soil water content.<br>2) Planting and growth of heavy vegetation, such as trees, at distances from the structure less than 1–1.5 times the height of mature trees, aggravates cyclic edge heave.<br>3) Drying of soil beneath heated areas of the foundation, such as furnace rooms, leads to soil shrinkage.<br>4) Leaking underground water and sewer lines can cause foundation heave and differential movement. |

## Table 14.2 Probable Expansion as Estimated from Classification Test Data[a] (from Holtz and Kovacs, 1981)

| Degree of Expansion | Probable Expansion as a % of the Total Volume Change (Dry to Saturated Condition)[b] | Colloidal Content (% – 1 μm) | Plasticity Index, PI | Shrinkage Limit, $w_S$ |
|---|---|---|---|---|
| Very high | >30 | >28 | >35 | <11 |
| High | 20–30 | 20–31 | 25–41 | 7–12 |
| Medium | 10–20 | 13–23 | 15–28 | 10–16 |
| Low | <10 | <15 | <18 | >15 |

[a] After Holtz (1959) and U.S.B.R. (1974).
[b] Under a surcharge of 6.9 kPa (1 psi).

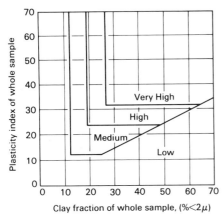

**Figure 14.1** Correlation between soil properties and swelling potential (from van de Merwe [1964]).

oedometer test results. In the oedometer methods, matric suction measurements are not required. A list of the various methods utilizing the oedometer test results is presented in Table 14.4. Three of these methods are briefly discussed herein.

The *direct model method* is based on a "free-swell" oedometer test on undisturbed samples (Fig. 14.2). The specimens are subjected to the overburden pressure (or the load that will exist at the end of construction) and allowed free access to water. The predicted heave is generally significantly below the actual heave experienced in the field. The stress path followed by the test procedure is shown in Fig. 14.2(b). The conventional two-dimensional manner for plotting the test data is shown in Fig. 14.2(a). The underestimation of the amount of heave appears to be primarily

**Table 14.3 Definitions of Volume Change Indices with Respect to Suction Changes (from Hamberg, 1985)**

| Reference | Symbol | Definition | Typical Values | Formation or Soil Type | Location |
|---|---|---|---|---|---|
| Fredlund, 1979 | $C_m$ | Slope of void ratio versus log matric suction: $C_m = \Delta e / \Delta \log(u_a - u_w)$ | 0.1–0.2 | Regina clay | Canada |
| Johnson, 1977 | $C_\tau$ | Slope of void ratio versus log matric suction, approximated by $C_\tau = \alpha G_s / 100 B$ where $\alpha$ = compressibility factor ($0 < \alpha < 1$), $B$ = slope of the suction versus water content relationship | 0.07 0.15–0.21 0.09–0.23 0.07–0.15 0.13–0.29 | Loess Yazoo clay Upper Midway Pierre Shale Marine clay | Mississippi Mississippi Texas Colorado Sicily |
| Lytton, 1977 McKeen, 1981 | $\gamma_h$ | Slope of the volumetric strain versus the log of total suction: $\gamma_h = \dfrac{\Delta e/(1 + e_o)}{\Delta \log h}$ where $\Delta e/(1 + e_o)$ = volumetric strain, $h$ = total suction | 0.02–0.18 0.02–0.20 0.05–0.22 | Engleford Yazoo clay Mancos | Texas Mississippi New Mexico |
| Aitchison and Martin (1973) | $I_{pt}''$ | Slope of vertical strain versus the log of total suction; $I_{pt}'' = \epsilon_v / \Delta \log h$ | 0–0.08 | | |
| Fargher et al. (1979) | $I_{pm}''$ | Slope of vertical strain versus the log of matric suction: $I_{pm}'' = \epsilon_v / \Delta \log(u_a - u_w)$ | 0–0.11 | Red–Brown Clay | Adelaide, S. Australia |
| | $I_{ps}''$ | Slope of vertical strain versus the log of solute (osmotic) suction: $I_{ps}'' = \epsilon_v / \Delta \log \pi$ | 0–0.20 | | |
| Grossman et al., 1968 U.S.D.A. Soil Conservation Service, 1971 | COLE | Value of linear strain corresponding to a suction change from 33 kPa to oven dry: $\text{COLE} = \Delta L / L_D = (\gamma_D / \gamma_W)^{1/3} - 1$ where $\Delta L / L_D$ = linear strain relative to dry dimensions, $\gamma_D$ = bulk density of oven dry sample, $\gamma_W$ = bulk density of sample at 33 kPa suction | 0–0.17 | Western and Midwestern U.S. soils | |

Table 14.4 Various Heave Predictions Methods Utilizing Oedometer Test Results

| Name of Method | Reference | | |
|---|---|---|---|
| | Year | Author | Country |
| 1) Double oedometer method | 1957 | Jennings, J.E. and Knight, K. | South Africa |
| | 1969 | Jennings, J.E.B. | |
| 2) Salas and Serratosa method | 1957 | Salas, J.A.J. and Serratos, J.M. | Spain |
| 3) Volumeter method | 1961 | DeBruijn, C.M.A. | South Africa |
| 4) Mississippi method | 1962 | Clisby, M.B. | U.S. |
| | 1972 | Teng, T.C., Mattox, R.M., and Clisby, M.B. | |
| | 1973 | Teng, T.C., Mattox, R.M., and Clisby, M.B. | |
| | 1975 | Teng, T.C. and Clisby, M.B. | |
| 5) Sampson, Schuster, and Budge's method | 1965 | Sampson, E., Schuster, R.L., and Budge, W.D. | U.S. |
| 6) Noble method | 1966 | Noble, C.A. | Canada |
| 7) Sullivan and McClelland method | 1969 | Sullivan, R.A., and McClelland, B. | U.S. |
| 8) Holtz method | 1970 | Holtz, W.G. | U.S. |
| 9) Navy method | 1971 | NAVFAC | U.S. |
| 10) Direct model method (Texas Highway Department) | 1973 | Smith, A.W. | U.S. |
| 11) Simple oedometer method | 1973 | Jennings, J.E., Firth, R.A., Ralph, T.K., and Nager, N. | South Africa |
| 12) U.S.B.R. method | 1973 | Gibbs, H.J. | U.S. |
| 13) Fredlund, Hasan and Filson's method | 1980 | Fredlund, D.G., Hasan, J.U., and Filson, H. | Canada |

due to a lack of consideration of disturbance which has been experienced by the soil during sampling.

The *Sullivan and McClelland method* is based on a "constant volume" oedometer test on an undisturbed sample initially subjected to the overburden pressure. Once the swelling pressure has been reached, the sample is rebounded. The stress path followed is shown in Fig. 14.3. The availability of published case histories is limited, but it is anticipated that this method would underestimate the amount of heave since sampling disturbance has not been taken into account.

The *double oedometer method* is based on the results of two oedometer tests, namely, a "free-swell" oedometer test and a "natural water content" oedometer test. The specimens are initially subjected to a token load of 1 kPa. No water is added to the oedometer pot during the "natural water content" test. The "natural water content" oedometer test data are adjusted vertically to match the "free-swell" test results at high applied loads. Various loading conditions and final pore–water pressures can be simulated in the analysis. The stress paths followed by the two tests are shown in Fig. 14.4. The predicted heave is generally satisfactory since the method of analyzing the data appears to compensate for the effects of sampling disturbance. In other words, the natural water content curve defines the effect of sampling disturbance. The stress paths of more recent, updated versions of the double oedometer method can also be visualized on similar three-dimensional plots in terms of net normal stress and matric suction.

Fredlund *et al.* (1980) proposed the use of "constant volume" oedometer test results in predicting total heave. It was suggested that the measured swelling pressure be corrected for sampling disturbance. A graphical technique for correcting the measured swelling pressure was proposed (see Chapter 13). The correction was similar in procedure to Casagrande's construction for determining the precon-

14.1 LITERATURE REVIEW   401

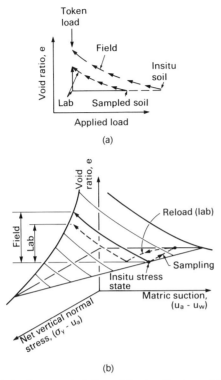

**Figure 14.2** Stress path followed in the direct model method. (a) Two-dimensional plot showing the stress path followed in the field and in the laboratory; (b) three-dimensional plot of the stress path.

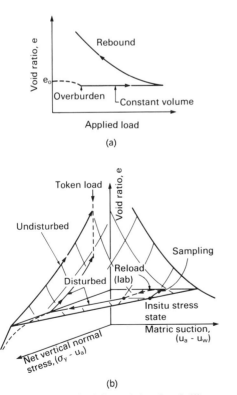

**Figure 14.3** Stress path followed in the Sullivan and McClelland method. (a) Two-dimensional plot of the stress path; (b) three-dimensional plot of the stress path.

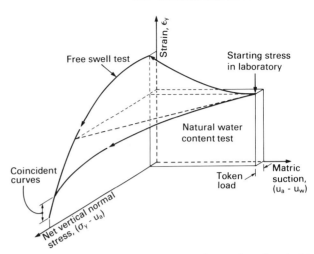

**Figure 14.4** Stress paths followed when using the double oedometer method (Jennings and Knight, 1957).

solidation pressure. The details of the procedure are explained later in this chapter.

The remainder of the procedures listed in Table 14.4 will not be explained. However, each procedure could be plotted on a three-dimensional plot with the stress state variables used for the horizontal axes. Later, a method for the prediction of heave will be outlined which is consistent with the fundamentals of the unsaturated soil theory of volume change (Fredlund et al., 1980).

### 14.1.1 Factors Affecting Total Heave

The preceding review of the literature on swelling soil behavior indicates that there are three primary factors controlling total heave, namely, the volume change indices and the present and future stress state variables. The properties of the soil have been shown to influence the volume change indices. For example, soils compacted at various densities and water contents produce different soil structures which have different volume change indices. In addition, various densities and water contents also affect the magnitude of the stress state of compacted soils. Soils compacted at low densities and high water contents (i.e., wet of optimum water content) have a low matric suction value as compared to the soils compacted dry of optimum water content. Therefore, it can be concluded that the density and water content conditions in a soil affect both the volume change indices and the stress state. These, in turn, control the amount of total heave.

Chen (1988) studied the effect of initial water content and dry density of compacted soils on the amount of total heave. The study was conducted using "free-swell" oedometer tests on expansive shales from Denver, CO. The shale had 63% silt and 37% clay, with a liquid limit and plasticity index of 44.4 and 24.4%, respectively. The results indicate that total heave increases with a decrease in the initial water content of specimens compacted at a con-

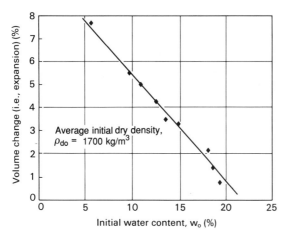

**Figure 14.5** Effect on total heave of varying the initial water content for specimens of constant initial dry density (from Chen, 1988).

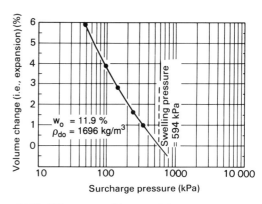

**Figure 14.7** Effect on total heave of the surcharge pressure for a specimen at a specified density and water content (from Chen, 1988).

stant initial dry density (Fig. 14.5). On the other hand, total heave increases with increasing initial dry density for specimens compacted at a constant initial water content (Fig. 14.6).

In addition to soil properties such as dry density and water content, the amount of total heave is also a function of the total stress applied to the soil specimen. Figure 14.7 demonstrates the influence of the surcharge pressure during an oedometer test on the amount of swelling. The results show that total heave decreases with an increasing surcharge pressure. In other words, it is possible to reduce the amount of swelling by applying a high total stress to the soil. Similar observations were reported by Dakshanamurthy (1979) as illustrated in Fig. 14.8. In this case, a commercial sodium montmorillonite was statically compacted and tested in a triaxial apparatus. The specimens were first subjected to various major and minor principal stresses, and then allowed to swell by imbibing water. The results in Fig. 14.8 indicate that total heave decreases with increasing mean normal stress, while the ratio of the principal stresses has an insignificant effect on the total heave.

Holtz and Gibbs (1956) summarized the effect of initial water content and dry density on the total heave of a compacted expansive clay (Fig. 14.9). The heave was obtained when the soil in an oedometer ring was submerged in water. The specimens were subjected to a surcharge load of 7 kPa. The diagram indicates an increasing total heave with respect to a decreasing initial water content or an increasing initial dry density. As an example, a clay compacted to optimum water content under standard AASHTO compaction will expand about 3% of its volume when saturated under a surcharge load of 7 kPa. On the other hand, expansion is reduced to zero at 3% wet of optimum water content, and is increased to 6% at a water content 3% dry of optimum. A similar diagram was also presented for the effect of dry density and water content on vertical swelling pressure (Fig. 14.10). The vertical swelling pressure was

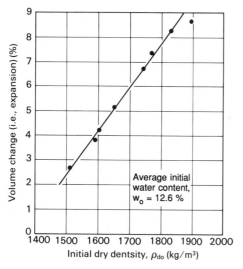

**Figure 14.6** Effect on total heave of the initial dry density for specimens, of constant initial water content (from Chen, 1988).

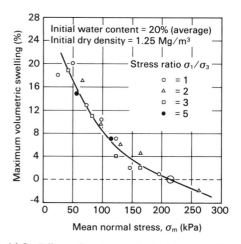

**Figure 14.8** Effect of various principal stress ratios on total heave (from Dakshanamurthy, 1979).

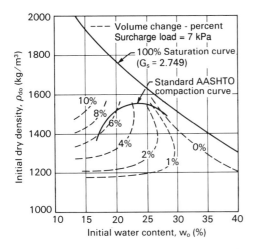

**Figure 14.9** Effect of initial water content and dry density on the expansion properties of compacted Porterville clay when wetted (from Holtz and Gibbs, 1956).

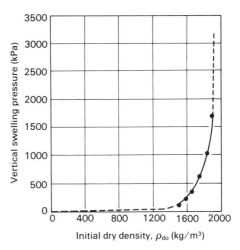

**Figure 14.11** Effect of initial dry density on swelling pressure (from Chen, 1988).

defined as the pressure developed in a specimen placed in an oedometer ring and saturated without allowing any volume change. The diagram shows that the vertical swelling pressure is more sensitive to variations in the initial dry density than it is to variations in initial water content.

Chen (1988) also studied the effect of volume–mass properties on vertical and lateral swelling pressures. The vertical swelling pressure was found to be essentially independent of the initial water content and the surcharge load. This finding is in agreement with the observations made by Holtz and Gibbs (1956), as shown in Fig. 14.10. The vertical swelling pressure, however, increases exponentially with an increasing dry density, as illustrated in Fig. 14.11. On the other hand, the lateral swelling pressure was found to be a function of the initial dry density, the degree of saturation, and the surcharge load. The experimental results indicate that the lateral swelling pressure increased rapidly to a peak value during saturation, and then decreased to an equilibrium state, as demonstrated in Fig. 14.12. This observation is in agreement with the results of extensive research into lateral swelling pressure behavior conducted by Katti *et al.* (1969).

## 14.2 PAST, PRESENT, AND FUTURE STATES OF STRESS

The prediction of volume change requires information on possible changes in the stress state and the soil property, known as a volume change index. Oedometer tests explained in Chapter 13 are commonly performed to determine the present *in situ* state of stress and the volume change indices. In this test, the present *in situ* state of stress is translated onto the net normal stress plane, and is referred to as the "corrected" swelling pressure. The mea-

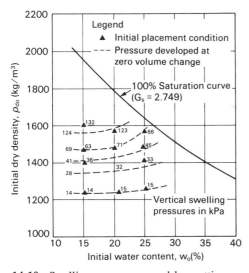

**Figure 14.10** Swelling pressure caused by wetting compacted Porterville clay at various placement conditions (from Holtz and Gibbs, 1956).

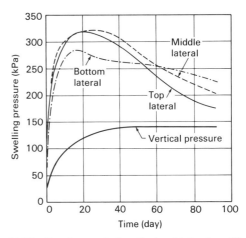

**Figure 14.12** Development of vertical and lateral swelling pressures with time (from Chen, 1988).

sured and corrected swelling pressure represents the sum of the overburden pressure and the matric suction equivalent. The matric suction equivalent is the matric suction of the soil translated onto the net normal stress plane. Thus, the oedometer test measures the *in situ* state of stress (on the total stress plane) without having to measure the individual components of stress.

### 14.2.1 Stress State History

The development of an expansive soil can be visualized in terms of changes in the stress state in the deposit. Changes in stress state occur during geological deposition, erosion, and during environmental changes resulting from precipitation, evaporation, and evapotranspiration. The following example illustrates the stress history of a preglacial lake sediment that has evolved to become an expansive soil over a period of time.

Consider a preglacial lake deposit that was initially consolidated under its own self weight. The drainage of the lake and the subsequent evaporation of water from the lake sediments result in desiccation of the deposit. The term "desiccation" refers to the drying of soils by evaporation and evapotranspiration. The water table is simultaneously drawn below the ground surface. As a result, the pore-water pressures above the water table decrease to negative values, while the total stress in the deposit remains essentially constant. In other words, the effective stress in the soil increases, and consolidation takes places. The negative pore-water pressures act in all directions (i.e., isotropically), resulting in a tendency for cracking and overall desaturation of the upper portion of the profile (Fig. 14.13).

The soil is further desaturated as a result of the growth of grass, trees, and other plants on the ground surface. Most plants are capable of applying as much as 10–20 atm of tension to the water phase prior to reaching their wilting point. A high tension in the water phase (i.e., high matric suction) causes the soil to have a high affinity for water [Fig. 14.13(a)].

The surface deposit can also be subjected to varying and changing environmental conditions. The changing water flux (i.e., wetting and drying) at the surface results in the swelling and shrinking of the upper portion of the deposit. Volume changes might extend to depths in excess of 3 m, causing the surface deposit to be highly desiccated.

Figure 14.14 illustrates the changes in the stress state of the surface deposit during wetting and drying due to infiltration and evaporation, respectively. Changes occur primarily on the matric suction plane at an almost constant net normal stress. The stress paths followed during the wetting and drying processes can be illustrated as hysteresis loops on the matric suction plane.

In arid and semi-arid regions, the natural water content in a deposit tends to decrease gradually with time. Low water content conditions in an unsaturated clay deposit indicate that the soil has a high swelling potential. Unsaturated soils with a high swelling index, $C_s$, in a changing environment are referred to as highly swelling and shrinking soils (i.e., expansive soils).

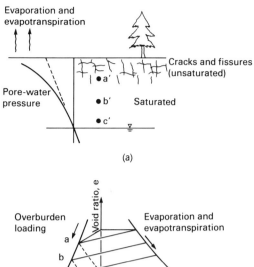

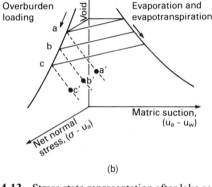

**Figure 14.13** Stress state representation after lake sediments are subjected to evaporation and evapotranspiration. (a) Pore-water pressures during drying of a lacustrine deposit; (b) stress-state path during drying of a lacustrine deposit.

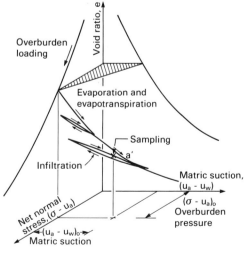

**Figure 14.14** Stress state representation when soil has undergone a complex stress history caused by drying and wetting.

### 14.2.2 In Situ Stress State

The prediction of heave requires information on the present *in situ* stress state and the possible future stress state. The difference between the present and future stress states is one of the main variables which indicates the amount of volume change or heave that can potentially occur. Therefore, it is important that the present *in situ* stress state be accurately assessed. When using laboratory oedometer tests for assessing the present *in situ* stress state, it is imperative to correct the results for sampling disturbance. In this book, most attention is given to the use of laboratory oedometer test results for the assessment of the *in situ* stress state. This procedure appears to be satisfactory as long as the oedometer test can be performed on specimens which adequately represent the *in situ* soil mass. When the soil mass is highly fissured and cracked, it is difficult to obtain representative samples. Under these conditions, it is prudent to give more attention to heave analysis procedures which measure the *in situ* suction and the three-dimensional volume change modulus for the soil.

When a soil is sampled for laboratory testing, its *in situ* state of stress is somewhere along either the wetting or the drying portion of the void ratio versus the stress state variable relationship (Fig. 14.14). In the field, the soil has been subjected to numerous cycles of wetting and drying. At the time of sampling, the soil has a specific net normal stress and a specific matric suction.

The laboratory information desired by the geotechnical engineer for predicting the amount of heave is an assessment of: 1) the *in situ* state of stress, and 2) the swelling properties with respect to changes in net normal stress [i.e., $(\sigma - u_a)$] and matric suction [i.e., $(u_a - u_w)$]. An understanding of the compressibility properties under loading conditions would also be useful. A demanding testing facility and program would be required to completely assess all of the above variables. For this reason, it is necessary to develop a simpler, more rapid, and economical procedure to obtain the information required to predict heave for practical problems.

Several laboratory testing procedures have been used in practice to obtain the necessary information. Among these procedures, the one-dimensional oedometer test is the most commonly used to determine the present *in situ* state of stress. The "free-swell" and the "constant volume" oedometer test methods have been explained in detail in Chapter 13, together with the necessary correction procedures. Other oedometer test procedures have also been used, but in general these procedures are variations of the "constant volume" and "free-swell" procedures.

The oedometer test translates the *in situ* stress state onto the net normal stress plane. The *in situ* stress state is referred to as the "corrected" swelling pressure, $P'_s$, which is equal to the sum of the overburden pressure and the matric suction equivalent (Chapter 13).

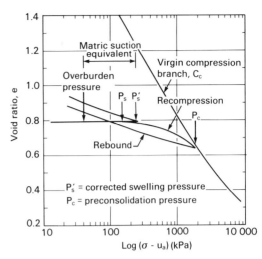

**Figure 14.15** Position of corrected swelling pressure relative to the preconsolidation pressure of the soil.

Figure 14.15 shows typical laboratory oedometer test data plotted to a scale which allows comparison to the virgin compression curve. Often, the entire laboratory loading curve is on the recompression portion, with the loading not even reaching the virgin compression branch. In other words, the preconsolidation pressure of many desiccated soils, $P_c$, may exceed the highest load applied during the test.

The preconsolidation pressure refers to the maximum stress producing a volume decrease that the soil has ever been subjected to in its history. Figure 14.15 illustrates the relative position of the corrected swelling pressure, $P'_s$, to the preconsolidation pressure, $P_c$. The corrected swelling pressure, $P'_s$, is located approximately at the intersection between the constant void ratio line and the recompression branch. On the other hand, the preconsolidation pressure, $P_c$, is located at the intersection between the recompression branch and the virgin compression curve.

Methods for determining the preconsolidation pressure have been described in many soil mechanics textbooks, and will not be repeated herein. However, it is appropriate and useful to redefine the overconsolidation ratio, OCR, as the ratio of the preconsolidation pressure, $P_c$, to the corrected swelling pressure, $P'_s$, where $P'_s$ is a representation of the *in situ* stress state of the soil:

$$\text{OCR} = \frac{P_c}{P'_s} \qquad (14.1)$$

where

OCR = overconsolidation ratio
$P'_s$ = corrected swelling pressure
$P_c$ = preconsolidation pressure.

Equation (14.1) is a modified definition for the overconsolidation ratio. At present, OCR is usually defined as the

**406** 14 VOLUME CHANGE PREDICTIONS

ratio of the preconsolidation pressure, $P_c$, to the *in situ* vertical effective overburden pressure, $\sigma'_v$ (i.e., $(\sigma_v - u_w)$, where $\sigma_v$ = total overburden pressure and $u_w$ = pore-water pressure). The use of the corrected swelling pressure, $P'_s$, in Eq. (14.1) eliminates the need for measuring the *in situ* negative pore-water pressure and assessing how it should be combined with the total stress when defining OCR. In addition, it seems more appropriate to define the degree of overconsolidation with respect to the *in situ* stress state on the net normal stress plane (i.e., $P'_s$).

### 14.2.3 Future Stress State and Ground Movements

Having determined the present *in situ* state of stress and the swelling indices from oedometer tests, the analysis can proceed to the prediction of possible changes in the stress state at some future time. The future state of stress corresponding to several years after construction must be estimated, based upon local experience and climatic conditions. Several procedures for estimating the final pore-water pressure profile are discussed in the next section.

Changes in total stress can be anticipated as a result of excavation, replacement with a relatively inert material (e.g., gravel), and other loadings. The effects of these changes can be taken into account using appropriate volume change indices for loading and unloading. However, it may be possible to assume that there is insufficient time for the soil to respond to each individual loading and unloading. The long-term volume change calculations will then be approximated as the net loading or unloading.

For discussions concerning possible future swelling, let us assume that the final pore-water pressures of the soil may go to zero under a constant net normal stress. Figure 14.16 shows the actual stress path that would be followed by a soil element at the depth from which the sample was retrieved. Swelling would follow a path from the initial void ratio, $e_0$, to the final void ratio, $e_f$, along the rebound surface on the matric suction plane. The entire rebound surface can be assumed to be unique since the direction of deformation is monotonic (Matyas and Radhakrishna, 1968; Fredlund and Morgenstern, 1976). Therefore, it is also possible to follow a stress path from the *in situ* stress state point over to the "corrected" swelling pressure, and then to proceed along the rebound curve on the net normal stress plane to the final stress condition. The advantage of the latter stress path is that the volume change indices determined on the net normal stress plane can be used to predict total heave.

### 14.3 THEORY OF HEAVE PREDICTIONS

Heave predictions should ideally be conducted using the volume change theories presented in Chapter 12 for various loading conditions. The volume change coefficients or in-

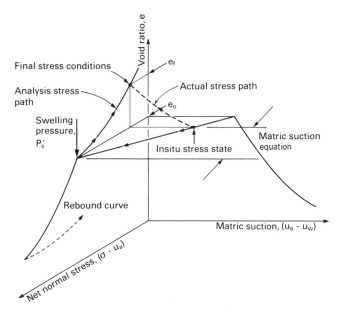

**Figure 14.16** "Actual" and "analysis" stress paths followed during the wetting of a soil.

dices and matric suction changes should be measured using the techniques presented in Chapters 13 and 4, respectively. These changes should then be used to compute the total heave in a manner consistent with Eq. (12.47) (i.e., $d\epsilon_y = m^s_{1k} d(\sigma_y - u_a) + m^s_2 d(u_a - u_w)$ where $d\epsilon_y$ = incremental strain in the y-direction, $(\sigma_y - u_a)$ = net normal stress in the y-direction, $(u_a - u_w)$ = matric suction, $m^s_{1k}$ = coefficient of volume change with respect to $(\sigma_y - u_a)$ for $K_0$ loading conditions, and $m^s_2$ = coefficient of volume change with respect to $(u_a - u_w)$). The stress state variable changes can be large and consequently, the soil properties can vary as the stress levels change. As a result, it would be necessary to integrate Eq. (12.47) between the initial and final stress states, while making the soil properties a function of the stress state. It is possible to circumvent this complexity by recognizing that the constitutive surface can be linearized by plotting the stress state variables on a logarithmic scale. In addition, it is possible to make a further simplification by transferring the matric suction variable onto the net total stress plane.

The total heave formulation will be presented with the initial and final stress states projected onto the net normal stress plane. The results from a one-dimensional oedometer test are plotted on a semi-logarithmic scale (i.e., the stress state) and the slope of the plot is used in the formulation for total heave. This method greatly simplifies the analysis for the prediction of total heave. The in situ stress state is equal to the vertical swelling pressure of the soil which is measured in a one-dimensional oedometer under $K_0$-loading conditions. As a result, only the vertical or one-dimensional heave is predicted. The vertical heave prediction is of importance in the design of shallow foundations for light structures. Two case histories dealing with highly expansive soils in Saskatchewan, Canada, are later analyzed and presented to illustrate the application of the total heave prediction theory.

For loading configurations other than $K_0$-conditions, volume change can also occur in the lateral directions. The swelling pressure in the lateral direction depends on several variables, such as the initial at rest earth pressure coefficient and the horizontal deformation moduli for the soil, as pointed out in Chapter 11. In a soil with wide desiccation cracks, substantial volume changes may occur in the horizontal direction prior to the development of the lateral swelling pressure. The ratio of the lateral to vertical swelling pressures can range from as low as the at rest earth pressure coefficient, which may be zero, to as high as the passive earth pressure coefficient (Pufahl et al., 1983; Headquarters, U.S. Department of the Army, 1983; Fourie, 1989). The lateral heave prediction is best analyzed using the volume change theory presented in Chapter 12, and it will not be further elaborated upon in this section.

### 14.3.1 Total Heave Formulations

The procedure for the calculation of total heave or swell is similar to that used for settlement calculations. The amount of total heave is computed from the changes in void ratios corresponding to the initial and final stress states and the swelling index. The formulation will be visualized on the void ratio versus the logarithm of the stress state. The following formulation assumes stress paths which have been projected onto the net normal stress plane, as shown in Fig. 14.17. The total heave stress path follows the rebound curve (i.e., $C_s$) from the initial stress state to the final stress state. The equation for the rebound portion of the oedometer test data can be written as,

$$\Delta e = C_s \log \left( \frac{P_f}{P_0} \right) \quad (14.2)$$

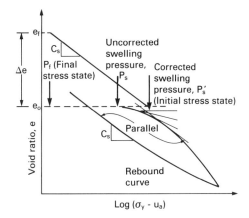

**14.17** One-dimensional oedometer test results showing the effect of sampling disturbance.

where

$\Delta e$ = change in void ratio between the initial and final stress states (i.e., $e_f - e_0$)
$e_0$ = initial void ratio
$e_f$ = final void ratio
$C_s$ = swelling index
$P_0$ = initial stress state, which is equivalent to the corrected swelling pressure (i.e., $P_0 = P_s'$)
$P_f$ = final stress state.

The initial stress state, $P_0$, or the corrected swelling pressure, $P_s'$, can be formulated as the sum of the overburden pressure and the matric suction equivalent (Fig. 14.15) as follows:

$$P_0 = (\sigma_y - u_a) + (u_a - u_w)_e \quad (14.3)$$

where

$\sigma_y$ = total overburden pressure
$\sigma_y - u_a$ = net overburden pressure
$u_a$ = pore-air pressure
$(u_a - u_w)_e$ = matric suction equivalent
$u_w$ = pore-water pressure.

Equation (14.3) defines the initial stress state, $P_0$. In practice, the value of $P_0$ is not calculated, but measured as the corrected swelling pressure, $P_s'$, in an oedometer test. The final stress state, $P_f$, must account for total stress changes and the final pore-water pressure conditions. The pore-air pressure in the field remains at atmospheric conditions. The final pore-water pressure conditions can be predicted or estimated as explained in the next section. Therefore, the final stress state, $P_f$, can be formulated as follows:

$$P_f = \sigma_y + \Delta \sigma_y - u_{wf} \quad (14.4)$$

where

$\Delta \sigma_y$ = change in total stress due to the excavation or placement of fill; the total stress change can have a positive or negative sign for either an increase or decrease in total stress, respectively
$u_{wf}$ = predicted or estimated final pore-water pressure.

The heave of an individual soil layer can be written in terms of a change in void ratio:

$$\Delta h_i = \frac{\Delta e_i}{1 + e_{0i}} h_i \quad (14.5)$$

where

$\Delta h_i$ = heave of an individual soil layer
$h_i$ = thickness of the layer under consideration
$\Delta e_i$ = change in void ratio of the layer under consideration (i.e., $e_{0i} - e_{fi}$)
$e_{0i}$ = initial void ratio of the soil layer
$e_{fi}$ = final void ratio of the soil layer.

The change in void ratio, $\Delta e_i$, in Eq. (14.5) can be rewritten, by incorporating the soil properties and the stress states [i.e., Eq. (14.2)], to give the following form for the heave of a soil layer:

$$\Delta h_i = \frac{C_s}{1 + e_{0i}} h_i \log \frac{P_{fi}}{P_{0i}} \quad (14.6)$$

where

$P_{fi}$ = final stress state in the soil layer
$P_{0i}$ = initial stress state in the soil layer.

The total heave from several layers, $\Delta H$, is equal to the sum of the heave for each layer:

$$\Delta H = \sum \Delta h_i. \quad (14.7)$$

### 14.3.2 Prediction of Final Pore–Water Pressures

The final pore–water pressures below a foundation or pavements can either be predicted or estimated. A prediction must take into consideration the surface flux boundary conditions (i.e., infiltration, evaporation, and evapotranspiration) and the fluctuation of the groundwater table. The surface flux boundary conditions can vary from one geographic location to another, depending upon the climatic conditions. Russam and Coleman (1961) related the equilibrium suction below asphaltic pavements to the Thornthwaite Moisture Index. On many smaller structures, however, it is often man-made causes such as leaky water lines and poor drainage which control the final pore–water pressures in the soil.

There are three possibilities for the estimation of final pore–water pressure conditions, as illustrated in Fig. 14.18. First, it can be assumed that the water table will rise to the ground surface, creating a hydrostatic condition. This assumption predicts the greatest amount of total heave. Second, it can be assumed that the pore–water pressure approaches a zero value throughout its depth. This may appear to be a realistic assumption; however, it should be noted that it is not an equilibrium condition. In many situations, this assumption may provide a suitable estimate for the final pore–water pressure state. Third, it can be assumed that under long-term equilibrium conditions, the pore–water pressure will remain slightly negative. This assumption predicts the smallest amount of total heave. It is also possible to have variations of the above assumptions with depth. Also, there may be a limit placed on the depth to which wetting will occur. Any of the above assumptions produces relatively similar predictions of heave in most situations. This is due to the fact that most of the heave occurs in the uppermost soil layer where the change in matric suction is largest.

### 14.3.3 Example of Heave Calculations

The following example problems are presented to illustrate the calculations associated with total heave. The first example considers a 2 m thick layer of swelling clay (Fig. 14.19). The initial void ratio of the soil is 1.0, the total unit weight is 18.0 kN/m$^3$, and the swelling index is 0.1. Only one oedometer test was performed on a sample taken from a depth of 0.75 m. The test data showed a corrected swelling pressure of 200 kPa. It is assumed that the corrected swelling pressure is constant throughout the 2 m layer.

Let us assume that the ground surface is to be covered with an impermeable layer such as asphalt. With time, the negative pore–water pressure in the soil below the asphalt will increase as a result of the discontinuance of evaporation and evapotranspiration. For analysis purposes, let us assume that the final pore–water pressure will increase to zero throughout the entire depth.

The 2 m layer is subdivided into three layers. The amount of heave in each layer is computed by considering the stress state changes at the middle of the layer. The initial stress state, $P_0$, will be equal to the corrected swelling pressure at all depths. The final stress state, $P_f$, will be the overburden pressure. Equation (14.6) is used to calculate the heave for each layer. The calculations in Fig. 14.19 show a total heave of 11.4 cm. Approximately 36% of the total heave occurs in the upper quarter of the clay strata. The calculations can also be used to show the amount of heave that would occur if each layer became wet from the surface downward.

The second example shows a more complex loading situation, and the results are presented in Fig. 14.20. Again, the clay layer is 2 m in thickness. The initial void ratio is 0.8, the total unit weight is 18.0 kN/m$^3$, and the swelling index is 0.21. Three oedometer tests were performed, which show a decrease in the corrected swelling pressure with depth (Fig. 14.20).

Suppose the engineering design suggests the removal of 1/3 m of swelling clay from the surface, prior to the placement of 2/3 m of gravel. The unit weight of the gravel is assumed to be equal to that of the clay. The $1\frac{2}{3}$ m of swell-

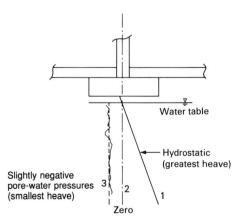

**Figure 14.18** Possibilities for the final pore–water pressure profile.

14.3 THEORY OF HEAVE PREDICTIONS 409

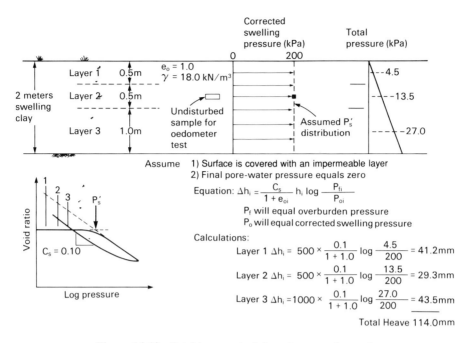

**Figure 14.19** Total heave calculations for example no. 1.

ing clay is subdivided into three strata. The thickness of each layer is shown in the table in Fig. 14.20.

The initial stress state, $P_0$, can be obtained by interpolation of the corrected swelling pressures at the midpoint of each layer. The final stress state, $P_f$, must take into account the final pore-water pressure and changes in the total stress. The final pore-water pressure is assumed to be $-7.0$ kPa. Equation (14.6) can be used to calculate the heave in each layer. The total heave is computed to be 22.1 cm.

Two assumptions are made during the heave analysis in the second example. First, it is assumed that the independent processes of excavation of the expansive soil and the placement of the gravel fill do not allow sufficient time for equilibrium to be established in the pore-water. Therefore, the soil responds only to the net change in total stress. Second, by estimating a final negative pore-water pressure, it is assumed that as saturation of the soil is approached, the slopes of the rebound curves on the matric suction and total stress planes approach the same value. This assumption is reasonable, provided the final pore-water pressure is relatively small.

A third example illustrates the amount of heave versus

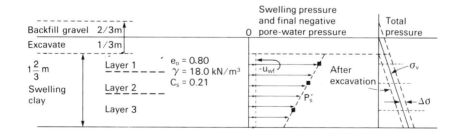

| | Initial stress state | | | | Final stress state | | |
|---|---|---|---|---|---|---|---|
| Layer no. | Thickness (mm) | $P_o = P_s'$ (kPa) | Initial overburden pressure $\sigma_v$ (kPa) | Change in total stress $\Delta\sigma$ (kPa) | Final pore-water pressure $u_{wf}$ (kPa) | $P_f = \sigma_v \pm \Delta\sigma - u_{wf}$ (kPa) | $\Delta h_i$ (mm) |
| 1 | 333 | 800 | 9.0 | +6.0 | -7.0 | 22.0 | 60.6 |
| 2 | 500 | 608 | 16.4 | +6.0 | -7.0 | 29.4 | 76.7 |
| 3 | 833 | 300 | 28.4 | +6.0 | -7.0 | 41.4 | 83.6 |

Total heave = 220.9mm

**Figure 14.20** Total heave calculations for example no. 2.

depth for a 4 m layer of expansive soil. The results show that the largest amount of heave occurs near the ground surface. In the zone near the ground surface, the matric suction is generally a maximum, and the overburden pressure has its lowest value. As a result, the ratio of the final stress state, $P_f$, to the initial stress state, $P_0$, in Eq. (14.6) approaches its minimum value, which results in the largest amount of heave.

Let us assume that the clay becomes wet, and the pore-water pressures go to zero throughout the deposit. Figure 14.21 shows the distribution of heave potential versus depth for the 4 m thick layer. Heave potential is defined as the heave in each layer of the profile divided by the maximum heave in any layer. Three initial swelling pressure profiles are assumed. The first profile assumes a constant swelling pressure of 720 kPa throughout the depth. The second profile assumes a constant value for the swelling pressure of 3600 kPa. The third profile assumes a linear decrease in swelling pressure from 720 kPa to a value equal to the overburden pressure at 4 m. All cases show that the heave potential decreases rapidly with depth. The swelling pressure generally decreases with depth *in situ*, and in this case, the decrease in heave potential is of an exponential form (Fig. 14.21).

The change in water content, $\Delta w$, can also be estimated as the soil swells provided the initial volume–mass soil properties are known:

$$\Delta w = \frac{S_f \Delta e}{G_s} + \frac{e_0 \Delta S}{G_s} \qquad (14.8)$$

where

$S_f$ = final degree of saturation
$G_s$ = specific gravity of soil solids
$\Delta S$ = change in degree of saturation.

The final degree of saturation can be assumed as some value approaching 100%. Changes in void, $\Delta e$, are obtained from the heave analysis.

### 14.3.4 Case Histories

Two case histories are briefly presented in order to demonstrate the application of the total heave prediction to engineering problems.

#### Slab-on-Grade Floor, Regina, Saskatchewan

In 1961, the Division of Building Research, National Research Council, monitored the performance of a light industrial building which has been constructed in north-central Regina. Details of the study were presented by Yoshida *et al.* (1983). Instrumentation was installed to monitor ground movements at various depths below the slab. Water content changes were monitored using a neutron moisture meter probe. Undisturbed samples were taken as part of the subsurface exploration prior to the construction of the building. "Constant volume" oedometer tests were performed on three samples, and the swelling pressure profiles are shown in Fig. 14.22. The average swelling index of the soil at this site was 0.09.

Approximately one year after construction, the owner

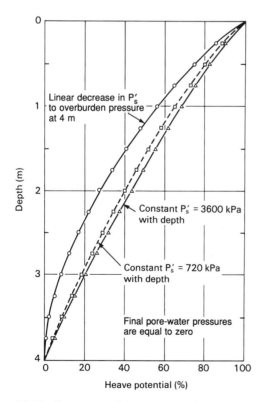

**Figure 14.21** Heave potential versus depth for various swelling pressure distributions.

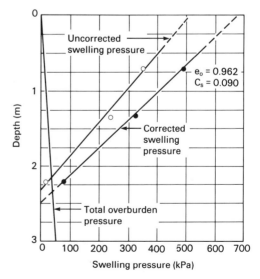

**Figure 14.22** Swelling pressure versus depth for Regina clay at the light industrial building site (from Yoshida *et al.* 1983).

noticed considerable cracking of the floor slab. Precise level surveys showed the maximum total heave to be 106 mm. The owner had also noticed a significant increase in water consumption (i.e., 35 000 L). It was discovered that a leak had occurred in the hot water line beneath the floor slab, near the location of the maximum heave. The leak was immediately repaired.

Total heave predictions were made on the basis of the laboratory oedometer test results. Various assumptions were made concerning the final pore-water pressure conditions. When it was assumed that the soil had become saturated and the water table rose to the base of the floor slab, the predicted heave was 141 mm. Assuming that the negative pore-water pressures were reduced to zero, the prediction of total heave was 118 mm. Assuming a final pore-water pressure profile of $-50$ kPa gave a total heave prediction of 66 mm. On the basis of the heave analysis, it appears that the assumption of zero pore-water pressure was probably the most realistic for this case history. It appears that further heave would likely have taken place had the leak had not been repaired. The prediction of heave at various depths showed close agreement with the actual measurements.

The importance of correcting the swelling pressure measurements can be illustrated by computing the predicted heave using the uncorrected swelling pressures. If the final pore-water pressures are assumed to be zero and the uncorrected swelling pressures are used in the computations, the predicted heave is 103 mm. This is less than the measured maximum heave and 13% less than the 118 mm predicted using the corrected swelling pressures.

### Eston School, Eston, Saskatchewan

Soils in the Eston area of Saskatchewan have long been known to be extremely high in swelling potential. The stratigraphy consists of approximately $7\frac{1}{2}$ m of highly plastic, brown clay overlying the glacial till. Many light structures have undergone serious distress. The building of particular interest is the Old Eston School constructed in the late 1920's.

The school building was constructed on concrete strip footings, and a wooden basement floor was supported by interior concrete footings. The school was a two-storey structure, with classrooms in both the lower and upper levels. The lower floor was approximately 1.2 m below grade. The exterior concrete walls were founded approximately 1.8 m below grade.

A substantial amount of heave took place below the interior footings. Although the record of performance was not precise, apparently the heave in one portion of the basement area had been quite extreme. On two occasions during the history of the school, 150–300 mm of soil was removed from below the interior footings. As much as 450–900 mm of total heave occurred during the life of the

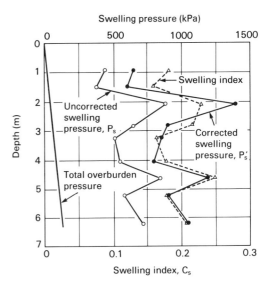

**Figure 14.23** Swelling index and swelling pressure versus depth for Eston clay.

school, according to maintenance records. Large amounts of differential heaving of the floor were recorded in 1960. The school was demolished in 1967.

In 1981, a subsurface investigation was conducted adjacent to the location of the old school. Undisturbed soil samples were taken, and "constant volume" oedometer tests were performed. The results are presented in Fig. 14.23. The average natural water content throughout the profile was 25%. The average plastic limit was 27%, and the average liquid limit was 100%. The average swelling index was 0.21. Due to a lack of detailed information on the soil and performance conditions of the school, it is not possible to do a precise total heave analysis. It is of interest, however, to perform an approximate analysis. Using the corrected swelling pressures from Fig. 14.23, and assuming that the negative pore-water pressures went to zero, the predicted heave would be approximately 990 mm. Using the uncorrected swelling pressures, the predicted heave would be about 86% of this value (i.e., in the order of 855 mm).

## 14.4 CONTROL FACTORS IN HEAVE PREDICTION AND REDUCTION

The following sections discuss the role of each variable that controls the magnitude of total heave (Rao *et al.*, 1988). The corrected swelling pressure and the swelling index are the main variables which are considered. The effect on the prediction of heave, of correcting the swelling pressure for sampling disturbance, is also demonstrated. These discussions utilize a simple case with a constant profile of soil properties, including the swelling pressure. As a result, the influence of each controlling variable on the prediction of heave can readily be identified.

# 14 VOLUME CHANGE PREDICTIONS

The same example is also used to illustrate the percentages of heave resulting from the wetting of an expansive soil to various depths. This analysis can be used to identify the critical zone where maximum heave will occur. Partial removal of expansive soils is often attempted in order to reduce the heaving potential. The effectiveness of this procedure in reducing the amount of heave is discussed using the same example.

## 14.4.1 Closed-Form Heave Equation when Swelling Pressure is Constant

An example is shown in Fig. 14.24 where the corrected swelling pressure, $P'_s$, is constant with depth. The soil is assumed to be homogeneous (i.e., $e_0$, $\rho$, and $C_s$ are constant with depth). These assumptions apply to all cases presented in the following sections. Sufficient water is supplied to the soil such that the pore–water pressures are assumed to go to zero throughout the entire soil profile. The final stress state, $P_f$, is assumed to be equal to the total overburden pressure.

The unsaturated, expansive soil swells upon wetting. Most of the heave will occur near ground surface where there is the largest difference between the corrected swelling pressure and the total overburden pressure. Heave continues to occur until a depth is reached where there is no difference between the corrected swelling pressure and the total overburden pressure. In this context, the depth at which the corrected swelling pressure equals the total overburden pressure can be defined as the *active depth* of swelling, $H$:

$$H = \frac{P'_s}{\rho g} \quad (14.9)$$

where

$H$ = active depth of swelling
$\rho$ = total density of the soil, which is assumed to remain constant with depth
$g$ = gravitational acceleration.

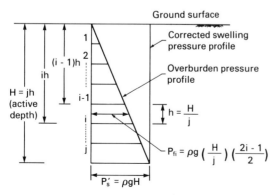

**Figure 14.24** Overburden and swelling pressure distributions versus depth for the case of constant swelling pressure.

Inherent in this definition of active depth is the assumption that the final pore–water pressures throughout the profile go to zero.

The heave analysis can be performed by first subdividing the computed active depth into $j$ number of layers of equal thickness (i.e., $h_i = h$):

$$h = \frac{H}{j} \quad (14.10)$$

where

$h$ = thickness of a soil layer.

An attempt is first made to determine the number of layers required to accurately predict total heave. The initial stress state, $P_{0i}$, is equal to the corrected swelling pressure $P'_s$, which can be defined in terms of the active depth in Eq. (14.9):

$$P_{0i} = \rho g H. \quad (14.11)$$

The final stress state, $P_{fi}$, in the "$i$"th layer from ground surface is computed as the average overburden pressure of the layer:

$$P_{fi} = \rho g \frac{(i-1)h + ih}{2} \quad (14.12)$$

where

$i$ = soil layer number (i.e., $1, 2, \cdots, j$).

The thickness of soil layer, $h$, in Eq. (14.12) can be written in terms of the total thickness divided by the number of layers [i.e., using Eq. (14.10)]:

$$P_{fi} = \frac{\rho g H (2i - 1)}{2j}. \quad (14.13)$$

The heave equation is written for each soil layer, and then summed to give the total heave. The amount of heave in each layer can be computed by substituting Eqs. (14.10), (14.11), and (14.13) for $h_i$, $P_{0i}$, and $P_{fi}$, respectively, in Eq. (14.6). The total heave for the entirely wetted active depth is the summation of the heave for each layer:

$$\Delta H = \frac{C_s H}{1 + e_0} \frac{1}{j} \sum_{i=1}^{j} \log\left(\frac{2i - 1}{2j}\right). \quad (14.14)$$

The computed total heave does not change significantly when using more than 35 layers (i.e., $j = 35$). Figure 14.25 shows the percentage of total heave, referenced to the case of 35 layers, versus the number of layers used in the analysis. If we assume the number of layers in the analysis to be 35, Eq. (14.14) can be simplified as follows:

$$\Delta H = -0.430 \frac{C_s H}{1 + e_0} \quad (14.15)$$

14.4 CONTROL FACTORS IN HEAVE PREDICTION AND REDUCTION    413

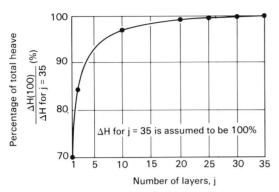

**Figure 14.25** Percentage of total heave as a function of the number of layers.

where:

$-0.430 = $ the computed value for the term:

$$\frac{1}{35} \sum_{i=1}^{35} \log\left(\frac{2i-1}{70}\right). \quad (14.16)$$

For an infinite number of soil layers, the constant, 0.430 in Eq. (14.15) approaches a value of 0.434, as obtained from a closed-form solution (Li, 1989).

The total heave expressed in Eq. (14.15) will later be used as a reference when comparing the amount of heave computed for different field conditions (i.e., partly wetted active depth, excavation, and backfill). In general, a total heave analysis can be performed using ten layers with an accuracy within practical limits.

Substituting Eq. (14.9) into Eq. (14.15) and assuming a unit weight (i.e., $\rho g$) of 20 kN/m³ and an initial void ratio, $e_0$, of 1.0 yields the following equation:

$$\Delta H = -0.01075 \, C_s P_s'. \quad (14.17)$$

$P_s'$ and $\Delta H$ have units of kPa and m, respectively. The corrected swelling pressure and the swelling index are the two primary variables involved in the prediction of heave. Figures 14.26 and 14.27 show the types of plots which can

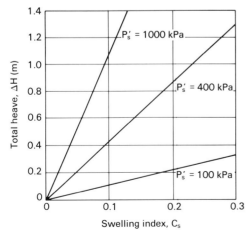

**Figure 14.26** Total heave versus swelling index for various corrected swelling pressures.

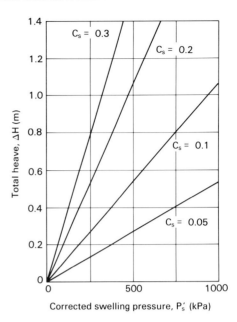

**Figure 14.27** Total heave versus corrected swelling pressure for various swelling indices.

be drawn from Eq. (14.17) to illustrate approximate amounts of total heave for various swelling indices and swelling pressures. The total heave varies linearly with both of the above variables. Viewed in another way, it is equally as important to accurately assess the swelling pressure as it is to assess the swelling index of an expansive soil.

### 14.4.2 Effect of Correcting the Swelling Pressure on the Prediction of Total Heave

Correcting the swelling pressure for the effect of sampling disturbance increases the change in void ratio, as shown in Fig. 14.28. As a result, the calculation of total heave also increases [see Eq. (14.5)]. The change in void ratio corresponding to the corrected swelling pressure can be computed from Eq. (14.2) by substituting $P_s'$ for $P_0$ (also see Fig. 14.28):

$$\Delta e' = C_s \log\left(\frac{P_f}{P_s'}\right). \quad (14.18)$$

The change in void ratio corresponding to the uncorrected swelling pressure, $P_s$, is written as (see Fig. 14.28)

$$\Delta e = C_s \log\left(\frac{P_f}{P_s}\right). \quad (14.19)$$

The difference in heave or in the change in void ratio can be obtained by subtracting Eq. (14.19) from Eq. (14.18):

$$\Delta e - \Delta e' = C_s \log\left(\frac{P_s}{P_s'}\right). \quad (14.20)$$

Referencing the difference between the two void ratio changes, to the change in void ratio when using the cor-

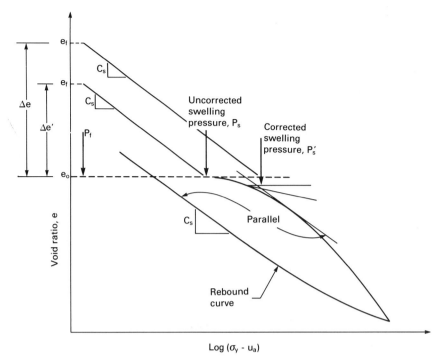

**Figure 14.28** Void ratio changes corresponding to the corrected and uncorrected swelling pressures.

rected swelling pressure gives

$$\frac{\Delta e - \Delta e'}{\Delta e'} = \frac{\log\left(\frac{P_s}{P'_s}\right)}{\log\left(\frac{P_f}{P_s}\right)}. \quad (14.21)$$

Eq. (14.21) is shown graphically in Fig. 14.29 for various ratios of swelling pressure, $(P'_s/P_s)$ and $(P_f/P_s)$. In

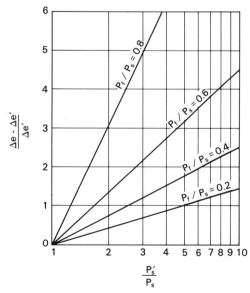

**Figure 14.29** Effect of correcting the swelling pressure on the computed change in void ratio for various overburden pressures.

other words, the initial and final stress states are referenced to the uncorrected swelling pressure. The plot shows that the relative difference between the two heave predictions increases as the difference between the corrected and uncorrected swelling pressures increases (i.e., as $P'_s/P_s$ increases). It is common for the corrected swelling pressure to be two–three times greater than the uncorrected swelling pressure [i.e., $P'_s/P_s$ can be 2–3 (Fredlund, 1983) (see Chapter 13)]. The relative difference in heave also increases as the overburden pressure increases (i.e., $P_f/P_s$ increases). The difference between the two predictions of total heave can be substantial.

### 14.4.3 Example with Wetting from the Top to a Specified Depth

Figure 14.30 shows the variables involved in studying the effect of wetting of the soil from the ground surface to a specified depth (e.g., by flooding the surface). For example, an insufficient amount of water infiltration into the ground may result in the wetting of only a portion of the active depth. Let $H_r$ be the portion of the active depth that has been wetted (i.e., $rH$). The wetted zone is also subdivided into "$j$" number of layers for the computation of total heave. The heave for any layer in the wetted zone can be calculated by substituting the variables shown in Fig. 14.30 into Eq. (14.6):

$$\Delta h_{ri} = \frac{C_s}{1 + e_0} \frac{rH}{j} \log\left\{\frac{\rho g r H(2i - 1)/2j}{\rho g H}\right\} \quad (14.22)$$

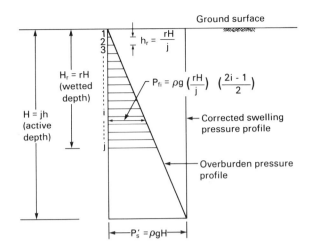

**Figure 14.30** Pressure distributions and definition of variables for the case of wetting a portion of the active depth.

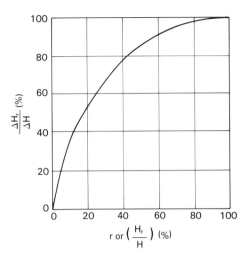

**Figure 14.31** Ratio of total heave predicted for wetting over a portion of the depth to the total heave for wetting of the entire active depth.

where

$\Delta h_{ri}$ = heave of an individual layer when a portion of the active depth is wetted

$r$ = the portion of the profile wetted (i.e., $H_r/H$).

Equation (14.22) reduces to the following form:

$$\Delta h_{ri} = \frac{C_s}{1 + e_0} \frac{rH}{j} \left\{ \log\left(\frac{2i - 1}{2j}\right) + \log r \right\}. \quad (14.23)$$

Equation (14.23) can be applied to all layers in the wetted zone to give the total heave:

$$\Delta H_r = \frac{C_s H}{1 + e_0} r \left\{ \frac{1}{j} \sum_{i=1}^{j} \log\left(\frac{2i - 1}{2j}\right) + \log r \right\}. \quad (14.24)$$

Substituting Eq. (14.15) into Eq. (14.24) for "$j$" equal to 35 gives

$$\Delta H_r = \frac{C_s H}{1 + e_0} r(-0.430 + \log r). \quad (14.25)$$

When "$r$" is equal to 1.0, the entire active depth is wetted, and Eq. (14.25) reverts to Eq. (14.15). A comparison between Eqs. (14.25) and (14.15) is as follows:

$$\frac{\Delta H_r}{\Delta H} = r(1 - 2.326 \log r). \quad (14.26)$$

Figure 14.31 shows a plot of Eq. (14.26) where various percentages of the profile are wetted. The relationship is unique for all values of swelling pressure and swelling index. For example, it can be seen that 80% of the total heave occurs if the depth of wetting is 40–50% of the active depth. This example can be used to illustrate that it may be possible to flood an area prior to construction, until a significant percentage of total heave has occurred. Whether this is a practical engineering solution may depend on other factors, such as the soil properties and the design of the structure.

### 14.4.4 Example with a Portion of the Profile Removed by Excavation and Backfilled with a Nonexpansive Soil

This section considers the possibility of excavating the upper portion of the profile. This is considered under the title of Case I. If the excavated portion is backfilled with a nonexpansive soil, the total heave calculations are considered as Case II. These two cases are shown in Fig. 14.32. In Case I, it is assumed that the active depth is unaffected by the removal of the soil. In both cases, the total heave calculations are made for the condition where the entire active depth is wetted.

For Case I, the heave for any layer in the expansive soil can be computed by substituting the variables for Case I in Fig. 14.32 into Eq. (14.6):

$$\Delta h_{li} = \frac{C_s}{1 + e_0} \frac{(1 - l)H}{j} \log \left\{ \frac{\rho g (1 - l) H (2i - 1)/2j}{\rho g H} \right\} \quad (14.27)$$

where

$\Delta h_{li}$ = heave in an expansive soil layer when the active depth is partially excavated

$l$ = the portion of the profile excavated (i.e., $H_l/H$).

Equation (14.27) can be rearranged to the following form:

$$\Delta h_{li} = \frac{C_s}{1 + e_0} \frac{(1 - l)H}{j} \left\{ \log\left(\frac{2i - 1}{2j}\right) + \log(1 - l) \right\}. \quad (14.28)$$

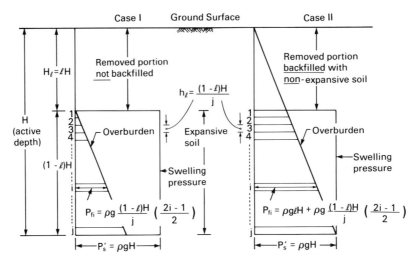

**Figure 14.32** Pressure distributions and definition of variables for the case of partial excavation and backfilling a nonexpansive soil within the active depth.

The total heave for Case I is computed by summing the individual magnitudes of heave in each soil layer:

$$\Delta H_l = \frac{C_s H}{1 + e_0} (1 - l) \left\{ \frac{1}{j} \sum_{i=1}^{j} \log \left( \frac{2i - 1}{2j} \right) \right. $$
$$\left. + \log (1 - l) \right\}. \quad (14.29)$$

Substituting Eq. (14.15) into Eq. (14.29) for "$j$" equal to 35 gives

$$\Delta H_l = \frac{C_s H}{1 + e_0} (1 - l) \{-0.430 + \log (1 - l)\}. \quad (14.30)$$

If $l = 0$, there is no removal, and Eq. (14.30) reverts to Eq. (14.15). If $l = 1$, the entire active depth has been excavated, and there is no more tendency for heaving. A comparison between Eqs. (14.30) and (14.15) can be written as

$$\frac{\Delta H_l}{\Delta H} = (1 - l)\{1 - 2.326 \log (1 - l)\}. \quad (14.31)$$

Equation (14.31) is illustrated in Fig. 14.33 for various percentages of soil removal (i.e., case of excavation without backfilling).

Let us now consider Case II, in which the excavated portion of the profile is backfilled with an inert or nonexpansive soil. The total density of the backfill is assumed to be equal to the density of the expansive soil. Similarly, the total heave, $\Delta H$, for Case II can be calculated using the variables given in Fig. 14.32:

$$\Delta H_b = \frac{C_s H}{1 + e_0} (1 - l)$$
$$\cdot \left[ \frac{1}{j} \sum_{i=1}^{j} \log \left\{ l + \frac{2i - 1}{2j} (1 - l) \right\} \right]. \quad (14.32)$$

If $l = 0$, the maximum heave is computed, and if $l = 1$, the heave is zero. A comparison between Eqs. (14.32) and (14.15) can be written as

$$\frac{\Delta H_b}{\Delta H} = -2.326(1 - l)$$
$$\cdot \left[ \frac{1}{j} \sum_{i=1}^{j} \log \left\{ l + \frac{2i - 1}{2j} (1 - l) \right\} \right]. \quad (14.33)$$

Dimensionless plots illustrating Cases I and II are shown in Fig. 14.33. These plots apply for all swelling indices and swelling pressures, and show the influence of excavating and backfilling on the computed total heave.

Figure 14.33 shows, for example, that if 50% of the active depth were removed, the total heave would still be 86% of that anticipated for the case of no excavation. However, if the excavated portion is backfilled with a nonexpansive soil, the total heave would be only 15% of

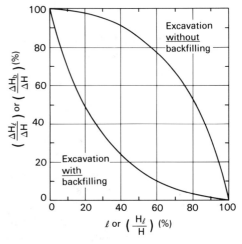

**Figure 14.33** Ratio of total heave for partial excavation and backfilling to total heave for wetting of the entire active depth.

## 14.5 NOTES ON COLLAPSIBLE SOILS

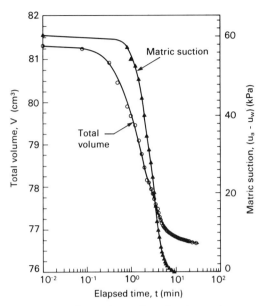

**Figure 14.34** Matric suction and total volume changes versus time after inundation of a compacted specimen of Indian Head silty sand (from Tadepalli *et al.*, 1992).

that anticipated if no material were excavated. In other words, partial excavation of the expansive soil depth does not significantly reduce the total heave unless there is backfilling with an inert soil.

Other possible boundary conditions could be assumed, but these examples illustrate the type of plots that can readily be generated using the equation for predicting total heave. The plots are of assistance in making engineering decisions when placing structures on expansive soils. Possible remedial measures for heave reduction by flooding, excavation, and backfilling can be studied by comparing these types of plots.

### 14.5 NOTES ON COLLAPSIBLE SOILS

An expansive soil exhibits a volume increase as a result of a reduction in matric suction, while a collapsible soil exhibits a volume decrease as a result of a reduction in matric suction. Collapsible soils have an open type of structure, with large void spaces which give rise to a metastable structure. Soils compacted dry of optimum at low densities often exhibit collapse behavior. In general, collapsible soils are unsaturated, and the reduction in matric suction is one of the major causes for the occurrence of collapse (Matyas and Radhakrishna, 1968; Escario and Sáez, 1973; Cox, 1978; Lloret and Alonso, 1980; Maswoswe, 1985).

Tadepalli (1990) conducted collapse tests in a specially designed oedometer with matric suction measurements. Soil specimens were statically compacted in the oedometer ring. Small-tip tensiometers were installed along the side of the specimen for the measurement of matric suction during the test. The specimen was then subjected to an applied total stress under constant water content conditions. At equilibrium, the soil specimen was inundated, with the result that the soil gradually decreased in volume as the suction went to zero. During inundation, the matric suction and volume decrease were measured simultaneously at various elapsed times.

Volume change and matric suction measurements taken during the inundation of compacted Indian Head silty sand are shown in Figs. 14.34 and 14.35 for small and large specimens, respectively. The results indicate that there is a unique relationship between the change in matric suction

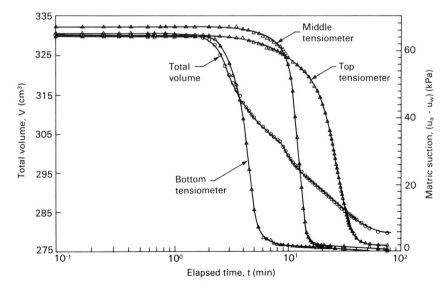

**Figure 14.35** Matric suction (measured using three tensiometers) and total volume changes versus time after inundation of a compacted specimen of Indian Head silty sand (from Tadepalli *et al.*, 1992).

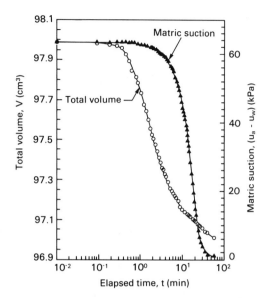

**Figure 14.36** Matric suction and total volume changes versus time after inundation of an undisturbed specimen of Mississippi Delta silt (from Tadepalli *et al.*, 1992).

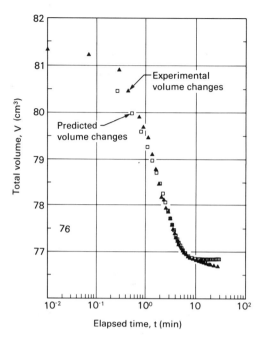

**Figure 14.37** Comparison between the predicted and measured volume changes during collapse of a compacted Indian Head silty sand (from Tadepalli and Fredlund, 1991).

and the total volume change during collapse. Figure 14.35 shows that the soil volume changed progressively with the reduction in matric suction, and the volume change ceased when the matric suction dropped to zero throughout the entire soil specimen. Similar types of observations were made on undisturbed specimens of Mississippi Delta silt, as shown in Fig. 14.36.

The volume changes during collapse can be predicted using the matric suction changes and volume change theory presented in Chapter 12. In this case, there is no change in the net normal stress, and the pore-air pressure remains essentially atmospheric. However, unlike the stable-structured soil, the metastable structure (i.e., collapsible soil) will have a positive sign for its volume change moduli. This means that a reduction in the matric suction variable will result in a volume decrease. Figure 14.37 shows a comparison between predicted and measured volume changes during the inundation of a compacted Indian Head silty sand. The predicted volume changes were computed using the measured matric suction changes.

# CHAPTER 15

# *One-Dimensional Consolidation and Swelling*

The application of a load to an unsaturated soil specimen will result in the generation of excess pore–air and pore–water pressures, as explained in Chapter 8. The excess pore pressures will dissipate with time, and will eventually return to their values prior to loading. The dissipation process of pore pressures is called consolidation, and the process results in a volume decrease or settlement.

In this chapter, the theory required for describing the consolidation and the swelling behavior of an unsaturated soil is derived. The derivation results in two transient flow equations which describe the air and water flow. The air and water flows are assumed to be interdependent, except in cases where the excess pore–air pressure dissipates instantaneously. In this case, only the water flow equation is required to analyze the consolidation (and the swelling) process.

The pore pressure changes at any time can be used to calculate the stress state variable changes. As a result, volume changes can also be computed by using the constitutive equations and the coefficients of volume change, as explained in Chapters 12 and 13, respectively.

## 15.1 LITERATURE REVIEW

The development of consolidation equations for saturated and unsaturated soils is summarized in this section. Terzaghi (1943) derived the classical theory for one-dimensional consolidation for *saturated* soils. Several assumptions were used in the derivations, such as: 1) the soil is homogeneous and saturated; 2) the strains are small; 3) the coefficients of volume change, $m_v$, and permeability, $k_s$, remain constant during consolidation; and 4) the water and soil particles are incompressible.

Terzaghi's classical derivations incorporated a constitutive equation for saturated soils and a flow law. The constitutive equation described the deformation of the soil structure with respect to changes in the stress state by using the soil property called the coefficient of volume change, $m_v$. Terzaghi (1936) proposed the effective stress, $(\sigma - u_w)$, as the stress variable to describe the behavior of a saturated soil. The flow rate of water during consolidation was calculated in accordance with Darcy's law. Darcy's law relates the flow rate of water to its hydraulic head gradient using the soil property called the coefficient of permeability, $k_s$.

The continuity of a saturated soil element requires that the volume change of a soil element be equal to the change in the volume of water in the element. Applying the continuity requirement results in the consolidation equation proposed by Terzaghi (1943):

$$\frac{\partial u_w}{\partial t} = c_v \frac{\partial^2 u_w}{\partial y^2} \qquad (15.1)$$

where

$c_v$ = coefficient of consolidation (i.e., $k_s/(\rho_w g m_v)$)
$k_s$ = coefficient of permeability with respect to water at saturation (i.e., $S = 100\%$)
$\rho_w$ = water density
$g$ = gravitational acceleration
$m_v$ = coefficient of volume change for saturated soils.

Equation (15.1) describes changes in pore–water pressures with respect to depth and time during the consolidation process. The changes in pore–water pressure result in changes in the effective stress, $(\sigma - u_w)$. The effective stress changes can be substituted into the constitutive equation in order to compute the volume change, which is equal to the volume of water flowing out of the saturated soils. The computed volume changes can, in turn, be used to compute the soil volume–mass properties, such as void ratio, water content, and density throughout the consolidation process.

In 1941, Biot proposed a general theory of consolidation for an unsaturated soil with occluded air bubbles. Two constitutive equations relating stress and strain were formulated in terms of the effective stress, $(\sigma - u_w)$, and the pore–water pressure, $u_w$. In other words, the need for separating the effects of total stress and pore–water pressure was recognized. One equation related the void ratio to the stress state, and the other equation related the water content to the stress state of the soil. Assumptions used in Biot's theory were similar to those used in Terzaghi's theory. For one-dimensional consolidation, Biot's theory resulted in an equation similar to Eq. (15.1), but the coefficient of consolidation, $c_v$, was modified to take into account the compressibility of the pore fluid. Larmour (1966), Hill (1967) and Olson (1986) showed that Terzaghi's equation with a modified coefficient of consolidation, $c_v$, can be used to describe the consolidation behavior of unsaturated soils with occluded air bubbles. Scott (1963) incorporated void ratio change and degree of saturation change into the formulation of the consolidation equation for unsaturated soils with occluded air bubbles.

Blight (1961) derived a consolidation equation for the air phase of a dry, rigid, unsaturated soil. Fick's law of diffusion, which relates mass transfer to the pressure gradient, was used in the derivation.

Barden (1965, 1974) presented an analysis of the one-dimensional consolidation of compacted, unsaturated clay. Darcy's law was used to describe flow for the air and water phases. Several independent types of analyses were proposed, depending upon the degree of saturation of the soil. The analyses remained indeterminate due to the lack of information on the stress state and the constitutive relations for the unsaturated soil.

Fredlund and Hasan (1979) presented two partial differential equations which could be solved for the pore–air and pore–water pressures during the consolidation process of an unsaturated soil. The air phase was assumed to be continuous. Darcy's and Fick's laws were applied to the flow of the water and air phases, respectively. The coefficients of permeability with respect to both the water and air phases were considered to be a function of matric suction or one of the volume–mass properties of the soil. In other words, the two partial differential equations contained terms which accounted for the variation in the coefficients of permeability. Both equations were solved simultaneously, and the method is commonly referred to as a two-phase flow approach.

The formulation by Fredlund and Hasan (1979) was similar in form to the conventional one-dimensional Terzaghi derivation [i.e., Eq. (15.1)]. The derivations also demonstrated a smooth transition between the unsaturated and saturated cases. Similar consolidation equations have also been proposed by Lloret and Alonso (1981).

The two partial differential equations have been used to simulate the total and water volume change behavior of compacted kaolin specimens during total stress and matric suction changes (Fredlund and Rahardjo, 1986). The pore pressure changes calculated from the two different equations resulted in changes in the stress state variables. The stress state variable changes were substituted into the soil structure and the water phase constitutive equations in order to compute the volume changes in the unsaturated soil. Comparisons between the predicted volume changes and experimental results indicate similar volume change behavior with respect to time. However, pore pressure changes in the specimens were not measured during the tests.

In 1984, Dakshanamurthy et al., extended the consolidation theory for unsaturated soils to the three-dimensional case. The continuity equations were coupled with the equilibrium equations in deriving the three-dimensional formulation.

Rahardjo (1990) conducted one-dimensional consolidation tests on an unsaturated silty sand in a specially designed $K_0$-cylinder. The cylinder was developed to accommodate $K_0$-loading, and allow for the simultaneous measurement of pore–air and pore–water pressures throughout the soil specimen. The total and water volume changes were measured independently during the consolidation tests. The results indicated an essentially instantaneous dissipation of the excess pore–air pressures for the particular soil used in the experiment. On the other hand, the excess pore–water pressure dissipation was found to be a time-dependent process which could be closely simulated using the water flow partial differential equation.

## 15.2 PHYSICAL RELATIONS REQUIRED FOR THE FORMULATION

The one-dimensional consolidation equation for saturated soils [i.e., Eq. (15.1)] can be derived by satisfying the continuity requirement for a saturated soil element. The continuity equation is satisfied by equating the time derivative of the constitutive equation to the divergence of the flow rate. The constitutive equation for a saturated soil relates the change in void ratio, $de$, to the change in effective stress, $d(\sigma - u_w)$. The rate of water flow through a soil mass is described by Darcy's law. The time derivative of the constitutive equation gives the rate of change of the total volume of the saturated soil element. The total volume change rate of a saturated soil element should be equal to the net rate of water flow from the element.

The one-dimensional consolidation equation for an unsaturated soil can similarly be formulated by satisfying the continuity requirement. In this case, the change in the total volume of the soil element is equal to the sum of the changes in the volume of water and the volume of air in the element. The constitutive equations for an unsaturated soil were formulated in Chapter 12. One-dimensional consolidation under a $K_0$-loading condition has the following constitutive equation for the soil structure:

$$\frac{dV_v}{V_0} = m_{1k}^s \, d(\sigma_y - u_a) + m_2^s d(u_a - u_w) \quad (15.2)$$

where

$dV_v/V_0$ = volume change of the soil element with respect to the initial volume of the element (also referred to as the volumetric strain change, $d\epsilon_v$)

$V_v$ = volume of soil voids in the element

$V_0$ = initial overall volume of the element

$m_{1k}^s$ = coefficient of volume change with respect to a change in net normal stress, $d(\sigma_y - u_a)$, for $K_0$-loading

$m_2^s$ = coefficient of volume change with respect to a change in matric suction, $d(u_a - u_w)$, during $K_0$-loading.

The water phase constitutive relation for $K_0$-loading can be written as

$$\frac{dV_w}{V_0} = m_{1k}^w \, d(\sigma_y - u_a) + m_2^w d(u_a - u_w) \quad (15.3)$$

where

$dV_w/V_0$ = change in the volume of water in the soil element with respect to the initial volume of the element

$V_w$ = volume of water in the element

$m_{1k}^w$ = coefficient of water volume change with respect to a change in the net normal stress, $d(\sigma_y - u_a)$, for $K_0$-loading

$m_2^w$ = coefficient of water volume change with respect to a change in matric suction, $d(u_a - u_w)$, during $K_0$-loading.

The air phase constitutive relation is given by the difference between the soil structure and the water phase constitutive equations. The air phase constitutive equation can also be expressed in a general form as a function of the stress state variables:

$$\frac{dV_a}{V_0} = m_{1k}^a d(\sigma_y - u_a) + m_2^a \, d(u_a - u_w) \quad (15.4)$$

where

$dV_a/V_0$ = change in the volume of air in the soil element with respect to the initial volume of the element

$V_a$ = volume of air in the element

$m_{1k}^a$ = coefficient of air volume change with respect to a change in net normal stress, $d(\sigma_y - u_a)$

$m_2^a$ = coefficient of air volume change with respect to a change in matric suction, $d(u_a - u_w)$.

Figure 15.1 shows the relationship among the three constitutive surfaces described by the above equations. At any stress point on the constitutive surfaces, the coefficients of volume change are related in accordance with the continuity requirement. That is,

$$m_{1k}^a = m_{1k}^s - m_{1k}^w \quad (15.5)$$

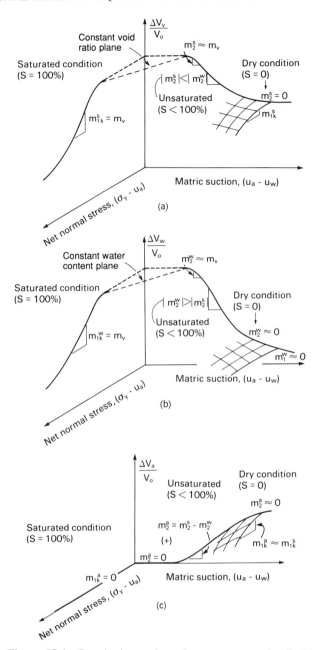

**Figure 15.1** Constitutive surfaces for an unsaturated soil. (a) Soil structure constitutive surface; (b) water phase constitutive surface; (c) air phase constitutive surface.

and

$$m_2^a = m_2^s - m_2^w. \quad (15.6)$$

When the soil is saturated (i.e., degree of saturation, $S$, at 100%), the four coefficients of volume change, $m_{1k}^s$, $m_2^s$, and $m_{1k}^w$, and $m_2^w$ are equal to the $m_v$ value used for a saturated soil. The $m_v$ coefficient can be obtained from a one-dimensional consolidation test on a saturated specimen. The $m_{1k}^a$ and $m_2^a$ values will be equal to zero under saturated conditions [Eqs. (15.5) and (15.6)]. As the soil becomes unsaturated (i.e., $S < 100\%$), the absolute values of the $m_{1k}^s$, $m_2^s$, $m_{1k}^w$, and $m_2^w$ coefficients decrease (Fig. 15.1). All of the above coefficients have a negative sign (Chapter 12), indicating that an increase in the stress state variable causes a volume decrease. In the unsaturated condition, however, the absolute magnitude of the $m_2^w$ coefficient will be greater than the absolute magnitude of the $m_2^s$ coefficient. In other words, an increase in matric suction causes a greater change in the volume of water in the soil than the overall volume change of the soil element. As a result, an increase in matric suction results in a decrease in the degree of saturation. On the other hand, the absolute magnitude of the $m_{1k}^s$ coefficient is greater than the absolute magnitude of the $m_{1k}^w$ coefficient because a net normal stress produces a greater change in the overall volume of the soil element than it produces for the change in water volume in the element.

As the soil approaches a dry condition (i.e., $S = 0$), the $m_2^s$ and $m_2^w$ coefficients approach zero (Fig. 15.1). Consequently the $m_2^a$ coefficient also approaches zero [Eq. (15.6)]. This means that a change in the matric suction when the degree of saturation of the soil approaches zero is no longer effective in producing a change in volume in the soil. Any volume change which takes place when the soil is in this dry condition will be due primarily to a change in the net normal stress. However, a net normal stress increase applied to a dry soil is unlikely to cause a change in the volume of water in the soil (i.e., $m_{1k}^w = 0$). Stated in another way, the total soil volume change in a dry soil is only associated with a change in the volume of air (i.e., $m_{1k}^s = m_{1k}^a$) due to a change in the net normal stress.

Darcy's and Fick's laws are used to describe the water and air flows through an unsaturated soil, respectively. These two flow laws have been presented in Chapter 5. Other required physical relations are the ideal gas law (Chapter 2) and the compressibility of air-water mixtures (Chapter 8).

## 15.3 DERIVATION OF CONSOLIDATION EQUATIONS

For a simple, isothermal, transient flow problem, two fluids can flow independently from an unsaturated soil, namely, water and air. Therefore, two independent partial differential equations are required to solve for the pore-water and pore-air pressures with respect to time (i.e., two-phase flow approach). The pore-water and pore-air partial differential equations must satisfy the continuity of the water and air phases, respectively. For two- and three-dimensional cases, it is necessary to couple the continuity equations with the equilibrium equations for a rigorous formulation (Dakshanamurthy et al., 1984). In most situations, it is satisfactory to use an uncoupled form of the continuity equation when solving problems in unsaturated soils.

For some problems, the excess pore-air pressure throughout the soil mass will be negligible or rapidly dissipated (Rahardjo, 1990). In these situations, only the water phase partial differential equation needs to be solved. This may be the case, for example, when predicting the swelling of an unsaturated soil as a result of a slow wetting from the boundary. For other problems, only the excess pore-air pressure may be of significance, and in these cases, only the pore-air partial differential equation needs to be solved.

The process associated with the application of a total stress to a soil gives the geotechnical engineer insight into the behavior of a soil. After applying a load, there is an immediate settlement and then a dissipation with time of the excess pore-water and pore-air pressures. In order to predict the pore-water and pore-air pressures as a function of time, it is necessary to simultaneously solve the pore-water and pore-air partial differential equations. The initial excess pore-air and pore-water pore pressures can be considered as initial conditions which can be evaluated using pore pressure parameters as described in Chapter 8. The application of a total load will always induce an excess pore-air pressure which is smaller than the excess pore-water pressure. The magnitude of each pressure depends upon the compressibility of the air-water mixture.

The one-dimensional transient flow equation for the water and air phases can be derived by equating the time derivative of the relevant constitutive equation to the divergence of the flow rate, as described by the flow law. The time derivative of the constitutive equation controls the amount of deformation that occurs under various stress conditions, while the divergence of the flow rates controls how fast the flow of air and water occurs.

The assumptions used in the derivation are similar to those proposed by Terzaghi for saturated soils, with some exceptions and additions.

1) The air phase is assumed to be continuous. However, even when the air phase becomes occluded, the coefficient of permeability with respect to the air phase approaches the diffusivity of air through water. At the other extreme, the excess pore-air pressure is assumed to remain at atmospheric pressure when the air voids are large. This negates the need for solving the air phase partial differential equation for this case.

2) The coefficients of volume change for the soil (i.e., $m_{1k}^w$, $m_2^w$, $m_{1k}^a$, and $m_2^a$) remain constant during the consolidation process. However, it is possible to make these coefficients a function of the stress state during the solution of the partial differential equations.

3) The coefficients of permeability with respect to the air and water phases are assumed to be a function of the stress state or the volume-mass soil properties during the consolidation process. However, the case where the coefficients of permeability are assumed to remain constant is also shown.

4) The effect of air diffusing through water, air dissolving in water, and the movement of water vapor are

ignored. In most practical problems, the initial and final states of stress in the air phase are the same (i.e., atmospheric). Therefore, although air may go into solution under increased pressure, it can also be assumed that it comes out of solution in response to a return to the boundary conditions prior to loading.

5) The soil particles and the pore–water are assumed to be incompressible.
6) Strains occurring during consolidation are assumed to be small.

The above assumptions are not completely accurate for all cases; however, these assumptions are reasonable as a first attempt in deriving a general consolidation theory for unsaturated soils.

### 15.3.1 Water Phase Partial Differential Equation

Consider a referential element of unsaturated soil, with air and water flow during one-dimensional consolidation (Fig. 15.2). The net flux of water through the element is computed from the volume of water entering and leaving the element within a period of time:

$$\frac{\partial V_w}{\partial t} = \left(v_w + \frac{\partial v_w}{\partial y} dy\right) dx\, dz - v_w dx\, dz \quad (15.7)$$

where

$dV_w$ = change in the volume of water in the soil element over a specific time, $dt$
$\partial V_w/\partial t$ = net flux of water through the soil element
$v_w$ = water flow rate across a unit area of the soil element in the $y$-direction
$dx, dy, dz$ = infinitesimal dimensions in the $x$-, $y$-, and $z$-directions, respectively.

Expressing the net flux of water *per unit volume* of the

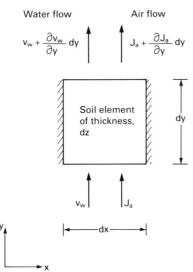

**Figure 15.2** Unsteady-state air and water flow during one-dimensional consolidation.

soil and rearranging Eq. (15.7) yields

$$\frac{\partial(V_w/V_0)}{\partial t} = \frac{\partial v_w}{\partial y} \quad (15.8)$$

where

$V_0$ = initial total volume of the soil element (i.e., $dx, dy, dz$)
$\partial(V_w/V_0)/\partial t$ = net flux of water per unit volume of the soil.

Substituting Darcy's law for the flow rate of water, $v_w$, into Eq. (15.8) gives

$$\frac{\partial(V_w/V_0)}{\partial t} = \frac{\partial\left(-k_w \dfrac{\partial h_w}{\partial y}\right)}{\partial y} \quad (15.9)$$

where

$k_w$ = coefficient of permeability with respect to water as a function of matric suction which varies with location in the $y$-direction [i.e., $k_w(u_a - u_w)$]
$h_w$ = hydraulic head (i.e., gravitational plus pore–water pressure head or $y + u_w/(\rho_w g)$)
$y$ = elevation
$u_w$ = pore–water pressures
$\rho_w$ = density of water
$g$ = gravitational acceleration
$\partial h_w/\partial y$ = hydraulic head gradient in the $y$-direction.

Rearranging Eq. (15.9) gives

$$\frac{\partial(V_w/V_0)}{\partial t} = -k_w \frac{\partial^2 h_w}{\partial y^2} - \frac{\partial k_w}{\partial y}\frac{\partial h_w}{\partial y}. \quad (15.10)$$

Let us substitute $(y + u_w/\rho_w g)$ for $h_w$ in Eq. (15.10):

$$\frac{\partial(V_w/V_0)}{\partial t} = -k_w \frac{\partial^2\left(y + \dfrac{u_w}{\rho_w g}\right)}{\partial y^2} - \frac{\partial k_w}{\partial y}\frac{\partial\left(y + \dfrac{u_w}{\rho_w g}\right)}{\partial y} \quad (15.11)$$

$$\frac{\partial(V_w/V_0)}{\partial t} = -\frac{k_w}{\rho_w g}\frac{\partial^2 u_w}{\partial y^2} - \frac{1}{\rho_w g}\frac{\partial k_w}{\partial y}\frac{\partial u_w}{\partial y} - \frac{\partial k_w}{\partial y}. \quad (15.12)$$

The water phase constitutive relation [Eq. (15.3)] defines the water volume change in a soil element caused by changes in the net normal stress, $d(\sigma_y - u_a)$, and the matric suction, $d(u_a - u_w)$. The flux of water *per unit volume* of the soil can be obtained by differentiating the water phase constitutive relation [Eq. (15.3)] with respect to time:

$$\frac{\partial(V_w/V_0)}{\partial t} = m_{1k}^w \frac{\partial(\sigma_y - u_a)}{\partial t} + m_2^w \frac{\partial(u_a - u_w)}{\partial t}. \quad (15.13)$$

The coefficients of volume change are assumed to be constant during the consolidation process. In other words, the coefficients of volume change are not a function of time, but only a function of the stress state variables. Their values can be updated in accordance with the stress state of the soil, and this process is explained later. The change in total stress with respect to time is generally set to zero during a consolidation process (i.e., $\partial \sigma_y / \partial t = 0$). Both equations for the flux of water [i.e., Eqs. (15.12) and (15.13)] can now be equated to give a partial differential equation for the water phase:

$$-m_{1k}^w \frac{\partial u_a}{\partial t} + m_2^w \frac{\partial u_a}{\partial t} - m_2^w \frac{\partial u_w}{\partial t}$$

$$= -\frac{k_w}{\rho_w g} \frac{\partial^2 u_w}{\partial y^2} - \frac{1}{\rho_w g} \frac{\partial k_w}{\partial y} \frac{\partial u_w}{\partial y} - \frac{\partial k_w}{\partial y}. \quad (15.14)$$

Rearranging Eq. (15.14) gives

$$m_2^w \frac{\partial u_w}{\partial t} = -(m_{1k}^w - m_2^w) \frac{\partial u_a}{\partial t} + \frac{k_w}{\rho_w g} \frac{\partial^2 u_w}{\partial y^2}$$

$$+ \frac{1}{\rho_w g} \frac{\partial k_w}{\partial y} \frac{\partial u_w}{\partial y} + \frac{\partial k_w}{\partial y}. \quad (15.15)$$

Equation (15.15) is a general form of the partial differential equation for the water phase. The equation can be simplified for special soil conditions, such as the fully saturated case, the dry case, or special cases of an unsaturated soil. The water phase partial differential equation for each condition is described in the following sections.

### Saturated Condition

The coefficients of water volume change, $m_{1k}^w$ and $m_2^w$, become equal to the coefficient of volume change, $m_v$, for a saturated soil (i.e., degree of saturation, $S = 100\%$) (Fig. 15.1). The coefficient of permeability, $k_w$, reverts to the coefficient of permeability at saturation, $k_s$. The saturated coefficient of permeability is usually assumed to remain constant during the consolidation process (i.e., $\partial k_s / \partial y = 0$). Substituting $m_v$ and making $\partial k_s / \partial y$ equal to 0 in Eq. (15.15) results in the Terzaghi form of the one-dimensional consolidation equation for saturated soils (Terzaghi, 1943) as presented in Eq. (15.1).

Equation (15.1) describes the pore-water pressure changes during one-dimensional consolidation of a saturated soil. The consolidation process of a saturated soil involves only the flow of water, which in turn produces a volume change.

### Dry soil condition

When the soil approaches a dry condition (i.e., $S$ approaches 0%), only a small amount of water is present around the soil particles. In this condition, changes in matric suction or net normal stress produce negligible changes in the volume of water. As a result, the coefficients of water volume change, $m_{1k}^w$ and $m_2^w$, and the coefficient of permeability, $k_w$, go to zero [Fig. 15.1(b)]. Any volume change that occurs in the dry soil is the result of the soil structure or the air phase volume change. Setting the coefficients of volume change and the coefficient of permeability to zero (i.e., $m_{1k}^w = m_2^w = 0$ and $k_w = 0$) in Eq. (15.15) causes the partial differential equation for the water phase to vanish. In other words, there is no water flow during the consolidation of a dry soil.

### Special Case of an Unsaturated Soil

In an unsaturated soil (i.e., $0 < S < 100\%$), air and water flows can take place simultaneously during consolidation. The consolidation equation for the water phase, presented in Eq. (15.15), can be rearranged as follows:

$$\frac{\partial u_w}{\partial t} = -C_w \frac{\partial u_a}{\partial t} + c_v^w \frac{\partial^2 u_w}{\partial y^2} + \frac{c_v^w}{k_w} \frac{\partial k_w}{\partial y} \frac{\partial u_w}{\partial y} + c_g \frac{\partial k_w}{\partial y}$$

$$(15.16)$$

where

$C_w$ = interactive constant associated with the water phase partial differential equation, i.e., $((1 - m_2^w / m_{1k}^w))/(m_2^w / m_{1k}^w))$
$c_v^w$ = coefficient of consolidation with respect to the water phase (i.e., $k_w / (\rho_w g \, m_2^w)$)
$c_g$ = gravity term constant (i.e., $1/m_2^w$).

The coefficient of permeability, $k_w$, can vary significantly with respect to matric suction, which in turn can vary in the $y$-direction. The permeability variation in the soil is taken into account by the last two terms in Eq. (15.16). Several relationships describing permeability functions, $k_w(u_a - u_w)$, were described in Chapter 5.

The inclusion of the gravitational component of the hydraulic head in the above derivation results in the last term of Eq. (15.16) [i.e., $(1/m_2^w)(\partial k_w / \partial y)$]. In some cases, this term can be considered negligible as compared to the other terms. If the gravity term is neglected and the coefficient of permeability, $k_w$, does not vary significantly with space (i.e., $(\partial k_w / \partial y)$ is negligible), a simplified form of the differential water flow equation can be written:

$$\frac{\partial u_w}{\partial t} = -C_w \frac{\partial u_a}{\partial t} + c_v^w \frac{\partial^2 u_w}{\partial y^2}. \quad (15.17)$$

The consolidation equation for the air phase is derived in the next section. However, there are many cases where the dissipation of the excess pore-air pressures occurs almost instantaneously (Rahardjo, 1990). In this case, only the water phase undergoes a transient process, and Eq. (15.17) can be further simplified to a form similar to Terzaghi's consolidation equation:

$$\frac{\partial u_w}{\partial t} = c_v^w \frac{\partial^2 u_w}{\partial y^2}. \quad (15.18)$$

where

$$c_v^w = k_w/(\rho_w g\, m_2^w).$$

### 15.3.2 Air Phase Partial Differential Equation

The air phase is compressible and flows in response to an air pressure gradient. Therefore, the flow of air through a referential element of unsaturated soil (Fig. 15.2) is computed in terms of the mass rate of air flow, $J_a$. The net mass rate of air flow across the element is obtained as the difference between the mass rates of air entering and leaving the element within a period of time:

$$\frac{\partial M_a}{\partial t} = \left(J_a + \frac{\partial J_a}{\partial y}dy\right)dx\,dz - J_a dx\,dz \quad (15.19)$$

where

$dM_a$ = change of air mass in the soil element for a specific time, $dt$
$\partial M_a/\partial t$ = net mass rate of air flowing through the soil element
$J_a$ = mass rate of air flowing across a unit area of the soil.

The net mass rate of air flow can be expressed for *a unit volume* of the soil element by rearranging Eq. (15.19) as follows:

$$\frac{\partial(M_a/V_0)}{\partial t} = \frac{\partial J_a}{\partial y} \quad (15.20)$$

where

$\partial(M_a/V_0)/\partial t$ = net mass rate of air flow per unit volume of the element
$V_0$ = initial total volume of the soil element (i.e., $dx\,dy\,dz$).

Substituting Fick's law for the mass rate of air flow, $J_a$, in Eq. (15.20), results in the following equation:

$$\frac{\partial(M_a/V_0)}{\partial t} = \frac{\partial\left(-D_a^*\dfrac{\partial u_a}{\partial y}\right)}{\partial y} \quad (15.21)$$

where

$D_a^*$ = coefficient of transmission for the air phase, which is a function of the volume–mass properties or matric suction of the soil; therefore, $D_a^*$ can vary with location in the $y$-direction
$u_a$ = pore-air pressure
$\partial u_a/\partial y$ = pore-air pressure gradient in the $y$-direction.

The mass of air, $M_a$, in Eq. (15.21) can be written in terms of the volume of air, $V_a$, and its density, $\rho_a$ (i.e., $M_a = V_a \rho_a$):

$$\frac{\partial(V_a \rho_a/V_0)}{\partial t} = -\left(D_a^* \frac{\partial^2 u_a}{\partial y^2} + \frac{\partial D_a^*}{\partial y}\frac{\partial u_a}{\partial y}\right) \quad (15.22)$$

where

$V_a$ = volume of air
$\rho_a$ = density of air.

Rearranging Eq. (15.22) gives

$$\rho_a \frac{\partial(V_a/V_0)}{\partial t} + \frac{V_a}{V_0}\frac{\partial \rho_a}{\partial t} = -D_a^* \frac{\partial^2 u_a}{\partial y^2} - \frac{\partial D_a^*}{\partial y}\frac{\partial u_a}{\partial y}. \quad (15.23)$$

The volume of air can be related to the volume–mass properties of the soil:

$$V_a = (1 - S)n\,V \quad (15.24)$$

where

$S$ = degree of saturation
$n$ = porosity
$V$ = current total volume of the soil element.

The total volume change of an unsaturated soil can be assumed to be small (i.e., small strains) during the consolidation process. This assumption is justifiable for an unsaturated soil that has a relatively rigid soil structure. Therefore, the current total volume of the soil, $V$, in Eq. (15.24) can be assumed to be equal to the initial total volume of the soil, $V_0$. Substituting Eq. (15.24) into Eq. (15.23) yields

$$\rho_a \frac{\partial(V_a/V_0)}{\partial t} + (1-S)n\frac{\partial \rho_a}{\partial t}$$

$$= -D_a^* \frac{\partial^2 u_a}{\partial y^2} - \frac{\partial D_a^*}{\partial y}\frac{\partial u_a}{\partial y}. \quad (15.25)$$

The density of air is a function of air pressure in accordance with the ideal gas law (Chapter 2):

$$\rho_a = \frac{\omega_a}{RT}\bar{u}_a \quad (15.26)$$

where

$\omega_a$ = molecular mass of air (kg/kmol)
$R$ = universal (molar) gas constant [i.e., 8.31432 J/(mol · K)]
$T$ = absolute temperature (i.e., $T = t° + 273.16$) (K)
$t°$ = temperature (°C)
$\bar{u}_a$ = absolute pore-air pressure (i.e., $\bar{u}_a = u_a + \bar{u}_{\text{atm}}$) (kPa)
$u_a$ = gauge pore-air pressure (kPa)
$\bar{u}_{\text{atm}}$ = atmospheric pressure (i.e., 101 kPa or 1 atm).

Replacing the air density, $\rho_a$, in Eq. (15.25) with Eq. (15.26) gives

$$\left(\frac{\omega_a}{RT}\right)\bar{u}_a \frac{\partial(V_a/V_0)}{\partial t} + (1-S)n\left(\frac{\omega_a}{RT}\right)\frac{\partial u_a}{\partial t}$$

$$= D_a^* \frac{\partial^2 u_a}{\partial y^2} - \frac{\partial D_a^*}{\partial y}\frac{\partial u_a}{\partial y}. \quad (15.27)$$

Equation (15.27) can be further rearranged to give the air flux *per unit volume* of the soil element:

$$\frac{\partial(V_a/V_0)}{\partial t} = -\frac{D_a^*}{(\omega_a/RT)\bar{u}_a}\frac{\partial^2 u_a}{\partial y^2}$$

$$-\frac{(1-S)n}{\bar{u}_a}\frac{\partial u_a}{\partial t} - \frac{1}{(\omega_a/RT)\bar{u}_a}$$

$$\cdot \frac{\partial D_a^*}{\partial y}\frac{\partial u_a}{\partial y}. \quad (15.28)$$

The air phase constitutive relation [Eq. (15.4)] defines the air volume change in the soil element due to changes in net normal stress, $d(\sigma_y - u_a)$, and matric suction, $d(u_a - u_w)$. The derivative of the air phase constitutive relation with respect to time is equal to the air flux *per unit volume* of the soil element:

$$\frac{\partial(V_a/V_0)}{\partial t} = m_{1k}^a \frac{\partial(\sigma_y - u_a)}{\partial t} + m_2^a \frac{\partial(u_a - u_w)}{\partial t}. \quad (15.29)$$

In differentiating Eq. (15.4), the coefficients of volume change, $m_{1k}^a$ and $m_2^a$, are assumed to be constant during consolidation. Their magnitudes are a function of the stress state variables and can be updated accordingly, if desired (see later sections). The total stress change with respect to time is set to zero during the consolidation process (i.e., $\partial \sigma_y / \partial t = 0$). Equating both equations for the flux of air [i.e., Eqs. (15.28) and (15.29)] gives rise to the partial differential equation for the air phase:

$$-m_{1k}^a \frac{\partial u_a}{\partial t} + m_2^a \frac{\partial u_a}{\partial t} - m_2^a \frac{\partial u_w}{\partial t}$$

$$= -\frac{D_a^*}{(\omega_a/RT)\bar{u}_a}\frac{\partial^2 u_a}{\partial y^2} - \frac{(1-S)n}{\bar{u}_a}\frac{\partial u_a}{\partial t}$$

$$-\frac{1}{(\omega_a/RT)\bar{u}_a}\frac{\partial D_a^*}{\partial y}\frac{\partial u_a}{\partial y}. \quad (15.30)$$

Rearranging Eq. (15.30) gives

$$-\left(m_{1k}^a - m_2^a - \frac{(1-S)n}{\bar{u}_a}\right)\frac{\partial u_a}{\partial t}$$

$$= m_2^a \frac{\partial u_w}{\partial t} - \frac{D_a^*}{(\omega_a/RT)\bar{u}_a}\frac{\partial^2 u_a}{\partial y^2}$$

$$-\frac{1}{(\omega_a/RT)\bar{u}_a}\frac{\partial D_a^*}{\partial y}\frac{\partial u_a}{\partial y}. \quad (15.31)$$

Equation (15.31) is a general form of the air phase partial differential equation. The equation can be simplified for special cases such as the fully saturated case, the dry soil case, or special cases of an unsaturated soil. The air phase partial differential equation for each of these cases is outlined in the following sections.

### Saturated Soil Condition

For the saturated condition, the degree of saturation goes to 100%, and the coefficients of air volume change, $m_{1k}^a$ and $m_2^a$, become equal to zero [Fig. 15.1(c)]. The coefficient of transmission, $D_a^*$, approaches zero, indicating the absence of air flow. Air may exist in the form of occluded bubbles which have a pressure equal to that of the water phase (i.e., $u_a - u_w) = 0$). At this point, air movement may occur in the form of air diffusion through the pore-water. In other words, only water flow takes place during consolidation [Eq. (15.1)].

When occluded air bubbles are present, the water phase becomes more compressible than that of pure water. In this case, a rigorous solution of Eq. (15.1) requires appropriate soil coefficient (i.e., $c_v$, $k_s$, $\rho_w$, and $m_v$), corresponding to a compressible pore fluid.

### Dry Soil Condition

In the dry condition (i.e., $S$ approaching 0), a change in matric suction produces negligible volume change. In other words, all coefficients of volume change with respect to matric suction approach zero (i.e., $m_2^s = m_2^w = m_2^a = 0$). For a compressible soil structure, a volume change may still occur due to a change in net normal stress. In this case, the soil volume change is equal to the air phase volume change [i.e., $m_{1k}^a = m_{1k}^s$ in Fig. 15.1(c)]. The coefficient of transmission, $D_a^*$, reverts to a constant coefficient, $D_d^*$, corresponding to a dry condition. Therefore, Eq. (15.31) can be simplified to the following form for a compressible soil under a dry condition:

$$\left(m_{1k}^a + \frac{n}{\bar{u}_a}\right)\frac{\partial u_a}{\partial t} = \frac{D_d^*}{(\omega_a/RT)\bar{u}_a}\frac{\partial^2 u_a}{\partial y^2}. \quad (15.32)$$

Rearranging Eq. (15.32) results in a consolidation equation similar in form to Terzaghi's one-dimensional consolidation equation:

$$\frac{\partial u_a}{\partial t} = \frac{D_d^*}{(\omega_a/RT)}\frac{1}{(m_{1k}^a \bar{u}_a + n)}\frac{\partial^2 u_a}{\partial y^2}. \quad (15.33)$$

Equation (15.33) describes the pore–air pressure changes during one-dimensional consolidation of a dry, compressible soil. Only air flow occurs during the consolidation process. The coefficient used in Eq. (15.33) is similar in form to the coefficient of consolidation, $c_v$. In the case where the soil is dry and incompressible (i.e., $m_{1k}^a = m_{1k}^s = 0$), Eq. (15.33) reverts to the equation proposed by Blight (1971):

$$\frac{\partial u_a}{\partial t} = \frac{D_d^*/n}{(\omega_a/RT)}\frac{\partial^2 u_a}{\partial y^2}. \quad (15.34)$$

### Special Case of an Unsaturated Soil

In the unsaturated soil condition, air and water may flow simultaneously during the consolidation process. The air phase consolidation equation [i.e., Eq. (15.31)] can be written in a simplified form:

$$\frac{\partial u_a}{\partial t} = -C_a \frac{\partial u_w}{\partial t} + c_v^a \frac{\partial^2 u_a}{\partial y^2} + \frac{c_v^a}{D_a^*} \frac{\partial D_a^*}{\partial y} \frac{\partial u_a}{\partial y} \quad (15.35)$$

where

$C_a$ = interactive constant associated with the air phase partial differential equation, i.e.,

$$\left[ \frac{m_2^a/m_{1k}^a}{1 - m_2^a/m_{1k}^a - (1 - S)n/(\bar{u}_a m_{1k}^a)} \right]$$

$c_v^a$ = coefficient of consolidation with respect to the air phase, i.e.,

$$\left[ \frac{D_a^*}{(\omega_a/RT)} \frac{1}{\bar{u}_a m_{1k}^a (1 - m_2^a/m_{1k}^a) - (1 - S)n} \right].$$

The coefficient of transmission, $D_a^*$, is a function of matric suction which can vary in the $y$-direction. The transmissivity variation of the air phase is taken into account by the last term in Eq. (15.35). Several relationships describing the air transmissivity variation are explained in Chapter 5. However, if the air transmissivity variation with space is negligible (i.e., $(\partial D_a^*/\partial y)$ is negligible), Eq. (15.35) can be simplified:

$$\frac{\partial u_a}{\partial t} = -C_a \frac{\partial u_w}{\partial t} + c_v^a \frac{\partial^2 u_a}{\partial y^2}. \quad (15.36)$$

## 15.4 SOLUTION OF CONSOLIDATION EQUATIONS USING FINITE DIFFERENCE TECHNIQUE

The use of the finite difference technique in solving a one-dimensional, steady-state water flow equation was described in Chapter 7. Similarly, the finite difference technique can be used to solve the unsteady-state (i.e., transient) air and water flow equations for one-dimensional consolidation. The air and water flow partial differential equations can be solved simultaneously using an explicit forward difference technique, as illustrated in Fig. 15.3. A special finite difference procedure can be used to linearize the nonlinear partial differential flow equations for the air and water phases (Dakshanamurthy and Fredlund, 1981).

Let us consider one-dimensional consolidation where air and water flows occur simultaneously. The spatial variation in the water coefficient of permeability and the air coefficient of transmission are assumed to be negligible. Therefore, Eqs. (15.17) and (15.36) are used as the partial differential equations for the water and air phases, respectively. The transient water flow equation [i.e., Eq. (15.17)] can be written in a finite difference form:

$$\frac{u_{w(i,j+1)} - u_{w(i,j)}}{\Delta t} = -C_w \frac{u_{a(i,j+1)} - u_{a(i,j)}}{\Delta t} + c_v^w \frac{u_{w(i+1,j)} - 2u_{w(i,j)} + u_{w(i-1,j)}}{\Delta y^2} \quad (15.37)$$

where

$i$ = space increment in the $y$-direction
$j$ = time increment.

The transient air flow equation [i.e., Eq. (15.36)] can be

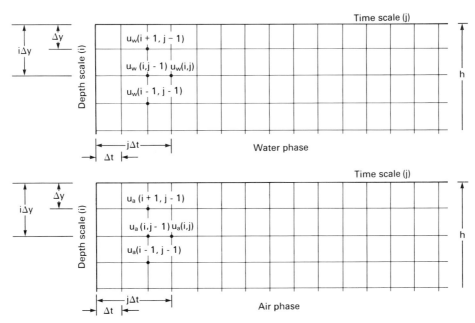

**Figure 15.3** Finite difference mesh for solving the transient air and water flow equations.

written in a finite difference form as follows:

$$\frac{u_{a(i,j+1)} - u_{a(i,j)}}{\Delta t}$$
$$= -C_a \frac{u_{w(i,j+1)} - u_{w(i,j)}}{\Delta t}$$
$$+ c_v^a \frac{u_{a(i+1,j)} - 2u_{a(i,j)} + u_{a(i-1,j)}}{\Delta y^2}. \quad (15.38)$$

The pore-water pressure is solved by multiplying Eq. (15.38) by $(-C_w)$ and substituting it into Eq. (15.37):

$$\frac{u_{w(i,j+1)} - u_{w(i,j)}}{\Delta t}$$
$$= C_a C_w \frac{u_{w(i,j+1)} - u_{w(i,j)}}{\Delta t}$$
$$- c_v^a C_w \frac{u_{a(i+1,j)} - 2u_{a(i,j)} + u_{a(i-1,j)}}{\Delta y^2}$$
$$+ c_v^w \frac{u_{w(i+1,j)} - 2u_{w(i,j)} + u_{w(i-1,j)}}{\Delta y^2}. \quad (15.39)$$

Simplifying and rearranging Eq. (15.39) allows the unknown pore-water pressure [i.e., at a given time step, $(j + 1)$] to be put on the left-hand side of the equation:

$$u_{w(i,j+1)} = u_{w(i,j)} + \frac{\beta_w g_1^w}{(1 - C_a C_w)}$$
$$- \left(\frac{C_w}{1 - C_a C_w}\right) \beta_a f_1^a \quad (15.40)$$

where

$$\beta_w = c_v^w \frac{\Delta t}{\Delta y^2}$$
$$\beta_a = c_v^a \frac{\Delta t}{\Delta y^2}$$
$$g_1^w = u_{w(i+1,j)} - 2u_{w(i,j)} + u_{w(i-1,j)}$$
$$f_1^a = u_{a(i+1,j)} - 2u_{a(i,j)} + u_{a(i-1,j)}.$$

Similarly, the pore-air pressure can be obtained by multiplying Eq. (15.37) by $(-C_a)$ and substituting this equation into Eq. (15.38):

$$\frac{u_{a(i,j+1)} - u_{a(i,j)}}{\Delta t}$$
$$= C_a C_w \frac{u_{a(i,j+1)} - u_{a(i,j)}}{\Delta t}$$
$$- c_v^w C_a \frac{u_{w(i+1,j)} - 2u_{w(i,j)} + u_{w(i-1,j)}}{\Delta y^2}$$
$$+ c_v^a \frac{u_{a(i+1,j)} - 2u_{a(i,j)} + u_{a(i-1,j)}}{\Delta y^2}. \quad (15.41)$$

Simplifying and rearranging Eq. (15.41) allow the unknown pore-air pressure [i.e., at a given time step, $(j + 1)$] to be put on the left-hand side and all known variables from the previous time step (i.e., $j$) to be put on the right-hand side of the equation:

$$u_{a(i,j+1)} = u_{a(i,j)} - \left(\frac{C_a}{1 - C_a C_w}\right) \beta_w g_1^w$$
$$+ \frac{\beta_a f_1^a}{(1 - C_a C_w)}. \quad (15.42)$$

The pore-water and pore-air pressures at a given time step are computed from the known values at the previous time step using Eqs. (15.40) and (15.42), respectively. When computations for both pore-water and pore-air pressures for all depth steps are completed, there can then be a march forward to the next time step. The above procedure can be repeated until equilibrium has been achieved for the air and water phases. The solution of the finite difference equations is obtained by ensuring that the values for $\beta_w$ and $\beta_a$ are less than 0.5 in order to satisfy stability conditions (Desai and Christian, 1977).

The initial and boundary conditions for the problem must be established prior to performing the above finite difference computations. The initial conditions for one-dimensional consolidation constitute the initial and excess pore pressures corresponding to the instant after the application of the total load. The excess pore-air and pore-water pressures can be calculated using the pore-air and pore-water pressure parameters explained in Chapter 8. The pore pressure parameters can be used to compute the magnitude of the excess pore pressures generated in response to the application of the total load.

In addition to the initial conditions, the boundary conditions for the problem must be specified at the boundaries during the consolidation process. At a free-draining boundary, the pore pressures must be set to the initial values at all times. At an impervious boundary, the pore pressure gradients must be set to zero since there is no flow across the boundary.

A mixed boundary condition can also be established for special applications such as occur in laboratory tests. As an example, laboratory equipment is commonly designed such that air flows upward, while water flows downward during a test. In this case, the top boundary is a free-draining boundary for the air phase and an impervious boundary for the water phase. The reverse condition exists at the bottom boundary.

Having set up the initial and final boundary conditions, the dissipation of the excess pore-air and pore-water pressures can be computed using the finite difference schemes shown by Eqs. (15.40) and (15.42), respectively.

The pore pressure changes can be used to compute the volume-mass changes in the soil through use of the soil constitutive equations, as explained in Chapter 12. In other words, changes in the water content, void ratio, and degree of saturation of the soil can be computed at any time during the consolidation process.

## 15.5 TYPICAL CONSOLIDATION TEST RESULTS FOR UNSATURATED SOILS

Several typical experimental results from consolidation tests on unsaturated soils are presented in this section to illustrate the consolidation behavior of an unsaturated soil. The equipment, the soil, and the test procedures used in each experiment are also described. The tests require the independent measurement or control of pore-air and pore-water pressures on the boundaries of the soil specimens during consolidation. The results were then best-fitted with the theory presented in this chapter.

### 15.5.1 Tests on Compacted Kaolin

Laboratory experiments were performed on several compacted kaolin specimens using a modified Anteus oedometer and a modified triaxial cell (Fredlund, 1973). The modified Anteus oedometer was used to perform one-dimensional consolidation tests. The modified Wykeham Farrance triaxial cell was used to perform isotropic volume change tests.

A high air entry disk was sealed to the lower pedestal. This disk allowed water to flow, but prevented the flow of air. Therefore, the measurement (or control) of the pore-air and pore-water pressures could be made independently. A low air entry disk was placed on the top of the specimen to facilitate the control of the pore-air pressure. This allowed the translation of the air and the water pressures to positive values in order to prevent cavitation of the water below the high air entry disk [i.e., the axis-translation technique (Hilf, 1956)]. The total, pore-air, and pore-water pressure conditions could therefore be controlled in the study of the volume change behavior of unsaturated soils.

Two rubber membranes separated by a slotted tin foil were placed around the specimen in order to reduce air diffusion. This composite membrane was found to be essentially impermeable to air and water over a long period of time.

### Presentation of Results

Five specimens of kaolin were compacted in accordance with the standard Proctor procedure. The initial volume-mass properties for each specimen are summarized in Table 15.1. Each experiment was performed by changing one of the components of the stress state variables [i.e., the total stress ($\sigma$), the pore-water pressure ($u_w$), or the pore-air pressure ($u_a$)]. The stress state component changes associated with each experiment are given in Table 15.2. In each case, the volume changes of the soil structure and water phase were monitored. However, the pore-air and pore-water pressure changes within the soil specimen, during the transient process, were not measured.

Table 15.2 indicates that only Test 1 was a consolidation test in the classical sense where the air and water flows were caused by an increase in the total stress. In other tests, the transient flows were caused by a change in either the pore-air or pore-water pressures at the boundary. In all tests, the air and water flowed one-dimensionally. Therefore, the one-dimensional consolidation theory can be used to describe the transient flows of air and/or water during each test.

Figure 15.4 shows a typical plot of volume changes due to an increment of total stress in Test 1. The plot shows a decrease in the volume of the soil structure, the air, and the water phases during the transient process. A large instantaneous volume decrease occurred at the time when the load was applied.

An increase in the pore-air pressure in Test 2 caused the soil structure to expand temporarily (Fig. 15.5). However, the increase in air pressure resulted in an increase in matric suction ($u_a - u_w$), which in turn caused a decrease in the volume of the soil structure at the end of the process.

The volume change processes associated with a metastable-structured soil are shown in Fig. 15.6. The decrease in air pressure (Test 3) reduced the matric suction ($u_a - u_w$) and allowed more water to flow into the specimen. The intake of water appears to have reduced the normal and shear stresses between the soil particles. As a result, the soil structure underwent a decrease in volume (i.e., a collapse phenomenon).

Table 15.1 Initial Volume-Mass Properties for Specimens Tested

| Test No.[a] | Diameter (cm) | Height (cm) | Total Volume (cm³) | Water Content, $w$ (%) | Void Ratio, $e$ | Dry Unit weight, $\gamma_d$ (kN/m³) | Degree of Saturation (%) |
|---|---|---|---|---|---|---|---|
| 1 | 10.006 | 11.815 | 929.09 | 34.32 | 1.0696 | 13.185 | 78.87 |
| 2 | 9.945 | 11.703 | 909.10 | 33.17 | 1.0251 | 13.298 | 80.61 |
| 3 | 10.543 | 5.867 | 503.53 | 29.62 | 1.2242 | 11.529 | 63.29 |
| 4 | 9.832 | 5.758 | 437.16 | 32.12 | 0.9310 | 13.281 | 90.25 |
| 5 | 6.350 | 2.283 | 72.29 | 31.18 | 1.1247 | 12.069 | 72.51 |

[a] All tests performed in a modified triaxial apparatus except Test 5, which was performed in a modified oedometer.

**Table 15.2 Change in Stress State Variable Components Associated with Each Test**

| Test No.[a] | Total Stress, $\sigma$ (kPa) | | Pore–Water Pressure, $u_w$ (kPa) | | Pore–Air Pressure, $u_a$ (kPa) | | Change (kPa) |
|---|---|---|---|---|---|---|---|
| | Initial | Final | Initial | Final | Initial | Final | |
| 1 | 358.7 | 560.9 | 163.8 | 164.4 | 214.4 | 215.6 | $\Delta\sigma = +202.2$ |
| 2 | 560.9 | 559.0 | 164.4 | 163.1 | 215.6 | 421.1 | $\Delta u_w = +205.5$ |
| 3 | 475.1 | 476.8 | 41.9 | 42.2 | 397.8 | 206.5 | $\Delta u_a = -191.3$ |
| 4 | 611.4 | 610.1 | 177.3 | 379.2 | 532.0 | 530.9 | $\Delta u_w = +201.9$ |
| 5 | 606.7 | 605.2 | 216.5 | 323.6 | 413.8 | 413.8 | $\Delta u_w = +107.1$ |

[a]All tests performed in a modified triaxial apparatus except Test 5, which was performed in a modified oedometer.

An increase in water pressure or a decrease in matric suction causes an increase in volume of the soil structure (Fig. 15.7 for Test 4) when the soil has a stable structure and is expansive in nature. On the other hand, when the soil structure is metastable, an increase in water pressures or a decrease in matric suction causes the soil structure to decrease in volume or collapse (Fig. 15.8).

*Theoretical analyses*

Attempts were made to best-fit the theoretical analyses of volume change with the results from laboratory experiments (Fredlund and Rahardjo, 1986). This was accomplished by approximating the coefficients of volume change for the soil structure, air, and water phases, based on the laboratory results. Table 15.3 summarizes the approximate coefficients of volume change for each of the five specimens.

The magnitudes of the coefficients of volume change for each phase were obtained by dividing the amount of deformation at the end of each process by the change in the stress state variable. A sign (i.e., positive or negative) is attached to each of the coefficients of volume change, based on the direction (i.e., increase or decrease) of the volume change associated with each phase and the change in the stress state variables.

Table 15.3 indicates that the coefficients of volume

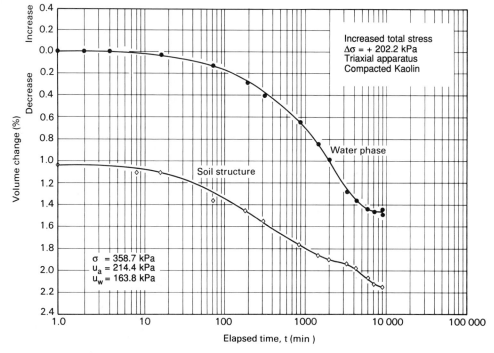

**Figure 15.4** Soil structure and water phase volume changes associated with an increase in total stress (Test 1).

15.5 TYPICAL CONSOLIDATION TEST RESULTS FOR UNSATURATED SOILS    431

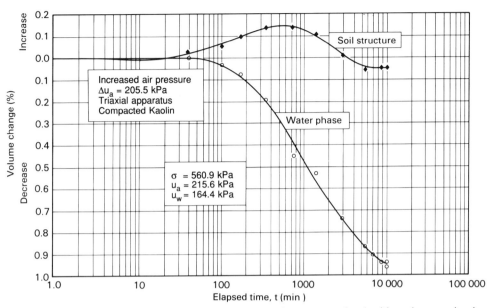

**Figure 15.5** Soil structure and water phase volume changes associated with an increase in air pressure (Test 2).

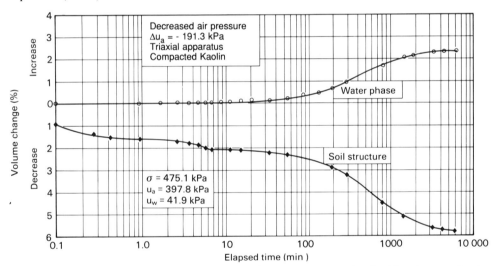

**Figure 15.6** Soil structure and water phase volume changes associated with a decrease in air pressure (Test 3).

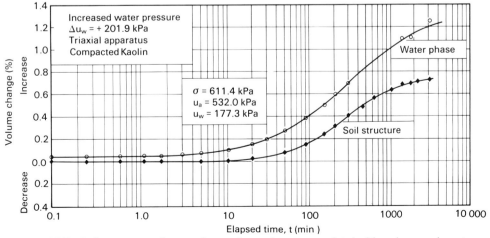

**Figure 15.7** Soil structure and water phase volume changes associated with an increase in water pressure (Test 4).

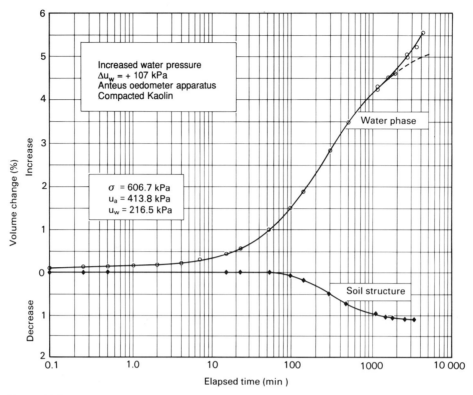

**Figure 15.8** Soil structure and water phase volume changes associated with an increase in water pressure (Test 5).

change for the soil structure have negative signs for a stable-structured soil (Tests 1, 2, and 4) and positive signs for a metastable-structured soil (Tests 3 and 5). A negative coefficient of volume change indicates an increase in volume for a decrease in the stress state variable, while a positive coefficient of volume change indicates the converse.

The theoretical analysis for each of the laboratory results was performed by assuming a one-dimensional transient flow process. The pore–water and pore–air pressures during the transient process were computed by simultaneously solving Eqs. (15.40) and (15.42) using the explicit central difference technique. The coefficients of permeability for air and water were assumed to be constant during the transient process.

The volume changes associated with the water and air phases during the transient process were computed using Eqs. (15.3) and (15.4). The coefficients of volume change used in the calculations were assumed to be constant throughout the process. The soil structure volume change was obtained by adding the volume changes associated with the water and air phases, in keeping with the continuity requirement.

**Table 15.3 Coefficients of Volume Change for Each Specimen**

| | Soil Structure | | Water Phase | | Air Phase | |
|---|---|---|---|---|---|---|
| Test No. | $m_1^s$ $\times 10^{-4} \text{kPa}^{-1}$ | $m_2^s$ $\times 10^{-4} \text{kPa}^{-1}$ | $m_1^w$ $\times 10^{-4} \text{kPa}^{-1}$ | $m_2^w$ $\times 10^{-4} \text{kPa}^{-1}$ | $m_1^a$ $\times 10^{-4} \text{kPa}^{-1}$ | $m_2^a$ $\times 10^{-4} \text{kPa}^{-1}$ |
| 1 | −1.17 | −3.52 | −0.80 | −2.41 | −0.37 | −1.11 |
| 2 | −0.006 | −0.032 | −0.131 | −0.657 | +0.125 | +0.625 |
| 3 | +0.76 | +3.80 | −0.30 | −1.49 | +1.06 | +5.29 |
| 4 | −0.09 | −0.35 | −0.15 | −0.61 | +0.06 | +0.26 |
| 5 | +0.26 | +1.04 | −1.20 | −4.81 | +1.46 | +5.85 |

## 15.5 TYPICAL CONSOLIDATION TEST RESULTS FOR UNSATURATED SOILS

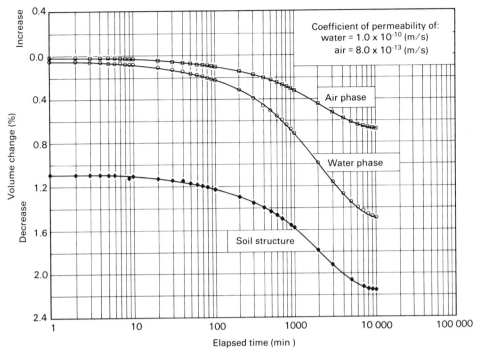

**Figure 15.9** Theoretical simulation of a consolidation test on an unsaturated compacted kaolin (Test 1).

Figures 15.9–15.13 show the theoretical analyses associated with Test 1–5. The fitting was accomplished by using different combinations for the coefficients of permeability for the water and air phases. The combinations of the water and air coefficients of permeability that gave the best-fit results for each test are shown in each figure. Comparisons between the theoretical analyses and laboratory results were made for Tests 4 and 3 (Figs. 15.14 and 15.15). The results indicate reasonably close agreement between the theoretical analyses and the results from the laboratory tests. However, some discrepancies can be observed during the transient process. The disagreements may be due to one or more reasons. For example, the assumption was made that the coefficients of permeability were constant throughout the process. In spite of this approximation, the theoretical analyses and the laboratory results show similar volume change behavior with respect to time.

### 15.5.2 Tests on Silty Sand

One-dimensional consolidation tests were conducted on a silty sand using a special $K_0$-cylinder designed by Rahardjo (1990). The $K_0$-cylinder allowed for the simultaneous measurement of pore-air and pore-water pressures at various depths along the soil specimen. The top and bottom of the soil specimen were connected to a low and a high air entry disk, respectively. Therefore, the air flowed upward and the water flowed downward during the consolidation test. The total and water volume changes were measured independently.

### Presentation of Results

Slurry specimens of the silty sand were subjected to a gradually increasing net normal stress and matric suction. The

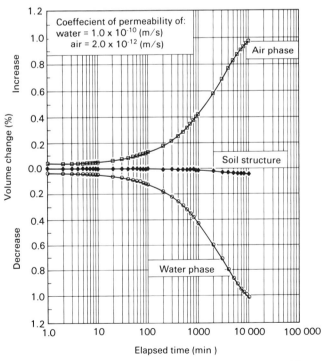

**Figure 15.10** Theoretical simulation of a laboratory test with an increase in air pressure (Test 2).

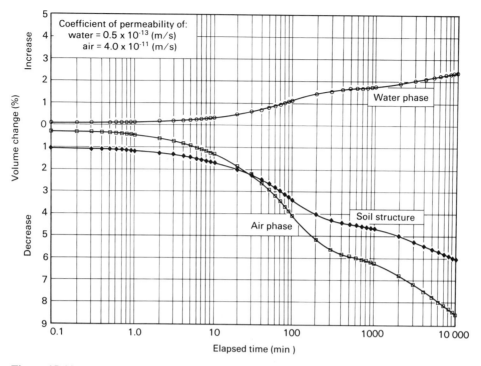

**Figure 15.11** Theoretical simulation of a laboratory test with a decrease in air pressure (Test 3).

specimens were then subjected to an increase in total stress under an undrained condition. Having fully developed the excess pore-air and pore-water pressures, the air and water phases were subsequently allowed to drain during a consolidation test. As a result, the pore-air and pore-water pressures reduced to their values prior to undrained loading. In other words, the net normal stress was increased, and the matric suction was returned (i.e., increased) to its original value at the end of the consolidation test.

Figure 15.16 presents the results of pore pressure and

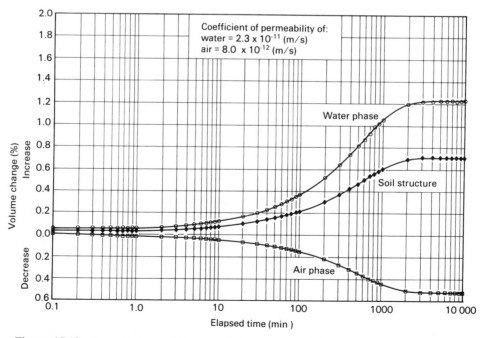

**Figure 15.12** Theoretical simulation of a laboratory test with an increase in water pressure (Test 4).

15.5 TYPICAL CONSOLIDATION TEST RESULTS FOR UNSATURATED SOILS    435

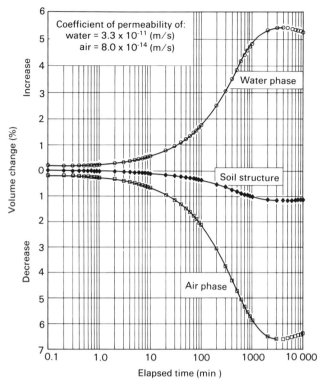

Figure 15.13 Theoretical simulation of a laboratory test with an increase in water pressure (Test 5).

**Table 15.4 Initial Soil Properties and Stress State Variables for the S1FG Consolidation Test on Silty Sand**

| Initial Volume-Mass | Properties |
| --- | --- |
| $S$ | 98.0% |
| $e$ | 0.578 |
| $w$ | 21.2% |
| $n$ | 0.366 |
| Stress State Variables (kPa) | |
| $\sigma$ | 184.3 |
| $u_a$ (top) | 119.5 |
| $u_w$ (bottom) | 106.7 |
| $\sigma - u_a$ | 64.8 |
| $d(\sigma - u_a)$ | |
| $u_a - u_w$ | 12.8 |
| $d(u_a - u_w)$ | |

volume change measurements during the consolidation test. The initial volume-mass properties of the soil specimen and its stress state variables are shown in Table 15.4.

The results shown in Fig. 15.16(a) indicate that the excess pore-air pressures dissipated instantaneously throughout the soil specimen. This characteristic was observed in every consolidation test performed on the silty sand (Rahardjo, 1990). On the other hand, the excess pore-water pressure dissipation occurred in a time-dependent manner (i.e., a transient process), as demonstrated in Fig. 15.16(a). During the transient process, the water volume change (i.e., 10.5 cm$^3$) exceeded the total volume change (i.e., 6 cm$^3$), as shown in Fig. 15.16(b). However, the soil structure underwent an immediate volume decrease of 9 cm$^3$ during the undrained loading, prior to the pore pressure dissipation process. This immediate volume change was caused by compression of the soil structure and the air-water mixture, as explained in Chapter 8. In total, the soil structure underwent a larger volume change than did the water phase in response to an increase in total stress.

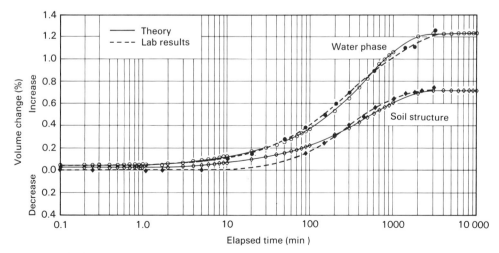

Figure 15.14 Comparisons between the theoretical simulation and the laboratory test results for Test 4 (i.e., an increase in water pressure).

**436** 15 ONE-DIMENSIONAL CONSOLIDATION AND SWELLING

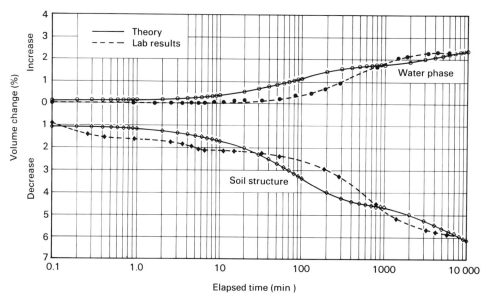

**Figure 15.15** Comparisons between the theoretical simulation and the laboratory test results for Test 3 (i.e., a decrease in air pressure).

### Theoretical Analysis

Another set of results from a consolidation test on the silty sand is presented in Fig. 15.17. The total stress was increased under a constant water content condition prior to consolidation. As a result, only excess pore-water pressures were developed during loading. The excess pore-water pressure dissipation during the consolidation process is shown in Fig. 15.17(a) for various depth and time intervals. The pore-water pressure isochrones were best-fitted with the water flow equation using a Terzaghi-type consolidation equation [i.e., Eq. (15.18)]. This equation was used since there was no air flow during the transient process. The coefficients of permeability and volume change for the water phase were assumed to be constant during consolidation, and this resulted in a best-fit, constant coef-

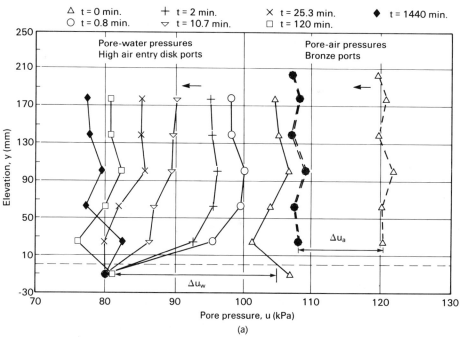

**Figure 15.16** Pore pressures and volume changes during a consolidation test on a silty sand (S1FG test). (a) Pore-air and pore-water pressure isochrones; (b) total and water volume changes (from Rahardjo, 1990).

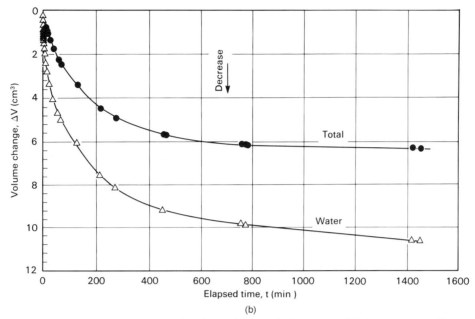

**Figure 15.16** Pore pressures and volume changes during a consolidation test on a silty sand (S1FG test). (a) Pore-air and pore-water pressure isochrones; (b) total and water volume changes (from Rahardjo, 1990). (*Continued*)

ficient of consolidation, $c_v^w$, of $1.2 \times 10^{-6}$ m²/s. There is good agreement between the theoretical and experimental results, as demonstrated in Fig. 15.17(a).

The total water volume changes during the consolidation process [i.e., Fig. 15.17(b)] are similar to those presented in Fig. 15.16(b). During the consolidation process following a constant water content loading, the net normal stress remained constant, while the matric suction increased due to the dissipation of the excess pore-water pressures. The total and water volume changes were simulated using Eqs. (15.2) and (15.3), respectively. The coefficients of volume change that give the best-fit results were found to be $8 \times 10^{-6}$ and $3.8 \times 10^{-5}$ kPa$^{-1}$ for the $m_2^s$ and $m_2^w$, respectively.

The above examples have been used to demonstrate the use of the theory presented in this chapter for describing the consolidation behavior of an unsaturated soil. In the use of the consolidation equation, or any transient process in an unsaturated soil, it is important to properly assess the coefficients of volume change (i.e., Chapter 13) and the coefficients of permeability (i.e., Chapter 6) of the soil under investigation.

## 15.6 DIMENSIONLESS CONSOLIDATION PARAMETERS

The simultaneous solution of the pore-water and pore-air partial differential equations [i.e., Eqs. (15.17) and (15.36)] can be generalized in terms of dimensionless numbers similar to those used for saturated soils. The dimensionless numbers are the average degree of consolidation and the time factor for the water and air phases. The average degree of consolidation for the water phase is defined as

$$U_w = 1 - \frac{\int_0^{2d} u_w \, dy}{\int_0^{2d} u_{w0} \, dy} \quad (15.43)$$

where

$U_w$ = average degree of consolidation with respect to the water phase
$u_{w0}$ = initial pore-water pressure
$u_w$ = pore-water pressure at any time
$d$ = length of drainage path; double drainage assumed.

The time factor for the water phase is defined as

$$T_w = \frac{c_v^w t}{d^2} \quad (15.44)$$

where

$t$ = elapsed time.

**438** 15 ONE-DIMENSIONAL CONSOLIDATION AND SWELLING

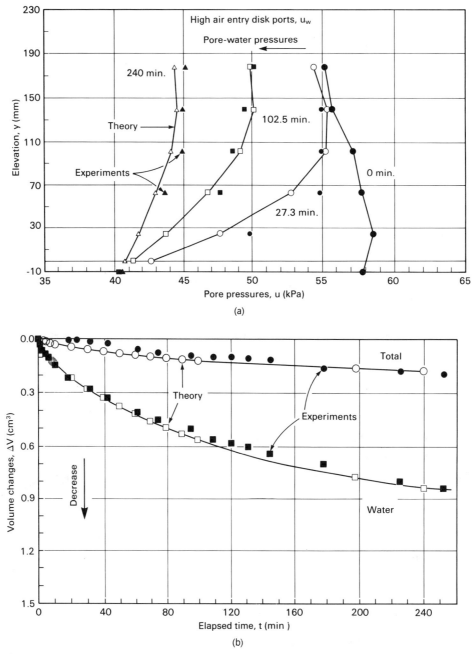

**Figure 15.17** Comparisons between a theoretical simulation and the experimental results from a consolidation test on a silty sand (S4Q4R test). (a) Pore–water pressure profiles at different times; (b) total and water volume changes during consolidation (from Rahardjo, 1990).

Similarly, the average degree of consolidation and the dimensionless time factor with respect to the air phase are defined as

$$U_a = 1 - \frac{\int_0^{2d} u_a \, dy}{\int_0^{2d} u_{a0} \, dy} \quad (15.45)$$

and

$$T_a = \frac{c_v^a t}{d^2} \quad (15.46)$$

where

$U_a$ = average degree of consolidation with respect to the air phase
$T_a$ = time factor with respect to the air phase

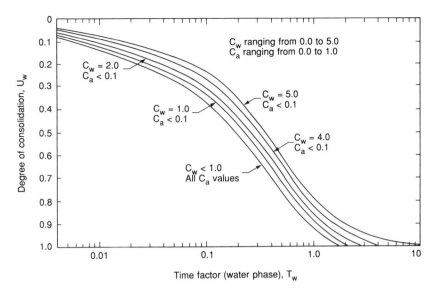

**Figure 15.18** Dimensionless time factor versus degree of consolidation curves for the water phase.

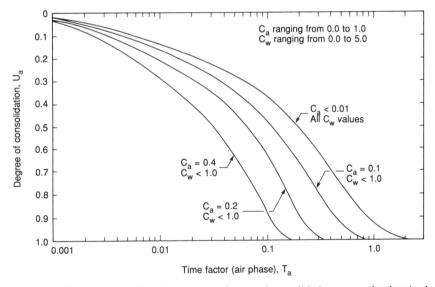

**Figure 15.19** Dimensionless time factor versus degree of consolidation curves for the air phase.

$u_{a0}$ = initial pore-air pressure
$u_a$ = pore-air pressure at any time.

Figure 15.18 shows the water phase degree of consolidation versus time factor curves for various air-water interaction constants. Similar curves for the air phase are shown in Figure 15.19. The interactive constant in the air phase partial differential equation was assumed to be constant for the calculation of pore-air pressure dissipation.

The curves show a smooth transition towards the case of a completely saturated soil (Terzaghi, 1943) and the case of a completely dry soil (Blight, 1971).

The relationships between these dimensionless numbers (Figs. 15.18 and 15.19) can be used as a general solution for one-dimensional consolidation. Appropriate curves corresponding to the air and water phases can be used for different values of consolidation coefficients and soil constants.

# CHAPTER 16

## *Two- and Three-Dimensional Unsteady-State Flow and Nonisothermal Analyses*

The types of analyses which can be performed by a geotechnical engineer have been strongly influenced by the available computing power. During the past few years, dramatic changes have been made in geotechnical engineering practice as a result of the computer becoming a part of the office equipment. A wide range of questions can be asked by clients, and the engineer can now perform parametric-type analyses that would not have been possible a few years ago.

One of the more recent changes in the practice of geotechnical engineering is related to types of unsteady-state or transient analyses which can be performed. This chapter sets forth the theoretical basis for two- and three-dimensional unsteady-state flow analysis in unsaturated soils. Also presented is the role of nonisothermal effects as it relates to the ground surface boundary and internal to the soil mass. In each case, the partial differential equations are presented for the analysis. Space does not permit the detailing of the finite element, numerical method formulations corresponding to each of the equations. The intent is rather to initiate the reader to the fundamentals of the theories involved.

Many situations encountered in practice have the ground surface as a flux boundary. In other words, the climatic conditions at a site give rise to thermal and moisture flux conditions at the ground surface. In particular, the conversion of the thermal boundary conditions into an actual evaporative flux becomes of primary interest in solving many geotechnical problems. This is an area of much needed research, and only recently have significant strides been made in providing the theoretical formulations for solving this problem. This book provides an introduction to the assessment of the moisture flux conditions at the ground surface.

## 16.1 UNCOUPLED TWO-DIMENSIONAL FORMULATIONS

Water flow through an earth dam during the filling of its reservoir is an example of two-dimensional, unsteady-state flow. Eventually, the flow of water through the dam will reach a steady-state condition, as illustrated in Chapter 7. Subsequent fluctuations of water level in the reservoir will again initiate unsteady-state water flow conditions. Furthermore, infiltration and evaporation cause almost a continuously changing flow condition. The transient analysis of seepage is strongly influenced by conditions in the unsaturated zone (Freeze, 1971).

The following uncoupled formulation independently satisfies the continuity equation for the water and air phases. The interaction between the fluid flows and the soil structure equilibrium condition is not considered in the uncoupled derivation. The two-dimensional derivation considers the plane strain case where fluid flows are assumed to occur only in two directions (i.e., the $x$- and $y$-directions). Fluid flow in the third direction (i.e., the $z$-direction) is assumed to be negligible.

### 16.1.1 Unsteady-State Seepage in Isotropic Soil

The term isotropy is used to refer to the coefficient of permeability (or any soil property), of the soil which does not vary with respect to direction. In other words, the coefficients of permeability in the $x$- and $y$-directions are equal at a point in the soil mass. The coefficient of permeability of an unsaturated soil, however, can vary with respect to space (i.e., heterogeneity), depending upon the magnitude of the matric suction at the point under consideration in the soil mass.

Consider a referential element of unsaturated soil subjected to unsteady-state air and water flow, as shown in Fig. 16.1. The flow equations for the air phase and the water phase can be derived by equating the time derivative of the relevant constitutive equation to the divergence of the flow rate as described by the flow law. The derivation for the uncoupled, two-dimensional case is essentially an extension of the one-dimensional consolidation equation derived in Chapter 15. The extension is associated with flow in the second direction. These terms are derived in a similar manner to the derivation for the one-dimensional case presented in Chapter 15. Therefore, the uncoupled

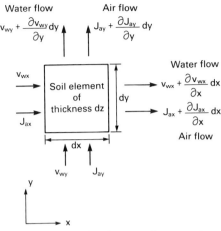

**Figure 16.1** Two-dimensional unsteady-state air and water flows.

two-dimensional flow equations are presented without repeating the details of their derivation.

*Water Phase Partial Differential Equation*

The partial differential equation for unsteady-state water flow can be written as follows:

$$\frac{\partial u_w}{\partial t} = -C_w \frac{\partial u_a}{\partial t} + c_v^w \frac{\partial^2 u_w}{\partial x^2} + \frac{c_v^w}{k_w}\frac{\partial k_w}{\partial x}\frac{\partial u_w}{\partial x}$$

$$+ c_v^w \frac{\partial^2 u_w}{\partial y^2} + \frac{c_v^w}{k_w}\frac{\partial k_w}{\partial y}\frac{\partial u_w}{\partial y} + c_g \frac{\partial k_w}{\partial y} \quad (16.1)$$

where

- $u_w$ = pore-water pressure
- $u_a$ = pore-air pressure
- $t$ = time
- $C_w = (1 - m_2^w/m_1^w)/(m_2^w/m_1^w)$; interaction constant associated with the water phase partial differential equation.
- $m_1^w$ = coefficient of water volume change with respect to a change in net normal stress, $(\sigma - u_a)$
- $\sigma$ = total stress
- $m_2^w$ = coefficient of water volume change with respect to a change in matric suction, $(u_a - u_w)$
- $c_v^w = k_w/(\rho_w g\, m_2^w)$; coefficient of consolidation with respect to the water phase for the x- and y-directions
- $k_w$ = coefficient of permeability with respect to the water phase for the x- and y-directions (i.e., isotropic soil); the permeability is a function of the matric suction at any point in the soil mass
- $\rho_w$ = density of water
- $g$ = gravitational acceleration
- $c_g = 1/m_2^w$; called the coefficient associated with the gravity term.

A comparison between Eq. (16.1) and the one-dimensional water phase partial differential equation for an unsaturated soil (Chapter 15) shows that the second and third terms of the right hand side of Eq. (16.1) are additional terms associated with the flow of water in the second direction (i.e., the x-direction). The gravity term, $c_g$ [i.e., the last term in Eq. (16.1)] is only applicable to the flow of water in the y-direction since this term is derived from the elevation head gradient (see Chapter 15).

*Air Phase Partial Differential Equation*

The partial differential equation for unsteady-state air flow is as follows:

$$\frac{\partial u_a}{\partial t} = -C_a \frac{\partial u_w}{\partial t} + c_v^a \frac{\partial^2 u_a}{\partial x^2} + \frac{c_v^a}{D_a^*}\frac{\partial D_a^*}{\partial x}\frac{\partial u_a}{\partial x}$$

$$+ c_v^a \frac{\partial^2 u_a}{\partial y^2} + \frac{c_v^a}{D_a^*}\frac{\partial D_a^*}{\partial y}\frac{\partial u_a}{\partial y} \quad (16.2)$$

where

- $C_a = \dfrac{m_2^a/m_1^a}{1 - m_2^a/m_1^a - (1-S)n/(\bar{u}_a m_1^a)}$; interaction constant associated with the air phase partial differential equation
- $m_1^a$ = coefficient of air volume change with respect to a change in net normal stress, $(\sigma - u_a)$
- $m_2^a$ = coefficient of air volume change with respect to a change in matric suction $(u_a - u_w)$
- $S$ = degree of saturation
- $n$ = porosity
- $\bar{u}_a$ = absolute pore-air pressure (i.e., $\bar{u}_a + \bar{u}_{atm}$)
- $\bar{u}_{atm}$ = atmospheric pressure (i.e., 101 kPa or 1 atm)
- $c_v^a = \dfrac{D_a^*}{\omega/(RT)} \dfrac{1}{\bar{u}_a m_1^a(1 - m_2^a/m_1^a) - (1-S)n}$; coefficient of consolidation with respect to the air phase for the x- and y-directions
- $D_a^*$ = coefficient of transmission with respect to the air phase for the x- and y-directions (i.e., isotropic soil); the coefficient is a function of the matric suction at a point in the soil mass
- $\omega_a$ = molecular mass of air
- $R$ = universal (molar) gas constant [i.e., 8.31432 J/(mol · K)]
- $T$ = absolute temperature (i.e., $T = t° + 273.16$) (K)
- $t°$ = temperature (°C).

The second and third terms in Eq. (16.2) are the terms associated with the air flow in the second direction.

### 16.1.2 Unsteady-State Seepage in an Anisotropic Soil

The term anisotropy is used to refer to the soil condition where the coefficient of permeability varies with respect to direction. Therefore, at any point in the soil mass, the coefficients of permeability in the x- and y-directions are assumed to be different. The conditions associated with anisotropy are discussed first, followed by the analysis of seepage. In addition, the coefficient of permeability varies with respect to space (i.e., heterogeneity) due to the variation in matric suction in the soil mass.

Unsteady-state water flow through an anisotropic soil is analyzed by considering the continuity of the water phase. The pore–air pressure is assumed to remain constant with time (i.e., $\partial u_a/\partial t = 0$). "The single-phase approach to unsaturated flow leads to techniques of analysis that are accurate enough for almost all practical purposes, but there are some unsaturated flow problems where the multiphase flow of air and water must be considered" (Freeze and Cherry, 1979).

The water phase partial differential equation can be obtained in a similar manner to the formulation of the water flow equation for an isotropic soil.

### Anisotropy in Permeability

Let us consider the general case of a variation in the coefficient of permeability with respect to space (heterogeneity) and direction (anisotropy) in an unsaturated soil, as illustrated in Fig. 16.2. At a particular point, the largest or major coefficient of permeability, $k_{w1}$, occurs in the direction $s_1$ which is inclined at an angle, $\alpha$, to the $x$-axis (i.e., horizontal). The smallest coefficient of permeability is in a perpendicular direction to the largest permeability (i.e., in the direction $s_2$), and is called the minor coefficient of permeability, $k_{w2}$. The ratio of the major to the minor coefficients of permeability is a constant not equal to unity at any point within the soil mass. The magnitudes of the major and minor coefficients of permeability, $k_{w1}$ and $k_{w2}$, can vary with matric suction from one location to another (i.e., heterogeneity), but their ratio is assumed to remain constant at every point.

The unsteady-state seepage analysis is generally derived with reference to the $x$- and $y$-directions. Therefore, it is necessary to write the coefficients of permeability in the $x$- and $y$-directions in terms of the major and minor coefficients of permeability. This relationship can be derived by first writing the water flow rates in the major and minor permeability directions (i.e., directions $s_1$ and $s_2$, respectively):

$$v_{w1} = -k_{w1}\frac{\partial h_w}{\partial s_1} \quad (16.3)$$

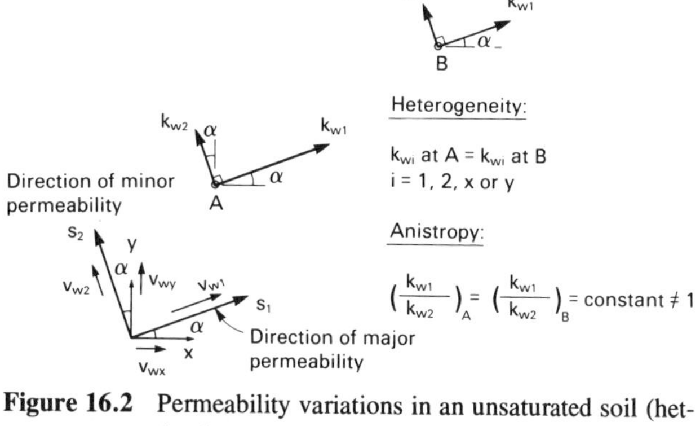

$$v_{w2} = -k_{w2}\frac{\partial h_w}{\partial s_2} \quad (16.4)$$

where

- $v_{w1}$ = water flow rate across a unit area of the soil element in the $s_1$-direction
- $v_{w2}$ = water flow rate across a unit area of the soil element in the $s_2$-direction
- $k_{w1}$ = major coefficient of permeability with respect to water as a function of matric suction which varies with location in the $s_1$-direction [i.e., $k_{w1}(u_a - u_w)$]
- $k_{w2}$ = minor coefficient of permeability with respect to water as a function of matric suction which varies with location in the $s_2$-direction [i.e., $k_{w2}(u_a - u_w)$]
- $h_w$ = hydraulic head (i.e., gravitational plus pore-water pressure head or $y + u_w/\rho_w g$)
- $y$ = elevation
- $s_1$ = direction of major coefficient of permeability, $k_{w1}$
- $s_2$ = direction of minor coefficient of permeability, $k_{w2}$
- $\partial h_w/\partial s_1$ = hydraulic head gradient in the $s_1$-direction
- $\partial h_w/\partial s_2$ = hydraulic head gradient in the $s_2$-direction.

The chain rule can be used to express the hydraulic head gradients in the $s_1$- and $s_2$-directions (i.e., $\partial h_w/\partial s_1$ and $\partial h_w/\partial s_2$, respectively) in terms of the gradients in the $x$- and $y$-directions (i.e., $\partial h_w/\partial x$ and $\partial h_w/\partial y$, respectively):

$$\frac{\partial h_w}{\partial s_1} = \frac{\partial h_w}{\partial x}\frac{\partial x}{\partial s_1} + \frac{\partial h_w}{\partial y}\frac{\partial y}{\partial s_1} \quad (16.5)$$

$$\frac{\partial h_w}{\partial s_2} = \frac{\partial h_w}{\partial x}\frac{\partial x}{\partial s_2} + \frac{\partial h_w}{\partial y}\frac{\partial y}{\partial s_2} \quad (16.6)$$

where

- $\partial h_w/\partial x$ = hydraulic head gradient in the $x$-direction
- $\partial h_w/\partial y$ = hydraulic head gradient in the $y$-direction.

From trigonometric relations, the following relationships can be obtained (see Fig. 16.2):

$$\frac{dx}{ds_1} = \cos\alpha \quad (16.7)$$

$$\frac{dy}{ds_1} = \sin\alpha \quad (16.8)$$

$$\frac{dx}{ds_2} = -\sin\alpha \quad (16.9)$$

$$\frac{dy}{ds_2} = \cos\alpha. \quad (16.10)$$

**Figure 16.2** Permeability variations in an unsaturated soil (heterogeneous and anisotropic).

Substituting Eqs. (16.5)–(16.10) into Eqs. (16.3) and

(16.4) gives

$$v_{w1} = -k_{w1}\left(\cos\alpha\frac{\partial h_w}{\partial x} + \sin\alpha\frac{\partial h_w}{\partial y}\right) \quad (16.11)$$

$$v_{w2} = -k_{w2}\left(-\sin\alpha\frac{\partial h_w}{\partial x} + \cos\alpha\frac{\partial h_w}{\partial y}\right). \quad (16.12)$$

The water flow rates in the $x$- and $y$-directions can be written by projecting the flow rates in the major and minor directions to the $x$- and $y$-directions (see Fig. 16.2):

$$v_{wx} = v_{w1}\cos\alpha - v_{w2}\sin\alpha \quad (16.13)$$

$$v_{wy} = v_{w1}\sin\alpha + v_{w2}\cos\alpha \quad (16.14)$$

where

$v_{wx}$ = water flow rate across a unit area of the soil element in the $x$-direction

$v_{wy}$ = water flow rate across a unit area of the soil element in the $y$-direction

$\alpha$ = angle between the direction of the major coefficient of permeability and the $x$-direction.

Substituting Eqs. (16.11) and (16.12) for $v_{w1}$ and $v_{w2}$, respectively, into Eqs. (16.13) and (16.14) results in the following relations:

$$v_{wx} = -k_{w1}\cos^2\alpha\frac{\partial h_w}{\partial x} - k_{w1}\sin\alpha\cos\alpha\frac{\partial h_w}{\partial y}$$

$$- k_{w2}\sin^2\alpha\frac{\partial h_w}{\partial x} + k_{w2}\sin\alpha\cos\alpha\frac{\partial h_w}{\partial y} \quad (16.15)$$

$$v_{wy} = -k_{w1}\sin\alpha\cos\alpha\frac{\partial h_w}{\partial x} - k_{w1}\sin^2\alpha\frac{\partial h_w}{\partial y}$$

$$+ k_{w2}\sin\alpha\cos\alpha\frac{\partial h_w}{\partial x} - k_{w2}\cos^2\alpha\frac{\partial h_w}{\partial y}. \quad (16.16)$$

Rearranging Eqs. (16.15) and (16.16) yields the expressions for the rate of water flow in the $x$- and $y$-directions:

$$v_{wx} = -\left(k_{wxx}\frac{\partial h_w}{\partial x} + k_{wxy}\frac{\partial h_w}{\partial y}\right) \quad (16.17)$$

$$v_{wy} = -\left(k_{wyx}\frac{\partial h_w}{\partial x} + k_{wyy}\frac{\partial h_w}{\partial y}\right) \quad (16.18)$$

where

$k_{wxx} = k_{w1}\cos^2\alpha + k_{w2}\sin^2\alpha$
$k_{wxy} = (k_{w1} - k_{w2})\sin\alpha\cos\alpha$
$k_{wyx} = (k_{w1} - k_{w2})\sin\alpha\cos\alpha$
$k_{wyy} = k_{w1}\sin^2\alpha + k_{w2}\cos^2\alpha$.

Equations (16.17) and (16.18) provide the flow rates in the $x$- and $y$-directions in terms of the major and minor coefficients of permeability. These flow rate expressions can then be used in the formulation for unsteady-state seepage analyses.

### Water Phase Partial Differential Equation

The water phase partial differential equation can be obtained by considering the continuity for the water phase. The net flux of water through an element of unsaturated soil (Fig. 16.1) can be computed from the volume rates of water entering and leaving the element within a period of time:

$$\frac{\partial(V_w/V_o)}{\partial t} = \frac{\partial v_{wx}}{\partial x} + \frac{\partial v_{wy}}{\partial y} \quad (16.19)$$

where

$V_w$ = volume of water in the element
$V_0$ = initial overall volume of the element (i.e., $dx$, $dy$, $dz$)
$dx, dy, dz$ = dimensions in the $x$-, $y$-, and $z$-directions, respectively
$\partial(V_w/V_0)/\partial t$ = rate of change in the volume of water in the soil element with respect to the initial volume of the element.

A summary of the constitutive equations required for the formulation of flow equations is given in Chapter 15. The net flux of water can be computed from the time derivative of the water phase constitutive equation. In this case, the time derivatives of the total stress and the pore–air pressure are equal to zero since both pressures are assumed to remain constant with time. In addition, the $m_2^w$ coefficient can be assumed to be constant for a particular time step during the transient process:

$$\frac{\partial(V_w/V_0)}{\partial t} = -m_2^w\frac{\partial u_w}{\partial t}. \quad (16.20)$$

The continuity condition can be satisfied by equating the divergence of the water flow rates [i.e., Eq. (16.19)] and the time derivative of the constitutive equation for the water phase [i.e., Eq. (16.20)]:

$$\frac{\partial v_{wx}}{\partial x} + \frac{\partial v_{wy}}{\partial y} = -m_2^w\frac{\partial u_w}{\partial t}. \quad (16.21)$$

Substituting Eqs. (16.17) and (16.18) for $v_{wx}$ and $v_{wy}$, respectively, into Eq. (16.21) yields

$$\frac{\partial}{\partial x}\left(k_{wxx}\frac{\partial h_w}{\partial x} + k_{wxy}\frac{\partial h_w}{\partial y}\right) + \frac{\partial}{\partial y}\left(k_{wyx}\frac{\partial h_w}{\partial x} + k_{wyy}\frac{\partial h_w}{\partial y}\right)$$

$$= m_2^w\rho_w g\frac{\partial h_w}{\partial t}. \quad (16.22)$$

Equation (16.22) is the governing partial differential equation for unsteady-state water seepage in an anisotropic soil when the pore–air pressure is assumed to remain constant with time. In many cases, the *major and minor coefficient of permeability directions* coincide with the $x$- and $y$-directions, respectively. In this case, the $\alpha$ angle is equal to zero, and the governing equation [Eq. (16.22)] can be simplified by setting the $\alpha$ angle to zero [i.e., Eqs. (16.17) and (16.18)]:

$$\frac{\partial}{\partial x}\left(k_{w1}\frac{\partial h_w}{\partial x}\right) + \frac{\partial}{\partial y}\left(k_{w2}\frac{\partial h_w}{\partial y}\right) = m_2^w \rho_w g \frac{\partial h_w}{\partial t}. \quad (16.23)$$

The $k_{w1}$ and $k_{w2}$ terms in Eq. (16.23) are the major and minor coefficients of permeability in the $x$- and $y$-directions, respectively. These permeability coefficients are a function of matric suction that can vary with location in the $x$- and $y$-directions [i.e., $k_{w1}(u_a - u_w)$ and $k_{w2}(u_a - u_w)$]. For *isotropic soil conditions*, the $k_{w1}$ and $k_{w2}$ terms are equal [i.e., $k_{w1} = k_{w2} = k_w(u_a - u_w)$] and Eq. (16.23) can be further simplified as

$$\frac{\partial}{\partial x}\left(k_w \frac{\partial h_w}{\partial x}\right) + \frac{\partial}{\partial y}\left(k_w \frac{\partial h_w}{\partial y}\right) = m_2^w \rho_w g \frac{\partial h_w}{\partial t}. \quad (16.24)$$

Rearranging Eq. (16.24) gives

$$k_w \frac{\partial^2 h_w}{\partial x^2} + \frac{\partial k_w}{\partial x}\frac{\partial h_w}{\partial x} + k_w \frac{\partial^2 h_w}{\partial y^2} + \frac{\partial k_w}{\partial y}\frac{\partial h_w}{\partial y} = m_2^w \rho_w g \frac{\partial h_w}{\partial t}. \quad (16.25)$$

Substituting $(y + u_w/\rho_w g)$ for the hydraulic head, $h_w$, in Eq. (16.25) and rearranging the equation results in the following form:

$$c_v^w \frac{\partial^2 u_w}{\partial x^2} + \frac{c_v^w}{k_w}\frac{\partial k_w}{\partial x}\frac{\partial u_w}{\partial x} + c_v^w \frac{\partial^2 u_w}{\partial y^2}$$
$$+ \frac{c_v^w}{k_w}\frac{\partial k_w}{\partial y}\frac{\partial u_w}{\partial y} + c_g \frac{\partial k_w}{\partial y} = \frac{\partial u_w}{\partial t} \quad (16.26)$$

where

$c_v^w = k_w/(\rho_w g m_2^w)$
$c_g = 1/m_2^w$.

Equation (16.26) is essentially equal to Eq. (16.1) without the interaction term, $C_w$ since the pore–air pressure is assumed to remain constant with time. Therefore, it has been shown that the general governing equation for an anisotropic soil [i.e., Eq. (16.22)] can be simplified for the anisotropic case where the $\alpha$ angle is equal to zero [i.e., Eq. (16.23)] and for the isotropic soil condition [i.e., Eq. (16.24) or (16.26)].

The general governing equation, Eq. (16.22), can therefore be used to solve water seepage problems through a saturated–unsaturated flow system. For the *saturated* portion, the water coefficient of permeability becomes equal to the saturated coefficient of permeability, $k_s$. The saturated coefficient of permeability may vary with respect to direction (i.e., anisotropy) or with respect to location (i.e., heterogeneity). In this case, both anisotropy and heterogeneity with respect to permeability are accounted for in Eq. (16.22). The coefficient of water volume change, $m_2^w$, in Eq. (16.22) approaches the value of the coefficient of volume change $m_v$, as the soil becomes saturated.

### Seepage Analysis Using the Finite Element Method

Unsteady-state water seepage through a saturated–unsaturated soil system can be analyzed by solving the general governing flow equation [i.e., Eq. (16.22)]. The analysis can be performed using the finite element method as described in Chapter 7 for steady-state seepage. A similar approach can be used for unsteady-state seepage, with the exception of some differences in the finite element formulation.

The finite element formulation for unsteady-state seepage in two dimensions can be derived using the Galerkin principle of weighted residual (Lam *et al.* (1988)). The Galerkin solution to the governing equation, Eq. (16.22), is given by the following integrals over the area and the boundary surface of a triangular element (Fig. 16.3):

$$\int_A [B]^T[k_w][B]\,dA\,\{h_{wn}\} + \int_A [L]^T\lambda[L]\,dA\,\frac{\partial\{h_{wn}\}}{\partial t}$$
$$- \int_S [L]^T \bar{v}_w\,dS = 0 \quad (16.27)$$

where

[$B$] = matrix of the derivatives of the area coordinates (Fig. 16.3), which can be written as

$$\frac{1}{2A}\left\{\begin{array}{c}(y_2 - y_3)(y_3 - y_1)(y_1 - y_2)\\(x_3 - x_2)(x_1 - x_3)(x_2 - x_1)\end{array}\right\}$$

$x_i, y_i (i = 1, 2, 3)$ = Cartesian coordinates of the three nodal points of an element
$A$ = area of the element
[$k_w$] = tensor of the water coefficients of permeability for the element, which

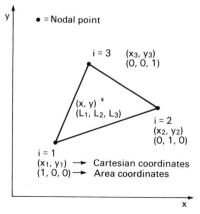

**Figure 16.3** Area coordinates in relation to the Cartesian coordinates for a triangular element.

can be written as

$$\begin{bmatrix} k_{wxx} & k_{wxy} \\ k_{wyx} & k_{wyy} \end{bmatrix}$$

$\{h_{wn}\}$ = matrix of hydraulic heads at the nodal points, that is,

$$\begin{Bmatrix} h_{w1} \\ h_{w2} \\ h_{w3} \end{Bmatrix}$$

$[L]$ = matrix of the element area coordinates (i.e., $\{L_1 L_2 L_3\}$)

$L_1 L_2 L_3$ = area coordinates of points in the element that are related to the Cartesian coordinates of nodal points as follows (Fig. 16.3):

$L_1 = (1/2A)\{(x_2 y_3 - x_3 y_2) + (y_2 - y_3)x + (x_3 - x_2)y\}$

$L_2 = (1/2A)\{(x_3 y_1 - x_1 y_3) + (y_3 - y_1)x + (x_1 - x_3)y\}$

$L_3 = (1/2A)\{(x_1 y_2 - x_2 y_1) + (y_1 - y_2)x + (x_2 - x_1)y\}$

$x, y$ = Cartesian coordinates of a point within the element

$\lambda = \rho_w g m_2^w$

$\bar{v}_w$ = external water flow rate in a direction perpendicular to the boundary of the element

$S$ = perimeter of the element.

Either the hydraulic head or the flow rate must be specified at the boundary nodal points. Specified hydraulic heads at the boundary nodes are called Dirichlet boundary conditions. A specified flow rate across the boundary is referred to as a Neuman boundary condition. The third term in Eq. (16.27) accounts for the specified flow rates across the boundary. The specified flow rates at the boundary must be projected to a direction normal to the boundary. As an example, a specified flow rate, $v_w$, in the vertical direction must be converted to a normal flow rate, $\bar{v}_w$, as illustrated in Fig. 16.4. The normal flow rate is in turn converted to a nodal flow, $Q_w$ (Segerlind, 1976). Figure 16.4 shows the computation of the nodal flows, $Q_{wi}$ and $Q_{wj}$, at the boundary nodes ($i$) and ($j$), respectively. A positive nodal flow signifies that there is infiltration at the node or that the node acts as a "source." A negative nodal flow indicates evaporation, evapotranspiration at the node, or that the node acts a "sink." When the flow rate across a boundary is zero (e.g., impervious boundary), the third term in Eq. (16.27) disappears.

The numerical integration of Eq. (16.27) results in a simpler expression of the equation:

$$[D]\{h_{wn}\} + [E]\{h_{wn}\} = [F] \quad (16.28)$$

where

$[D]$ = stiffness matrix, that is,

$$[B]^T [k_w] [B] A$$

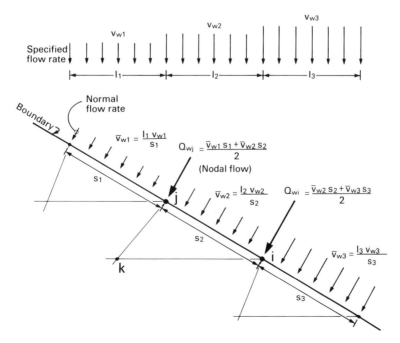

**Figure 16.4** Applied flow rate across the boundary expressed as nodal flows.

$[E]$ = capacitance matrix, that is,

$$\frac{\lambda A}{12} \begin{bmatrix} 2 & 1 & 1 \\ 1 & 2 & 1 \\ 1 & 1 & 2 \end{bmatrix}$$

$\{h_{wn}\}$ = matrix of the time derivatives of the hydraulic heads at the nodal points, i.e.,

$$\frac{\partial \{h_{wn}\}}{\partial t}$$

$[F]$ = flux vector reflecting the boundary conditions, i.e.,

$$\int_S [L]^T \bar{v}_w \, dS$$

The time derivative in Eq. (16.28) can be approximated using a finite difference technique. The relationship between the nodal heads of an element at two successive time steps can be expressed using the central difference approximation:

$$\left([D] + \frac{2[E]}{\Delta t}\right) \{h_{wn}\}_{t+\Delta t}$$
$$= \left(\frac{2[E]}{\Delta t} - [D]\right) \{h_{wn}\}_t + 2[F] \quad (16.29)$$

or the backward difference approximation:

$$\left([D] + \frac{[E]}{\Delta t}\right) \{h_{wn}\}_{t+\Delta t} = \frac{[E]}{\Delta t} \{h_{wn}\}_t + [F] \quad (16.30)$$

The above time derivative approximations are considered to be unconditionally stable. The central difference approximation generally gives a more accurate solution than that obtained from the backward difference approximation. However, the backward difference approximation is found to be more effective in reducing numerical oscillations commonly encountered in highly nonlinear systems of flow equations (Neuman and Witherspoon (1971), Neuman, 1973).

The finite element flow equation [i.e., Eq. (16.28)] can be written for each element and assembled to form a set of global flow equations. This is performed while satisfying nodal compatibility (Desai and Abel, 1972). Nodal compatibility requires that a particular node shared by the surrounding elements have the same hydraulic head in all the elements (Zienkiewicz, 1971, Desai, 1975b).

The set of global flow equations for the whole system is solved for the hydraulic heads at the nodal points, $\{h_{wn}\}$. However, Eq. (16.28) is nonlinear because the coefficients of permeability are a function of matric suction, which is related to the hydraulic head at the nodal points.

The hydraulic heads are the unknown variables in Eq. (16.28). Therefore, Eq. (16.28) must be solved using an iterative procedure that involves a series of successive approximations. In the first approximation, the coefficients of permeability are estimated in order to calculate the first set of hydraulic heads at the nodal points. The computed hydraulic heads are used to calculate the average matric suction within an element. In the subsequent approximations, the coefficient of permeability is adjusted to a value depending upon the average matric suction in the element. The adjusted permeability value is then used to calculate a new set of nodal hydraulic heads. The above procedure is repeated until both the hydraulic head and the permeability differences within each element at two successive iterations are smaller than a specified tolerance.

The above iterative procedure causes the global flow equations to be linearized and solved simultaneously using a Gaussian elimination technique. The convergency rate is highly dependent on the degree of nonlinearity of the permeability function and the spatial discretization of the problem. A steep permeability function requires more iterations and a larger convergency tolerance. A finer discretization in both element size and time step will assist in obtaining convergence faster with a smaller tolerance. Generally, the solution will converge to a tolerance of less than 1% in ten iterations.

The unsteady-state seepage equation, Eq. (16.28), is considered solved for one time step once the converged nodal hydraulic heads of the system have been obtained. Having reached convergence at a particular time step, other secondary quantities, such as pore–water pressures, hydraulic head gradients, and water flow rates, can then be calculated using the converged nodal hydraulic heads. The equation for nodal pore–water pressures is

$$\{u_{wn}\} = (\{h_{wn}\} - \{y_n\})\rho_w g \quad (16.31)$$

where

$\{u_{wn}\}$ = matrix of pore–water pressures at the nodal points, i.e.,

$$\begin{Bmatrix} u_{w1} \\ u_{w2} \\ u_{w3} \end{Bmatrix}$$

$\{y_n\}$ = matrix of elevation heads at the nodal points, i.e.,

$$\begin{Bmatrix} y_1 \\ y_2 \\ y_3 \end{Bmatrix}.$$

The hydraulic head gradients in the $x$- and $y$-directions can be computed for an element by taking the derivative of the element hydraulic heads with respect to $x$ and $y$, respectively:

$$\begin{Bmatrix} i_x \\ i_y \end{Bmatrix} = [B]\{h_{wn}\} \quad (16.32)$$

where

$i_x$, $i_y$ = hydraulic head gradient within an element in the $x$- and $y$-directions, respectively.

The element flow rates, $v_w$, can be calculated from the hydraulic head gradients and the coefficients of permeability in accordance with Darcy's law:

$$\begin{Bmatrix} v_{wx} \\ v_{wy} \end{Bmatrix} = [k_w][B]\{h_{wn}\} \tag{16.33}$$

where

$v_{wx}$, $v_{wy}$ = water flow rates within an element in the $x$- and $y$-directions, respectively.

The hydraulic head gradient and the flow rate at nodal points are computed by averaging the corresponding quantities from all elements surrounding the node. The weighted average is computed in proportion to the element areas.

### Examples of Two-Dimensional Problems and Their Solutions

Three example problems are presented in this section to illustrate the unsteady-steady state seepage analysis using the finite element method. Lam (1984) has solved several classical problems of seepage through saturated–unsaturated soil systems. The three problems which will be discussed in this section deal with water flows through an earth dam, flow below a lagoon, and flow through a layered hill slope. In all cases, the flow regime of the problem must first be defined. This includes the spatial dimension of the soil boundaries, the boundary conditions of the flow system, and the determination of soil properties.

The boundaries of different soil layers can be determined through a subsurface soil investigation. The boundary conditions of the flow system can be obtained from piezometric records and hydrological data. Soil properties, such as coefficients of permeability and coefficients of volume change (i.e., $k_w$ and $m_2^w$), can be measured using *in situ* or laboratory tests (Chapters 6 and 13).

### Example of Water Flow Through an Earth Dam

The example problem involving water flow through an earth dam is discussed first. The cross section and discretization of the earth dam are shown in Fig. 16.5. The soil is assumed to be isotropic with respect to its coefficient of permeability, and the permeability function used in the analysis is shown in Fig. 16.6. The saturated coefficient of permeability, $k_s$, is $1.0 \times 10^{-7}$ m/s. The pore-air pressure is assumed to be atmospheric. Therefore, the matric suction values in Fig. 16.6 are numerically equal to the pore-water pressures, and can be expressed as a pore-water pressure head, $h_p$. The base of the dam is selected as the datum. In addition, a coefficient of water volume change, $m_2^w$, of $1.0 \times 10^{-3}$ kPa$^{-1}$ is used in the analysis.

The dam is initially at a steady-state condition, with the reservoir water level 4 m above the datum. At a time assumed to be equal to zero, the water level in the reservoir is instantaneously raised to a level of 10 m above the datum. The water level remains constant at 10 m during the transient process. The rising of the phreatic line from the initial steady-state condition (i.e., at time, $t$, equal to zero) to the final steady-state condition (i.e., at time, $t$, equal to 19 656 h or 819 days) is illustrated in Fig. 16.7.

The development of equipotential lines, phreatic surface, and water flow rates across the dam are illustrated in Figs. 16.8–16.11 for four different times during the transient process. The increase in the reservoir level results in an increase in pore-water pressures with time. This is demonstrated by the advancement of equipotential lines from the upstream to the downstream of the dam with increasing times. It should also be noted that the equipotential lines extend from the saturated to the unsaturated zones, as shown in Figs. 16.8(a), 16.9(a), 16.10(a), and 16.11(a). In other words, water flows in both the saturated and the unsaturated zones as a result of the hydraulic head differences between the equipotential lines. The flow of water in both zones can be observed directly from the flow rate vectors that exist in both the saturated and the unsaturated zones, as shown in Figs. 16.8(b), 16.9(b), 16.10(b), and 16.11(b). The amount of water flowing in the unsaturated zone depends on the rate of change in the coefficient of permeability with respect to the matric suction changes.

### Example of Groundwater Seepage Below a Lagoon

The second example problem illustrates unsteady-state groundwater seepage below a lagoon. The lagoon is placed on top of a 1 m thick soil linear, as shown in Fig. 16.12. The geometry of the problem is symmetrical, and the liner and the surrounding soil are assumed to be isotropic with respect to their permeability. Therefore, the problem can be analyzed by only considering half of the geometry. The discretized cross section of the soil liner and its surrounding soil are depicted in Fig. 16.12. The permeability functions for the soil liner and the surrounding soil are shown in Fig. 16.13. The saturated coefficients of permeability are equal to $5.0 \times 10^{-6}$ and $1.0 \times 10^{-5}$ m/s for the liner and the surrounding soil, respectively. A coefficient of water volume change, $m_2^w$, of $2.0 \times 10^{-3}$ kPa$^{-1}$ is used in the analysis for both the liner and the surrounding soil.

An initial steady-state condition with a groundwater table located 5 m below the ground surface is assumed (Fig. 16.14). "No flow" boundary conditions are assumed along the ground surface, the bottom boundary, and the axis of symmetry. On the right-hand boundary, a hydrostatic and "no-flow" condition is assumed to exist below and above the groundwater table, respectively. At a time assumed to be equal to zero, the lagoon is filled with water to the 1 m height, which gives rise to a 1 m pore-water pressure head. As a result, water seepage occurs from the lagoon, causing

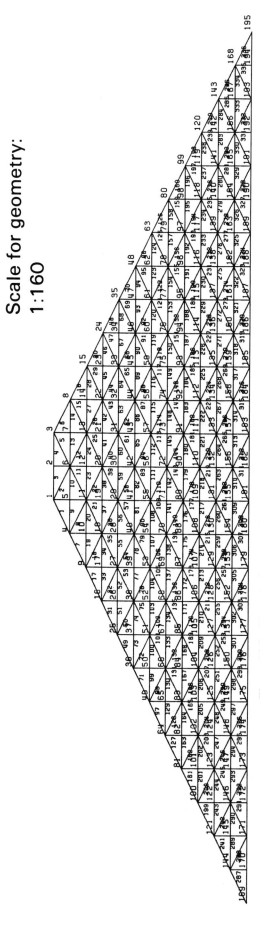

**Figure 16.5** Discretized cross section of a dam for a finite element analysis involving unsteady-state seepage.

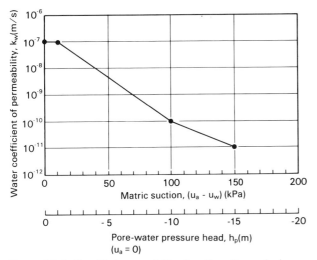

**Figure 16.6** Specified permeability functions for analyzing unsteady-state seepage through an isotropic dam.

the groundwater to gradually mound to its final steady-state condition. Figure 16.14 illustrates the transient positions of the water table at various times during the mounding process. The water level in the lagoon is assumed to remain constant at a 1 m height throughout the transient process.

The seepage flow pattern and the development of the pore–water pressures in the soil below the lagoon are presented in Figs. 16.15–16.18 for various times during the transient process. In the beginning of the process, water seeps downward from the lagoon, while the position of the groundwater table is not yet affected (Fig. 16.15). With time, the wetting front moves deeper into the soil mass, while the groundwater begins to mound towards the lagoon (Fig. 16.16). Eventually, the downward wetting front from the lagoon joins the rising groundwater table, as shown in Fig. 16.17. The final steady-state condition is achieved at an elapsed time equal to 189 h or approximately 8 days for the designated coefficients of permeability. After 8 days, the water seepage from the lagoon is balanced by the seepage leaving the right-hand boundary.

*Example of Seepage within a Layered Hill Slope*

The third example problem illustrates unsteady-state seepage within a layered hill slope under constant infiltration conditions. Rulon and Freeze (1985) studied this problem using a sandbox model of the layered hill slope equipped with pore–water pressure measurements. The slope consisted of medium sand with a horizontal layer of fine sand, as shown in Fig. 16.19. The fine sand had a lower coefficient of permeability than the medium sand in order to impede water flow and create a seepage face on the slope. A constant rate of infiltration was applied to the crest of the hill slope. The results of measurements were then calibrated using Neuman's finite element model, as depicted in Fig. 16.20. A close comparison between the observed measurements and the finite element analysis for a steady-

state condition was obtained when using calibrated saturated permeabilities of $1.4 \times 10^{-3}$ and $5.5 \times 10^{-5}$ m/s for the medium and fine sands, respectively. In addition, a constant infiltration rate of $2.1 \times 10^{-4}$ m/s was used in the analysis.

The unsteady-state seepage analysis of this third problem can also be performed using the finite element method described in the previous section. The discretized cross section of the layered hill slope is shown in Fig. 16.21. The same coefficient of permeability values and infiltration rate as designated by Rulon and Freeze (1985) are used in this unsteady-state analysis. The permeability functions for both the medium and fine sands are given in Fig. 16.22. In addition, a coefficient of water volume change, $m_2^w$, of $2.0 \times 10^{-3}$ kPa$^{-1}$ is used in the analysis for both sands.

An initial steady-state condition is assumed, with the water table located at a height of 0.3 m from the toe of the slope. At an elapsed time equal to zero, a constant infiltration rate of $2.1 \times 10^{-4}$ m/s is applied to the top portion of the hill slope. The development of the groundwater table from the initial steady-state condition (i.e., at time equal to zero) to the final steady-state condition (i.e., at an elapsed time equal to 208 s) is illustrated in Fig. 16.23. The water level at the toe of the slope and the infiltration rate at the top of the slope are assumed to remain constant throughout the transient process. No flow boundary conditions are assumed along the remainder of the slope boundaries.

The seepage flow pattern and the development of the equipotential lines within the slope are presented in Figs. 16.24–16.27 for various times during the transient process. At the start of the infiltration process, water infiltrates vertically towards the impeding layer, while the position of the groundwater table is not yet affected (Fig. 16.24). As infiltration progresses, water seeps through the impeding layer, causing the groundwater table to rise, as shown in Fig. 16.25. Eventually, a perched water table on the impeding layer develops (Fig. 16.26) and moves towards the slope face (Fig. 16.27). At steady-state conditions (Fig. 16.27), a wedge-shaped unsaturated zone is formed, and two seepage faces develop, one near the toe of the slope, and the other just above the impeding layer. In other words, the presence of the impeding layer results in a complex configuration for the groundwater table and the equipotential line positions. A comparison between the above results (Fig. 16.27) and the model presented by Rulon and Freeze (1985) (Fig. 16.20) indicates a close agreement in the positions of the developed water table and seepage faces.

The analyses presented from the above three example problems have demonstrated the usefulness of the general water flow equation [i.e., Eq. (16.22)] in solving typical saturated–unsaturated flow problems. The application of a saturated flow model to each of these problems is difficult. Furthermore, water flows primarily in the unsaturated zone in the early stage of the infiltration process, as indicated by

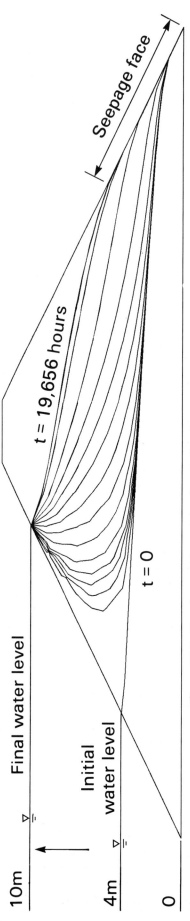

**Figure 16.7** Transient positions of the phreatic line from an elapsed time equal to zero to an elapsed time equal to 19 656 h (i.e., 819 days).

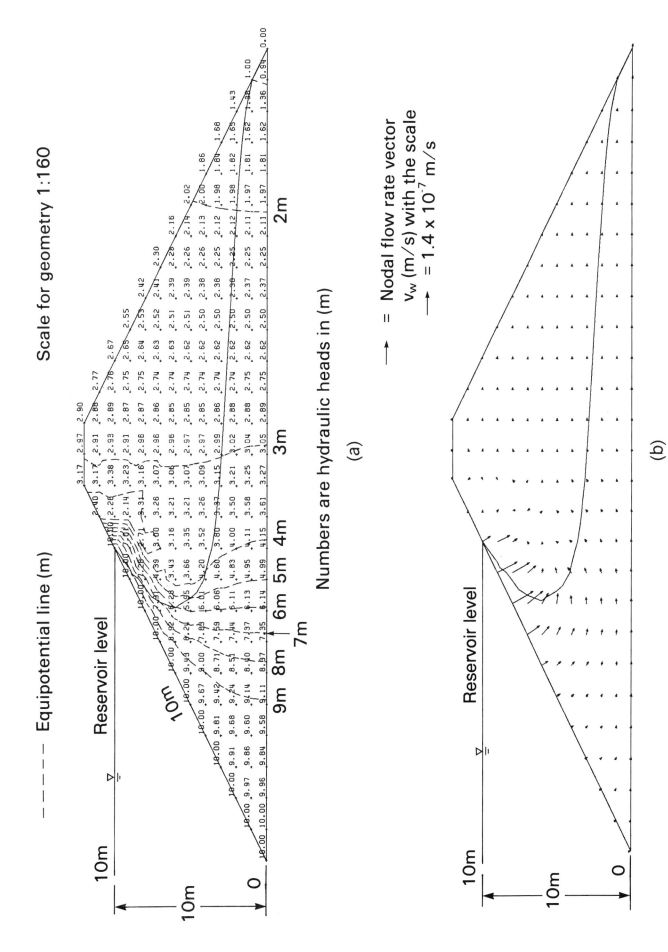

**Figure 16.8** Unsteady-state seepage through an isotropic dam at an elapsed time equal to 36 min. (a) Equipotential lines; (b) nodal flow rate vectors throughout the dam.

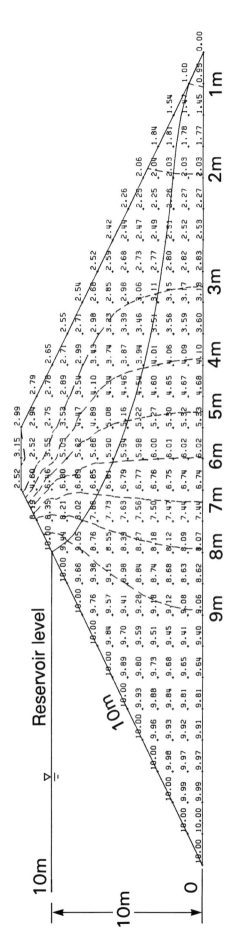
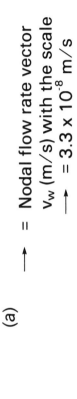
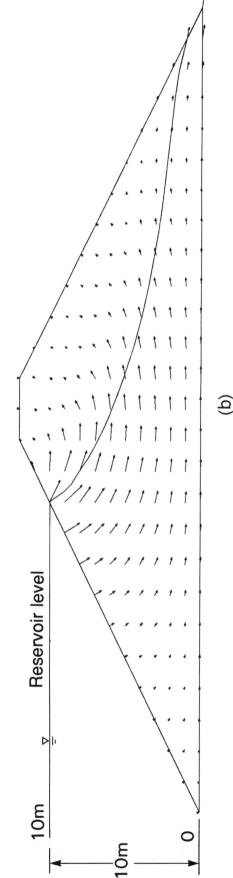

**Figure 16.9** Unsteady-state seepage through an isotropic dam at an elapsed time equal to 1032 h (i.e., 43 days). (a) Equipotential lines; (b) nodal flow rate vectors throughout the dam.

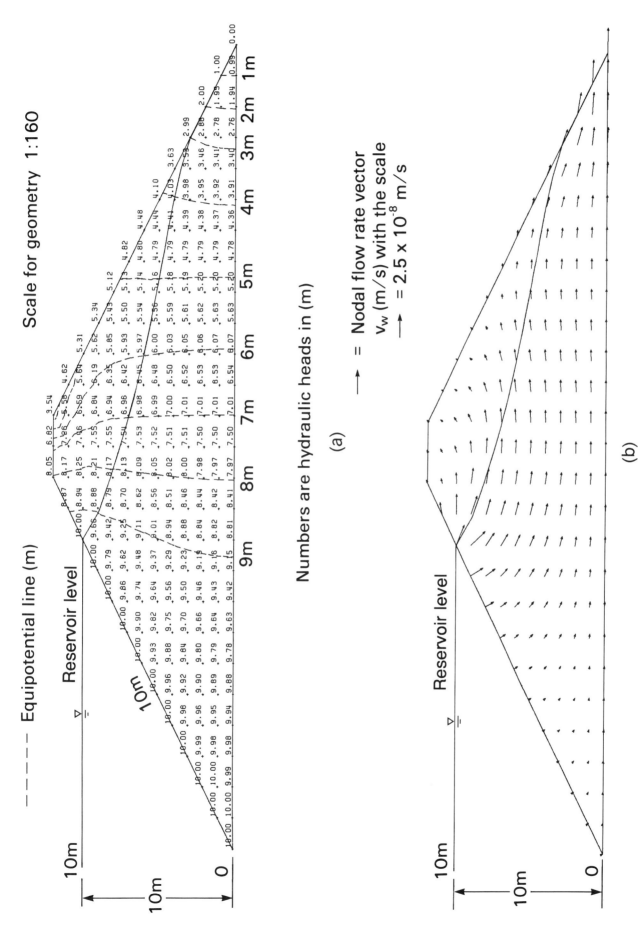

**Figure 16.10** Unsteady-state seepage through an isotropic dam at an elapsed time equal to 3147 h (i.e., 131 days). (a) Equipotential lines; (b) nodal flow rate vectors throughout the dam.

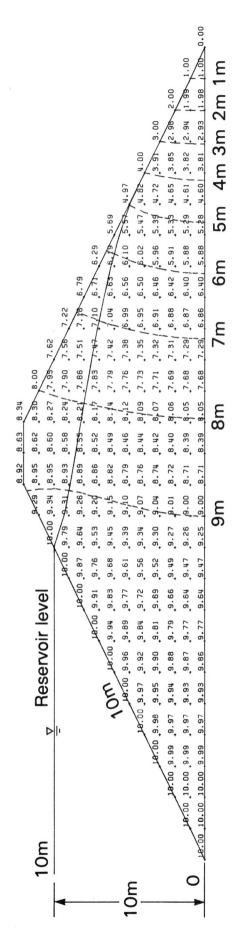
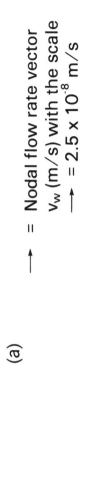
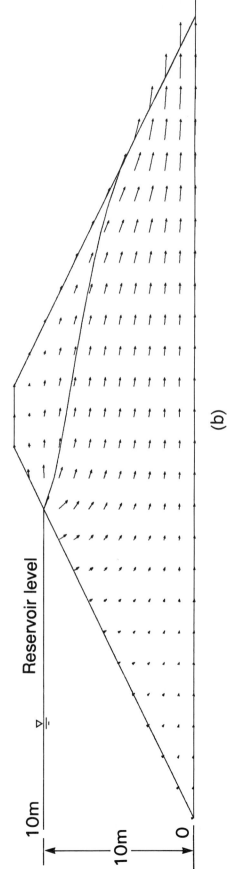

**Figure 16.11** Steady-state seepage through an isotropic dam at an elapsed time equal to 19 656 h (i.e., 819 days). (a) Equipotential lines; (b) nodal flow rate vectors throughout the dam.

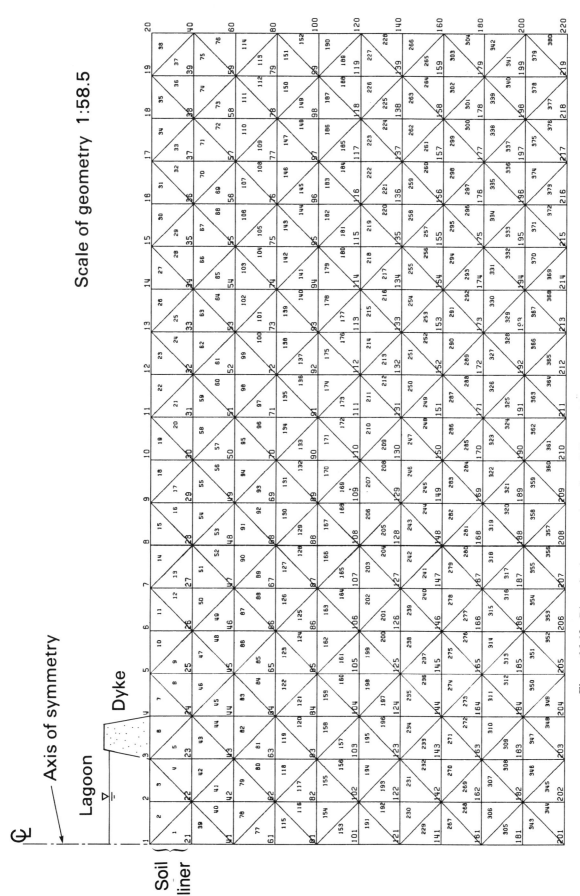

**Figure 16.12** Discretized cross section of a soil liner and its surrounding soil for unsteady-state seepage analysis below a lagoon.

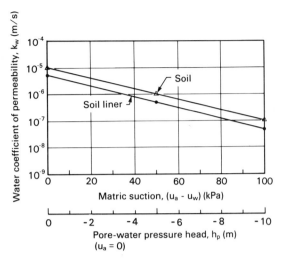

**Figure 16.13** Specified permeability functions for analyzing unsteady-state groundwater seepage below a lagoon.

the computed seepage flow rates. The flow system developed in this type of problem can be complex, depending upon the contrast in the coefficients of permeability values for the different soils involved in the system.

## 16.2 COUPLED FORMULATIONS OF THREE-DIMENSIONAL CONSOLIDATION

A rigorous formulation of two- and three-dimensional consolidation requires that the continuity equation be coupled with the equilibrium equations. This method was proposed by Biot (1941) to analyze the consolidation process for a special case of an unsaturated soil. The derivations were based on the assumption that occluded bubbles of air existed in the soil during the consolidation process.

A summary of the coupled formulation for three-dimensional consolidation of an unsaturated soil with a continuous air phase is presented in this section. Reference is made to Dakshanamurthy *et al.* (1984) for detailed explanations on the coupled consolidation equations. The constitutive relations for the soil structure, the water phase, and the air phase are required for formulating the equilibrium and continuity equations. Therefore, these constitutive relations are first summarized in their elasticity forms prior to the formulation of the consolidation equations.

### 16.2.1 Constitutive Relations

The stress state and deformation state variables can be linked by suitable constitutive relations which incorporate soil properties in the form of coefficients. For an unsaturated soil, there are three available constitutive relations for an unsaturated soil, namely, one for the soil structure, one for the water phase, and one for the air phase. In each constitutive equation, the deformation state variable can be the total, water, or air volume change, while the stress state variables are $(\sigma - u_a)$ and $(u_a - u_w)$. These constitutive equations are similar to those suggested by Biot (1941) and Coleman (1962), and have been presented in Chapter 12. The following sections summarize the constitutive equations for unsaturated soils.

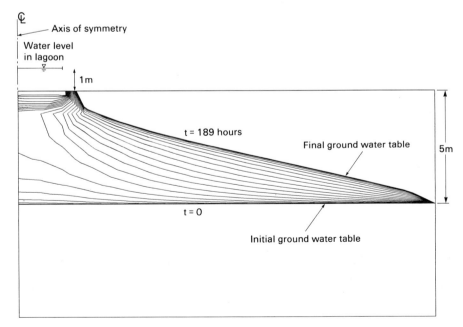

**Figure 16.14** Transient positions of the water table from an elapsed time equal to zero to an elapsed time equal to 189 h (i.e., 8 days).

16.2 COUPLED FORMULATIONS OF THREE-DIMENSIONAL CONSOLIDATION 457

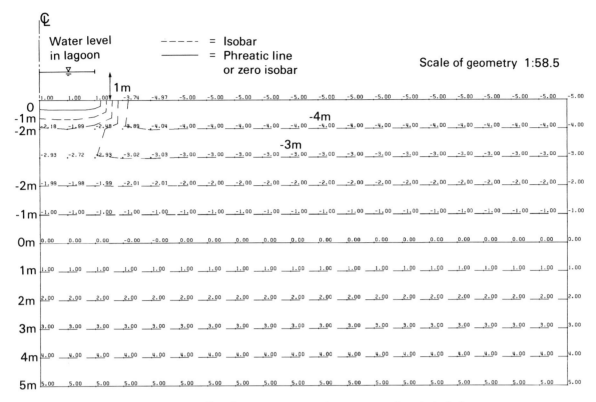

(a)

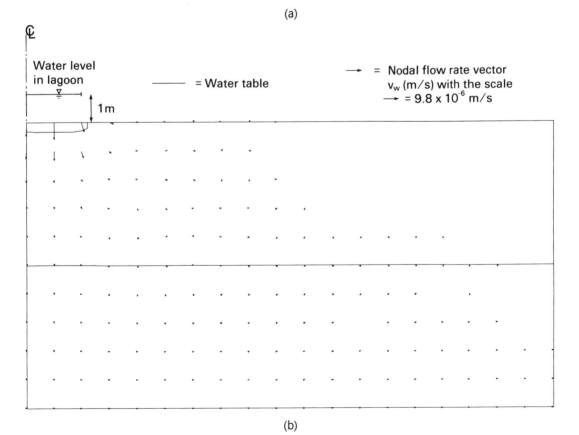

(b)

**Figure 16.15** Unsteady-state groundwater seepage below a lagoon at an elapsed time equal to 73 min. (a) Contours of pore-water pressure heads (i.e., isobars); (b) nodal flow rate vectors.

# 16 TWO- AND THREE-DIMENSIONAL UNSTEADY-STATE FLOW AND NONISOTHERMAL ANALYSES

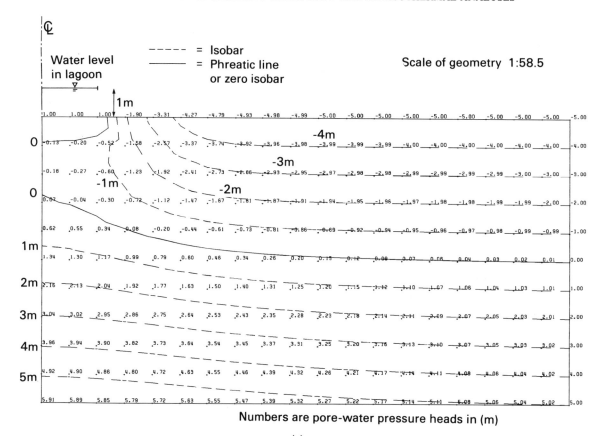

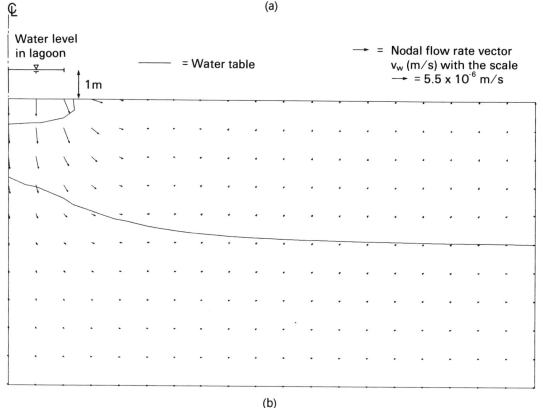

**Figure 16.16** Unsteady-state groundwater seepage below a lagoon at an elapsed time equal to 416 min. (i.e., 7 hours). (a) Contours of pore–water pressure heads (i.e., isobars); (b) nodal flow rate vectors.

16.2 COUPLED FORMULATIONS OF THREE-DIMENSIONAL CONSOLIDATION 459

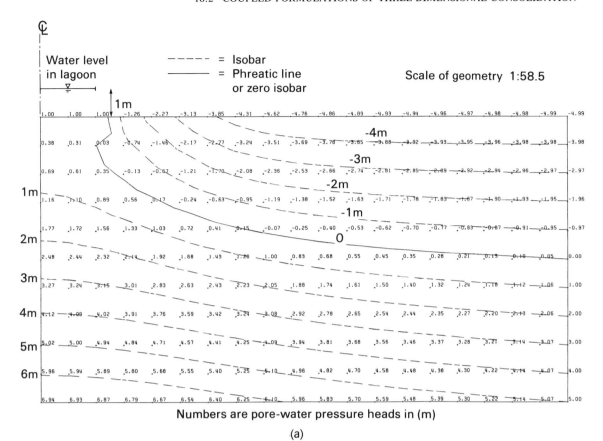

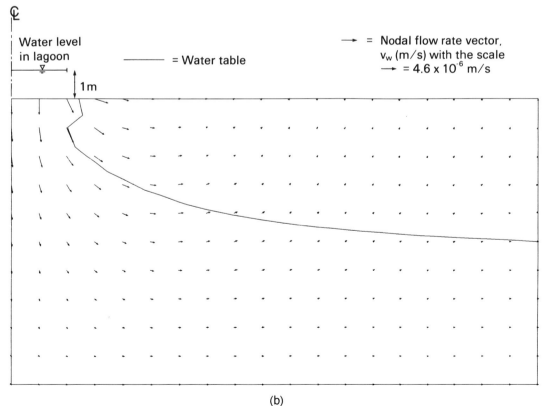

**Figure 16.17** Unsteady-state groundwater seepage below a lagoon at an elapsed time equal to 792 min. (i.e., 13 hours). (a) Contours of pore-water pressure heads (i.e., isobars); (b) nodal flow rate vectors.

460  16 TWO- AND THREE-DIMENSIONAL UNSTEADY-STATE FLOW AND NONISOTHERMAL ANALYSES

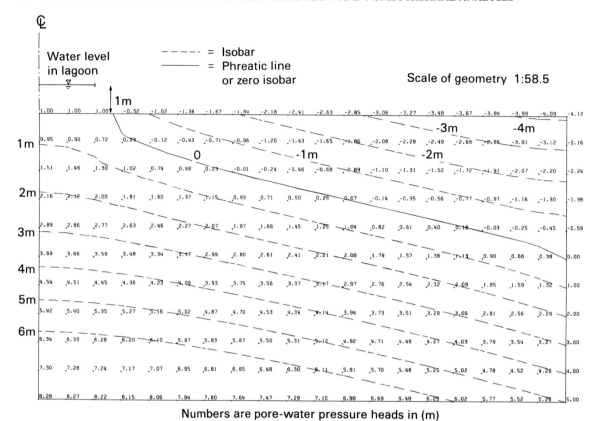

**Figure 16.18** Steady-state groundwater seepage below a lagoon at an elapsed time equal to 189 h (i.e., 8 days). (a) Contours of pore–water pressure heads (i.e., isobars); (b) nodal flow rate vectors.

## 16.2 COUPLED FORMULATIONS OF THREE-DIMENSIONAL CONSOLIDATION

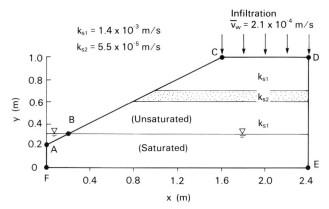

**Figure 16.19** Geometry and boundary conditions for example 3, seepage in layered hill slope.

### Soil Structure

The constitutive equation for the soil structure is derived by assuming that the soil behaves as an isotropic, linear elastic material. This assumption is acceptable in an incremental sense. The constitutive relations can be developed in a semi-empirical manner as an extension of the elasticity formulation used for saturated soils. The soil structure constitutive relations associated with the normal strains in the $x$-, $y$-, and $z$-directions can be written in an incremental form as follows:

$$d\epsilon_x = \frac{d(\sigma_x - u_a)}{E} - \frac{\mu}{E}d(\sigma_y + \sigma_z - 2u_a) + \frac{d(u_a - u_w)}{H} \quad (16.34)$$

$$d\epsilon_y = \frac{d(\sigma_y - u_a)}{E} - \frac{\mu}{E}d(\sigma_x + \sigma_z - 2u_a) + \frac{d(u_a - u_w)}{H} \quad (16.35)$$

$$d\epsilon_z = \frac{d(\sigma_z - u_a)}{E} - \frac{\mu}{E}d(\sigma_x + \sigma_y - 2u_a) + \frac{d(u_a - u_w)}{H} \quad (16.36)$$

and the constitutive equations associated with shear deformations are

$$d\gamma_{xy} = \frac{d\tau_{xy}}{G} \quad (16.37)$$

$$d\gamma_{yz} = \frac{d\tau_{yz}}{G} \quad (16.38)$$

$$d\gamma_{zx} = \frac{d\tau_{zx}}{G} \quad (16.39)$$

where

- $\epsilon_x$ = normal strain in the $x$-direction
- $\epsilon_y$ = normal strain in the $y$-direction
- $\epsilon_z$ = normal strain in the $z$-direction
- $\sigma_x$ = total normal stress in the $x$-direction
- $\sigma_y$ = total normal stress in the $y$-direction
- $\sigma_z$ = total normal stress in the $z$-direction
- $(\sigma_x - u_a)$ = net normal stress in the $x$-direction
- $(\sigma_y - u_a)$ = net normal stress in the $y$-direction
- $(\sigma_z - u_a)$ = net normal stress in the $z$-direction
- $\mu$ = Poisson's ratio
- $E$ = modulus of elasticity or Young's modulus for the soil structure with respect to a change in $(\sigma - u_a)$
- $H$ = modulus of elasticity for the soil structure with respect to a change in $(u_a - u_w)$
- $\gamma_{xy}$ = shear strain on the $z$-plane (i.e., $\gamma_{xy} = \gamma_{yx}$)
- $\gamma_{yz}$ = shear strain on the $x$-plane (i.e., $\gamma_{yz} = \gamma_{zy}$)
- $\gamma_{zx}$ = shear strain on the $y$-plane (i.e., $\gamma_{zx} = \gamma_{xz}$)
- $\tau_{xy}$ = shear stress on the $x$-plane in the $y$-direction (i.e., $\tau_{xy} = \tau_{yx}$)
- $\tau_{yz}$ = shear stress on the $y$-plane in the $z$-direction (i.e., $\tau_{yz} = \tau_{zy}$)
- $\tau_{zx}$ = shear stress on the $z$-plane in the $x$-direction (i.e., $\tau_{zx} = \tau_{xz}$)
- $G$ = shear modulus.

The change in volumetric strain, $d\epsilon_v$, can be obtained

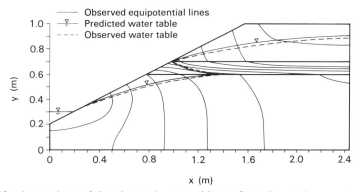

**Figure 16.20** Comparison of the observed water table configuration in the experimental sand tank with that predicted by Neuman's model at steady-state conditions (from Rulon and Freeze 1985).

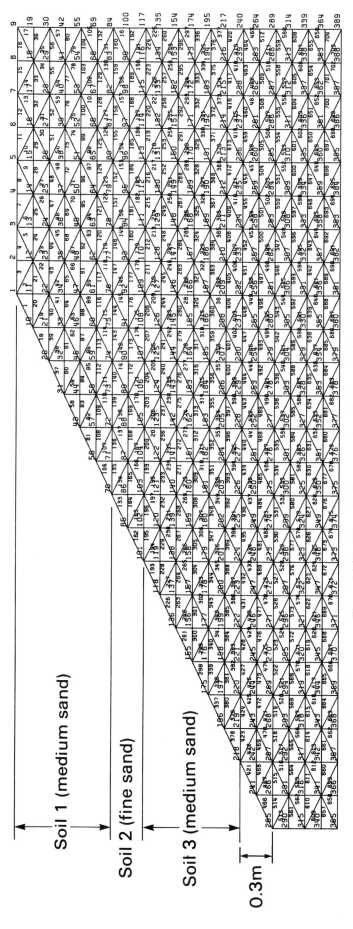

**Figure 16.21** Discretized cross section of a layered hill slope.

16.2 COUPLED FORMULATIONS OF THREE-DIMENSIONAL CONSOLIDATION 463

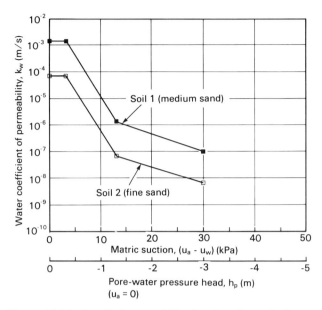

**Figure 16.22** Specified permeability functions for analyzing unsteady-state seepage through a layered hill slope.

from the summation of the normal strain changes:

$$d\epsilon_v = d\epsilon_x + d\epsilon_y + d\epsilon_z \quad (16.40)$$

where

$d\epsilon_v$ = volumetric strain change for each increment (i.e., $dV_v/V_0$)
$dV_v$ = total volume change of a soil element
$V_0$ = initial total volume of a soil element.

Rearranging Eqs. (16.34), (16.35), and (16.36), the change in net normal stresses can be expressed as functions of the normal strain and matric suction changes:

$$d(\sigma_x - u_a) = 2G(d\epsilon_x + \alpha d\epsilon_v) - \beta d(u_a - u_w) \quad (16.41)$$

$$d(\sigma_y - u_a) = 2G(d\epsilon_y + \alpha d\epsilon_v) - \beta d(u_a - u_w) \quad (16.42)$$

$$d(\sigma_z - u_a) = 2G(d\epsilon_z + \alpha d\epsilon_v) - \beta d(u_a - u_w) \quad (16.43)$$

where

$$G = \frac{E}{2(1 + \mu)} \quad (16.44)$$

$$\alpha = \frac{\mu}{1 - 2\mu} \quad (16.45)$$

$$\beta = \frac{E}{H(1 - 2\mu)} \quad \text{or} \quad \frac{2G}{H} \frac{(1 + \mu)}{(1 - 2\mu)}. \quad (16.46)$$

*Water Phase*

The constitutive equation for the water phase defines the water volume change in the soil element for any change in the total, pore-air, and pore-water pressures. The constitutive equation for the water phase in a referential element can be written as a linear combination of the stress state variable changes, and has the following incremental form:

$$\frac{dV_w}{V_0} = \frac{d(\sigma_x - u_a)}{E_w} + \frac{d(\sigma_y - u_a)}{E_w} + \frac{d(\sigma_z - u_a)}{E_w} + \frac{d(u_a - u_w)}{H_w} \quad (16.47)$$

where

$dV_w$ = water volume change in the soil element
$E_w$ = water volumetric modulus associated with a change in $(\sigma - u_a)$
$H_w$ = water volumetric modulus associated with a change in $(u_a - u_w)$.

Substituting Eqs. (16.41), (16.42), and (16.43) into Eq. (16.47), the water volume change in the soil element can

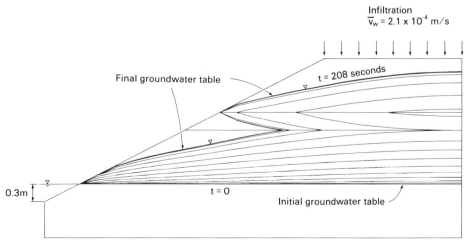

**Figure 16.23** Transient positions of the water table for elapsed times ranging from zero to 208 s (i.e., 3.5 min).

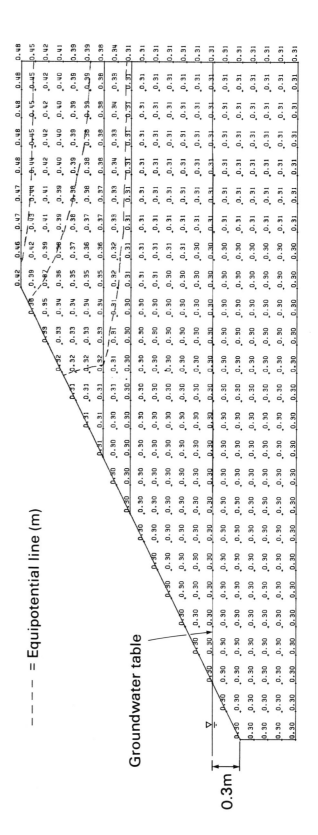

(a) Numbers are hydraulic heads in (m)

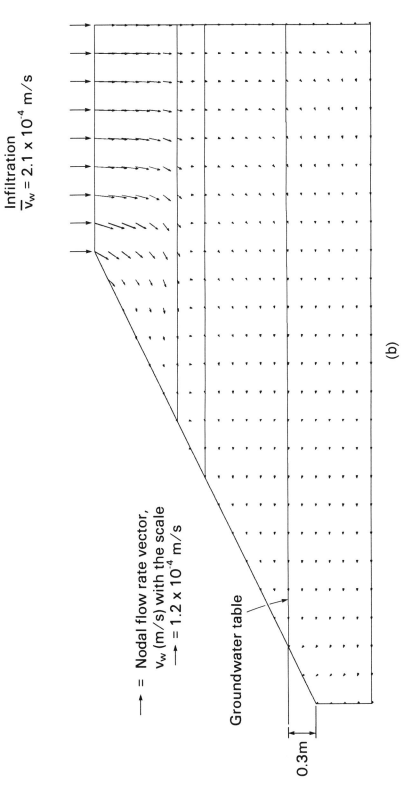

**Figure 16.24** Unsteady-state seepage through a layered hill slope at an elapsed time equal to 4.6 s. (a) Equipotential lines; (b) nodal flow rate vectors throughout the slope.

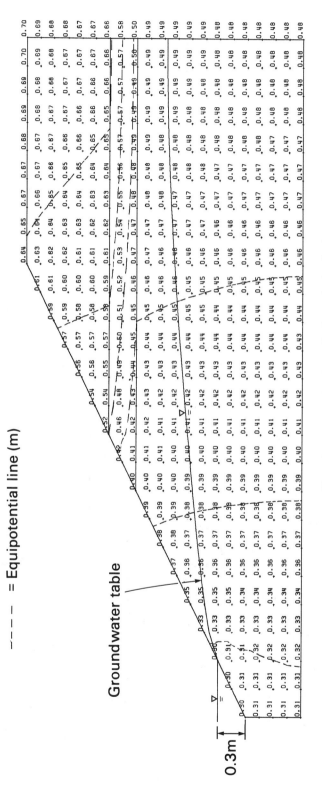

(a)

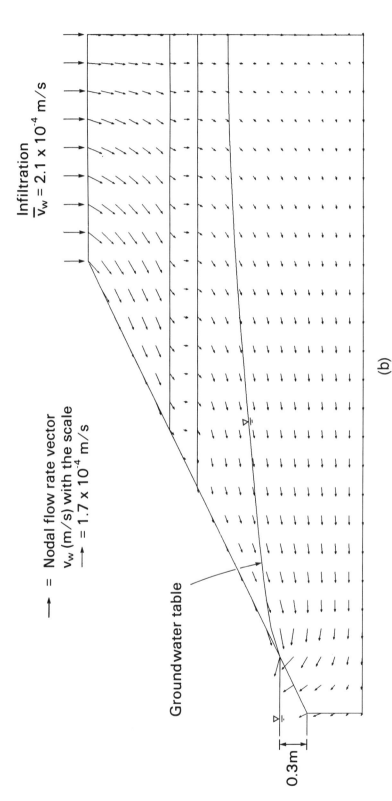

**Figure 16.25** Unsteady-state seepage through a layered hill slope at an elapsed time equal to 31 s. (a) Equipotential lines; (b) nodal flow rate vectors throughout the slope.

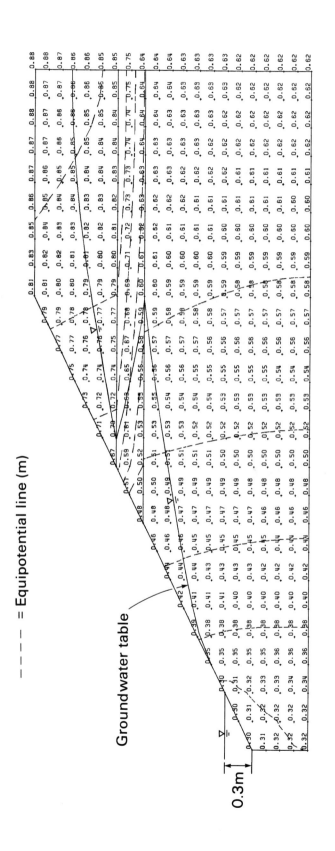

(a) Numbers are hydraulic heads in (m)

468

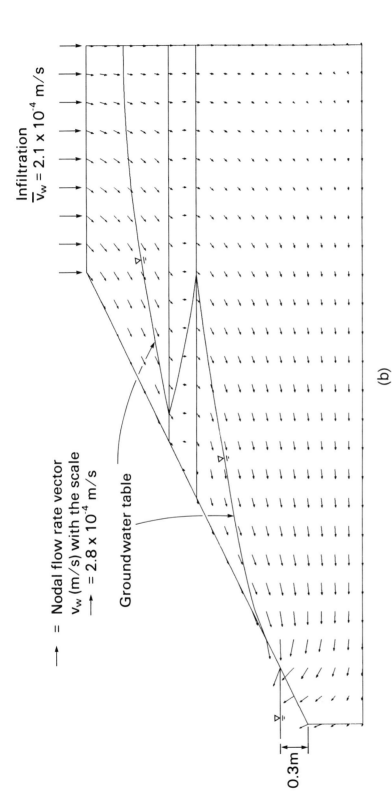

**Figure 16.26** Unsteady-state seepage through a layered hill slope at an elapsed time equal to 65.1 s. (a) Equipotential lines; (b) nodal flow rate vectors throughout the slope.

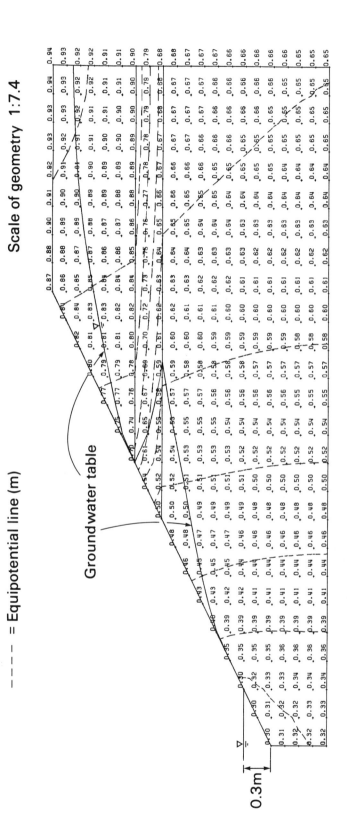

(a)

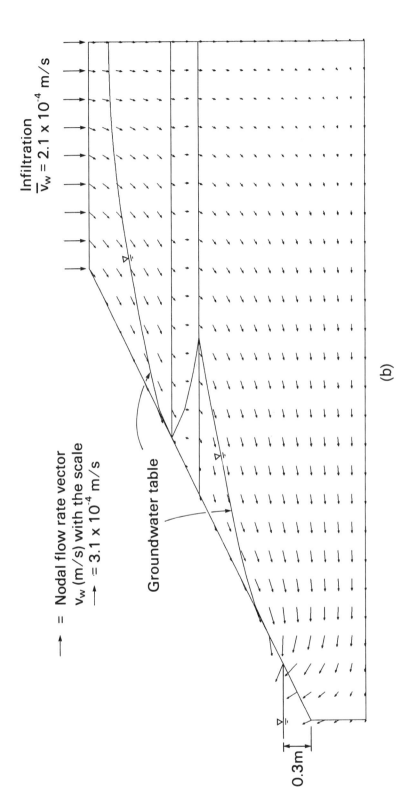

**Figure 16.27** Steady-state seepage through a layered hill slope at an elapsed time equal to 208 s. (a) Equipotential lines; (b) nodal flow rate vectors throughout the slope.

be written as

$$\frac{dV_w}{V_0} = \beta_{w1} d\epsilon_v + \beta_{w2} d(u_a - u_w) \quad (16.48)$$

where

$$\beta_{w1} = \frac{E}{E_w(1 - 2\mu)}$$

$$\beta_{w2} = \frac{1}{H_w} - \frac{3\beta}{E_w}.$$

### Air Phase

The constitutive equation for the air phase defines the air volume change in the soil element for any change in the total, pore–air, and pore–water pressures. The constitutive equation for the air phase in a referential element can be written as a linear combination of the stress state variable changes, and it has the following incremental form:

$$\frac{dV_a}{V_0} = \frac{d(\sigma_x - u_a)}{E_a} + \frac{d(\sigma_y - u_a)}{E_a} + \frac{d(\sigma_z - u_a)}{E_a}$$

$$+ \frac{d(u_a - u_w)}{H_a} \quad (16.49)$$

where

$dV_a$ = air volume change in the soil element
$E_a$ = air volumetric modulus associated with a change in $(\sigma - u_a)$
$H_a$ = air volumetric modulus associated with a change in $(u_a - u_w)$.

Substituting Eqs. (16.41), (16.42), and (16.43) into Eq. (16.49), the air volume change in the soil element can be written as

$$\frac{dV_a}{V_0} = \beta_{a1} d\epsilon_v + \beta_{a2} d(u_a - u_w) \quad (16.50)$$

where

$$\beta_{a1} = \frac{E}{E_a(1 - 2\mu)}$$

$$\beta_{a2} = \frac{1}{H_a} - \frac{3\beta}{E_a}.$$

### 16.2.2 Coupled Consolidation Equations

The coupled consolidation equations presented herein assume that the air phase is continuous. Several other assumptions used in the derivation are similar to those proposed by Terzaghi (1943) and Biot (1941). An outline of the assumptions used is as follows: 1) material is isotropic, 2) reversibility of stress–strain relations, 3) linearity of stress–strain relations, 4) small strains, 5) pore-water is incompressible, 6) coefficients of permeability of water and air phases are functions of the volume–mass soil properties during the consolidation process, and 7) the effects of air diffusing through water, air dissolving in the water phase, and the movement of water vapor are ignored.

In a three-dimensional consolidation problem, there are five unknowns of deformation and volumetric variables to be solved, namely, the displacements in the $x$-, $y$-, and $z$-directions (i.e., $u$, $v$, and $w$, respectively), and the water and air volume changes (i.e., $dV_w$ and $dV_a$). The displacements in the $x$-, $y$-, and $z$-directions are used to compute the total volume change. The five unknowns can be obtained from three equilibrium equations and two continuity equations (i.e., the water and air phase continuities). These equations are outlined in the following sections. The continuity equations for a three-dimensional consolidation are essentially an extension of the continuity applied to a one-dimensional consolidation problem, as shown in Chapter 15. Therefore, detailed derivations of the continuity equations are not presented in the following sections, but reference is made to Chapter 15. In summary, the solution to the three-dimensional consolidation problem can be obtained by solving the three equilibrium equations and two continuity equations.

### Equilibrium Equations

The stress state for an unsaturated soil element should satisfy the following equilibrium conditions (see Appendix B):

$$\frac{\partial \sigma_x}{\partial x} + \frac{\partial \tau_{yx}}{\partial y} + \frac{\partial \tau_{zx}}{\partial z} = 0 \quad (16.51)$$

$$\frac{\partial \tau_{xy}}{\partial x} + \frac{\partial \sigma_y}{\partial y} + \frac{\partial \tau_{zy}}{\partial z} = 0 \quad (16.52)$$

$$\frac{\partial \tau_{xz}}{\partial x} + \frac{\partial \tau_{yz}}{\partial y} + \frac{\partial \sigma_z}{\partial z} = 0. \quad (16.53)$$

The effect of the body force [i.e., the gravity force in Eq. (16.52)] is assumed to be negligible.

Substituting Eqs. (16.41), (16.42), and (16.43) into Eqs. (16.51), (16.52), and (16.53) gives the following form for the equilibrium equations:

$$G\nabla^2 u + \frac{G}{1 - 2\mu} \frac{\partial \epsilon_v}{\partial x} - \beta \frac{\partial (u_a - u_w)}{\partial x} + \frac{\partial u_a}{\partial x} = 0$$

$$(16.54)$$

$$G\nabla^2 v + \frac{G}{1 - 2\mu} \frac{\partial \epsilon_v}{\partial y} - \beta \frac{\partial (u_a - u_w)}{\partial y} + \frac{\partial u_a}{\partial y} = 0$$

$$(16.55)$$

$$G\nabla^2 w + \frac{G}{1 - 2\mu} \frac{\partial \epsilon_v}{\partial z} - \beta \frac{\partial (u_a - u_w)}{\partial z} + \frac{\partial u_a}{\partial z} = 0$$

$$(16.56)$$

where

$$\nabla^2 = \frac{\partial^2}{\partial x^2} + \frac{\partial^2}{\partial y^2} + \frac{\partial^2}{\partial z^2}.$$

## Water Phase Continuity

The continuity equation for the water phase can be obtained by equating the time derivative of the water phase constitutive equation to the divergence of the flow rate, as described by Darcy's law (see Chapter 15):

$$\beta_{w1} \frac{\partial \epsilon_v}{\partial t} + \beta_{w2} \frac{\partial (u_a - u_w)}{\partial t}$$
$$= -\left[ \frac{k_w}{\rho_w g} \nabla^2 u_w + \frac{1}{\rho_w g} \left( \frac{\partial k_w}{\partial x} \frac{\partial u_w}{\partial x} + \frac{\partial k_w}{\partial y} \frac{\partial u_w}{\partial y} \right. \right.$$
$$\left. \left. + \frac{\partial k_w}{\partial z} \frac{\partial u_w}{\partial z} \right) + \frac{\partial k_w}{\partial y} \right]. \qquad (16.57)$$

## Air Phase Continuity

The continuity equation for the air phase can be obtained from the time derivative of the air phase constitutive equation and the net mass rate of air flow, as described by Fick's law (see Chapter 15):

$$\beta_{a1} \frac{\partial \epsilon_v}{\partial t} + \beta_{a2} \frac{\partial (u_a - u_w)}{\partial t}$$
$$= -\left[ \frac{D_a^*}{\rho_a \bar{u}_a} \nabla^2 u_a + \frac{1}{\rho_a \bar{u}_a} \left( \frac{\partial D_a^*}{\partial x} \frac{\partial u_a}{\partial x} + \frac{\partial D_a^*}{\partial y} \frac{\partial u_a}{\partial y} \right. \right.$$
$$\left. \left. + \frac{\partial D_a^*}{\partial z} \frac{\partial u_a}{\partial z} \right) + \frac{(1-S)n}{\bar{u}_a} \frac{\partial \bar{u}_a}{\partial t} \right] \qquad (16.58)$$

where

$D_a^*$ = coefficient of transmission for the air phase
$\rho_a$ = density of air
$\bar{u}_a$ = absolute pore–air pressure (i.e., $\bar{u}_a = u_a + \bar{u}_{atm}$)
$u_a$ = gauge pore–air pressure
$\bar{u}_{atm}$ = atmospheric pressure (i.e., 101 kPa)
$S$ = degree of saturation
$n$ = porosity.

The solution of a coupled analysis for two-phase flow through an unsaturated soil involves the simultaneous solution of Eqs. (16.54), (16.55), (16.56), (16.57), and (16.58). It is beyond the scope of this book to illustrate the solution to such a coupled analysis.

## 16.3 NONISOTHERMAL FLOW

Air and water flow not only is a result of applied loads, but also is a result of temperature gradients within a soil mass. Nonisothermal conditions are closely related to microclimatic changes in the field. As an example, seasonal temperature changes can cause air and water flow, and consequently volume changes within a soil mass. This nonisothermal condition often occurs under highway and airfield pavements or under the shallow foundations of lightweight structures.

The following sections present a one-dimensional flow formulation for nonisothermal conditions in an unsaturated soil. Temperature gradients are incorporated into the formulation of flow. In addition, water vapor flow is also incorporated into the formulation since vapor flow can be a significant factor under nonisothermal conditions. Surface boundary conditions for the air, liquid water, water vapor, and heat flow equations are also reviewed. The soil–atmospheric boundary conditions incorporate the microclimatic conditions at a site.

### 16.3.1 Air Phase Partial Differential Equation

The differential equation for one-dimensional air flow during consolidation (see Chapter 15) can be used for nonisothermal conditions, with the addition of one term. The additional term accounts for the effect of temperature changes on air flow, and consequently on pore–air pressures. The one-dimensional air flow equation in the y-direction can be expressed as follows (Dakshanamurthy and Fredlund, 1981):

$$\frac{\partial u_a}{\partial t} = -C_a \frac{\partial u_w}{\partial t} + c_v^a \frac{\partial^2 u_a}{\partial y^2} + C_{at} \frac{\partial T}{\partial t} \qquad (16.59)$$

where

$C_a$ = interaction coefficient associated with the air phase partial differential equation, i.e.,

$$\frac{m_2^a/m_{1k}^a}{1 - m_2^a/m_{1k}^a - (1-S)n/(\bar{u}_a m_{1k}^a)}$$

$m_{1k}^a$ = coefficient of air volume change with respect to a change in net normal stress, $d(\sigma_y - u_a)$, for $K_0$-loading

$c_v^a$ = coefficient of consolidation with respect to the air phase, i.e.,

$$\frac{D_a^*}{\omega_a/RT} \frac{1}{\bar{u}_a m_{1k}^a (1 - m_2^a/m_{1k}^a) - (1-S)n}$$

$C_{at}$ = interaction temperature coefficient associated with the air phase partial differential equation, i.e.,

$$\frac{1}{T} \left[ \frac{(1-S)n \bar{u}_a}{(1 - m_2^a/m_{1k}^a)\bar{u}_a m_{1k}^a + (1-S)n} \right]$$

$T$ = absolute temperature (i.e., $T = t° + 273.16$) (K)
$t°$ = temperature (°C)
$t$ = time.

Equation (16.59) does not take into account any variation which might occur in the coefficient of transmission, $D_a^*$. The variation in the coefficient of transmission with respect to space (i.e., $\partial D_a^*/\partial y$) is assumed to be negligible.

### 16.3.2 Fluid and Vapor Flow Equation for the Water Phase

The differential equation for fluid and vapor flow under nonisothermal conditions is similar to the water flow equation for one-dimensional consolidation (see Chapter 15), with the addition of one term. The additional term accounts for the water vapor flow due to diffusion and advection processes (Wilson, 1990). The water vapor flow can be significant under nonisothermal conditions. The one-dimensional fluid and vapor flow equation for the water phase in the $y$-direction can be expressed as follows:

$$\frac{\partial u_w}{\partial t} = -C_w \frac{\partial u_a}{\partial t} + \frac{1}{\rho_w g m_2^w} \frac{\partial}{\partial y}\left(k_w \frac{\partial u_w}{\partial y}\right)$$
$$+ \frac{\bar{u}_a + \bar{u}_v}{\bar{u}_a} \frac{1}{\rho_w m_2^w} \frac{\partial}{\partial y}\left(D_v \frac{\partial \bar{u}_v}{\partial y}\right) \quad (16.60)$$

where

- $C_w$ = interaction coefficient associated with the water phase partial differential equation [i.e., $(1 - m_2^w/m_{1k}^w)/(m_2^w/m_{1k}^w)$]
- $u_v$ = partial pressure of water vapor in air
- $D_v$ = diffusion coefficient of the water vapor through the soil (kg · m/kN · s).

The coefficient of water vapor diffusion, $D_v$, can be calculated as follows (Philip and de Vries, 1957; de Vries, 1975; Dakshanamurthy and Fredlund, 1981; Wilson, 1990):

$$D_v = \alpha\beta\left(D_{vm}\frac{\omega_v}{RT}\right) \quad (16.61)$$

where

- $\alpha$ = tortuosity factor for the soil (i.e., $\epsilon = \beta^{2/3}$)
- $\beta$ = cross-section area of the soil available for water vapor flow (i.e., $(1 - S)n$)
- $D_{vm}$ = molecular diffusivity of water vapor in air (i.e., $0.229[1 + T/273]^{1.75} \times 10^{-4}$ (m$^2$/s) (from Kimball et al., 1976)
- $\bar{\omega}_v$ = molecular mass of water vapor (i.e., 18.016 kg/kmol).

Equation (16.61) indicates that the diffusion coefficient, $D_v$, is a function of soil properties (i.e., $S$ and $n$), which in turn are a function of matric suction, $(u_a - u_w)$. In addition, $D_v$ is also a function of temperature, $T$ [Eq. (16.61)]. Similarly, the coefficient of permeability, $k_w$, is also a function of soil properties (see Chapter 5) that may vary with respect to location in the soil mass. If the variations in the $k_w$ and $D_v$ coefficients with respect to space are considered to be negligible, Eq. (16.60) can be simplified as

$$\frac{\partial u_w}{\partial t} = -C_w \frac{\partial u_a}{\partial t} + c_v^w \frac{\partial^2 u_w}{\partial y^2} + c_v^{wv} \frac{\partial^2 \bar{u}_v}{\partial y^2} \quad (16.62)$$

where

- $c_v^w$ = coefficient of consolidation with respect to the water phase (i.e., $k_w/(\rho_w g m_2^w)$)
- $c_v^{wv}$ = coefficient of consolidation with respect to the water vapor phase, i.e.,

$$\frac{(\bar{u}_a + \bar{u}_v)}{\bar{u}_a} \frac{D_v}{\rho_w m_2^w}.$$

### 16.3.3 Heat Flow Equation

A Fourier diffusion equation can be used to describe heat transfer in soils (Jame and Norum, 1980; Fredlund and Dakshanamurthy, 1982; Wilson, 1990). The form of the heat flow equation proposed by Wilson (1990) is as follows:

$$\zeta\frac{\partial T}{\partial t} = \frac{\partial}{\partial y}\left(\lambda\frac{\partial T}{\partial t}\right) - L_v\left(\frac{\bar{u}_a + \bar{u}_v}{\bar{u}_a}\right)\frac{\partial}{\partial y}\left(D_v \frac{\partial \bar{u}_v}{\partial y}\right)$$
$$(16.63)$$

where

- $\zeta$ = volumetric specific heat of the soil as a function of water content (J/m$^3$/°C)
- $\lambda$ = thermal conductivity of the soil as a function of water content (W/m/°C)
- $L_v$ = latent heat of vaporization of water (i.e., 2 418 000 J/kg).

Equation (16.63) describes the heat flow due to conduction and the latent heat transfer caused by phase changes. Convective heat flow is considered to be negligible (Jame and Norum, 1980; Wilson, 1990). The volumetric specific heat of the soil, $\zeta$, can be calculated using the relationship given by de Vries (1963):

$$\zeta = \zeta_s\theta_s + \zeta_w\theta_w + \zeta_a\theta_a \quad (16.64)$$

where

- $\zeta_s$ = volumetric specific heat capacity of the soil solids (i.e., a typical value is $2.235 \times 10^6$ J/m$^3$/°C for fine sands (de Vries, 1963))
- $\theta_s$ = volumetric solid content (i.e., $V_s/V$)
- $V_s$ = volume of soils solids in the soil
- $V$ = total volume of the soil
- $\zeta_w$ = volumetric specific heat capacity for the water phase (i.e., $4.154 \times 10^6$ J/m$^3$/°C for water at 35°C (Wilson, 1990))
- $\theta_w$ = volumetric water content (i.e., $V_w/V$)
- $V_w$ = volume of water in the soil
- $\zeta_a$ = volumetric specific heat capacity for the air phase
- $\theta_a$ = volumetric air content (i.e., $V_a/V$).

The third term on the right-hand side of Eq. (16.64) is small and can be considered negligible (de Vries, 1963). The thermal conductivity of the soil, $\lambda$, can be obtained from the following expression (de Vries, 1963):

$$\lambda = \frac{f_s \theta_s \lambda_s + f_w \theta_w \lambda_w + f_a \theta_a \lambda_a}{f_s \theta_s + f_w \theta_w + f_a \theta_a} \quad (16.65)$$

where

$f_s, f_w, f_a$ = weighting factors for the solid, water, and air phases, respectively

$\lambda_s, \lambda_w, \lambda_a$ = thermal conductivities of the solid, water, and air phases, respectively.

Typical thermal conductivities for sand particles and water are 6.0 and 0.57 W/m/°C, respectively (de Vries, 1963; Jame, 1977; Wilson, 1990). The thermal conductivity of air, $\lambda_a$, consists of two components (Jame, 1977; Wilson, 1990):

$$\lambda_a = \lambda_{\text{dry air}} + \lambda_{\text{water vapor}} \quad (16.66)$$

where

$\lambda_{\text{dry air}} = 0.025$ W/m/°C

$\lambda_{\text{water vapor}}$ = varies linearly between 0 and 0.0736 W/m/°C for volumetric water content, $\theta_w$, between 0 and 12.1%.

The weighting factors, $f$, are calculated using the assumption that the soil particles are ellipsoidal in shape. The weighting factor for the continuous medium (i.e., air or water) is equal to 1.0. Water can be selected as the continuous medium, with the $f_w$ value equal to 1.0. The weighting factors for the solid and air phases can then be calculated in accordance with the following relationship:

$$f = \frac{1}{3} \sum_{i=1}^{3} \left[ 1 + \left( \frac{\lambda}{\lambda_w} - 1 \right) g_i \right]^{-1} \quad (16.67)$$

where

$f = f_a$ or $f_s$
$\lambda = \lambda_a$ or $\lambda_s$
$g$ = depolarization factors for the ellipsoid (i.e., $g_1, g_2,$ and $g_3$ where $g_1 + g_2 + g_3 = 1$); the values of $g_1, g_2,$ and $g_3$ are independent of particle size, and are dependent only on the ratio of the length of the ellipsoid axes.

Wilson (1990) used equal depolarization factors of $\frac{1}{3}$ in computing the weighting factors for a sandy soil. In this case, the sand particles were assumed to have spherical shapes. The depolarization factors, $g_1$ and $g_2$, for the air phase were assumed to decrease linearly from 0.333 to 0.105 for volumetric water content, $\theta_w$, ranging from 23.6 to 12.1% respectively, and from 0.105 to 0.015 for $\theta_w$ values ranging from 12.1 and 0%, respectively (Jame, 1977). Figure 16.28 illustrates the thermal conductivity, $\lambda$, variation for Beaver Creek sand with respect to volumetric water contents as computed using Eqs. (16.65), (16.66), and (16.67).

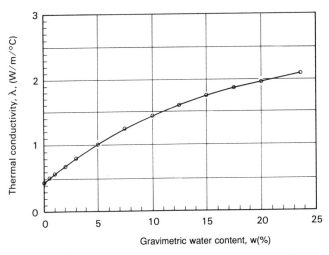

**Figure 16.28** Thermal conductivity versus water content for the Beaver Creek sand (from Wilson, 1990).

### 16.3.4 Atmospheric Boundary Conditions

The solution for nonisothermal flows can be obtained by solving the differential equations for the air phase [Eq. (16.59)], the liquid water and water vapor phases [Eqs. (16.60) or (16.62)], and the heat flow equation [Eq. (16.63)] simultaneously. In some cases, the pore–air pressure changes during a transient process can be considered negligible (Rahardjo, 1990; Wilson, 1990). As a result, the air flow differential equation, Eq. (16.59), can be dropped and the $(\partial u_a / \partial t)$ term in the water flow equation [i.e., Eq. (16.60) or (16.62)] is set to zero. In other words, only two differential equations need to be solved simultaneously. These are Eqs. (16.60) [or (16.62)] and (16.63).

*Surface Boundary Conditions for Air and Fluid Water Flow*

Solving the three differential flow equations requires information on the upper or atmospheric boundary conditions with respect to air, liquid water, water vapor and temperature. Atmospheric air pressure can be used as a head boundary condition at the soil surface when solving the air flow equation, Eq (16.59). The net water infiltration at the ground surface can be used as a flux boundary condition at the soil surface when solving the liquid water flow component in Eq. (16.60) or (16.62). The net infiltration at the ground surface can be computed as the difference between the total rainfall and the amount of runoff.

*Surface Boundary Conditions for Water Vapor Flow*

Net water exfiltration at the ground surface can be written as a water vapor flow or evaporation. The evaporative flux at the ground surface can be used as a flux boundary condition for the water vapor flow component in Eq. (16.60) or (16.62). The evaporative flux is a function of the water vapor pressure gradient between the soil surface and the atmosphere. The maximum evaporative rate corresponds to the condition where the soil surface is fully saturated. This rate is commonly referred to as the potential evaporation (Thornthwaite, 1948; Penman, 1948). The actual rate of evaporation decreases significantly as the soil surface becomes unsaturated and desiccated (Hillel, 1980; Wilson, 1990). In addition, the actual evaporative rate is a function of climatic conditions, as illustrated in Fig. 16.29.

The evaporation rate from a soil surface is directly related to the water vapor pressure gradient between the surface of the soil and the air immediately above the surface. The evaporation rate can be written using Dalton's mass diffusion equation (Wilson, 1990):

$$E_v = f_v(\bar{u}_v - \bar{u}_v^a) \quad (16.68)$$

where

$E_v$ = vertical evaporative flux
$f_v$ = a function dependent on wind speed, surface roughness, and eddy diffusion
$\bar{u}_v$ = water vapor pressure of the evaporating soil surface
$u_v^a$ = water vapor pressure in the air above the evaporating surface.

The use of Eq. (16.68) in solving field problems requires climatic data in order to evaluate the $f_v$ coefficient. Penman (1948) presented a formulation for computing the potential evaporation from saturated surfaces by combining Eq. (16.68) and the energy balance at the ground surface. The formulation was later extended by Wilson (1990) to calculate the actual evaporation rate from unsaturated soil surfaces:

$$E_v = \frac{\Gamma Q_n + \eta E}{\Gamma + \eta A} \quad (16.69)$$

where

$E_v$ = vertical evaporative flux (mm/day)
$\Gamma$ = slope of the saturation water vapor pressure versus temperature curve at the mean temperature of the air (see Chapter 2)

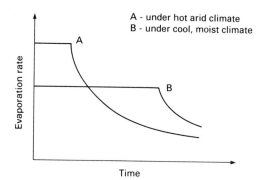

**Figure 16.29** Evaporation rates versus time for a desiccating soil surface under varying climatic conditions (from Wilson, 1990).

$Q_n$ = heat budget or all net radiation (mm/day)
$\eta$ = psychrometric constant (i.e., 0.27 mmHg/°F)
$E_{va} = f_v u_v^a (B - A)$
$f_v = 0.35(1 + 0.146 W_w)$
$W_w$ = wind speed (km/h)
$\bar{u}_v^a$ = water vapor pressure in the air above the evaporating surface (mmHg)
$B$ = inverse of the relative humidity in the air
$A$ = inverse of the relative humidity at the soil surface; the relative humidity, RH, at the soil surface can be related to soil suction, as shown in Chapter 4.

The heat budget, $Q_n$, can be determined using the empirical formula given by Penman (1948):

$$Q_n = R_c(1 - r) - \tau_{sb} T_a^4 (0.56 - 0.092 \sqrt{\bar{u}_v^a})$$
$$\cdot \left(0.10 + \frac{0.90 n_{as}}{N_{ps}}\right) \quad (16.70)$$

where

$R_c$ = shortwave radiation measured at the site (i.e., $0.95 R_a (0.18 + 0.55 n_{as}/N_{ps})$)
$R_a$ = solar radiation (from charts) for a completely transparent atmosphere
$r$ = reflectance coefficient
$\tau_{sb}$ = Stefan–Boltzmann constant (i.e., $5.67 \times 10^{-8}$ W/m²/K⁴)
$T_a$ = air temperature (K)
$n_{as}/N_{ps}$ = ratio of actual/possible sunshine hours.

Wilson (1990) demonstrated the use of Eq. (16.69) for predicting a typical evaporation rate for Saskatoon, Sask., Canada in the month of July. The soil used in the analyses was a fine uniform aeolian sand. A potential evaporation rate of 7.7 mm/day was calculated using the following typical climatic data provided by Gray (1970):

1) Air temperature = 23.6°C
2) Relative humidity = 47.5%
3) Wind speed = 4.4 m/s
4) Solar radiation = 39 MJ/m² day or 16.2 mm/day
5) Insolation = 30 MJ/m² day or 12.5 mm/day
6) Reflection coefficient = 0.05
7) Sunshine ratio, $n_{as}/N_{ps} = 65\%$.

Figure 16.30 shows the predicted evaporation rates for three positions of the water table. The high water table condition (i.e., curve 1) gives an evaporation rate equal to the potential rate of evaporation of 7.7 mm/day for a duration of 10 days. In other words, the soil surface remains as a saturated evaporating surface for the entire 10 day period since the high water table provides adequate recharge to the soil surface for full evaporation. However, the lower water table conditions (i.e., curves 2 and 3) cannot provide sufficient recharge to the soil surface during the evapora-

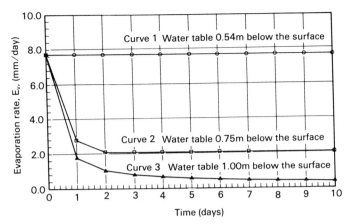

**Figure 16.30** Evaporation rate versus time using the Modified Penman Equation with the water table at various depths in the Beaver Creek sand (from Wilson, 1990).

tion period. As a result, the initially saturated sand surface quickly desaturates.

The soil suction at the ground surface increases significantly, and evaporation decreases. The evaporation reduces from 7.7 to approximately 2 mm/day. It appears that a drying front develops near the soil surface, which controls the evaporative flux. The high soil suction at the surface causes the coefficient of permeability with respect to the water phase to approach zero (or a low value). The liquid water becomes discontinuous or absent. Therefore, the upward liquid water flux from the water table only extends to a short distance, below ground surface. Then water vapor diffusion through the dry surface soil layer becomes the dominant process.

### Surface Boundary Conditions for Heat Flow

The soil temperature at the ground surface can be used as the surface boundary condition when solving the heat flow equation, Eq. (16.63). The surface temperature can be calculated as follows (Wilson, 1990):

$$T_s = T_a + \frac{1}{\eta f_v}(Q_n - E_v) \qquad (16.71)$$

where

$T_s$ = soil temperature at the surface (K).

The $Q_n$ and $E_v$ values in Eq. (16.71) are computed using Eqs. (16.70) and (16.69), respectively.

The solution of further example problems related to the prediction of surface evaporative flux based on microclimatic conditions is beyond the scope of this book. However, the prediction of evaporative flux is becoming an increasingly important problem. There appears to be considerable need for further research and verification of soil surface evaporative models.

# APPENDIX A

# *Units and Symbols*

The universal and consistent system of units, "The International System of Units," (SI) is used throughout the book. This version of the metric system is described in the ASTM (1966) Metric Practice Guide and in the ASTM Standard for Metric Practice. The SI system which was formally recognized by the General Conference on Weights and Measures, Rome, in 1960, has now been essentially adopted in many countries (e.g., Great Britain, Australia, New Zealand, Canada, and most European countries).

The SI system is based on seven basic units to cover the entire spectrum of science and engineering. The units are comprehensive and coherent. These basic units are length (meter), mass (kilogram), time (second), electric current (ampere), thermodynamic temperature (Kelvin), luminous intensity (candela) and the amount of substance (mole). All units are independent of the nature of the physical process being considered. Only the first three units, as well as temperature, are used in this book.

Other physical quantities can be derived from the basic units. For example, the unit of force is a newton. The unit of force is derived from Newton's second law, $F = Ma$, where mass "M" is in kilograms and the acceleration, "a" is in $m/s^2$. The correct unit to express the weight of an object is newtons and weight is a function of gravitational acceleration. It is recommended that a mass designation be used in engineering practice whenever possible.

Confusion is generated when the word "weight" is used to refer to obtaining the "mass" of an object. A laboratory balance, for example, is used to compare two masses with the objective of obtaining an unknown "mass." This process is commonly referred to as "weighing the object."

The unit weight of a soil is another derived unit used in engineering practice. It is preferable, however, to use the density of a soil multiplied by acceleration due to gravity. Acceleration due to gravity can generally be taken as 9.807 $m/s^2$.

Table A.1 contains a list of derived units of interest to geotechnical engineering. Pressure or stress has the derived units of pascals or newtons per square meter. This unit is relatively small and is often used in conjunction with a prefix. Table A.2 lists prefixes which can be used to indicate multiples and submultiples of basic or derived units. The prefixes are used to render numbers lying within the range of $10^{-9}$ to $10^9$.

In soil science, the pF unit has been used to designate negative pore-water pressures (Glossary of Soil Science Terms, 1975). The pF unit was defined as the logarithm of the negative pore-water pressure expressed as the height of a column of water in centimeters. This unit is now considered to be obsolete (Johnson, 1981) and is not used in this book. Rather, negative pore-water pressures are expressed in standard SI units (e.g., kPa).

Although the book is written using the SI designation of units, there is often need to convert variables to other units. Table A.3 contains some conversion factors useful in making conversions between the SI (metric) system and the British Engineering system of units.

The symbols used throughout this book are, for the most part, consistent with those used in soil and rock mechanics. The extensive list of symbols published by the International Society for Soil Mechanics and Foundation Engineering (ISSMFE, 1977) has been used as a guide. However, it has been necessary to extend some of the symbols for unsaturated soil mechanics. For example, the soil has two fluid phases and it becomes necessary to carry a subscript to differentiate the air and water phases (e.g., $u_a$ and $u_w$).

All symbols are defined upon their first usage in the book. Some of the definitions are repeated from one chapter to another. Some of the main symbols used in the book are summarized in Table A.4.

## Table A.1 Derived Units Common to Geotechnical Engineering

| Physical Quantity/Property | Unit | SI Symbol |
|---|---|---|
| Mass | kilogram | kg |
| Length | meter | m |
| Time | seconds | s |
| Area | square meter | $m^2$ |
| Volume | cubic meter | $m^3$ |
| Velocity | meter per second | m/s |
| Acceleration | meter per square second | $m/s^2$ |
| Discharge, Flux | cubic meter per second | $m^3/s$ |
| Weight, Force | newton | N |
| Density | kilogram per cubic meter | $kg/m^3$ |
| Unit weight | newton per cubic meter | $N/m^3$ |
| Pressure or Stress | newton per square meter (i.e., pascal) | Pa |
| Moment or Torque | newton meter | Nm |
| Permeability | meter per second | m/s |
| Viscosity (dynamic) | newton second per square meter | $Ns/m^2$ |
| Temperature | Kelvin | K |

## Table A.2 Prefixes for Basic and Derived Units

| Factor | Prefix | Symbol |
|---|---|---|
| $10^9$ | giga | G |
| $10^6$ | mega | M |
| $10^3$ | kilo | k |
| $10^2$ | hecto | h |
| $10^1$ | deca | da |
| $10^{-1}$ | deci | d |
| $10^{-2}$ | centi | c |
| $10^{-3}$ | milli | m |
| $10^{-6}$ | micro | $\mu$ |
| $10^{-9}$ | nano | n |

## Table A.3 Conversion Factors for the British Engineering and SI Systems

| British System | SI |
|---|---|
| *Length* | |
| 1 inch, in | = 25.4 mm = 0.0254 m |
| 1 foot, ft | = 0.3048 m |
| 1 yard, yd | = 0.9144 m |
| *Mass* | |
| 1 pound mass, 1 bm (avoirdupois) | = 0.4536 kg |
| 1 slug (1 lb-force/$fts^2$) | = 14.59 kg |
| *Force* | |
| 1 lb-force | = 4.448 N |
| *Stress and Pressure* | |
| 1 psi (lb-force/$in.^2$) | = $6.895 \times 10^3$ Pa or 6.895 kPa |
| 1 atmosphere | = $1.013 \times 10^5$ Pa or 101.3 kPa |
| 1 bar | = $1 \times 10^5$ Pa or 100 kPa |
| *Density* | |
| 1 lb-mass/$ft^3$ | = 16.018 $kg/m^3$ |

**Table A.4 Principal Notations for Unsaturated Soils**

| Soil Property | Symbol | Unit | Dimensions |
|---|---|---|---|
| Void ratio | $e$ | — | — |
| Porosity | $n$ | % | — |
| Degree of saturation | $S$ | % | — |
| Water content (gravimetric) | $w$ | % | — |
| Volumetric water content | $\theta_w$ | decimal or % | — |
| Specific gravity of solids | $G_s$ | — | — |
| Total density | $\rho$ | kg/m³ | $ML^{-3}$ |
| Dry density | $\rho_d$ | kg/m³ | $ML^{-3}$ |
| Density of water | $\rho_w$ | kg/m³ | $ML^{-3}$ |
| Density of air | $\rho_a$ | kg/m³ | $ML^{-3}$ |
| Universal gas constant | $R$ | J/mol K | — |
| Relative Humidity | RH | % | — |
| Temperature | $t°$ | °C[a] | — |
| Time | $t$ | s, min., hour, day | — |
| Normal stress | $\sigma$ | kPa | $ML^{-1}T^{-2}$ |
| Effective normal stress | $\sigma'$ | kPa | $ML^{-1}T^{-2}$ |
| Pore-air pressure | $u_a$ | kPa | $ML^{-1}T^{-2}$ |
| Pore-water pressure | $u_w$ | kPa | $ML^{-1}T^{-2}$ |
| Net normal stress | $(\sigma - u_a)$ | kPa | $ML^{-1}T^{-2}$ |
| Matric suction | $(u_a - u_w)$ | kPa | $ML^{-1}T^{-2}$ |
| Coefficient of permeability at saturation | $k_s$ | m/s | $LT^{-1}$ |
| Coefficient of permeability (water phase) | $k_w$ | m/s | $LT^{-1}$ |
| Coefficient of permeability (air phase) | $k_a$ | m/s | $LT^{-1}$ |
| Hydraulic or Total head | $h_w$ | m | L |
| Hydraulic head gradient | $i_w$ | — | — |
| Total suction | $\Psi$ | kPa | $ML^{-1}T^{-2}$ |
| Osmotic suction | $\pi$ | kPa | $ML^{-1}T^{-2}$ |
| Surface tension | $T_s$ | mN/m | $MT^{-2}$ |
| Total cohesion | $c$ | kPa | $ML^{-1}T^{-2}$ |
| Effective cohesion | $c'$ | kPa | $ML^{-1}T^{-2}$ |
| Undrained shear strength | $c_u$ | kPa | $ML^{-1}T^{-2}$ |
| Angle of internal friction | $\phi$ | degree | — |
| Effective angle of internal friction | $\phi'$ | degree | — |
| Angle indicating the rate of increase in shear strength related to matric suction | $\phi^b$ | degree | — |
| Friction angle associated with the matric suction when using the $(\sigma - u_w)$ and $(u_a - u_w)$ stress state variable combination | $\phi''$ | degree | — |
| Shear stress | $\tau$ | kPa | $ML^{-1}T^{-2}$ |
| Major principal stress | $\sigma_1$ | kPa | $ML^{-1}T^{-2}$ |
| Minor principal stress | $\sigma_3$ | kPa | $ML^{-1}T^{-2}$ |
| Net major normal stress | $\sigma_1 - u_a$ | kPa | $ML^{-1}T^{-2}$ |
| Net minor normal stress | $\sigma_3 - u_a$ | kPa | $ML^{-1}T^{-2}$ |
| Deviator stress | $\sigma_1 - \sigma_3$ | kPa | $ML^{-1}T^{-2}$ |
| $(\sigma_1 + \sigma_3)/2$ | $p$ | kPa | $ML^{-1}T^{-2}$ |
| $(\sigma_1 - \sigma_3)/2$ | $q$ | kPa | $ML^{-1}T^{-2}$ |
| $(u_a - u_w)$ | $r$ | kPa | $ML^{-1}T^{-2}$ |

**Table A.4** (*Continued*)

| Soil Property | Symbol | Unit | Dimensions |
|---|---|---|---|
| Coefficient of air volume change with respect to change in net normal stress | $m_1^a$ | 1/kPa | $M^{-1}LT^2$ |
| Coefficient of air volume change with respect to change in matric suction | $m_2^a$ | 1/kPa | $M^{-1}LT^2$ |
| Coefficient of soil structure volume change with respect to change in net normal stress | $m_1^s$ | 1/kPa | $M^{-1}LT^2$ |
| Coefficient of soil structure volume change with respect to change in matric suction | $m_2^s$ | 1/kPa | $M^{-1}LT^2$ |
| Coefficient of water volume change with respect to change in net normal stress | $m_1^w$ | 1/kPa | $M^{-1}LT^2$ |
| Coefficient of water volume change with respect to change in matric suction | $m_2^w$ | 1/kPa | $M^{-1}LT^2$ |
| Coefficient of volume change | $m_v$ | 1/kPa | $M^{-1}LT^2$ |
| Coefficient of consolidation with respect to the water phase | $c_v^w$ | m²/s | $L^2T^{-1}$ |
| Coefficient of consolidation with respect to the air phase | $c_v^a$ | m²/s | $L^2T^{-1}$ |
| Tangent pore-air pressure parameter (isotropic loading) | $B_a$ | — | — |
| Tangent pore-water pressure parameter (isotropic loading) | $B_w$ | — | — |
| Isothermal compressibility of air | $C_a$ | 1/kPa | $M^{-1}LT^2$ |
| Compressibility of water | $C_w$ | 1/kPa | $M^{-1}LT^2$ |
| Volumetric coefficient of solubility | $h$ | — | — |
| Coefficient of lateral earth pressure at rest | $K_o$ | — | — |

[a] °Celsius = K − 273.15

# APPENDIX B

# *Theoretical Justification for Stress State Variables*

Appendix B contains a detailed stress analysis on an unsaturated soil element. The purpose of the analysis is to demonstrate that the stress state variables $(\sigma - u_a)$ and $(u_a - u_w)$ appear as surface tractions in the equilibrium equations for the soil structure and the contractile skin. The inclusion of this stress analysis is to re-affirm the central and improtant role of the independent stress state variables.

## B.1 EQUILIBRIUM EQUATIONS FOR UNSATURATED SOILS

The state of stress at a point in an unsaturated soil can be analyzed using a cubical element of infinitesimal dimensions. Figure B.1 illustrates the total normal and shear stresses that act on the boundaries of the soil element. The gravitational force, $\rho g$, (i.e., soil density, $\rho$, times gravitational acceleration, $g$) is a body force. The gravitational force acts through the centroid of the element, but is not shown in Fig. B.1 in order to maintain simplicity.

The equilibrium analysis of a soil element is based upon the conservation of linear momentum. The conservation of linear momentum can be applied to the soil element in Fig. B.1 by summing forces first in the y-direction.

$$\left(\frac{\partial \tau_{xy}}{\partial x} + \frac{\partial \sigma_y}{\partial y} + \frac{\partial \tau_{zy}}{\partial z} + \rho g\right) dx\, dy\, dz = \left(\rho \frac{Dv_y}{Dt}\right) \quad \text{(B.1)}$$

where

$\tau_{xy}$ = shear stress on the x-plane in the y-direction
$\sigma_y$ = total normal stress on the y-plane
$\tau_{zy}$ = shear stress on the z-plane in the y-direction
$\rho$ = total density of the soil
$g$ = gravitational acceleration
$dx, dy, dz$ = dimension of the element in the x-, y-, and z-directions respectively

$\frac{Dv_y}{Dt} = \frac{\partial v_y}{\partial t} + \frac{\partial v_y}{\partial y}\frac{\partial y}{\partial t}$; acceleration in the y-direction

$v_y$ = velocity in the y-direction

Since the soil element does not undergo acceleration, the right hand side of Eq. (B.1) becomes zero

$$\left(\frac{\partial \tau_{xy}}{\partial x} + \frac{\partial \sigma_y}{\partial y} + \frac{\partial \tau_{zy}}{\partial z} + \rho g\right) dx\, dy\, dz = 0 \quad \text{(B.2)}$$

Equation B.2 is commonly referred to as the equilibrium equation for the y-direction. Similarly, equilibrium equations can be derived for the x-direction,

$$\left(\frac{\partial \sigma_x}{\partial x} + \frac{\partial \tau_{yx}}{\partial y} + \frac{\partial \tau_{zx}}{\partial z}\right) dx\, dy\, dz = 0 \quad \text{(B.3)}$$

and for the z-direction.

$$\left(\frac{\partial \tau_{xz}}{\partial x} + \frac{\partial \tau_{yz}}{\partial y} + \frac{\partial \sigma_z}{\partial z}\right) dx\, dy\, dz = 0 \quad \text{(B.4)}$$

## B.2 TOTAL OR OVERALL EQUILIBRIUM

Total equilibrium refers to the force equilibrium of a complete soil element with its four phases (i.e., air, water, contractile skin, and soil particles). The total stress fields of an unsaturated soil element in the x-, y-, and z-directions are presented in Fig. B.1 with the exception that no body forces are shown (i.e., gravitational force, $\rho g$). Only the equilibrium in the y-direction will be analyzed and presented. However, the same principles apply to equilibrium in the x- and z-directions. Figure B.2 depicts the total stress fields in the y-direction. The force equilibrium equation associated with Fig. B.2 is given in Eq. (B.2).

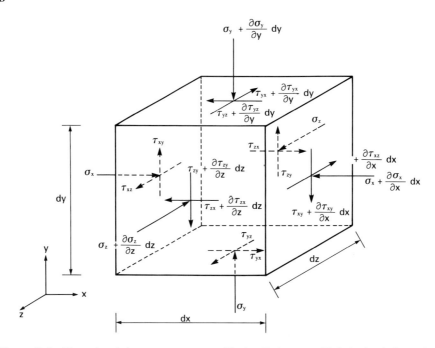

**Figure B.1** Normal and shear stresses on a cubical soil element of infinitesimal dimensions.

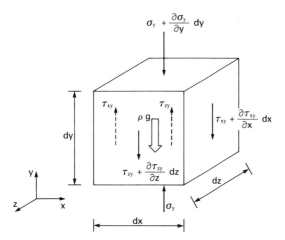

**Figure B.2** Components for total equilibrium in the *y*-direction for an unsaturated soil element.

## B.3 INDEPENDENT PHASE EQUILIBRIUM

The soil particles and the contractile skin are assumed to behave as solids in an unsaturated soil. In other words, it is assumed that these phases come to equilibrium under applied stress gradients. The arrangement of soil particles is referred to as the soil structure. The water and air phases are fluids that flow under applied stress gradients. In the equilibrium analysis, each phase is assumed to behave as an independent, linear, continuous, and coincident stress field in each direction. Therefore, an independent equilibrium equation can be written for each phase. The principle of superposition can be applied to the equilibrium equations for each of the phases because the stress fields are linear. The sum of the equilibrium equations for the individual phases is equal to the total equilibrium of the soil element (Westergaard, 1952; Green and Naghdi, 1965; Truesdell, 1966; and Faizullaev, 1969).

The equilibrium equations for the water phase, air phase, and contractile skin can be written independently. The three individual phase equilibrium equations together with the total equilibrium equation form the basis for fomulating the equilibrium equation for the soil structure.

The areas or volumes associated with each phase can be written relative to the dimensions of the soil element (i.e., $dx$, $dy$, and $dz$). This is done using a porosity designation relative to the individual phases. The porosity, $n$, of a phase is defined in Chapter 2 as the ratio of the volume of the phase relative to the total volume. For a homogeneous soil, the porosity, $n$, is also equal to the area ratio of each phase relative to the total cross-sectional area. This concept is referred to as the "theorem of the equality of volume and surface porosities in a homogeneous porous medium" (Faizullaev, 1969). Therefore, the same porosity term can be applied to the body force as to the surface tractions. The sum of the porosities relative to each phase is equal to one.

$$n_a + n_w + n_c + n_s = 1 \tag{B.5}$$

where

$n_a$ = porosity relative to the air phase
$n_w$ = porosity relative to the water phase
$n_c$ = porosity relative to the contractile skin
$n_s$ = porosity relative to the soil particles

Similarly the total density of a soil, $\rho$, can be expressed in terms of the densities of the individual phases.

$$\rho = \frac{M_a + M_w + M_c + M_s}{V} \quad (B.6)$$

where

$M_a$ = mass of the air phase
$M_w$ = mass of the water phase
$M_c$ = mass of the contractile skin
$M_s$ = mass of the soil particles
$V$ = total volume of the soil

Equation B.6 can be written as,

$$\rho = \frac{V_a \rho_a}{V} + \frac{V_w \rho_w}{V} + \frac{V_c \rho_c}{V} + \frac{V_s \rho_s}{V} \quad (B.7)$$

or

$$\rho = n_a \rho_a + n_w \rho_w + n_c \rho_c + n_s \rho_s \quad (B.8)$$

where

$V_a$ = volume of the air phase
$V_w$ = volume of the water phase
$V_c$ = volume of the contractile skin
$V_s$ = volume of the soil particles
$\rho_a$ = air density
$\rho_w$ = water density
$\rho_c$ = contractile skin density
$\rho_s$ = soil particle density

### B.3.1 Water Phase Equilibrium

The water phase equilibrium in the $y$-direction is shown in Fig. B.3. Summing forces in the $y$-direction gives the equilibrium equation for the water phase.

$$\left( n_w \frac{\partial u_w}{\partial y} + n_w \rho_w g + F_{sy}^w + F_{cy}^w \right) dx\, dy\, dz = 0 \quad (B.9)$$

where

$u_w$ = pore–water pressure
$F_{sy}^w$ = interaction force (i.e., body force) between the water phase and the soil particles in the $y$-direction
$F_{cy}^w$ = interaction force (i.e., body force) between the water phase and the contractile skin in the $y$-direction

### B.3.2 Air Phase Equilibrium

The air phase equilibrium in the $y$-direction is shown in Fig. B.4. The summation of forces in the $y$-direction gives the equilibrium equation for the air phase.

$$\left( n_a \frac{\partial u_a}{\partial y} + n_a \rho_a g + F_{sy}^a + F_{cy}^a \right) dx\, dy\, dz = 0 \quad (B.10)$$

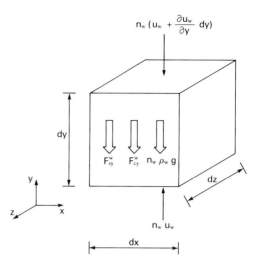

**Figure B.3** Components for force equilibrium of the water phase in the $y$-direction.

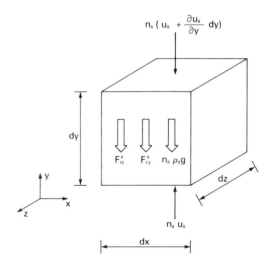

**Figure B.4** Components for force equilibrium of the air phase in the $y$-direction.

where

$u_a$ = pore–air pressure
$F_{sy}^a$ = interaction force (i.e., body force) between the air phase and the soil particles in the $y$-direction
$F_{cy}^a$ = interaction force (i.e., body force) between the air phase and the contractile skin in the $y$-direction.

### B.3.3 Contractile Skin Equilibrium

The contractile skin is only a few molecular layers in thickness. However, its presence affects the equilibrium conditions in an unsaturated soil. This is due to its ability to exert a surface tension, $T_s$.

The magnitude of the unit surface tension is constant at

a particular temperature (see Chapter 2). However, the direction in which the surface tension acts depends primarily on two factors, namely, the magnitude of the matric suction, and the arrangement of the soil particles. On the other hand, the equilibrium of the soil particles (i.e., soil structure) is affected by the surface tension. This interdependency causes the equilibrium of the contractile skin to be somewhat complex.

The following illustrations demonstrate the relationship between matric suction, the equilibrium of the soil structure and the magnitude and direction of the surface tension. Consider a curved contractile skin between soil particles. The contractile skin is in equilibrium with the air and water pressures as shown in Fig. B.5. The horizontal surface through the contractile skin has an area, $A$, and a periphery, $L$. The vertical cross-section through the contractile skin has a radius of curvature, $R_s$, and a thickness $t$, as shown in Fig. B.5a. The surface tension $T_s$, acts at an angle, $\alpha$, to the horizontal in response to the initial matric suction, $(u_a - u_w)$. The surface tension is defined as the tensile force per unit length of the contractile skin. This force can be divided vectorially into a normal component (i.e., vertical), $T_v$, and a component parallel to the horizontal surface, $T_h$. In general, the horizontal components, $T_h$ will not cancel. However, for illustrative purposes the assumption will be made that the horizontal components cancel. This is an idealized representation of the more general case. The vertical resultant force associated with all the normal components, $T_v$, acting along the periphery of the contractile skin is called $X_v$.

$$X_v = (u_a - u_w)A \qquad (B.11)$$

$X_v$ acts on the cross-sectional area of the contractile skin cut by the horizontal plane, which can be written as the product of $L$ (periphery) and $t$ (thickness) (i.e., $Lt$). Therefore, the normal stress acting on the cross-sectional area of the contractile skin, $\sigma_v^c$, can be written as follows:

$$\sigma_v^c = \frac{A}{Lt}(u_a - u_w) \qquad (B.12)$$

or

$$\sigma_v^c = f(u_a - u_w) \qquad (B.13)$$

where

$$f = A/(Lt)$$

When the matric suction is changed to $\{(u_a - u_w) + \Delta(u_a - u_w)\}$, the radius of curvature of the contractile skin changes to $R_s^*$ (Fig. B.5b). However, the magnitude of the surface tension, $T_s$, remains unchanged. The direction of the surface tension changes to an angle $\alpha'$, from the horizontal. This means that the magnitude of the shear and the normal components of the surface tension vary as matric

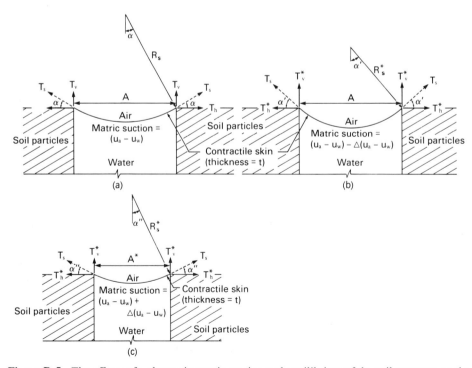

**Figure B.5** The effects of a change in matric suction and equilibrium of the soil structure on the normal and shear stresses associated with the contractile skin. (a) Initial matric suction conditions, $(u_a - u_w)$; (b) effect of a change in matric suction for an incompressible soil structure (i.e., $A$ constant); (c) effect of a change in matric suction for a compressible soil structure (i.e., $A$ changes to $A^*$).

suction changes. Consequently, the equilibrium of the soil structure is altered since the shear and normal components of the surface tension are transferred to the soil particles.

If the soil structure is incompressible, the stress change does not cause any deformation in the soil. In this case, the area of the horizontal surface of the contractile skin remains constant at, $A$, after a change in the matric suction (Fig. B.5b). The length of the periphery, $L$, also does not change. The thickness of the contractile skin, $t$, also remains constant. The resultant force of the normal surface tension, $X_v^*$, can again be computed by considering the vertical equilibrium of the contractile skin in Fig. B.5b.

$$X_v^* = \{(u_a - u_w) + \Delta(u_a - u_w)\}A \quad (B.14)$$

The normal stress acting on the cross-sectional area of the contractile skin, $\sigma_v^{c*}$, can be written as follows:

$$\sigma_v^{c*} = \frac{A}{Lt}(u_a - u_w) + \frac{A}{Lt}\Delta(u_a - u_w) \quad (B.15)$$

or

$$\sigma_v^{c*} = f(u_a - u_w) + f\Delta(u_a - u_w) \quad (B.16)$$

Substituting Eq. (B.13) into Eq. (B.16) gives,

$$\sigma_v^{c*} = \sigma_v^c + f\Delta(u_a - u_w) \quad (B.17)$$

A comparison of Eqs. (B.17) and (B.13) shows that the change in the normal stress associated with the contractile skin is a function of the change in the matric suction, $\Delta(u_a - u_w)$. This function is expressed in terms of, $f$, and is a constant (i.e., $A/Lt$) for the case of an incompressible soil structure.

In the case of a compressible soil structure, any alteration in stress equilibrium results in deformation. Figure B.5c demonstrates the deformation in the soil structure due to a change in matric suction. The equilibrium of the soil structure can also be altered by changing the total normal and/or shear stresses of the soil.

For the case shown in Fig. B.5c, the area of the horizontal surface of the contractile skin changes, for example, from $A$ to $A^*$. Similarly the length of the periphery becomes $L^*$. As a result, the direction of the surface tension is different from its direction for the incompressible soil structure case. The surface tension, $T_s$, now acts at an angle, $\alpha''$, from the horizontal, although the radius of curvature of the contractile skin will be the same as shown in Fig. B.5b (i.e., $R_s^*$).

The resultant force for the nornal surface tension, $X_v^*$, is obtained by considering the vertical force equilibrium of the contractile skin (Fig. B.5c).

$$X_v^* = \{(u_a - u_w) + \Delta(u_a - u_w)\}A^* \quad (B.18)$$

The normal stress on the cross-sectional area of the contractile skin, $\sigma_v^{c*}$, is written as follows:

$$\sigma_v^{c*} = \frac{A^*}{L^*t}(u_a - u_w) + \frac{A^*}{L^*t}\Delta(u_a - u_w) \quad (B.19)$$

Rearranging Eq. (B.19) gives,

$$\sigma_v^{c*} = \frac{A}{Lt}(u_a - u_w) + \left\{\frac{A^*}{L^*t} - \frac{A}{Lt}\right\}(u_a - u_w)$$

$$+ \frac{A^*}{L^*t}\Delta(u_a - u_w) \quad (B.20)$$

The term $(A^*/L^*t)$ is referred to as the interaction function, $f^*$, between the equilibrium of the soil structure and the equilibrium of the contractile skin.

$$\sigma_v^{c*} = f(u_a - u_w) + (f^* - f)(u_a - u_w)$$

$$+ f^*\Delta(u_a - u_w) \quad (B.21)$$

Substituting the initial normal stress (i.e., Eq. B.13) into Eq. (B.21) gives,

$$\sigma_v^{c*} = \sigma_v^c + \Delta f^*(u_a - u_w) + f^*\Delta(u_a - u_w) \quad (B.22)$$

where

$\sigma_v^{c*}$ = final normal stress in the contractile skin
$\sigma_v^c$ = initial normal stress in the contractile skin
$(u_a - u_w)$ = initial matric suction
$\Delta(u_a - u_w)$ = change in the matric suction
$f^*$ = final interaction between the contractile skin and the soil structure equilibriums
$\Delta f^*$ = change in the interaction function, $f^*$

Equation B.22 describes the variation in the normal stress in the contractile skin for a compressible soil structure. In this case, the function, $f^*$, is no longer a constant. The interaction function, $f^*$, and its change, $\Delta f^*$, depend upon the stress state and the compressibility of the soil structure. Therefore, the stress state variables that govern the equilibrium of the soil structure also influence the equilibrium of the contractile skin in compressible soils. If the soil structure is incompressible, Eq. (B.22) reverts to Eq. (B.17), and the interaction function, $f^*$, becomes a constant, $f$.

A change in the matric suction always affects the equilibrium of the contractile skin for both compressible and incompressible soils as shown in Eqs. (B.22) and (B.17), respectively. However, a change in the total stress of a soil only influences the equilibrium of the contractile skin for a compressible soil. This occurs through the change in the interaction function, $\Delta f^*$. If a compressible soil is subjected only to a change in the total stress, the term, $(f^*\Delta(u_a - u_w))$, of Eq. (B.22) can be omitted. In this case, the effect of a total stress change on the contractile skin is taken into account only by the term, $(\Delta f^*(u_a - u_w))$. On the other hand, if the soil is subjected only to a change in matric suction, both terms (i.e., $f^*\Delta(u_a - u_w)$ and $\Delta f^*(u_a - u_w)$) of Eq. (B.22) are applicable. This can be used to explain why a change in matric suction is more effective than a change in total stress for changing the degree of saturation or the water content of a soil.

All soils can be considered to have compressible structures to varying degrees. Equation B.22 is therefore used

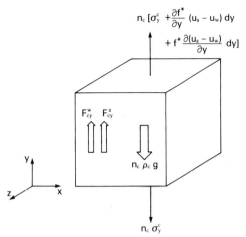

**Figure B.6** Components for force equilibrium of the contractile skin in the y-direction.

in the equilibrium formulation for the contractile skin in an unsaturated soil element. The interaction function, $f^*$, cannot be mathematically derived because the equilibrium of the soil structure is unknown. However, more important is the fact that the stress state variables for the equilibrium of the soil structure are also the stress state variables for the contractile skin.

Figure B.6 shows the components for the contractile skin equilibrium in the y-direction according to Eq. (B.22). The variation in the normal stress associated with the contractile skin in the y-direction is written in a differential form, where $\sigma_y^c$ is the normal component of the stress in the contractile skin on the y-plane. The equilibrium equation for the contractile skin is as follows:

$$\left\{ -n_c \frac{\partial f^*}{\partial y}(u_a - u_w) - n_c f^* \frac{\partial (u_a - u_w)}{\partial y} \right. $$
$$\left. + n_c \rho_c g - F_{cy}^w - F_{cy}^a \right\} dx\, dy\, dz = 0 \quad (B.23)$$

## B.4 EQUILIBRIUM OF THE SOIL STRUCTURE (i.e., ARRANGEMENT OF SOIL PARTICLES)

The total equilibrium of an unsaturated soil element (i.e., Eq. B.2) is equivalent to the resultant of the equilibrium equations for the individual phases (i.e., Eqs. B.9, B.10, B.23 and the soil structure equilibrium). Therefore, the equilibrium of the soil structure in the y-direction can be written as the difference between the total equilibrium equation and the sum of the water, air and contractile skin equilibrium equations.

$$\left\{ \frac{\partial \tau_{xy}}{\partial x} + \frac{\partial \sigma_y}{\partial y} - n_a \frac{\partial u_a}{\partial y} - n_w \frac{\partial u_w}{\partial y} + n_c f^* \frac{\partial(u_a - u_w)}{\partial y} \right.$$
$$+ \frac{\partial \tau_{zy}}{\partial z} + \rho g - n_a \rho_a g - n_w \rho_w g - n_c \rho_c g - F_{sy}^w$$
$$\left. - F_{sy}^a + n_c(u_a - u_w)\frac{\partial f^*}{\partial y} \right\} dx\, dy\, dz = 0 \quad (B.24)$$

Substituting Eq. (B.8) into (B.24) gives,

$$\left\{ \frac{\partial \tau_{xy}}{\partial x} + \frac{\partial \sigma_y}{\partial y} - n_a \frac{\partial u_a}{\partial y} - n_w \frac{\partial u_w}{\partial y} \right.$$
$$+ n_c f^* \frac{\partial(u_a - u_w)}{\partial y} + \frac{\partial \tau_{zy}}{\partial z} + n_s \rho_s g - F_{sy}^w$$
$$\left. - F_{sy}^a + n_c(u_a - u_w)\frac{\partial f^*}{\partial y} \right\} = 0 \quad (B.25)$$

The last term, $(n_c(u_a - u_w)\partial f^*/\partial y)$, is the interaction force between the soil structure and the contractile skin. Equation B.25 can be rearranged using the pore-air pressure, $u_a$, as a reference. The term, $(n_a \partial u_a/\partial y)$, in Eq. (B.25) can be written using Eq. (B.5) as,

$$n_a \frac{\partial u_a}{\partial y} = (1 - n_w - n_c - n_s) \frac{\partial u_a}{\partial y} \quad (B.26)$$

Substituting Eq. (B.26) into Eq. (B.25) produces another form for the equilibrium equation for the soil structure in the y-direction.

$$\frac{\partial \tau_{xy}}{\partial x} + \frac{\partial (\sigma_y - u_a)}{\partial y} + (n_w + n_c f^*)\frac{\partial(u_a - u_w)}{\partial y}$$
$$+ \frac{\partial \tau_{zy}}{\partial z} + (n_c + n_s)\frac{\partial u_a}{\partial y} + n_s \rho_s g - F_{sy}^w - F_{sy}^a$$
$$+ n_c(u_a - u_w)\frac{\partial f^*}{\partial y} = 0 \quad (B.27)$$

Similar equilibrium equations can be written for the x-direction.

$$\frac{\partial (\sigma_x - u_a)}{\partial x} + (n_w + n_c f^*)\frac{\partial(u_a - u_w)}{\partial x} + \frac{\partial \tau_{yx}}{\partial y} + \frac{\partial \tau_{zx}}{\partial z}$$
$$+ (n_c + n_s)\frac{\partial u_a}{\partial x} - F_{sx}^w - F_{sx}^a$$
$$+ n_c(u_a - u_w)\frac{\partial f^*}{\partial x} = 0 \quad (B.28)$$

and for the z-direction,

$$\frac{\partial \tau_{xz}}{\partial x} + \frac{\partial \tau_{yz}}{\partial y} + \frac{\partial (\sigma_z - u_a)}{\partial z} + (n_w + n_c f^*)\frac{\partial(u_a - u_w)}{\partial z}$$
$$+ (n_c + n_s)\frac{\partial u_a}{\partial z} - F_{sz}^w - F_{sz}^a$$
$$+ n_c(u_a - u_w)\frac{\partial f^*}{\partial z} = 0 \quad (B.29)$$

The stress variables controlling the equilibrium of the soil structure are the stress state variables which control the mechanical behavior of soils. There are three independent sets of normal stresses (i.e., surface tractions) that can be extracted from the equilibrium equations for the soil structure (i.e., Eqs. B.27, B.28 and B.29) to form the stress state variables. The three variables are $(\sigma - u_a)$, $(u_a - u_w)$, and $(u_a)$. The stress variable, $u_a$, can be eliminated if the soil particles are assumed to be incompressible. Therefore, the stress state variables for the soil structure and contrac-

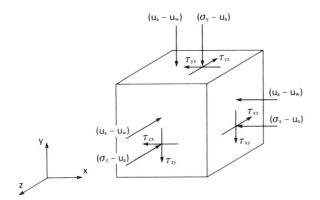

**Figure B.7** The stress state variables for an unsaturated soil using the combination of $(\sigma - u_a)$ and $(u_a - u_w)$.

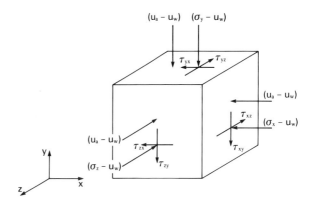

**Figure B.8** The stress state variables for an unsaturated soil using the combination of $(\sigma - u_w)$ and $(u_a - u_w)$.

tile skin in an unsaturated soil are $(\sigma - u_a)$ and $(u_a - u_w)$. The complete form of the stress state for an unsaturated soil can be written as two independent stress tensors. Figure B.7 illustrates the two independent stress tensors at a point in an unsaturated soil.

$$\begin{bmatrix} (\sigma_x - u_a) & \tau_{yx} & \tau_{zx} \\ \tau_{xy} & (\sigma_y - u_a) & \tau_{zy} \\ \tau_{xz} & \tau_{yz} & (\sigma_z - u_a) \end{bmatrix} \quad (B.30)$$

and

$$\begin{bmatrix} (u_a - u_w) & 0 & 0 \\ 0 & (u_a - u_w) & 0 \\ 0 & 0 & (u_a - u_w) \end{bmatrix} \quad (B.31)$$

## B.5 OTHER COMBINATIONS OF STRESS STATE VARIABLES

The equilibrium equation for the soil structure (i.e., Eq. (B.25)) can be rewritten using the pore-water pressure, $u_w$, or the total normal stress, $\sigma_y$, as a reference. If the pore-water pressure, $u_w$, is used as a reference, the term $(n_w \partial u_w / \partial y)$ in Eq. (B.25) can be written as,

$$n_w \frac{\partial u_w}{\partial y} = (1 - n_a - n_c - n_s) \frac{\partial u_w}{\partial y} \quad (B.32)$$

Substituting Eq. (B.32) into Eq. (B.25) yields,

$$\frac{\partial \tau_{xy}}{\partial x} + \frac{\partial (\sigma_y - u_w)}{\partial y} - (n_a - n_c f^*) \frac{\partial (u_a - u_w)}{\partial y}$$

$$+ \frac{\partial \tau_{zy}}{\partial z} + (n_c + n_s) \frac{\partial u_w}{\partial y}$$

$$+ n_s \rho_s g - F_{sy}^w - F_{sy}^a + n_c (u_a - u_w) \frac{\partial f^*}{\partial y} = 0$$

(B.33)

The combination of stress state variables that can be extracted from Eq. (B.33) is $(\sigma_y - u_w)$, $(u_a - u_w)$, and $(u_w)$.

The stress variable, $u_w$, is only of relevance when dealing with compressible soil particles. Figure B.8 illustrates the two independent stress tensors at a point in an unsaturated soil using the combination of $(\sigma - u_w)$ and $(u_a - u_w)$.

Equation B.25 can be rearranged using the total normal stress, $\sigma_y$, as a reference, resulting in the following equation:

$$\frac{\partial \tau_{xy}}{\partial x} + \frac{\partial \sigma_y}{\partial y} - (n_a - n_c f^*) \frac{\partial u_a}{\partial y} - (n_w + n_c f^*) \frac{\partial u_w}{\partial y}$$

$$+ \frac{\partial \tau_{zy}}{\partial z} + n_s \rho_s g - F_{sy}^w - F_{sy}^a$$

$$+ n_c (u_a - u_w) \frac{\partial f^*}{\partial y} = 0$$

(B.34)

Equation B.34 can be rearranged as follows:

$$\frac{\partial \tau_{xy}}{\partial x} + (n_a - n_c f^*) \frac{\partial (\sigma_y - u_a)}{\partial y}$$

$$+ (n_w + n_c f^*) \frac{\partial (\sigma_y - u_w)}{\partial y} + (n_c + n_s) \frac{\partial \sigma_y}{\partial y}$$

$$+ n_s \rho_s g - F_{sy}^w - F_{sy}^a + n_c (u_a - u_w) \frac{\partial f^*}{\partial y} = 0$$

(B.35)

Three independent stress state variables can be extracted from Eq. (B.35). These are $(\sigma_y - u_a)$, $(\sigma_y - u_w)$, and $(\sigma_y)$. Again, the stress variable, $\sigma_y$, can be ignored when assuming the soil particles are incompressible.

Equations (B.25), (B.27), (B.33), and (B.35) are all acceptable equilibrium equations for the soil structure in the $y$-direction. The soil structure equilibrium equations can have three different forms depending upon the selected stress reference. Each form of the equilibrium equations contains a combination of two stress state variables. In other words, any two of three possible stress variables (i.e., $(\sigma - u_a)$, $(\sigma - u_w)$, and $(u_a - u_w)$) can be used to describe the stress state for the soil structure and contractile skin in an unsaturated soil.

# References

Y. M. Abelev, "The Stabilization of Foundations of Structures on Loess Soils," *Proc. of the Fourth Int. Conf. on Soil Mech. and Found. Eng.* (London), vol. I, 1967, pp. 259–263.

G. D. Aitchison, "The Strength of Quasi-Saturated and Unsaturated Soils in Relation to the Pressure Deficiency in the Pore Water," in *Proc. 4th Int. Conf. Soil Mech. Found. Eng.* (London), 1959, pp. 135–139.

G. D. Aitchison, "Relationship of Moisture and Effective Stress Functions in Unsaturated Soils," in *Pore Pressure and Suction in Soils Conf.*, organized by British Nat. Soc. of Int. Soc. Soil Mech. Found. Eng. at Inst. Civil Eng. London, England: Butterworths, 1961, pp. 47–52.

G. D. Aitchison, "Engineering Concepts of Moisture Equilibria and Moisture Changes in Soils," Statement of the Review Panel, Ed., published in *Moisture Equilibria and Moisture Changes in Soils Beneath Covered Areas*, A Symp.-in-print (Australia), Butterworths, 1964, pp. 7–21.

G. D. Aitchison, Ed., *Moisture Equilibria and Moisture Changes in Soils Beneath Covered Areas, A Symp. in Print*, G. D. Aitchison, Ed. Australia: Butterworths, 1965, 278 pp.

G. D. Aitchison, "Soil Properties, Shear Strength, and Consolidation," in *Proc. 6th Int. Conf. Soil Mech. Found. Eng.* (Montreal, Canada), vol. 3, 1965, pp. 318–321.

G. D. Aitchison, "Separate Roles of Site Investigation, Quantification of Soil Properties, and Selection of Operational Environment in the Determination of Foundation Design on Expansive Soils," in *Proc. 3rd Asian Reg. Conf. Soil Mech. Found. Eng.* (Haifa, Israel), vol. 3, 1967, pp. 72–77.

G. D. Aitchison, "The Quantitative Description of the Stress-Deformation Behavior of Expansive Soils—Preface to Set of Papers," in *Proc. 3rd Int. Conf. Expansive Soils* (Haifa, Israel), vol. 2, 1973, pp. 79–82.

G. D. Aitchison and I. B. Donald, "Effective Stresses in Unsaturated Soils," in *Proc. 2nd Australia–New Zealand Conf. Soil Mech.*, 1956, pp. 192–199.

G. D. Aitchison and R. Martin, "A Membrane Oedometer for Complex Stress-Path Studies in Expansive Clays," in *Proc. 3rd Int. Conf. Expansive Soils* (Haifa, Israel), vol. 2, 1973, pp. 83–88.

G. D. Aitchison and J. A. Woodburn, "Soil Suction in Foundation Design," in *Proc. 7th Int. Conf. Soil Mech. Found. Eng.* (Mexico), vol. 2, 1969, pp. 1–8.

G. D. Aitchison, P. Peter, and R. Martin, "The Instability Indices $I_{pm}$ and $I_{ps}$ in Expansive Soils," in *Proc 3rd Int. Conf. Expansive Soils* (Haifa, Israel), vol. 2, 1973, pp. 101–104.

S. Al-Khafaf and R. J. Hanks, "Evaluation of the Filter Paper Method for Estimating Soil Water Potential," *Soil Sci.*, vol. 117, pp. 194–199, 1974.

E. E. Alonso and A. Lloret, "Behaviour of Partially Saturated Soil in Undrained Loading and Step by Step Embankment Construction," in *Proc. IUTAM Conf. Deformation and Failure of Granular Mater.* (Delft, The Netherlands), 1982, pp. 173–180.

M. G. Anderson, "Prediction of Soil Suction for Slopes in Hong Kong," Geotech. Control Office, Eng. Development Dep., Hong Kong, GCO Publ. No. 1/84, 243 pp., 1983.

M. Anderson and T. Burt, "Automatic Monitoring of Soil Moisture Conditions in A Hillslope Spur and Hollow," *J. Hydro.*, vol. 33, pp. 27–36, 1977.

A. Arbhabhirama and C. Kridakorn, "Steady Downward Flow to a Water Table," *Water Resources Res.*, vol. 4, 1968.

ASTM, *ASTM Metric Practice Guide*, 2nd ed., 46 pp., 1966.

ASTM, "Standard Test Method for Capillary-Moisture Relationships for Coarse and Medium Textured Soils by Porous-Plate Apparatus," ASTM D-2325-68- (1981), *1985 Annual Book of ASTM Standards*, vol. 4, no. 8, pp. 363–369, 1985.

ASTM D2325-68, "Standard Test Method for Capillary-Moisture Relationships for Coarse and Medium Textured Soils by Porous Plate Apparatus," *1985 Annual Book of ASTM Standards, Vol. 04.08, Soil and Rock; Building Stones*, ASTM, Philadelphia, PA, 1981.

ASTM D2435, "Standard Test Method for One-Dimensional Consolidation Properties of Soils," *1985 Annual Book of ASTM Standards, Vol. 04.08, Soil and Rock; Building Stones*, sect. 4, ASTM, Philadelphia, PA, 1985.

ASTM D427, "Standard Test Method for Shrinkage Factors of Soils," *1985 Annual Book of ASTM Standards, Vol. 04.08, Soil and Rock; Building Stones*, sect. 4, ASTM, Philadelphia, PA, 1985.

ASTM D4546, "Standard Test Method for One-Dimensional Swell or Settlement Potential of Cohesive Soils," *1986 Annual Book of ASTM Standards, Vol. 04.08, Soil and Rock; Building Stones*, ASTM, Philadelphia, PA, 1986.

S. F. Averjanov, "About Permeability of Subsurface Soils in Case of Incomplete Saturation," *Eng. Collect.*, vol. 7, 1950.

W. A. Bailey, "The Effect of Salt Content on the Consolidation

Behavior of Saturated Remoulded Clays," U.S. Corps of Army Eng., Waterways Experimental Station, Vicksburg, MS, Contract Report 3-101, 1965.

S. L. Barbour, "Osmotic Flow and Volume Change in Clay Soils." Ph.D. Thesis, Department of Civil Engineering, Saskatoon, SK, Canada, 1987.

S. L. Barbour and D. G. Fredlund, "Mechanics of Osmotic Flow and Volume Change in Clay Soils," *Can. Geot. Journ.*, vol. 26, pp. 551-562, 1989.

S. L. Barbour and D. G. Fredlund, "Physico-chemical State Variables for Clay Soils," *Proc. of the XII Int. Conf. on Soil Mech. and Fdtn Eng.* (Rio de Janeiro, Brazil), 1989, pp. 1839-1843.

S. L. Barbour, D. G. Fredlund, J. K-M. Gan, and G. W. Wilson, "Prediction of Moisture Movement in Highway Subgrade Soils," *Proc of the Can. Geot. Conf.* (Toronto), October 26-28, 1992, pp. 41A.1-41A.13.

S. L. Barbour, D. G. Fredlund, and D. E. Pufahl, "The Osmotic Role in the Behavior of Swelling Clay Soils," *NATO Advanced Research Workshop* (Korfu, Greece), July 1-6, 1991, pp. 97-139.

L. Barden, "Consolidation of Compacted and Unsaturated Clays," *Geotechnique*, vol. 15, no. 3, pp. 267-286, 1965.

L. Barden, "Consolidation of Clays Compacted 'Dry' and 'Wet' of Optimum Water Content," *Geotechnique*, vol. 24, no. 4, pp. 605-625, 1974.

L. Barden and G. Pavlakis, "Air and Water Permeability of Compacted Unsaturated Cohesive Soil," *J. Soil Sci.*, vol. 22, no. 3, pp. 302-317, 1971.

L. Barden and G. R. Sides, "The Diffusion of Air Through the Pore Water of Soils," in *Proc. 3rd Asian Reg. Conf. Soil Mech. Found. Eng.* (Israel), vol. 1, 1967, pp. 135-138.

L. Barden and G. R. Sides, "Engineering Behavior and Structure of Compacted Clay," *ASCE J. Soil Mech. Found. Eng.*, vol. 96, pp. 1171-1200, July 1970.

L. Barden, A. O. Madedor, and G. R. Sides, "Volume Change Characteristics of Unsaturated Clay," *ASCE J. Soil Mech. Found. Div.*, vol. 95, SM1, pp. 33-52, 1969.

L. Barden, A. O. Madedor, and G. R. Sides, "The Flow of Air and Water in Partly Saturated Clay Soil," presented at the Int. Symp. Fundamentals of Transport Phenomena in Flow Through Porous Media, Haifa, Israel, 1969.

F. B. J. Barends, "The Compressibility of an Air-Water Mixture in a Porous Medium," *Igm-Mededelingen*, publ. by the Delft Soil Mech. Lab., The Netherlands, part 20, no. 1, pp. 49-66, May 1979.

N. S. Beal, "Direct Determination of Linear Dimension Versus Moisture Content Relationship in Expansive Clays," in *Proc. 5th Int. Conf. Expansive Soils* (Adelaide, Australia), 1984, pp. 62-66.

J. Bear, *Hydraulic of Groundwater*. New York: McGraw-Hill (McGraw-Hill Series in Water Resources and Environmental Eng.), 1979, 567 pp.

W. Bernatzik, "The Determination of the Capillary Rise in Sand by Means of Prism Pressure Test," in *Proc. 2nd Int. Conf. Soil Mech. Found. Eng.* (Rotterdam, The Netherlands), vol. 5, 1948, p. 28.

M. A. Biot, "General Theory of Three-Dimensional Consolidation," *J. Appl. Phys.*, vol. 12, no. 2, pp. 155-164, 1941.

M. A. Biot, "Theory of Elasticity and Consolidation for a Porous Anisotropic Solid," *J. Appl. Phys.*, vol. 26, no. 2, pp. 182-185, 1955.

A. W. Bishop, "The Use of Pore Pressure Coefficients in Practice," *Geotechnique*, vol. 4, no. 4, pp. 148-152, 1954.

A. W. Bishop, "The Use of the Slip Circle in the Stability Analysis of Slopes," *Geotechnique*, vol. 5, pp. 7-17, 1955.

A. W. Bishop, "Some Factors Controlling the Pore Pressures Set Up During the Construction of Earth Dams," in *Proc. 4th Int. Conf. Soil Mech. Found. Eng.* (Paris, France), vol. 2, 1957, pp. 294-300.

A. W. Bishop, "The Principle of Effective Stress," lecture delivered in Oslo, Norway, in 1955; published in *Teknisk Ukeblad*, vol. 106, no. 39, pp. 859-863, 1959.

A. W. Bishop, "The Measurement of Pore Pressure in Triaxial Test," in *Proc. Conf. Pore Pressure and Suction in Soils*. London: Butterworths, 1960, pp. 38-46.

A. W. Bishop, "Discussion on General Principles and Laboratory Measurements," in *Proc. Conf. Pore Pressure and Suction in Soils*. London: Butterworths, 1961, pp. 63-66.

A. W. Bishop, "The Measurement of Pore Pressure in the Triaxial Test," in *Proc. Conf. Pore Pressure and Suction in Soils*. London: Butterworths, 1961, pp. 38-46.

A. W. Bishop, "Pore Pressure Measurements in the Field and in the Laboratory," in *Proc. 7th Int. Conf. Soil Mech. Found. Eng., Specialty Sessions* (Mexico), 1969, pp. 427-441.

A. W. Bishop, "Test Requirements of Measuring the Coefficient of Earth Pressure at Rest," *Proc. of the Conf. on Earth Pressure Problems* (Brussels), vol. I, 1985, pp. 2-14.

A. W. Bishop and G. E. Blight, "Some Aspects of Effective Stress in Saturated and Unsaturated Soils," *Geotechnique*, vol. 13, no. 3, pp. 177-197, 1963.

A. W. Bishop and I. B. Donald, "The Experimental Study of Partly Saturated Soil in the Triaxial Apparatus," in *Proc. 5th Int. Conf. Soil Mech. Found. Eng.* (Paris, France), vol. 1, 1961, pp. 13-21.

A. W. Bishop and A. K. G. Eldin, "Undrained Triaxial Tests on Saturated Sands and Their Significance in the General Theory of Shear Strength," *Geotechnique*, vol. 2, pp. 13-32, 1950.

A. W. Bishop and R. E. Gibson, "The Influence of the Provisions for Boundary Drainage on Strength and Consolidation Characteristics of Soils Measured in the Triaxial Apparatus," in *Laboratory Shear Testing of Soils*, ASTM, STP No. 361, pp. 435-458, 1963.

A. W. Bishop and D. J. Henkel, *The Measurement of Soil Properties in the Triaxial Test*, 2nd ed. London, England: Edward Arnold, 1962, 227 pp.

A. W. Bishop and N. R. Morgenstern, "Stability Coefficients for Earth Slopes," *Geotechnique*, vol. 10, no. 4, pp. 129-147, Dec. 1960.

A. W. Bishop, I. Alpan, G. E. Blight, and I. B. Donald, "Factors Controlling the Shear Strength of Partly Saturated Cohesive Soils," in *ASCE Res. Conf. Shear Strength of Cohesive Soils* (Univ. of Colorado, Boulder), 1960, pp. 503-532.

D. K. Black and K. L. Lee, "Saturating Laboratory Samples by Back Pressure," *ASCE J. Soil Mech. Found. Eng. Div.*, vol. 99, SM1, pp. 75-95, 1973.

W. P. M. Black and D. Croney, "Pore Water Pressure and Moisture Content Studies Under Experimental Pavements," in *Proc. 4th Int. Conf. Soil Mech. Found. Eng.*, vol. 2, 1957, pp. 94-103.

W. P. M. Black, D. Croney, and J. C. Jacobs, "Field Studies of the Movement of Soil Moisture," *Tech. Paper*, Road Research Board, no. 41, H.M.S.O., London, 1959.

G. E. Blight, "Strength and Consolidation Characteristics of Compacted Soils," Ph.D. dissertation, Univ. of London, London, England, 217 pp., 1961.

G. E. Blight, "A Study of Effective Stresses for Volume Change," in *Moisture Equilibria and Moisture Changes in Soils Beneath Covered Areas, A Symp. in Print*, G. D. Aitchison, Ed. Australia: Butterworths, 1965, pp. 259-269.

G. E. Blight, "Strength Characteristics of Desiccated Clays," *ASCE J. Soil Mech. Found. Eng. Div.*, vol. 92, SM6, pp. 19-37, 1966.

G. E. Blight, "Effective Stress Evaluation for Unsaturated Soils," *ASCE J. Soil Mech. Found. Eng. Div.*, vol. 93, SM2, pp. 125-148, 1967.

G. E. Blight, "Flow of Air through Soils," *ASCE J. Soil Mech. Found. Eng. Div.*, vol. 97, SM4, pp. 607-624, 1971.

G. E. Blight, "The Mechanics of Unsaturated Soils," notes for a series of lectures delivered as part of Course 270C at the Univ. of California, Berkeley, 1980.

M. E. Bloodworth and J. B. Page, "Use of Thermistor for the Measurement of Soil Moisture and Temperature," *Soil Sci. Soc. of Amer. Proc.*, vol. 21, pp. 11-15, 1957.

K. A. Bocking and D. G. Fredlund, "Use of the Osmotic Tensiometer to Measure Negative Pore Water Pressure," *Geotech. Testing J., GTJODJ*, vol. 2, no. 1, pp. 3-10, Mar. 1979.

K. A. Bocking and D. G. Fredlund, "Limitations of the Axis Translation Technique," in *Proc. 4th Int. Conf. Expansive Soils*, Denver, CO, pp. 117-135, June 1980.

G. H. Bolt and R. D. Miller, "Calculation of Total and Component Potentials of Water in Soil," *Amer. Geophys. Union Transportation*, vol. 39, pp. 917-928, 1958.

V. Boulichev, "Apparatus for Testing Compressibility and Capillary Properties of Soils," in *Proc. 1st Int. Conf. Soil Mech. Found. Eng.*, vol. 2, 1936, pp. 37-38.

I. J. A. Brackley, "Partial Collapse in Unsaturated Expansive Clay," in *Proc. 5th Reg. Conf. Soil Mech. Found. Eng.* (South Africa), 1971, pp. 23-30.

E. W. Brand, "Analysis and Design in Residual Soils," in *Proc. ASCE Specialty Conf. Eng. and Construction in Tropical and Residual Soils* (Honolulu, HI), 1982, pp. 89-143.

E. W. Brand, "Geotechnical Engineering in Tropical Residual Soils," in *Proc. 1st Int. Conf. Geomech. in Tropical Lateritic and Saprolitic Soils* (Brasilia, Brazil), vol. 3, Feb. 1985, pp. 23-91.

B. B. Broms, "Effect of Degree of Saturation on Bearing Capacity of Flexible Pavements," presented at the 43rd Annual Meeting of the Committee on Flexible Pavement Design, Ithaca, NY, 1965.

E. W. Brooker and H. O. Ireland, "Earth Pressures at Rest Related to Stress History," *Can. Geotech. J.*, vol. 2, no. 1, pp. 1-15, 1965.

R. H. Brooks and A. T. Corey, "Hydraulic Properties of Porous Media," Colorado State Univ. Hydrol. Paper, No. 3, 27 pp., Mar. 1964.

R. H. Brooks and A. T. Corey, "Properties of Porous Media Affecting Fluid Flow," *J. Irrigation and Drainage Div.*, ASCE IR 2, vol. 92, pp. 61-88, 1966.

R. W. Brown, "Measurement of Water Potential with Thermocouple Psychrometers: Construction and Applications," U.S.D.A. Forest Service Res. Paper INT-80, 27 pp., 1970.

R. W. Brown and D. L. Bartos, "A Calibration Model for Screen-Caged Peltier Thermocouple Psychrometers," U.S. Dept. Agriculture, Research Paper INT-293, July 1982.

R. W. Brown and J. M. Collins, "A Screen-Caged Thermocouple Psychrometer and Calibration Chamber for Measurements of Plant and Soil Water Potential," *Agron. J.*, vol. 72, pp. 851-854, 1980.

E. Buckingham, "Studies of The Movement of Soil Moisture," *U.S.D.A. Bur. of Soils, Bulletin No. 38*, 1907.

N. T. Burdine, "Relative Permeability Calculations from Pore-Size Distribution Data," *Trans. AIME*, 1952.

J. B. Burland, "Effective Stresses in Partly Saturated Soils," discussion of "Some Aspects of Effective Stress in Saturated and Partly Saturated Soils," by G. E. Blight and A. W. Bishop, *Geotechnique*, vol. 14, pp. 65-68, Feb. 1964.

J. B. Burland, "Some Aspects of the Mechanical Behavior of Partly Saturated Soils," in *Moisture Equilibria and Moisture Changes in the Soils Beneath Covered Areas, A Symp. in Print*, G. D. Aitchison, Ed. Australia: Butterworths, 1965, pp. 270-278.

G. S. Campbell and W. H. Gardner, "Psychrometric Measurement of Soil Potential: Temperature and Bulk Density Effects," *Soil Sci. Soc. Amer.*, vol. 35, pp. 8-12, 1971.

J. D. Campbell, "Pore Pressures and Volume Changes in Unsaturated Soils," Ph.D. dissertation, Univ. of Illinois at Urbana-Champaign, 104 pp., 1973.

Canada Department of Agriculture, "Glossary of Terms in Soil Science," Publication No. 1459, published by Research Branch, Canada Dept. of Agric., Ottawa, Canada, 1976, 44 pp.

Canadian Geotechnical Society, *Canadian Foundation Engineering Manual*, 2nd ed., G. G. Meyeroff and B. H. Fellenius, Eds., 1985, 456 pp.

H. S. Carslaw and J. C. Jaeger, *Conduction of Heat in Solids*, 2nd ed. Oxford: Clarendon, 1959, 510 pp.

A. Casagrande, "The Determination of the Preconsolidation Load and Its Practical Significance," Discussion D-34, in *Proc. 1st Int. Conf. Soil Mech. Found. Eng.* (Cambridge, MA), vol. 3, 1936, pp. 60-64.

A. Casagrande, "Seepage Through Dams," *J. New England Water Works*, vol. 51, no. 2, pp. 295-336, 1937.

A. Casagrande and R. C. Hirschfeld, "Stress-Deformation and Strength Characteristics of Clay Compacted to a Constant Dry Unit Weight," in *Proc. Res. Conf. Strength of Cohesive Soils* (ASCE, Boulder, CO), 1960, pp. 359-417.

L. Casagrande, "Electro-Osmotic Stabilization of Soils," *Boston Soc. Civ. Eng., Contributions to Soil Mech., 1941-1953*, pp. 285-317, 1952.

D. K. Cassel and A. Klute, "Water Potential: Tensiometry," in *Methods of Soil Analysis, Part 1, Physical and Mineralogical Methods*, 2nd ed., A. Klute, Ed., Amer. Soc. of Agronomy, Soil Sci. Soc. of Amer., Madison, WI, 1986.

H. R. Cedergren, *Seepage Drainage and Flow Nets*. New York: Wiley, 1989, 465 pp.

R. J. Chandler and C. I. Gutierrez, "The Filter-Paper Method of Suction Measurement," *Geotechnique*, vol. 36, pp. 265-268, 1986.

C. S. Chang, C. W. Adegoke, and E. T. Selig, "GEOTRACK Model for Railroad Track Performance," *J. Geotech. Eng. Div.* (ASCE), vol. 106, no. GT11, pp. 1201–1218, 1980.

K. Chantawarangul, "Comparative Study of Different Procedures to Evaluate Effective Stress Strength Parameters for Partially Saturated Soils," M. Sc. thesis, Asia Inst. of Tech., Bangkok, Thailand, 1983.

P. K. Chattopadhyay, "Residual Shear Strength of Some Pure Clay Minerals," Ph.D. dissertation, Univ. of Alberta, Edmonton, Alta., Canada, 1972, 340 pp.

F. H. Chen, *Foundations on Expansive Soils*, 1st ed. New York: Elsevier, 1975, 280 pp.

F. H. Chen, *Foundations on Expansive Soils*, 2nd ed. Amsterdam: Elsevier, 1988, 463 pp.

F. H. Chen and D. Huang, "Lateral Expansion Pressure on Basement Walls," *Proc. of the 6th Int. Conf. on Expansive Soils* (Delhi, India), vol. 1, Dec. 1–4, 1987, pp. 55–60.

E. C. Childs, *An Introduction to the Physical Basis of Soil Water Phenomena*. London: Wiley-Interscience, 1969, 493 pp.

E. C. Childs and N. Collis-George, "Soil Geometry and Soil-Water Equilibria," *Faraday Soc.* (Discussion), no. 3, pp. 78–85, 1948.

E. C. Childs and N. Collis-George, "The Permeability of Porous Materials," *Proc. Royal Soc.*, vol. 201A, pp. 392–405, 1950.

R. K. H. Ching, "A Theoretical Examination and Practical Applications of the Limit Equilibrium Methods to Slope Stability Problems," M.Sc. thesis, Univ. of Saskatchewan, Saskatoon, Sask., Canada, 1981, 427 pp.

R. K. H. Ching and D. G. Fredlund, "Some Difficulties Associated with the Limit Equilibrium Method of Slices," *Can. Geotech. J.*, vol. 20, no. 4, pp. 661–672, 1983.

R. K. H. Ching and D. G. Fredlund, "A Small Saskatchewan Town Copes with Swelling Clay Problems," in *Proc. 5th Int. Conf. Expansive Soils*, May 1984, pp. 306–310.

R. K. H. Ching, J. Sweeney, and D. G. Fredlund, "Increase in Factor of Safety Due to Soil Suction for Two Hong Kong Slopes," in *Proc. 4th Int. Symp. Landslides*, Toronto, Ont., Canada, vol. 1, Sept. 1984, pp. 617–623.

P. C. Chou and N. J. Pagano, *Elasticity*. Toronto: Van Nostrand, 1967, 290 pp.

R. N. Chowdhury, "A Reassessment of Limit Equilibrium concepts in Geotechnique," in *Proc. Symp. Limit Equilibrium, Plasticity, and Generalized Stress Strain Applications in Geotech. Eng.*, ASCE, 1980 Annu. Conv., Hollywood, FL, 1980.

T. F. Christie, "Secondary Compression Effects During One-Dimensional Consolidation Tests," in *Proc. 6th Int. Conf. Soil Mech. Found. Eng.* (Montreal, Canada), vol. 1, 1965, pp. 198–202.

A. W. Clifton, R. Yoshida, D. G. Fredlund, and R. W. Chursinoff, "Performance of Darke Hall, Regina, Canada, Constructed on a Highly Swelling Clay," *Proc. of the 5th Int. Conf. on Expansive Soils*, May 21–23 (Adelaide, Australia), 1984, pp. 197–201.

M. B. Clisby, "An Investigation of the Volumetric Fluctuation of Active Clay Soils," Ph.D. dissertation, Univ. of Texas, Austin, 1962, 108 pp.

D. M. Cole, "A Technique for Measuring Radial Deformation During Repeated Load Triaxial Testing," *Can. Geotech. J.*, vol. 15, pp. 426–429, 1978.

J. D. Coleman, "Stress/Strain Relations for Partly Saturated Soils," *Geotechnique* (Correspondence), vol. 12, no. 4, pp. 348–350, 1962.

A. T. Corey, "The Interrelation between Gas and Oil Relative Permeabilities," *Producer's Monthly*, vol. 19, no. 1, 1954.

A. T. Corey, "Measurement of Water and Air Permeability in Unsaturated Soil," *Proc. Soil Sci. Soc. Amer.*, vol. 21, no. 1, pp. 7–10, 1957.

A. T. Corey, "Mechanics of Heterogeneous Fluids in Porous Media," Water Resources Publ., Fort Collins, CO, 259 pp., 1977.

A. T. Corey and W. D. Kemper, "Concept of Total Potential in Water and Its Limitations," *Soil Sci.*, vol. 91, no. 5, pp. 299–305, May 1961.

A. T. Corey, R. O. Slayter, and W. D. Kemper, "Comparative Terminologies for Water in the Soil-Plant-Atmosphere System," in *Irrigation in Agricultural Soils*, R. M. Hagan et al., Eds., Amer. Soc. Agron., Mono. No. 11, ch. 22, 1967.

D. W. Cox, "Volume Change of Compacted Clay Till," in *Proc. ICE Conf. Clay Tills*, 1978, pp. 79–86.

M. C. Crilly, R. M. C. Driscoll, and R. J. Chandler, "Seasonal Ground and Water Movement Observations from an Expansive Clay Site in the UK," Proc. 7th Int. Conf. on Expansive Soils, (Dallas, TX), Aug. 3–5, vol. 1, pp. 313–318, 1992.

D. Croney, "The Movement and Distribution of Water in Soils," *Geotechnique*, vol. 3, pp. 1–16, 1952.

D. Croney and J. D. Coleman, "Soil Thermodynamics Applied to the Movement of Moisture in Road Foundations," in *Proc. 7th Int. Cong. Appl. Mech.*, vol. 3, 1948, pp. 163–177.

D. Croney and J. D. Coleman, "Soil Structure in Relation to Soil Suction (pF)," *J. Soil Sci.*, vol. 5, no. 1, pp. 75–84, 1954.

D. Croney and J. D. Coleman, "Pore Pressure and Suction in Soil," in *Proc. Conf. Pore Pressure and Suction in Soils*. London: Butterworths, 1961, pp. 31–37.

D. Croney, J. D. Coleman, and W. A. Lewis, "Calculation of the Moisture Distribution Beneath Structures," *Cov. Eng. L.*, vol. 45, pp. 524, 1950.

D. Croney, J. D. Coleman, and W. P. M. Black, "Movement and Distribution of Water in Soil in Relation to Highway Design and Performance," *Water and Its Conduction in Soils*, Highway Res. Board, Special Report, Washington, DC, no. 40, pp. 226–252, 1958.

D. Croney, J. D. Coleman, and P. M. Bridge, "The Suction of Moisture Held in Soils and Other Porous Materials," *Tech. Paper*, Road Research Board, no. 24, H.M.S.O., London, 1952.

A. A. Curtis and C. D. Johnston, "Monitoring Unsaturated Soil Water Conditions in Groundwater Recharge Studies," in *Proc. Int. Conf. Measurement of Soil and Plant Water Status* (Utah State Univ., Logan), vol. 1, 1987, pp. 267–274.

V. Dakshanamurthy, V. "A Stress-Controlled Study of Swelling Characteristics of Compacted Expansive Clays," *Geotech. Testing J., GTJODJ*, vol. 2, no. 1, pp. 57–60, Mar. 1979.

V. Dakshanamurthy, V. and D. G. Fredlund, "Moisture and Air Flow in an Unsaturated Soil," in *Proc. 4th Int. Conf. Expansive Soils*, ASCE, vol. 1, 1980, pp. 514–532.

V. Dakshanamurthy and D. G. Fredlund, "Transient Flow Processes in Unsaturated Soils (Temperature, Relative Humidity, Evaporation and Infiltration)," CD-16.4 Transportation and Geotechnical Group, Univ. of Saskatchewan, Saskatoon, Sask., Canada, 1981, 92 pp.

V. Dakshanamurthy, V., D. G. Fredlund, and H. Rahardjo, "Coupled Three-Dimensional Consolidation Theory of Unsaturated Porous Media," *Preprint of Papers: 5th Int. Conf. Expansive Soils* (Adelaide, South Australia), Institute of Engineers, Australia, May 1984, pp. 99-104.

F. N. Dalton, "Development of Time-Domain Reflectometry for Measuring Soil Water Content and Bulk Soil Electrical Conductivity," published in *Advances in Measurement of Soil Physical Properties: Bringing Theory into Practice*, SSSA Special Publication no. 30, Soil Science Society of America, Madison, WI, 1992, pp. 143-168.

D. E. Daniel, "Permeability Test for Unsaturated Soil," *Geotech. Testing J.* (ASTM), vol. 6, no. 2, pp. 81-86, 1983.

D. E. Daniel, J. M. Hamilton, and R. E. Olson, "Suitability of Thermocouple Psychrometers for Studying Moisture Movement in Unsaturated Soils," in *Permeability and Groundwater Contaminant Transport*, ASTM STP 746, T. F. Zimmie and C. O. Riggs, eds., ASTM 1981, pp. 84-100.

H. Darcy, *Histoire Des Foundataines Publique de Dijon*. Paris: Dalmont, 1856, pp. 590-594.

J. M. Davidson, L. R. Stone, D. R. Nielsen, and M. E. Larue, "Field Measurement and Use of Soil-Water Properties," *Water Resources Res.*, vol. 5, pp. 1312-1321, 1969.

J. T. Davies and E. K. Rideal, *Interfacial Phenomena*, 2nd ed. New York: Academic, 1963, 747 pp.

C. M. A. de Bruijn, "Swelling Characteristics of a Transported Soil Profile at Leeuhof Vereeniging (Transvaal)," in *Proc. 5th Int. Conf. Soil Mech. Found. Eng.*, vol. 1, 1961, pp. 43-49.

C. M. A. de Bruijn, "Some Observations on Soil Moisture Conditions Beneath and Adjacent to Tarred Roads and Other Surface Treatments in South Africa," in *Moisture Equilibria and Moisture Changes in Soils Beneath Covered Areas, A Symp. in Print*, G. D. Aitchison, Ed. Australia: Butterworths, 1965, pp. 135-142.

D. A. de Vries, "Thermal Properties of Soils," in *The Physics of Plant Environment*, W. R. Van Wijk, Ed. Amsterdam: North-Holland, 1963, p. 382.

D. A. de Vries, "Heat Transfer in Soils," in *Heat and Mass Transfer in the Biosphere, 1. Transfer Processes in Plant Environment*. Washington, DC: Scripta, 1975, pp. 5-28.

D. U. Deere and F. D. Patton, "Slope Stability in Residual Soils," in *Proc. 4th Pan-Am Conf. Soil Mechanics Found. Eng.* (San Juan, Puerto Rico), vol. 1, 1971, pp. 87-170.

B. V. Derjaguin, "Recent Research into the Properties of Water in Thin Films and in Micro-Capillaries," in *Proc. Soc. for Experimental Biology Symp., XIX, The State and Movement of Water in Living Organisms*. London: Cambridge Univ. Press, 1965.

C. S. Desai, "Finite Element Methods for Flow in Porous Media," in *Finite Elements in Fluids*. New York: Wiley, 1975.

C. S. Desai, "Finite Element Residual Schemes for Unconfined Flow," *Int. J. Numer. Methods Eng.*, vol. 10, pp. 1415-1418, 1975.

C. S. Desai and J. F. Abel, *Introduction to the Finite Element Method; A Numerical Method for Engineering Analysis*. New York: Van Nostrand Reinhold, 1972, 477 pp.

C. S. Desai and J. T. Christian, *Numerical Methods in Geotechnical Engineering*. New York: McGraw-Hill, 1977, 783 pp.

C. S. Desai and W. C. Sherman, "Unconfined Transient Seepage in Sloping Banks," *Proc. ASCE, J. Soil Mech. Found. Eng. Div.*, vol. 97, Soil Mech. 2, pp. 357-373, 1971.

I. B. Donald, "Shear Strength Measurements in Unsaturated Non-Cohesive Soils with Negative Pore Pressures," in *Proc. 2nd Australia-New Zealand Conf. Soil Mech. Found. Eng.* (Christchurch, New Zealand), 1956, pp. 200-205.

I. B. Donald, "The Mechanical Properties of Saturated and Partly Saturated Soils with Special Reference to Negative Pore Water Pressure," Ph.D. dissertation, Univ. of London, London, England, 1961.

I. B. Donald, "Effective Stress Parameters in Unsaturated Soils," in *Proc. 4th Australia-New Zealand Conf. Soil Mech. Found. Eng.* (Adelaide, South Australia), 1963, pp. 41-46.

N. E. Dorsey, *Properties of Ordinary Water-Substances*, Amer. Chemical Soc., Mono. Series. New York: Reinhold, 1940, 673 pp.

H. E. Dregne, *Soils in Arid Regions*. New York: American Elsevier, 1976, 5 pp.

E. Drumright, "Shear Strength for Unsaturated Soils," Ph.D. dissertation, Univ. of Colorado, Fort Collins, 1987.

C. S. Dunn, "Developments in the Design of Triaxial Equipment for Testing Compacted Soils," in *Proc. Symp. Economic Use of Soil Testing in Site Investigation* (Birmingham, AL), vol. 3, 1965, pp. 19-25.

T. B. Edil, S. E. Motan, and F. X. Toha, "Mechanical Behavior and Testing Methods of Unsaturated Soils," in *Laboratory Shear Strength of Soil*, ASTM STP No. 740, R. N. Yong and F. C. Townsend, Eds., 1981, pp. 114-129.

N. E. Edlefsen and A. B. C. Anderson, "Thermodynamics of Soil Moisture," *Hilgardia*, vol. 15, p. 31-298, 1943.

D. E. Elrick and D. H. Bowman, "Improved Apparatus for Soil Moisture Flow Measurements," *Proc. Soil Sci. Soc. Amer.*, vol. 28, pp. 450-453, 1964.

A. Elzeftawy and K. Cartwright, "Evaluating the Saturated and Unsaturated Hydraulic Conductivity of Soils," in *Permeability and Groundwater Contaminant Transport*, ASTM STP 746, T. F. Zimmie and C. D. Riggs, Eds., Amer. Soc. Testing and Materials, 1981, pp. 168-181.

V. Escario, "Swelling of Soils in Contact with Water at a Negative Pressure," in *Proc. 2nd Int. Conf. Expansive Soils* (Texas A & M Univ., College Station), 1969, pp. 207-217.

V. Escario, "Suction Controlled Penetration and Shear Tests," in *Proc. 4th Int. Conf. Expansive Soils* (Denver, CO), ASCE, vol. 2, 1980, pp. 781-797.

V. Escario and J. Sáez, "Measurement of the Properties of Swelling and Collapsing Soils Under Controlled Suction," in *Proc. 3rd Int. Conf. Expansive Soils* (Haifa, Israel), vol. 1, 1973, pp. 195-200.

V. Escario and J. Sáez, "The Shear Strength of Partly Saturated Soils," *Geotechnique*, vol. 36, no. 3, pp. 453-456, 1986.

V. Escario and J. Sáez, "Shear Strength of Partly Saturated Soil Versus Suction," *Proc. of the 6th Int. Conf. on Expansive Soils* (Delhi, India), vol. 2, Dec. 1-4, 1987, pp. 602-604.

V. Escario, J. F. T. Jucá, and M. S. Coppe, "Strength and Deformation of Partly Saturated Soils," in *Proc. 12th Int. Conf.*

*Soil Mech. Found. Eng.* (Rio de Janeiro, Brazil), Aug. 1989, pp. 43-46.

D. F. Faizullaev, "Laminar Motion of Multiphase Media in Conduits," Special Res. Report, translated from Russian by Consultants Bureau, New York, NY, 144 pp., 1969.

K. Fan, "Evaluation of the Interslice Forces for Lateral Earth Force and Slope Stability Problems," M.Sc. thesis, Univ. of Saskatchewan, Saskatoon, Sask., Canada, 369 pp., 1983.

K. Fan, D. G. Fredlund, and G. W. Wilson, "An Interslice Force Function for Limit Equilibrium Slope Stability Analysis," *Can. Geotech. J.*, vol. 23, no. 3, pp. 287-296, 1986.

P. J. Fargher, J. A. Woodburn, and J. Selby, "Footings and Foundations for Small Buildings in Arid Climates," Inst. of Eng. Aust., S. A. Div. and Univ. Adelaide Adult Educ. Dep., Adelaide, Australia, 1979.

R. G. Fawcett and N. Collis-George, "A Filter-Paper Method for Determining the Moisture Characteristics of Soil," *Australian J. Exp. Agriculture and Animal Husbandry*, vol. 7, pp. 162-167, Apr. 1967.

A. Fick, "Ueber Diffusion," *Ann. der Phys.* (Leipzig), vol. 94, pp. 59-86, 1855.

H. L. Filson, "A Study of Heave Analysis Using the Oedometer Test," M.Sc. thesis, Univ. of Saskatchewan, Saskatoon, Sask., Canada, 1980.

A. B. Fourie, "Laboratory Evaluation of Lateral Swelling Pressure," *ASCE J. Geotech. Eng.*, vol. 115, no. 10, pp. 1481-1486, 1989.

D. G. Fredlund, "Comparison of Soil Suction and One-Dimensional Consolidation Characteristics of a Highly Plastic Clay," Nat. Res. Council Tech. Report No. 245. v, Div. of Building Res., Ottawa, Ont., Canada, 26 pp., July 1964.

D. G. Fredlund, "Consolidometer Test Procedural Factors Affecting Swell Properties," in *Proc. 2nd Conf. Expansive Clay Soils.* College Station, TX: Texas A and M Press, 1969, pp. 435-456.

D. G. Fredlund, "Manual of Volume Change Test Procedures for Unsaturated Soils," Internal Note SM12, Univ. of Alberta, Edmonton, Alta., Canada, 1972.

D. G. Fredlund, "Volume Change Behavior of Unsaturated Soils," Ph.D. dissertation, Univ. of Alberta, Edmonton, Alta., Canada, 490 pp., 1973.

D. G. Fredlund, Discussion at "The Second Technical Session, Div. Two (Flow and Shear Strength)," *Proc. of the 3rd Int. Conf. on Expansive Soils*, vol. II, Jeruselam Academic Press, Haifa, Israel, 1973, pp. 71-76.

D. G. Fredlund, "Engineering Approach to Soil Continua," in *Proc. 2nd Symp. Application of Solid Mech.*, Hamilton, Ont., Canada, 1974, pp. 46-59.

D. G. Fredlund, "A Diffused Air Volume Indicator for Unsaturated Soils," *Can. Geotech. J.*, vol. 12, no. 4, pp. 533-539, 1975.

D. G. Fredlund, "Prediction of Heave in Unsaturated Soils," *Proc. of the 5th Asian Regional Conf.* (Bangalore, India), December, 1975, pp. 19-22.

D. G. Fredlund, "Density and Compressibility Characteristics of Air-Water Mixtures," *Can. Geotech. J.*, vol. 13, no. 4, pp. 386-396, 1976.

D. G. Fredlund, "Two-Dimensional Finite Element Program Using Constant Strain Triangles (FINEL)," Univ. of Saskatchewan Transportation and Geotech. Group, Internal Report CD-2, 1978.

D. G. Fredlund, "Second Canadian Geotechnical Colloquium: Appropriate Concepts and Technology for Unsaturated Soils," *Can. Geotech. J.*, vol. 16, no. 1, pp. 121-139, 1979.

D. G. Fredlund, "Use of Computers for Slope Stability Analysis," State-of-the-Art Paper, in *Proc. Int. Symp. Landslides* (Delhi, India), vol. 2, 1980, pp. 129-138.

D. G. Fredlund, "Seepage in Saturated Soils," Panel Discussion: Ground Water and Seepage Problems, in *Proc. 10th Int. Conf. Soil Mech. Found. Eng.* (Stockholm, Sweden), vol. 4, 1981, pp. 629-641.

D. G. Fredlund, Discussion: "Consolidation of Unsaturated Soils including Swelling and Collapse Behavior," by Lloret, A. and Alonso, E. E. (1981) *Geotechnique* 30, no. 4, pp. 449-477. Discussion in *Geotechnique*, 1981.

D. G. Fredlund, "The Shear Strength of Unsaturated Soils and its Relationship to Slope Stability Problems in Hong Kong," *The Hong Kong Engineer*, vol. 9, no. 4, pp. 37-45, 1981.

D. G. Fredlund, "Consolidation of Unsaturated Porous Media," in *Proc. NATO Advanced Study Inst. Mech. of Fluids in Porous Media* (Newark, DE), July 1982.

D. G. Fredlund, "Prediction of Ground Movements in Swelling Clays," presented at the 31st Annu. Soil Mech. Found. Eng. Conf., ASCE, Invited Lecture, Minneapolis, MN, 1983.

D. G. Fredlund, *SLOPE-II User's Manual, A Slope Stability Program to Calculate the Factor of Safety*, Doc. No. S-20 GEO-SLOPE Programming Ltd., Calgary, Alta., Canada, 850 pp., 1984.

D. G. Fredlund, "Theory Formulation and Application for Volume Change and Shear Strength Problems in Unsaturated Soils," in *Proc. 11th Int. Conf. Soil Mech. Found. Eng.* (San Francisco, CA), Aug. 1985.

D. G. Fredlund, "Soil Mechanics Principles that Embrace Unsaturated Soils," in *Proc. 11th Int. Conf. Soil Mech. Found. Eng.* (San Francisco, CA), vol. 2, Aug. 1985, pp. 465-473.

D. G. Fredlund, "PC-SLOPE Microcomputer Slope Stability Program to Calculate Factor of Safety," *User's Manual S-30*, GEO-SLOPE Programming Ltd., Calgary, Alta., Canada, 550 pp., 1985.

D. G. Fredlund, "Slope Stability Analysis Incorporating the Effect of Soil Suction," chapter 4 in *Slope Stability*, M. G. Anderson and K. S. Richards, Eds. New York: Wiley, 1987, pp. 113-144.

D. G. Fredlund, "The Stress State for Expansive Soils," in *Proc. 6th Int. Conf. Expansive Soils* (Delhi, India), Dec. 1987, pp. 1-9.

D. G. Fredlund. "Negative Pore-Water Pressures in Slope Stability," in *Simposio Suramericano de Deslizamiento* (Paipa, Colombia), 1989, pp. 1-31.

D. G. Fredlund, "The Character of the Shear Strength Envelope for Unsaturated Soils," The Victor de Mello Volume, the *XII Int. Conf. on Soil Mech. and Found. Eng.* (Rio de Janeiro, Brazil), 1989, pp. 142-149.

D. G. Fredlund, "Soil Suction Monitoring for Roads and Airfields," *Symposium on the State-of-the-Art of Pavement Response Monitoring Systems for Roads and Airfields*, Sponsored by the U.S. Army Corps of Engineers (Hanover, NH), March 6-9, 1989.

D. G. Fredlund, "How Negative Can Pore-Water Pressure Get?" *Geotechnical News*, vol. 9, no. 3, Can. Geot. Society, September, 1991, pp. 44–46.

D. G. Fredlund, "Background, Theory and Research Related to the Use of Thermal Conductivity Sensors of Matric Suction Measurements," published in *Advances in Measurement of Soil Physical Properties; Bringing Theory into Practice*, SSSA Special Publication No. 30, Soil Science of America, Madison, WI, 1992, pp. 249–262.

D. G. Fredlund and S. L. Barbour, "Integrated Seepage Modelling and Slope Stability Analysis—A Generalized Approach for Saturated/Unsaturated Soils," chapter in *Geomechanics and Water Engineering in Environmental Management*, R. Chowdhury, Ed., 1990.

D. G. Fredlund and V. Dakshanamurthy, "Transient Flow Theory for Geotechnical Engineering," Special Session: Impact of Richard's Equation: Semi-Centennial Session, Fall Meeting, Amer. Geophys. Union, 1981.

D. G. Fredlund and V. Dakshanamurthy, "Prediction of Moisture Flow and Related Swelling or Shrinking in Unsaturated Soils," *Geotechnical Engineering*, vol. 13, pp. 15–49, 1982.

D. G. Fredlund and J. U. Hasan, "Statistical Geotechnical Properties of Lake Regina Sediments," Transportation and Geotech. Group, Dep. of Civil Eng., Univ. of Saskatchewan, Saskatoon, Sask., Canada, Internal Report IR-9, 102 pp., Jan. 1979.

D. G. Fredlund and J. U. Hasan, "One-Dimensional Consolidation Theory: Unsaturated Soils," *Can. Geot. Jour.*, vol. 16, no. 3, pp. 521–531, 1979.

D. G. Fredlund and J. Krahn, "Comparison of Slope Stability Methods Analysis," *Can. Geotech. J.*, vol. 14, no. 3, pp. 429–439, 1977.

D. G. Fredlund and N. R. Morgenstern, "Pressure Response Below High Air Entry Discs," in *Proc. 3rd Int. Conf. Expansive Soils*. Haifa, Israel: Jerusalem Academic Press, vol. 1, 1973, pp. 97–108.

D. G. Fredlund and N. R. Morgenstern, "Constitutive Relations for Volume Change in Unsaturated Soils," *Can. Geotech. J.*, vol. 13, no. 3, pp. 261–276, 1976.

D. G. Fredlund and N. R. Morgenstern, "Stress State Variables for Unsaturated Soils," *ASCE J. Geotech. Eng. Div. GT5*, vol. 103, pp. 447–466, 1977.

D. G. Fredlund and N. R. Morgenstern, Closure to Paper Entitled, "Stress State Variables for Unsaturated Soils," ASCE, *Geot. Div.*, GT11, pp. 1415–1416, 1978.

D. G. Fredlund and Ning Yang, "The Mechanical Properties of Unsaturated Soil and their Engineering Application," Yantu Gongcheng Xuebao, *Journal of China Academy of Building Research*, Published in the Chinese Journal of Geotechnical Engineering (Beijing, China), vol. 13, no. 5, pp. 24–35, September 1991.

D. G. Fredlund and H. Rahardjo, "Theoretical Context for Understanding Unsaturated Residual Soil Behavior," in *Proc. 1st Int. Conf. Geomech. in Tropical Lateritic and Saprolitic Soils* (Sao Paulo, Brazil), Feb. 1985, pp. 295–306.

D. G. Fredlund and H. Rahardjo, "Unsaturated Soil Consolidation Theory and Laboratory Experimental Data," *Consolidation of Soils: Testing and Evaluation*, ASTM STP 892, R. N. Yong and F. C. Townsend, Eds., ASTM, Philadelphia, PA, pp. 154–169, 1986.

D. G. Fredlund and H. Rahardjo, "Soil Mechanics Principles for Highway Engineering in Arid Regions," *Transportation Res. Record 1137*, pp. 1–11, 1987.

D. G. Fredlund and H. Rahardjo, "State-of-Development in the Measurement of Soil Suction," in *Proc. Int. Conf. Eng. Problems on Regional Soils* (Beijing, China), Aug. 1988, pp. 582–588.

D. G. Fredlund and H. Rahardjo, "The Role of Unsaturated Soil Mechanics in Geotechnical Engineering," Invited Lecture, Proc. of the 11th S.E.A.G.C (Singapore), May 4–8, 1993.

D. G. Fredlund and D. K. H. Wong, "Calibration of Thermal Conductivity Sensors for Measuring Soil Suction," *ASTM Geotech. J.*, vol. 12, no. 3, pp. 188–194, 1989.

D. G. Fredlund, J.K-M. Gan, and H. Rahardjo, "Measuring Negative Pore-Water Pressures in a Freezing Environment," *Transportation Research Record 1307*, pp. 291–299, 1992.

D. G. Fredlund, J. U. Hasan, and H. Filson, "The Prediction of Total Heave," *Proc. of the 4th Int. Conf. on Expansive Soils* (Denver, CO), 1980, vol. 1, pp. 1–17.

D. G. Fredlund, J. Krahn, and J. U. Hasan, "Variability of an Expansive Clay Deposit," *Proc. of the 4th Int. Conf. on Expansive Soils* (Denver, CO), 1980, vol. 1, pp. 322–338.

D. G. Fredlund, N. R. Morgenstern, and R. A. Widger, "The Shear Strength of Unsaturated Soils," *Can. Geotech. J.*, vol. 15, no. 3, pp. 313–321, 1978.

D. G. Fredlund, J. Krahn, and J. U. Hasan, "Variability of an Expansive Clay Deposit," in *Proc. 4th Int. Conf. Expansive Soils* (Denver, CO), vol. 1, 1980, pp. 322–338.

D. G. Fredlund, J. U. Hasan, and H. Filson, "The Prediction of Total Heave," in *Proc. 4th Int. Conf. Expansive Soils* (Denver, CO), vol. 1, 1980, pp. 1–17.

D. G. Fredlund, J. Krahn, and D. Pufahl, "The Relationship Between Limit Equilibrium Slope Stability Methods," in *Proc. 10th Int. Conf. Soil Mech. Found. Eng.* (Stockholm, Sweden), vol. 3, 1981, pp. 409–416.

D. G. Fredlund, H. Rahardjo, and J. Gan, "Nonlinearity of Strength Envelope for Unsaturated Soils," in *Proc. 6th Int. Conf. Expansive Soils* (New Delhi, India), vol. 1, pp. 49–54, Dec. 1987.

D. G. Fredlund, H. Rahardjo, and J.K-M. Gan, Discussion on the paper entitled, "A Framework for Unsaturated Soils Behavior," by D. G. Toll (1990), *Geotechnique*, 40, no. 1, pp. 31–44, 1990.

D. G. Fredlund, H. Rahardjo, and T. Ng, "Effect of Pore-Air and Negative Pore-Water Pressures on Stability at the End-of-Construction," Proc. Int. Conf. on Dam Engineering, Johor Bahru, Malaysia, pp. 43–51, Jan. 12–13, 1993.

R. A. Freeze, "Influence of the Unsaturated Flow Domain on Seepage Through Earth Dams," *Water Resources Res.*, vol. 7, no. 4, pp. 929–940, 1971.

R. A. Freeze and J. A. Cherry, *Groundwater*. Englewood Cliffs, NJ: Prentice-Hall, 1979, 604 pp.

R. A. Freeze and P. A. Witherspoon, "Theoretical Analysis of Regional Groundwater Flow, 1: Analytical and Numerical Solutions to the Mathematical Model," *Water Resources Res.*, vol. 2, no. 4, pp. 641–656, 1966.

Y. C. Fung, *Foundations of Solid Mechanics*. Englewood Cliffs, NJ: Prentice-Hall, 1965, 525 pp.

Y. C. Fung, *A First Course in Continuum Mechanics*, 2nd ed. Englewood Cliffs, NJ: Prentice-Hall, 1977, 340 pp.

R. H. Galagher, J. T. Oden, C. Taylor, and O. C. Zienkiewicz, *Finite Element in Fluids, Vol. 1: Viscous Flow and Hydrodynamics.* New York: Wiley, 1975.

P. M. Gallen, "Measurement of Soil Suction Using the Filter Paper Methods: A Literature Review; Final Report," Internal Report, Transportation and Geotech. Group, Dep. of Civil Eng., Univ. of Saskatchewan, Saskatoon, Sask., Canada, 32 pp., 1985.

J. K. M. Gan, "Direct Shear Strength Testing of Unsaturated Soils." M.Sc. thesis, Dep. of Civil Eng., Univ. of Saskatchewan, Saskatoon, Sask., Canada, 587 pp., 1986.

J. K. M. Gan and D. G. Fredlund, "Multistage Direct Shear Testing of Unsaturated Soils," *ASTM, Geotech. Testing J.*, vol. 11, no. 2, pp. 132–138, 1988.

J. K. M. Gan, D. G. Fredlund, and H. Rahardjo, "Determination of the Shear Strength Parameters of an Unsaturated Soil Using the Direct Shear Test," *Can. Geotech. J.*, vol. 25, no. 3, pp. 500–510, 1988.

R. Gardner, "A Method of Measuring the Capillary Tension of Soil Moisture over a Wide Moisture Range," *Soil Sci.*, vol. 43, no. 4, pp. 277–283, 1937.

W. R. Gardner, "Some Steady State Solutions of the Unsaturated Moisture Flow Equation with Application to Evaporation from a Water-Table," *Soil Sci.*, vol. 85, no. 4, 1958.

W. R. Gardner, "Laboratory Studies of Evaporation from Soil Columns in the Presence of a Water-Table," *Soil Sci Amer.*, vol. 85, p. 244, 1958.

W. R. Gardner, "Soil Suction and Water Movement," *Pore Pressure and Suction in Soils.* London: Butterworths, pp. 137–140, 1961.

W. R. Gardner and M. Fireman, "Laboratory Studies of Evaporation from Soil Columns in the Presence of a Water-Table," *Soil Sci.*, vol. 85, pp. 244–249, 1958.

W. Gardner and J. A. Widtsoe, "The Movement of Soil Moisture," *Soil Sci.*, vol. 11, pp. 215–232, 1921.

H. J. Gibbs, "Pore Pressure Control and Evaluations for Triaxial Compressions," Amer. Soc. for Testing and Materials, Special Tech. Publ. No. 361, pp. 212–221, 1963.

H. J. Gibbs, "Discussion, Proceedings of the Specialty Session No. 3 on Expansive Soils and Moisture Movement in Partly Saturated Soils," *7th Int. Conf. on Soil Mec. and Found. Eng.* (Mexico City, Mexico), 1969, pp. 424–424.

H. J. Gibbs, "Use of a Consolidometer for Measuring Expansion Potential of Soils," in *Proc. Workshop on Expansive Clays and Shales in Highway Design and Construction*, Univ. of Wyoming, Laramie, vol. 1, May 1973, pp. 206–213.

H. J. Gibbs and C. T. Coffey, "Application of Pore Pressure Measurements to Shear Strength of Cohesive Soils," Report No. EM-761, U.S. Dep. of the Interior, Bureau of Reclamation, Soils Eng. Branch, Div. of Research, Boulder, CO, 1969.

H. J. Gibbs and C. T. Coffey, "Techniques for Pore Pressure Measurements and Shear Testing of Soil," in *Proc. 7th Int. Conf. Soil Mech. Found. Eng.* (Mexico), vol. 1, 1969, pp. 151–157.

H. J. Gibbs, J. W. Hilf, W. G. Holtz, and F. C. Walker, "Shear Strength of Cohesive Soils," in *Proc. ASCE Res. Conf. Shear Strength of Cohesive Soils*, Boulder, CO, June 1960, pp. 33–162.

R. E. Gibson, "An Analysis of System Flexibility and its Effect on Time-Lag in Pore Water Pressure Measurements," *Geotechnique*, vol. 13, pp. 1–11, 1963.

R. E. Gibson and D. J. Henkel, "Influence of Duration of Tests at Constant Rate of Strain on Measured 'Drain' Strength," *Geotechnique*, vol. 4, pp. 6–15, 1954.

H. G. Gilchrist, "A Study of Volume Change of a Highly Plastic Clay," M.Sc. thesis, Univ. of Saskatchewan, Saskatoon, Sask., Canada, 141 pp., 1963.

*Glossary of Soil Science Terms*, Soil Science Society of America, Madison, WI, 1979, 37 pp.

P. A. Gonzalez and B. J. Adams, "Mine Tailings Disposal: I. Laboratory Characterization of Tailings," Dep. of Civil Eng., Univ. of Toronto, Toronto, Ont., Canada, pp. 1–14, 1980.

D. M. Gray, "Handbook on the Principles of Hydrology," *Canadian National Committee for the International Hydrological Decade*, 1970, 631 pp.

H. Gray, "Simultaneous Consolidation of Contiguous Layers of Unlike Compressible Soils," *Trans. ASCE*, vol. 110, pp. 1327–1344, 1945.

A. E. Green and P. M. Naghdi, "A Dynamical Theory of Interacting Continua," *Int. J. Eng. Sci.* (Pergamon, London), vol. 3, pp. 231–241, 1965.

R. E. Green and J. C. Corey, "Calculation of Hydraulic Conductivity: A Further Evaluation of Some Predictive Methods," *Proc. Soil Sci. Soc. Amer.*, vol. 35, pp. 3–8, 1971.

R. E. Green and J. C. Corey, "Calculation of Hydraulic Soils, Prediction, Laboratory Test and In-Situ Measurements," in *Proc. 8th Int. Conf. Soil Mech. Found. Eng.* (Moscow, Russia), 1971, pp. 163–170.

G. J. Gromko, "Review of Expansive Soils," *Journal of the Geotechnical Engineering Division*, ASCE, vol. 100, no. GT6, pp. 667–687, 1974.

R. B. Grossman, B. R. Brashner, D. P. Franzmeier, and J. L. Walker, "Linear Extensibility as Calculated from Natural-Clod Bulk Density Measurements," *Proc. Soil Sci. Soc. Amer.*, vol. 32, no. 4, pp. 570–573, 1968.

S. K. Gulhati, "Shear Behaviour of Partially Saturated Soil," in *Proc. 5th Asian Reg. Conf. Soil Mech. Found. Eng.* (Bangalore, India), 1975, pp. 87–90.

S. K. Gulhati and D. J. Satija, "Shear Strength of Partially Saturated Soils," in *Proc. 10th Int. Conf. Soil Mech. Found. Eng.* (Stockholm), 1981, pp. 609–612.

A. B. Gureghian, "A Two-Dimensional Finite Element Solution Scheme for the Saturated-Unsaturated Flow with Applications to Flow Through Ditch-Drained Soils," *J. Hydrol.*, vol. 50, pp. 333–353, 1981.

S. A. Habib, T. Kato, and D. Karube, "One Dimensional Swell Behavior of Unsaturated Soil," *Proceedings of the 7th Int. Conf. on Expansive Soils* (Dallas, Texas), August 3–5, 1992, vol. 1, pp. 222–226.

W. B. Haines, "The Volume Changes Associated with Variations of Water Content in Soil," *J. Agric. Sci.*, vol. 13, pp. 296–310, 1923.

W. B. Haines, "A Further Contribution to the Theory of Capillary Phenomena in Soils," *J. Agric.Sci.*, vol. 17, pp. 264–290, 1927.

D. J. Hamberg, "A Simplified Method for Predicting Heave in Expansive Soils," M.Sc. thesis, Colorado State Univ., Fort Collins, 275 pp., 1985.

A. P. Hamblin, "Filter-Paper Method for Routine Measurement of Field Water Potential," *Journal of Hydrology*, vol. 53, no. 3/4, pp. 355–360, 1984.

J. J. Hamilton, "Foundations on Swelling or Shrinking Subsoils," *Can. Building Digest*, CBD 184, Div. of Building Research, Nat. Res. Council, Ottawa, Ont., Canada, 4 pp., 1977.

J. M. Hamilton, "Measurement of Permeability of Partially Saturated Soils," M.Sc. thesis, Univ. of Texas, Austin, 1979.

J. M. Hamilton, D. E. Daniel, and R. E. Olson, "Measurement of Hydraulic Conductivity of Partially Saturated Soils," in *Permeability and Groundwater Contaminant Transport*, ASTM Special Tech. Publ. 746, T. F. Zimmie and C. O. Riggs, Eds., ASTM, pp. 182–196, 1981.

R. J. Hanks and S. A. Bowers, "Numerical Solution of the Moisture Flow Equation into Layered Soil," *Proc. Soil Sci. Amer.*, vol. 26, no. 6, 1962.

A. M. Hanna and G. G. Meyerhof, "Design Charts for Ultimate Bearing Capacity of Foundations on Sand Overlying Soft Clay," *Can. Geotech. J.*, vol. 17, pp. 300–303, 1980.

F. Hardy, "The Shrinkage Coefficient of Clays and Soils," *J. Agric. Sci.*, vol. 13, pp. 244–264, 1923.

L. P. Harrison, "Fundamental Concepts and Definitions Relating to Humidity," in *Humidity and Moisture, Fundamentals and Standards*, A. Wexler and W. A. Wildhack, Eds. New York: Reinhold, vol. 3, 1965, pp. 3–69.

J. U. Hasan and D. G. Fredlund, "Pore Pressure Parameters for Unsaturated Soils," *Can. Geotech. J.*, vol. 17, no. 3, pp. 395–404, 1980.

K. H. Head, *Manual of Soil Laboratory Testing, Vol. 1, Soil Classification and Compaction Tests.* London, Plymouth: Pentech, 1984.

K. H. Head, *Effective Stress Tests, Manual of Soil Laboratory Testing.* New York: Wiley, vol. 3, pp. 743–1238, 1986.

Headquarters, U.S. Dep. of the Army, "Foundations in Expansive Soils," Tech. Man., TM5-818-7, Washington, DC, 1983.

E. Heyer, A. Cass, and A. Mauro, "A Demonstration of the Effect of Permeant and Impermeant Solutes and Unstirred Boundary Layers on Osmotic Flow," *Yale J. Biol. Med.*, vol. 42, pp. 139–153, Dec. 1969–Feb. 1970.

J. W. Hilf, "Estimating Construction Pore Pressures in Rolled Earth Dams," in *Proc. 2nd Int. Conf. Soil Mech. Found. Eng.* (Rotterdam, The Netherlands), vol. 3, 1948, pp. 234–240.

J. W. Hilf, "An Investigation of Pore-Water Pressure in Compacted Cohesive Soils," Ph.D. dissertation, Tech. Memo. No. 654, U.S. Dep. of the Interior, Bureau of Reclamation, Design and Construction Div., Denver, CO, 654 pp., 1956.

C. B. Hill, "Consolidation of Compacted Clay," M.Sc. thesis, Univ. of Manchester, Manchester, England, 1967.

D. Hillel, *Introduction to Soil Physics.* New York: Academic, 1982, 364 pp.

D. Hillel, V. D. Krentos, and Y. Stylianou, "Procedure and Test of an Internal Drainage Method for Measuring Soil Hydraulic Characteristics In-Situ," *Soil Sci.*, vol. 114, pp. 295–400, 1972.

D. Y. F. Ho, "Measurement of Soil Suction Using the Filter Paper Technique," Internal Report, IR-11, Transportation and Geotech. Group, Dep. of Civil Eng., Univ. of Saskatchewan, Saskatoon, Sask., Canada, 90 pp., 1979.

D. Y. F. Ho, "The Relationship Between the Volumetric Deformation Moduli of Unsaturated Soils," Ph.D. dissertation, Dep. of Civil Eng., Univ. of Saskatchewan, Saskatoon, Sask., Canada, 379 pp., 1988.

D. Y. F. Ho and D. G. Fredlund, Discussion of "Shear Strength of Partially Saturated Soils," by S. K. Gulhati and B. S. Satija, Session No. 4, *10th Int. Conf. on Soil Mech. and Found. Eng.* (Stockholm, Sweden), June 1981, vol. 4, pp. 672–674.

D. Y. F. Ho and D. G. Fredlund, "Increase in Shear Strength due to Soil Suction for Two Hong Kong Soils," in *Proc. ASCE Geotech. Conf. Eng. and Construction in Tropical and Residual Soils* (Honolulu, HI), Jan. 1982, pp. 263–295.

D. Y. F. Ho and D. G. Fredlund, "A Multi-Stage Triaxial Test for Unsaturated Soils," *ASTM Geotech. Testing J.*, vol. 5, no. 1/2, pp. 18–25, June 1982.

D. Y. F. Ho and D. G. Fredlund, "Strain Rates for Unsaturated Soil Shear Strength Testing," in *Proc. 7th Southeast Asian Geotech. Conf.* (Hong Kong), Nov. 1982.

D. Y. F. Ho and D. G. Fredlund, "Laboratory Measurement of the Volumetric Deformation Moduli for Two Unsaturated Soils," in *Proc. 42nd Can. Geo. Conf.* (Winnipeg, Man., Canada), Oct. 1989, pp. 50–60.

D. Y. F. Ho, D. G. Fredlund and H. Rahardjo, "Volume Change Indices During Loading and Unloading of an Unsaturated Soil," *Can. Geot. Jour.*, vol. 29, no. 2, pp. 195–207, 1992.

C. A. Hogentogler and E. S. Barber, "Discussion in Soil Water Phenomena," *Proc. Hwy. Research Board*, vol. 21, pp. 452–465, 1941.

W. G. Holtz, "Expansive Clays—Properties and Problems," in *Theoretical and Practical Treatment of Expansive Soils, 1st Conf. Mech. Soils*, Colorado School of Mines, Golden, vol. 54, no. 4, 1959, pp. 89–117.

W. G. Holtz, "Suggested Method of Test for One-Dimensional Expansion and Uplift Pressure of Clay Soils," in *Special Procedures for Testing Soil and Rock for Engineering Purposes*, 5th ed., ASTM STP 479, 1970.

W. G. Holtz and H. J. Gibbs, "Engineering Properties of Expansive Clays," *Trans. ASCE*, vol. 121, pp. 641–663, 1956.

R. D. Holtz and W. G. Kovacs, *An Introduction to Geotechnical Engineering.* Englewood Cliffs, NJ: Prentice-Hall, 733 pp. 1981.

G. M. Hornberger, I. Remson, and A. A. Fungaroli, "Numeric Studies of a Composite Soil Moisture Ground-Water System," *Water Resources Res.*, vol. 5, no. 4, pp. 797–802, 1969.

B. K. Hough, *Basic Soils Engineering.* New York: Ronald Press, 1969, 634 pp.

M. K. Hubbert and W. W. Rubey, "Role of Fluid Pressure in Mechanics of Overthrust Faulting," *Bull. Geol. Soc. Amer.*, vol. 70, pp. 115–166, Feb. 1959.

J. E. Ingersoll, "Method for Coincidentally Determining Soil Hydraulic Conductivity and Moisture Retention Characteristics," Special Report 81-2, U.S. Army Cold Regions Res. and Eng. Lab., Hanover, NH, 1981.

International Society for Soil Mechanics and Foundation Engineering, "List of Symbols, Units, and Definitions," Subcommittee on Symbols, Units, and Definitions, *Proc. 9th Inter. Conf. Soil Mech. Found. Eng.* (Tokyo), 1977, vol. 3, pp. 156–170.

International Society of Soil Science, "News of the Commission,

Commission I (Soil Physics)," *Soil Physics Terminology*, Bulletin No. 23, 7, published by the Soil Science Society of America, Madison, WI, 1963.

International Society of Soil Science, "Glossary of Soil Science Terms," published by the Soil Science Society of America, Madison, WI, May, 1970.

International Society of Soil Science, "Glossary of Soil Science Terms," published by the Soil Science Society of America, Madison, WI, October, 1979, 37 pp.

H. O. Ireland, "Design and Construction of Retaining Walls," in *Soil Mechanics Lecture Series: Design of Structures to Resist Earth Pressure*, Amer. Soc. of Civil Eng., Chicago, IL, 1964.

S. Irmay, "On the Hydraulic Conductivity of Unsaturated Soils," *Trans. Amer. Geophys. Union*, vol. 35, 1954.

J. Jaky, "The Coefficient of Earth Pressure at Rest," *J. Soc. Hungarian Architects and Eng.*, pp. 355-358, 1944.

Y. W. Jame, "Heat and Mass Transfer in Freezing Soil," Ph.D. dissertation, Univ. of Saskatchewan, Saskatoon, Sask., Canada, 1977.

Y. W. Jame and D. I. Norum, "Heat and Mass Transfer in Freezing Unsaturated Porous Medium," *Water Resources Res.*, vol. 16, no. 4, pp. 811-819, 1980.

D. J. Janssen and B. J. Dempsey, "Soil-Moisture Properties of Subgrade Soils," presented at the 60th Annu. Transportation Res. Board Meeting, Washington, DC, Aug. 1980.

I. Javandel and P. A. Witherspoon, "Application of the Finite Element Method to Transient Flow in Porous Media," *Trans. Soc. Petro. Eng.*, vol. 243, 1968.

J. E. Jennings, "A Revised Effective Stress Law for Use in the Prediction of the Behavior of Unsaturated Soils," in *Pore Pressure and Suction in Soils*, conf. organized by the British Nat. Soc. of Int. Soc. Soil Mech. Found. Eng. at the Inst. of Civil Eng. London: Butterworths, 1961, pp. 26-30.

J. E. Jennings, "The Prediction of Amount and Rate of Heave Likely to be Experienced in Engineering Construction on Expansive Soils," in *Proc. 2nd Int. Conf. Expansive Clay Soils* (Texas A&M Univ., College Station), 1969, pp. 99-109.

J. E. Jennings and J. B. Burland, "Limitations to the Use of Effective Stresses in Partly Saturated Soils," *Geotechnique*, vol. 12, no. 2, pp. 125-144, 1962.

J. E. Jennings and K. Knight, "The Prediction of Total Heave from the Double Oedometer Test," in *Proc. Symp. Expansive Clays* (South African Inst. of Civil Engineers, Johannesberg), vol. 7, no. 9, 1957, pp. 13-19.

J. E. Jennings, R. A. Firth, T. K. Ralph, and N. Nagar, "An Improved Method for Predicting Heave Using the Oedometer Test," in *Proc. 3rd Int. Conf. Expansive Soils* (Haifa, Israel), vol. 2, 1973, pp. 149-154.

G. D. Johnpeer, "Land Subsidence Caused by Collapsible Soils in Northern New Mexico," in *Ground Failure, No. 3*, Nat. Res. Council on Ground Failure Hazards, 1986, pp. 1-22.

A. I. Johnson, "Glossary," Permeability and Groundwater Contaminant Transport, ASTM STP 746, T. F. Zimmie and C. O. Riggs, Eds., American Society for Testing and Materials, pp. 3-17, 1981.

L. D. Johnson, "Evaluation of Laboratory Suction Tests for Prediction of Heave in Foundation Soils," Tech. Reports S-77-7, U.S. Army Engineer Waterways Experiment Station, Vicksburg, MS, 118 pp., 1977.

L. N. Johnston, "Water Permeable Jacketed Thermal Radiators as Indicators of Field Capacity and Permanent Wilting Percentage in Soils," *Soil Sci.*, vol. 54, pp. 123-126, 1942.

D. E. Jones and W. G. Holtz, "Expansive Soils—The Hidden Disaster," *Civil Eng.*, ASCE, New York, NY pp. 87-89, Aug. 1973.

J. L. Justo and A. Delgado, "Stress-Strain Paths in the Collapse-Swelling-Shrinkage of Soils Insitu," *Proc. of the 6th Int. Conf. on Expansive Soils* (Delhi, India), vol. 2, Dec. 1-4, 1987, pp. 596-601.

J. L. Justo, A. Delgado, and J. Ruiz, "The Influence of Stress-Path in the Collapse-Swelling of Soils at the Laboratory," in *Proc. 5th Int. Conf. Expansive Soils* (Adelaide, Australia), 1984, pp. 67-71.

G. Kassiff, "Compaction and Shear Characteristics of Remoulded Negev Loess," in *Proc. 4th Int. Conf. Soil Mech. Found. Eng.* (London), vol. 1, 1957, pp. 56-61.

R. K. Katti, S. K. Kulkarni, and S. K. Fotedar, "Shear Strength and Swelling Pressure Characteristics of Expansive Soils," in *Proc. 2nd Int. Res. Eng. Conf. Expansive Clay Soils* (College Station, TX), 1969, pp. 334-342.

M. Katzir, "Design Considerations and Performance of Anchored, Cast in Place Retaining Piles in Swelling Clays," in *Proc. 3rd Int. Conf. Expansive Soils* (Haifa, Israel), vol. 2, 1973, pp. 185-188.

G. W. C. Kaye and T. H. Laby, *Tables of Physical and Chemical Constants*, 14th ed. Longman, 1973, 386 pp.

T. C. Kenney and G. H. Watson, "Multiple-Stage Triaxial Test for Determining $c'$ and $\phi'$ of Saturated Soils," in *Proc. 5th Int. Conf. Soil Mech. Found. Eng.* (Paris, France), vol. 1, 1961, pp. 191-195.

A. H. Khan, "Evaluation of Laboratory Suction Tests by Filter Paper Technique for Prediction of Heave in Expansive Soils," Report No. GM-81-001, Dep. of Civil Eng., Univ. of Western Australia, 1981.

M. H. Khan and D. L. Hoag, "A Noncontacting Transducer for Measurement of Lateral Strains," *Can. Geotech. J.*, vol. 16, pp. 409-411, 1979.

W. E. I. Khogali, K. O. Anderson, J.K-M. Gan and D. G. Fredlund, "Installation and Monitoring of Moisture Suction Sensors in a Fine-grained Subgrade Soil Subjected to Seasonal Frost," Presented to the *2nd International Symposium*, "State-of-the-Art of Pavement Response-Monitoring Systems for Roads and Airfields," sponsored by U.S. Army Cold Regions, Research and Engineering Laboratory, Hanover, New Hampshire and Federal Aviation Administration, Washington, D.C., West Levanon, N.H., Sept. 10-13, 1991.

B. A. Kimball, R. D. Jackson, R. J. Reginato, F. S. Nakayama, and S. B. Idso, "Comparison of Field-Measured and Calculated Sol-Heat Fluxes," *Proc. Soil Sci. Soc. Amer.*, vol. 40, no. 1, pp. 18-25, 1976.

B. Kjaernsli and N. Simons, "Stability Investigations of the North Bank of the Drammen River," *Geotechnique*, vol. 12, pp. 147-167, 1962.

Klute, A. "Laboratory Measurement of Hydraulic Conductivity of Unsaturated Soil," in *Methods of Soil Analysis*, C. A. Black, D. D. Evans, J. L. White, L. E. Ensminger, and F. E. Clark, Eds., Mono. 9, Part 1, Amer. Soc. of Agronomy, Madison, WI, 1965, pp. 253-261.

A. Klute, "Water Diffusivity," in *Methods of Soil Analysis*, C. A. Black, D. D. Evans, J. L. White, L. E. Ensminger, and

F. E. Clark, Eds., Mono. 9, Part 1, Amer. Soc. of Agronomy, Madison, WI, 1965, pp. 262–272.

A. Klute, "The Determination of the Hydraulic Conductivity and Diffusivity of Unsaturated Soils," *Soil Sci.*, vol. 113, pp. 264–276, 1972.

P. C. Knodel and C. T. Coffey, "Measurement of Negative Pore Pressure of Unsaturated Soils—Shear and Pore Pressure Research—Earth Research Program," U.S.B.R., Lab. Report No. EM-738, Soil Eng. Branch, Bureau of Reclamation, Denver, CO, 20 pp., June 1966.

P. G. Kohn, "Tables of Some Physical and Chemical Properties of Water," S.E.B. Symp., XIX, *The State and Movement of Water in Living Organisms* (Cambridge), 1965.

A. Komornik, M. Livneh, and S. Smucha, "Shear Strength and Swelling of Clays Under Suction," in *Proc. 4th Int. Conf. Expansive Soils* (CO), vol. 1, 1980, pp. 206–226.

H. L. Koning, "Some Observations on the Modulus of Compressibility of Water," in *Proc. Conf. Settlement and Compressibility of Soils*, Weisbaden, Germany, 1963, pp. 33–36.

P. Koorevaar, G. Menelik, and C. Dirksen, *Elements of Soil Physics*. Amsterdam, The Netherlands: Elsevier, 1983, 228 pp.

J. Krahn, "Comparison of Soil Pore Water Potential Components," M.Sc. thesis, Univ. of Saskatchewan, Saskatoon, Sask., Canada, 125 pp., 1970.

J. Krahn and D. G. Fredlund, "On Total Matric and Osmotic Suction," *J. Soil Sci.*, vol. 114, no. 5, pp. 339–348, 1972.

J. Krahn, D. G. Fredlund, and M. J. Klassen, "Effect of Soil Suction on Slope Stability at Notch Hill," *Can. Geotech. J.*, vol. 26, no. 2, pp. 269–278, 1989.

J. Krahn, D. G. Fredlund, L. Lam and S. L. Barbour, "PC-SEEP: A Finite Element Program for Modeling Seepage," *Proc. of the 1st Canadian Symposium on Microcomputer Applications to Geotechnique* (Regina, Canada), October 22–23, 1987, pp. 243–251.

J. P. Krohn and J. E. Slosson, "Assessment of Expansive Soils in the United States," in *Proc. 4th Int. Conf. Expansive Soils* (Denver, CO), vol. 1, 1980, pp. 596–608.

D. P. Krynine, "Analysis of the Latest American Tests on Soil Capillarity," in *Proc. 2nd Int. Conf. Soil Mech. Found. Eng.* (Rotterdam, The Netherlands), vol. 3, 1948, pp. 100–104.

R. J. Kunze, G. Uehara, and K. Graham, "Factors Important in the Calculation of Hydraulic Conductivity," *Proc. Soil Sci. Soc. Amer.*, vol. 32, pp. 760–765, 1968.

C. C. Ladd, "Mechanisms of Swelling in Compacted Clay," *Highway Res. Board Bull.*, Nat. Research Council, No. 245, pp. 10–26, 1960.

C. C. Ladd, "Settlement Analysis for Cohesive Soils," Research Report R71-2, Soils Publication 272, Department of Civil Engineering, Massachusetts Institute of Technology, 1971, 107 pp.

G. E. Laliberte and A. T. Corey, "Hydraulic Properties of Disturbed and Undisturbed Soils," ASTM, STP, No. 417, 1966.

L. Lam, "Transient Finite Element Seepage Program TRASEE," *User's Manual*, CD-18, Dep. of Civil Eng., Univ. of Saskatchewan, Saskatoon, Sask., Canada, 1984.

L. Lam and D. G. Fredlund, "Saturated-Unsaturated Transient Finite Element Seepage Model for Geotechnical Engineering," *Adv. Water Resources*, vol. 7, pp. 132–136, 1984.

L. Lam, D. G. Fredlund and S. L. Barbour, "Transient Seepage Model for Saturated-Unsaturated Soil Systems: A Geotechnical Engineering Approach," *Can. Geot. Jour.*, vol. 24, no. 4, pp. 565–580, 1988.

T. W. Lambe, *Soil Testing for Engineers*. New York: Wiley, 1951, 165 pp.

T. W. Lambe, "Capillary Phenomena in Cohesionless Soils," *Trans. ASCE*, vol. 116, pp. 401–423, 1951.

T. W. Lambe, "The Engineering Behavior of Compacted Clay," *ASCE J. Soil Mech. Found Div.*, vol. 84, SM2, Paper No. 1655, pp. 1–35, 1958.

T. W. Lambe, "Stress Path Method," *ASCE J. Soil Mech. Found. Eng. Div.*, vol. 93, SM6, pp. 309–331, 1967.

T. W. Lambe and R. V. Whitman, *Soil Mechanics*. New York: Wiley, 1979, 553 pp.

K. S. Lane and S. E. Washburn, "Capillarity Tests by Capillarimeters and by Soil Filled Tubes," *Proc. Highway Res. Board*, vol. 26, pp. 460–473, 1946.

A. R. G. Lang, "Osmotic Coefficients and Water Potentials of Sodium Chloride Solutions from 0 to 40 Degrees C," *Australian J. Chem.*, vol. 20, pp. 2017–2023, 1967.

L. J. Langfelder, C. F. Chen, and J. A. Justice, "Air Permeability of Compacted Cohesive Soils," *ASCE J. Soil Mech. Found. Eng. Div.*, vol. 94, SM4, pp. 981–1001, 1968.

D. Y. Larmour, "One-Dimensional Consolidation of Partially Saturated Clay," M.Sc. thesis, Univ. of Manchester, Manchester, England, 1966.

J. T. K. Lau, "Desiccation Cracking of Soils," M.Sc. thesis, Department of Civil Engineering, University of Saskatchewan, Saskatoon, Sask., Canada, 1987, 208 pp.

A. S. Laughton, "The Compaction of Ocean Sediments," Ph.D. dissertation, Univ. of Cambridge, Cambridge, England, 1955.

B. Leach and R. Herbert, "The Genesis of a Numerical Model for the Study of the Hydrogeology of a Steep Hillside in Hong Kong," *Quart. J. Eng. Geol.*, vol. 15, pp. 253–259, 1982.

H. C. Lee and W. K. Wray, "Evaluation of Soil Suction Instruments," *Proc. 7th Int. Conf. on Expansive Soils* (Dallas, Texas), August 3–5, 1992, vol. 1, pp. 307–312.

K. L. Lee, R. A. Morrison, and S. C. Haley, "A Note on the Pore Pressure Parameter B," in *Proc. 7th Int. Conf. Soil Mech. Found. Eng.*, vol. 1, 1969, pp. 231–238.

R. K. C. Lee, "Measurement of Soil Suction Using the MCS Sensor," M.Sc. thesis, Univ. of Saskatchewan, Saskatoon, Sask., Canada, 162 pp., 1983.

R. K. C. Lee and D. G. Fredlund, "Measurement of Soil Suction Using the MCS 6000 Gauge," in *Proc. 5th Int. Conf. Expansive Soils*, Inst. of Eng., Adelaide, Australia, 1984, pp. 50–54.

K. S. Li, Discussion on "Closed-Form Heave Solutions for Expansive Soils," *ASCE J. Geotech. Eng.*, vol. 115, no. 12, pp. 1819–1823, 1989.

A. C. Liakopoulos, "Theoretical Solution of the Unsteady Unsaturated Flow Problems in Soils," *Bull. Int. Assoc. Sci. Hydrol.*, vol. 10, pp. 5–39, 1965.

A. C. Liakopoulos, "Retention and Distribution of Moisture in Soils after Infiltration has Ceased," *Bull. Int. Assoc. Sci. Hydrol.*, vol. 10, pp. 58–69, 1965.

W. D. L. Finn, "Finite Element Analysis of Seepage Through Dams," *ASCE J. Soil Mech. Found. Eng. Div.*, SM6, pp. 41–48, 1967.

R. A. Lidgren, "Volume Change Characteristics of Compacted Till," M.Sc. thesis, Univ. of Saskatchewan, Saskatoon, Sask., Canada, 158 pp., 1970.

A. L. Little, "The Engineering Classification of Residual Tropical Soils," in *Proc. 7th Int. Conf. Soil Mechanics Found. Eng.* (Mexico City, Mexico), vol. 1, 1969, pp. 1–10.

A. Lloret and E. E. Alonso, "Consolidation of Unsaturated Soils Including Swelling and Collapse Behavior," *Geotechnique*, vol. 30, no. 4, pp. 449–477, 1980.

A. Lloret and E. E. Alonso, "State Surfaces for Partially Saturated Soils," in *Proc. 11th Int. Conf. Soil Mech. Found. Eng.* (San Francisco, CA), vol. 2, 1985, pp. 557–562.

J. Loi, J. Gan, and D. G. Fredlund, "An Evaluation of AGWA-II Sensors in Measuring Soil Suction from Full Scale Testing at the Regina Test Track," Geotech. Group of the Dep. of Civil Eng., Univ. of Saskatchewan, Saskatoon, Sask., Canada, 118 pp., 1989.

J. Loi, D. G. Fredlund, J.K-M. Gan, and R. A. Widger, "Monitoring Soil Suction in an Indoor Test Track Facility," presented to 71st Annual Meeting of the Transportation Board, Session #228 on Environmental Factors Except Frost, Washington, DC, January 12–16, 1992.

J. Lowe, III and T. C. Johnson, "Use of Back Pressure to Increase Degree of Saturation of Triaxial Test Specimens," in *Proc. Res. Conf. Shear Strength of Cohesive Soils*, ASCE, Boulder, CO, 1960, pp. 819–836.

P. Lumb, "The Residual Soils of Hong Kong," *Geotechnique* (Correspondence), vol. 16, no. 4, pp. 359–360, 1966.

A. V. Lykov, *Heat and Mass Transfer in Capillary-Porous Bodies (Teoreticheskiye Omorey Stroitel'noli Teplofiziki)*, translated by P. N. B. Harrison, W. M. Pun, Ed. New York: Pergamon, 1961, pp. 201–277.

R. L. Lytton, "Isothermal Water Movement in Clay Soils," Ph.D. dissertation, University of Texas, Austin, 231 pp., 1967.

R. L. Lytton, "Expansive Soils—Effect of Environmental Changes and Related Engineering Requirements." Session 66, ASCE Annual Meeting, Denver, Colorado, November 1975.

R. L. Lytton, "Foundations in Expansive Soils," in *Numerical Methods in Geotechnical Engineering*, Chapt. 13, C. S. Desai and J. T. Christian (Eds.), New York: McGraw-Hill, 1977, pp. 427–458.

R. L. Lytton, "The Design of Foundations and Pavements on Expansive Clays," *Proc. of the 4th Regional Geotechnical Conference*, Baranquilla, Colombia, November 1981.

R. L. Lytton and J. A. Woodburn, "Design and Performance of Mat Foundations on Expansive Clay," *Proc. of the 3rd Int. Conf. on Expansive Soils*, Haifa, Israel, January, 1973, vol. 1, pp. 301–307.

M. Maksimovic, "Limit Equilibrium for Nonlinear Failure Envelope and Arbitrary Slip Surface," in *Proc. 3rd Int. Conf. Numer. Methods in Geomech.* (Aachen, Germany), Apr. 1979, pp. 769–777.

F. T. Manheim, "A Hydraulic Squeezer for Obtaining Interstitial Water from Consolidated and Unconsolidated Sediment," U.S. Geological Survey Prof. Paper 550-C, pp. 256–261, 1966.

T. J. Marshall, "A Relation Between Permeability and Size Distribution of Pores," *J. Soil Sci.*, vol. 9, pp. 1–8, 1958.

G. E. Mase, *Continuum Mechanics.* New York: McGraw-Hill, 1970.

M. L. Maslia and M. M. Aral, "Evaluation of a Chimney Drain Design in an Earthfill Dam," *Groundwater*, vol. 20, no. 1, pp. 22–31, 1982.

Massachusetts Institute of Technology, "Engineering Behavior of Partially Saturated Soils," Publ. No. 134 presented to U.S. Army Eng., Waterways Experimental Station, Vicksburg, MS, Soil Eng. Dep., Dep. of Civil Eng., M.I.T., Contract No. DA-22-079-eng-288, Boston, MA, 1963.

J. Maswoswe, "Stress Paths for a Compacted Soil During Collapse due to Wetting," Ph.D. dissertation, Imperial College, London, England, 1985.

J. Maswoswe, J. B. Burland, and P. R. Vaughan, "Stress Path Method of Predicting Compacted Soil Collapse," *Proc. of the 7th Int. Conf. on Expansive Soils* (Dallas, Texas), August 3–5, 1992, vol. 1, pp. 61–66.

E. L. Matyas, "Air and Water Permeability of Compacted Soils," in *Permeability and Capillary of Soils*, ASTM STP 417 Amer. Soc. Testing and Materials, 1967, pp. 160–175.

E. L. Matyas and H. S. Radhakrishna, "Volume Change Characteristics of Partially Saturated Soils," *Geotechnique*, vol. 18, no. 4, pp. 432–448, Dec. 1968.

R. G. McKeen, "Suction Studies: Filter Paper Method," Design of Airport Pavements for Expansive Soils: Final Report (No. DOT/FAA/RD-81/25), U.S. Dep. of Transportation, Federal Aviation Administration, Systems Research and Development Service, Washington, DC, 1981.

R. G. McKeen, "Validation of Procedures for Pavement Design on Expansive Soils," Final Report, U.S. Dep. of Transportation, Washington, DC, 1985.

I. S. McQueen and R. F. Miller, "Calibration and Evaluation of a Wide Range Method of Measuring Moisture Stress," *J. Soil Sci.*, vol. 106, no. 3, pp. 225–231, 1968.

D. B. McWhorter and J. D. Nelson, "Unsaturated Flow Beneath Tailings Impoundments," *ASCE J. Geotech. Eng. Div.*, vol. 105, pp. 1317–1334, 1979.

R. O. Meeuwig, "A Low-Cost Thermocouple Psychrometer Recording System," in *Psychrometry in Water Relations Res., Proc. Symp. Thermocouple Psychrometers*, R. W. Brown and B. P. Van Haveren, Eds., published by Utah Agricultural Experiment Station, Utah State Univ., Logan, 1972.

P. Meigs, "World Distribution of Arid and Semi-Arid Homoclimates," in *UNESCO, Reviews of Res. on Arid Zone Hydrology, Arid Zone Res.*, 1953, pp. 203–210.

H. A. C. Meintjes, "Suction-Load-Strain Relations in Expansive Soil," *Proc. of the 7th Int. Conf. on Expansive Soils* (Dallas, Texas), August 3–5, 1992, pp. 51–54.

W. C. Michels, editor-in-chief, *The International Dictionary of Physics and Electronics*, 2nd ed. Toronto: Van Nostrand, 1961.

B. W. Mickleborough, "An Experimental Study of the Effects of Freezing on Clay Subgrades," M.Sc. thesis, Univ. of Saskatchewan, Saskatoon, Sask., Canada, 1970.

R. F. Miller and I. S. McQueen, "Moisture Relations in Rangelands, Western United States," in *Proc. 1st Int. Rangeland Congress*, 1978, pp. 318–321.

R. J. Millington and J. P. Quirk, "Permeability of Porous Media," *Nature*, vol. 183, pp. 387–388, 1959.

R. J. Millington and J. R. Quirk, "Permeability of Porous Solids," *Trans. Faraday Soc.*, vol. 57, pp. 1200–1207, 1961.

L. J. Milne and M. Milne, "Insects of the Water Surface," *Sci. Amer.*, pp. 134–142, Apr. 1978.

J. K. Mitchell, *Fundamentals of Soil Behavior*. New York: John Wiley and Sons, 1976, 422 pp.

P. W. Mitchell, "A Simple Method of Design of Shallow Footings on Expansive Soil," *Proc. of the Fifth Conf. on Expansive Soils*, National Conference Publication No. 84/3 (Adelaide, Australia), May 21–23, 1984, pp. 159–164.

J. K. Mitchell and N. Sitar, "Engineering Properties of Tropical Residual Soils," in *Proc. ASCE Geotech. Eng. Div. Specialty Conf.: Engineering and Construction in Tropical and Residual Soils*, Honolulu, HI, 1982, pp. 30–57.

P. W. Mitchell and D. L. Avalle, "A Technique to Predict Expansive Soils Movements," in *Proc. 5th Int. Conf. Expansive Soils*, Adelaide, Australia, 1984, pp. 124–130.

R. E. Moore, "Water Conduction From Shallow Water Tables," *Hilgardia*, vol. 12, no. 6, pp. 383–426, 1939.

O. Moretto, A. J. L. Bolonesi, A. O. Lopez, and E. Nunez, "Propiedades Y Comportamiento De Un Suelo Limoso De Baja Plasticidad," in *Proc. 2nd Panamerican Conf. Soil Mech. Found. Eng.* (Brazil), vol. 2, 1963, pp. 131–146.

N. R. Morgenstern, "Properties of Compacted Soils," Contribution to Panel Discussion, Session IV, *Proc. of the 6th Panamerican Conf. Soil Mech. Found. Eng.* (Lima, Peru), vol. 3, December 1979, pp. 349–354.

N. R. Morgenstern and Z. Eisenstein, "Methods of Estimating Lateral Loads and Deformations," in *Proc. 1970 Specialty Conf. Lateral Stresses in the Ground and Design of Earth Retaining Structures*, ASCE, New York, NY, 1970, pp. 51–102.

N. R. Morgenstern and V. E. Price, "The Analysis of the Stability of General Slip Surfaces," *Geotechnique*, vol. 15, pp. 79–93, 1965.

N. R. Morgenstern and V. E. Price, "A Numerical Method for Solving the Equations of Stability of General Slip Surfaces," *Computer J.*, vol. 9, pp. 388–393, 1965.

N. R. Morgenstern and V. E. Price, "A Numerical Method for Solving the Equation of General Slip Surfaces," *British Computer J.*, vol. 9, pp. 338–393, 1967.

R. D. Morrison, "Ground Water Monitoring Technology: Procedures, Equipment, and Applications," published by Timco Mfg., Inc., Prairie Du Sac, WI, 111 pp., 1983.

J. Morrison and D. G. Fredlund, Discussion: "Emploi d'une methode psychromitrique dans des assais triaxiaux sur un limon remanii non saturi," *Can. Geot. Journ.*, vol. 16, no. 3, pp. 614–615, 1979.

C. H. Mou and T. Y. Chu, "Soil Suction Approach for Swelling Potential Evaluation," *Transportation Res. Rec. 790*, Washington, DC, pp. 54–60, Jan. 1981.

R. R. Nachlinger and R. L. Lytton, "Continuum Theory of Moisture Movement and Swell in Expansive Clay," University of Texas Center for Highway Research, Report 118-2, 1969.

N. K. Nagpal and L. Boersma, "Air Entrapment as a Possible Source of Error in the Use of a Cylindrical Heat Probe," *Proc. Soil Sci. Soc. Amer.*, vol. 37, pp. 828–832, 1973.

T. N. Narasimhan and P. A. Witherspoon, "Numerical Model for Saturated-Unsaturated Flow in Deformable Porous Media, 1. Theory," *Water Resources Res.*, vol. 13, no. 3, pp. 657–664, 1977.

NAVFAC, *Soil Mechanics Foundation and Earth Structures, Design Manual*, NAVFACDM-7, Naval Facilities Engineering Command, Department of the Navy, Bureau of Yards and Docks, Washington, DC, 1971.

J. D. Nelson and D. J. Miller, "Expansive Soils," Wiley and Sons, New York, NY, 295p, 1992.

S. P. Neuman, "Finite Element Computer Programs for Flow in Saturated-Unsaturated Porous Media," Hydro. and Hydraulic Lab., Project No. A10-SWC-77, Technion, Haifa, Israel, 1972.

S. P. Neuman, "Saturated-Unsaturated Seepage by Finite Elements," *Proc. ASCE, Hydraulics Div.*, vol. 99, no. 12, pp. 2233–2250, 1973.

S. P. Neuman, *Galerkin Approach to Saturated-Unsaturated Flow in Porous Media Finite Elements in Fluids*. New York: Wiley, 1975, pp. 201–217.

S. P. Neumann and P. A. Witherspoon, "Analysis of Nonsteady Flow with a Free Surface using the Finite Element Method," *Water Resources Res.*, vol. 7, no. 33, pp. 611–623, 1971.

E. M. D. Neves, "Influencis das Tensoes Negativas Nas Caracteristicas Estruturais Dos Solos Compactados," Ministerio das Ofras Publicas, Laboratorio Nacional de Engenharia Civil, Memo. No. 386, 1971.

T. N. Ng, "The Effect of Negative Pore-Water Pressures on Slope Stability Analysis," M.Sc. thesis, Univ. of Saskatchewan, Saskatoon, Sask., Canada, 245 pp., 1988.

B. Nicolae, S. Tudor, and S. Anghel, "A New Method to Estimate the Behavior of Expansive Clay," *Proc. of the 6th Int. Conf. on Expansive Soils* (Delhi, India), Dec. 1–4, 1987, vol. 2, pp. 591–595.

D. R. Nielsen and Y. W. Biggar, "Measuring Capillary Conductivity," *Soil Sci.*, vol. 92, pp. 192–193, 1961.

D. R. Nielsen, J. M. Davidson, J. W. Biggar, and R. J. Miller, "Water Movement Through Panoche Clay Loam Soil," *Hilgardia*, vol. 35, pp. 491–506, 1964.

D. R. Nielsen, R. D. Jackson, J. W. Cary, and D. D. Evans, "Soil Water," Amer. Soc. Agronomy and Soil Sci. Amer., Madison, WI, 1972.

D. R. Nielsen, J. W. Biggar, and K. T. Erb, "Spatial Variability of Field-Measured Soil-Water Properties," *Hilgardia*, vol. 42, no. 7, pp. 215–359, 1973.

C. A. Noble, "Swelling Measurements and Prediction of Heave for Lacustrine Clay," *Can. Geotech. J.*, vol. 3, no. 1, pp. 32–41, 1966.

D. Northey and F. Thomas, "Consolidation Test Pore Pressure," in *Proc. 6th Int. Conf. Soil Mech. Found. Eng.* (Montreal, Canada), vol. 1, 1965, pp. 323–327.

G. Ogata and L. A. Richards, "Water Content Changes Following Irrigation of Bare-Field Soil That is Protected from Evaporation," *Proc. Soil Sci. Soc. Amer.*, vol. 21, pp. 355–356, 1957.

R. E. Olson, "Effective Stress Theory of Compaction," *Proc. Amer. Soc. Civil Eng.*, vol. 89, pp. 27–44, 1963.

R. E. Olson, "State-of-the-Art: Consolidation Testing," in *Consolidation of Soils; Testing and Evaluation*, ASTM, Philadelphia, PA, 1986, pp. 7–70.

R. E. Olson and D. E. Daniel, "Measurement of the Hydraulic

Conductivity of Fine-Grained Soils," in *Permeability and Groundwater Contaminant Transport*, ASTM Special Tech. Publ. 746, T. F. Zimmie and C. O. Riggs, 1981, pp. 18–64.

R. E. Olson and L. J. Langfelder, "Pore-Water Pressures in Unsaturated Soils," *J. Soil Mech. Found. Div., Proc. Amer. Soc. Civil Eng.*, vol. 91, SM4, pp. 127–160, 1965.

N. A. Ostashev, "The Law of Distribution of Moisture in Soils and Methods for the Study of Same," in *Proc. 1st Int. Conf. Soil Mech. Found. Eng.*, vol. 1, 1936, pp. 227–228.

A. T. Papagiannakis, "A Steady State Model for Flow in Saturated-Unsaturated Soils," M.Sc. thesis, Univ. of Saskatchewan, Saskatoon, Sask., Canada, 196 pp., 1982.

A. T. Papagiannakis and D. G. Fredlund, "A Steady State Model for Flow in Saturated-Unsaturated Soils," *Can. Geotech. J.*, vol. 21, no. 13, pp. 419–430, 1984.

A. T. Papagiannakis, L. Lam, and D. G. Fredlund, "Steady State and Transient Mass Transport Models for Saturated-Unsaturated Soils," *Proc. of the 1st Int. Potash Tech. Conf.*, Oct. 3–5, 1983, pp. 741–747.

Papers and Discussion from the First Soil Mechanics Conference, "Theoretical and Practical Treatment of Expansive Soils," Colorado School of Mines, Golden, CO, published in *Quart. Colorado School of Mines*, vol. 54, no. 4, 1959.

A. J. Peck and R. M. Rabbidge, "Direct Measurement of Moisture Potential: A New Technique," in *Proc. UNESCO-Neth. Gov. Symp. Water in the Unsaturated Zone* (Paris, France), 1966, pp. 165–170.

A. J. Peck and R. M. Rabbidge, "Design and Performance of an Osmotic Tensiometer for Measuring Capillary Potential," *Proc. Soil Sci. Soc. Amer.*, vol. 33, no. 2, pp. 196–202, Mar.–Apr. 1969.

H. L. Penman, "Natural Evapotranspiration from Open Water, Bare Soil and Grass," *Proc. Roy. Soc., London*, ser. A. no. 193, pp. 120–145, 1948.

W. H. Perloff, K. Nair, and J. G. Smith, "Effect of Measuring System on Pore Water Pressures in the Consolidation Test," in *Proc. 6th Int. Conf. Soil Mech. Found. Eng.* (Montreal, Canada), vol. 1, 1965, pp. 338–341.

C. J. Phene, G. J. Hoffman, and S. L. Rawlins, "Measuring Soil Matric Potential in Situ by Sensing Heat Dissipation with a Porous Body: Theory and Sensor Construction," *Proc. Soil Sci. Soc. Amer.*, vol. 35, pp. 27–32, 1971.

C. J. Phene, D. A. Clark, G. E. Cardon, and R. M. Mead, "The Soil Matric Potential Sensor Research and Applications," published in *Advances in Measurement of Soil Physical Properties: Bringing Theory into Practice*, SSSA Special Publication No. 30, Soil Science Society of America, Madison WI, 1992, pp. 263–280.

J. R. Philip and D. A. de Vries, "Moisture Movement in Porous Materials under Temperature Gradients," *Trans. Amer. Geophys. Union*, vol. 38, no. 2, pp. 222–232, 1957.

M. Picornell and R. L. Lytton, "Detection and Sizing of Surface Cracks in Expansive Soil Deposits," presented at the Transportation Research Board Annual Meeting, Washington, DC, January, 1988.

M. Picornell and R. L. Lytton, "Modelling the Heave of a Heavily Loaded Foundation," *5th Int. Conf. on Expansive Soils* (Adelaide, Australia), May, 1984, pp. 104–108.

M. Picornell, R. L. Lytton, and M. Steinberg "Matrix Suction Instrumentation of a Vertical Moisture Barrier," *J. Transportation*, ASCE, 1983.

M. Picornell, R. L. Lytton and M. L. Steinberg, "Assessment of the Effectiveness of the Effectiveness of a Moisture Barrier," *5th Int. Conf. on Expansive Soils* (Adelaide, Australia), May, 1984, pp. 354–358.

M. Popescu, "Behavior of Expansive Soils with a Crumb Structure," in *Proc. ASCE 4th Int. Conf. Expansive Soils* (Denver, CO), vol. 1, 1980, pp. 158–171.

S. J. Poulos, "Control of Leakage in the Triaxial Test," Soil Mech. Series No. 71, Massachusetts Institute of Technology, Cambridge, MA, 1964.

L. Prandtl, "On the Penetrating Strength (Hardness) of Plastic Construction Materials and the Strength of Cutting Edges," *Zeit. Agnew. Mathematical Mech.*, vol. 1, no. 1, pp. 15–20, 1921.

D. E. Pufahl, "Evaluation of Effective Stress Components in Non-Saturated Soils," M.Sc. thesis, Univ. of Saskatchewan, Saskatoon, Sask., Canada, 88 pp., 1970.

D. E. Pufahl and R. L. Lytton, "Temperature and Suction Profiles Beneath Highway Pavements: Computed and Measured," Transportation Research Record No. 1307, pp. 268–276, 1991.

D. E. Pufahl, D. G. Fredlund, and H. Rahardjo, "Lateral Earth Pressures in Expansive Clay Soils," *Can. Geotech. J.*, vol. 20, no. 2, pp. 228–241, 1983.

D. E. Pufahl, R. L. Lytton, and H. S. Liang, "An Integrated Computer Model to Estimate Moisture and Temperature Effects Beneath Pavements," Transportation Research Record No. 1286, pp. 259–269, 1990.

H. Rahardjo, "The Study of Undrained and Drained Behavior of Saturated Soils," Ph.D. dissertation, Univ. of Saskatchewan, Saskatoon, Sask., Canada, 385 pp. 1990.

H. Rahardjo and D. G. Fredlund, "Calculation Procedures for Slope Stability Analyses Involving Negative Pore-Water Pressures," in *Proc. Int. Conf. Slope Stability Eng., Development, Applications* (Isle of Wight, U.K.), 1991.

H. Rahardjo and D. G. Fredlund, "Mechanics of Soil with Matric Suction," in *Proc. Int. Conf. Geotech. Eng. (GEOTROPIKA'92)*, Universiti Teknologi Malaysia, Johor Bahru, Malaysia, 1992.

H. Rahardjo and D. G. Fredlund, "Stress Paths for Shear Strength Testing of Unsaturated Soils," *Proc. of the 11th S.E.A.G.C.* (Singapore), May 4–8, 1993.

H. Rahardjo, D. G. Fredlund, and S. K. Vanapalli, "Use of Linear and Non-Linear Shear Strength versus Matric Suction Relations in Slope Stability Analysis," *Proc. of the 6th Int. Symp. on Landslides* (Christchurch, New Zealand), February 10–14, 1992, pp. 531–537.

H. Rahardjo, J. Loi, and D. G. Fredlund, "Typical Matric Suction Measurements in the Laboratory and in the Field Using Thermal Conductivity Sensors," in *Proc. Indian Geotech. Conf. Geotechniques of Problematic Soils and Rocks: Characterisation, Design, and Ground Improvements* (Sarita Prakashan, Meerut), vol. 1, 1989, pp. 127–133.

H. Rahardjo, D. Y. F. Ho, and D. G. Fredlund, "Testing Procedures for Obtaining Volume Change Indices During Loading of an Unsaturated Soil," in *Proc. 1990 CSCE Annual Conf.* (Hamilton, Canada), vol. II-2, pp. 558–573, May, 1990.

R. R. Rao, H. Rahardjo, and D. G. Fredlund, Closure on "Closed-Form Heave Solutions for Expansive Soils," for Discussion by K. S. Li, Binnie and Partners, Hong Kong, ASCE *Geot. Journ.*, vol. 115, no. 12, pp. 1822–1833, 1989.

R. R. Rao, H. Rahardjo, and D. G. Fredlund, "Closed-Form Heave Solution for Expansive Soils," *ASCE J. Geotech. Eng.*, vol. 114, no. 5, pp. 573–588, May 1988.

M. Reeves, and J. O. Duguid, "Water Movement through Saturated-Unsaturated Porous Media: A Finite Element Galerkin Model," U.S. Dep. of Commerce, Nat. Tech. Information Service, Oak Ridge Nat. Laboratory, TN, 1975.

L. Rendulic, "Relation Between Void Ratio and Effective Principal Stresses for a Remoulded Silty Clay," in *Proc. 1st Int. Conf. Soil Mech. Found. Eng.* (Cambridge, MA), vol. 3, 1936, pp. 48–51.

B. G. Richards, "Measurement of the Free Energy of Soil Moisture by the Psychrometric Technique Using Thermistors," in *Moisture Equilibria and Moisture Changes in Soils Beneath Covered Areas, A Symp. in Print.* Australia: Butterworths, 1965, pp. 39–46.

B. G. Richards, "The Significance of Moisture Flow and Equilibria in Unsaturated Soils in Relation to the Design of Engineering Structures Built on Shallow Foundations in Australia," presented at the Symp. on Permeability and Capillary, Amer. Soc. Testing Materials, Atlantic City, NJ, 1966.

B. G. Richards, "A Review of Methods for the Determination of the Moisture Flow Properties of Unsaturated Soils," Tech. Memo. No. 5, Commonwealth Sci. and Industrial Res. Organization, Australia, 1967.

B. G. Richards, "Behavior of Unsaturated Soils," in *Soil Mechanics—New Horizons*, I. K. Lee, Ed. New York: American Elsevier, 1974, pp. 112–157.

B. G. Richards, P. Peter, and R. Martin, "The Determination of Volume Change Properties in Expansive Soils," in *Proc. 5th Int. Conf. Expansive Soils* (Adelaide, Australia), 1984, pp. 179–186.

L. A. Richards, "The Usefulness of Capillary Potential to Soil Moisture and Plant Investigators," *J. Agric. Res.*, vol. 37, pp. 719–742, 1928.

L. A. Richards, "Capillary Conduction of Liquids Through Porous Medium," *J. Physics*, vol. 1, pp. 318–333, 1931.

L. A. Richards and M. Fireman, "Pressure-Plate Apparatus for Measuring Moisture Sorption and Transmission by Soils," *Soil Sci.*, vol. 56, pp. 395–404, 1943.

L. A. Richards and G. Ogata, "A Thermocouple for Vapour Pressure Measurement in Biological and Soil Systems at High Humidity," *Science*, vol. 128, pp. 1089–1090, 1958.

L. A. Richards and B. D. Wilson, "Capillary Conductivity Measurements in Peat Soils," *J. Amer. Soc. Agronomy*, vol. 28, pp. 427–436, 1936.

L. A. Richards, W. R. Gardner, and G. Ogata, "Physical Processes Determining Water Loss from Soil," *Proc. Soil Sci. Soc. Amer.*, vol. 20, pp. 310–314, 1956.

H. H. Rieke and G. V. Chilinicarian, "Compaction of Argillaceous Sediments," in *Developments in Sedimentology, No. 16.* Amsterdam: Elsevier, 1974, pp. 31–86.

W. H. Rodebush and A. M. Buswell, "Properties of Water Substances," Highway Res. Board Special Report No. 40, 1958.

C. W. Rose and A. Krishnan, "A Method of Determining Hydraulic Conductivity Characteristics for Non-Swelling Soils Insitu, and of Calculating Evaporation from Bare Soil," *Soil Sci.*, vol. 103, pp. 369–373, 1967.

C. W. Rose, W. R. Stern, and J. Drummond, "Determination of Hydraulic Conductivity as a Function of Depth and Water Content for Soil Insitu," *Australian J. Soil Res.*, vol. 3, pp. 1–9, 1965.

J. Rubin, "Theoretical Analysis of Two-Dimensional Transient Flow of Water in Unsaturated and Partly Unsaturated Soils," *Proc. Soil Sci. Soc. Amer.*, vol. 32, pp. 607–615, 1968.

E. C. Ruddock, Discussion on "The Residual Soils of Hong Kong," *Geotechnique*, vol. 16, no. 1, pp. 78–81, 1966.

J. J. Rulon and R. A. Freeze, "Multiple Seepage Faces on Layered Slopes and Their Implications for Slope Stability Analysis," *Can. Geotech. J.*, vol. 22, pp. 347–356, 1985.

K. Russam and C. D. Coleman, "The Effect of Climatic Factors on Subgrade Moisture Conditions," *Geotechnique*, vol. XI, no. 1, pp. 22–28, 1961.

J. A. J. Salas and M. Roy, "Preliminary Evaluation of the Swelling Danger—Overview on the Spanish Practice," *Proc. of the 6th Int. Conf. on Expansive Soils* (Delhi, India), Dec. 1–4, 1987, vol. 1, pp. 217–224.

J. A. J. Salas and J. M. Serratosa, "Foundations on Swelling Clays," in *Proc. 4th Int. Conf. Soil Mech. Found. Eng.* (London, England), vol. I, 1957, pp. 424–428.

E. Sampson, R. L. Schuster, and W. D. Budge, "A Method of Determining Swell Potential of an Expansive Clay," in *Proc. Int. Res. Eng. Conf. Expansive Clays* (College Station, TX), 1965, pp. 255–275.

B. S. Satija, "Shear Behavior of Partly Saturated Soils," Ph.D. dissertation, Indian Inst. of Technol., Delhi, 327 pp., 1978.

B. S. Satija and S. K. Gulhati, "Strain Rate for Shearing Testing of Unsaturated Soil," in *Proc. 6th Asian Reg. Conf. Soil Mech. Found. Eng.* (Singapore), 1979, pp. 83–86.

P. J. Sattler, "Numerical Modelling of Vertical Ground Movements," M.Sc. thesis, Univ. of Saskatchewan, Saskatoon, Sask., Canada, 425 pp., 1989.

P. J. Sattler and D. G. Fredlund, "Use of Thermal Conductivity Sensors to Measure Matric Suction in the Laboratory," *Can. Geotech. J.*, vol. 26, no. 3, pp. 491–498, 1989.

P. J. Sattler and D. G. Fredlund, "Modeling Vertical Ground Movements Using Surface Climatic Flux," ASCE, *Geot. Eng. Congress* (Boulder, Colorado), June 9–12, 1991.

P. J. Sattler and D. G. Fredlund, "Numerical Modeling of Vertical Ground Movements in Expansive Soils," *Can. Geot. Journ.*, vol. 28, no. 2, pp. 189–199, 1991.

P. J. Sattler, D. G. Fredlund, L. W. Lam, A. W. Clifton, and M. J. Klassen, "Implementation of a Bearing Capacity Design Procedure for Railway Subgrades: A Case Study," Transportation Res. Board, Washington, DC, 1990.

E. K. Sauer, "Application of Geotechnical Principles to Road Design Problems," Dr. Eng. dissertation, Univ. of California, Berkeley, 1967.

J. M. Schmertmann, "The Undisturbed Consolidation Behavior of Clay," *ASCE Trans.*, vol. 120, pp. 1201–1227, 1955.

R. Schofield, "The pF of the Water in Soil," *Trans. 3rd Int. Congress Soil Sci.*, vol. 2, pp. 37–48, 1935.

E. Schuurman, "The Compressibility of an Air/Water Mixture

and a Theoretical Relation Between the Air and Water Pressures,'' *Geotechnique*, vol. 16, no. 4, pp. 269–281, 1966.

R. F. Scott, ''Principles of Soil Mechanics,'' Addison-Wesley Publishing Company, Inc., Reading, MA, 1963.

H. B. Seed, ''A Modern Approach to Soil Compaction,'' *Proc. 11th California Street and Highway Conference*, Reprint No. 69, The Institute of Transportation and Traffic Engineering, University of California, 1959, 93 pp.

H. B. Seed, R. J. Woodward, and R. Lundgren, ''Prediction of Swelling Potential for Compacted Clays,'' *Journ. of the Soil Mech. Found. Div.*, ASCE, vol. 88, no. SM4, pp. 57–59, 1962.

L. G. Segerlind, *Applied Finite Element Analysis*, 2nd ed. New York: Wiley, 1984.

E. Seker, ''The Study of Deformation for Unsaturated Soils,'' Ph.D. dissertation, Ecole Polytecnique Federale, Lausanne, Switzerland, 224 pp., 1983.

R. A. Sevaldson, ''The Slide at Lodalen, Oct. 6th, 1954,'' *Geotechnique*, vol. 6, pp. 167–182, 1956.

B. Shaw and L. D. Baver, ''An Electrothermal Method for Following Moisture Changes of the Soil Insitu,'' *Proc. Soil Sci. Soc. Amer.*, vol. 4, pp. 78–83, 1939.

N. Simons, ''Discussion: Test Requirements for Measuring the Coefficient of Earth Pressure at Rest,'' in *Proc. Brussels Conf. Earth Pressure Problems*, vol. 3, 1958, pp. 50–53.

H. H. Sisler, C. A. Vanderwerf, and A. W. Davidson, *General Chemistry—A Systematic Approach*. New York: Macmillan, 1953.

M. Sitz, ''Discussion on Terzaghi's Ideas on: Surface Tension of Water and the Rise of Water in Capillaries,'' in *Proc. 2nd Int. Conf. Soil Mech. Found. Eng.* (Rotterdam, The Netherlands), vol. 5, 1948, 289–292.

W. Skaling, ''TRASE: A Product History,'' published in Adv. Meas. Soil Phy. Prop.: Bringing Theory into Practice, SSSA Special Publication No. 30, Soil Science Society of America, Madison, WI., 1992, pp. 169–186.

A. W. Skempton, ''The Bearing Capacity of Clays,'' Building Res. Congress, England, 1951.

A. W. Skempton, ''The Pore Pressure Coefficients, A and B,'' *Geotechnique*, vol. 4, no. 4, pp. 143–147, 1954.

A. W. Skempton, ''Effective Stress in Soils, Concrete and Rocks,'' in *Proc. Conf. Pore Pressure*. London: Butterworths, 1961, pp. 4–16.

A. W. Skempton and A. W. Bishop, ''Building Materials, Their Elasticity and Inelasticity,'' in *Soils*. Amsterdam: North-Holland, 1954, ch. 10.

A. W. Skempton and J. Hutchinson, ''Stability of Natural Slopes and Embankment Foundations,'' *State-of-the-Art Report, 7th Int. Conf. Soil Mech. Found. Eng.* (Mexico City, Mexico), State-of-the-Art Vol., 1969, pp. 291–340.

A. W. Smith, ''Method for Determining the Potential Vertical Rise, PVR,'' Texas Test Method Tex-126-E, *Proc. Workshop Expansive Clays and Shales in Highway Design and Construction* (Univ. of Wyoming, Laramie), vol. 1, 1973, pp. 189–205.

G. N. Smith, *Elements of Soil Mechanics for Civil and Mining Engineers*, 4th ed. London: Granada Publ. Ltd., 1978, 424 pp.

D. R. Snethen, ''Expansive Soils: Where Are We?,'' in *Ground Failure, No. 3*, Nat. Res. Council Comm. on Ground Failure Hazards, 1986, pp. 12–16.

Soilmoisture Equipment Corp., ''Commercial Publications,'' P.O. Box 30025, Santa Barbara, CA, 1985.

D. C. Spanner, ''The Peltier Effect and Its Use in the Measurement of Suction Pressure,'' *J. Exp. Bot.*, vol. 11, pp. 145–168, 1951.

E. Spencer, ''A Method of Analysis of the Stability of Enbankments Assuming Parallel Interslice Forces,'' *Geotechnique*, vol. 17, pp. 11–26, 1967.

E. Spencer, ''Effect of Tension on Stability on Embankments,'' *ASCE J. Soil Mech. Found. Eng. Div.*, vol. 94, SM5, pp. 1159–1173, 1968.

E. Spencer, ''Thrust Line Criterion in Embankment Stability Analysis,'' *Geotechnique*, vol. 23, pp. 85–100, 1973.

D. I. Stannard, ''Tensiometers—Theory, Construction, and Use,'' in *Groundwater and Vadose Zone Monitoring*, ASTM STP 1053, D. M. Neilson and A. I. Johnson, Eds., ASTM, Philadelphia, PA, 1990, pp. 34–51.

V. L. Streeter and B. Wylie, *Fluid Mechanics*, 6th ed. New York: McGraw-Hill, 1975, 717 pp.

R. A. Sullivan and B. McClelland, ''Predicting Heave of Buildings on Unsaturated Clays,'' in *Proc. 2nd Int. Res. Eng. Conf. Expansive Clay Soils*. College Station, TX: Texas A&M Press, 1969, pp. 404–420.

D. J. Sweeney, ''Some Insitu Soil Suction Measurements in Hong Kong's Residual Soil Slopes,'' in *Proc. 7th Southeast Asia Geotech. Conf.* (Hong Kong), vol. 1, 1982, pp. 91–106.

D. J. Sweeney and P. Robertson, ''A Fundamental Approach to Slope Stability in Hong Kong,'' *Hong Kong Eng.*, pp. 35–44, 1979.

B. J. Szafron and D. G. Fredlund, ''Monitoring Matric Suction in the Subgrade of Gravel Roads,'' presented to the 45th *Can. Geot. Conf.* (Toronto, Canada), October 26–28, 1992, pp. 52.1–52.10.

R. Tadepalli, ''Study of Collapse Behavior During Inundation,'' M.Sc. thesis, Univ. of Saskatchewan, Saskatoon, Sask., Canada, 276 pp., 1990.

R. Tadepalli and D. G. Fredlund, ''The Collapse Behavior of a Compacted Soil During Inundation,'' *Can. Geot. Jour.*, vol. 28, pp. 477–488, 1991.

R. Tadepalli, D. G. Fredlund, and H. Rahardjo, ''Soil Collapse and Matric Suction Change,'' *Proc. of the 7th Int. Conf. on Expansive Soils* (Dallas, Texas), August 3–5, 1992, vol. 1, pp. 286–291.

R. Tadepalli, H. Rahardjo, and D. G. Fredlund, ''Measurements of Matric Suction and Volume Changes During Inundation of Collapsible Soils,'' *Geotech. Test. J.*, ASTM, GTJODJ, Vol. 15, No. 2, pp. 115–122, June, 1992.

R. K. W. Tang, ''Measurement of Soil Suction by Filter Paper Method, Phases I–V,'' Laboratory Reports, Dep. of Civil Eng., Univ. of Saskatchewan, Saskatoon, Sask., Canada, 1978.

D. W. Taylor, *Fundamentals of Soil Mechanics*. New York: Wiley, 1948, 700 pp.

S. A. Taylor and G. L. Ashcroft, *Physical Edaphology*. San Francisco, CA: 1972, 533 pp.

R. L. Taylor and C. B. Brown, ''Darcy Flow Solutions with a

Free Surface," *ASCE Hydraulics Div.*, vol. 93, no. HY2, pp. 25–33, Mar. 1967.

G. S. Taylor and A. Luthin, "Computer Methods for Transient Analysis of Water-Table Aquifers," *Water Resources Res.*, vol. 5, no. 1, 1969.

T. C. P. Teng and M. B. Clisby, "Experimental Work for Active Clays in Mississippi," *Transportation Eng. J., ASCE*, vol. 101, no. TE1, Proc. Paper No. 11105(6798), pp. 77–95, 1975.

T. C. P. Teng, R. M. Mattox, and M. B. Clisby, "A Study of Active Clays as Related to Highway Design," Res. Report 72-045, Mississippi State Highway Dep., Jackson, 134 pp., 1972.

T. C. P. Teng, R. M. Mattox, and M. B. Clisby, "Mississippi Experimental Work on Active Clays," in *Proc Workshop Expansive Clays and Shales in Highway Design and Construction*, D. R. Lamb and S. J. Hanna, Eds., vol. 2, 1973, pp. 1–27.

K. Terzaghi, *Erdbaumechanik* (in German). Vienna: Franz Deuticke, 1925, 399 pp.

K. Terzaghi, "Der Spannungszustand im Porenwasser Trocknender Betonkoerper (Pressure due to Capillarity in Drying Concrete Bodies)," *Bauingenieur*, vol. 15, pp. 303–306, 1934.

K. Terzaghi, "The Shear Resistance of Saturated Soils," in *Proc. 1st Int. Conf. Soil Mech. Found. Eng.* (Cambridge, MA), vol. 1, 1936, pp. 54–56.

K. Terzaghi, *Theoretical Soil Mechanics*. New York: Wiley, 1943, 510 pp.

K. Terzaghi, "Anchored Bulkheads," *Trans. ASCE*, vol. 119, pp. 1234–1280, 1954.

K. Terzaghi and R. B. Peck, *Soil Mechanics in Engineering Practice*. New York: Wiley, 1967, 729 pp.

The Commission of Inquiry, *Rainstorm Disasters 1972*, published by the Government Printer, Hong Kong, 1972.

C. W. Thornthwaite, "An Approach Toward a Rational Classification of Climate," *Geographical Rev.*, vol. 38, pp. 55–94, 1948.

G. R. Tomlin, "Seepage Analysis through Zoned Anisotropic Soils by Computer," *Geotechnique*, vol. 16, no. 3, pp. 220–230, 1966.

G. C. Topp and E. E. Miller, "Hysteretic Moisture Characteristics and Hydraulic Conductivities for Glass-Bead Media," *Proc. Soil Sci. Soc. Amer.*, vol. 30, pp. 156–162, 1966.

C. Truesdell, *The Elements of Continuum Mechanics*. New York: Springer-Verlag, 1966.

J. J. Tuma, *Handbook of Physical Calculations*. New York: McGraw-Hill, 1976, 370 pp.

R. E. Tuncer and R. A. Lohnes, "An Engineering Classification of Certain Basalt Derived Lateritic Soils," *Eng. Geology*, vol. 11, 1977.

U.S.D.A. Agricultural Handbook No. 60, *Diagnosis and Improvement of Saline and Alkali Soils, 1950*.

U.S. Department of Agriculture, U.S.D.A., Soil Conservation Service, *Guide for Interpreting Engineering Uses of Soils*, Washington, DC, 1971, 87 pp.

U.S. Department of Interior, *Earth Manual*, Bureau of Reclamation, Denver, CO, 1974.

U.S. Department of Interior Bureau of Reclamation, U.S.B.R., "Measurement of Negative Pore Pressure of Unsaturated Soil Shear and Pore Pressure Research Earth Research Program," Report No. EM-738, Soils Eng. Branch, Div. of Res., Denver, CO, 1966.

U.S. Research Council, *The International Critical Tables*. New York: McGraw-Hill, 1933, vol. 1, no. 7.

R. Valle-Rodas, "Capillarity in Sands," in *Proc. Highway Res. Board*, vol. 24, pp. 389–396, 1944.

G. J. van Amerongen, "Permeability of Different Rubbers to Gases and Its Relation to Diffusivity and Solubility," *J. Appl. Phys.*, vol. 17, no. 11, pp. 972–985, 1946.

C. H. M. van Bavel, G. B. Stirk, and K. J. Brust, "Hydraulic Properties of a Clay Loam Soil and the Field Measurement of Water Uptake by Roots: I. Interpolation of Water Content and Pressure Profiles," *Proc. Soil Sci. Soc. Amer.*, vol. 32, pp. 310–317, 1968.

D. H. Van der Merwe, "The Prediction of Heave from the Plasticity Index and Percentage Fraction of Soils," *Civil Eng. in South Africa*, vol. 6, no. 6, 1964, pp. 103–107.

P. van der Raadt, "Field Measurement of Soil Suction Using Thermal Conductivity Matric Potential Sensors," M.Sc. thesis, Univ. of Saskatchewan, Saskatoon, Sask., Canada, 1988, 210 pp.

P. van der Raadt, D. G. Fredlund, A. W. Clifton, M. J. Klassen, and W. E. Jubien, "Soil Suction Measurement at Several Sites in Western Canada," *Transportation Res. Rec. 1137, Soil Mech. Considerations in Arid and Semi-Arid Areas*, Transportation Res. Board, Washington, DC, pp. 24–35, 1987.

B. P. Van Haveren and R. W. Brown, "The Properties and Behavior of Water in the Soil-Plant-Atmosphere Continuum," in *Psychrometry in Water Relations Research*, R. W. Brown and B. P. Van Haveren, Eds., Utah Agric. Experimental Station, Utah State Univ., Logan, pp. 1–27, 1972.

M. Vanclin, G. Vachand, and J. Khanji, "Two Dimension Numerical Analysis of Transient Water Transfer in Saturated-Unsaturated Soils," in *Modelling and Simulation of Water Resources Systems*. Amsterdam: North-Holland, 1975, pp. 299–323.

M. Vargas, "The Concept of Tropical Soils," in *Proc. 1st Int. Conf. Geomech. in Tropical Lateritic and Saprolitic Soils* (Brasilia, Brazil), vol. 3, pp. 11–14, 1985.

A. Verruijt, "Elastic Storage of Aquifers," in *Flow Through Porous Media*, R. J. M. De Wiest, Ed. New York: Academic, 1969, ch. 8.

K. K. Watson, "An Instantaneous Profile Method for Determining the Hydraulic Conductivity of Unsaturated Porous Materials," *Water Resources Res.*, vol. 2, pp. 709–715, 1966.

R. C. Weast, M. J. Astle, and W. H. Beyer, *CRC Handbook of Chemistry and Physics*, 65th ed. Boca Raton, FL: CRC Press, 1984, 370 pp.

H. F. Weimer, "The Strength, Resilience, and Frost Durability Characteristics of a Lime-Stabilized Till," M.Sc. thesis, Univ. of Saskatchewan, Saskatoon, Sask., Canada, 134 pp., 1972.

H. M. Westergaard, *Theory of Elasticity and Plasticity*. New York: Dover, 1952, 176 pp.

S. J. Wheeler and V. Sivakumar, "Critical State Concepts for Unsaturated Soil," *Proc. 7th Int. Conf. on Expansive Soils* (Dallas, Texas), August 3–5, 1992, pp. 167–172.

R. V. Whitman and W. A. Bailey, "Use of Computer for Slope Stability Analysis," *ASCE J. Soil Mech. Found. Eng. Div.*, vol. 93, SM4, pp. 519–542, 1967.

R. V. Whitman, A. M. Richardson, and K. A. Healy, "Time-Lags in Pore Pressure Measurements," in *Proc. 5th Int. Conf. Soil Mech. Found. Eng.* (Paris, France), vol. 1, 1961, pp. 407–411.

R. A. Widger, "Slope Stability in Unsaturated Soils," M.Sc. thesis, Univ. of Saskatchewan, Saskatoon, Sask., Canada, 71 pp., 1976.

R. A. Widger and D. G. Fredlund, "Stability of Swelling Clay Embankments," *Can. Geot. Journ.*, vol. 16, no. 1, pp. 140–151, 1979.

A. A. B. Williams, "Studies of Shear Strength and bearing Capacity of Some Partially Saturated Sands," in *Proc. 4th Conf. Soil Mech. Found. Eng.* (London, England), vol. 3, 1957, pp. 453–456.

A. A. B. Williams, "Discussion of the Prediction of Total Heave from Double Oedometer Test," by J. E. B. Jennings and K. Knight, *Trans. South African Inst. Civil Eng.*, vol. 8, no. 6, 1958.

A. A. B. Williams, "Some Regional Aspects of Soil Mechanics in South Africa," *Proc. of the 2nd Regional Conf. African Soil Mech. and Found. Eng.*, 1959, Vol. 1, pp. 11–16.

J. Williams and C. F. Shaykewich, "The Influence of Soil Water Matric Potential on the Strength Properties of Unsaturated Soil," *Proc. Soil Sci. Soc. Amer.*, vol. 34, no. 6, Div. S-1, pp. 835–840, 1970.

G. W. Wilson, "Soil Evaporative Fluxes for Geotechnical Engineering Problems," Ph.D. dissertation, University of Saskatchewan, Saskatoon, Sask., Canada, 464 pp., 1990.

R. G. Wilson, "Methods of Measuring Soil Moisture," in *Int. Field Year for the Great Lakes Tech. Manual Series, No. 1*, J MacDowall, Ed., The Secretariat Can. Nat. Committee for the Int. Hydrol. Decade, 1971.

G. W. Wilson, S. L. Barbour, and D. G. Fredlund, "The Evaluation of Evaporative Fluxes from Soil Surfaces for the Design of Dry Covers and the Abatement of Acid Drainage," *Proc. of the 2nd Int. Conf. on the Abatement of Acid Drainage* (Montreal, Quebec), September 16–18, 1991.

G. W. Wilson, D. G. Fredlund, and S. L. Barbour, "The Evaluation of Evaporative Fluxes from Soil Surfaces for Problems in Geotechnical Engineering," *Proc. of the Can. Geot. Conf.* (Calgary, AB), Sept. 29–Oct. 2, 1991, pp. 68.1–68.9.

H. Y. Wong, "Soil Strength Parameter Determination," *Hong Kong Eng.*, pp. 33–39, Mar. 1978.

D. K. H. Wong, D. G. Fredlund, E. Imre, and G. Putz, "Evaluation of AGWA-II Thermal Conductivity Sensors for Soil Suction Measurement," presented at the Transportation Res. Board Meeting, Washington, DC, 1989.

F. L. D. Wooltorton, *The Scientific Basis of Road Design*. Edward Arnold, 1954.

W. K. Wray, "The Principle of Soil Suction and its Geotechnical Engineering Applications," in *Proc. 5th Int. Conf. Expansive Soils* (Adelaide, South Australia), May 1984, pp. 114–119.

W. K. Wray, "Evaluation of Static Equilibrium Soil Suction Envelopes for Predicting Climate-Induced Soil Suction Changes Occurring Beneath Covered Surfaces," *Proc. of the 6th Int. Conf. on Expansive Soils* (Delhi, India), vol. 1, Dec. 1–4, 1987, pp. 235–240.

S. G. Wright, "SSTAB1—A General Computer Program for Slope Stability Analysis," Res. Report No. GE-74-1, Dep. of Civil Eng., Univ. of Texas at Austin, 1974.

R. D. Wyckoff and H. G. Botset, "The Flow of Gas-Liquid Mixtures through Unsaturated Sands," *Physics*, vol. 7, Sept. 1936.

D. Yang and Z. J. Shen, "Generalized Nonlinear Constitutive Theory of Unsaturated Soils," *Proc. 7th Int. Conf. on Expansive Soils* (Dallas, Texas), August 3–5, 1992, vol. 1, pp. 158–162.

R. N. Yong and B. P. Warkentin, "Introduction to Soil Behavior," Macmillan, New York, 451 pp., 1966.

R. N. Yong and B. P. Warkentin, "Soil Properties and Behaviour," Elsevier Scientific Publishing Co., New York, 1975, 449 pp.

R. Yoshida, D. G. Fredlund, and J. J. Hamilton, "The Prediction of Total Heave of a Slab-on-Ground Floor on Regina Clay," *Can. Geo. J.*, vol. 20, no. 1, pp. 69–81, 1983.

S. J. Zegelin, Ian White, and G. F. Russell, "A Critique of the Time Domain Reflectometry Technique for Determining Field Soil-Water Content," published in *Advances in Measurement of Soil Physical Properties: Bringing Theory into Practice*, SSSA Special Publication No. 30, Soil Science Society of America, Madison, WI, 1992, pp. 187–208.

O. C. Zienkiewicz, *The Finite Element Method in Engineering Science*. London: McGraw-Hill, 1971.

O. C. Zienkiewicz and Y. K. Cheung, "Finite Elements in the Solution of Field Problems," *The Engineer*, Sept. 24, 1965.

O. C. Zienkiewicz and C. J. Parekh, "Transient Field Problems: Two-Dimensional and Three-Dimensional Analysis by Isoparametric Finite Elements," *Int. J. Numer. Method for Eng.*, vol. 2, pp. 61–67, 1970.

O. C. Zienkiewicz, P. Mayer, and Y. K. Cheung, "Solution of Anisotropic Seepage by Finite Elements," *ASCE Eng. Mech. Div., Eng. Mech. 1*, vol. 92, pp. 111–120, 1966.

# About the Authors

**Delwyn G. Fredlund** attended the University of Saskatchewan, Saskatoon, Canada, where he obtained his B.Sc. degree in 1962. He worked as a research assistant for the Division of Building Research, National Research Council, Saskatoon, and then went for graduate studies at the University of Alberta, Edmonton, Canada. He obtained his M.Sc. degree in 1964 and later returned to obtain his Ph.D. degree in 1973, both at the University of Alberta.

In 1964 he started work as a geotechnical engineer for R. M. Hardy and Associates, Edmonton, where he was involved in studies associated with a wide range of geotechnical problems. In 1966 he went to the University of Saskatchewan to teach in geotechnical engineering. He developed an active research program which attracted graduate students from many countries of the world.

Over the years, Professor Fredlund has concentrated his research program on the study of the behavior of unsaturated soils and geotechnical problems in arid regions of the world. At the same time, there has been a diversity to the subjects studied. Some of these are: slope stability analyses and case studies, expansive soils, collapsing soils, measurement of soil suction, shear strength and volume change in unsaturated soils, permeability functions and flow through unsaturated soils, and probability theory applied to geotechnical engineering. He has been extensively involved in numerical modelling and the computer programming of various geotechnical problems. This has resulted in the establishment of a geotechnical software company called GEO-SLOPE International Ltd., of Calgary, Canada. This company has extended many of the classic saturated soils analyses to also embrace unsaturated soils.

Professor Fredlund has published more than 150 research papers in geotechnical engineering; most of them related to unsaturated soil behavior and slope stability analyses. He has been actively involved in the geotechnical community on the national level (e.g., Canadian Geotechnical Society) and the international level (e.g., International Society for Soil Mechanics and Foundation Engineering). He is presently chairman of the TC6 subcommittee of ISSMFE on Expansive Soils.

Professor Fredlund has travelled to many countries of the world, both to give technical presentations and to view firsthand, geotechnical problems in arid regions. These endeavours have resulted in International Development and Research Centre (IDRC) cooperative programs with China and Kenya. These programs promote technology transfer amongst all countries involved with the objective of being mutually supportive in solving geotechnical problems.

Professor Fredlund is presently Head of the Department of Civil Engineering at the University of Saskatchewan. Over the years he has acted as a consultant to numerous consulting firms and government agencies on problems related to unsaturated soil behavior and slope stability. He is presently a staff consultant to Clifton Associates, Regina, Canada. He has held research contracts from numerous private and government agencies in Canada, United States and other countries.

**Harianto Rahardjo** obtained his civil engineering degree, Ir., from Bandung Institute of Technology, Bandung, Indonesia in 1979. He worked as a consulting engineer for Wiratman and Associates in Indonesia and was involved in various geotechnical investigations and studies. He then went to the University of Saskatchewan, Saskatoon, Canada for his graduate studies, obtaining his M.Sc. degree in 1982. He subsequently worked as a research engineer with the geotechnical group in the Department of Civil Engineering, University of Saskatchewan. His research work was primarily on the behavior of unsaturated soils, slope stability and lateral earth pressure analyses. In 1988 he was awarded a Postgraduate Scholarship from the National Science and Engineering Research Council of Canada for pursuing his Ph.D. degree which he obtained in 1990.

In 1990 Dr. Rahardjo received a post doctoral fellowship from the University of Saskatchewan to continue his research work on unsaturated soil behavior. In 1991, he took up a teaching position in the School of Civil and Structural Engineering, Nanyang Technological University, (NTU), Singapore. He is presently a senior lecturer in the Division of Geotechnics and Surveying at NTU.

Dr. Rahardjo has published over thirty research papers in the area of unsaturated soil mechanics, slope stability analyses and lateral earth pressure studies. He is a member of the Canadian and South East Asian Geotechnical Societies. He is also a member of the TC6 subcommittee of the International Society for Soil Mechanics and Foundation Engineering on Expansive Soils. He has served as a consultant to several engineering firms in Singapore and Indonesia. He is presently conducting research programs at the Nanyang Technological University, Singapore, on the application of the unsaturated soil mechanics theories to slope stability, seepage and bearing capacity problems in residual soils.

# INDEX

*A* pore pressure parameter, *see* Pore pressure parameter
$A_a$ parameter, *see* Pore pressure parameter
$A_w$ parameter, *see* Pore pressure parameter
Active depth of swelling, 412, 415, 416
Active earth pressure, *see also* Coefficient of active earth pressure; Earth pressure
 coefficient of, *see* Earth pressure
 distribution of, 304, 306
 effect of matric suction, 304, 305, 306, 310, 311, 312
 effect of surcharge, 306
 effect of tension cracks, 305, 311
 effect of wall friction, 309–310
 orientation of failure plane, 304
 Rankine, 304, 309
 strain required, 308–309
 with tension cracks, 306
Active zone (for expansive soils), 8
Activity, typical values, 300
Actual evaporation rate, *see* Evaporation
Air coefficient of permeability, 199, 140, 141, 143, 175, 251, *see also* Coefficient of transmission; Permeability
 direct methods of measurement, 138
 effect of (relationship with) degree of saturation, 120
 effect of (relationship with) matric suction, 120–121
 function, 120
 indirect methods of measurement, 120
 measurement, 138–143, 175–176, 275
Air compressibility, *see* Compressibility, of air
Air entry value, *see also* High air entry disks
 determination, 112
Air phase, 21, 34, 43
 continuous, 20, 31, 108, 117, 420, 456, 472
 occluded, 31, 108, 117, 120, 251, 261, 419, 426, 456
Air pressure:
 absolute, 22, 23, 37, 44
 gauge, 22, 44, 52, 221, 256
Air volume change, 350, 351, 353
 coefficient of, 421, 426, 441, 473
 measurement, 275, 281
Air volumetric modulus, 472
Air-water interphase, 1, 14, 15, 17, 20, 24, 39, 63, 263, 275. *See also* Contractile skin
Allowable bearing pressure, *see* Bearing capacity

Angle of internal friction, typical values, 238. *See also* Shear strength
Anisotropic consolidation, *see* Shear strength tests; Triaxial test
Anisotropy, 130, 151, 190, 441, 442
At-rest condition, *see* $K_0$
Atmospheric boundary condition, 475
Atterberg limits:
 liquid limit, 300
 plastic limit, 300
 plasticity index, 300, 398, 399
 shrinkage limit, 398
Axis-translation:
 apparatus, 82
 concept, 47, 379
 technique, 47, 48, 54, 91–93, 94, 221, 223, 224, 238, 239, 241, 247, 248, 253, 256, 260, 261, 262, 277, 290, 291, 379, 429

*B* pore pressure parameters, *see* Pore pressure parameter
$B_a$ parameter, *see* Pore pressure parameter
$B_w$ parameter, *see* Pore pressure parameter
Backpressure (backpressuring), 273, 277, 278, 279, 281
Basic volume-mass relationship, *see* Volume-mass relations
Bearing capacity, 7, 217, 247, 297, 315, 318
 allowable, 317
 effect of embedment, 315
 effect of footing size, 316–317
 effect of matric suction, 317–318
 effect of shape, 316–317
 equation, 315–316, 318
 factors, 315, 316, 317
 layered systems, 319–320
 relation to cohesion, 315
 relation to friction angle, 316
 stress state variable approach, 317
 total stress approach, 317, 318
 ultimate, 315, 316, 317, 318, 319
Boyle's law, 22, 27, 35, 179, 180, 183, 184, 192, 193, 277, 279
Bubble collapse theory, 183–184
Bubble pump, 146–147, 221
Bubbling pressure, 112
Bulk modulus, 354

Capillarimeter, 16
Capillarity (also Capillary), 16, 65, 67–70.
 height (of rise), 67–68, 69–70

 model, 69–70
 pressure, 68–69
 radius, 69–70, *see also* Radius of curvature
 rise, 17, 18, 69–70
 tube, 16, 65, 66, 69
 zone, 17
Cavitation, of water, 44, 47, 54, 67, 83, 86, 90, 92, 219, 221, 238, 260, 261, 429
CD tests (*see* Triaxial test)
Cell pressure, *see* Confining pressure
Chemical diffusion, *see* Diffusion
Clays, swelling, 11
Climate, 1–3, 398
Coefficient of active earth pressure, definition, 304
Coefficient of air volume change, *see* Air volume change
Coefficient of compressibility, *see* Compressibility
Coefficient of consolidation, *see* Consolidation
 measurement of, *see* Oedometer test
Coefficient of diffusion (diffusivity), 28, 29, 30, 122, 146, 269, 279, 474
 measurement of, 144–146, 177
Coefficient of earth pressure at-rest, $K_0$, 298, 299, 300, 302, 312, 407
 typical values, 299, 300
Coefficient of lateral earth pressure, 52
Coefficient of passive earth pressure, 299, 307, 407
Coefficient of permeability, 110, 151, 163, 247, 251, 252, 253, 261, 266, 341, 437, 440, 441, 442, 446, 447, 449, 456, 474. *See also* Permeability
 air phase, 119–121, 175, 251, 420, 433, 472.
 determination of, *see* Permeability, measurement
 effect of degree of saturation, *see* Degree of saturation
 effect of water content, 111
 effect of void ratio, 111
 function, 111, 113, 341, 424, *see also* Permeability function
 (equivalent) layered systems, 145–146
 saturated, 111, 138, 153, 164, 340, 341, 419
 unsaturated, 110, 114, 116, 127, 133, 151, 251, 341
 water phase, *see* Water coefficient of permeability
Coefficient of transmission, for air flow, 118,

119, 121, 175, 176, 177, 425, 426, 427, 441, 473
Coefficient of volume change, 187, 188, 194, 213, 357, 358, 359, 361, 362, 364, 365, 367, 376, 377, 396, 406, 419, 421, 424, 426, 430, 432, 436, 437, 447, 449
Coefficient of swelling, 375, 376
Coefficient of water content, *see* Water content
Coefficient of water content change, 358, 360, 375, 376
Coefficient of water volume change, 341, 424, 441, 444
Cohesion:
  apparent, 230
  use in bearing capacity analysis, 315, 318
  effective, *see* Shear strength
  in terms of total stress, 230
  intercept, 220, 228, 229, 230, 233, 235, 236, 238, 240, 242, 334
  measurement, *see* Direct shear test; Triaxial test
  use in retaining wall analysis, 302–303, 307
  use in slope stability analysis, 323, 333–354
  total, *see* Shear strength
  typical values, 229
  use in slope stablity analysis, 323, 333–334
Collapsible soils, 417–418
Collapsing soils, 9, 13, 361
  laboratory test for, 417–418
Compacted soils, 13, 217, 219, 238, 295, 368, 371
  use in earth dams, 219
  use in pavement design, 319–320
  strength, 219
Compaction:
  definition, 33
  line of optimums, 34
Compaction tests, typical curves, 34
Compactive effort, 225, 238, 239
Compartment:
  below high air entry disk, 144, 261, 264, 265, 266, 268, 272, 274, 276, 379, 393
  water, 81, 261, 264, 265, 266, 273, 274, 380
Compliance factor, 270, 271
Composite slip surface, *see* Slip surface
Compressibility, *see also* Isothermal compressibility
  of air, 179, 183, 330
  of air-water mixture, 179–184, 266, 268, 269, 271, 272, 347, 422
  coefficient of, 358, 366, 369, 375
  form, 357, 358, 361, 365, 367
  measurement, *see* Oedometer test; Triaxial test
  soil (structure), 191, 194, 206, 266, 330, 363
  of water, 179, 206, 269, 272
Compressibility equations:
  air-water mixture, 180–181, 183, 191, 195, 196, 269
  soil structure, 191
Compressibility of pore fluid, 178–184
  air compressibility, *see* Compressibility of air
  air-water mixtures, *see* Compressibility of air-water mixtures
  effect of dissolved air, 179, 182
  effect of free air, 182
  water compressibility, *see* Compressibility of water

Compression index, 369, 370, 371, 375, 378, 392,
Compressive strength, *see* Undrained strength
Confined compression, *see* $K_0$-condition, Oedometer test, 8
Confining pressure, triaxial tests, 51, 224, 225, 226, 236, 238, 239, 240, 241, 242, 243, 246, 255, 260, 262, 277, 280
Conservation of mass, 36, 349. *See also* Continuity equation, Continuity requirement
Consolidated drained shear test, *see also* Triaxial test
  test procedure, 229
  typical results, 229
Consolidated undrained shear test, *see* Triaxial test
Consolidation, 239, 240, 264, 266, 280, 281, 283, 290
  coefficient of, 252, 253, 254, 419, 420, 424, 426, 427, 435, 473, 474
  defined, 236, 419
  degree of, 437, 438, 439
  equation derivation, 422–427
  process, 238, 241, 251
  secondary, 48
  theory of, 9, 39, 250, 419
  three-dimensional, 346, 420
  time factor, 437, 438, 439
Consolidation tests, 429–437
  data presentation, 430–436
Consolidation theory:
  assumptions, 422
  boundary conditions, 428
  derivations, 422–427
  solution, 427–429
Constant water content shear test, *see also* Triaxial test
  test procedure, 229
  typical results, 229
Constant-volume test, *see* Oedometer test
Constitutive relations (or equations), 15, 39, 347, 348, 349, 351–360
  air phase, 348, 353, 421, 426, 456, 472
  single-valued equations, 18, 38, 39, 41, 346
  soil structure, 298, 346, 348, 352, 353, 356, 358, 420, 456, 461
  unsaturated soils, 351, 352
  volume change, 346
  water phase, 346, 348, 353, 356, 358, 421, 423, 456, 463
Constitutive surfaces, 347, 349, 354, 358, 360, 361, 363, 364, 372, 406
  air phase, 421
  degree of saturation, 348, 363, 364, 365
  loading, 374, 377
  soil structure, 348, 349, 354, 357, 360, 362, 363, 364, 365, 367, 368, 369, 421
  unloading, 374, 377, 392, 395
  void ratio, 348, 359, 363, 364, 365, 366, 371, 372, 373, 374, 378, 382, 385
  water content, 359, 365, 366, 371, 373, 374, 378
  water phase, 348, 349, 354, 357, 368, 362, 363, 367, 368, 370, 386, 421
Constrained modulus, *see also* Compressibility; Modulus
  defined, 357

Continuity equation, 349, 420, 422, 456, 472
Continuity requirement, 349, 350, 351, 353, 419, 420, 421, 432
Continuum mechanics, 15, 16, 19, 41, 56
  multiphase, 348, 349
Contractile skin, 1, 14, 15, 20, 21, 24, 25, 26, 29, 39, 43, 44, 63, 67, 69, 81, 263, 350, 360. *See also* Air-water interphase
Coupled equations, 472–473
  three-dimensional formulation, 420, 422, 456
  two-dimensional formulation, 422
Crack depth, 298, 300, 301, 302, 304, 305
Critical height, 298, 313, 314, 315
Critical slip surface, *see* Slip surface
CU tests, *see* Triaxial test
Cuttings, *see* Excavations, Slopes

Dalton's, law, 26, 476
Darcy's law, 117, 123, 140, 153, 159, 164, 268, 419, 422, 447, 473
  for air phase, 119, 122
  definition, 110, 124
  in theory of consolidation, 420, 423
  validity for unsaturated soils, 110
Deformation, 255, 297, 308, 355
  distortion, 351
  types of, 355
Deformation state variables, *see* State variables
Degree of saturation, 27, 31, 32, 34, 35, 40, 47, 111, 114, 117, 223, 228, 239, 244, 257, 258, 259, 269, 277, 278, 279, 290, 366, 368, 389
  constitutive surface, *see* Constitutive surfaces
  critical, 346
  definition, 30
  effect on compressibility, 34, 279, 366
  effect of permeability, 54, 111–114
  effect on pore pressures generation, 204–210, 223
  effect on undrained strength, 223
  effective, 112, 113, 120
  residual, 112
Density, 21, 22, 23, 27, 35, 225, 228
  of air, 21, 35
  of air-water mixtures, 37
  buoyant or submerged, definition, 32
  dry, 225
  dry, definition, 32, 33
  saturated, definition, 32
  solids, definition, 37
  total (wet), definition, 32, 36, 37, 305
  typical values, 31
  of water, 21, 22, 27, 33, 35
  water phase, 36
Depth of cracking, *see* Crack depth
Desaturation, 1, 257, 387, 397, 404
Desiccated soil, 53, 300, 301, 317, 390
Desiccation, 2, 298, 404
  causing preconsolidtion, 404
Deviator stress, *see* Stress
"Different" soil, 217, 225, 259, 288
Differential equations of equilibrium, 43
Differential movement, 92
Diffused air, 144, 253, 261, 264, 274, 281
  method of measurement, 127, 146, 265, 280, 281, 282, 283

Diffused air volume indicator, (DAVI), 146,
    261, 266, 267, 274, 279, 280, 281, 283
  accuracy of, 149
  apparatus, 147
Diffusion, 48, 107, 121–123, 275, 474
  of air, 28, 107, 177, 221, 262, 276, 282
  of air through water, 29, 30, 108, 120, 121–
    123, 127, 242, 251, 426
  chemical, 123
  Fick's law for, 28, 121
  of gases through water, 28, 123
  measurement, 143–149
  properties, 145
Diffusion process:
  air through ceramic disk, 121, 143, 144, 177,
    253, 261, 262
  air through water, 121, 143, 177
  chemical concentration, 121
  osmotic, 110
Diffusivity, 28, 251, 474
Direct model method, 399, 400
Direct shear test, 217, 220, 223, 224, 225,
    227, 247, 248, 254, 256, 275, 277, 284,
    289, 290, 295, 296
  consolidated drained, 223, 224, 229, 247,
    254, 282, 283
  consolidated drained multistage, 229, 254, 290
Dirichlet boundary condition, 163, 445
Displacement rate, 254, 256
  direct shear, 254, 290, 291
Dissolved air, 27, 34, 86, 277
  volume of, 28, 35, 36, 35, 180
Disturbance, 255, 285, 363, 388, 389
  sampling, 130, 389, 390, 391, 400, 407, 411
Double layer, 63
Double oedometer method, 400, 401
Drainage path:
  double drainage, 252
  single drainage, 252, 253, 254
Drained loading, 188–189
Driving potential, 107, 108, 110, 119, 121
  air phase, 117
  water phase, 108–110
Dry soils, 30, 34, 45, 46

Earth dam, 3, 219, 329
Earth pressure, 297, 298, 301
  active, 7, 298, 301, 302, 303, 305, 306, 310,
    311, 314
  at-rest, 298
  coefficient of, 308, 309
  passive, 7, 298, 301, 302, 303, 314
  Rankine's theory, 301
Effective stress, 38, 45, 52, 237, 317
  concept, 38, 39, 217
  equation, 18, 39, 40, 41, 346
  law, 38, 39, 45
  single-valued equation, see Constitutive relations
Elastic equilibrium, 298
Elastic material:
  pore pressure coefficients for, see Pore
    pressure coefficients
  stress-strain relationships, 298
Elastic modulus, 298, 352
Elasticity:
  linear elastic material, 351, 461
  theory (of), 52, 185, 297, 320

Elasticity form, 351, 357, 365, 456
  constitutive relations, 298
Elevation head, see Head
Equilibrium:
  force, 328, 330, 331
  moment, 328, 330, 331
Equilibrium equations, 43, 297, 456, 472
Equipotential (lines), 154, 156, 165, 167–170,
    329, 330, 331, 447, 449, 451–454, 461,
    464–471
Evaporation, 1, 4, 45, 48, 130, 131, 152, 156,
    157, 163, 275, 317, 318, 398, 404, 408,
    440, 445
  actual, 476
  potential, 476
  rate, 53, 476, 477
Evapo-transpiration, 1, 2, 53, 163, 404, 408,
    445
Excavation, 6, 52, 406, 407, 409, 415, 417
Excess pore pressure, 236, 237, 239, 241, 243,
    245, 251, 254, 428
  pore-air, 238, 241, 243, 251, 260, 419, 422,
    428, 434
  pore-water, 238, 241, 243, 251, 252, 260,
    419, 428, 434
Expansion, see Swell
Expansive soils, 297, 300, 388, 397, 404, 407,
    409, 412, 416, 417
Extended Mohr diagram (circle), 49, 54, 56,
    57–58, 60, 301, 312
  construction of, 56
Extended Mohr-Coulomb failure, envelope,
    227, 228, 229, 230, 231, 232, 233, 234,
    235, 248, 255, 258, 259, 260, 318
Extended Rankine theory, 301
"Extended shear strength" method, 340

Factor of safety, 5, 6, 317, 321, 323, 326, 330,
    332, 333, 334, 335, 337, 338, 339, 340,
    343, 344, 345
  force equilibrium, 324, 325, 327, 328, 332
  moment equilibrium, 324, 327, 328, 332, 336
Failure criteria, 225, 226, 227, 297, 303, 307
  maximum deviator stress, 226
  maximum principal stress ratio, 221, 226, 227
  Mohr-Coulomb criterion, 217, 218
  strain (or displacement) limit, 227, 289
Failure envelope, 217, 218, 220, 223, 224,
    225, 228, 229, 230, 231, 232, 233, 234,
    235, 238, 239, 240, 241, 242, 247, 248,
    258, 259, 284, 285, 287, 288, 293, 296
  Mohr-Coulomb, 218, 223, 255, 259, 290,
    303, 307
Failure plane, 217, 218, 227, 228, 229, 230,
    247, 302, 313
  angle of inclination, 228, 313
Fick's law, 117, 121, 123, 422
  application to air phase, 117–119, 175, 420,
    425, 473
  definition, 117
Filter paper, 54, 67, 77–80, 81
  calibration, 77–79
  contact, 77, 78, 79, 80
  measure matric suction, 77
  measure total suction, 76, 77
  non-contact, 77, 78, 79, 80
  test procedure, 78–79

Finite difference method, 155–158, 427–428,
    446
  backward difference, 446
  central difference, 155, 432, 446
  explicit, 158, 432
  implicit, 156
Finite element method, 161, 313, 326, 342,
    440, 444, 447, 449
Flow, see also Seepage
  capillary, 16, 18
  one-dimensional, 152–158
  three-dimensional, 173–174
  two-dimensional, 159–173
  velocity head, see Head
Flow laws, 107–123, 347
  air phase, 117
  water phase, 107
Flownet, 160, 162
  equipotential lines, 160
  flow channels, 171
  flow lines, 160, 171
Flux boundary condition, 5, 9, 156, 157, 342,
    475
Foundations, see also Bearing capacity
  shallow, 7, 12
Free air, 27, 28, 34, 35, 180, 277, 278
Free swell test, see Oedometer test
Free (water) surface, 160, 161, 166
Friction angle, see Shear strength

Gaussian elimination technique, 163, 446
Geostatic stress (condition), see Stress
Gradient, 107
  (chemical) concentration, 107, 117, 118, 121,
    123
  electrical, 107
  elevation, 108, 117
  hydraulic head, 107, 108, 110, 119, 124, 126,
    127, 128, 129, 130, 132, 153, 163, 164,
    268, 419, 423, 442, 446, 447
  matric suction, 107, 108
  osmotic, 110
  osmotic suction, 121, 123, 127
  pressure, 107, 108, 117, 119
  total head, 109
  thermal (or temperature), 107, 473
  water content, 107
Grain size, 257
Ground water table, see Water table

Head, 107, 109
  elevation, 109
  gravitational, 109, 130, 132
  hydraulic, 109, 129, 130, 132, 153, 154, 171,
    423, 442, 444, 445, 446, 451–454
  matric suction, 115
  pressure, 109, 127, 129, 166
  velocity, 109
Head boundary condition, 155, 160
Heat flow, equation, 474
Heave, 8, 49. See also Swell
  methods of prediction, 400
  potential, 397, 410
  prediction, 397, 401, 405, 411
Heave calculations, 407, 408–410
  case histories, 410–411
  closed-form equation, 412
  effect of correcting swelling pressure, 413

Henry's law, 27, 28, 35, 180, 183, 192, 279
  application, 184
High air entry disk, 48, 80, 81–82, 126, 127, 133, 134, 140, 141, 142, 143, 218, 219, 221, 223, 224, 225, 248, 250, 251, 252, 253, 254, 260, 263, 264, 265, 266, 267, 268, 269, 270, 272, 273, 275, 277, 279, 282, 283, 284, 363, 379, 429, 433
  air entry value, 48, 81, 82, 83, 84, 112, 218, 221, 257, 258, 260, 261, 263, 264, 265, 266, 272, 273, 275, 380, 386, 387
  air passage through, 82, 84, 221, see also Diffusion process
  (water) coefficient of permeability of, 82, 83, 84, 264, 268, 270, 272, 276
  properties of, 83, 263, 264
Highway pavement, 473
Hilf's analysis, 186, 192–194, 201–204, 213, 219
Hooke's law, generalized, 352
Hydraulic (total) head, 152
Hydraulic gradient, see Gradient
Hydrostatic condition (profile), 298, 301, 302, 305, 314, 317, 335, 343
Hydrostatic line, 305
Hysteresis, 389
  in permeability function, 116–117, 126
  in soil structure relationships, 348, 360, 361, 404
  in water content relationships, 69, 112, 117, 136, 364, 369, 376, 386, 387, see also Soil water characteristic

Ideal gas law, 22, 121, 122, 422, 425
"Identical" soil (or specimens), 126, 217, 225, 243, 259, 284, 347, 361, 362, 364, 388
Immiscible mixtures, 25
  of air and water, 25
Impedance factors, 251, 252, 253, 254
Impeded flow, 250, 251, 252, 253
Impervious (or impermeable) membrane, 9, 315
Infiltration, 4, 53, 117, 130, 131, 152, 156, 157, 158, 163, 166, 334, 338, 404, 408, 414, 440, 449, 461, 463, 465, 467, 469, 471, 475
Infinite slope, 171
In situ stress state, 389, 397, 404, 406
  determination of, 388, 403, 404, 405–406
Instantaneous profile method, see Water coefficient of permeability
Intermediate principal stress, see Stress
Interslice force, 321, 326, 330, 332
Interslice force function, 324, 325, 326, 327, 328, 333
Intrinsic permeability, see Permeability
Isobar(s), 165, 167–170, 172, 457–460
Isochrones, 436
Isothermal compressibility, 178, 179
  air, 179, 180, 182
  air-free water, 179
  air-saturated water, 179
  water, 179, 182
Isotropic compression, 347, 354, 356
Isotropic loading, 194–196, 204–215, 348, 349, 354, 355, 356, 359, 363, 366

Jaky's equation, 300

$K_0$ loading, 61, 62, 191–192, 196, 347, 348, 349, 355, 356, 359, 362, 363, 364, 394, 406, 420
$K_0$ condition, 298, 309
$K_f$-line, 233, 309
  Mohr-Coulomb, relationship with, 233
  parameters, $p_f$, $q_f$, 231, 233
Kelvin's equation, 25, 81, 183–184, 263

Laplace (or Laplacian) equation, 25, 160, 161
Latent heat, 474
Lateral earth pressure, 7, 52, 217, 297, 313
  active, see Earth pressure
  at-rest, 52, 54
  passive, see Earth pressure
Limit equilibrium, 297, 320, 321, 326, 332
Loading, drained, see Drained loading
  undrained, see Undrained loading

Matric suction, 40, 41, 43, 45, 46, 48, 49, 52, 53, 55, 58, 59, 60, 61, 63, 65, 217, 218, 219, 220, 222, 223, 224, 225, 226, 228, 229, 230, 233, 234, 238, 239, 241, 242, 244, 245, 246, 247, 248, 252, 258, 259, 261, 262, 263, 264, 276, 283, 290, 291, 293, 294, 295, 296, 297, 299, 300, 301, 302, 303, 304, 306, 307, 308, 309, 311, 313, 315, 317, 332, 333, 334, 363, 364, 366. See also Soil suction
  definition, 25, 66, 69, 183, 379
  direct measurement, 82–93
  envelope, see Shear strength versus matric suction
  equivalent, 389, 404, 405, 406, 407
  indirect measurement, 93–106
  in situ, 54, 62, 317, 318, 336, 337, 338, 340, 397
  measurement, 63, 67, 80–104, 94, 100, 101, 102, 103, 104, 105, 336, 337, 338, 340
  profiles, 53, 54, 89, 90, 300, 301, 305, 334, 336, 337, 338, 339, 340, 342, 343, 344
Matric suction versus shear strength, see Shear strength versus matric suction
Matric suction versus water volume change (see Water volume change versus matric suction
Maximum past stress, see Preconsolidation pressure
Measurement of:
  pore-air pressure, see Pore-air pressure
  pore-water pressure, see Pore-water pressure
Meniscus, 273, 305
  defined, 66
Meta-stable structured soil, 9, 361, 418, 429, 432
Microclimate, 317
Microclimatic conditions, 9, 54, 473, 477
Miscible mixture, of air and water, 26
Mixtures, 34
  air-water, 37
  soil particles-water-air, 37
  theory of, 19, 36
Mobilized shear force, see Shear force mobilized
Modified direct shear apparatus, 219, 223, 224, 260, 266, 267, 272, 282, 290, 291
Modified Penman equation, 477
Modified triaxial apparatus (cell), 48, 220, 222, 260, 265, 272, 279, 280, 363

Modulus, see also Compressibility; Elastic modulus; Shear modulus; Young's modulus
  bulk, see Bulk modulus
  constrained, see Constrained modulus
  deformation, 313
Mohr circle, 57, 59, 218, 220, 226, 228, 231, 233, 235, 236, 238, 242, 244, 247, 255
  construction, 303, 307
  equation, 55–56
  envelope, 223
  failure envelope, 223
  pole point method, 56, 57, 58, 60, 218
Mohr-Coulomb:
  envelope, 220, 231, 318
  failure criterion, see Failure criteria
Molecular diffusivity:
  water vapor, 474
Monotonic (loading), 40, 49, 51, 221, 262, 349, 360, 363, 364, 368, 370, 406
Multiphase system, 349
Multistage tests, 255, 276, 281, 283, 285
  cyclic loading procedure, 255, 284, 285
  direct shear, 224, 229, 255, 256, 289, 292, 293
  sustained loading procedure, 255, 284, 285
  triaxial, 223, 229, 251, 255, 284, 285

Negative pore-water pressure, 1, 3, 6, 7, 8, 9, 10, 13, 17, 18, 40, 47, 53, 83, 219, 245, 257, 297, 298, 300, 304, 305, 311, 315, 317, 318, 320, 323, 324, 329, 330, 332, 333, 334, 336, 340, 343, 344, 345, 404, 408, 409
Net horizontal stress, see Stress
Net normal stress, see Stress
Net vertical stress, see Stress
Neuman boundary condition, 163, 445
Neutral stress, see Pore-water pressure
Non-contacting radial deformation transducer, 275–276
Non-isothermal, 440, 473, 474, 475
Nonlinearity of failure envelope, 217, 224, 228, 255, 257, 258
Normal force, 327
  base of a slice, 322, 323, 324, 332
  interslice, 322, 325, 328
Normal strain, 219, 298, 350, 351, 352, 353, 461
Normal stress, see Stress
Null tests, 47, 48, 49, 50

Oedometer test, 378, 382, 399, 400
  compressibility of apparatus, 389–390
  constant volume oedometer tests, 313, 374, 375, 381, 384, 386, 388, 389, 392, 394, 395, 400, 405, 410, 411
  free swell oedometer tests, 374, 375, 388, 389, 390, 392, 393, 394, 396, 399, 400, 401, 405
  procedure, 378
  stress path, 389
  typical results, 375
One-dimensional consolidation, 356
  assumptions, 422–423
  derivation, 423–427
  solution of, 427–428
  theory, 419, 420
  typical results, 429–436

## 514  INDEX

One-dimensional swelling, 406, *see also* One-dimensional heave
Osmotic pressure, 54, 91, 105, 106
Osmotic suction, 65, 66, 105, 106, 110, 225, *see also* Soil suction
   measurement of, 67, 104
   role of, 63
Osmotic tensiometers, *see* Tensiometer
Overall (total) volume change, 346, 349, 350, 352, 362
   measurement, 275
Overburden pressure, *see* Stress
Overconsolidated soils, 52
Overconsolidation ratio, OCR, 406
   defined, 405
   effect on $K_0$, 300

Partial pressure, 23
Passive earth pressure, 300
   coefficient, *see* Coefficient of passive earth pressure
   distribution, 307, 308, 309
   effect of matric suction, 307, 308, 312
   effect of wall friction, 309–310
   failure, 301
   orientation of failure planes, 308
   Rankine, 308, 309
   strain required, 308–309
Peltier effect, *see* Psychrometer
Permeability, 54. *See also* Coefficient of permeability; Water coefficient of permeability
   coefficient of, 6
   effect of degree of saturation, *see* Degree of saturation
   effect of void ratio, *see* Coefficient of permeability
   intrinsic, 110–111
   measurement of, 124–149
   relative, *see* Relative water coefficient of permeability
Permeability function, 111, 113, 114, 115, 116, 137, 151, 159, 164, 166, 341, 446, 447, 449, 456, 463
Permeability tests:
   in situ measurements, 130–133
   laboratory measurements, 124–130
Permeameter, 124, 125, 127, 128, 129
   for air permeability, 140–143
   constant head, 125–127, 142
   for water permeability, 140–143
Phase, 14, 20
   four-phase system, 15, 20, 41
   multiphase, 41
   properties of a phase, 20
Phase diagram, 20, 21, 30, 32
Phase relations, 21
Phenomenological approach, 15
Phreatic surface (line), 5, 160, 161, 165, 167–170, 171, 172, 173, 329, 447, 450, 457–460
Piezometric line, 330, 331
Plane strain loading, 355, 357, 359
Plane stress loading, 355, 357, 359
Plastic equilibrium, 297, 301, 303, 316
   stress state approach, 297
   total stress approach, 297

Plastic limit, *see* Atterberg limits
Plasticity index, *see* Atterberg limits
Poiseuille's equation, 115
Poisson's ratio, 298, 299, 301, 302, 352, 356, 364
Pore-air pressure, 4, 40, 43, 44, 45, 47, 48, 51, 52, 61, 208, 209, 211, 212, 213, 218, 219, 221, 222, 223, 227, 228, 229, 230, 237, 239, 240, 241, 245, 246, 264, 277, 280, 298, 366, 379, 422, 428, 430
   coefficient, *see* Pore pressure coefficient
   control of, 48, 218, 221, 224, 238, 247, 248, 260, 263, 265, 266, 272
   excess, *see* Excess pore pressure
   measurement of, 218, 219, 221, 222, 243, 248, 260, 263, 265, 272, 273
   parameters, *see* Pore pressure parameters
Pore fluid, 251, 260
Pore fluid squeezer, 54, 67, 105, *see also* Squeezing technique
Pore pressure coefficient, 328, 329, 330
   air pore pressure coefficient, 329
   water pore pressure coefficient, 328, 329, 330
Pore pressure parameters, 62, 178–216, 239, 347, 422, 428. *See also* Tangent Pore pressure parameter
   definition, 181, 198
   $A$ parameter, 246, 329
   $A_a$ parameter, 198, 330
   $A_w$ parameter, 197–198, 330
   $B$-parameter, 198, 204, 216, 243, 329
   $B_a$ parameter, 181, 183, 184, 186, 199, 200, 205, 206, 208, 209, 210, 211, 212, 214, 330
   $B_w$ parameter, 181, 183, 184, 186, 197, 199, 200, 204, 205, 206, 208, 209, 210, 211, 212, 214, 244, 278, 279, 330
   $D$ parameter, 198, 215, 216, 239, 241, 243, 245
   $D_a$ parameter, 197, 214, 215
   $D_w$ parameter, 197, 214, 215
Pore pressure response, below high air entry disk, 266, 271
Pore size distribution, 111, 120
Pore size distribution index:
   definition, 112
   determination, 112
   suggested values, 114
Pore-water pressure, 4, 39, 43, 44, 47, 48, 51, 53, 61, 62, 208, 209, 211, 212, 213, 217, 218, 219, 221, 222, 223, 224, 227, 228, 229, 230, 237, 238, 239, 240, 241, 245, 246, 247, 250, 256, 298, 366, 379, 408, 422, 428, 430
   backpressure, *see* Backpressure
   coefficient, *see* Pore pressure coefficient
   excess, *see* Excess pore pressure
   measurement, 218, 219, 221, 222, 225, 238, 243, 248, 260, 261, 262, 263, 265, 266, 267, 268, 283
   negative, *see* Negative pore-water pressure
   parameters, *see* Pore pressure parameters
   profile, 317, 342, 344, 406, 411, 438
   response, 278, 279
Porosity, 32, 33, 43, 44, 264
   defined, 29, 30
   typical values, 31

Potential heave, 8, 397
Precipitation, 1, 6, 318, 404
Preconsolidation pressure, 366, 386, 388, 405
   Casagrande construction, 390–391
   determination of, 391, 405
   effect of sample disturbance, 389
   factors affecting determination of, 389
Pressure, *see* Stress
Pressure head, *see* Head
Pressure membrane extractor, 379–380
Pressure plate, 63, 223
   equipment, 379–380
   extractor, 379, 380
   null type, 48, 67, 92
Pressure plate tests, 47, 368, 379, 381, 382, 393
   drying portion, 393
   typical results, 382, 386–388
   wetting portion, 393
Principal planes, *see* Stress
Principal stress, *see* Stress
   at failure, 226
   intermediate, *see* Stress
   major, *see* Stress
   minor, *see* Stress
Principal stress ratio, 227, 248, 250
   definition, 226
Profiles:
   of residual soils (and lateritic soils), 10
   of swelling or expansive soils, 11, 12
   of unsaturated soils, 9
Psychrometers, 54, 63, 67, 70–77, 126, 128, 129
   calibration of, 73–74
   Peltier effect, 70–71, 72
   Peltier-type, 70, 71–73, 76, *see also* Spanner psychrometer
   performance of, 74–77
   Seebeck effect, 70, 71
   Spanner psychrometer, 71, *see also* Peltier-type
   thermocouple, 70–71

Radius of curvature, 25, 66, 67, 68, 69, 81, 183, 184
   of air-water interphase, 184
   of contractile skin, 263
Rainfall, 5, 61, 63, 312, 336, 340, 344, 345
Relative air phase coefficient of permeability, 120
Relative humidity, 23, 27, 64, 65, 66, 70, 71, 72, 476
Relative water phase coefficient of permeability, 113, 114
Residual degree of saturation, *see also* Degree of Saturation
   determination
Residual shear strength, 255
Residual soil, 1, 9, 13, 322, 340
Retaining structures (or wall), 297, 303, 305, 309, 310, 312

Sampling disturbance, *see* Disturbance
Salt content, 63
Saturated-unsaturated flow (seepage), 5, 151, 161, 162, 164, 166, 171, 444, 449

INDEX  515

Saturation procedure:
  high air entry disk, 266
  for an unsaturated soil, 277
Secant pore pressure parameter, 185–186, 192, 201–202
  pore-air pressure, 185–186, 194
  pore-water pressure, 185–186, 194
Seebeck effect, *see* Psychrometer
Seepage:
  analysis, 342, 344, 444
  flownets, steady state, *see* Steady state seepage
  transient analysis, 342, 344, 440, 442, 443, 444, 447
Semi-permeable membrane, 110, 123, 263
Shallow foundations, 317, 406, 473
Shape factor, 316
Shear force mobilized, 322, 323, 324, 325, 328, 333, 345
  base of slice, 322, 323
  interslice, 322, 324, 325, 327, 328, 331, 333, 336
Shear modulus, 352, 461
Shear strain, 351, 461
Shear strength, 38, 39, 41, 46, 48, 49, 62, 63, 217, 218, 219, 222, 224, 226, 227, 230, 235, 237, 238, 244, 247, 248, 249, 260, 315, 317, 332, 333, 344, *see also* Cohesion; Friction angle
  angle of internal friction, 217, 218, 223, 224, 227, 235, 238, 239, 247, 284, 285, 286, 289, 300, 310, 316, 317, 341
  cohesion, $c$, 304, 315
  effective cohesion, $c'$, 217, 218, 222, 227, 230, 235, 239, 285, 286, 289, 302, 304, 305, 307, 308, 309, 310, 315, 317, 341
  envelope, *see* Mohr circle
  equation, 227, 230, 257, 258, 301, 302, 346
  failure criteria:
    maximum principal stress difference, *see* Failure criteria
    maximum principal stress ratio, *see* Failure criteria
  formulation, 227–235
  measurement, 236, 260
  parameters, $c'$, $\phi'$, 220, 227, 228, 229, 235, 238, 256, 260, 277, 284, 287, 289, 310, 320, 334, 339
  theory, 217, 255
  total cohesion, 302, 303, 304, 307, 315, 317, 323
  total stress approach, 225, 237, 238
Shear strength tests, *see* Triaxial test
  CD, consolidated-drained, *see* Triaxial test
  CU, consolidated-undrained, *see* Triaxial test
  UU, unconsolidated-undrained, *see* Triaxial test
Shear (stress) strength versus matric suction, 223, 224, 225, 229, 230, 234, 235, 240, 255, 258, 259, 284, 287, 291
  failure envelope, 256, 290, 291, 293, 294
  nonlinear envelope, 255, 257, 284, 285, 286, 288, 291, 296
Shrinkage test, 369, 380, 381, 382
  procedure, 380
  typical results, 369, 377, 382, 388
Shrinking soils, 12, 397

Slip surface, 6, 297, 310, 315, 320, 322, 323, 326, 331, 332, 333, 345
  circular, 321, 322, 324, 325, 327, 332, 333, 336, 338, 342
  composite, 321, 322, 324, 325, 327
  critical, 320, 321, 322, 336, 337, 345
Slopes:
  stability, 217, 247, 320, 321
  stability at end-of-construction, *see* Stability analysis
Soil mechanics:
  definition, 1
  historical development, 16–18
Soil profiles, residual soils, 10
Soil structure, 43, 44, 278, 285, 401
  metastable structure, 348, 417, 430
  microstructure, 10
  stable structure, 430
Soil structure volume change, 350, 351
Soil suction, 39, 54, 64
  components of, 65–66, 67
  matric suction, *see* Matric suction
  measurement of, 64–106
  osmotic suction, *see* Osmotic suction
  theory of, 64
  total suction, *see* Total suction
Soil-structure interaction, 313
Soil-water characteristic curve, 45, 67, 74, 98, 111, 114, 115, 116, 124, 126, 128, 129, 130, 134, 136–138, 257, 341, 368, 369, 370, 371, 372, 375, 376, 377, 378, 381, 382, 386, 387, 388, 393
  drying portion, 136, 369
  wetting portion, 136, 369
Solubility:
  of air in water, 27, 28, 330
  coefficient of, 28, 29
  of gases in water, 28, 29
Specific gravity, 21, 33, 284, 366, 375, 378, 381, 392
Specific volume, 21
Squeezing technique, 63, 105, 106
  osmotic suction measurement, 63
Stability analysis, 297, 320, 329, 333, 334, 336, 337, 344
  Bishop simplified method, 330, 331, 332, 335
  Corps of Engineers method, 330, 331, 332
  General Limit Equilibrium, GLE method, 321, 324, 330, 331, 332, 343
  incorporating negative pore-water pressures, 333–345, *see also* Total cohesion method
  Janbu Simplified method, 330, 331, 332, 335
  Janbu Generalized method, 330, 331, 332
  Lowe-Karafiath method, 330, 331, 332
  Morgenstern-Price method, 330, 331, 332, 335
  Ordinary method, 327, 330, 331, 332
  Spencer method, 330, 331, 332
Stable structured soil, 361, 418, 432
State, 15
State variables:
  deformation state variables, 15, 346, 348, 349, 350, 351, 353, 354, 358, 359, 456
  stress state varibales, 15, 18, 38, 39, 40, 41, 42, 43, 44, 45, 46, 48, 51, 217, 218, 222, 223, 227, 228, 230, 233, 237, 238, 243, 256, 257, 259, 346, 347, 348, 349, 351,

352, 353, 354, 358, 359, 360, 361, 362, 364, 370, 371, 373, 377, 401, 406, 421, 424, 437, 456
Steady state air diffusion, 177
Steady state air flow, 118, 119, 175–177
  one-dimensional, 175–176
  two-dimensional, 176–177
Steady state seepage (flow), 5, 150–177, 264, 265, 268, 329, 342
  analyses, 150
  heterogeneous anisotropic, 159, 173
  heterogeneous isotropic, 151, 159, 174
  homogeneous anisotropic, 151
  homogeneous isotropic, 160
  one-dimensional water flow, 152–158
  three-dimensional water flow, 173–174
  two-dimensional water flow, 159–173
Strain rate, 248, 249, 250, 251, 252, 281
  consolidated drained, 249
  constant water content, 249
  drained shear, 248
  direct shear, *see* Displacement rate
  effect, 222, 228, 248, 249
  triaxial, 222, 223, 248, 250, 251, 253, 284, 286, 287, 288
  undrained shear, 248, 249, 250, 253, 282
  unsaturated soils, 251
Stress, *see also* Pressure
  analysis, 38, 49
  at a point, 54, 56
  deviator, 51, 57, 59, 219, 222, 223, 226, 227, 231, 232, 238, 239, 241, 243, 244, 245, 246, 248, 249, 250, 255, 260, 263, 281, 284, 285, 286, 289, 297, 347
  effective, *see* Effective stress
  geostatic, 52
  history, 52, 217, 225, 228, 250, 254, 298, 299, 301, 347, 397, 404
  horizontal, 298, 299, 300, 305, 307
  invariants, 49, 58, 59
  lateral, 300
  net normal, 43, 45, 46, 54, 55, 56, 57, 58, 59, 60, 62, 217, 222, 223, 224, 225, 226, 227, 228, 230, 231, 232, 233, 234, 235, 236, 239, 240, 242, 243, 244, 246, 248, 256, 257, 258, 259, 286, 290, 295, 296, 303, 307
  normal, 42
  overburden, 246, 298, 299, 302, 303, 305, 307, 308, 309, 329, 330, 400, 404, 405
  paths, 49, 59, 60, 61, 62, 231, 238, 239, 240, 241, 242, 243, 244, 245, 246, 255, 288, 299, 309, 312, 313, 318, 347, 399, 400, 401, 404, 406, 407, 408
  point, 49, 56, 59, 60, 221, 228, 231, 238, 239
  point envelope, 231, 232, 233, 234, 235
  point surface, 231
  preconsolidation, *see* Preconsolidation pressure
  principal:
    intermediate, 58, 236
    major, 56, 58, 231, 232, 236, 238, 239, 241, 244, 245, 303, 307, 402
    minor, 58, 223, 231, 232, 236, 241, 303, 307, 402
  princpial planes, 54, 57, 58, 59, 301

principal stresses, 55, 226, 301, 329
shear, 42, 43, 54, 55, 56, 57, 58, 59, 60, 217, 218, 220, 223, 224, 225, 226, 227, 228, 229, 231, 232, 233, 234, 235, 236, 239, 240, 242, 243, 244, 245, 246, 247, 248, 256, 257, 258, 284, 286, 287, 292, 293, 294, 295, 296, 303, 307
state variables, see Stress state variables
total, 219, 225, 238, 240, 242, 243, 245, 247, 303, 323
total normal, 52, 54, 244, 245, 246, 298, 302, 352
vertical, 298, 303, 307
Stress history, see Stress
Stress path, see Stress
Stress point, see Stress
Stress state variables, see State variables
Stress-strain relationships:
 elastic, 297, 298
 plastic, 297
Structure of soil:
 compacted soils, 225
 dispersed, 225, 364
 flocculated, 225, 364
Subgrade, 16
Suction, see Soil suction
Surface boundary conditions, 473, 475, 476
 heat flow, 473
 vapor flow, 473, 475
Surface tension, 21, 24, 25, 65, 67, 69, 81, 183, 263, 273
Swell, see Heave
Swelling, 46, 63, 361, 388, 402, 406, 419
 compacted clays, 301, 408, 410
 potential, 399, 404, 411
 soils, 397, 401
Swell (or Swelling) index, 369, 375, 392, 393, 394, 398, 402, 404, 406, 407, 408, 410, 411, 413, 415
 correlation to plasticity index, 399
Swelling pressure, 7, 313, 374, 375, 389, 391, 400, 409, 415
 corrected swelling pressure, 374, 384, 389, 390, 391, 403, 404, 405, 406, 407, 408, 409, 410, 411, 412, 413, 414
 correction for sampling disturbance, 390–391, 400, 407
 lateral, 403
 profile, 410
 relationship with earth pressure, 313, 407
 uncorrected swelling pressure, 389, 390, 391, 407, 410, 411, 413, 414

Tangent pore pressure parameter, 185, 195, 196, 204–216. See also Pore pressure parameter
 pore-air pressure, 186, 197
 pore-water pressure, 186, 197
Tempe pressure cell, 126, 379
 apparatus, 133–134
 test procedure, 133–134
Tensiometers, 54, 67, 82, 83–80, 109, 125, 126, 127, 128, 129, 130, 131, 315, 334, 336, 338, 417
 installation of, 338

Jet-fill, 86, 87
osmotic, 90–91
performance of, 88
Quick-draw, 88–90
servicing of, 84–87
scanning valve system, 90
small tip, 86–88, 417
Tension crack (zone), 6, 305, 306, 308, 310, 311, 312, 315, 323, 325, 326, 332, 333
 filled with water, 322, 323
Tension zone, 304, 305, 306, 310, 314, 322
 depth, 304, 305, 306, 311, 314
Thermal conductivity, 95, 97, 474, 475
Thermal conductivity sensors, 54, 67, 104
 AGWA-II, 97, 98, 99, 100
 calibration of, 97,
 MCS 6000, 96, 97, 99
 response times of, 99
 theory of operation, 97
 typical results, 99–104
Thermocouple psychrometer, see Psychrometer
Thornthwaite moisture index, 2, 408
Three-dimensional loading, 194–195, 199–200, 355, 358, 359
Time factor, see Consolidation
Time to failure, 295
Tortuosity, 111, 123, 474
Total cohesion intercept, 230
"Total cohesion" method, 333, 334, 344
Total heave, 397, 402, 408, 411, 413, 415, 416
 factors affecting, 401–403
 formulation of, 406, 407–408
 prediction, 411, 413
 theory of heave prediction, 406, 407
Total stress analysis, 317
Total suction, (see also Soil suction), 54, 63, 64, 65,
 measurement, 63, 67, 70–80, 76, 80
Transient analysis, see Seepage
Triaxial test, 40, 51, 217, 219, 236, 237, 279, 284
 consolidated drained, CD, 220, 221, 222, 223, 224, 225, 226, 229, 236, 237, 238, 239, 240, 251, 255, 263, 272, 280, 281, 283, 284, 285, 287, 288
 consolidated undrained, CU, 220, 221, 222, 225, 236, 237, 240, 241, 251, 255, 263, 272, 281, 282, 285, 287
 constant water content, CW, 220, 221, 222, 223, 229, 236, 237, 238, 239, 240, 241, 248, 249, 251, 255, 262, 263, 272, 278, 281, 282, 284, 285, 287, 288
 drained, 226, 253, 266
 equipment, 222, 223
 loading, 197–199, 215–216, 346, 354, 355, 356, 359
 measurement of pore pressures, 204–206, 212–216
 unconfined compression, UC, 220, 225, 236, 237, 245, 246, 247, 251, 262, 278, 282, 284, 288
 undrained, 219, 220, 227, 235, 236, 237, 243, 251, 266, 282, 284, 288, 289
Two-dimensional loading, see Plane strain; Plane stress

Ultimate bearing capacity, see Bearing capacity
Unconfined compression strength, 246, 247, 318
Unconfined compression test, 237
Unconsolidated undrained shear test, UU see Triaxial test
Uncoupled formulations, 440–472
 anisotropic soil, 441
 isotropic soil, 440
 plane strain, 440
 two-dimensional, 159, 440, 447
 unsteady state seepage, 440
Undrained loading, 188–189, 238, 239, 260, 347, 349, 434
Undrained shear strength: 144, 244, 246, 247, 318, 319
 use in bearing capacity analysis, 318
 $\phi = 0$ concept, 244
Uniaxial loading, 196–197, 354, 355, 356, 359
Uniqueness (also Nonuniqueness), 287, 288, 346, 349, 351, 360, 361, 362, 363, 364
Unit weight, 21, 341
 total, 52, 341
Unsaturated soil, 12, 13, 16, 21, 39, 42, 43, 44, 45, 46, 48, 49, 56, 63, 220, 222, 225, 230, 236, 237, 244, 246, 247, 248, 388
 definition, 1, 30, 31, 349
 permeability, 54
 strength, 218, 224, 225, 227
Unsaturated soil mechanics, 12, 15, 18, 19
 definition, 13
Unsteady-state flow (or seepage), 150, 440, 441, 442, 449
Unsupported excavations, 310, 313, 314, 315
 effect of tension cracks, 314

Vegetation, 53, 54
Velocity head, see Head
Vertical excavations, 6, 311
 near vertical excavations, 6
Virgin compression curve, 405
Viscosity, 23, 24
 absolute, 23, 24
Void ratio, 32, 33, 34, 46, 228, 290, 291, 358
 convergence, 384, 385
 definition, 30, 35
 typical values, 31
Void ratio constitutive surface, see Constitutive surfaces
Volume change, 8, 38, 39, 41, 46, 48, 49, 50, 62, 63, 187–189
 analysis, 349
 coefficient of, see Coefficient of volume change
 measurement of, 274
 prediction, 403
 theory, 374, 401, 406, 418
Volume change indices, 374, 377, 378, 380, 381, 382, 384, 385, 392, 395, 397, 399, 401, 403, 406
 determination, 395
 measurement of, 374–396
 typical values, 399
Volume-mass relations, 20, 29, 30, 32–33, 34, 35, 366
 density, 32, 33

Volumetric coefficient of solubility, 29, 35, 36, 37, 122, 146, 180, 206, 277, 278
Volumetric deformation coefficients, 346, 351, 358, 359, 360, 361, 365, 366, 367, 369, 370, 374
 defined, 354
 laboratory tests, 367–369
 loading surface, 369–370
 relationship between, 365–373
 unloading surface, 369–370
 for void ratio, 366–367
 for water content, 366–367
Volumetric deformation index, 370, 371, 384
Volumetric deformation modulus, 361, 366
Volumetric modulus, 353
Volumetric pressure plate extractor, 379
 apparatus, 134–135, 146
 test procedure, 134–136
Volumetric specific heat, 474
Volumetric strain, 352, 353, 357, 361, 421, 461
 definition, 351
Volumetric water content, *see* Water content

Water, density, 33
Water coefficient of permeability, 110–117, 127, 128, 130, 131, 133, 139, 143, 152, 153, 161, 164, 251, 252, 263, 264, 420, 427, 433, 435, 444, 449, 456, 463, 472
 direct methods, 124–133
 indirect methods, 124, 133–138
 in situ field method, 130–133
 in situ instantaneous profile method, 130–133
 instantaneous profile method, 124, 127–130, 131
 laboratory test, 124–130
 measurement, 124–138
 (and) matric suction relationship, 113, 114, 126, 341
 steady state method, 124–127
 (and) volumetric water content relationship, 113–117
Water compressibility, *see* Compressibility of water
Water content, 46, 225, 239, 249, 290
 coefficient, 376
 definition, 30, 31–32
 gravimetric, 32, 33, 34, 35, 120, 238, 358
 volumetric, 30, 32, 33, 46, 110, 111, 113, 114, 116, 128, 129, 132, 133, 137, 151, 341, 474, 475
Water content constitutive surface, *see* Constitutive surfaces
Water content index, 371, 372, 376, 378
 with respect to matric suction, 371
 with respect to net normal stress, 371
Water table, 6, 54, 130, 152, 300, 301, 302, 304, 305, 311, 317, 320, 322, 334, 336, 338, 339, 340, 342, 343, 398, 408, 456, 461–471, 476
Water vapor, 26, 473
 partial pressure of, 23, 27, 474
 saturation pressure of, 23, 26, 27
 vaporization curve, 26
Water volume change, 48, 266, 273, 281, 290, 350, 351, 352, 353, 361, 362, 423, 430, 463
 coefficient, *see* Coefficient of water volume change
 measurement, 265, 267, 273, 284, 362, 380
Water volume change versus matric suction, 252
Water volumetric modulus, 353, 361, 463
Wilting point, 404

Young's modulus, 352, 361, 364, 461

Zero air voids curve, 33

$\alpha$ parameter, 200–201, 216
$\phi^b$ angle, 223, 224, 227, 228, 229, 230, 231, 232, 233, 234, 235, 238, 240, 242, 244, 247, 256, 257, 258, 259, 260, 284, 285, 286, 287, 288, 290, 291, 293, 294, 302, 310, 313, 317, 318, 319, 323, 325, 334, 335, 336, 339, 340
$\chi$ parameter, 40, 41, 258, 259